The Late Eocene Earth—Hothouse, Icehouse, and Impacts

edited by

Christian Koeberl
Department of Lithospheric Research
Center for Earth Sciences
University of Vienna, Althanstrasse 14
A-1090 Vienna, Austria

Alessandro Montanari
Osservatorio Geologico di Coldigioco
62020 Frontale di Apiro (MC)
Italy

THE GEOLOGICAL SOCIETY OF AMERICA®

Special Paper 452

3300 Penrose Place, P.O. Box 9140 ▪ Boulder, Colorado 80301-9140, USA

2009

Published by The Geological Society of America, Inc.
3300 Penrose Place, P.O. Box 9140, Boulder, Colorado 80301-9140, USA
www.geosociety.org

Printed in U.S.A.

GSA Books Science Editors: Marion E. Bickford and Donald I. Siegel

Library of Congress Cataloging-in-Publication Data

The late Eocene Earth : hothouse, icehouse, and impacts / edited by Christian Koeberl, Alessandro Montanari.
p. cm. — (Special paper ; 452)
Includes bibliographical references.
ISBN 978-0-8137-2452-2 (pbk.)
1. Paleoclimatology—Eocene. 2. Paleoclimatology—Oligocene. 3. Geology, Stratigraphic—Eocene. 4. Geology, Stratigraphic—Oligocene. I. Koeberl, Christian, 1959– II. Montanari, Alessandro.
QC884.L386 2009
551.609'012—dc22

2008053543

Cover photograph by A. Montanari and using parts of the original Penrose meeting logo by D. Jalufka (University of Vienna).

10 9 8 7 6 5 4 3 2 1

Contents

Preface

The Late Eocene and the Eocene-Oligocene (E-O) transition mark the most profound oceanographic and climatic changes of the past 50 million years of Earth history, with cooling beginning in the middle Eocene and culminating in the major earliest Oligocene Oi-1 isotopic event. The Late Eocene is characterized by an accelerated global cooling, with a sharp temperature drop of about 2°C near the Eocene/Oligocene (E/O) boundary, and significant stepwise floral and faunal turnovers. These global climate changes, which are reflected by a gradual increase of marine oxygen isotope values and biotic crises, are commonly attributed to the expansion of the Antarctic ice cap following its gradual isolation from other continental masses. However, multiple bolide impact events, possibly related to a comet shower over a duration of over 2 million years, may have played an important role related to the deterioration of the global climate at the end of the Eocene Epoch.

One of the best and most famous exposures of the Late Eocene, as well as the site of the Eocene-Oligocene Global Stratotype Section and Point (GSSP), is at Massignano, at the Cònero Riviera near Ancona in East-Central Italy; this was the subject and location of an earlier meeting in 1987. A Geological Society of America (GSA) Penrose Conference was held nearby at the Hotel Monteconero, a converted monastery, from October 3–6, 2007. The meeting included three days of oral and poster presentations, as well as a one-day field trip (visiting the E-O transition, the Late Eocene impact layers, K-T boundaries, and rock units in between, of the Umbria-Marche pelagic sequence). The oral and poster sessions covered the following topics: integrated stratigraphy of the Late Eocene-Early Oligocene transition and re-evaluation of the GSSP; paleoecology and paleoclimate through the critical period of transition from hothouse to icehouse; and Late Eocene impact processes and impact stratigraphy. Contributions in the present volume were largely, but not exclusively, derived from presentations at the Penrose Conference. The book gives an excellent overview of what happened on Earth during the last few million years of the Eocene and around the time of the Eocene-Oligocene boundary, from hot to cold conditions, and the role that impact events played. The contributions provide an updated and enhanced understanding of the Eocene-Oligocene boundary transition, using different climate proxies, improved time control, and climate models.

Following a general historical discussion by Alvarez and colleagues, the book is divided into three main parts. First, the late Eocene impact events are discussed, and then the actual E/O boundary conditions, split into marine and terrestrial locations, and their consequences (e.g., for the early Oligocene glaciations) are evaluated. A review paper by C. Koeberl provides an overview of the impact structures that are currently known to have a Late Eocene age (mainly the 100-km-diameter Popigai crater in Russia and the 85-km-diameter Chesapeake Bay structure in the USA, but also a few smaller structures), as well as of impact ejecta deposits of this age around the world. K. Farley discusses the pros and cons of the comet shower versus the asteroid shower hypotheses that have been proposed to explain both the ^{3}He anomaly and the multiple impacts. S. Liu and colleagues give a detailed discussion and present abundant new data on the late Eocene impact ejecta layers. Then B. Schmitz and colleagues compare the extensive evidence for a large Mid-Ordovician asteroid and meteorite shower with the situation at the end of the Eocene. Next is a paper in which R. Coccioni and co-authors review the evidence for multiple extraterrestrial signatures at the Massignano quarry. This is followed by a report by A. Pusz and colleagues on how the Late Eocene impact events influenced the global carbon cycle. Finally, the last contribution in the impact-related section is by R. Brown and colleagues, who demonstrate a disturbance in the Milankovic cyclicity at Massignano that coincided with the timing of the major impact events (Chesapeake Bay, Popigai), implying a causal relationship.

The next section begins with a contribution by F. Hilgen and K. Kuiper, who review research aimed at numerical dating of the boundary and show that current ages using totally independent astronomical and radio-isotopic dating methods essentially reveal the same age of ~33.7 Ma, but that this is not necessarily the case if attempts to intercalibrate these methods are taken into account. A contribution by L. Jovane and colleagues integrates previous multidisciplinary studies with recent data from the Massignano section (Umbria-Marche basin) and summarizes interpretations of studies from the past twenty years that have been based on the Eocene/Oligocene boundary GSSP. Next is a paper by K. Miller and co-authors, who discusses the geologic evidence for the E-O greenhouse to icehouse transition from the point of view of the Antarctic icesheet. Then J. Hooker and colleagues summarize studies on the continental Eocene-Oligocene transition in the Hampshire Basin (UK), indicating no major paleo-temperature decrease across the boundary. Next, A. Zanazzi and colleagues describe a very interesting study on the biostratigraphy and implications for paleoclimatology from the Eocene-Oligocene boundary section at Toadstool Park in northwestern Nebraska. Finally in this section, B. McGowran presents the Auversian Facies Shift, which marks a fundamental shift in oceanic facies, spanning the Bartonian and Priabonian stages and marking the changeover from the Eocene Climatic Optimum to the major Antarctic glaciation of Oi-1.

The section on terrestrial boundaries and synthesis of the E-O work starts with two papers by N. Sheldon and colleagues, who presents nonmarine records of climate change across the E-O boundary. He focuses in particular on the application of paleosols as climate proxies and showed that, contrary to expectation, the data revealed relatively steady annual climate conditions across the E-O transition, suggesting that other climate variables such as seasonality may have driven faunal and floral turnovers. Furthermore, paleosol derived records of chemical weathering are inconsistent with the timing of the Himalayan uplift hypothesis and reveal a reduction in chemical weathering associated with the E-O transition. Then R. Fluegeman et al. show that the boundary between the Jackson and Vicksburg Groups is marked by an erosional unconformity and a prominent benthic foraminiferal fauna turnover that do not coincide with the biostratigraphic E-O boundary but slightly higher, probably coincident with the Oi-1 shift. M. Aubry and R. Bord discuss the long-term evolutionary turnover in the coccolithophorids that took place from the late middle Eocene through the early Oligocene. The E-O boundary itself appears a drastic event at low latitudes, but is essentially unmarked in assemblages from high-latitudes. The final contribution is by E. Hyland and co-authors, who report on the Eocene-Oligocene transition in the Monte Cagnero section in central Italy, how it correlates with the GSSP of Massignano, and that it might contain an even more complete record of the E-O transition than Massignano.

Clearly, much has happened since the first meeting over 20 years ago. A basic conclusion is that for the time being, the location of the GSSP at Massignano, and its placement at meter level 19, is sufficient, despite some suggestions that the disturbance of the global carbon signature might be a more suitable time marker. Concerns were raised that some effort needs to be put into the correlation between the marine and terrestrial boundary time markers. The past 20 years of research also led to the discovery of several major impact events during the Late Eocene—if as a result of a comet or asteroid shower still requires more work—and also that these impacts caused only minor, but distinctly noticeable disturbances in the global carbon cycle and thus global climate, and must have had severe, but not catastrophic, implications for the Earth System at the time.

We would like to thank the meeting participants for their excellent contributions, both at the meeting and as authors and reviewers of the current volume: Laia Alegret, Walter Alvarez, Milly Alvarez, Katerina Bartosowa, Steven Bohaty, Henk Brinkhuis, Rachel Brown, Philippe Claeys, Rodolfo Coccioni, Daniel Condon, Cameron Davidson, Robert DeConto, Ken Farley, Kristen Farley, Raquel Fenero, Chiara Fioroni, Richard Fluegemann, Simone Galeotti, Steven Goderis, David Griscom, Scott Harris, Frits Hilgen, Jerry Hooker, Dona Jalufka, Patricia Jannett, Simon Kelley, Christian Koeberl, Paul Kopsick, Brian McGowran, Paula Metallo, Simonetta Monechi, Alessandro Montanari, Silvia Ortiz Sainz-Aja, Mark Pagani, Francois Paquay, Isabella Premoli Silva, Aimee Pusz, Greg Ravizza, Stephen Schellenberg, Birger Schmitz, Nathan Sheldon, Jan Smit, Selena Smith, Dennis Terry, Flavia Tori, Simone Ulmer, Bridget Wade, Paul Wilson, and Alessandro Zanazzi.

The meeting was punctuated by an art exhibit in which two American artists living in Europe, Dona Jalufka (Vienna, Austria) and Paula Metallo (Coldigioco, Italy), used paintings, objects, and installations to provide an artistic interpretation of topics related to the Penrose conference (e.g., impacts, craters, evolution, and climate change). This ended up being an unusual show at the intersection of art and science, with the appropriate title "Rimanere Colpiti" (Awestruck). We would like to thank both artists for giving the scientists something to think about. In addition, we thank Dona Jalufka for designing the logo and the meeting identity.

In addition, we thank the reviewers of the various chapters of this book: Robert Aunger, David Bice, Henk Brinkhuis, David Christian, Philippe Claeys, Laura Cleaveland, Steve Driese, Lucy Edwards, Ludovic Ferriere, Bill Glass, Alessandro Grippo, Frits Hilgen, Linda Hinnov, David Hodell, David T. Dockery III, Gerald Johnson, Luigi Jovane, David T. King Jr., Andrew A. Kulpecz, Frank Kyte, Qianyu Li, Jovane Luigi, Iain McDonald, Brian McGowran, Ken Miller, Ted Moore, Jared Morrow, C. Wylie Poag, Chris Poulsen, David Powars, Isabella Premoli, Donald Prothero, Pat Quilty, Greg Retallack, Birger Schmitz, Philip Sexton, Nathan Sheldon, John Steinmetz, Etienne Steurbaut, Neil Tabor, Dennis Terry, and Bridget Wade.

Christian Koeberl
Department of Lithospheric Research
University of Vienna, Althanstrasse 14
A-1090 Vienna, Austria
christian.koeberl@univie.ac.at

Alessandro Montanari
Osservatorio Geologico di Coldigioco
I-62020 Frontale di Apiro (MC)
Italy
sandro.ogc@fastnet.it

Figure 1. Group picture of participants in front of the church at the Hotel Monteconero, Italy (including some visiting undergraduate students from Carleton College, USA). The editors are in the front row on the left.

Figure 2. Some of the participants inspecting the impact layers at the Massignano quarry at the Cònero Riviera, Italy (the GSSP for the Eocene-Oligocene boundary is in the very left front of the image.

The Geological Society of America
Special Paper 452
2009

Time-scale construction and periodizing in Big History: From the Eocene-Oligocene boundary to all of the past

Walter Alvarez
Department of Earth and Planetary Science, University of California, Berkeley, California 94720-4767, USA, and
Osservatorio Geologico di Coldigioco, 62020 Frontale di Apiro, Macerata, Italy

Philippe Claeys
Department of Geology, Vrije Universiteit Brussel, Pleinlaan 2, B-1050 Brussels, Belgium

Alessandro Montanari
Osservatorio Geologico di Coldigioco, 62020 Frontale di Apiro, Macerata, Italy

ABSTRACT

The Ancona Penrose Conference of October 2007 dealt with the current state of understanding of the late Eocene and the Eocene-Oligocene boundary, ~34 million years ago, a critical but very brief interval in Earth history. In this paper, we place that brief interval and the lessons from the conference in the broadest possible context by viewing them in the light of "Big History." This new intellectual concept maintains that there may be value in considering the entire past, from the big bang until today, as a single unit of study. At this very early stage in the study of Big History, not even the most fundamental questions have been well formulated, let alone answered. As a first cut, Big History can be divided into regimes by considering the disciplines that study it: cosmic history (studied by cosmology and astronomy), Earth history (studied by geology), life history (studied by paleontology and evolutionary biology), and human history (studied by archaeology and historiography). These disciplines differ in terms of problems, techniques, and intellectual traditions. If we seek a common basis for a finer subdivision, the changes in utilization of the energy that have driven historical changes would seem like a good candidate. In two thought-provoking papers in 2007, Robert Aunger proposed "periodizing" all of history by placing divisions between periods when new methods of utilizing concentrated energy came into being, e.g., cellular metabolism or human agriculture. Aunger draws a parallel between this periodization of Big History and the establishment of the geological time scale. In this paper, we carefully consider this parallel and conclude that periodizing history on the basis of energy use or any other conceptual scheme is quite different from the division of Earth history into the intervals that yielded the geological time scale. Both are important but they have different purposes. Time-scale construction is a procedure that ties history to the rocks that record the history. It is a necessary step in reconstructing Earth and life history but is neither necessary nor possible in studying

Alvarez, W., Claeys, P., and Montanari, A., 2009, Time-scale construction and periodizing in Big History: From the Eocene-Oligocene boundary to all of the past, *in* Koeberl, C., and Montanari, A., eds., The Late Eocene Earth—Hothouse, Icehouse, and Impacts: Geological Society of America Special Paper 452, p. 1–15, doi: 10.1130/2009.2452(01). For permission to copy, contact editing@geosociety.org.

the history of cosmos or humanity. In contrast, periodizing history into intervals provides a conceptual framework on which to hang a growing understanding of history. We conclude that it is important to differentiate between (1) time-scale construction, (2) correlating events, (3) dating events, and (4) periodizing history. In this light, Aunger's focus on changes in energy use remains an instructive way of periodizing history, but it must be clearly differentiated from time-scale construction.

Keywords: Earth history, big history, time scale, periodization, Eocene, Oligocene.

INTRODUCTION: THE ANCONA PENROSE CONFERENCE

The Ancona Penrose Conference of October 2007 dealt with the current state of geological and paleontological understanding of the late Eocene and of the Eocene-Oligocene (E-O) boundary, ~34 million years before present (Ma). This was a critical time in Earth history, in which a long-enduring warm, "hothouse" climatic phase shifted over into the cold, "icehouse" glacial phase that has persisted until the present time. The beginning of the shift is marked by the "Oi-1" oxygen-isotope shift, which reflected that climate change and occurred shortly after at least two large extraterrestrial impacts and a marked increase in the flux of extraterrestrial dust particles reaching Earth. Many technical questions were considered at the conference—how to read the stratigraphic record of these events, how to divide up the chronology of this part of Earth history, and how to interpret the causal relationships.

THE EOCENE-OLIGOCENE INTERVAL IN THE CONTEXT OF BIG HISTORY

We do not intend to debate the specific questions considered at the conference, but rather to show how they fit into a very much broader context. Earth history, of which the Eocene-Oligocene interval is a significant detail, deals with only a small part of everything that has happened in the past. Taking the broadest possible viewpoint, there is currently an effort by a small community of scholars and scientists to develop a coherent history of everything that has ever happened. Pioneered by the historian David Christian, who gave it the name "Big History," this approach deals with the combined history of cosmos, earth, life, and humanity. It is clear from this subdivision of Big History that, in practice, it cannot really deal with everything that has ever happened, but only with the parts that we have the practical possibility of studying—mostly what has happened in the visible parts of the cosmos and on this specific planet.

Christian (2004) and Brown (2007) offered narrative accounts of all of Big History, while Spier (1996) provided a thoughtful introduction to some of the fundamental questions and concepts in this emerging field. Big History is an intellectually rich concept that is seen differently by investigators from different fields. The Christian, Brown, and Spier books are centered on a human focus. Chaisson (2001, 2006) and other astronomers see the field from a cosmological perspective, in which Big History has been called "cosmic evolution."

Earth and life history, the middle levels in the Big History spectrum, have long been seen as unified by geologists and paleontologists. The findings of these two fields are closely related, and they combine to present an integrated view of the biological and geological evolution of planet Earth. Furthermore, geologists and paleontologists have long recognized that their combined field of earth and life history is conditioned by cosmic evolution and provides the setting for human history, so the concept of Big History has long been inherent, if seldom specifically recognized, in the work of geologists and paleontologists, for example, Cloud (1978).

The findings of geologists and paleontologists come primarily from the historical record preserved in rocks, and largely from sedimentary rocks, because solids remember, while liquids and gases forget. Reading the rock record of Earth and life history is dependent on the construction of the geologic time scale. This task is not necessary in other regimes of Big History, where the information is not extracted from rocks. In the present paper, we explain in some detail this central effort required to understand Earth and life history.

The geologic time scale has been constructed over the last two hundred years, and its subdivisions are continually being improved and refined, while the ages in years that have been attached to it over the last half-century are constantly being made more accurate and precise. The Ancona Penrose Conference dealt both with the most recent improvements in the Eocene-Oligocene portion of the geologic time scale, and with the growing understanding of the trends and events in the history of Earth and life in that interval.

Some of the points made in this paper will seem obvious to any geologist or paleontologist. They are included and discussed at length because the paper is a summary of the lessons learned by Earth scientists during the setting up and refining of the geologic time scale, and it is intended to be useful to scholars from other fields who may be trying to bridge the disciplinary boundaries that compartmentalize Big History.

We wish to stress here a critical point that will be developed in this paper: the ongoing task we refer to as "time-scale construction" is essential to the entire enterprise of reading the rock record of Earth history and of life on Earth, but it is neither possible nor necessary in cosmic history, or in human history based on written documents. Time-scale development is completely different from what we shall call "periodization," which is the procedure of identifying and naming coherent intervals within any of the regimes of Big History. Periodization gives rise to names like "Renaissance" and "Hadron epoch," which serve entirely different purposes than do the intervals in the geologic time scale. The situation is confusing because "periods," which are one of the kinds of subdivision in the geologic time scale, are not established by "periodizing" history. This kind of confusion caused by the differing usage of the same word in different fields of study is a typical problem in the early stages of developing an interdisciplinary field like Big History (Alvarez, 1991).

DIVIDING UP BIG HISTORY

In an infant discipline like Big History, a critical task is simply to formulate the questions that are to be asked. Without research-grade questions, Big History could turn out to be no more than a useful way to organize an overview course for undergraduate students. Whether it also becomes a vibrant scholarly field will depend on whether challenging questions can be asked that will attract the attention and activity of scholars and scientists from its component disciplines. Spier (1996) made an initial effort in this direction.

One such question concerns how Big History as a chronological record might be divided up into coherent intervals to clarify understanding and guide research. At this point we see two potentially useful ways to divide and subdivide the time scale of Big History. The coarser division is based on the major historical entities and thus on the scholarly disciplines that study them. A finer subdivision might be based on innovations in the way concentrated energy is used to drive the processes that generate history.

The Major Regimes of History

Spier (1996) recommended using the usefully vague word "regime" as a term with which to divide up Big History in various ways. He defined regime (p. 4) as "a more or less regular but ultimately unstable pattern that has a certain temporal permanence." It is worth noting that a particular regime may occur at different times in different places, e.g., biological regimes in different parts of the Earth or on different planets. We suggest that a reasonable first-cut organization of Big History would be to divide it into a few major regimes that correspond to historical entities and, equivalently, to scientific and scholarly fields of study.

One such organization would have four major regimes, reflecting the historical objects of interest: *cosmic history* would deal with the history of the Universe and the Milky Way Galaxy from the big bang to the present, as studied by cosmology and astronomy. *Earth history* would study the planet Earth, and perhaps the rest of the solar system, from its origin to the present time, corresponding to the fields of geology, the other earth sciences, and planetary science. *Life history* would treat the evolution of organisms from the origin of life to the present, thus including the subject matter of paleontology and evolutionary biology. *Human history* would have a poorly defined beginning and correspond to the subject matter of archaeology and of historiography (history in the strict sense). This division is slightly problematic, because human history and archaeology have similar subject matter but different methods, whereas geology and paleontology have different subject matter but similar methods.

In other cases, perhaps it would be useful to identify six major regimes, to better correspond to the disciplines involved. In this arrangement, cosmic history could be divided into *big bang history*, where theoretical physics is of central importance, and *post–big bang history*, which is in the purview of astronomy, while human history could be separated into archaeology-based *preliterate human history* and document-based *literate human history*. Clearly, there is no "correct" division, simply divisions that are useful for certain purposes. In a table presented later in this paper, we use the six-regime division because we are interested in the kinds of techniques used by the scholarly and scientific disciplines that study these regimes.

It is interesting to note that although the history of the big bang ended after only a few minutes or a few hundred thousand years, depending on how its end is defined, none of the other major historical regimes has ended. Cosmic history continued after the appearance in our solar system of solid planets; Earth history continued after the origin of life, and all continue now along with human history.

The subdivision into four or six major regimes leads to the question of what changes mark the transition from each stage of Big History to the next. In each regime, different sets of rules and information govern historical change. The fundamental rules are the laws of physics, but with variations. In cosmic history, subsequent to the high-energy physics conditions of the big bang, change is governed by the physical laws of gravity and of gases and plasmas. In Earth history, the laws of condensed-matter physics and of chemistry come into play. In life history, the rules of organic chemistry, the complexity of biochemistry, and the information stored in deoxyribonucleic acid (DNA) are central. However, as mentioned already, Earth and life history are closely related and mutually interacting, dealing with the combined geological and biological evolution of Earth. In human history, new kinds of information are stored in brains and documents, while social interactions, the laws of which, if any, seem still to be obscure, enter the picture. The complexity of organization and of interactions seems to increase along this sequence (Chaisson, 2001, 2006; Christian, 2008), although both have proven difficult to quantify.

Periodizing History According to Changes in the Use of Energy

In two recent and thought-provoking papers, Aunger (2007a, 2007b) suggested a coherent way to periodize Big History all the way back to the big bang at ca. 13.7 Ga, at a finer level than our four or six major regimes. Noting that Big

History has been marked by a set of transitions in the ways in which concentrated energy is utilized in driving historical processes, Aunger chose these transitions as the boundaries between regimes that he called "periods." This is an important concept: stars, planets, organisms, and human societies require very different material resources, but they all require sources of concentrated energy, so energy use provides a possible unified basis for subdividing all of Big History. Taking the hierarchical structure of the geological time scale as a model, Aunger proposed a set of "eons," "eras," and "periods." The resulting periodization is shown in Figure 1.

Although as geologists we are flattered by the choice of the geologic time scale as a model, we find three aspects of this approach that deserve comment. First, it is ironic that although the hierarchical units in this periodization are based on the units in the geologic time scale, from which the names of the three levels of units are taken, geology and the evolution of the Earth itself are omitted, with the periods passing directly from "Galaxy" to "Cell." Judging by the subdivision of the "Terrestrial Eon," Earth is seen as a relatively inert stage upon which the drama of life and humanity are played out. As a result, the long and fascinating history of the Earth itself that geologists have uncovered is not considered.

In a second, related observation, this compilation omits several geological events of the type used by Aunger to periodize the rest of Big History—it omits the innovations in energy use made possible by the formation of a solid planet like the Earth in orbit around a star like the Sun. We prefer to include these innovations in the interest of portraying a richer Earth history, although Aunger's approach of reducing the absolute number of events in the interest of simplicity and clarity is equally valid. Our additional thermodynamic innovations include (1) conversion of gravitational potential energy to heat through the accretion of the planetesimals, and the sinking of its iron component to the center of the planet, forming the core; (2) the driving of most of the internal processes within Earth, first by the heat released by accretion and core formation, then by radioactive decay, and then with additional latent heat from freezing of the liquid iron outer core to form the solid iron inner core; (3) heating of the Earth's surface by solar radiation, which drives most of the surface processes on Earth, such as erosion, sediment transportation and deposition, and the ocean-atmosphere-climate system; (4) the use of solar energy to produce biomass through photosynthesis, and the use of that stored solar energy in the complex food chains that power a large part of the biosphere; and (5) tides within the oceans and the solid Earth. We recognize that Earth and the Solar System are not the first places in the universe where these innovations will have occurred, but since Big History can only deal with the history for which we have a record, the energy innovations made possible by formation of Earth should be considered in periodizing all of history.

Third and most fundamental, this energy-based subdivision of Big History entangles two different procedures involved in reconstructing and understanding the past.

The first procedure involves choosing or developing a chronological framework, a way of ordering historical events in a time sequence. With events or objects ordered sequentially in a chronological framework, it is possible to judge proposed causal relationships. The ideal chronological framework would be simply to have all events dated in years (Christian, 2008). In cosmic history, it is often possible to date an observed event like a distant supernova in years, and in human history, many documents carry dates in years. However, this is not the case in Earth and life history, where events are most often recoded in rocks. Consequently, it has been necessary to go through the intermediate step of constructing a time scale tied to the rock record. Now, with many kinds of sophisticated radiometric dating available, the geologic time scale is being calibrated in years.

A second procedure commonly used for understanding the past involves identifying historical intervals that have some kind of coherence. This appears to be what Aunger has done in identifying his energy use–based intervals (Fig. 1). Following Aunger and accepted usage in human history, we call this procedure "periodization." For the interdisciplinary communication that will be necessary if Big History is to emerge as a valid field, it is important to keep separate the procedures of time-scale construction and periodization that are entangled in Figure 1. The rest of the present paper explores these themes in depth.

	EON	ERA	PERIOD
older ↑	Cosmological	Material	Atomic
			Stellar
			Galaxy
	Terrestrial	Biological	Cell
			Complex-cell
			Multi-cell
		Cultural	Tool
			Fire
			Multi-tool
			Complex tool
			Agriculture
		Technological	Machine
			Steam
			Electric
			Engine
younger ↓			Nuclear

Figure 1. The periodization of Big History suggested by Aunger (2007b, his Table 2) on the basis of changes in the way concentrated energy is used to drive the processes that produce history. The hierarchy of units (eon, era, period) is based on the units used in the geological time scale, which is tied directly to the rock record. However, the periodizations (i.e., Aunger's "periods") are intellectual concepts, not rock units. Note that the top is older, opposite to the way geologic time scales are drawn.

HISTORICAL PROCEDURES IN SCHOLARSHIP AND SCIENCE

To make possible the kind of interdisciplinary conversation needed in Big History, it is critical for participants in each of the relevant disciplines to understand the special conditions and terminology of the other disciplines. In this discussion, we will consider each of these main concepts as they apply to each of the four main regimes of Big History, drawing lessons mainly from Earth and life history, where our expertise lies. It will be seen that students of each of the regimes face complex problems that may be quite different from regime to regime. Table 1 provides an overview of the considerations presented here.

Chronological Framework and Time-Scale Construction

In trying to understand what has happened in the past and in seeking causal relationships and a general understanding of history, one first needs a chronological framework for ordering, correlating, and then comparing the ages of different events. Each historical discipline has developed such a framework, but they are not all the same.

Human History

In historiography based on written documents, it is often possible to use calendar years as a chronological framework, but this entails the problem of choosing the starting point. Counting backward from the present is not an option in human history, for the present is changing as the research is being done (in contrast to Earth history, where these changes in the present are smaller than the resolution of the dating). Dates in human history are thus counted backward and forward from a standard starting point—B.C. (or BCE) and A.D. (or CE) in the most widespread system. These dates are essentially conventional, since the birth date of Jesus is not known, and since other traditions such as the Islamic world use other starting points. In addition, for early written documents, it can be difficult to date these sources accurately, so, for example, one sees Egyptian history based on the "low chronology" or the "high chronology."

Because individual years are so closely spaced, historians commonly refer events to centuries or millennia. Thus, historiographical treatments are full of references to the second millennium B.C., the mid–twelfth century A.D., or the late nineteenth century. Anticipating a problem that will arise in Earth history, these millennium and century intervals are generally not bounded by significant events, so the French Revolution occurs in the late eighteenth century, but not at the end of that century. Millennia and centuries are thus neither time-scale units like those of geology, nor conceptual periodizations; they are simply coarser divisions, or approximations, of the chronological framework in years.

The preliterate history of humans is of necessity based on archaeology, yielding situations quite analogous to geology and paleontology, where history is recorded in stratified sedimentary deposits, and where artifacts play the role of fossils. Just as the early geologists could only divide up Earth history on the basis of characteristic fossils, archaeologists were forced to divide the history they were uncovering into units based on the type of artifacts—units like the Stone Age, the Bronze Age, or subdivisions thereof, for example, the Paleolithic or the Late Helladic IIA, an interval in the Greek Bronze Age.

It was clear early on that this was an unsatisfactory procedure because new technologies diffused slowly through the ancient world, so that the Bronze Age began earlier in Turkey than it did in Italy. This is inherent to fossil- or artifact-based chronologies and analogous to the diachronous first appearance of species of fossil organisms. Establishing the sequence of artifacts is an additional procedure that is aided when archaeologists encounter situations where long intervals of time are recorded by successive cities at the same site, such as Troy or the Mesopotamian tells, or in trash dumps that were in long-continued use and are full of pottery fragments (Potter, 1976). These situations are analogous to the stratigraphic sections studied by geologists, but often archaeologists deal with sites representing only brief time intervals. The development of radiocarbon dating has made it possible to put numerical dates on the succession of observed archaeological events, although radiocarbon years require a not-yet-fully determined correction for conversion to absolute years.

Earth and Life History

Although human history merges insensibly back into prehuman Earth history, the development of a chronological framework for Earth history has been completely different from that of written human history. The effort to understand Earth's past in a scientific way began with the discovery by Steno (1669) that the sequence of rock layers deposited one on top of the other carries a stratigraphic record of Earth history (Cutler, 2003). Steno recognized in Tuscany that the sea had twice covered parts of what is now dry land (Alvarez, 2009, ch. 5), but in the absence of any other way to date these inundations, he could only attribute them to the universal ocean of Genesis before the creation of the land, and to Noah's flood.

The first serious attempt to develop a time scale corresponding to the rocks seen in the field was made by Giovanni Arduino (1714–1795), who used the terms Primary, Secondary, Tertiary, and Quaternary, from oldest to youngest (Cita, 2009). Primary and Secondary fell into disuse long ago, there is currently debate whether to discontinue the use of Tertiary (Salvador, 2006), and the definition of Quaternary is under review (Gradstein et al., 2004, p. 28, 411).

The development of the modern time scale for the stratigraphic record of Earth history was made possible, beginning around 1800, by the careful study of fossils. Even then, it was a clear observational fact that the shapes of fossils have changed through the time represented by layers of rock. The changes in fossils allow the time correlation of rocks in distant places and thus the placing of rocks in their correct chronological order, even if they do not happen to lie in contiguous stratigraphic

TABLE 1. PROCEDURES AND UNITS USED IN STUDYING SIX MAJOR REGIMES OF BIG HISTORY

Regime	Discipline	Time-scale construction based on solid rocks or artifacts	Chronology in units of time (years, etc.)	Methods for determining age in units of time	Periodization
Big bang history	Cosmology, physics	Not needed—dates are computed	Years, minutes, seconds, or negative log seconds, after instant of big bang. (Time runs forward from $t = 0$ = instant of big bang.)	Calculations based on principles of physics.	Terms are apparently used informally and inconsistently (e.g., Primordial era, Hadron epoch *or* Hadron era).
Cosmic history	Astronomy, cosmology	Not needed—dates are based on distance or computation	Billions of years ago, back to big bang at ca. 13.7 billion years ago (Time runs backward from the present.)	Age is related to distance, which can be determined by red shift of light. History of stars is inferred from observing stars in many stages of evolution and from physical calculations.	Terms are apparently used informally and inconsistently (e.g., Primordial dark age, Stelliferous era).
Earth history	Geology	Phanerozoic Eon/Eonothem* Cenozoic Era/Erathem* Paleogene Period/System* Eocene Epoch/Series* Priabonian Age/State* (An example of the hierarchical structure.)	Millions or billions of years ago (Ma, Ga), back to origin of Earth at 4.56 Ga. (Time runs backward from the present.)	Various methods based on radioactive decay (K/Ar, U/Pb, U/Th, Rb/Sr, etc.); counting of orbital cycles, varves, etc.	Geologists and paleontologists generally use the well-defined, agreed upon time-scale divisions, finding them more precise than periodizations like "late heavy bombardment," "time of banded iron formations," or "assembly of Pangea."
Life history	Paleontology, evolutionary biology, paleoanthropology	Same as for Earth history; relative dating is based on fossils.	Same as for Earth history.	Same as for Earth history.	Same as for Earth history.
Human archaeology	Archaeology	Paleolithic, Late Bronze Age, etc., based on materials used for, and styles represented by, artifacts recovered in excavations. No global system because of diachronous appearances.	Years, centuries, or millennia before the present. Sometimes in uncorrected radiocarbon years. (Time runs backward from the present.)	Various methods based on radioactive decay (especially radiocarbon); counting of tree rings; dated inscriptions on archaeological materials.	Sometimes periodized in terms of building styles, agricultural practices, or social organization, but none of these can be global.
Human history	Historiography	Not needed—dates are based on documents.	Years before or since an arbitrary starting point with labels like B.C. (=BCE), A.D. (=CE), AH, Hebrew, or Roman calendars, etc. (Time may run forward or backward.)	Dates, or references to events, in written documents, which may be difficult to calibrate in older documents.	Historians commonly use numbered centuries, which are in fact precisely defined units of time, chronologically specified terms such as dynasties, or poorly defined, regional terms, like World War II, Renaissance, or Baroque.

*The first term refers to a time interval, and the second refers to the rock deposited during that time interval.

sequences. The philosophical context of these discoveries was reviewed by Allègre (1992); this context emerged through debates over whether geologic time was infinite or finite, whether Earth history has been cyclic or trend-like, and whether it has always changed slowly and gradually, or has been punctuated by occasional catastrophes (Eldridge and Gould, 1972; Gould, 1987).

The pioneering stratigraphic geologists in the early nineteenth century found that different parts of the geologic record of Earth history—the geologic column—are preserved in different places in the world. As they worked out the sequence of fossils, they named parts of the geologic column in ways that reflect those places, but with no guiding principles or system. Thus, we have the Devonian named for Devonshire, the Permian for ancient Perm in Russia, the Lutetian for the Roman town that is now Paris, and the Jurassic for the Jura Mountains. In a more whimsical mood, the Ordovician and Silurian are named for pre-Roman British tribes that lived where rocks of those ages are exposed. Sometimes intervals were named for the dominant kind of rock in a region where that time interval was studied; this is the origin of Carboniferous for a coal-rich interval, and the Cretaceous, from the Latin name for chalk, or even Triassic, for a German region where rocks of that age occur in a triad of distinctive layers. Some units were named in a semisystematic way, including the rocks that were the subject of the Ancona Penrose Conference—Eocene meaning the dawn of the recent, Oligocene meaning less recent.

The geologic time scale was thus built up in a disorderly way, which is understandable, since those who built it were exploring deep time with no guiding principles because no one had done this before. The time scale was gradually refined during the nineteenth and twentieth centuries (Fig. 2).

The intense research in many parts of the world that produced the time scale made it clear that the fossil organisms upon which the time scale was based have changed, or evolved, over geologic time, and this made it possible to correlate rocks over long distances. Evolution in this sense is an observable fact. In 1859, the geologist Charles Darwin explained the evolutionary changes in organisms as the result of descent with modification resulting from natural selection. Thus, the fundamental paradigm in modern biology was largely an outcome of research aimed at building a geologic time scale, although insights from other fields also contributed.

Fossils allowed the division of the most recent part of Earth history into the Paleozoic, Mesozoic, and Cenozoic Eras, along with a hierarchy of finer subdivisions. Below the oldest Paleozoic layers—a period called the Cambrian—were rocks lacking fossils, which therefore could not be dated or fitted into a detailed succession. These rocks were called Precambrian, and their history remained intractable until the development of radiometric age dating in the mid–twentieth century.

By the 1960s, it had become clear that different stratigraphic traditions in different countries were making it hard to compare and integrate the findings of geologists and paleontologists around the world. Further progress in dating and interpreting

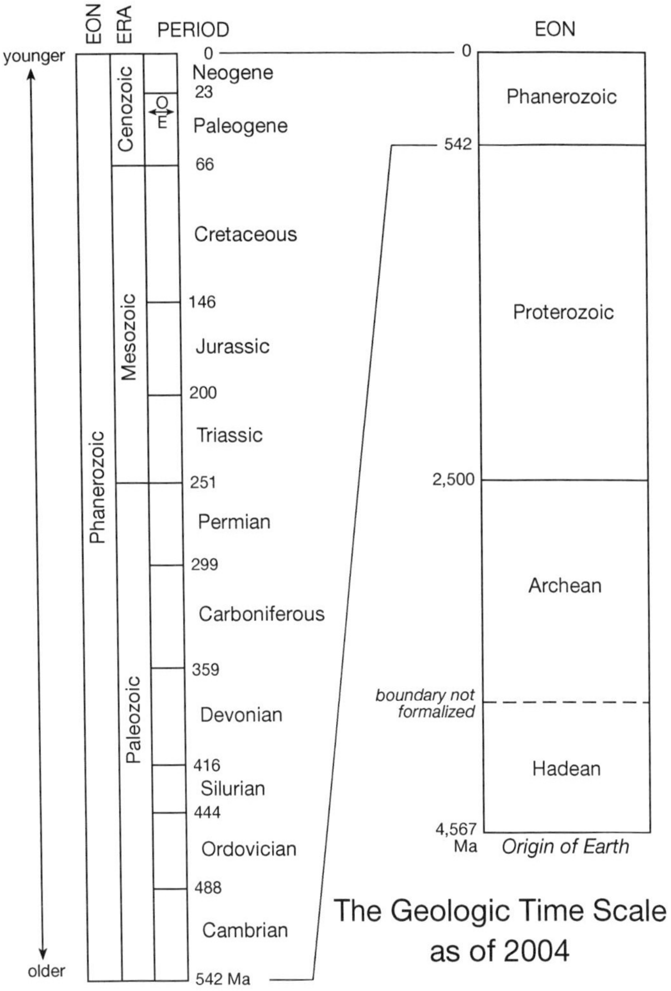

Figure 2. The current geologic time scale with age calibration in Ma (million years before present). The Phanerozoic has abundant fossils, and boundaries are defined by global stratotype sections and points (GSSPs). Pre-Phanerozoic rocks have very little fossil record, and boundaries are defined by round-number ages in Ma. O/E is the Eocene-Oligocene boundary, discussed in this paper. "Hadean" is widely used for the time represented by little or no rock record, but it is not currently a formally accepted unit. Top is younger in geologic time scale drawings, such as Figures 2, 3, and 5 here.

the rock record of Earth history would require standard procedures to be adopted on a global basis. A major effort in geological diplomacy, in which Hollis D. Hedberg repeatedly circulated questionnaires among prominent stratigraphers around the world and gradually widened the areas of agreement, eventually produced the International Stratigraphic Guide (Hedberg, 1976; Salvador, 1994).

A critical problem at that time was the way in which to tie the intervals in the time scale to the rock record itself. In nineteenth-century work, an investigator would find a region with abundant, fossil-rich rocks representing some part of geologic time and would use those fossils to define a time interval—the

Triassic Period, for example. Rocks anywhere else that carried those typical fossils would be identified as Triassic in age. As work progressed, the order of these periods became evident, for in undeformed areas, rocks with Jurassic fossils rest on top of those with Triassic fossils. As the time scale became more detailed, the problem arose of defining the boundaries between adjacent time intervals. It was clear that the time intervals needed to be tied to the rock record, and so the concept of a type section, or stratotype, emerged. However, if the top and the bottom of the Triassic were picked in a type section in Germany and the top and bottom of the Jurassic were established in Switzerland, for example, it was unlikely that the top of the type Triassic and the base of the type Jurassic would be exactly the same age; probably there would be either a gap or an overlap in time.

Largely through the work of Hedberg and others interested in the problem (Ager, 1973), a solution was developed that has become the basis of the modern time scale. In this procedure, only the lower boundary of a time interval is tied to the rock record. This boundary is formally established at a specific level in a specific exposure of stratified sedimentary rocks. The top of a time unit is not defined anywhere in the rock record, but is considered to be the same as the base of the next younger unit. The defined position of the lower boundary of a unit is often informally called a "golden spike," but the formal term is "global stratotype section and point" (GSSP), where the stratotype section is the exposure of rocks chosen for defining the lower boundary of a time unit, and the point is the specific level in that section where the formal boundary is placed (Fig. 3).

International working groups were set up to identify possible GSSPs and choose the best one for defining each stage, where a stage is defined as a subdivision of an epoch (Fig. 3). The base of a time interval is placed at the base of its lowest stage. This was the context of the Ancona meeting. The GSSP for the base of the Rupelian stage, and the equivalent Eocene-Oligocene boundary, had been established in an abandoned quarry at nearby Massignano in 1992 (Fig. 4). The 2007 meeting was held to consider the accumulated research on this part of the time scale since the boundary had been formalized (Fig. 5).

Obviously, there is a great deal of investigation, argument, advocacy, and negotiation involved in choosing each of the dozens of GSSPs. This crucial work is now well advanced, although not yet complete. For an online list of the GSSPs for all recognized stages and their current status, see http://www.stratigraphy.org/gssp.htm. For an example of the intensely detailed work that goes into proposing a GSSP, see Coccioni et al. (2008), who documented a candidate GSSP for the base of the Chattian, the upper of the two Oligocene stages, in the same part of Italy where the base of the Oligocene was previously established at Massignano. Formal ratification of several GSSPs is anticipated, based on discussions at the International Geological Congress at Oslo in August 2008.

The history of this critical work in geological time-scale construction has been reviewed in general by Walsh et al. (2004) and for the Neogene stages of the Mediterranean area by Cita (2009). Generally, it has worked extremely well, although technical debates continue, for example, whether a GSSP should

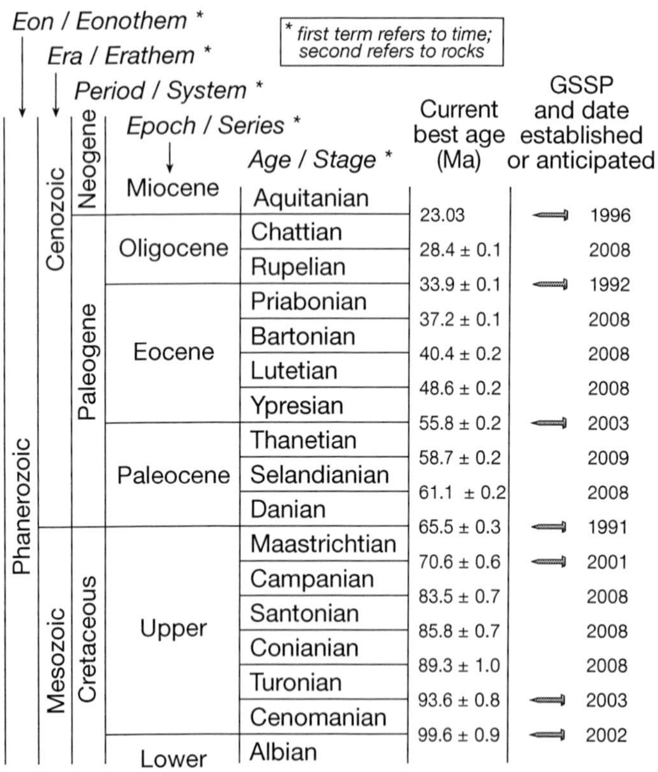

Figure 3. Current status of the geologic time scale and the global stratotype sections and points (GSSPs) for the Upper Cretaceous and Paleogene. This diagram is not scaled with time; instead, each stage has an equal-sized box. This figure combines information from Gradstein et al. (2004) and the web site of the International Commission on Stratigraphy (http://www.stratigraphy.org/gssp.htm) as of April 2008.

Figure 4. View of the Massignano quarry near Ancona, Italy, with bedding dipping gently to the left. The position of the global stratotype section and point (GSSP) that defines the base of the Oligocene, and thus the Eocene-Oligocene boundary, is marked. The inset close-up shows the actual physical spike driven into the rock at the GSSP.

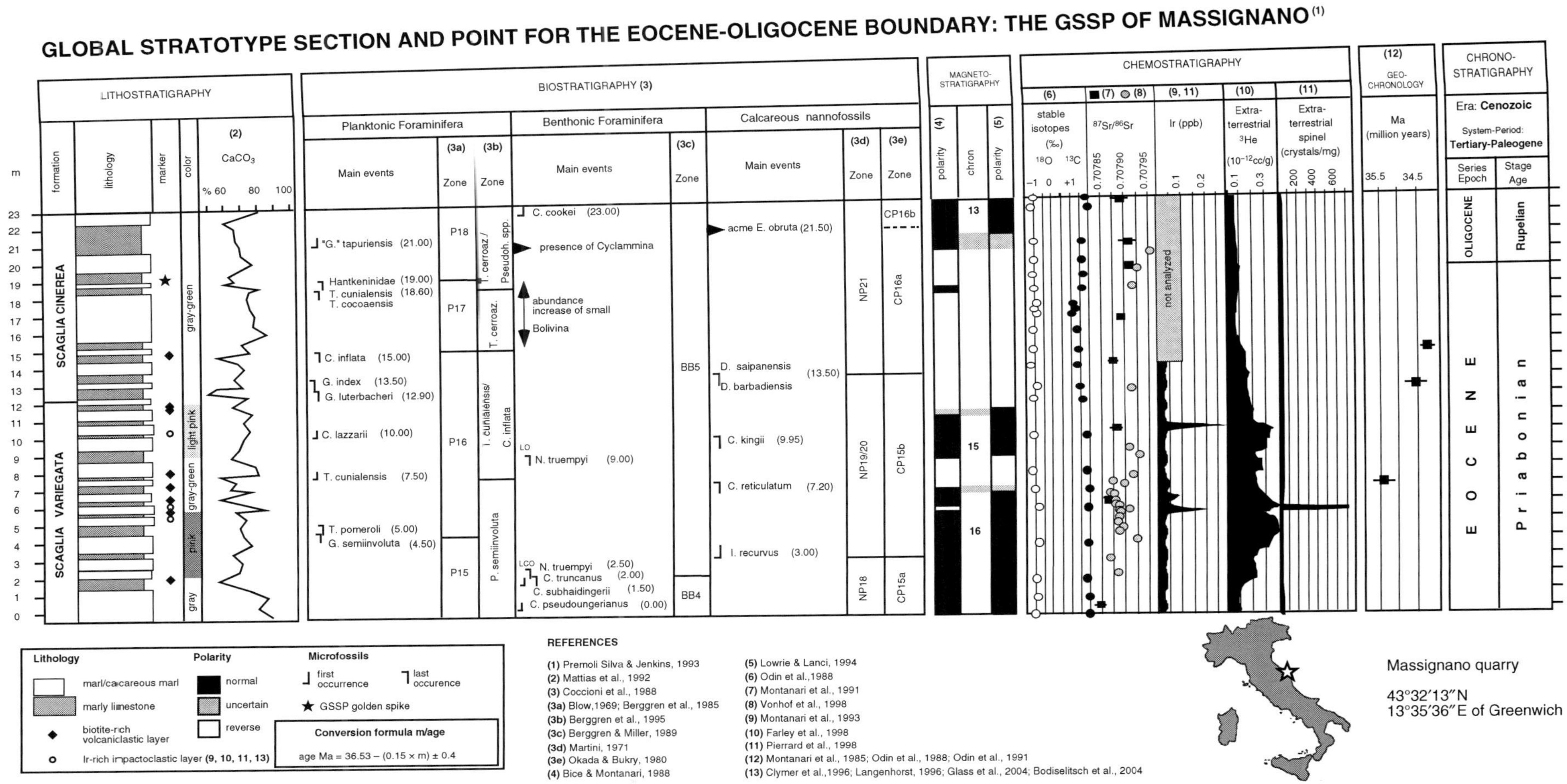

Figure 5. Detailed stratigraphy of the Massignano global stratotype section for the Eocene-Oligocene boundary (after Montanari and Koeberl, 2000, p. 260–261), which may be consulted for the genus names and the references cited in the figure. The star marks the global stratotype section and point (GSSP).

be placed at a level corresponding to a major event in Earth history (a "golden event"), or at a minor but easily correlated event like the appearance or disappearance of a widespread and abundant fossil species, which must then be considered to have occurred globally at the same time.

The critical point from the foregoing discussion is that, unlike the human history based on often-datable written documents, Earth and life history are based on the rock record, which cannot be directly calibrated in years. The subdivision of rock-based Earth history has been carried to a detailed and sophisticated level, and currently these intervals of the rock record are being dated in years using isotopic methods. A similar situation prevails in the human prehistory based on archaeological excavations, and yet this time-scale construction has no analogue in either document-based historiography or in cosmic history.

Cosmic History

Cosmic history was the most recent major regime for which it was determined that there actually is a significant history to be understood. The expansion of the universe was discovered in the 1920s, and through the middle of the twentieth century this was explained in two mutually contradictory ways—the big bang hypothesis and the steady-state hypothesis. Both hypotheses were developed theoretically until the discovery of the cosmic microwave background radiation in the 1960s definitively confirmed a prediction of the big bang theory, demolished the steady-state hypothesis, and rendered cosmic history significant and interesting. Ferris (1988) has reviewed these developments.

Much is now understood about cosmic history, and the sources of that information are quite unlike the sources of information about the history of Earth, life, and humanity. In part, cosmic history can simply be observed with telescopes. Observations of galaxies billions of light years away show us the conditions of those objects billions of years ago. Although there is no way to observe our own planet or solar system far back in time, the distant observations illustrate what the universe in general was like in remote times.

Telescopes make it possible, however, to infer the early history of our sun and solar system, by observing nearby protostars and very young stars. Spectacular images from the Hubble Space Telescope and other observatories (http://apod.nasa.gov/apod/astropix.html) show brand-new stars accreting from giant clouds of dust and gas, lighting up from thermonuclear fusion, and blowing away the residual dust and gas. Geological understanding of the very early history of Earth in the context of the solar system has been greatly enhanced by these astronomical images.

Our knowledge of the big bang itself has come from an interplay between theoretical calculations based on high-energy physics and astronomical observations, especially of the cosmic microwave background radiation.

None of these sources of information on cosmic history requires or is susceptible to the kind of time-scale construction that geologists, paleontologists, and archaeologists have had to use to understand the rock record. Cosmic history has always been dated, as well as possible, in years. Even when it sounds like cosmic history is being divided into units like those of the geologic time scale (for example, the Radiation Era), this is a completely different kind of historical procedure, which we treat later under "Periodizing history."

Correlating Events

In rock- and sediment-based history, it is often useful to correlate events, that is, to determine that they are the same age, or that one is older than the other, whether or not the events can be dated in years. Clearly, this is important in trying to establish causal relationships. Correlation is one of the two main uses of fossils in geology; the other being environmental reconstruction.

Beginning in the 1960s and 1970s, a new correlation technique was developed based on the measurement of remnant magnetization in rocks. Paleomagnetists discovered that Earth's magnetic field has reversed many times in the past, and it has been possible gradually to work out the sequence of geomagnetic reversals recorded in rock sequences. Earth's magnetic field collapses during a geomagnetic reversal and then regrows in the opposite direction, so these events must be synchronous worldwide, making them an ideal basis for correlating distant rock sequences. If an event is recorded in one place in rocks with normal magnetic polarity (normal is the polarity of today's Earth) and an event is recorded elsewhere in reversely polarized rocks, they cannot be the same age. The "fingerprint" of long and short, normal and reversed polarity zones has been recovered from the continuous magnetic record in ocean crust (Glen, 1982), so if enough polarity zones can be identified in a section of layered rocks to recognize the fingerprint, the rocks can be correlated to the seafloor record and thus to other rock sequences.

The modern time scale (Gradstein et al., 2004) shows the history of geomagnetic reversals along with the fossil-based subdivisions, and both are dated in years as well as possible. This development (Lowrie and Alvarez, 1981; Glen, 1982) has no obvious analogue in either human or cosmic history.

Dating Events

At the beginning of the twenty-first century, we find ourselves in a situation never before experienced, in which all the major regimes of Big History can place their findings in a chronology based on years. Years form the ideal basis for chronology because the orbital period of Earth circling the Sun has been essentially constant since nearly the beginning of the solar system, unlike the length of the day, which progressively slows because of tidal friction, and because they can be understood intuitively.

Christian (2008) writes of a First Chronometric Revolution marked by the invention of writing, which allowed events of the past to be preserved with more or less accurate dates for longer than was possible using only memory and legend. He points out that as a result, history based on written documents came to be perceived as the only way there was to know and date the past.

Christian calls the current situation, in which all of the past can be known and studied, the Second Chronometric Revolution. This has led to the slightly touchy problem that various sciences now claim to be doing history, which may distress some historians. In cases where this distinction may cause discomfort, it may be useful to think of history based on written documents as history *sensu stricto*, and the broader view of the past to which various sciences are contributing, as history *sensu lato*.

Christian (2008) gives a useful overview of the various methods by which ages in years are determined for geological and paleontological materials. The most important, of course, is radioactive decay. Many different radioactive isotopes have been exploited because their half-lives and useful materials have made them applicable to a wide range of geological time scales and problems. For age calibration of the geological time scale in cases like the Eocene and Oligocene under consideration at the Ancona Penrose Conference, probably the most useful is $^{40}Ar/^{39}Ar$. This system is based on the decay of potassium to argon and is measured on volcanic mineral grains that were erupted explosively and carried by wind, in this case, to the (then submerged) part of Italy where the Eocene and Oligocene sediments were accumulating. These mineral grains were too hot before eruption to retain any daughter argon, but after eruption, they cooled, began to accumulate radiogenic argon, fell to the ocean, and sank to be incorporated in the accumulating sediment, all in a geological instant. Their radiometric age thus gives the age of the fossils in the layer where they are found or of any other stratigraphic feature, such as magnetic polarity reversals or chemostratigraphic signatures. This approach, which has been referred to as "integrated stratigraphy," was pioneered in the Tertiary portion of the northeastern Apennine pelagic succession, providing unprecedented precision and accuracy in the calibration of the geological time scale (Montanari et al., 1985, 1988, 1991, 1997).

Christian (2008) also lists other techniques by which ages in years can be determined to calibrate Earth and life history: dendrochronology, counting of annual layers in glacial ice and in the sediments of periglacial lakes (varves), thermoluminescence, electron-spin resonance, fission-track dating, and genetic dating methods. Yet another technique that is having a major impact on the dating of the geological time scale is cyclostratigraphy (Kuiper et al., 2008). This method is based on finding rhythms in pelagic sediments, such as cyclic oscillations in the proportions of detrital clay and biogenic calcium carbonate, which reflect plankton productivity controlled by cyclic climate changes. In turn, these climate changes are responses to oscillations in the three main parameters of Earth's orbit and spin, with periods of ~20,000 yr (precession of the spin axis), 40,000 yr (obliquity of the spin axis), and 100,000 yr (eccentricity of Earth's orbit). These orbital rhythms, known as Milankovitch cycles after the Serbian engineer and mathematician who first calculated them in the late 1930s (Milankovitch, 1941), control the changing amount of solar energy that reaches Earth's surface at any given latitude and time. These changes in energy have an effect on the production of primary biogenic carbonate and, ultimately, on the type of sediment deposited in marine pelagic basins (especially marl versus limestone). In addition to calcium carbonate, other proxies are used today to resolve orbitally forced climate cycles, including stable carbon and oxygen isotopes, magnetic susceptibility, and the relative abundances of different types of microfossils, e.g., rhythmically alternating radiolarian-rich layers and foraminifera-rich layers (Mitchell et al., 2008).

The history of the orbital parameters of Earth was recently calculated astronomically with extreme precision at 1000 yr intervals back in time to 100 Ma by Laskar et al. (2004). This reference curve makes it possible to calibrate with high precision the orbital cyclicity recorded in pelagic sediments. The marine carbonate succession of the northeastern Apennines of Italy, which represents a nearly 200 m.y. time interval from the Early Jurassic to the Pliocene, provides an ideal situation for applying high-resolution cyclostratigraphy. The Miocene succession exposed on the cliffs along the Conero Riviera, near Ancona, has been subjected to high-resolution cyclostratigraphic analysis (Cleaveland et al., 2002; Mader et al., 2004; Hilgen et al., 2005), leading to a highly precise astrochronological calibration of this portion of the geological time scale. The Eocene-Oligocene succession exposed at Massignano and elsewhere in this region (Premoli Silva et al., 1988; Montanari and Koeberl, 2000) is also being dated by the cyclostratigraphic method, providing not only an ever more accurate calibration of the geologic time scale, but insights into the response of the marine biota to climate changes and extraterrestrial events such as impacts and comet showers as well (Jovane et al., 2006; Brown et al., this volume; Hyland et al., this volume).

Thus, the cyclostratigraphic approach, combined with integrated stratigraphy, is leading toward a realistic goal of a geologic time scale with events calibrated to an accuracy of around 1000 yr—a time resolution previously almost inconceivable.

Periodizing History

The whole purpose of time-scale development and of correlating and dating the past must be to understand history. In human history, we would like to know the social and environmental factors that drove historical change, as well as the human choices that were made and their historical consequences. In Earth and life history, we want to understand the evolution of Earth and of life on Earth that led to the world we know today. In cosmic history, we would like to know how the initial conditions at the time of the big bang led to a universe that developed as it has. In Big History, we hope that it may be possible to discern the structure and character of all of the past, and to recognize the nature and timing of the first-order changes that have produced the major regimes of history.

It would not be easy to discern the structure underlying a continuous written narrative with no chapter breaks, no headings or subheadings, and no separation into paragraphs. It is thus imperative for writers to provide these markers and breaks, to make the reader's job easier. Analogously, when historians study the past, they generally impose breaks or separations on their narrative. Within the field of human history, there is a long tradition of making these separations, and it is called "periodizing" history. However, as David Christian (2008, personal commun.) points out, "periodization is much more than a matter of making life easier for the reader. It is fundamental to understanding the past as it entails an attempt (which may or may not succeed) to separate the trivial from the significant, and to identify the large changes. It has to be done because history without periodization is a meaningless jumble of information." Periodization is intimately related to the questions that are chosen for study. It is, however, a controversial procedure to which philosophically minded historians have given a great deal of thought (Gerhard, 1973–1974).

Examples of periodization are everywhere at hand. Ancient Egyptian history is conventionally divided into the Old, Middle, and New Kingdoms, separated by the First and Second Intermediate Periods. These are our divisions, not theirs; surely Senusret III did not think of himself as a Pharaoh of the late Middle Kingdom. Indian history is summarized as a sequence of Kingdoms and Empires. Chinese history is conventionally presented as a chronology of dynasties. European history is usually divided into a sequence something like the following: Roman Empire, Early Middle Age, High Middle Age, Late Middle Age, Renaissance, Reformation, and so on.

Periodization is probably unavoidable if we are to gain any understanding of history. The problem is that the choice of periods and breaks between periods is very subjective. Where should one place the division between the Late Middle Age and the Renaissance? A historian of Italy with a particular interest in art and literature might place the break at the beginning of the Florentine humanist movement around 1400. A historian of Iberia interested in the end of the Late Middle Age and in exploration might place the break at some critical discovery of the Portuguese, perhaps Gil Eannes finally passing the longtime barrier of Cape Bojador on the Saharan coast of Africa in 1434, or at Spain's miracle year of 1492. Historians of Germany, France, or England would make other choices, but to historians of China, India, Africa, Mexico, or Peru, the transition from the Middle Ages to the Renaissance is completely irrelevant. It seems clear that such periodizations are useful but rather subjective, and they are virtually impossible to correlate worldwide. In view of this inherent subjectivity, it is difficult to imagine the world's historians undertaking to agree on a standard periodization, nor would it be either necessary or desirable.

A central point of the present paper is that periodization in this sense is completely different from the time-scale development that has been necessary in order for geologists and paleontologists to decipher the rock record of Earth and life history, and yet they are commonly confused because, at first glance, the geological periods and other ranks of time-scale division look like periodized intervals. For example, the Wikipedia article on "Periodization," as of April 2008, states, "Periodization from the sciences includes the geologic Cretaceous or Jurassic periods, while examples from human history include the Baroque Period, the Age of Anxiety, and the Harlem Renaissance." This ignores the difference that Cretaceous and Jurassic are objective time-scale divisions formally tied to the rock record and agreed to at the highest level of international scientific cooperation—or will be, when the bounding GSSPs are agreed upon, probably in the next couple of years (http://www.stratigraphy.org/gssp.htm), whereas the Baroque Period, the Age of Anxiety, and the Harlem Renaissance are subjective terms, although they may be very useful and may reflect a great deal of scholarship. We seek to make this difference clear with the following examples.

EXAMPLES

Historiography: The Second World War

In his thoughtful and influential little book, *The Nature of the Stratigraphical Record*, Ager (1981, p. 71–73) drew an analogy between establishing a unit in the geological time scale and trying to define the Edwardian Era. This analogy clearly illustrated the problems in choosing the beginning and end of a geological time interval and in recognizing it from the material remains that may be preserved. However, we would note two critical differences. First, defining geological periods like the Devonian or Permian on the basis of the rock record was an essential step in establishing the chronology of Earth history; nothing could be understood until that was done. Defining the Edwardian Era can be an aid in understanding what was happening at that time—it is an example of periodizing history—but it was not necessary in order to establish a chronology, because the dates in years were already there, although the archaeological study of earlier human history would be more analogous to the geological procedure. Second, once the GSSPs that define the limits of the Devonian and Permian were placed in the rock record, those geological periods were precisely defined. It is difficult to imagine an analogous procedure for precisely defining the beginning and end of the Edwardian Era.

For another example, consider the Second World War, for which the age of almost every critical event is known not just to the year, but to the day or even the hour. Time-scale construction is not necessary for establishing chronology, but periodizing is helpful in understanding what happened. Periodization will be arbitrary and designed for the purpose of a particular study. For example, did the Second World War begin in 1939, when Germany invaded Poland, or was it then a regional conflict only? Did it begin when the conflict widened to include Britain and later the Soviet Union, or when it became truly global with the entry of Japan and the United States? Did it end with the surrender of Germany, or of Japan? Or did it continue seamlessly into the Cold War? Perhaps each historian will structure the Second World War differently, for the purposes of a particular study, as Winston Churchill did in the titles of the six volumes in his study of the war: "The Gathering Storm," "Their Finest Hour," "The Grand Alliance," "The Hinge of Fate," "Closing the Ring," and "Triumph and Tragedy."

Cosmic History: The Early Galaxies

As an example of how astronomers can literally *see* ancient history, consider the Hubble Ultra Deep Field (Beckwith et al., 2006), an image of a tiny portion of the sky, covering 11 arcmin^2, made with a 1,000,000 s exposure using the Hubble Space Telescope. The HUDF image (www.aip.de/groups/galaxies/sw/udf/swudfV1.0.html) contains at least 10,000 objects, almost all of them galaxies, with the wavelength of their light shifted toward the red because of motion away from us. The red shifts increase with distance—which is the signature of the expansion of the universe—and the greater the distance, the longer it has taken the light to reach us. Thus, in the most distant objects astronomers are seeing galaxies as they were when the universe was very young. Red shift is measurable; converting it to age is somewhat model dependent, but in some objects in the HUDF image, we are clearly seeing galaxies as they were in the first billion years after the big bang, 13.7 billion years ago. Beckwith et al. (2006) observed that distant, young-universe galaxies have ragged, irregular shapes, contrasting with the elegant spirals and ellipses of closer galaxies, which we see as they were after galactic shapes had had time to evolve, pulling together gravitationally into spirals and ellipses.

The HUDF early galaxies represent an intermediate step between the original slight fluctuations in the density of the universe immediately after the big bang, which are another example of observed history, seen as patchiness in the cosmic background radiation (http://antwrp.gsfc.nasa.gov/apod/ap050925.html), and the present universe of discrete, gravitationally organized galaxies.

Earth and Life History: Mass Extinctions

The geologic record bears witness to mass extinctions of fossil organisms. Six great extinctions are recorded since the appearance of an abundant fossil record at the beginning of the Cambrian, ca. 540 Ma, and there have been several minor extinctions as well. The early stratigraphers used these extinctions as natural breaks between periods when they set up the time scale. They did this work well, and four of the six great extinctions currently recognized (end Ordovician, end Permian, end Triassic, end Cretaceous) are at or very close to modern geological period boundaries. In the absence of a method for dating rocks in years, it was possible to interpret these extinctions as due to missing rock record, since an originally gradual evolutionary change will appear sudden where a hiatus has removed the record. Darwin, influenced by the then-current doctrine of uniformitarianism, chose this interpretation—*Natura non fecit saltum* (Nature does not make jumps); it was perhaps his greatest mistake, for we now know from places with demonstrably continuous sedimentation that at least some of the mass extinctions took place suddenly (Alvarez, 1997).

A brief review of three of the extinctions throws light on the interplay between time-scale development and periodization. A minor extinction can be seen around the Eocene-Oligocene boundary on a graph of the number of marine fossil genera recorded in the literature, plotted to substage (Sepkoski, 1997), but it is not easy to locate with precision in the rock record, even in an intensely studied section like the Massignano stratotype. The clearest event at Massignano is the extinction of the Hantkeninidae, a family of planktic foraminifera, and this level was selected as the GSSP for the Eocene-Oligocene boundary because of the utility of these floating microorganisms for correlation. However, there is no corresponding extinction in two other well-represented fossil groups, the benthic foraminifera and the calcareous nannofossils (Fig. 5). Furthermore, this GSSP does not correlate with either of two indicators of extraterrestrial influences—a level of impact debris or an anomaly in helium isotopes that marks an input of extraterrestrial dust (Farley et al., 1998), nor does it correlate with the striking cooling event and increase in global ice volume recorded by oceanic isotopic event 1, labeled Oi1 (Miller et al., 2008). Thus, for the Eocene-Oligocene boundary, the time scale is well defined, but it does not quite match any of the interesting events that might be taken as periodization boundaries.

The most recent of the great mass extinctions, the Cretaceous-Tertiary (K-T) boundary, at 65.5 Ma in the 2004 time scale, is unique in being extremely abrupt and synchronous in a number of groups of organisms, and in correlating precisely in time with evidence for a very large impact (Smit and Romein, 1985; Alvarez, 1997; Arenillas et al., 2006). Hundreds of excellent stratigraphic sections on land and in deep-sea cores are available for studying this event worldwide (Claeys et al., 2002). The GSSP for the K-T boundary has been placed at the base of a thin clay layer carrying the geochemical evidence for the impact, at a stratotype at El Kef in Tunisia. This is a unique case of a precise match between the time-scale boundary and a critically important periodization boundary.

The greatest of the mass extinctions, close to the Permian-Triassic boundary (251 Ma in the 2004 time scale) has been harder to study because useful stratigraphic sections are very rare, occurring almost exclusively in China. The extinction is fairly abrupt, yet at the well-studied Meishan section in China, it is spread over ~25 m of section, corresponding to more than a million years (Jin et al., 2000). The extinction is recorded primarily in marine invertebrate fossils, which are less useful for long-range correlation than are conodonts—the microscopic jaws of otherwise soft-bodied, swimming vertebrates, which largely escaped the extinction (Sepkoski, 2002). The gradual character of the mass extinction and the greater utility of conodonts for correlation have led to a surprising situation in which the GSSP for the base of the Triassic has been placed at the first appearance of the fossil conodont *Hindeodus parvus*, slightly above the level of what has long been called the Permian-Triassic extinction. Thus, by definition, the greatest mass extinction in life history took place very late in the Permian, but not at the Permian-Triassic boundary. Nothing could show more clearly the difference between time-scale construction and periodization!

CONCLUSION

We conclude that it is important for historians of all kinds to distinguish carefully between dating events and establishing time scales, on the one hand, and periodizing, or dividing the past into manageable and useful chunks, on the other. Certainly, some of the concepts and distinctions we have discussed are subtle and slippery, and we hope this paper may generate discussion and exchange between the component disciplines of Big History and will lead to the sharpening and clarification of these important differences.

ACKNOWLEDGMENTS

We thank Robert Aunger and David Christian for very helpful reviews and comments, and Christian Koeberl for his work as editor of the volume. The Ancona Penrose Conference, convened by Alessandro Montanari, Christian Koeberl, Frits Hilgen, and Rodolfo Coccioni, provided an excellent stimulus for this paper.

REFERENCES CITED

Ager, D.V., 1973, The Nature of the Stratigraphical Record: New York, Wiley, 114 p.

Ager, D.V., 1981, The Nature of the Stratigraphical Record (2nd edition): New York, Wiley, 122 p.

Allègre, C., 1992, From Stone to Star: A View of Modern Geology: Cambridge, Massachusetts, Harvard University Press, 287 p.

Alvarez, W., 1991, The gentle art of scientific trespassing: GSA Today, v. 1, p. 29–34.

Alvarez, W., 1997, *T. rex* and the Crater of Doom: Princeton, New Jersey, Princeton University Press, 185 p.

Alvarez, W., 2009, The Mountains of Saint Francis: New York, W.W. Norton, 304 p.

Arenillas, I., Arz, J.A., Grajales-Nishimura, J.M., Murillo-Muñetón, G., Alvarez, W., Camargo-Zanoguera, A., Molina, E., and Rosales-Domínguez, C., 2006, Chicxulub impact event is Cretaceous/Paleogene boundary in age: New micropaleontological evidence: Earth and Planetary Science Letters, v. 249, p. 241–257, doi: 10.1016/j.epsl.2006.07.020.

Aunger, R., 2007a, Major transitions in 'big' history: Technological Forecasting and Social Change, v. 74, p. 1137–1163, doi: 10.1016/j.techfore.2007.01.006.

Aunger, R., 2007b, A rigorous periodization of 'big' history: Technological Forecasting and Social Change, v. 74, p. 1164–1178, doi: 10.1016/j.techfore.2007.01.007.

Beckwith, S.V.W., Stiavelli, M., Koekemoer, A.M., Caldwell, J.A.R., Ferguson, H.C., Hook, R., Lucas, R.A., Bergeron, L.E., Corbin, M., Jogee, S., Panagia, N., Robberto, M., Royle, P., Somerville, R.S., and Sosey, M., 2006, The Hubble Ultra Deep Field: The Astronomical Journal, v. 132, p. 1729–1755, doi: 10.1086/507302.

Brown, C.S., 2007, Big History: From the Big Bang to the Present: New York, New Press, 288 p.

Brown, R.E., Koeberl, C., Montanari, A., and Bice, D.M., 2009, this volume, Evidence for a change in Milankovitch forcing caused by extraterrestrial events at Massignano, Italy, *in* Koeberl, C., and Montanari, A., eds., The Late Eocene Earth—Hothouse, Icehouse, and Impacts: Geological Society of America Special Paper 452, doi: 10.1130/2008.2452(08).

Chaisson, E., 2001, Cosmic Evolution: The Rise of Complexity in Nature: Cambridge, Massachusetts, Harvard University Press, 274 p.

Chaisson, E., 2006, Epic of Evolution: Seven Ages of the Cosmos: New York, Columbia University Press, 478 p.

Christian, D., 2004, Maps of Time. An Introduction to Big History: Berkeley, University of California Press, 642 p.

Christian, D., 2008, History, complexity and the chronometric revolution: Revista de Occidente (Madrid, Spain), no. 323, p. 27–57.

Cita, M.B., 2009, Mediterranean Neogene stratigraphy: Development and evolution through the centuries, *in* Bernoulli, D., Cita, M.B., and McKenzie, J., eds., Major Discoveries in Sedimentary Geology: International Association of Sedimentologists (in press).

Claeys, P., Kiessling, W., and Alvarez, W., 2002, Distribution of Chicxulub ejecta at the Cretaceous-Tertiary boundary, *in* Koeberl, C., and MacLeod, K.G., eds., Catastrophic events and mass extinctions: Impacts and Beyond: Geological Society of America Special Paper 356, p. 55–68.

Cleaveland, L.C., Jensen, J., Goese, S., Bice, D.M., and Montanari, A., 2002, Cyclostratigraphic analysis of pelagic carbonates at Monte dei Corvi (Ancona, Italy) and astronomical correlation of the Serravallian-Tortonian boundary: Geology, v. 30, p. 931–934, doi: 10.1130/0091-7613(2002)030<0931:CAOPCA>2.0.CO;2.

Cloud, P., 1978, Cosmos, Earth, and Man: A Short History of the Universe: New Haven, Yale University Press, 372 p.

Coccioni, R., Marsili, A., Montanari, A., Bellanca, A., Neri, R., Bice, D.M., Brinkhuis, H., Church, N., Macalady, A., McDaniel, A., Deino, A., Lirer, F., Sprovieri, M., Maiorano, P., Monechi, S., Nini, C., Nocchi, M., Pross, J., Rochette, P., Sagnotti, L., Tateo, F., Touchard, Y., Van Simaeys, S., and Williams, G.L., 2008, Integrated stratigraphy of the Oligocene pelagic sequence in the Umbria-Marche Basin (northeastern Apennines, Italy): A potential global stratotype section and point (GSSP) for the Rupelian/Chattian boundary: Geological Society of America Bulletin, v. 120, p. 487–511, doi: 10.1130/B25988.1.

Cutler, A., 2003, The Seashell on the Mountaintop: A Story of Science, Sainthood, and the Humble Genius Who Discovered a New History of the Earth: New York, Dutton, 228 p.

Eldridge, N., and Gould, S.J., 1972, Punctuated equilibria: An alternative to phyletic gradualism, *in* Schopf, T.J.M., ed., Models in Paleobiology: San Francisco, Freeman, Cooper and Co., p. 82–115.

Farley, K.A., Montanari, A., Shoemaker, E.M., and Shoemaker, C.S., 1998, Geochemical evidence for a comet shower in the late Eocene: Science, v. 280, p. 1250–1253, doi: 10.1126/science.280.5367.1250.

Ferris, T., 1988, Coming of Age in the Milky Way: New York, Morrow, 495 p.

Gerhard, D., 1973–74, Periodization in history, *in* Wiener, P.P., ed., The Dictionary of the History of Ideas, Volume 3: New York, Charles Scribner's Sons, p. 477–482.

Glen, W., 1982, The Road to Jaramillo: Stanford, Stanford University Press, 459 p.

Gould, S.J., 1987, Time's Arrow, Time's Cycle: Cambridge, Massachusetts, Harvard University Press, 222 p.

Gradstein, F., Ogg, J., and Smith, A., eds., 2004, A Geologic Time Scale 2004: Cambridge, Massachusetts, Cambridge University Press, 589 p.

Hedberg, H.D., 1976, International Stratigraphic Guide: New York, John Wiley and Sons, 200 p.

Hilgen, F., Iaccarino, S., Krijgsman, W., Montanari, A., Raffi, I., Turco, E., and Zachariasse, W.-J., 2005, The global boundary stratotype section and point (GSSP) of the Tortonian Stage (Upper Miocene) at Monte dei Corvi: Episodes, v. 28, p. 1–12.

Hyland, E., Murphy, B., Varela, P., Marks, K., Colwell, L., Tori, F., Monechi, S., Cleaveland, L., Brinkhuis, H., van Mourik, C.A., Coccioni, R., Bice, D., and Montanari, A., 2009, this volume, Integrated stratigraphy and astrochronologic calibration of the Eocene-Oligocene transition in the Monte Cagnero section (northeastern Apennines, Italy): A potential parastratotype for the Massignano GSSP, *in* Koeberl, C., and Montanari, A., eds., The Late Eocene Earth—Hothouse, Icehouse, and Impacts: Geological Society of America Special Paper 452, doi: 10.1130/2009.2452(19).

Jin, Y.G., Wang, Y., Wang, W., Shang, Q.H., Cao, C.Q., and Erwin, D.H., 2000, Pattern of marine mass extinction near the Permian-Triassic boundary in South China: Science, v. 289, p. 432–436, doi: 10.1126/science.289.5478.432.

Jovane, L., Florindo, F., Sprovieri, M., and Pälike, H., 2006, Astronomic calibration of the late Eocene/early Oligocene Massignano section (central Italy): Geochemistry, Geophysics, Geosystems, v. 7, doi: 10.1029/2005GC001195.

Kuiper, K.F., Deino, A., Hilgen, F.J., Krijgsman, W., Renne, P.R., and Wijbrans, J.R., 2008, Synchronizing rock clocks of Earth history: Science, v. 320, p. 500–504, doi: 10.1126/science.1154339.

Laskar, J., Robutel, P., Joutel, F., Gastineau, M., Correia, A.C.M., and Levrard, B., 2004, A long term numerical solution for the insolation quantities of the Earth: Astronomy & Astrophysics, v. 428, p. 261–285, doi: 10.1051/0004-6361:20041335.

Lowrie, W., and Alvarez, W., 1981, One hundred million years of geomagnetic polarity history: Geology, v. 9, p. 392–397, doi: 10.1130/0091-7613(1981)9<392:OHMYOG>2.0.CO;2.

Mader, D., Cleaveland, L., Koeberl, C., Bice, D., and Montanari, A., 2004, High-resolution geochemical proxies of a short Langhian pelagic sequence at the Cònero Riviera, Ancona (Italy): Some palaeoenvironmental and cyclostratigraphic considerations: Palaeogeography, Palaeoclimatology, Palaeoecology, v. 211, p. 325–344, doi: 10.1016/j.palaeo.2004.06.001.

Milankovitch, M., 1941, Kanon der Erdbestrahlung und seine Anwendung auf das Eiszeitenproblem: Belgrade, Serbian Academy of Science, 633 p.

Miller, K.G., Browning, J.V., Aubry, M.-P., Wade, B.S., Katz, M.E., Kulpecz, A.A., and Wright, J.D., 2008, Eocene–Oligocene global climate and sea-level changes: St. Stephens Quarry, Alabama: Geological Society of America Bulletin, v. 120, p. 34–53, doi: 10.1130/B26105.1.

Mitchell, R.N., Bice, D.M., Montanari, A., Cleaveland, L.C., Christianson, K.T., Coccioni, R., and Hinnov, L.A., 2008, Oceanic anoxic events? Orbital prelude to the Bonarelli Level (OAE2): Earth and Planetary Science Letters, v. 267, p. 1–16, doi: 10.1016/j.epsl.2007.11.026.

Montanari, A., and Koeberl, C., 2000, Impact Stratigraphy: The Italian Record (Lecture notes in Earth science, v. 93): Berlin, Springer, 364 p.

Montanari, A., Langenheim, V.E., and Coccioni, R., 1988, Stratigraphy and geochronologic potential of the pelagic and hemipelagic sequence of the northeastern Apennines: A research note: IGCP Project No. 196, Bulletin de Liason et Informations, v. 7, p. 17–23.

Montanari, A., Drake, R., Bice, D.M., Alvarez, W., Curtis, G.H., Turrin, B.D., and DePaolo, D.J., 1985, Radiometric time scale for the Upper Eocene and Oligocene based on K/Ar and Rb/Sr dating of volcanic biotites from the pelagic sequence of Gubbio, Italy: Geology, v. 13, p. 596–599, doi: 10.1130/0091-7613(1985)13<596:RTSFTU>2.0.CO;2.

Montanari, A., Deino, A., Coccioni, R., Langenheim, V.E., Capo, R., and Monechi, S., 1991, Geochronology, Sr isotope stratigraphy, magnetostratigraphy, and plankton stratigraphy across the Oligocene-Miocene boundary in the Contessa section (Gubbio, Italy): Newsletters in Stratigraphy, v. 23, p. 151–180.

Montanari, A., Beaudoin, B., Chan, L.S., Coccioni, R., Deino, A., DePaolo, D.J., Emmanuel, L., Fornaciari, E., Kruge, M., Lindblad, S., Mozzato, C., Portier, E., Renard, M., Rio, D., Sandroni, P., and Stankiewicz, A., 1997, Integrated stratigraphy of the middle to upper Miocene pelagic sequence of the Cònero Riviera (Marche Region, Italy), *in* Montanari, A., Odin, G.S., and Coccioni, R., eds., Miocene Stratigraphy: An Integrated Approach: Amsterdam, Elsevier, p. 409–450.

Potter, T.W., ed., 1976, A Faliscan Town in South Etruria: Excavations at Narce, 1966–71: Rome, British School of Rome, 352 p.

Premoli Silva, I., Coccioni, R., and Montanari, A., eds., 1988, The Eocene-Oligocene Boundary in the Marche-Umbria Basin: International Union of Geological Sciences (IUGS) Special Publication: Ancona, F.lli Aniballi Publishers, 268 p.

Salvador, A., 1994, International Stratigraphic Guide: A Guide to Stratigraphic Classification, Terminology, and Procedure (2nd edition): Boulder, Colorado, Geological Society of America, 214 p.

Salvador, A., 2006, A stable Cenozoic geologic time scale is indispensable: Episodes, v. 29, p. 43–45.

Sepkoski, J.J., Jr., 1997, Biodiversity: Past, present, and future: Journal of Paleontology, v. 71, p. 533–539.

Sepkoski, J.J., Jr., 2002, A Compendium of Fossil Marine Animal Genera, edited by D. Jablonski and M. Foote: Bulletin of American Paleontology, no. 363, 560 p.

Smit, J., and Romein, A.J.T., 1985, A sequence of events across the Cretaceous-Tertiary boundary: Earth and Planetary Science Letters, v. 74, p. 155–170, doi: 10.1016/0012-821X(85)90019-6.

Spier, F., 1996, The Structure of Big History: Amsterdam, Amsterdam University Press, 113 p.

Steno, N., 1669, De solido intra solidum naturaliter contento dissertationis prodromus: Florence, *in* Winter, J.G., trans., 1916, The Prodromus of Nicholas Steno's Dissertation: New York, Macmillan, University of Michigan Studies, Humanistic Series, v. 11, pt. 2, p. 169–283.

Walsh, S., Gradstein, F., and Ogg, J., 2004, History, philosophy, and application of the global stratotype section and point (GSSP): Lethaia, v. 37, p. 201–218, doi: 10.1080/00241160410006500.

Manuscript Accepted by the Society 16 September 2008

The Geological Society of America
Special Paper 452
2009

Late Eocene impact craters and impactoclastic layers—An overview

Christian Koeberl*
Department of Lithospheric Research, University of Vienna, Althanstrasse 14, A-1090 Vienna, Austria, and
Osservatorio Geologico di Coldigioco, I-62020 Frontale di Apiro (MC), Italy

ABSTRACT

Multiple bolide impact events, possibly related to a comet or asteroid shower over a duration of ~2–3 m.y., may have played an important role in the deterioration of the global climate at the end of the Eocene. Upper Eocene marine sediments around the world contain evidence for at least two closely spaced impactoclastic layers, i.e., layers containing impact debris such as tektites and microtektites, shocked minerals, and rock fragments. The upper layer correlates with the North American tektite strewn field (mostly on the eastern side of North America), and the 85-km-diameter Chesapeake Bay crater (USA) has been suggested as its source crater, whereas the lower, microkrystite layer (with clinopyroxene-bearing spherules) was most likely derived from the 100-km-diameter Popigai impact crater (Russia). In summary, at least five impact structures with late Eocene ages are known. Disturbances in the climate at that time are documented, and connection with the impact events is likely. This contribution provides a short review of late Eocene impact craters and ejecta layers.

Keywords: late Eocene impacts, ejecta, Chesapeake Bay crater, Popigai crater, tektites.

INTRODUCTION

The late Eocene was a time interval of major changes characterized by an accelerated global cooling, with a sharp temperature drop of ~2 °C near the Eocene-Oligocene (E-O) boundary, and significant stepwise floral and faunal turnovers (see papers in this volume). These global climate changes, which are recorded by a gradual increase of marine oxygen isotope values and biotic crises, are commonly attributed to the expansion of the Antarctic ice cap following its gradual isolation from other continental masses. However, multiple bolide impact events documented from this time (e.g., Farley et al., 1998; Montanari and Koeberl, 2000, and references therein) may have played an important role in the deterioration of the global climate during the end of the Eocene Epoch (e.g., Vonhof et al., 2000; Bodiselitsch et al., 2004). It is clear, however, that the major climate change at the Eocene-Oligocene boundary, ~1.5 m.y. later, was most likely unrelated to the late Eocene impact events (e.g., Zanazzi et al., 2007; Katz et al., 2008; Miller et al., this volume). Also, the eruption of the Ethiopian Traps happened even later, in the early Oligocene (Touchard et al., 2002), and is thus completely unrelated to the late Eocene climate events.

There is abundant evidence for multiple impact events in the late Eocene, in terms of both impact craters and ejecta layers (Figs. 1 and 2). The first ejecta layer to be recognized was the North American tektite strewn field, with locations on or near the eastern part of the North American continent, which contains both tektites and microtektites (see following). Later, other (mostly recrystallized) microspherules of almost the same age were found in deep-sea sediments at a variety of locations around the world. Both layers were originally thought be in the middle to the early part of magnetic chron C15, although it now appears

*christian.koeberl@univie.ac.at.

Koeberl, C., 2009, Late Eocene impact craters and impactoclastic layers—An overview, *in* Koeberl, C., and Montanari, A., eds., The Late Eocene Earth—Hothouse, Icehouse, and Impacts: Geological Society of America Special Paper 452, p. 17–26, doi: 10.1130/2009.2452(02). For permission to copy, contact editing@geosociety.org.

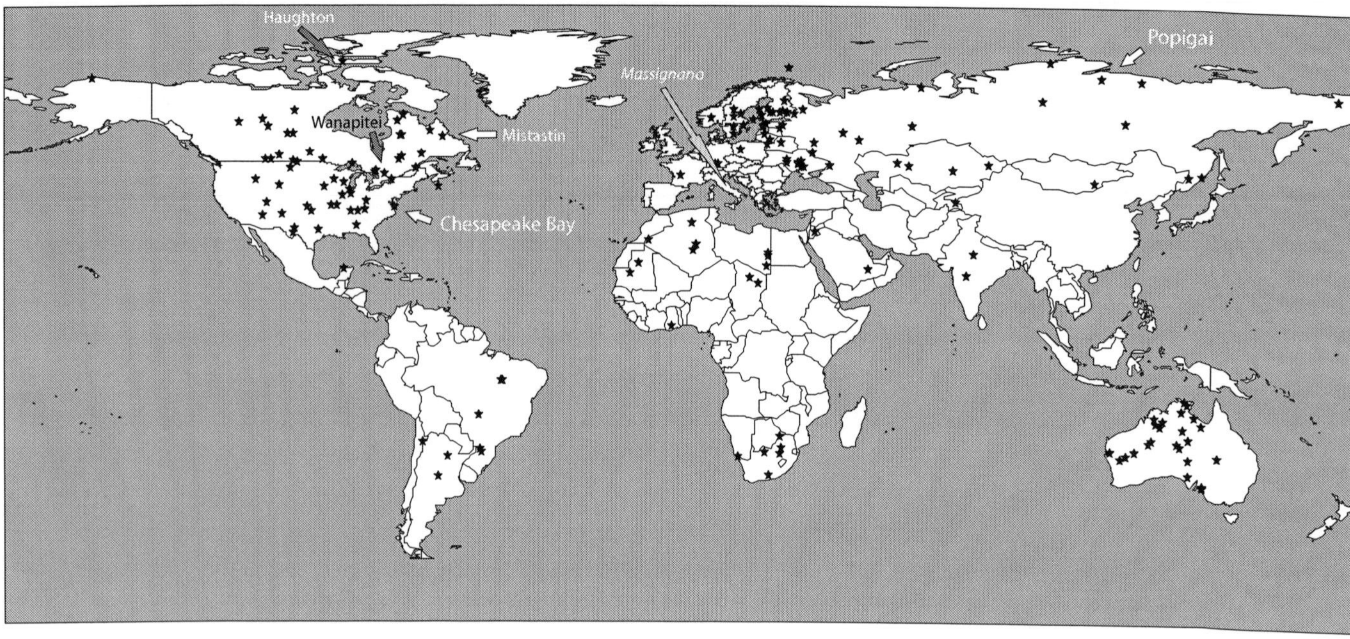

Figure 1. Late Eocene impact craters and structures (marked) on a global map of all well-documented craters and impact structures (crater coordinates modified from the Earth Impact Database at http://www.unb.ca/passc/ImpactDatabase/). Arrow points to the Eocene-Oligocene global stratotype section and point (GSSP) at Massignano, Italy.

that the older layer is in chron C16n. Some of these deposits have been known for much of the twentieth century but were only confirmed as being of impact origin within the past 40 yr or so. Research over the past 20 yr has led to the identification of source craters for these ejecta deposits. Thus, there are currently at least five impact craters known that are of late Eocene age, representing a relatively large number within a short time span. The two largest impact structures are Popigai, Russia, and Chesapeake Bay, USA, with respective diameters of 100 and 85 km, and there are several smaller craters with similar ages (see Fig. 3 for a time line). The present contribution provides a brief review of the late Eocene ejecta layers and coeval impact structures, and possible connections to environmental changes.

TEKTITES, MICROTEKTITES, AND MICROKRYSTITES

The North American tektite strewn field includes tektites found on land and microtektites found in deep-sea sediments. In this strewn field, tektites, which are glassy distal impact ejecta (cf. Koeberl, 1986, 1994, 2007) found on land, occur in two main areas, namely the bediasites in Texas, named after the town of Bedias, in the northeast corner of Grimes County, ~20 miles west of Huntsville, and the georgiaites in Georgia (e.g., Albin et al., 2000). In addition, two samples have been found in Cuba and Martha's Vineyard, respectively, but it is uncertain if these are genuine locations. In addition, tektite fragments were found underwater at Deep Sea Drilling Project (DSDP) Site 612 (see following), and microtektites were found on land at Barbados. Microtektites have been found in late Eocene marine sediments on Barbados and in sediment cores recovered from the Gulf of Mexico, Caribbean Sea, and northwestern Atlantic Ocean (e.g., Glass and Zwart, 1979; Glass et al., 1985, 1998; Sanfilippo et al., 1985; D'Hondt et al., 1987; Keller et al., 1987; Glass, 2002). On the basis of stratigraphic, compositional, isotopic, and age data, they were identified to be part of the North American tektite field. Chemical analyses (Koeberl and Glass, 1988), Sr and Nd isotopic studies (Ngo et al., 1985), and age data (Glass et al., 1986) showed that, without doubt, the Barbados tektites are North American tektites. From their Rb-Sr/Sm-Nd isotopic studies, Shaw and Wasserburg (1982) concluded that the source rocks from which the North American tektites were derived was crustal material that formed very late in the Precambrian (Sm-Nd model age = 0.62–0.67 Ga), which excludes most of the Precambrian shield areas of North America as well as sediments derived from these areas. The Rb-Sr data are in agreement with derivation from a sedimentary source rock.

The North American microtektites are associated with unmelted impact ejecta including shocked quartz and feldspar with multiple sets of planar deformation features, coesite, stishovite, and reidite (a high-pressure polymorph of zircon) (Glass, 1989; Glass and Wu, 1993; Glass et al., 1998; Glass and Liu, 2001). In 1987, microtektites, tektite fragments, and impact debris were discovered in a relatively thick layer in a core at DSDP Site 612 (e.g., Thein, 1987). The chemical composition of the DSDP Site 612 tektites and microtektites is similar to that of other North American tektites, but there are some important differences, e.g., lower Na and higher K contents in the 612 tektites (Koeberl and

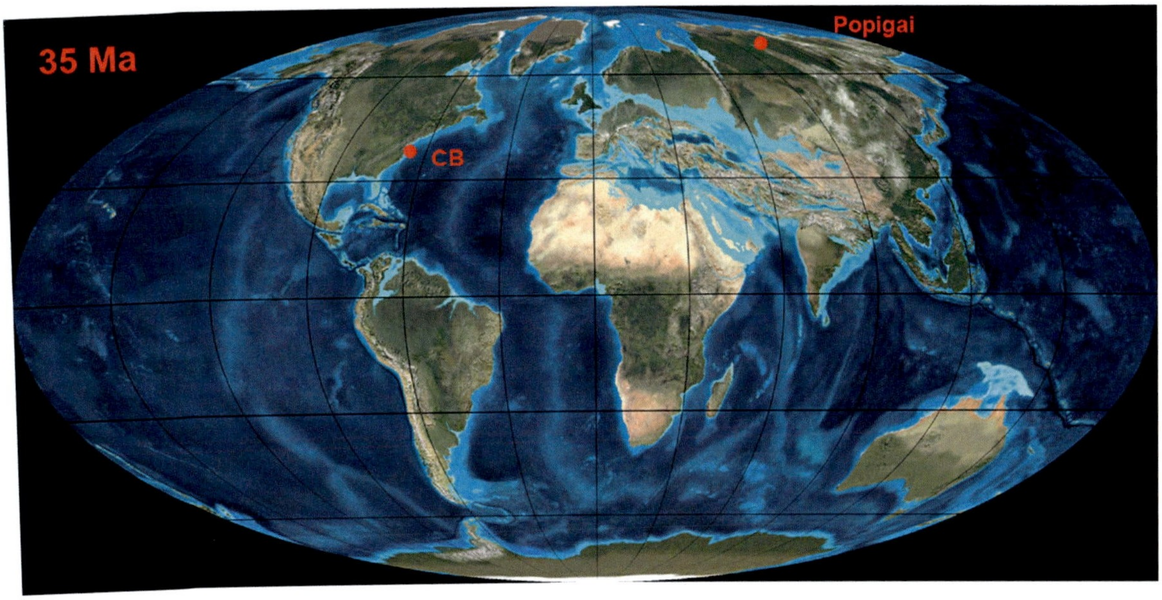

Figure 2. Paleogeographic reconstruction of the plate tectonic setting at 35 Ma (from http://jan.ucc.nau.edu/~rcb7/35moll.jpg; for explanation: http://jan.ucc.nau.edu/~rcb7/globaltext2.html), around the time of formation of the late Eocene impact craters. The locations of the Chesapeake Bay (CB) and Popigai impact structures are indicated.

Glass, 1988). Stecher et al. (1989) found that the Rb-Sr/Sm-Nd isotopic composition of the 612 tektites shows a much wider scatter than that of the North American tektites (but they may still be on a mixing line with the other tektites). Radiometric dating of the DSDP 612 and Barbados tektite fragments indicated an age of ca. 35 Ma for this layer (Glass et al., 1986; Obradovich et al., 1989). This age is consistent with the fission-track, K-Ar, and ^{40}Ar-^{39}Ar ages obtained for the North American tektites (e.g., Zähringer, 1963; Storzer and Wagner, 1971; Albin and Wampler, 1996). The Chesapeake Bay structure has been suggested to be the source of this strewn field based on geographic location, age, composition, and isotopic data (Poag et al., 1994; Koeberl et al., 1996) (see next section for more detailed discussion).

Clinopyroxene-bearing spherules (cpx spherules; microkrystites) are closely associated with the North American microtektites at some locations. Thus, occurrences of cpx spherules in the Indian Ocean and equatorial Pacific were originally interpreted to indicate a major extension of the North American strewn field. However, at a few sites, it is clear that the cpx spherules are slightly older than the North American microtektites, and it is now accepted that they belong to a different impact event. By now, it is clear that the late Eocene microkrystite layer occurs at numerous sites around the world and may be global in geographic extent (Glass et al., 1985, 1998; Keller et al., 1987; Vonhof and Smit, 1999; Glass and Koeberl, 1999a, 1999b; Glass, 2002; Liu et al., 2006). In locations where the North American microtektite layer is present, the cpx spherule layer occurs just below it, but in many cases, the two layers overlap. Based on biostratigraphic data, the cpx spherule layer appears to be ~10–20 k.y. older than the North American microtektite layer (Glass et al., 1998), although a shorter time interval of ~5000 yr is also possible (Glass and Koeberl, 1999a). In the cpx spherules, the most abundant crystalline phase is clinopyroxene, but Ni- and Cr-rich spinel crystals are also present in many of the spherules (e.g., Glass et al., 1985; Glass and Burns, 1987). The cpx spherule layer is, in some cases, associated with an enrichment if the element Ir (e.g., Glass et al., 1985; Keller et al., 1987; Vonhof and Smit, 1999; Kyte and Liu, 2002) and with the extinction of several species of radiolaria (Glass et al., 1985; Sanfilippo et al., 1985). Both the microtektites from the North American strewn field and microkrystites from the cpx-strewn field have major- and trace-element compositions that are close to those of the upper continental crust (e.g., Koeberl and Glass, 1988; Glass et al., 2004a).

Based on similarity in age and Sr and Nd isotopic compositions, it has been suggested (e.g., Whitehead et al., 2000; Deutsch and Koeberl, 2006) that the source crater for the cpx spherule layer may be the 35.7 ± 0.8 Ma (Bottomley et al., 1997), 100-km-diameter Popigai impact crater in northern Siberia. Additional evidence for Popigai as the source of the cpx spherules is found at the Eocene-Oligocene global stratotype section and point (GSSP) at Massignano, Italy, where at the 5.61 m level, shocked quartz and pancake-shaped smectite spherules containing Ni- and Cr-rich spinel crystals are associated with a positive Ir anomaly in late Eocene deposits that have the same age, as far as can be determined, as the cpx spherule layer (Montanari et al., 1993; Clymer et al., 1996; Pierrard et al., 1998). The pancake spherules appear to be diagenetically altered and flattened cpx spherules (Glass et al., 2004b). Langenhorst (1996) suggested that the nature of the shocked quartz is consistent with derivation of this layer (and thus the cpx spherules) from the Popigai

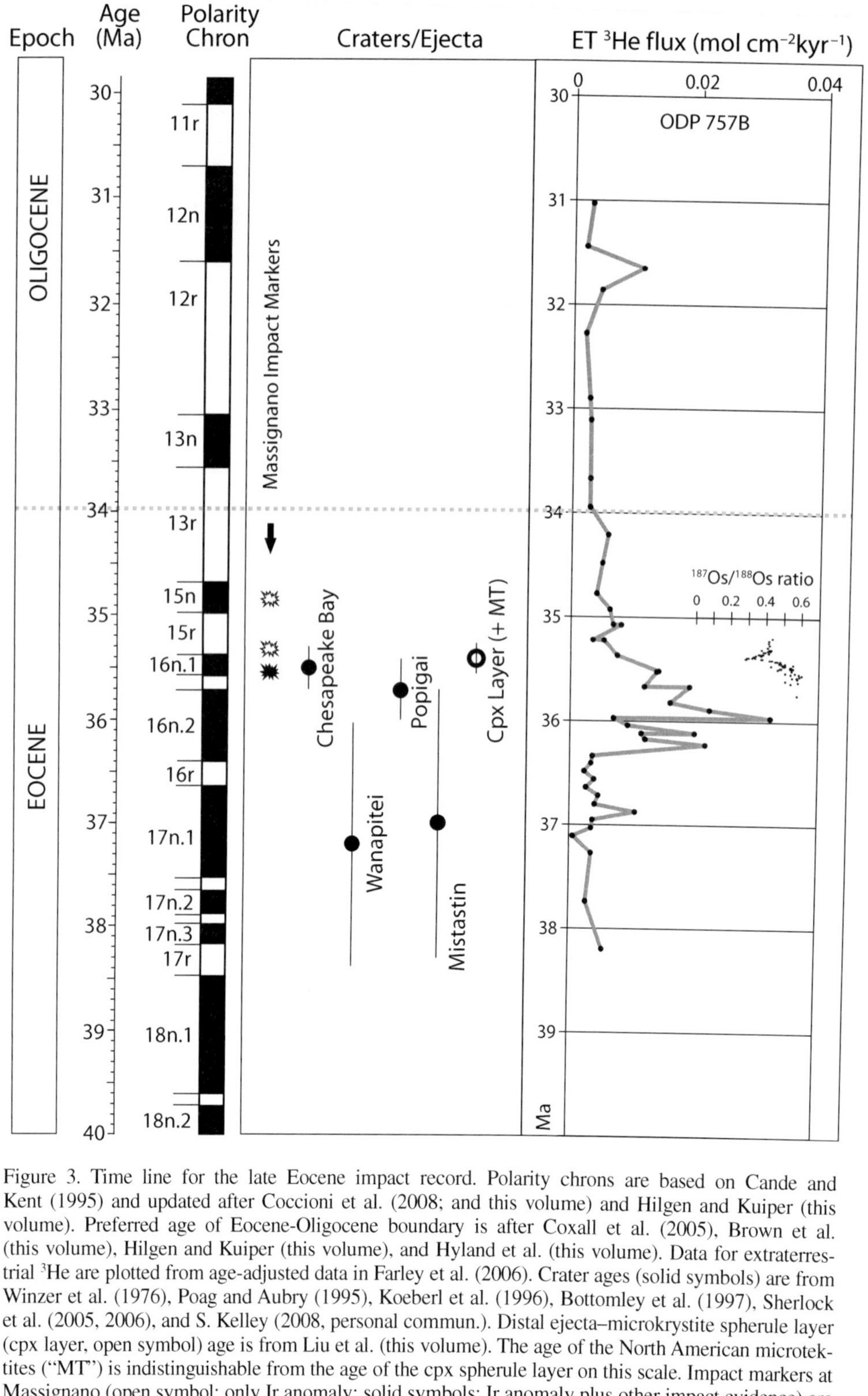

Figure 3. Time line for the late Eocene impact record. Polarity chrons are based on Cande and Kent (1995) and updated after Coccioni et al. (2008; and this volume) and Hilgen and Kuiper (this volume). Preferred age of Eocene-Oligocene boundary is after Coxall et al. (2005), Brown et al. (this volume), Hilgen and Kuiper (this volume), and Hyland et al. (this volume). Data for extraterrestrial ^{3}He are plotted from age-adjusted data in Farley et al. (2006). Crater ages (solid symbols) are from Winzer et al. (1976), Poag and Aubry (1995), Koeberl et al. (1996), Bottomley et al. (1997), Sherlock et al. (2005, 2006), and S. Kelley (2008, personal commun.). Distal ejecta–microkrystite spherule layer (cpx layer, open symbol) age is from Liu et al. (this volume). The age of the North American microtektites ("MT") is indistinguishable from the age of the cpx spherule layer on this scale. Impact markers at Massignano (open symbol: only Ir anomaly; solid symbols: Ir anomaly plus other impact evidence) are from Bodiselitsch et al. (2004) and Coccioni et al. (this volume). In addition, the Os isotopic compositions of deep-sea sediments from Ocean Drilling Program (ODP) Site 1090 are shown (Paquay et al., 2008). Due to differences in the polarity chron measurements at different locations, it is not easy to correlate the stratigraphic positions of the various impact markers across the world. In the case of Massignano, the temporal location of the closely spaced impacts is in lowermost part of foraminiferal zone E15, across the short transitional interval between C16n2 and C16n1 (A. Montanari, 2008, personal commun.).

crater. At Massignano, a second Ir anomaly exists just above the one that is correlative with the pancake spherules, which might be associated with the North American strewn field (Montanari et al., 1993; Montanari and Koeberl, 2000). This second Ir anomaly at Massignano was confirmed by Bodiselitsch et al. (2004). The age of the 5.61 m Ir anomaly was determined to be 35.7 ± 0.4 Ma, by interpolation from several dated volcanic ashes found in the same section (cf. Montanari and Koeberl, 2000). According to the astrochronologic calibration by Brown et al. (this volume), which indicates a mean sedimentation rate for the Massignano section of 8.7 m/m.y., the second Ir peak at 6.19 m would have followed the previous by some 90,000 yr (i.e., close to a short eccentricity cycle), which is a significantly larger time difference than the 5000–20,000 yr of separation estimated between the North American microtektites and the cpx-spherule layer at North American deep-sea sediment sites.

The number of late Eocene microtektite/spherule layers has been a matter of debate, with estimates ranging from two (Glass et al., 1985; Glass and Burns, 1987), to three (Keller et al., 1987), to six or more (Hazel, 1989). Most of the debate is concerned with the number of cpx spherule layers, but Miller et al. (1991) proposed that the microtektite layer at Site 612, off New Jersey, is older than the microtektite layer on Barbados. However, most authors now agree that there are just two layers: (1) the North American microtektite layer and (2) the cpx spherule layer (Glass, 1990; Wei, 1995; Vonhof and Smit, 1999; Montanari and Koeberl, 2000; Whitehead et al., 2000; Marchand and Whitehead, 2002). A detailed review is provided by Glass (2002).

LATE EOCENE IMPACT STRUCTURES: CHESAPEAKE BAY, POPIGAI, AND OTHERS

As noted already, the source of the North American tektite strewn field has now been linked, with a certain degree of confidence, to the 85-km-diameter Chesapeake Bay impact structure (Koeberl et al., 1996; Deutsch and Koeberl, 2006). The size of the Chesapeake Bay structure is generally given as 85 km (Poag et al., 2004); however, some authors suggest that the diameter of the crater before resurge may have been only ~40 km (e.g., Collins and Wünnemann, 2005). Nevertheless, this does not change the fact that the current, measurable crater diameter is 85 km. An impact event that created a crater of this size would be capable of globally distributing its distal ejecta. The late Eocene Chesapeake Bay impact structure is among the largest and best preserved of the known impact craters on Earth (Poag et al., 2004), and it is the largest impact structure currently identified in the United States (Fig. 3). The Chesapeake Bay structure is unique among subaerial and submarine impact craters on Earth because: (1) it is a relatively young structure and, in comparison to other known impact structures of such size, very well preserved; (2) its location on a passive continental margin has prevented the tectonic or orogenic disruption or distortion that is typical of many large terrestrial craters; (3) its original location on a relatively deep continental shelf allowed marine deposition to resume immediately following the impact, which buried the crater rapidly and completely, thereby preventing subsequent erosion; (4) the upper part of the breccia section inside the crater was derived from resurge currents and impact-generated tsunami waves; (5) the breccia body contains a large volume of impact-generated brine; and (6) the crater underlies a densely populated urban corridor whose two million citizens are still affected by crater-related phenomena, such as freshwater availability. Besides glasses and impact debris of the North American tektite strewn field, Chesapeake Bay impact ejecta have also been identified (in the form of shocked minerals) in a quarry in Georgia (Harris et al., 2004).

There is a second large crater with a radiometric age that is indistinguishable from that of the Chesapeake Bay structure and the two ejecta layers, namely the 100-km-diameter Popigai impact structure in Siberia, which has been dated at 35.7 ± 0.8 Ma. The Popigai structure is exposed in Archean crystalline rocks of the Anabar Shield, with overlying Proterozoic to Mesozoic sedimentary sequences, and it is the largest Cenozoic crater on Earth. As noted previously, it is now commonly assumed that the global late Eocene microkrystite layer originated from the Popigai impact event. A detailed geochemical and Sr/Nd isotope survey of target and impact melt lithologies finally established that the very broad range of Popigai target lithologies indeed represents the precursor material for melanocratic and leucocratic microkrystites as well as for the associated microtektites (Kettrup et al., 2003). Also, the isotope compositions and model parameters of the ejecta material allowed Kettrup et al. (2003) to constrain their origin from the uppermost layers at lithologically different and geographically defined parts of the Popigai target area (cf. Whitehead et al., 2002). This explains why the microtektites and microkrystites of the cpx spherule layer display a wider variation in compositions than tektites from other strewn fields.

OTHER LATE EOCENE–AGE CRATERS

There are several smaller late Eocene craters. The locations of the craters are shown in Figure 1, the ages—in comparison to distal ejecta markers and polarity chrons—are plotted in Figure 3, and satellites images of the craters are shown in Figure 4. Mistastin Crater in central Labrador, Canada, is dated at 38 ± 4 Ma (Mak et al., 1976), although Grieve (2006) cited a recalculated age of 36 ± 4 Ma; it is unclear how this age was recalculated. In Figure 3, I use a value of 37.0 ± 2.6 Ma (based on preliminary new Ar-Ar data; S. Kelley, 2008, personal commun.). Mistastin is a heavily eroded complex structure with a diameter of 28 km. Glacial erosion has imparted an eastward elongation to the crater that is particularly evident in the shape of the lake that occupies the central 10 km of the structure. Horseshoe Island, in the center of the lake, is part of the central uplift and contains shocked Precambrian crystalline target rocks. Outside the margins of the lake, there are remnants of the impact melt sheet (with columnar jointing), which contains evidence of shock metamorphism (e.g., Grieve, 1975). The age needs to be redetermined with more modern methods.

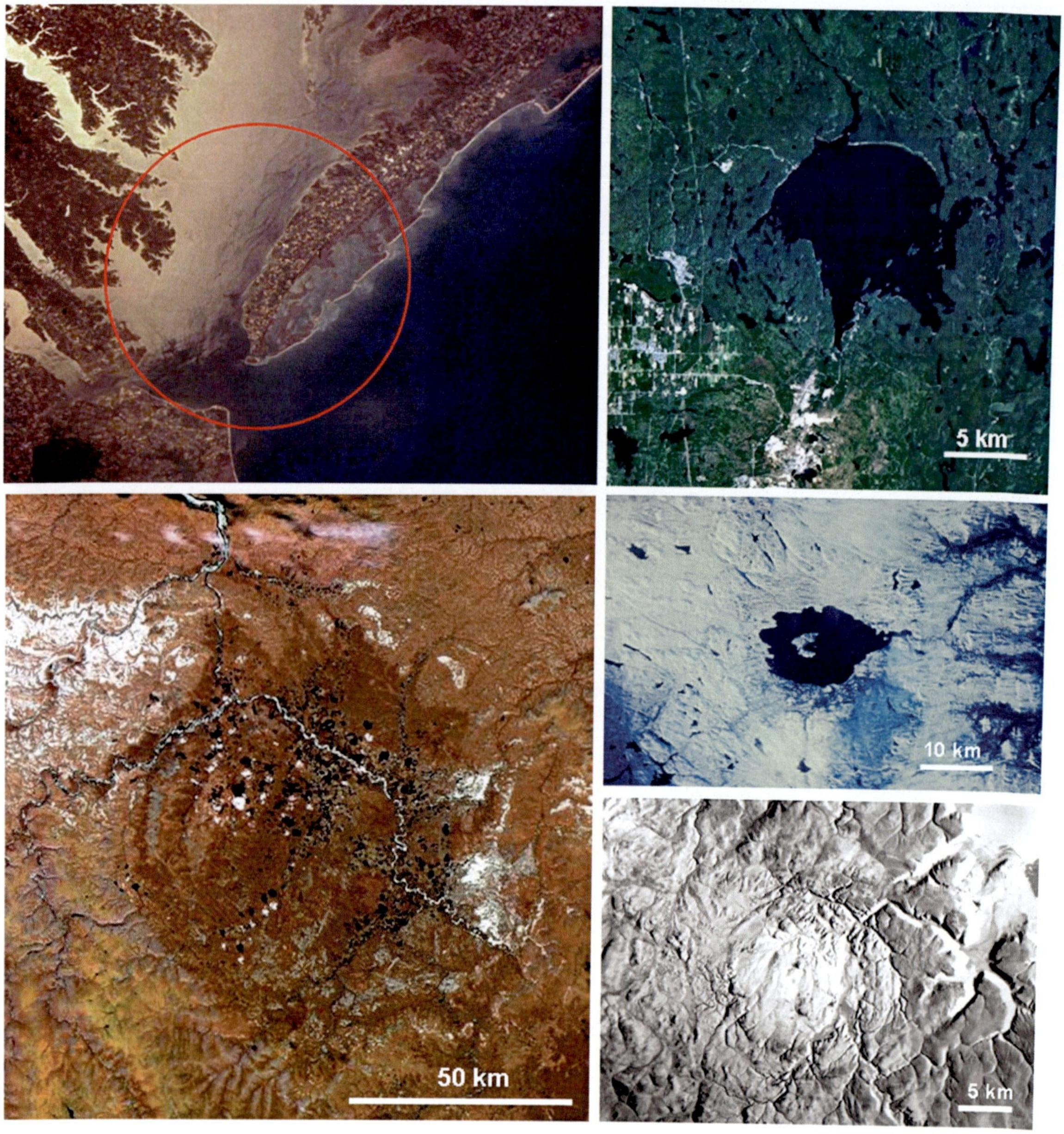

Figure 4. Overview of impact structures of (possibly) late Eocene age. Top left: Chesapeake Bay, USA (Space Shuttle image; circle indicates approximate dimension [85-km diameter] of the structure). Top right: Wanapitei, Canada (Landsat image). Bottom left: Popigai, Russia (Landsat image). Center right: Mistastin, Canada (Space Shuttle image). Bottom right: Haughton, Canada (Radarsat image).

Wanapitei near Sudbury in Canada is dated at 37.2 ± 1.2 Ma (Winzer et al., 1976). The crater is usually listed to have a diameter of 7.5 km. However, it has recently been reclassified as a simple crater because there is no evidence of a central uplift in the submerged crater, and geophysical work by L'Heureux et al. (2005) indicates that if the observed circular structure is due to a meteorite impact, it is at most a 3–4-km-diameter simple crater. This is still an ongoing debate. Wanapitei Lake, which probably fully encompasses the impact structure, has a semicircular shape along the north coast and a highly indented southern margin (Fig. 3). Target rocks are Precambrian crystalline and metasedimentary units of the Southern Structural Province Archean gneisses, Huronian Gowganda (metasedimentary rocks) and Mississagi Formations. They are intruded by Nipissing and younger diabase dikes. The outline of the lake has been enlarged and modified by erosion (e.g., Lazorek et al., 2006; Ugalde et al., 2006).

Petrographic evidence for an impact origin at Wanapitei comes from rock samples that demonstrate shock metamorphic effects (quartzite fragments and the presence of glass) found in glacial drift on the southern shores of the lake (Grieve and Ber,

1994). These include boulders of suevite and glassy breccia, as well as samples of coesite (e.g., Dence et al., 1974; Dressler et al., 1997). There are only a few samples that indicate these shock metamorphic features, none of which were found in their natural or original position or place (Dressler et al., 1997). Shatter cones have been found on certain islands in the southern part of the lake as well as on the shore, but these cannot be unequivocally attributed to the Wanapitei structure due to the close proximity of the Sudbury impact structure.

Sherlock et al. (2005) reevaluated the published age information for the Haughton impact structure, which was believed to have formed close to 23 Ma, during the Miocene age, and reported new late Eocene Ar/Ar laser probe data from shocked basement clasts that are at odds with the published Miocene stratigraphic, apatite fission-track, and Ar/Ar data. Sherlock et al. (2005) found that the age of the Haughton impact structure is close to 39 Ma. This age is not yet confirmed, as the reason for the younger fission-track ages and stratigraphic ages (which agree with each other) are not clear yet, but it is at least conceivable that Haughton is another late Eocene–age impact structure.

The 17-km-diameter Logoisk (Belarus) impact structure, which was earlier dated at 40 ± 5 Ma, has now been provisionally redated at 42.3 ± 1.1 Ma (Sherlock et al., 2006) and thus may not be part of the same late Eocene event.

A CLUSTER OF LATE EOCENE IMPACTS?

Significantly enhanced levels of ^{3}He were found to coincide with the two Upper Eocene impactoclastic layers. This isotope is a proxy for the influx of extraterrestrial dust and is interpreted as indicating that, during the late Eocene, there was a time of enhanced collision activity in the inner solar system, probably resulting in a higher impact rate than usual (e.g., Farley et al., 1998). The data were interpreted to represent a 2.2 m.y. interval of increased flux of interplanetary dust particles (IDPs). The observed higher impact rate and the enhanced flux of interplanetary dust particles during that time can be explained by (1) the arrival of long-period comets to the center of the solar system, resulting from a perturbation of the Oort cloud (Farley et al., 1998), or (2) the occurrence of an asteroid shower, triggered by a major collision in the asteroid belt (Tagle and Claeys, 2004).

High-resolution studies (δ^{13}C, δ^{18}O, and elemental abundances) were conducted by Bodiselitsch et al. (2004) in rocks at and below the E-O boundary GSSP boundary at Massignano, Italy. In addition to an earlier known Ir anomaly at 5.61 m, which is possibly linked to the Popigai impact event, they confirmed the presence of two additional Ir anomalies in the intervals from 6.00 to 6.40 m and from 10.00 to 10.50 m, with maximum values of 259 ± 32 ppt at 6.17 m, and 149 ± 24 ppt at 10.28 m, respectively. The lower Ir anomaly might be derived from the Chesapeake Bay impact event, whereas no impact event is known for the upper one. Similar δ^{13}C and δ^{18}O trends related to the two Ir anomalies indicate that the Ir anomaly at 10.28 m might be also derived from an impact into a continental shelf, similar to the Chesapeake Bay impact event. The δ^{18}O values decrease in the high-Ir layers to –1.16‰ and –1.17‰, respectively, and, together with negative shifts in δ^{13}C in the Ir-rich levels, they indicate a warm pulse superimposed on a general late Eocene cooling trend that is characterized by δ^{18}O values ranging between –0.6‰ and –0.4‰. The release of methane hydrate after an impact in a continental shelf or seafloor, or impacts of ^{12}C-rich comets during a 2.2 m.y. comet shower, respectively, could have produced these more negative carbon and oxygen excursions compared to the continuously decreasing trend documented through the whole late Eocene Massignano section (Bodiselitsch et al., 2004).

As discussed already, there is abundant evidence for the presence of an extraterrestrial component in late Eocene impact ejecta layers. In addition, the melt rocks of the Popigai impact structure show enrichments in characteristic siderophile trace elements, the ratios of which point toward an ordinary chondrite projectile, possibly of the L-chondrite type (e.g., Tagle and Claeys, 2005). Tagle and Claeys (2005) used regression techniques on platinum group element (PGE) data to derive evidence of an L-chondritic projectile for Popigai. Evans et al. (1993) used PGE data to indicate a nonspecified chondrite type for Wanapitei. Goderis et al. (2007) confirmed that the PGE characteristics of melt rocks from the 7.5 km Wanapitei structure point toward ordinary chondrites (possibly even L-chondrites), suggesting that the projectiles for these two impact structures were of the same type. It should be mentioned, however, that the exact projectile identification at Popigai (and possibly at Wanapitei) depends on not only on the quality and number of comparison data for the various meteorite groups, but also on the type of regression calculation used in interelement plots (see Farley, this volume, for a detailed discussion).

It is, therefore, in agreement with the currently available data that two of the late Eocene impact craters were thus formed by the same type of projectile. This hypothesis agrees with the chromium isotope analyses of several samples from Ocean Drilling Program (ODP) Site 709C (Kyte et al., 2004). The positive δ^{53}Cr results exclude carbonaceous chondrites, which make cometary sources less likely. Collisions in the main asteroid belt would result in an increase of the terrestrial impact rate in the form of asteroid showers lasting 2–30 m.y. (Zappalà et al. 1998). This would also be consistent with an asteroid shower between ca. 37 and 34 Ma. However, the association of the layers studied by Kyte et al. (2004) with either Popigai or Chesapeake Bay is not exactly clear. Independent confirmation of a meteoritic component in the late Eocene ejecta layers comes from a study by Paquay et al. (2008), who used the Os isotopic record of marine sediments to derive the projectile sizes. In sediments from ODP Sites 1090 and 1219, these authors also found a decrease in the ^{187}Os/^{188}Os ratio, and a correlated increase in the Ir concentrations, in the age range 35.3–35.6 Ma, with strong peaks in both signals at around 35.43 Ma, and possible secondary peaks at 35.5 Ma. These signals could conceivably be from the Popigai and Chesapeake Bay impacts, respectively.

At the Chesapeake Bay impact structure, the situation is not clear yet. The North American tektites do not seem to contain any distinct enrichment in siderophile elements, and analyses of drill core samples from within the structure have yielded a range of results. Most analyses (e.g., Poag et al., 2004) were below the 1 ppb detection limit for Ir. The highest content of Ir found by Lee et al. (2006) in a clast of impact melt rock from the Sustainable Technology Park (STP) test hole was 0.466 ppb. Lee et al. (2006) reported on osmium isotope ratios and PGE concentrations of impact-melt rocks. They found that the $^{187}Os/^{188}Os$ ratios of impact-melt rocks ranged from 0.151 to 0.518, and that PGE concentrations of some of these rocks were much higher than concentrations in basement gneiss. Together with the osmium isotopes ratios, these data indicate a measurable meteoritic component (0.01%–0.1%) in some impact-melt rocks. However, because the PGE abundances in the impact-melt rocks are dominated by the target materials, interelemental ratios of the impact-melt rocks are highly variable and nonchondritic (Lee et al., 2006). Due to the limitations of the Os isotopic method, the projectile type for the Chesapeake Bay impact structure cannot be constrained by these analyses.

CONCLUSIONS

There is abundant evidence for two large and a number of still poorly constrained (at least two, possibly three or four) smaller impact events during the late Eocene. This coincides with He isotopic evidence of an increased flux of extraterrestrial dust onto Earth over an ~2 m.y. period during this time. The source of the enhanced impact rate and accretion flux is not yet clear, even though the He isotope data were initially interpreted to indicate a comet shower. However, chemical data on the projectile compositions for several of the late Eocene impact craters indicate L-chondritic compositions, which would be difficult to reconcile with a comet source. Disturbances in the climate and, possibly, in cyclostratigraphy coinciding with the impact events have been documented (Brown et al., this volume). At least two important conclusions can be drawn from the study of the late Eocene impact events. First, it is now possible, with the increased resolution of stratigraphic studies and detailed geochemical, isotopic, paleontologic, and petrophysical studies of the rock strata, to determine the global and regional effects of impact events on the Earth system. Second, even a series of fairly large impact events did not lead to any significant mass extinctions, constraining the lower threshold for global biological extinction events above the combined energy of the cluster of late Eocene impacts.

ACKNOWLEDGMENTS

The present work was supported by the Austrian Science Foundation (FWF) project P18862-N10 (to Koeberl). I am grateful to my Penrose co-organizers Sandro Montanari, Rodolfo Coccioni, and Frits Hilgen for lots of help and the opportunity to work together on a fascinating topic, and to J. Morrow and D.T. King for very helpful and rapid reviews.

REFERENCES CITED

Albin, E., and Wampler, J.M., 1996, New potassium-argon ages for georgiaites and the Upper Eocene Dry Branch Formation (Twiggs Clay Member): Inferences about tektite stratigraphic occurrence: Lunar and Planetary Science, v. 27, p. 5–6.

Albin, E.F., Norman, M.D., and Roden, M., 2000, Major and trace element compositions of georgiaites: Clues to the source of North American tektites: Meteoritics & Planetary Science, v. 35, p. 795–806.

Bodiselitsch, B., Montanari, A., Koeberl, C., and Coccioni, R., 2004, Delayed climate cooling in the late Eocene caused by multiple impacts: High-resolution geochemical studies at Massignano, Italy: Earth and Planetary Science Letters, v. 223, p. 283–302, doi: 10.1016/j.epsl.2004.04.028.

Bottomley, R., Grieve, R., York, D., and Masaitis, V., 1997, The age of the Popigai impact event and its relation to events at the Eocene/Oligocene boundary: Nature, v. 388, p. 365–368, doi: 10.1038/41073.

Brown, R.E., Koeberl, C., Montanari, A., and Bice, D.M., 2009, this volume, Evidence for a change in Milankovitch forcing caused by extraterrestrial events at Massignano, Italy, Eocene-Oligocene boundary GSSP, *in* Koeberl, C., and Montanari, A., eds., The Late Eocene Earth—Hothouse, Icehouse, and Impacts: Geological Society of America Special Paper 452, doi: 10.1130/2008.2452(08).

Cande, S.C., and Kent, D.V., 1995, Revised calibration of the geomagnetic polarity time scale for the Late Cretaceous and Cenozoic: Journal of Geophysical Research, v. 100, p. 6093–6095, doi: 10.1029/94JB03098.

Clymer, A.K., Bice, D.M., and Montanari, A., 1996, Shocked quartz from the late Eocene: Impact evidence from Massignano, Italy: Geology, v. 24, p. 483–486, doi: 10.1130/0091-7613(1996)024<0483:SQFTLE>2.3.CO;2.

Coccioni, R., Marsili, A., Montanari, A., Bellanca, A., Neri, R., Bice, D.M., Brinkhuis, H., Church, N., Macalady, A., McDaniel, A., Deino, A., Lirer, F., Sprovieri, M., Maiorano, P., Monechi, S., Nini, C., Nocchi, M., Pross, J., Rochette, P., Sagnotti, L., Tateo, F., Touchard, Y., Van Simaeys, S., and Williams, G.L., 2008, Integrated stratigraphy of the Oligocene pelagic sequence in the Umbria-Marche basin (northeastern Apennines, Italy): A potential global stratotype section and point (GSSP) for the Rupelian/Chattian boundary: Geological Society of America Bulletin, v. 120, p. 487–511, doi: 10.1130/B25988.1.

Collins, G.S., and Wünnemann, K., 2005, How big was the Chesapeake Bay impact? Insight from numerical modeling: Geology, v. 33, p. 925–928, doi: 10.1130/G21854.1.

Coxall, H.K., Wilson, P.A., Pälike, H., Lear, C.H., and Backman, J., 2005, Rapid stepwise onset of Antarctic glaciation and deeper calcite compensation in the Pacific Ocean: Nature, v. 433, p. 53–57, doi: 10.1038/nature03135.

Dence, M.R., Robertson, P.B., and Wirthlin, R.L., 1974, Coesite from the Lake Wanapitei crater, Ontario: Earth and Planetary Science Letters, v. 22, p. 118–122, doi: 10.1016/0012-821X(74)90071-5.

Deutsch, A., and Koeberl, C., 2006, Establishing the link between the Chesapeake Bay impact structure and the North American tektite strewn field: The Sr-Nd isotopic evidence: Meteoritics & Planetary Science, v. 41, p. 689–703.

D'Hondt, S.L., Keller, G., and Stallard, R.F., 1987, Major element compositional variation within and between different late Eocene microtektite strewn fields: Meteoritics, v. 22, p. 61–79.

Dressler, B.O., Crabtree, D., and Schuraytz, B.C., 1997, Incipient melt formation and devitrification at the Wanapitei impact structure, Ontario, Canada: Meteoritics & Planetary Science, v. 32, p. 249–258.

Evans, N.J., Gregoire, D.C., Grieve, R.A.F., Goodfellow, W.D., and Veizer, J., 1993, Use of platinum-group elements for impactor identification: Terrestrial impact craters and Cretaceous-Tertiary boundary: Geochimica et Cosmochimica Acta, v. 57, p. 3737–3748, doi: 10.1016/0016-7037(93)90152-M.

Farley, K.A., 2009, this volume, Late Eocene and late Miocene cosmic dust events: Comet showers, asteroid collisions, or lunar impacts?, *in* Koeberl, C., and Montanari, A., eds., The Late Eocene Earth—Hothouse, Icehouse, and Impacts: Geological Society of America Special Paper 452, doi: 10.1130/2009.2452(03).

Farley, K.A., Montanari, A., Shoemaker, E.M., and Shoemaker, C.S., 1998, Geochemical evidence for a comet shower in the late Eocene: Science, v. 280, p. 1250–1253, doi: 10.1126/science.280.5367.1250.

Farley, K.A., Vokrouhlicky, D., Bottke, W.F., and Nesvorny, D., 2006, A late Miocene dust shower from the break-up of an asteroid in the main belt: Nature, v. 439, p. 295–297, doi: 10.1038/nature04391.

Glass, B.P., 1989, North American tektite debris and impact ejecta from DSDP Site 612: Meteoritics, v. 24, p. 209–218.

Glass, B.P., 1990, Chronostratigraphy of Upper Eocene microspherules: Comment and Reply: Palaios, v. 5, p. 387–390, doi: 10.2307/3514897.

Glass, B.P., 2002, Upper Eocene impact ejecta/spherule layers in marine sediments: Chemie der Erde, v. 62, p. 173–196, doi: 10.1078/0009-2819-00017.

Glass, B.P., and Burns, C.A., 1987, Late Eocene crystal-bearing spherules: Two layers or one?: Meteoritics, v. 22, p. 265–279.

Glass, B.P., and Koeberl, C., 1999a, Ocean Drilling Project Hole 689B spherules and Upper Eocene microtektite and clinopyroxene-bearing spherule strewn fields: Meteoritics & Planetary Science, v. 34, p. 197–208.

Glass, B.P., and Koeberl, C., 1999b, North American microtektites in the Indian Ocean? (abstract): Meteoritics & Planetary Science, v. 34, p. A42–A43.

Glass, B.P., and Liu, S., 2001, Discovery of high-pressure $ZrSiO_4$ polymorph in naturally occurring shock-metamorphosed zircons: Geology, v. 29, p. 371–373, doi: 10.1130/0091-7613(2001)029<0371:DOHPZP>2.0.CO;2.

Glass, B.P., and Wu, J., 1993, Coesite and shocked quartz discovered in the Australasian and North American microtektite layers: Geology, v. 21, p. 435–438, doi: 10.1130/0091-7613(1993)021<0435:CASQDI>2.3.CO;2.

Glass, B.P., and Zwart, M.J., 1979, North American microtektites in Deep Sea Drilling Project cores from the Caribbean Sea and Gulf of Mexico: Geological Society of America Bulletin, v. 90, p. 595–602, doi: 10.1130/0016-7606(1979)90<595:NAMTID>2.0.CO;2.

Glass, B.P., Burns, C.A., Crosbie, J.R., and DuBois, D.L., 1985, Late Eocene North American microtektites and clinopyroxene-bearing spherules, *in* Proceedings of the 16th Lunar and Planetary Science Conference, Part 1: Journal of Geophysical Research, v. 90, supplement, p. D175–D196.

Glass, B.P., Hall, C.M., and York, D., 1986, $^{40}Ar/^{39}Ar$ laser-probe dating of North American tektite fragments from Barbados and the age of the Eocene-Oligocene boundary: Chemical Geology, v. 5, p. 181–186.

Glass, B.P., Koeberl, C., Blum, J.D., and McHugh, C.M.G., 1998, Upper Eocene tektite and impact ejecta layer on the continental slope off New Jersey: Meteoritics & Planetary Science, v. 33, p. 229–241.

Glass, B.P., Huber, H., and Koeberl, C., 2004a, Geochemistry of Cenozoic microtektites and clinopyroxene-bearing spherules: Geochimica et Cosmochimica Acta, v. 68, p. 3971–4006, doi: 10.1016/j.gca.2004.02.026.

Glass, B.P., Liu, S.B., and Montanari, A., 2004b, Impact ejecta in Upper Eocene deposits at Massignano, Italy: Meteoritics & Planetary Science, v. 39, p. 589–597.

Goderis, S., Tagle, R., and Claeys, P., 2007, Current situation on the characterization of the late Eocene impact structures using platinum group element (PGE) ratios, *in* Montanari, A., and Koeberl, C., eds., "The Late Eocene Earth: Hothouse, Icehouse, and Impacts": Ancona, Italy, Penrose Conference Abstract volume, p. 27–28.

Grieve, R.A.F., 1975, Petrology and chemistry of the impact melt at Mistastin Lake crater, Labrador: Geological Society of America Bulletin, v. 86, p. 1617–1629, doi: 10.1130/0016-7606(1975)86<1617:PACOTI>2.0.CO;2.

Grieve, R.A.F., 2006, Impact Structures in Canada: St. Johns, Canada, Geological Association of Canada, 210 p.

Grieve, R.A.F., and Ber, T., 1994, Shocked lithologies at the Wanapitei impact structure, Ontario, Canada: Meteoritics, v. 29, p. 621–631.

Harris, R.S., Roden, M., Schroeder, P., Holland, S., Duncan, M., and Albin, E., 2004, Upper Eocene impact horizon in east-central Georgia: Geology, v. 32, p. 717–720, doi: 10.1130/G20562.1.

Hazel, J.E., 1989, Chronostratigraphy of Upper Eocene microspherules: Palaios, v. 4, p. 318–329, doi: 10.2307/3514556.

Hilgen, F.J., and Kuiper, K.F., 2009, this volume, A critical evaluation of the numerical age of the Eocene-Oligocene boundary, *in* Koeberl, C., and Montanari, A., eds., The Late Eocene Earth—Hothouse, Icehouse, and Impacts: Geological Society of America Special Paper 452, doi: 10.1130/2009.2452(09).

Hyland, E., Murphy, B., Varela, P., Marks, K., Colwell, L., Tori, F., Monechi, S., Cleaveland, L., Brinkhuis, H., van Mourik, C.A., Coccioni, R., Bice, D., and Montanari, A., 2009, this volume, Integrated stratigraphy and astrochronologic calibration of the Eocene-Oligocene transition in the Monte Cagnero section (northeastern Apennines, Italy): A potential parastratotype for the Massignano GSSP, *in* Koeberl, C., and Montanari, A., eds., The Late Eocene Earth—Hothouse, Icehouse, and Impacts: Geological Society of America Special Paper 452, doi: 10.1130/2009.2452(19).

Katz, M.E., Miller, K.G., Wright, J.D., Wade, B.S., Browning, J.V., Cramer, B.S., and Rosenthal, Y., 2008, Stepwise transition from greenhouse to icehouse climates: Nature Geoscience, v. 1, p. 329–334, doi: 10.1038/ngeo179.

Keller, G., D'Hondt, S.L., Orth, C.J., Gilmore, J.S., Oliver, P.Q., Shoemaker, E.M., and Molina, E., 1987, Late Eocene impact microspherules: Stratigraphy, age, and geochemistry: Meteoritics, v. 22, p. 25–60.

Kettrup, B., Deutsch, A., and Masaitis, V.L., 2003, Homogeneous impact melts produced by a heterogeneous target? Sr-Nd isotopic evidence from the Popigai crater, Russia: Geochimica et Cosmochimica Acta, v. 67, p. 733–750, doi: 10.1016/S0016-7037(02)01143-2.

Koeberl, C., 1986, Geochemistry of tektites and impact glasses: Annual Review of Earth and Planetary Sciences, v. 14, p. 323–350, doi: 10.1146/annurev.ea.14.050186.001543.

Koeberl, C., 1994, Tektite origin by hypervelocity asteroidal or cometary impact: Target rocks, source craters, and mechanisms, *in* Dressler, B.O., Grieve, R.A.F., and Sharpton, V.L., eds., Large Impacts and Planetary Evolution: Geological Society of America Special Paper 293, p. 133–151.

Koeberl, C., 2007, The geochemistry and cosmochemistry of impacts, *in* Davis, A., ed., Treatise of Geochemistry (Online Edition), Volume 1: Amsterdam, Elsevier, p. 1.28.1–1.28.52, doi: 10.1016/B978-008043751-4/00228-5.

Koeberl, C., and Glass, B.P., 1988, Chemical composition of North American microtektites and tektite fragments from Barbados and DSDP Site 612 on the continental slope off New Jersey: Earth and Planetary Science Letters, v. 87, p. 286–292, doi: 10.1016/0012-821X(88)90016-7.

Koeberl, C., Poag, C.W., Reimold, W.U., and Brandt, D., 1996, Impact origin of Chesapeake Bay structure and the source of North American tektites: Science, v. 271, p. 1263–1266, doi: 10.1126/science.271.5253.1263.

Kyte, F.T., and Liu, S., 2002, Iridium and spherules in late Eocene impact deposits: Houston, Texas, Lunar and Planetary Institute, Lunar and Planetary Science, v. 33, abs. 1981 (CD-ROM).

Kyte, F.T., Shukolyukov, A., Hildebrand, A.R., and Lugmair, G.W., 2004, Initial Cr-isotopic and iridium measurements of concentrates from Late-Eocene cpx-spherule deposits: Houston, Texas, Lunar and Planetary Institute, Lunar and Planetary Science, v. 35, abs. 1824 (CD-ROM).

Langenhorst, F., 1996, Characteristics of shocked quartz in late Eocene impact ejecta from Massignano (Ancona, Italy): Clues to shock conditions and source crater: Geology, v. 24, p. 487–490, doi: 10.1130/0091-7613(1996)024<0487:COSQIL>2.3.CO;2.

Lazorek, M., Eyles, N., Eyles, C., Doughty, M., L'Heureux, E., and Milkereit, B., 2006, Late Quaternary seismo-stratigraphy of Lake Wanapitei, Sudbury, Ontario, Canada: Arguments for a possible meteorite impact origin: Sedimentary Geology, v. 192, p. 231–242, doi: 10.1016/j.sedgeo.2006.04.010.

Lee, S.R., Horton, J.W., and Walker, R.J., 2006, Osmium-isotope and platinum-group element systematics of impact melt rocks, Chesapeake Bay impact structure, Virginia, USA: Meteoritics & Planetary Science, v. 41, p. 819–833.

L'Heureux, E., Ugalde, H., Milkereit, B., Boyce, J., Morris, W., Eyles, N., and Artemieva, N., 2005, Using vertical dikes as a new approach to constraining the size of buried craters: An example from Lake Wanapitei, Canada, *in* Kenkmann, T., Hörz, F., and Deutsch, A., eds., Large Meteorite Impacts Volume III: Geological Society of America Special Paper 384, p. 43–50.

Liu, S., Papanastassiou, D.A., Ngo, H.H., and Glass, B.P., 2006, Sr and Nd analyses of Upper Eocene spherules and their implications for target rocks: Meteoritics & Planetary Science, v. 41, p. 705–714.

Liu, S., Glass, B.P., Kyte, F.T., and Bohaty, S.M., 2009, this volume, The Upper Eocene clinopyroxene-bearing spherule layer: New sites and discovery of coesite and shocked quartz, *in* Koeberl, C., and Montanari, A., eds., The Late Eocene Earth—Hothouse, Icehouse, and Impacts: Geological Society of America Special Paper 452, doi: 10.1130/2009.2452(04).

Mak, E.K.C., York, D., Grieve, R.A.F., and Dence, M.R., 1976, The age of the Mistastin Lake crater, Labrador, Canada: Earth and Planetary Science Letters, v. 31, p. 345–357, doi: 10.1016/0012-821X(76)90116-3.

Marchand, E., and Whitehead, J., 2002, Statistical Evaluation of Compositional Differences between Upper Eocene Impact Ejecta Layers: Mathematical Geology, v. 34, p. 555–572.

Miller, K.G., Berggren, W.A., Zhang, J., and Palmer-Julson, A.A., 1991, Biostratigraphy and isotope stratigraphy of Upper Eocene microtektites at Site 612: How many impacts?: Palaios, v. 6, p. 17–38, doi: 10.2307/3514951.

Miller, K.G., Wright, J.D., Katz, M.E., Wade, B.S., Browning, J.V., Cramer, B.S., and Rosenthal, Y., 2009, this volume, Climate threshold at the Eocene-Oligocene transition: The Antarctic ice sheet influences on ocean circulation, *in* Koeberl, C., and Montanari, A., eds., The Late Eocene

Earth—Hothouse, Icehouse, and Impacts: Geological Society of America Special Paper 452, doi: 10.1130/2009.2452(11).

Montanari, A., and Koeberl, C., 2000, Impact Stratigraphy: Heidelberg, Springer, 364 p.

Montanari, A., Asaro, F., Michel, E., and Kennett, J.P., 1993, Iridium anomalies of late Eocene age at Massignano (Italy), and in ODP Site 689B (Maud Rise, Antarctica): Palaios, v. 8, p. 420–437, doi: 10.2307/3515017.

Ngo, H.H., Wasserburg, G.J., and Glass, B.P., 1985, Nd and Sr isotopic compositions of tektite material from Barbados and their relationship to North American tektites: Geochimica et Cosmochimica Acta, v. 49, p. 1479–1485, doi: 10.1016/0016-7037(85)90297-2.

Obradovich, J.D., Snee, L.W., and Izett, G.A., 1989, Is there more than one glassy impact layer in the late Eocene?: Geological Society of America Abstracts with Programs, v. 21, no. 6, p. A134.

Paquay, F.S., Ravizza, G.E., Dalai, T.K., and Peucker-Ehrenbrink, B., 2008, Determining chondritic impactor size from the marine osmium isotope record: Science, v. 320, p. 214–218, doi: 10.1126/science.1152860.

Pierrard, O., Robin, E., Rocchia, R., and Montanari, A., 1998, Extraterrestrial Ni-rich spinel in Upper Eocene sediments from Massignano, Italy: Geology, v. 26, p. 307–310, doi: 10.1130/0091-7613(1998)026<0307:ENRSIU>2.3.CO;2.

Poag, C.W., and Aubry, M.-P., 1995, Upper Eocene impactites of the U.S. East Coast: Depositional origins, biostratigraphic framework, and correlation: Palaios, v. 10, p. 16–43, doi: 10.2307/3515005.

Poag, C.W., Powars, D.S., Poppe, L.J., and Mixon, R.B., 1994, Meteoroid mayhem in Ole Virginny: Source of the North American tektite strewn field: Geology, v. 22, p. 691–694, doi: 10.1130/0091-7613(1994)022<0691:MMIOVS>2.3.CO;2.

Poag, C.W., Koeberl, C., and Reimold, W.U., 2004, Chesapeake Bay Crater: Geology and Geophysics of a Late Eocene Submarine Impact Structure: Heidelberg, Springer, 522 p.

Sanfilippo, A., Riedel, W.R., Glass, B.P., and Kyte, F.T., 1985, Late Eocene microtektites and radiolarian extinctions on Barbados: Nature, v. 314, p. 613–615, doi: 10.1038/314613a0.

Shaw, H.F., and Wasserburg, G.J., 1982, Age and provenance of the target materials for tektites and possible impactites as inferred from Sm-Nd and Rb-Sr systematics: Earth and Planetary Science Letters, v. 60, p. 155–177, doi: 10.1016/0012-821X(82)90001-2.

Sherlock, S.C., Kelley, S.P., Parnell, J., Green, P., Lee, P., Osinski, G.R., and Cockell, C.S., 2005, Re-evaluating the age of the Haughton impact event: Meteoritics & Planetary Science, v. 40, p. 1777–1787.

Sherlock, S.C., Kelley, S.P., and Glazovskaya, L., 2006, A new age for the Logoisk impact structure, Berlarus and implications for the late Eocene comet shower, *in* First International Conference on Impact Cratering in the Solar System abstract volume: Noordwijk, Holland, European Space Agency (CD-ROM).

Stecher, O., Ngo, H.H., Papanastassiou, D.A., and Wasserburg, G.J., 1989, Nd and Sr isotopic evidence for the origin of tektite material from DSDP Site 612 off the New Jersey Coast: Meteoritics, v. 24, p. 89–98.

Storzer, D., and Wagner, G.A., 1971, Fission track ages of North American tektites: Earth and Planetary Science Letters, v. 10, p. 435–440, doi: 10.1016/0012-821X(71)90093-8.

Tagle, R., and Claeys, P., 2004, Comet or asteroid shower in the late Eocene?: Science, v. 305, p. 492, doi: 10.1126/science.1098481.

Tagle, R., and Claeys, P., 2005, An ordinary chondrite impactor for the Popigai crater, Siberia: Geochimica et Cosmochimica Acta, v. 69, p. 2877–2889, doi: 10.1016/j.gca.2004.11.024.

Thein, J., 1987, A tektite layer in Upper Eocene sediments of the New Jersey continental slope (Site 612, Leg 95), *in* Poag, C.W., Watts, A., et al., Initial Reports of the Deep Sea Drilling Project, Volume 95: Washington, D.C., U.S. Government Printing Office, p. 565–574.

Touchard, Y., Rochette, P., Aubry, M.-P., and Michard, A., 2003, High-resolution magnetostratigraphic and biostratigraphic study of Ethiopian traps-related products in Oligocene sediments from the Indian Ocean: Earth and Planetary Science Letters, v. 206, p. 493–508, doi: 10.1016/S0012-821X(02)01084-1.

Ugalde, H.A., L'Heureux, E., Lachapelle, R., and Milkereit, B., 2006, Measuring gravity on ice: An example from Wanapitei Lake, Ontario, Canada: Geophysics, v. 71, p. J23–J29.

Vonhof, H.B., and Smit, J., 1999, Late Eocene microkrystites and microtektites at Maud Rise (Ocean Drilling Project Hole 689B; Southern Ocean) suggest a global extension of the approximately 35.5 Ma Pacific impact ejecta strewn field: Meteoritics & Planetary Science, v. 34, p. 747–755.

Vonhof, H.B., Smit, J., Brinkhuis, H., Montanari, A., and Nederbragt, A.J., 2000, Global cooling accelerated by early late Eocene impacts?: Geology, v. 28, p. 687–690, doi: 10.1130/0091-7613(2000)28<687:GCABEL>2.0.CO;2.

Wei, W., 1995, How many impact-generated microspherule layers in the Upper Eocene?: Palaeogeography, Palaeoclimatology, Palaeoecology, v. 114, p. 101–110, doi: 10.1016/0031-0182(95)00075-W.

Whitehead, J., Papanastassiou, D.A., Spray, J.G., Grieve, R.A.F., and Wasserburg, G.J., 2000, Late Eocene impact ejecta: Geochemical and isotopic connections with the Popigai impact structure: Earth and Planetary Science Letters, v. 181, p. 473–487, doi: 10.1016/S0012-821X(00)00225-9.

Whitehead, J., Grieve, R.A.F., and Spray, J.G., 2002, Mineralogy and petrology of melt rocks from the Popigai impact structure, Siberia: Meteoritics & Planetary Science, v. 37, p. 623–647.

Winzer, S.R., Lum, R.K.L., and Schuhmann, S., 1976, Rb, Sr and strontium isotopic composition, K/Ar age and large ion lithophile trace element abundances in rock and glasses from the Wanapitei Lake impact structure: Geochimica et Cosmochimica Acta, v. 40, p. 51–57, doi: 10.1016/0016-7037(76)90193-9.

Zähringer, J., 1963, K-Ar measurements of tektites, *in* Radioactive Dating: Proceedings IAEA Symposium: Vienna, International Atomic Energy Agency (IAEA), p. 289–305.

Zanazzi, A., Kohn, J.J., MacFadden, B.J., and Terry, D.O., Jr., 2007, Large temperature drop across the Eocene-Oligocene transition in central North America: Nature, v. 445, p. 639–642, doi: 10.1038/nature05551.

Zappalà, V., Cellino, A., Gladman, B.J., Manley, S., and Migliorini, F., 1998, Asteroid showers on Earth after family breakup events: Icarus, v. 134, p. 176–179, doi: 10.1006/icar.1998.5946.

Manuscript Accepted by the Society 16 September 2008

Printed in the USA

The Geological Society of America
Special Paper 452
2009

Late Eocene and late Miocene cosmic dust events: Comet showers, asteroid collisions, or lunar impacts?

K.A. Farley
Division of Geological and Planetary Sciences, MS 170-25, California Institute of Technology, Pasadena, California 91125, USA

ABSTRACT

Two long-lived peaks in extraterrestrial ^{3}He flux have been identified in the sedimentary record of the Cenozoic Era: at 8.2 Ma (late Miocene) and 35.8 Ma (late Eocene). These peaks document the occurrence of important events in the recent history of the solar system. The timing and temporal evolution of the dust event in the late Miocene as well as the absence of major terrestrial impact craters are consistent with an origin in the catastrophic collision that produced the Veritas asteroid family at this time. In contrast, there is no known asteroid collision corresponding to the late Eocene peak. Instead, the late Eocene event has elements consistent with a comet shower produced by a close stellar encounter. Both the rise time and the fall time of the enhanced dust flux and the occurrence of two major terrestrial impacts at the peak of the event are predicted by considerations of comet shower dynamics. However, debris from one of these impacts has a Cr isotope ratio that appears to exclude a carbonaceous chondrite impactor. If the presumption that comets have a carbonaceous chondrite–like composition is accurate, then an alternative mechanism for the late Eocene event may be required. The recent suggestion of an asteroid shower, where the ^{3}He-bearing dust resulted from lunar impacts, is one such possibility, but it too fails to account fully for existing observations. The cause of the late Eocene event thus remains uncertain.

Keywords: cosmic dust, extraterrestrial ^{3}He, impacts, comet shower, Veritas.

INTRODUCTION

Earth currently sweeps from space ~25–40 million kg/yr of interplanetary dust particles (IDPs) that range in size from a few micrometers to a few hundred micrometers in diameter (Love and Brownlee, 1993; Karner et al., 2003). Because small extraterrestrial particles migrate into the sun on time scales from ~10^3 to 10^5 yr, the abundance of IDPs in the solar system reflects their production from comets and asteroids with very limited temporal buffering. Occasional solar system events such as asteroid collisions and comet showers will temporarily dominate the production of these particles, so their flux to Earth must vary over geologic time. Models suggest that large dust-producing asteroid collisions have a recurrence interval of 10^7–10^8 yr (Durda and Dermott, 1997); comet showers produced by close encounters with passing stars or molecular clouds or due to galactic tidal forces will have a similar recurrence period (Weissman, 1990). Thus, a record of the IDP flux over the last tens to hundreds of millions of years can potentially identify significant events in the recent history of the solar system.

Farley, K.A., 2009, Late Eocene and late Miocene cosmic dust events: Comet showers, asteroid collisions, or lunar impacts?, *in* Koeberl, C., and Montanari, A., eds., The Late Eocene Earth—Hothouse, Icehouse, and Impacts: Geological Society of America Special Paper 452, p. 27–35, doi: 10.1130/2009.2452(03). For permission to copy, contact editing@geosociety.org.

Under most circumstances, these particles are too small, too rare, and too indistinctive to be observed directly in terrestrial materials such as sedimentary rocks. However, their presence can be deduced, and their relative abundance can be quantified using ^{3}He as a tracer (Merrihue, 1964; Ozima et al., 1984; Farley, 1995). As the dust particles migrate from their source bodies toward the sun, they rapidly become saturated with solar wind He. Compared to terrestrial dust, IDPs are enormously enriched in He, and especially in the rare isotope ^{3}He (Nier and Schlutter, 1990). Using isotopic systematics, the He in many marine sediments can be unequivocally deconvolved into terrestrial and extraterrestrial components (Takayanagi and Ozima, 1987; Farley, 2001). By estimating the sedimentation rate, the concentration of extraterrestrial ^{3}He can be converted into a ^{3}He flux (Takayanagi and Ozima, 1987; Farley, 1995). Because particles larger than ~10 μm are intensely heated and outgassed during atmospheric entry, the ^{3}He flux records delivery of only the smallest particles in the IDP population (Ozima et al., 1984; Farley et al., 1997). Particles in this size class have solar system lifetimes <10^4 yr. Importantly, large objects such as those that produce impact craters are so intensely heated that they are not expected to leave a record in terms of ^{3}He. This is nicely demonstrated by the absence of a ^{3}He peak in association with the Ir anomaly at the Cretaceous-Tertiary boundary (Mukhopadhyay et al., 2001b).

Temporal variations in the flux of IDP-hosted ^{3}He have been documented in pelagic sediments spanning the last 70 m.y. (Farley et al., 1998, 2006; Mukhopadhyay et al., 2001a). In this interval, two episodes of highly elevated flux have been detected (Fig. 1): in the late Eocene (36 Ma; Farley et al., 1998) and the late Miocene (8 Ma; Farley et al., 2006). The late Miocene dust event is thought to record a collision between asteroids (Farley et al., 2006). The origin of the late Eocene dust event is more controversial. Although originally ascribed to a comet shower (Farley et al., 1998), more recent work has called on an asteroid shower in which the enhanced ^{3}He flux is attributed to dust lofted by impacts on the moon (Tagle and Claeys, 2004; Fritz et al., 2007). Distinguishing among possible causes of these dust events is important for understanding the history of the solar system and for the origin of cosmic dust particles and possibly for the cratering history of rocky bodies in the solar system.

Here, I contrast expectations for the asteroidal and cometary models and then compare these expectations with observations of the two events. The result is a fairly consistent body of evidence that the late Miocene peak was caused by an asteroid collision. While most aspects of the late Eocene event are consistent with the original comet shower hypothesis, at least one line of evidence may contradict it. The asteroid shower–lunar impacts model similarly fails to explain some observations of this event.

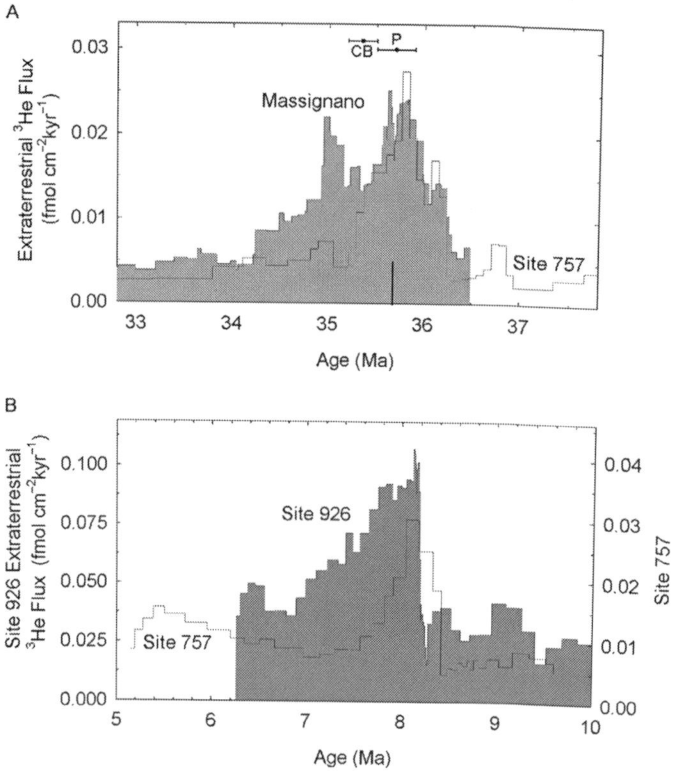

Figure 1. A comparison of the late Eocene and late Miocene extraterrestrial ^{3}He flux peaks. (A) The late Eocene event recorded in Tethyan limestones of the Massignano site (shaded; Farley et al., 1998) and at Ocean Drilling Program (ODP) Site 757 in the Indian Ocean (Farley et al., 2006). The ages and uncertainties of the Chesapeake Bay (CB) (Koeberl et al., 1996) and Popigai (P) (Bottomley et al., 1997) craters are indicated; the short vertical line is the 5.6 m level at Massignano, which has the highest concentrations of Ir in the section. (B) The late Miocene event at ODP Site 926 (Brazil margin) and Site 757 (Indian Ocean) from Farley et al. (2006). Note the vertical scale difference between the two late Miocene sites, consistent with sedimentary focusing at Site 926. For both events, the shaded records are preferred; Site 757 has low temporal resolution due to very low sedimentation rate and poor chronostratigraphy.

ASSIGNING DUST EVENTS TO SPECIFIC SOURCES

Attribution of the interplanetary dust events to a specific solar system process relies on several types of observations. These include: (1) the temporal evolution of the dust event as recorded by the ^{3}He flux; (2) the temporal relationship between the enhanced ^{3}He flux and terrestrial impact craters (if any); and (3) the geochemical fingerprint of any impactors associated with the event. In addition, an asteroid-collision origin can be given support if the event age matches either: (4) a peak in the distribution of cosmic-ray exposure ages of meteorites or (5) the age of formation of asteroid families obtained by orbital "backtracking" of family members (Nesvorny et al., 2003). In the remainder of this section, I describe some general considerations relating to the use of these observations for identifying the sources of the dust events.

The ^{3}He Proxy

It is important to recognize that ^{3}He does not provide a direct measure of the dust production associated with specific solar system events. First, there are myriads of dust sources in the solar system, so the enhanced ^{3}He flux is superimposed on a more or less constant baseline value. Second, there are strong biases in the way that ^{3}He records solar system dustiness. Obvious factors include grain size and velocity. Particle capture by Earth and He preservation during atmospheric entry (Farley et al., 1997) both favor small particles and those with low geocentric encounter velocities. Asteroid collisions, lunar impacts, and comet showers all likely change the mass flux as well as the size and velocity distribution of infalling particles, but these effects are integrated into just a single observable value, the ^{3}He flux. Additional information may be obtained from replicate ^{3}He analyses of specific samples, which can provide insight to the size distribution of ^{3}He-bearing particles (Farley et al., 1997). For example, a process which significantly increases the flux of low-velocity large particles might be detectable as unusually scattered replicates reflecting an increase in the "nugget effect."

Impactor Identification

Efforts have been made to determine the origin of impacting bodies in several ways. In some cases, actual impactor fragments or impactor spinels have been found in the crater or in correlative sediments (Kyte, 1998, 2002; Maier et al., 2006; Alwmark and Schmitz, 2007). The chemistry and mineralogy of these fragments can be used to assign the impactor to known meteorite types and potentially to specific asteroidal sources identified spectroscopically. In the absence of such fragments, the geochemical signature of the impactor can sometimes be determined from analyses of impact glasses or distal outfall, and then tied to specific meteorite types. Cr isotopes and platinum group element (PGE) concentrations have both been used for this purpose.

Several recent studies have documented a systematic difference in ^{54}Cr/^{52}Cr between carbonaceous and ordinary chondrites (Shukolyukov and Lugmair, 2006; Trinquier et al., 2007), probably reflecting incomplete mixing and radial heterogeneity of these isotopes within the solar nebula. In some cases, there is sufficient extraterrestrial Cr in impact debris that the Cr isotope composition of the impactor and an associated meteorite class can be inferred (Trinquier et al., 2006; Koeberl et al., 2007). Similarly, platinum group interelement ratios such Ru/Rh and Ru/Ir can apparently discriminate between various meteorite classes (McDonald et al., 2001; Tagle and Claeys, 2005; Tagle and Hecht, 2006). Figure 2A, based on data compiled in Tagle (2004), shows that carbonaceous chondrites may have different Ru-Rh-Ir systematics than ordinary chondrites, and that ordinary chondrite varieties may also be distinctive in this space. However, it is important to note that there is substantial compositional overlap in PGE ratios when analytical uncertainties are considered (especially the possibility of interlaboratory biases). Because many of the classes are documented by just one or two analyzed meteorites, it is unclear how well the data in Figure 2 characterize each entire class and differences among classes.

For present purposes, one essential limitation of these approaches is that the chemical and isotopic characteristics of

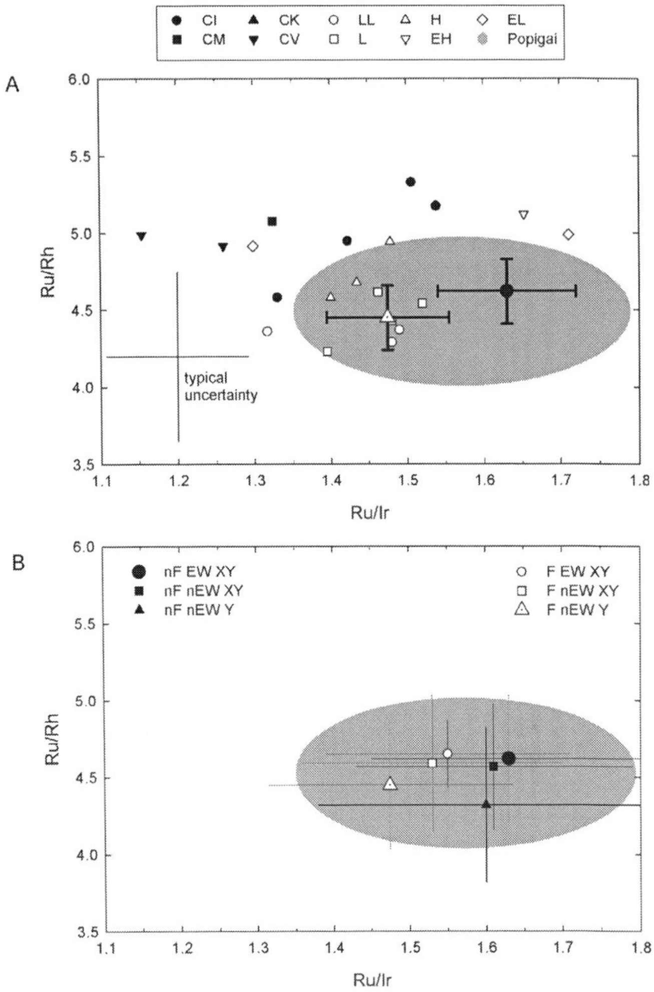

Figure 2. Platinum group element (PGE) fingerprint from Popigai and comparison to meteorite observations. This pair of PGE ratios was shown (Tagle, 2004; Tagle and Claeys, 2005) to provide the best discrimination among meteorite classes (indicated by letters). (A) Meteorite data from the new data and tabulation of Tagle (2004) supplemented with recent data of Koiwa et al. (2006). To eliminate nugget effects, only individual samples analyzed for all three elements are included. Unlike the class means used by Tagle and Claeys (2005) to discriminate among impactors, the points plotted here are for individual meteorites. The error bar labeled "typical" represents the analytical uncertainties on the meteorite values propagated from the uncertainties reported by the original authors. The two bold points with error bars are the Popigai impactor value computed by Tagle and Claeys (2005; open symbol), and the one preferred here (closed symbol). The gray ellipse encompasses the 2σ spread of various estimates of the Popigai impactor composition from panel B. Many different meteorite classes are consistent with the Popigai impactor composition, depending on how the composition is estimated (see text), especially when the uncertainties are considered. (B) Several different estimates of the PGE fingerprint of the Popigai impactor, with 2σ uncertainty. These estimates were computed from regression of Tagle and Claeys' (2005) Popigai data, plotted in Figure 4. The different regression techniques are indicated (F—filtered to remove data point marked in Fig. 4; nF—not filtered; EW—error weighted regression; nEW—not error-weighted; *y*—standard regression; *x*-*y*—*x*-*y* or York regression). The regression values proposed by Tagle and Claeys (2005; F nEW *y*) and the one preferred here (nF EW *x*-*y*) are highlighted with the larger symbols.

comets are essentially unknown. While it may be reasonable to assume that because comets are primitive carbon-rich bodies, they will be compositionally similar to carbonaceous chondrites, this remains unproven. Indeed, recent analyses of dust particles captured from Comet Wild 2 by the *Stardust* mission illustrate the complex nature of cometary composition. Contrary to common expectations, the Wild 2 particles document at least some high-temperature processing and presumably radial transfer of constituent particles from the inner to the outer solar system (Ciesla, 2007; Ishii et al., 2008; Marty et al., 2008). It is not yet clear how this process might influence either Cr isotopes or PGE ratios of entire comets.

Meteorite Cosmic-Ray Exposure Ages

Meteorites are generally thought to be fragments from the interiors of larger bodies, in which they were shielded from cosmic-ray bombardment until released by collision. Over 1600 meteorites have been analyzed for cosmogenic nuclide concentrations, and histograms of the resulting cosmic-ray exposure ages suggest the occurrence of perhaps a dozen meteorite-forming asteroid collisions over the last ~40 m.y. (Graf and Marti, 1994, 1995; Eugster, 2003). Some of these events are restricted to a single meteorite class, while others seem to occur in multiple classes. Each of these events is a possible source of dust to Earth, with the dust arriving nearly synchronously with the collision and hence with the peak in cosmic-ray exposure ages. However, connecting specific extraterrestrial dust peaks to specific cosmic-ray age peaks is complicated by the fact that (1) there are a large number of candidate collisions, separated by on average ~3–4 m.y. (Eugster, 2003); (2) the age maxima are fairly broad, making it hard to assign an age precisely; and (3) the accuracy of these ages is no better than the uncertainty on the cosmogenic nuclide production rate, which is ~10% (Eugster, 2003). For recent events (<10 m.y.), the ages of individual events may be sufficiently precise, accurate, and resolved to be diagnostic of a match to the terrestrial sediment record. However by the Eocene, these uncertainties compound such that almost any terrestrial event age may be mapped to one or more cosmic-ray age peaks.

It is possible that some meteorites are derived from comets, but, in general, comets have shorter lifetimes than most asteroid fragments owing to their distinct orbital properties and possibly also due to their lesser mechanical strength. Thus, peaks in cosmic-ray ages of modern meteorites are not expected to record a comet shower, at least not one that occurred more than a few million years ago.

Orbital Dynamics

Like cosmic-ray exposure ages of meteorites, the orbital characteristics of asteroid families document the collisional history of the asteroid belt. By integrating family member orbits backward in time, it is possible to estimate both the age of collision and the sizes of the bodies involved. The precision of the collisional age estimates is highly variable depending on family location in the asteroid belt, and it generally becomes much poorer with age. Ages for about a dozen families have been estimated, spanning 0.5–500 Ma (Nesvorny et al., 2003, 2006, 2007; Bottke et al., 2007). For example, the orbits of Datura family asteroids suggest the catastrophic disruption of a 15-km-diameter body at 450 ± 50 ka (Nesvorny et al., 2006), while the 170 km Baptistina collision event occurred in a very large time window between 140 and 190 Ma (Bottke et al., 2007). Thus, collisional age estimates are much like cosmic-ray exposure age peaks: for young events, it may be possible to link a specific asteroid collision to the terrestrial record, but at older ages, the uncertainties become very permissive.

Astrometric observations and orbital backtracking can also document the timing and proximity of close stellar encounters and resulting comet showers in the past. Available data have not identified any shower-producing encounters (Garcia-Sanchez et al., 2001). However, the data are observationally very incomplete and are restricted to about the last 10 m.y., and so they are of limited value for current purposes.

EVIDENCE FOR THE ORIGIN OF THE TWO DUST EVENTS

Late Miocene Event

Figure 1B shows the Ocean Drilling Program (ODP) Site 926 (Atlantic Ocean) ^{3}He record that documents the late Miocene dust event at 8.2 Ma. A second record of this event was obtained at ODP Site 757 (Indian Ocean), but this core has a very low accumulation rate and poor chronostratigraphy and so serves mainly to confirm the global nature of the event. To emphasize the timing and magnitude of the ^{3}He enhancement, the Site 926 data were recast in Figure 3 in terms of time relative to the maximum ^{3}He flux (t = 0), where the flux was normalized to the value prior to the event. The event is characterized by an extremely rapid onset, just ~30 k.y. from baseline to maximum. The true onset may have been even more rapid given the inevitable downward mixing by bioturbation. The flux rises to about five times higher than the baseline value and then declines steadily over at least 2 m.y. The Site 926 record does not extend sufficiently upsection to establish precisely when the flux returns to a baseline value.

Coincident with the study that revealed this peak, Nesvorny et al. (2003) discovered a major collision in the asteroid belt at the appropriate time. Using the orbital backtracking method, they found that the Veritas family was produced by the catastrophic disruption of a 140-km-diameter parent body at 8.3 ± 0.5 Ma, and they correctly predicted a "^{3}He spike" of this age. In terms of parent body mass, this is the largest collision yet discovered to have occurred in the Cenozoic era.

These observations are consistent with the source of the late Miocene ^{3}He peak in an asteroid collision, the event that simultaneously produced the Veritas family. The steep rise in the terrestrial IDP flux is expected from its rapid delivery from the initial

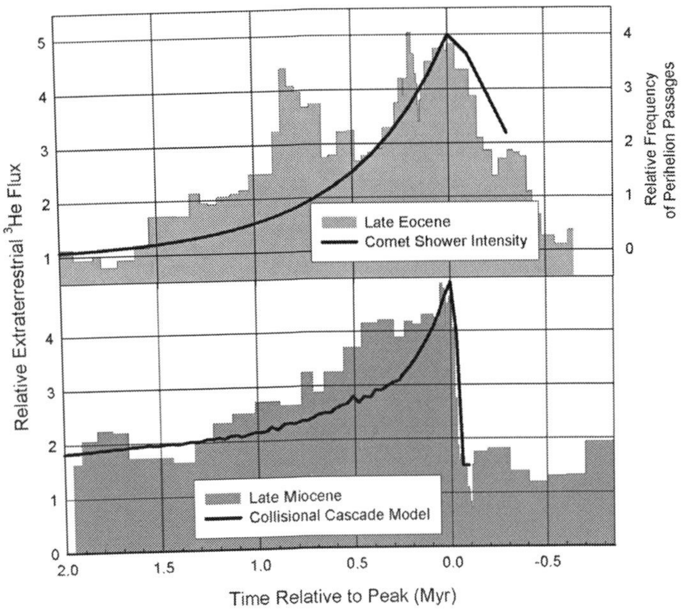

Figure 3. Normalized records of the ^{3}He peaks shown in Figure 1. The *x*-axis has been converted into time relative to the highest ^{3}He flux observed during each event, and the *y*-axis is normalized to the ^{3}He flux observed prior to the onset of each event. These normalizations emphasize the temporal evolution and magnitude of the two peaks, and the surprising similarity between the two events. For comparison, the upper panel also shows the frequency of cometary perihelion passages after an impulsive perturbation of the Oort cloud (Hut et al., 1987), a reasonable proxy for the IDP flux during the shower. The peak of this curve was aligned with the ^{3}He record. The lower panel shows a model of the IDP flux expected from a collisional cascade following the Veritas asteroid collision (Farley et al., 2006). Unlike the comet shower model, this model was developed specifically to match the ^{3}He data. Note in particular the distinction in rise time between the two models (longer for the comet shower), also observed in the two ^{3}He data sets.

massive collision with little dispersion during transit to Earth. The temporal decay of the peak is not a priori predicted because it depends on further collisions between fragments following the collision, but a model of this collisional cascade has been made to fit the ^{3}He observations (Farley et al., 2006). This fit is shown by the solid curve in the lower panel of Figure 3.

Large terrestrial impact craters are unknown in the 6–8 Ma time period. Just two small craters (~10 km diameter) have ages (5 Ma) that overlap this period within uncertainties (http://www.unb.ca/passc/ImpactDatabase/; Grieve, 2001). The apparent absence of major terrestrial impacts during the dust event is consistent with expectations for the Veritas collision. To become Earth-crossing, collisional fragments in the asteroid belt must have their orbits modified by a variety of different processes, especially resonant interactions with Jupiter, Saturn, and Mars. Ordinarily, this requires millions of years or more, although objects fortuitously delivered by collision directly into appropriate resonances may become Earth-crossing in as little as a few hundred thousand years (Gladman et al., 1997; Zappala et al., 1998; Nesvorny et al., 2007). The Veritas family is not located favorably for such rapid delivery (Farley et al., 2006), so impacts within the ~2 m.y. duration of the dust event would not be anticipated.

Histograms of meteorite cosmic-ray exposure ages show a large peak at ca. 7 Ma; indeed, this is the most prominent peak in the age distribution for all meteorite classes (Eugster, 2003). Although a peak of this age seems a sensible consequence of the Veritas breakup, this interpretation is problematic (Farley et al., 2006). The 7 Ma cosmogenic age peak occurs in H chondrites, yet spectral and mineralogic data indicate that Veritas is not likely to be the parent body of the H chondrites (Burbine et al., 2002). In addition, as mentioned already, the Veritas family is not well-positioned to deliver fragments to Earth-crossing orbits. If there was a relationship between the H chondrites and the Veritas breakup, it may have arisen from a secondary collision involving a Veritas fragment and an appropriately positioned body of H chondrite composition (Farley et al., 2006). This ad hoc mechanism is difficult to disprove.

Late Eocene Event

Figures 1 and 3 show the ^{3}He record that defines the late Eocene dust event, which peaks at 35.8 Ma. The most detailed record of this event is from Tethyan limestones exposed at Massignano, Italy (Farley et al., 1998). A lower-resolution and generally similar record has been obtained from ODP Site 757 (Farley et al., 2006), and a very crude record has been obtained from an extremely slowly accumulating Pacific pelagic clay core (Farley, 1995). In the high-resolution Massignano record, the maximum flux is about five times pre-event levels, and it decays back to baseline over a period of ~2 m.y. The general decline is interrupted by a prominent secondary maximum ~750 k.y. after the first peak, but this peak is not present in the Site 757 record. In terms of duration and magnitude, the late Eocene event is remarkably similar to that in the late Miocene (Fig. 3). However, unlike the nearly instantaneous onset of the late Miocene event, the late Eocene event builds up over ~500 k.y. This is far longer than can be attributed to bioturbation in the Massignano section, which would probably smear the signal over <20 k.y.

As shown in Figure 1A, the peak of the ^{3}He episode is coincident with the occurrence of the two largest impact craters of the Cenozoic Era: Popigai (100 km) and Chesapeake Bay (90 km). The peculiar temporal clustering of these impacts has been noted previously, and a comet shower origin has been considered (Hut et al., 1987). Shortly after the realization that IDP-hosted ^{3}He is preserved in the geologic record, the late Gene Shoemaker proposed that ^{3}He could provide a unique test of the comet shower model, an idea that set the stage for the initial interpretation of the event.

Several studies have examined the temporal evolution of cometary activity produced by an impulsive perturbation of the Oort cloud, for example, by a close stellar encounter (Fernandez and Ip, 1987; Hut et al., 1987; Dybczynski, 2002). After the impulse, the abundance of active comets initially rises as new comets arrive in the inner solar system from their initially distant orbits. Eventually, the flux of new comets declines, so the ~0.5 m.y. mean comet lifetime begins to dominate, and the shower dissipates. A quantitative model of the intensity of the shower as indicated by the frequency of perihelion passages (Hut et al., 1987) is shown in the upper panel of Figure 3. The frequency is asymmetric: it rises over ~400 k.y. and then decreases to near background values over ~2 m.y. More recent modeling has yielded generally similar results, and it further indicates the temporal evolution of the shower depends on the geometry of the encounter and the size of the perturber (Dybczynski, 2002).

The frequency of perihelion passages curve was originally generated as a proxy for the relative probability of a comet impact with Earth during a shower. The otherwise improbable coincidence of the two late Eocene impacts was deemed compatible with expectations for a comet shower (Hut et al., 1987) but of course could not prove its occurrence.

Making the assumption that the release rate of dust during a comet shower is proportional to the frequency of perihelion passages and recognizing the short dynamical lifetime of dust particles, the perihelion passage curve is also a reasonable estimate of the relative flux of shower-derived IDPs to Earth during the event. When scaled and positioned appropriately, this model matches remarkably well the ^{3}He observations (Fig. 3). Unlike the collisional cascade model, the comet shower predictions were derived solely from orbital mechanics and were published long before the ^{3}He data. Note in particular that the comet shower model predicts the observed slow rise, which is not seen in the asteroid collision model. The only feature not well matched by the model is the secondary peak 750 k.y. after the first, but, as noted above, this peak may not be global.

Based on the temporal coincidence of the large impacts and the dust peak and the match between modeled and observed temporal evolution, the late Eocene event was originally ascribed to a comet shower (Farley et al., 1998). However, this conclusion has been challenged (Tagle and Claeys, 2004; Fritz et al., 2007). Tagle and Claeys (2005) presented evidence that the Popigai crater was produced by an impactor chemically distinct from carbonaceous chondrites, and therefore possibly not a comet. Based on this conclusion, Tagle and Claeys (2004) proposed the occurrence of an asteroid shower in the late Eocene. To account for the temporal coincidence of impacts and cosmic dust, Fritz et al. (2007) made the novel suggestion that asteroid impacts with the moon during the shower could loft ^{3}He-rich dust that would then be captured by Earth. In the next section, I present and evaluate evidence that bears on this proposed model.

Mixed Evidence against a Cometary Impactor at Popigai

The main motivation for the "lunar impact" alternative to the comet shower model is observations of the impactor signature in the Popigai crater and associated ejecta, both PGEs (Tagle and Claeys, 2005) and Cr isotopes (Kyte et al., 2004). Tagle and Claeys (2005) analyzed the PGE composition of impact glasses at Popigai and attributed the varying abundances to mixing of terrestrial and impactor components. By plotting one PGE against another, they were able to identify apparent PGE ratios of the impactor. An example for the key ratio Ru/Ir is shown in Figure 4, replotted from the data in Tagle and Claeys (2005).

To obtain the slope of the mixing relationship and thus establish the signature of the impactor, Tagle and Claeys (2005) performed a non–error-weighted linear regression. For reasons never specified, they excluded one data point from the Ru-Ir regression (circled in Fig. 4). The result is the slope shown in the figure, implying an impactor Ru/Ir of 1.47 ± 0.08 (1σ). As shown in Figure 2B, this ratio plots broadly with the L chondrites and away from carbonaceous chondrites. Tagle and Claeys (2004, 2005) proposed that since the Popigai signature is distinct from carbonaceous chondrites, the impactor was not a comet. Instead they argued for an L-chondrite impactor.

This conclusion is remarkably sensitive to precisely how the regression is done and the ways in which uncertainties are treated. I illustrate this in several ways in Figures 2B and 4. The approach of Tagle and Claeys (2005) is arguably flawed in three ways. (1) The standard linear regression technique they used is not appropriate when both *x* and *y* variables have associated uncertainties. One can demonstrate this conclusion simply by swapping *x* for *y* and recomputing the regression—the new slope is very different from the inverse of the original slope. One proper technique to be used on such data is the York or "*x-y*" regression, for example, as provided in Isoplot (Ludwig, 2003). (2) It is important to error-weight the individual points in the regression. (3) It seems arbitrary to disregard a data point without clear justification; while the discarded point lies far from the "best-fit" array (Fig. 4), it is apparent that even the remaining data do not conform

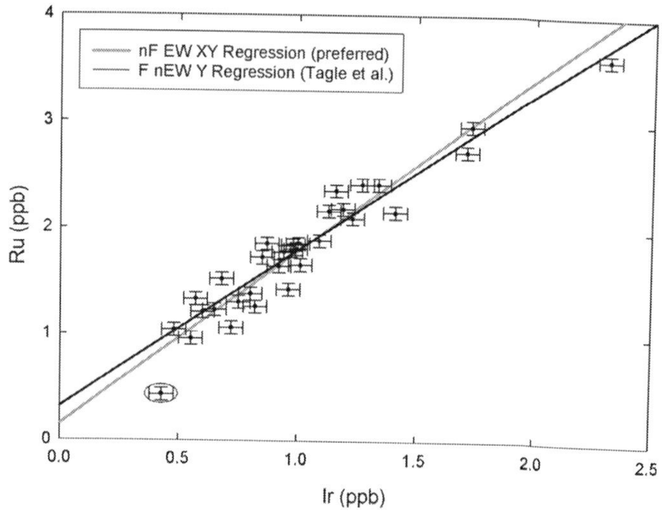

Figure 4. Regressions of the Popigai impact glass platinum group element (PGE) data of Tagle and Claeys (2005). Two possible fits to the data are shown, the one proposed by Tagle and Claeys (2005) and the one preferred here. Using the notation in the caption to Figure 2, these are the (F nEW *y*) and (nF EW *x-y*) regressions, respectively. The circled data point was ignored in the regression performed by Tagle and Claeys (2005) without explanation.

to the assumed two-component model. This discarded data point thus has equal validity to all others when fitting the slope.

To demonstrate the sensitivity of the impactor signature to the assumptions and methods used in the regression, using Isoplot I have recomputed the slopes for the ratios Ru/Ir and Ru/Rh from the data of Tagle and Claeys (2005) in six different ways that reflect the permutations of the three variables: (1) data point discarded or not discarded; (2) error-weighted or not error-weighted; (3) *x-y* regression or *y*-only regression. The range of resulting Ru/Ir ratios is indicated in Figure 2 and defines the gray 2σ field. Compared with the point computed by Tagle and Claeys (2005) (*y* regression, not error-weighted, filtered), the point that seems most defensible (*x-y* regression, error-weighted, unfiltered) lies at much higher Ru/Ir and slightly higher Ru/Rh. Within error, this point overlaps individual members of almost all of the different meteorite classes, including carbonaceous chondrites. The point of these calculations is not to assert what the "true" slope is, but simply to demonstrate that the results are very dependent on the assumptions made, and, in that sense, the conclusion cannot be considered robust.

In contrast, Kyte et al. (2004) presented evidence from Cr isotopes that argues strongly against a carbonaceous chondrite impactor for at least one large late Eocene impact. Debris presumably derived from the Popigai impact has $^{54}Cr/^{52}Cr$ ratios that plot within the field of ordinary chondrites and far removed from that of carbonaceous chondrites. This conclusion is analytically robust and based on a fairly extensive and consistent database of Cr isotopes in meteorites (Trinquier et al., 2007). Thus, if comets *must* be chemically similar to carbonaceous chondrites in terms of Cr isotopes, then a key argument in favor of the late Eocene comet shower model—the time-correlated large impact at Popigai—falls.

The Lunar Impact Alternative

The model proposed by Fritz et al. (2007) to account for an ordinary chondrite (L-class) impactor during the dust event invokes a transient shower of asteroid fragments. An intriguing feature of the model is that significant amounts of dust and hence ^3He accumulating on Earth may be derived from lunar impacts. This possibility seems very likely because the lunar surface is rich in ^3He, and the lofted dust would likely encounter Earth at comparatively low velocity, favoring capture and reducing the heating and He degassing associated with atmospheric entry. Indeed, transfer of ^3He rich dust from moon to Earth after a lunar impact seems almost inescapable.

However, there are several observations that are either problematic or not predicted by the asteroid shower–lunar dust model. Perhaps the biggest problems are that (1) there is no specific model for the proposed asteroid shower, and (2) there is no independent evidence from asteroid orbits for a major collision at the appropriate time from which such a model might be developed. As evidence for the asteroid shower scenario, Fritz et al. (2007) noted a peak in cosmic-ray exposure ages of L-chondrites suggesting a collision at 40 Ma. Interestingly, there is no dust record supporting this event (Farley et al., 2006).

In the absence of a quantitative model for the asteroid shower, there is no natural temporal evolution with which to compare the ^3He data. Why does the ^3He (or lunar impact frequency) follow the 2.5 m.y. temporal pattern documented in Figure 3, a pattern that closely matches the comet shower model? Existing models of asteroid showers are all much longer than this duration (Zappala et al., 1998). Because lunar dust is likely removed from the Earth-moon system on a very short time scale (<10–20 k.y.), it is necessary to call on literally hundreds of lunar impacts to account for the ~2.5 m.y. duration of the event. Why does the lunar dust from the asteroid shower rise and fall in such a characteristic pattern? Why does dust from the individual impacts coalesce so smoothly in the ^3He record, such that only one (possible) subsidiary peak rises above the main trend in Figure 3? Finally, as noted by Fritz et al. (2007), the lunar impacts scenario may have transferred relatively large particles to Earth at low velocities. If so, one might expect an increase in the "nugget effect" in terms of ^3He reproducibility. This has not been observed (Farley et al., 1998).

CONCLUSIONS

A nearly complete extraterrestrial ^3He flux record spanning the last ~70 m.y. (Mukhopadhyay et al., 2001b; Farley et al., 2006) indicates two periods of enhanced solar system dustiness, in the late Eocene and the late Miocene. These two events are remarkably similar in duration and magnitude—a 2.5 m.y. period of enhanced flux rising to a maximum of about five times pre-event levels. Despite their similarities, the two events probably have different causative mechanisms.

The strong temporal match between the onset of dustiness and one of the largest recent collisions in the asteroid belt convincingly implicates the Veritas-family forming event as the source of the late Miocene dust. A collisional cascade model can account for the nearly instantaneous rise of the flux and its ~2 m.y. decay. Interestingly, a large fraction of the dust during the peak of the event should have come from the Veritas parent body; if cosmic dust particles, melt spherules, or cosmic spinels (Pierrard et al., 1998; Schmitz et al., this volume) of this age can be isolated, it may be possible to map them to a specific extraterrestrial source. The apparent absence of large terrestrial impact craters during this dust event is consistent with expectations that it would take millions of years (if ever) for large asteroid fragments produced by this collision to cross Earth's orbit. The only dissatisfying element in this story is that the prominent 7 Ma peak in cosmic-ray exposure ages, which seems a natural outcome for such a large asteroid collision event, appears inconsistent with a Veritas source. The two may be completely unrelated despite their similar timing, or, conceivably, the H chondrite parent body may have been destroyed by a fragment of the Veritas collision.

The late Eocene event consists of two major components: a period of enhanced dust accumulation, and, near its peak, the almost simultaneous occurrence of the two by-far largest impact craters in the last 65 m.y. It is almost inconceivable that these events are unrelated. A model involving an impulsive perturbation of the Oort cloud by close stellar encounter nicely accounts for the impacts and the temporal evolution of the dust event. However, Cr isotope data indicate that at least one of the two impacts was produced by an object that was chemically distinct from carbonaceous chondrites. To the extent that comets must have the same fingerprint as carbonaceous chondrites, this substantially weakens the comet shower model. A similar conclusion was reached based on the PGE fingerprint of the Popigai impactor, though these data are not as robust.

A recently proposed alternative (Fritz et al., 2007), which states that the late Eocene event was an asteroid shower where dust was delivered by impacts on the moon, is plausible. However, it lacks a clear causative mechanism, does not naturally predict several key observations, and makes predictions that seem to be inconsistent with observations.

Clearly, further work is required to establish the cause of the late Eocene event. Several types of observations might settle the issue in favor of one or the other of these mechanisms. For example, a quantitative model that explains how the asteroid shower occurs, how long it lasts, and how lunar dust is sampled by Earth could support or invalidate this scenario. Because the comet shower model is consistent with all observations except the presumption that comets have carbonaceous-chondrite–like chemistry, further work on (or motivated by) *Stardust* samples from Comet Wild 2 may help. Finally, upcoming astrometric observations might be sufficiently complete and detailed to directly observe the "smoking gun" of a comet shower—a star that passed close to our solar system at the appropriate time.

ACKNOWLEDGMENTS

I would like to thank the conveners of the 2007 Penrose Conference at which this work was presented: Alessandro Montanari, Christian Koeberl, Fritz Hilgen, and Rodolfo Coccioni. I appreciate the helpful reviews of this manuscript by B. Schmitz and I. McDonald.

REFERENCES CITED

Alwmark, C., and Schmitz, B., 2007, Extraterrestrial chromite in the resurge deposits of the early Late Ordovician Lockne crater, central Sweden: Earth and Planetary Science Letters, v. 253, p. 291–303, doi: 10.1016/j.epsl.2006.10.034.

Bottke, W.F., Vokrouhlicky, D., and Nesvorny, D., 2007, An asteroid breakup 160 Myr ago as the probable source of the K/T impactor: Nature, v. 449, p. 48–53, doi: 10.1038/nature06070.

Bottomley, R., Grieve, R., York, D., and Masaitis, V., 1997, The age of the Popigai impact event and its relation to events at the Eocene/Oligocene boundary: Nature, v. 388, p. 365–368, doi: 10.1038/41073.

Burbine, T.H., McCoy, T.J., Meibom, A., Gladman, B., and Keil, K., 2002, Meteoritic parent bodies: Their number and identification, *in* Bottke, W.F., ed., Asteroids III: Tucson, Arizona, University of Arizona Press, p. 653–667.

Ciesla, F.J., 2007, Outward transport of high-temperature materials around the midplane of the solar nebula: Science, v. 318, p. 613–615, doi: 10.1126/science.1147273.

Durda, D., and Dermott, S., 1997, The collisional evolution of the asteroid belt and its contribution to the zodiacal cloud: Icarus, v. 130, p. 140–164, doi: 10.1006/icar.1997.5803.

Dybczynski, P.A., 2002, Simulating observable comets: I. The effects of a single stellar passage through or near the Oort cometary cloud: Astronomy & Astrophysics, v. 396, p. 283–292, doi: 10.1051/0004-6361:20021400.

Eugster, O., 2003, Cosmic-ray exposure ages of meteorites and lunar rocks and their significance: Chemie der Erde–Geochemistry, v. 63, p. 3–30, doi: 10.1078/0009-2819-00021.

Farley, K.A., 1995, Cenozoic variations in the flux of interplanetary dust recorded by ^{3}He in a deep sea sediment: Nature, v. 376, p. 153–156, doi: 10.1038/376153a0.

Farley, K.A., 2001, Extraterrestrial helium in seafloor sediments: Identification, characteristics, and accretion rate over geologic time, *in* Peucker-Ehrenbrink, B., and Schmitz, B., eds., Accretion of Extraterrestrial Matter throughout Earth's History: Dordrecht, The Netherlands, Kluwer/Plenum, p. 179–204.

Farley, K.A., Love, S.G., and Patterson, D.B., 1997, Atmospheric entry heating and helium retentivity of interplanetary dust particles: Geochimica et Cosmochimica Acta, v. 61, p. 2309–2316, doi: 10.1016/S0016-7037(97)00068-9.

Farley, K.A., Montanari, A., Shoemaker, E.M., and Shoemaker, C.S., 1998, Geochemical evidence for a comet shower in the late Eocene: Science, v. 280, p. 1250–1253, doi: 10.1126/science.280.5367.1250.

Farley, K.A., Vokrouhlicky, D., Bottke, W.F., and Nesvorny, D., 2006, A late Miocene dust shower from the break-up of an asteroid in the main belt: Nature, v. 439, p. 295–297, doi: 10.1038/nature04391.

Fernandez, J.A., and Ip, W.H., 1987, Time-dependent injection of Oort cloud comets into Earth-crossing orbits: Icarus, v. 71, p. 46–56, doi: 10.1016/0019-1035(87)90161-8.

Fritz, J., Tagle, R., and Artemieva, N., 2007, Lunar helium-3 in marine sediments: Implications for a late Eocene asteroid shower: Icarus, v. 189, p. 591–594, doi: 10.1016/j.icarus.2007.03.026.

Garcia-Sanchez, J., Weissman, P.R., Preston, R.A., Jones, D.L., Lestrade, J.F., Latham, D.W., Stefanik, R.P., and Paredes, J.M., 2001, Stellar encounters with the solar system: Astronomy & Astrophysics, v. 379, p. 634–659, doi: 10.1051/0004-6361:20011330.

Gladman, B.J., Migliorini, F., Morbidelli, A., Zappalà, V., Michel, P., Cellino, A., Froeschlé, C., Levison, H.F., Bailey, M., and Duncan, M., 1997, Dynamical lifetimes of objects injected into asteroid belt resonances: Science, v. 277, p. 197–201, doi: 10.1126/science.277.5323.197.

Graf, T., and Marti, K., 1994, Collisional records in LL-chondrites: Meteoritics, v. 29, p. 643–648.

Graf, T., and Marti, K., 1995, Collisional history of H-chondrites: Journal of Geophysical Research–Planets, v. 100, p. 21,247–21,263, doi: 10.1029/95JE01903.

Grieve, R.A.F., 2001, The terrestrial cratering record, *in* Peucker-Ehrinbrink, B., and Schmitz, B., eds., Accretion of Extraterrestrial Matter throughout Earth's History: New York, Kluwer, p. 379–402.

Hut, P., Alvarez, W., Elser, W.P., Hansen, T., Kauffman, E.G., Keller, G., Shoemaker, E.M., and Weissman, P.R., 1987, Comet showers as a cause of mass extinctions: Nature, v. 329, p. 118–127, doi: 10.1038/329118a0.

Ishii, H.A., Bradley, J.P., Dai, Z.R., Chi, M.F., Kearsley, A.T., Burchell, M.J., Browning, N.D., and Molster, F., 2008, Comparison of comet 81P/Wild 2 dust with interplanetary dust from comets: Science, v. 319, p. 447–450, doi: 10.1126/science.1150683.

Karner, D.B., Levine, J., Muller, R.A., Asaro, F., Ram, M., and Stolz, M.R., 2003, Extraterrestrial accretion from the GISP2 ice core: Geochimica et Cosmochimica Acta, v. 67, p. 751–763, doi: 10.1016/S0016-7037(02)01145-6.

Koeberl, C., Poag, C.W., Reimold, W.U., and Brandt, D., 1996, Impact origin of the Chesapeake Bay structure and the source of the North American tektites: Science, v. 271, p. 1263–1266, doi: 10.1126/science.271.5253.1263.

Koeberl, C., Shukolyukov, A., and Lugmair, G.W., 2007, Chromium isotopic studies of terrestrial impact craters: Identification of meteoritic components at Bosumtwi, Clearwater East, Lappajarvi, and Rochechouart: Earth and Planetary Science Letters, v. 256, p. 534–546, doi: 10.1016/j.epsl.2007.02.008.

Koiwa, Y., Shirai, N., and Ebihara, M., 2006, Fractionation of platinum group elements in carbonaceous chondrites: Houston, Texas, Lunar and Planetary Sciences Conference XXXVII, abs. 1928 (CD-ROM).

Kyte, F.T., 1998, A meteorite from the Cretaceous/Tertiary boundary: Nature, v. 396, p. 237–239, doi: 10.1038/24322.

Kyte, F.T., 2002, Unmelted meteoritic debris collected from Eltanin ejecta in Polarstern cores from expedition ANT XII/4: Deep-Sea Research Part I–Topical Studies in Oceanography, v. 49, p. 1063–1071, doi: 10.1016/S0967-0645(01)00141-2.
Kyte, F.T., Shukolyukov, A., Hildebrand, A.R., Lugmair, G.W., and Hanova, J., 2004, Initial Cr isotopic and iridium measurements of concentrates from late Eocene cpx-spherule deposits: Houston, Texas, Lunar and Planetary Sciences Conference XXXV, abs. 1824 (CD-ROM).
Love, S.G., and Brownlee, D.E., 1993, A direct measurement of the terrestrial mass accretion rate of cosmic dust: Science, v. 262, p. 550–553, doi: 10.1126/science.262.5133.550.
Ludwig, K.R., 2003, Isoplot 3.0: Berkeley Geochronology Center Special Publication 4.
Maier, W.D., Andreoli, M.A.G., McDonald, I., Higgins, M.D., Boyce, A.J., Shukolyukov, A., Lugmair, G.W., Ashwal, L.D., Graser, P., Ripley, E.M., and Hart, R.J., 2006, Discovery of a 25-cm asteroid clast in the giant Morokweng impact crater, South Africa: Nature, v. 441, p. 203–206, doi: 10.1038/nature04751.
Marty, B., Palma, R.L., Pepin, R.O., Zimmermann, L., Schlutter, D.J., Burnard, P.G., Westphal, A.J., Snead, C.J., Bajt, S., Becker, R.H., and Simones, J.E., 2008, Helium and neon abundances and compositions in cometary matter: Science, v. 319, p. 75–78, doi: 10.1126/science.1148001.
McDonald, I., Andreoli, M.A.G., Hart, R.J., and Tredoux, M., 2001, Platinum-group elements in the Morokweng impact structure, South Africa: Evidence for the impact of a large ordinary chondrite projectile at the Jurassic-Cretaceous boundary: Geochimica et Cosmochimica Acta, v. 65, p. 299–309, doi: 10.1016/S0016-7037(00)00527-5.
Merrihue, C., 1964, Rare gas evidence for cosmic dust in modern Pacific red clay: Annals of the New York Academy of Sciences, v. 119, p. 351–367, doi: 10.1111/j.1749-6632.1965.tb47445.x.
Mukhopadhyay, S., Farley, K.A., and Montanari, A., 2001a, A 35 Myr record of helium in pelagic limestones: Implications for interplanetary dust accretion from the early Maastrichtian to the middle Eocene: Geochimica et Cosmochimica Acta, v. 65, p. 653–669, doi: 10.1016/S0016-7037(00)00555-X.
Mukhopadhyay, S., Farley, K.A., and Montanari, A., 2001b, A short duration of the Cretaceous-Tertiary boundary event: Evidence from extraterrestrial helium-3: Science, v. 291, p. 1952–1955, doi: 10.1126/science.291.5510.1952.
Nesvorny, D., Bottke, W.F., Levison, H.F., and Dones, L., 2003, Recent origin of the solar system dust bands: The Astrophysical Journal, v. 591, p. 486–497, doi: 10.1086/374807.
Nesvorny, D., Vokrouhlicky, D., and Bottke, W.F., 2006, The breakup of a main-belt asteroid 450 thousand years ago: Science, v. 312, p. 1490, doi: 10.1126/science.1126175.
Nesvorny, D., Vokrouhlicky, D., Bottke, W.F., Gladman, B., and Haggstrom, T., 2007, Express delivery of fossil meteorites from the inner asteroid belt to Sweden: Icarus, v. 188, p. 400–413, doi: 10.1016/j.icarus.2006.11.021.
Nier, A.O., and Schlutter, D.J., 1990, Helium and neon isotopes in stratospheric particles: Meteoritics, v. 25, p. 263–267.
Ozima, M., Takayanagi, M., Zashu, S., and Amari, S., 1984, High $^3He/^4He$ ratios in ocean sediments: Nature, v. 311, p. 448–450, doi: 10.1038/311448a0.
Pierrard, O., Robin, E., Rocchia, R., and Montanari, A., 1998, Extraterrestrial Ni-rich spinel in Upper Eocene sediments from Massignano, Italy: Geology, v. 26, p. 307–310, doi: 10.1130/0091-7613(1998)026<0307:ENRSIU>2.3.CO;2.
Schmitz, B., Cronholm, A., and Montanari, A., 2009, this volume, Relict extraterrestrial minerals in the late Eocene Massignano section, central Italy: Preliminary results, *in* Koeberl, C., and Montanari, A., eds., The Late Eocene Earth—Hothouse, Icehouse, and Impacts: Geological Society of America Special Paper 452, doi: 10.1130/2009.2452(05).
Shukolyukov, A., and Lugmair, G.W., 2006, Manganese-chromium isotope systematics of carbonaceous chondrites: Earth and Planetary Science Letters, v. 250, p. 200–213, doi: 10.1016/j.epsl.2006.07.036.
Tagle, R., 2004, Platingruppenelemente in Meteoriten und Gesteinen irdischer Impaktkratar: Identifizierung der Einschlagskörper [Ph.D. thesis]: Berlin, Humboldt-Universität, 159 p.
Tagle, R., and Claeys, P., 2004, Comet or asteroid shower in the late Eocene?: Science, v. 305, p. 492, doi: 10.1126/science.1098481.
Tagle, R., and Claeys, P., 2005, An ordinary chondrite impactor for the Popigai crater, Siberia: Geochimica et Cosmochimica Acta, v. 69, p. 2877–2889, doi: 10.1016/j.gca.2004.11.024.
Tagle, R., and Hecht, L., 2006, Geochemical identification of projectiles in impact rocks: Meteoritics & Planetary Science, v. 41, p. 1721–1735.
Takayanagi, M., and Ozima, M., 1987, Temporal variation of $^3He/^4He$ in deep-sea sediment cores: Journal of Geophysical Research, v. 92, p. 12,531–12,538, doi: 10.1029/JB092iB12p12531.
Trinquier, A., Birck, J.L., and Allegre, C.J., 2006, The nature of the KT impactor: A Cr-54 reappraisal: Earth and Planetary Science Letters, v. 241, p. 780–788, doi: 10.1016/j.epsl.2005.11.006.
Trinquier, A., Birck, J.L., and Allegre, C.J., 2007, Widespread Cr-54 heterogeneity in the inner solar system: The Astrophysical Journal, v. 655, p. 1179–1185, doi: 10.1086/510360.
Weissman, P., 1990, The Oort cloud: Nature, v. 344, p. 825–830, doi: 10.1038/344825a0.
Zappala, V., Cellino, A., Gladman, B.J., Manley, S., and Migliorini, F., 1998, Asteroid showers on Earth after family breakup events: Icarus, v. 134, p. 176–179, doi: 10.1006/icar.1998.5946.

Manuscript Accepted by the Society 16 September 2008

The Geological Society of America
Special Paper 452
2009

The late Eocene clinopyroxene-bearing spherule layer: New sites, nature of the strewn field, Ir data, and discovery of coesite and shocked quartz

Shaobin Liu*
B.P. Glass
Department of Geological Sciences, University of Delaware, Newark, Delaware 19716, USA

Frank T. Kyte
Institute of Geophysics and Planetary Physics, University of California, Los Angeles, California 90095-1567, USA

Steven M. Bohaty
School of Ocean and Earth Science, University of Southampton, European Way, Southampton SO14 3ZH, UK

ABSTRACT

In order to better define the late Eocene clinopyroxene-bearing (cpx) spherule layer and to determine how the ejecta vary with distance from the presumed source crater (Popigai), we searched for the layer at 23 additional sites. We identified the layer at six (maybe seven) of these sites: Deep Sea Drilling Project (DSDP) and Ocean Drilling Program (ODP) Holes 592, 699A, 703A, 709C, 786A, 1090B, and probably 738B. The cpx spherule layer occurs in magnetochron 16n.1n, which indicates an age of ca. 35.4 ± 0.1 Ma for the layer. We found the highest abundance of cpx spherules and associated microtektites in Hole 709C in the northwest Indian Ocean, and we found coesite and shocked quartz in the cpx spherule layer at this site. We also found coesite in the cpx spherule layer at Site 216 in the northeast Indian Ocean. This is the first time that coesite has been found in the cpx spherule layer, and it provides additional support for the impact origin of this layer. In addition, the discovery of coesite and shocked quartz grains (with planar deformation features [PDFs]) supports the conclusion that the pancake-shaped clay spherules associated with quartz grains exhibiting PDFs are diagenetically altered cpx spherules. An Ir anomaly was found associated with the cpx spherule layer at all four of the new sites (699A, 709C, 738B, 1090B) for which we obtained Ir data. The geometric mean of the Ir fluence for the 12 sites with Ir data is 5.7 ng/cm^2, which is ~10% of the fluence estimated for the Cretaceous-Tertiary boundary. Based on the geographic distribution of the 23 sites now known to contain the cpx spherule layer, and 12 sites where we have good chronostratigraphy but the cpx spherule layer is apparently absent, we propose that

*Liu_Shaobin@cat.com

Liu, S., Glass, B.P., Kyte, F.T., and Bohaty, S.M., 2009, The late Eocene clinopyroxene-bearing spherule layer: New sites, nature of the strewn field, Ir data, and discovery of coesite and shocked quartz, *in* Koeberl, C., and Montanari, A., eds., The Late Eocene Earth—Hothouse, Icehouse, and Impacts: Geological Society of America Special Paper 452, p. 37–70, doi: 10.1130/2009.2452(04). For permission to copy, contact editing@geosociety.org.

the cpx spherule strewn field may have a ray-like distribution pattern. Within one of the rays, the abundance of spherules decreases and the percent microtektites increases with distance from Popigai. Shocked quartz and coesite have been found only in this ray at the two sites that are closest to Popigai. At several sites in the Southern Ocean, an increase in $\delta^{18}O$ in the bulk carbonate occurs immediately above the cpx spherule layer. This increase may indicate a drop in temperature coincident with the impact that produced the cpx spherule layer.

Keywords: impact spherules, iridium, coesite, shocked quartz, oxygen isotopes.

INTRODUCTION

The late Eocene is marked by a series of stepwise faunal and floral turnovers, a deterioration of climate (Prothero et al., 2003), and several high-energy impacts, including the two largest known impacts in the Cenozoic, i.e., the impacts that produced the 100-km-diameter Popigai (Kettrup et al., 2003) and 85-km-diameter Chesapeake Bay (Poag et al., 2004) impact structures. Distal impact ejecta layers have been found in late Eocene marine deposits around the globe, but there has been much controversy regarding the number of late Eocene ejecta layers (Glass et al., 1979; Keller et al., 1983, 1987; Glass et al., 1985; Glass and Burns, 1987; Hazel, 1989; Miller et al., 1991; Wei, 1995; Whitehead et al., 2000; Glass, 2002; Marchand and Whitehead, 2002; Liu et al., 2006). However, since the early 1990s, most researchers have agreed that there are probably only two widespread impact ejecta layers in late Eocene deposits: the North American microtektite layer, and the slightly older (~15 k.y. older) clinopyroxene-bearing (cpx) spherule layer. The North American microtektite layer is believed to have been derived from the Chesapeake Bay impact structure (Poag et al., 1994; Koeberl et al., 1996; Glass et al., 1998; Deutsch and Koeberl, 2006), whereas the cpx spherule layer is thought to have been derived from the Popigai impact structure (Whitehead et al., 2000; Glass et al., 2004a; Liu et al., 2006).

In earlier studies, the clinopyroxene-bearing spherules were referred to as microtektites (e.g., Keller et al., 1983), crystal-bearing microspherules (Keller et al., 1987), microspherules (Hazel, 1989), or microkrystites (e.g., Wei, 1995; Vonhof and Smit, 1999). Although the clinopyroxene-bearing spherule layer does contain microtektites, the clinopyroxene-bearing spherules are not microtektites, because they contain crystals. We prefer the term "spherule" rather than "microspherule," since spherule refers to small spherical bodies (Gary et al., 1972). The term "microkrystite" refers to all crystal-bearing impact spherules (Glass and Burns, 1988), and, therefore, the clinopyroxene-bearing spherules are microkrystites; however, we prefer to call them clinopyroxene-bearing spherules because the term is more precise. Throughout most of this paper, we use cpx as the abbreviation for clinopyroxene and refer to the clinopyroxene-bearing spherules as cpx spherules.

The North American microtektites are found closer to their proposed source crater than the cpx spherules are to theirs, and the North American microtektites are associated with shocked mineral grains and rock fragments (Thein, 1987; Glass, 1989; Glass and Wu, 1993; Glass et al., 1998; Glass and Liu, 2001). The cpx spherules contain Ni- and/or Cr-rich spinel crystallites (John and Glass, 1974; Glass et al., 1985; Glass and Burns, 1987) and the cpx spherule layer is associated with an Ir anomaly (Alvarez et al., 1982; Glass et al., 1985; Keller et al., 1987; Kyte and Liu, 2002). The cpx spherule layer may be nearly global in extent and at a few sites appears to be associated with the last occurrence (extinction?) of several radiolarian species (Maurrasse and Glass, 1976; Glass et al., 1985).

Montanari et al. (1993) reported the presence of a positive Ir anomaly at the 5.61 m level in late Eocene deposits at Massignano, Italy. Clymer et al. (1996) reported finding shocked quartz grains at the same level as this Ir anomaly. Pierrard et al. (1998) found flattened spherules (pancake spherules), some with Ni-rich spinel crystals on their surfaces, at approximately the same level as the Ir anomaly and shocked quartz. It seemed possible that the pancake spherules with Ni-rich spinel crystals on their surfaces are altered cpx spherules. However, since shocked quartz had been found associated with the North American microtektite layer, but not with the cpx spherule layer, Glass et al. (2004b) re-examined the Ir-enriched layer at Massignano to determine if the shocked quartz grains were above the pancake spherules, rather than associated with them as previously proposed. They found that the peak abundance of the shocked quartz does indeed correlate with the peak abundance of the pancake spherules. Assuming that the pancake-shaped spherules are altered cpx spherules, Massignano is the first site where shock-metamorphosed quartz grains have been reported to occur in the cpx spherule layer.

The cpx spherules have not been radiometrically dated, but biostratigraphic studies of sites where both layers occur indicate that the cpx spherule layer is ~10–15 k.y. older than the North American microtektite layer (Glass et al., 1998), which has been radiometrically dated. Tektite fragments from the North American microtektite layer on Barbados were determined to have an age of 35.4 ± 0.6 Ma (2σ error) using $^{40}Ar/^{39}Ar$ laser-probe dating (Glass et al., 1986). North American tektite fragments from Deep Sea Drilling Project (DSDP) Site 612 on the continental slope off New Jersey gave a $^{40}Ar/^{39}Ar$ plateau age of 35.5 ± 0.3 Ma (Obradovich et al., 1989). Within the quoted errors, these ages agree with the more recently reported K-Ar age for Georgia tektites of 35.2 ± 0.2 Ma (Albin and Wampler, 1996) and weighted mean

$^{40}Ar/^{39}Ar$ age of North American tektites (five from Georgia and 14 from Texas) of 35.3 ± 0.1 Ma (Horton and Izett, 2005). Thus, the cpx spherule layer appears to have an age of ca. 35.4 ± 0.3 Ma. This age is in good agreement with the age of 35.7 ± 0.8 Ma for the Popigai impact structure in northern Siberia (Bottomley et al., 1997), which is believed to be the source crater for the cpx spherule layer (Langenhorst, 1996; Poag, 1997; Vonhof and Smit, 1999; Whitehead et al., 2000; Glass et al., 2004a; Liu et al., 2006).

At equatorial sites, the cpx spherule layer is associated with a sharp drop in abundance (or extinction) of several radiolarian species: *Thyrsocyrtis bromia*, *T. tetracantha*, *T. triacantha*, *T. finalis*, and *Calocyclas turris* (Maurrasse and Glass, 1976; Glass et al., 1985; Sanfilippo et al., 1985). In earlier studies, the North American microtektite–cpx spherule layer was found in what had been identified as the *Globigerinatheka semiinvoluta* (P15) or the *Gl. cerroazulensis* (P16) planktonic foraminifera zone and the *Discoaster barbadiensis* (CP15) or *Isthmolithus recurvus–Sphenolithus pseudoradians* (NP19/20) calcareous nannofossil zone (Glass et al., 1985; Glass, 1986; Keller et al., 1987). In these earlier studies, it was also determined that the cpx spherule layer is in a normal magnetochron, but it was not known how this polarity interval correlated with the polarity time scale (Glass et al., 1985; Glass and Burns, 1987). However, the cpx spherule layer at Ocean Drilling Program (ODP) Hole 689B and at Massignano (assuming that the pancake spherules are altered cpx spherules) appears to be within magnetochron 16n.1n (Funakawa and Nishi, 2005; Jovane et al., 2006).

The purpose of the present research was to search for additional sites containing the cpx spherule layer in an attempt to better define this layer and to look for shocked quartz and other evidence of shock-metamorphosed grains associated with the cpx spherules. During this research, we searched cores from 23 Deep Sea Drilling Project (DSDP) and Ocean Drilling Program (ODP) sites for microtektites, cpx spherules, and unmelted impact ejecta (e.g., quartz with planar deformation features [PDFs], coesite) in late Eocene sediments. We found cpx spherules, microtektites, and/or unmelted ejecta at nine sites where they previously had not been found. In addition, we searched for the cpx spherule layer at ODP Site 689 (Hole 689D), which was taken ~24 m from Hole 689B, where a late Eocene Ir anomaly and cpx spherule layer had previously been found (Vonhof and Smit, 1999; Glass and Koeberl, 1999). We also report here, for the first time, the discovery of coesite associated with the cpx spherule layer. This paper is based primarily on the Ph.D. dissertation of the senior author.

METHODS

Twenty-three DSDP and ODP sites were selected for this study based on biostratigraphic and magnetostratigraphic data and >50% core recovery (Table 1). Initially, samples of 2 or 3 cm^3 were requested at 20 or 50 cm intervals in order to narrow down the interval of interest based on biostratigraphy and/or by finding impact spherules. After the initial search, the section of interest

TABLE 1. DEEP SEA DRILLING PROJECT (DSDP)/OCEAN DRILLING PROGRAM (ODP) SITES SEARCHED IN THIS STUDY

Site/Hole*	Latitude	Longitude	Water depth (mbsl)†	Core-section§	Sample interval (cm)	Number of samples
DSDP 111A	50°25.572′N	46°22.050′W	1797	6R-3	10	5
DSDP 522A	26°06.840′S	5°07.782′W	4441	29H-CC–30H-CC	20	14
DSDP 543	15°42.738′N	58°39.222′W	5633	27R-4–27R-5	1–20	29
DSDP 592	36°28.40′N	165°26.53′E	1088	38-2–38-6	10–50	33
ODP 647A	53°19.876′N	45°15.717′W	3868	35R-1–36R-4	20	59
ODP 672A	15°32.40′N	58°38.46′W	4982	39X-1–40X-4	20	59
ODP 674A	15°32.29′N	58°51.09′W	4550	36X-1–36X-2	20	16
ODP 689D	64°31.020′S	3°06.024′E	2080	11H-1–12H-CC	1–40	65
ODP 699A	51°32.520′S	30°40.620′W	3705	33X-7–36X-5	2–20	101
ODP 703A	47°03.042′S	07°53.679′E	1796	14H-6–15H-CC	20	38
ODP 709C	03°54.9′S	60°33.1′E	3039	31X-4–32-1	3–20	25
ODP 738B	62°42.54′S	82°47.25′E	2252	5H-2–5H3	5–10	22
ODP 744A	61°34.66′S	80°35.46′E	2307	18H-5–20H-CC	20	124
ODP 757B	17°01.458′S	88°10.889′E	1652	13H-1–15H-6	20	138
ODP 762C	19°53.220′S	112°15.24′E	1360	7X-1–8X-2	20	45
ODP 763B	20°35.19′S	112°12.52′E	1368	3X-1–4X-3	20	56
ODP 782A	30°51.66′N	141°18.85′E	2959	40X-4–41X-2	20	16
ODP 786A	31°52.500′N	141°13.560′E	3058	11X-1–11X-2	1–10	39
ODP 884B	51°27.026′N	168°20.228′E	3825	75X, 83X-3	10	100
ODP 900A	46°40.994′N	11°36.252′W	5037	53R-1–54R-CC	20	77
ODP 1053A	29°59.538′N	76°31.434′W	1630	7H-1–8H-CC	50	47
ODP 1073A#	39°13.522′N	72°16.546′W	639	72X-1–72X-CC	2–20	53
ODP 1090B	42°54.822′S	8°53.982′E	3699	30X-1–31X-CC	20	55

*Impact spherules/ejecta were found at the sites/holes in bold.
†mbsl—meters below sea level.
§For example, 6R-3 means section 3 in core 6R; 27R-4–27R-5 means core 27R, section 4 through core 27R, section 5; CC—core catcher.
#Only unmelted ejecta thought to belong to the North American microtektite layer were found at this site; no cpx spherules or microtektites were observed.

was then sampled at closer intervals (e.g., 10 cm or less). When impact spherules and/or unmelted impact ejecta were found, additional samples were requested in order to better define the vertical distribution and the peak abundance. In total, over 1200 samples were analyzed during this study. We also added unpublished data from DSDP Site 592 and data from Hole 738B where glass spherules were discovered by one co-author (Bohaty).

The samples were disaggregated in water and sieved into three size fractions: <63, 63–125, and >125 μm. The 63–125 and >125 μm size fractions were searched for impact spherules (cpx spherules and microtektites) and possible unmelted ejecta using a binocular microscope with up to 50× magnification. For some samples, heavy liquid or magnetic separation was used to help concentrate the cpx spherules and microtektites from the lighter or nonmagnetic grains, respectively (e.g., nonmagnetic mineral grains, microfossils). Translucent quartz grains were studied using a petrographic microscope in order to determine whether or not they exhibited planar deformation features. White opaque mineral grains were recovered and studied using X-ray diffraction (XRD) to determine if they showed evidence of shock metamorphism (e.g., the presence of high-pressure polymorphs). Each grain was mounted at the end of a glass rod using water-soluble glue. A holder with the glass rod inserted in it was put into a camera, and the grain was positioned to intercept the X-ray beam aligned. Film was put into the camera, and the grain was subjected to X-rays for 10–20 h. After developing the film, the d-spacing of each line was determined using a template. The intensity of each line was estimated visually. The patterns were identified using selected powder diffraction data for minerals (Berry, 1974).

Selected cpx spherules and microtektites were mounted on 2.56 cm glass discs using epoxy, ground down to exposed an interior surface, polished, and coated with carbon. Major oxide compositions were determined using an energy dispersive X-ray (EDX) analyzer (Princeton Gamma-Tech System 4) mounted on a Cambridge Steroscan 90 scanning electron microscope at the University of Delaware. Cr and Ni contents were determined for the cpx spherules and most of the microtektites using EDX. The acceleration voltage was 20 kV, the beam current was adjusted to give a count rate of ~2000/s, and the spectra were each collected for 200 s. The spectra were corrected for background, absorption, X-ray fluorescence, atomic number effects, and normalized to 100 wt% using software provided by Princeton Gamma-Tech. A glass of australite tektite composition made by Corning Glass Company and analyzed by the U.S. Geological Survey was used as a standard. A fragment of this glass was mounted on each slide and analyzed at the same time and under the same conditions as the samples. The compositions obtained for the glass standard were used to standardize the sample compositions. The corrections were generally quite small. For additional information on the method of chemical analysis, see Glass (1989), Glass et al. (2004a), and Liu (2006). Most compositions are an average of five spot analyses.

We report Ir analyses on sediments from five ODP Holes: 689D, 699A, 709C, 738B, and 1090B. The Ir analyses were performed at the University of California–Los Angeles. This work used a separate set of samples from those used for spherule extraction. Initial Ir measurements for samples from Hole 1090B were made independently from the spherule study, and details of this work are presented by Kyte (2001). We have since analyzed an additional 29 samples from Hole 1090B, and all of these data are summarized here. The samples from Holes 699A and 709C were selected based on results of the spherule work. In addition to the bulk sediment samples used to define the spherule/ejecta distribution at Hole 709C, we analyzed splits of spherules and fine fractions from a large-volume sample at Hole 709C, Core 31X, Section 4, at a depth of 145–150 cm (709C-31X-4, 145–150 cm). We also measured Ir concentrations in Holes 689D and 738B, which were studied in a separate collaboration with S. Bohaty and are now combined into this study.

Sediment samples of ~1 cm^2 were heated to 200 °C overnight, ground in high-purity alumina mortars, and splits of ~100 mg were sealed in quartz-glass tubes. Samples were irradiated at the University of Missouri research reactor for 30 or 40 h at a neutron flux of 7×10^{13} n cm^{-2} s^{-1}. Prior to counting, radiochemical purification of Ir was performed following the procedures described in Schellenberg et al. (2004). Typical 1σ counting errors for the Ir data are <10 rel% for samples with elevated Ir, but samples with concentrations <50 pg/g can have errors as large as ~30%. Iridium was detected at concentrations as low as 7 pg/g. For a few samples, upper limits are given a value of four standard deviations above the counting background.

We also calculated an estimate of the total amount of Ir deposited at each site and for other published localities. This "fluence," given in ng Ir/cm^2, is an attempt to integrate the amount of excess Ir in the anomaly relative to background concentrations. This estimate assumes that the Ir concentration of a given sample is representative of the interval extending halfway up or down to the next adjacent sample. This calculation also requires an estimate of the sediment dry bulk density (DBD), which is the weight of dry sediment divided by the corresponding volume of wet sediment. Where possible, we took DBD values from physical property data published in DSDP or ODP *Scientific Results* volumes for each site. For some samples, no data were available, and we had to assume DBD values (which we took to be 2.5 g/cm^3 for lithified sediments in Barbados and Massignano, and 1 g/cm^3 for RC9–58 and DSDP 149). Some Ir data (i.e., 689B, Massignano from Montanari et al., 1993) were estimated from published figures.

Oxygen stable isotope data were obtained across the spherule layer and associated Ir anomaly at ODP Hole 738. Small samples (~0.1 g) from this hole were freeze-dried and ground with an agate mortar and pestle in order to fully homogenize the sediment. The analyses were performed using a VG Optima dual-inlet isotope-ratio mass spectrometer at the University of California–Santa Cruz. Approximately 100 μg of sample was reacted in a common acid bath maintained at 90 °C using an automated carbonate preparation system with a carousel device. NBS-19, Atlantis II, and an in-house Carrara marble standard were included in all sample runs. All values are reported in standard delta notation (δ) in parts per mil (‰) relative to VPDB (Vienna Peedee belemnite). Analytical precision is estimated at 0.06‰ (1σ) for $\delta^{18}O$.

We sent microtektites and cpx spherules to the California Institute of Technology for Sr-Nd isotopic analysis. The results of these analyses were published in Liu et al. (2006). Liu et al. (2006) concluded that the Sr-Nd data support the hypothesis that there is only one cpx spherule layer in the late Eocene, and that the cpx spherules were derived from the 100-km-diameter Popigai impact structure in northern Siberia.

RESULTS

New Late Eocene Sites Containing Impact Ejecta

We found impact ejecta (microtektites, cpx spherules, and/or shock-metamorphosed mineral grains) at seven new sites: Deep Sea Drilling Project (DSDP) Site 592, and Ocean Drilling Program (ODP) Holes 699A, 703A, 709C, 786A, 1073A, and 1090B (Table 1; Fig. 1). We also found cpx spherules and associated microtektites in ODP Hole 689D. In addition, glass spherules (microtektites?) were found in association with an Ir anomaly in late Eocene sediments at ODP Hole 738B. We conclude that the glass spherules are probably microtektites belonging to the cpx spherule layer.

DSDP Site 592

This site is located on the Lord Howe Rise in the Tasman Sea at a depth of 1098 m below sea level (mbsl) (Shipboard Scientific Party, 1986). An Ir anomaly was discovered at this site in late Eocene sediments at a depth between 39 and 69 cm in Section 5 of Core 38 (Frank Asaro, 1987, personal commun.). We found no microtektites or cpx spherules associated with the Ir anomaly, but we did find some pyrite spherules in the 63–125 µm size fraction. Scanning electron microscope images of polished sections

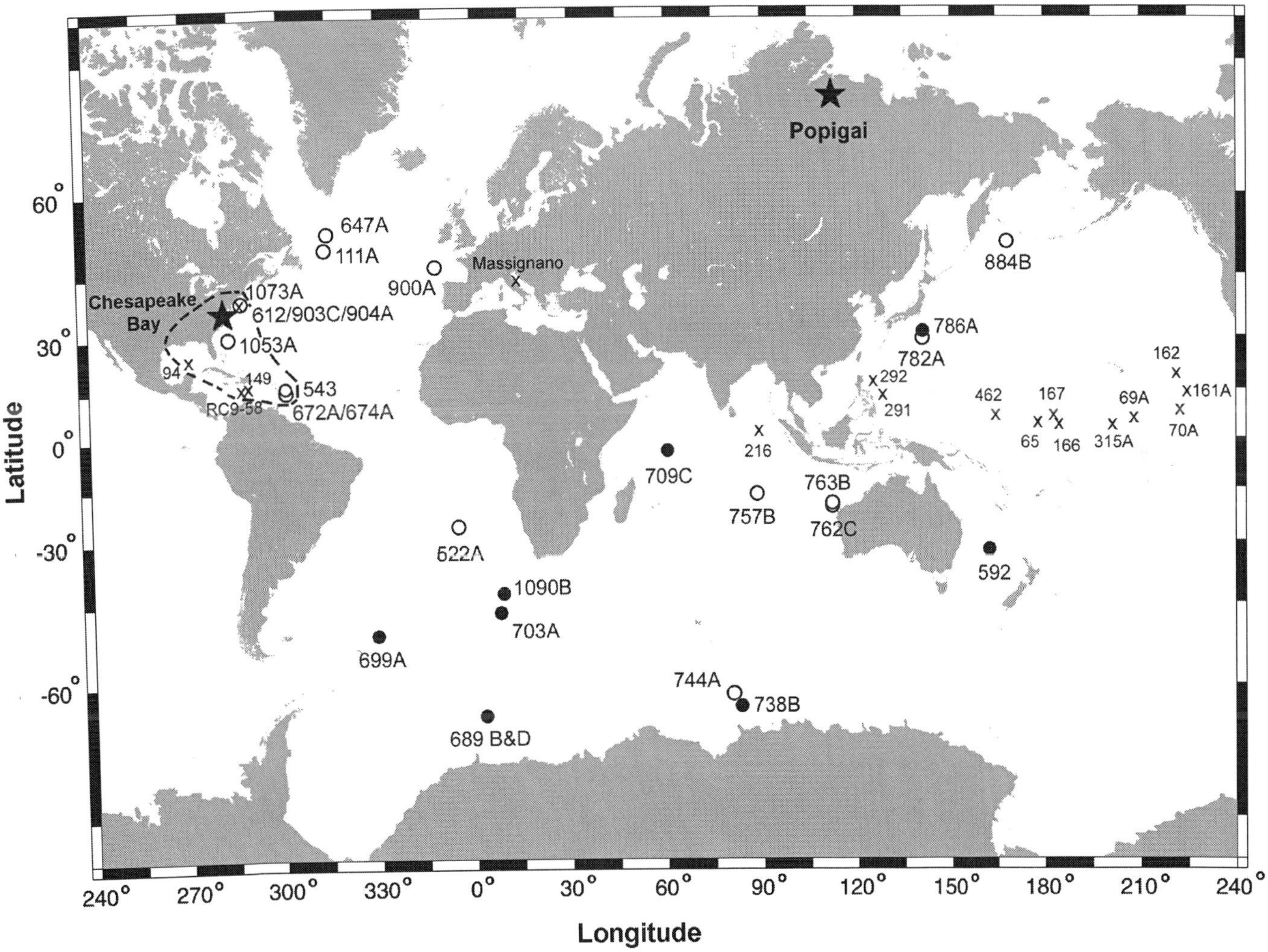

Figure 1. Map showing core sites and locations of the Chesapeake Bay and Popigai impact structures (stars). Sites where the clinopyroxene-bearing spherule (cpx spherule) layer was previously found are indicated with an "x"; solid circles indicate where the cpx spherule layer was found in this study; open circles indicate sites where the spherule layer was not found. No cpx spherules or microtektites were found at Ocean Drilling Program Hole 1073A in the northwest Atlantic off New Jersey. However, coesite was found there, which we believe is ejecta from the Chesapeake Bay impact structure.

of these spherules revealed the presence of crystalline textures (Fig. 2A) similar to those observed in cpx spherules. The Site 592 pyrite spherules are similar to cpx spherules found at DSDP Site 612 and ODP Sites 903 and 904, which were partly to completely replaced by pyrite (Glass, 1989; Glass et al., 1998). These three sites are in the northwest Atlantic Ocean on the upper continental rise off New Jersey in water depths between 446 and 1386 m. Like Sites 612, 903, and 904, Site 592 was drilled in relatively shallow water (1088 m). The water at these shallow depths was probably undersaturated with silica, and the bottom conditions would have been more reducing. No spherules were found in the >125 µm size fraction at Site 592. At the peak abundance, we recovered only 1 spherule (63–125 µm) per gram of sediment. We speculate that the pyrite spherules at this site are diagenetically altered (pyritized) cpx spherules. This spherule layer occurs within the NP19/NP20 calcareous nannofossil zone (Shipboard Scientific Party, 1986).

ODP Hole 689D

Hole 689D is located on the Maud Rise, in the South Atlantic Ocean, ~24 m from Hole 689B, where an Ir anomaly and cpx spherules had previously been documented (Montanari et al., 1993; Vonhof and Smit, 1999; Glass and Koeberl, 1999). The water depth at this hole is 2091 mbsl (Shipboard Scientific Party, 1988a). The purpose of searching for the cpx spherule layer at this hole was to determine if the abundance of impact ejecta in a distal impact ejecta layer varies considerably over distances of a few tens of meters or if the abundance, and therefore original thickness of the layer, remains fairly constant over short distances. We found microtektites associated with the cpx spherules at Hole 689B, and the peak abundance of the microtektites appeared to be slightly above the peak abundance of the cpx spherules, suggesting that they might belong to the North American microtektite layer (Glass and Koeberl, 1999). We wanted to determine if the same relationship was present at Hole 689D.

In Hole 689D, 345 microtektites and 1458 cpx spherules (>63 µm diameter) were found in 17 samples between 72 and 149 cm in Section 2 of Core 12 at 126.32–127.09 m below the seafloor (mbsf). This is the highest-latitude site where the cpx spherule layer has been found, and some of the biostratigraphic zonations differ from those of lower-latitude cores. The cpx spherule layer at this site occurs within the *Isthmolithus recurvus* calcareous nannofossil zone (Wei and Wise, 1990) and the AP12 planktonic foraminifera zone (near the AP12/13 zonal boundary), which is equivalent to the upper part of the lower-latitude P15 zone (Stott and Kennett, 1990). The layer is also in the radiolarian *Eucyrtidium spinosum* zone (Funakawa and Nishi, 2005).

The cpx spherules found in Hole 689D range in color from translucent yellow to opaque brown. Crystalline textures are visible on the surfaces of some of the spherules. Most of the cpx spherules are broken, and only fragments of cpx spherules were found in the >125 µm size fraction. Unbroken cpx spherules are all spherical. The majority of the microtektites are transparent colorless, but some are transparent pale brown like at Hole 689B.

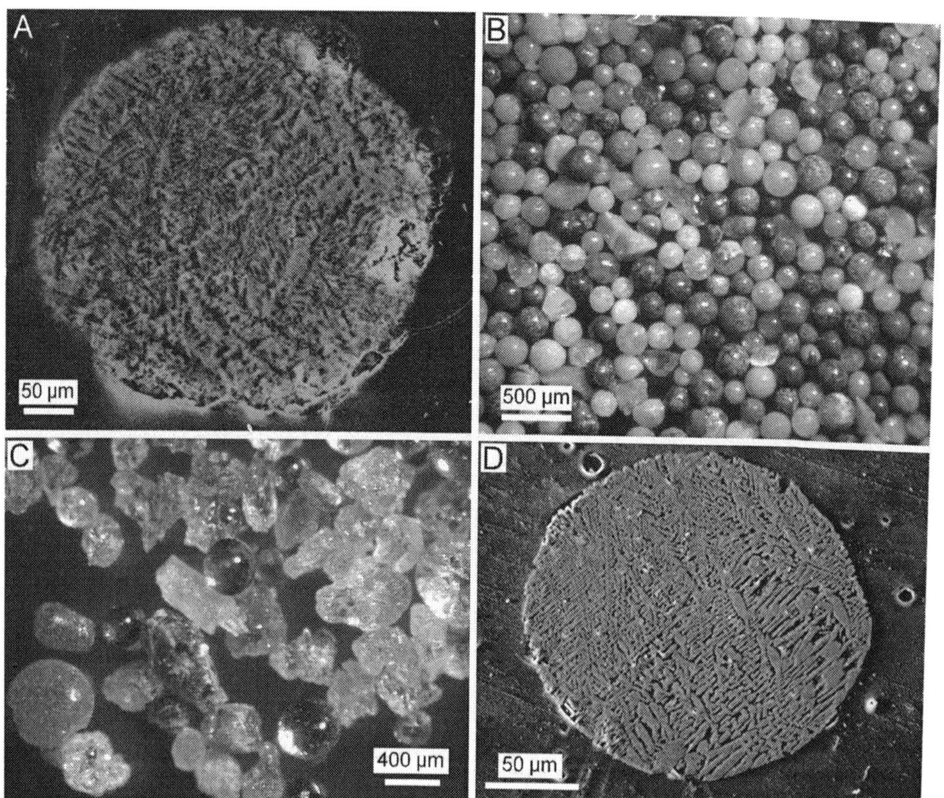

Figure 2. Cpx spherules and microtektites recovered from the cpx spherule layer. (A) Scanning electron microscope (SEM) image of a polished section of a pyritized cpx spherule from Deep Sea Drilling Project Site 592 in the Tasman Sea. (B) Optical microscope image of cpx spherules from Ocean Drilling Program (ODP) Hole 709C taken in the northeastern Indian Ocean. (C) Optical microscope image of microtektites from the cpx spherule layer at ODP Hole 709C in the northwestern Indian Ocean. (D) SEM image of a polished section of a cpx spherule from ODP Hole 786A in the northwestern Pacific Ocean (Izu-Bonin forearc basin).

Approximately 60% (by number) of the microtektites are spherical, 20% disc-shaped, and the remainder are cylindrical-, dumbbell-, or teardrop-shaped, or fragments of microtektites. The three largest microtektites (760, 860, 1100 μm in diameter) are oblate spheres.

We determined the major oxide contents for three cpx spherules and nine microtektites from Hole 689D (Table 2). The cpx spherules and microtektites from Hole 689D have major oxide compositions similar to the major oxide compositions previously obtained (Vonhof and Smit, 1999; Glass and Koeberl, 1999) for cpx spherules and microtektites from Hole 689B. The microtektites have high SiO_2 (70–81 wt%), low MgO (<1.2 wt%) and CaO (<0.45 wt%), and very low Na_2O (<0.1 wt%) contents and can be divided into an iron-poor group (FeO < 1.5 wt%) and an iron-rich group (FeO > 4 wt%) (Table 2).

ODP Hole 699A

This site is located on the northeastern slope of the Northeast George Rise in the South Atlantic (Shipboard Scientific Party, 1988b). Microtektites were found concentrated in an interval between 70 and 90 cm in Section 4 of Core 35, ~321 mbsf (Fig. 3). The peak abundance of microtektites (>125 μm), ~2/g, was at a depth of ~321.45 mbsf. The microtektite layer was found in the calcareous nannofossil zone NP19/20, zone CP15b, and the *Reticulofenestra oamaruensis* zone (Madile and Monechi, 1991). However, Crux (1991) placed the core section containing the microtektites in the *Isthmolithus recurvus* zone rather than the *Reticulofenestra oamaruensis* zone.

The microtektites at Hole 699A are transparent colorless. Most are spheroids; teardrop- and dumbbell-shaped microtektites are rare. Fragments of microtektites are also present. Approximately half of the microtektites have smooth surfaces, the rest have grooved and/or pitted surfaces. With one exception, the microtektites from this site are silica rich (>84 wt%) and have Al_2O_3 contents between ~7 and 12 wt%; FeO, MgO, CaO, Na_2O, K_2O, and TiO_2 contents are all <1 wt% (Table 2). We observed no cpx spherules or shock-metamorphosed grains at this site. Ir data were obtained for this site and are discussed later.

ODP Hole 703A

This site is on the Meteor Rise in the southwest Atlantic Ocean at a depth of 1796.1 mbsl (Shipboard Scientific Party, 1988c). Two microtektites and one cpx spherule were recovered from the 63–125 μm size fraction from 35 to 36 cm in Section 3, Core 15 (~131.75 mbsf). The section of Core 15, which contains the microtektites and cpx spherule, lies near the boundary between calcareous nannofossil subzones *Reticulofenestra oamaruensis* and *Isthmolithus recurvus* and in the N19/20 and CP15b zones (Madile and Monechi, 1991) and in the P16 or P17 planktonic foraminifera zone (Nocchi et al., 1991). The section of core containing the spherules was identified as likely being in magnetochron 15n (Hailwood and Clement, 1991), but biostratigraphic control on the magnetochron boundaries is poor at this site, and correlation of magnetochrons with the geomagnetic polarity time scale is equivocal. No other microtektites, cpx spherules, or shock-metamorphosed grains were recovered from this site. We cannot rule out the possibility that the two microtektites and one cpx spherule found in one sample from this site are due to contamination from the 63 μm sieve. No compositional data were obtained for the microtektites and cpx spherule, and no Ir measurements were made at this site.

ODP Hole 709C

This site is located in the western Indian Ocean in a water depth of 3041 mbsl (Shipboard Scientific Party, 1988d). Cpx spherules and microtektites were found in the interval between Section 4, 100 cm, and Section 5, 40 cm, in Core 31X (291.30–292.20 mbsf) (Fig. 3). In the >125 μm size fraction, cpx spherules are ~25 times more abundant than microtektites at this site. A peak abundance of cpx spherules (>125 μm) of ~854/g was found at a depth of 291.74 mbsf. The peak abundance of microtektites (>125 μm), ~47/g, was found at the same depth. The microtektite–cpx spherule layer at this site is associated with the last appearance datum (LAD) of the radiolarian *Thyrsocyrtis tetracantha* (as well as *T. bromia* and *T. rhizodon*) (Johnson, 1990) and thus occurs near the *Calocyclas bandyca–Cryptopora ornata* zonal boundary. It is also in the CP15 zone, perhaps near the CP15a/CP15b zonal boundary, but the indicator taxa are rare (Okada, 1990). The core containing the cpx spherule layer is in an undifferentiated interval belonging to planktonic foraminifera zones P15 through P17 (Premoli Silva and Spezzaferri, 1990).

The cpx spherules range from translucent yellow or bluish to white opaque to mottled with dark brown to reddish brown to tan markings. There are more light-colored cpx spherules than dark ones (Fig. 2B). About half of the light-colored cpx spherules are fragments, while <5% of the dark-colored cpx spherules are fragments. Most of the microtektites (Fig. 2C) are transparent colorless, but several brownish, deeply corroded microtektites or tektite fragments (>500 μm) were recovered, as well as a few green microtektites. Most of the glass occurs as fragments. Spherical microtektites are common, but other splash forms are also present (teardrop and oval shapes).

The major oxide compositions of six microtektites were determined. A greenish microtektite was found to have relatively low SiO_2 (~56 wt%), Al_2O_3 (~10 wt%), Na_2O (~1 wt%), K_2O (~0.4 wt%) contents and high MgO and CaO (~12 wt%) contents (Table 2). The green microtektite also appeared to have detectable Cr_2O_3 and NiO contents (Table 2). A brown cylindrical microtektite had the highest FeO content (~7.8 wt%). The remaining glass particles (two fragments, a vesicular teardrop, and an oval-shaped microtektite) had similar compositions: SiO_2 contents between ~76 and 78 wt%, Al_2O_3 contents between ~12 and 13 wt%, MgO, CaO, and Na_2O contents generally <1 wt%, and K_2O contents between 2.8 and 3.2 wt%. The major oxide compositions of 20 cpx spherules were determined (Table 2). Compared with the microtektites, the cpx spherules generally had lower SiO_2 (average of ~67 wt%), Al_2O_3 (average ~7 wt%), and K_2O (average ~2 wt%) contents and higher MgO, CaO, Cr_2O_3, and NiO contents.

TABLE 2. MAJOR OXIDE COMPOSITIONS (WT%) OF CPX SPHERULES AND ASSOCIATED MICROTEKTITES*

Sample	Hole	SiO_2	Al_2O_3	FeO†	MgO	CaO	Na_2O	K_2O	TiO_2	Cr_2O_3	NiO	Total§
Cpx spherules												
833-3a	689D	71.5	10.1	7.5	1.5	3.5	1.0	3.4	0.4	0.1	<0.1	99.0
833-3c	689D	67.1	6.3	9.5	3.8	7.8	1.2	3.1	0.3	0.1	<0.1	99.3
833-3b	689D	64.3	10.8	10.9	3.6	4.9	1.3	2.9	0.4	<0.1	<0.1	99.0
838-09	709C	79.7	3.4	6.6	5.0	1.9	0.5	1.8	0.2	0.1	0.1	99.4
839-05	709C	72.2	6.9	2.4	6.0	11.3	<0.1	1.5	0.3	0.1	0.1	100.7
839-07	709C	70.5	5.7	4.7	6.9	9.4	0.3	2.2	0.3	0.1	0.1	100.2
838-8	709C	69.7	5.7	6.6	6.7	8.3	0.5	1.9	0.3	0.1	0.1	99.9
839-09	709C	68.6	6.2	6.2	6.9	8.1	0.5	2.3	0.3	0.1	0.3	99.4
839-02	709C	67.9	8.5	4.2	6.6	9.7	0.3	1.8	0.4	<0.1	0.1	99.4
839-08	709C	67.6	5.8	6.7	6.5	8.8	0.5	2.8	0.3	0.1	0.2	99.2
838-3	709C	67.2	7.1	5.1	7.6	9.9	0.5	1.6	0.3	0.1	0.1	99.4
839-01	709C	66.8	8.8	3.8	6.7	10.2	0.6	1.9	0.5	0.1	0.1	99.3
850-B	709C	66.7	8.7	3.0	5.7	11.3	1.1	1.9	0.4	<0.1	<0.1	98.7
838-2	709C	66.6	7.2	4.5	7.4	11.3	0.4	1.7	0.3	0.1	0.1	99.5
838-6	709C	66.0	11.1	4.7	4.6	8.3	1.6	2.5	0.6	0.1	<0.1	99.4
839-03	709C	65.8	8.3	4.6	7.0	10.6	0.6	1.9	0.5	0.1	0.1	99.4
839-11	709C	65.7	5.8	7.8	7.9	8.9	0.9	1.9	0.3	0.2	0.2	99.4
839-10	709C	65.0	9.6	6.7	8.0	4.9	1.8	2.7	0.4	0.2	0.2	99.5
838-9	709C	65.0	9.3	8.2	7.8	4.9	1.3	2.1	0.4	0.1	0.2	99.5
838-05	709C	64.8	6.3	2.8	8.0	16.6	0.3	1.3	0.3	0.1	<0.1	100.4
839-12	709C	63.8	6.9	5.1	8.0	11.2	1.3	2.5	0.3	0.1	0.3	99.4
838-10	709C	63.4	8.9	9.0	6.9	7.3	1.0	2.1	0.4	0.2	0.1	99.3
839-06	709C	61.2	5.4	2.6	10.4	17.5	<0.1	2.0	0.2	0.1	0.1	99.5
839-04	709C	58.5	7.2	2.8	9.5	18.9	0.2	1.7	0.3	0.1	0.1	99.2
851-A01	709C	58.2	7.4	2.9	7.8	18.2	1.0	1.3	0.3	0.1	0.1	97.1
839-12	709C	54.2	2.8	3.0	15.5	22.7	0.2	0.4	0.1	0.3	0.4	99.5
839-08	709C	50.0	5.0	6.6	12.3	23.3	0.3	0.5	0.3	0.1	0.4	98.7
849-2A	786A	53.8	2.4	2.3	14.4	25.6	0.9	0.0	0.0	0.1	0.1	99.5
849-1A	786A	52.9	7.0	4.4	11.8	21.8	0.8	0.6	0.1	<0.1	<0.1	99.4
849-3A	786A	52.5	3.9	2.5	13.2	26.4	0.7	<0.1	<0.1	<0.1	<0.1	99.3
GL07	786A	52.4	2.7	3.2	18.2	21.5	1.4	0.3	0.1	0.1	0.1	100.0
849-2B	786A	51.7	3.7	4.4	12.4	26.1	1.0	<0.1	<0.1	<0.1	0.1	99.3
849-1B	786A	51.1	3.9	3.9	12.6	26.8	0.9	<0.1	0.1	<0.1	0.1	99.4
GL05	786A	51.0	7.0	4.1	15.0	16.9	4.1	1.8	<0.1	0.1	<0.1	99.9
GL04	786A	50.5	3.0	6.1	16.1	21.3	1.2	1.1	0.2	0.2	0.2	100.0
841-04	786A	50.3	3.2	3.3	17.1	25.5	<0.1	<0.1	<0.1	<0.1	<0.1	99.4
831-5-2	1090B	78.3	2.0	0.1	3.9	6.5	8.0	0.5	0.1	<0.1	<0.1	99.3
831-5-1	1090B	74.8	2.2	0.1	4.2	6.0	11.6	0.5	<0.1	<0.1	<0.1	99.3
833-5a	1090B	73.2	7.7	6.5	5.0	2.5	1.1	2.8	0.3	0.1	<0.1	99.2
831-B1	1090B	72.2	7.5	15.0	1.8	1.0	<0.1	1.4	0.3	<0.1	<0.1	99.3
831-D1	1090B	72.2	7.7	10.8	2.8	3.3	0.3	1.9	0.3	<0.1	0.1	99.4
831-A3	1090B	71.0	7.9	8.8	4.1	4.6	0.3	2.2	0.3	<0.1	0.1	99.3
831-C1	1090B	69.4	7.1	8.1	5.3	6.4	0.9	1.5	0.3	0.1	0.3	99.4
831-A1	1090B	68.9	7.1	10.6	4.3	6.2	0.3	1.5	0.3	0.1	0.2	99.5
831-4-1	1090B	67.9	6.7	3.9	7.6	11.0	0.2	1.8	0.2	<0.1	0.1	99.4
833-6a	1090B	66.9	5.9	6.2	8.2	9.0	0.7	2.0	0.2	<0.1	0.3	99.4
833-4b	1090B	66.9	7.0	4.6	6.9	10.9	0.1	2.6	0.2	0.1	<0.1	99.4
833-5b	1090B	66.0	8.9	9.0	3.3	8.2	0.8	2.7	0.4	<0.1	<0.1	99.3
833-2b	1090B	65.7	7.9	4.6	7.4	10.8	0.6	1.8	0.3	<0.1	<0.1	99.2
833-5c	1090B	65.4	8.8	14.0	3.1	4.6	0.7	2.0	0.4	0.1	<0.1	99.0
831-4-2	1090B	64.9	7.0	2.5	8.4	14.5	0.2	1.7	0.3	<0.1	0.1	99.6
833-4a	1090B	63.8	6.2	3.3	9.0	15.0	0.2	1.5	0.2	<0.1	<0.1	99.3
833-6b	1090B	63.8	6.7	6.7	8.2	11.0	0.6	2.0	0.2	<0.1	<0.1	99.2
833-2c	1090B	62.0	5.9	3.7	9.5	16.2	0.2	1.5	0.3	<0.1	<0.1	99.3
831-6-1	1090B	61.8	5.2	3.2	10.2	16.7	0.3	1.6	0.2	<0.1	0.1	99.4
833-2a	1090B	52.8	4.8	1.9	13.4	25.2	0.2	0.8	0.2	<0.1	<0.1	99.4
Microtektites												
832-5-B	689D	81.3	13.4	0.8	0.6	0.3	<0.1	2.6	0.6	<0.1	<0.1	99.4
832-1-A	689D	79.4	14.6	1.5	0.7	0.3	<0.1	2.4	0.6	<0.1	<0.1	99.3
832-2-B	689D	78.3	12.9	4.2	0.6	0.4	<0.1	2.4	0.6	<0.1	<0.1	99.4
832-5-A	689D	78.0	10.8	7.1	0.5	0.4	<0.1	2.1	0.4	<0.1	<0.1	99.3
832-6-B	689D	78.0	10.2	7.0	0.5	0.4	<0.1	2.8	0.3	<0.1	0.1	99.3
832-7-A	689D	77.9	13.4	4.3	0.5	0.3	<0.1	2.4	0.5	<0.1	<0.1	99.4
832-6-A	689D	76.9	19.4	0.3	0.5	0.1	<0.1	1.0	0.7	0.1	0.1	99.0
832-3-A	689D	73.5	20.1	1.3	0.8	0.2	<0.1	2.7	0.7	<0.1	0.1	99.4

(*continued*)

TABLE 2. MAJOR OXIDE COMPOSITIONS (WT%) OF CPX SPHERULES AND ASSOCIATED MICROTEKTITES*
(*continued*)

Sample	Hole	SiO_2	Al_2O_3	FeO†	MgO	CaO	Na_2O	K_2O	TiO_2	Cr_2O_3	NiO	Total§
Microtektites												
832-2-A	689D	70.2	19.0	6.0	1.2	0.4	<0.1	1.8	0.8	<0.1	<0.1	99.4
848-4B	699A	90.8	7.1	0.1	0.5	<0.1	0.4	0.2	0.2	<0.1	<0.1	99.3
848-4C	699A	90.4	7.2	0.2	0.6	0.1	0.4	0.2	0.3	<0.1	<0.1	99.3
848-4D	699A	87.4	9.9	0.2	0.6	0.1	0.4	0.5	0.3	<0.1	0.1	99.4
848-2B	699A	86.9	10.4	0.3	0.6	<0.1	0.5	0.5	0.3	<0.1	<0.1	99.4
848-2A	699A	86.5	0.4	0.0	0.0	<0.1	2.8	9.7	0.0	<0.1	<0.1	99.4
848-1B	699A	86.1	9.8	0.5	0.5	0.1	0.4	1.8	0.3	<0.1	<0.1	99.5
848-1C	699A	85.0	10.5	0.5	0.9	0.6	0.4	1.1	0.3	<0.1	<0.1	99.3
848-4F	699A	84.8	12.0	0.3	0.6	0.1	0.4	0.7	0.4	<0.1	<0.1	99.3
848-1A	699A	84.5	12.0	0.3	0.7	0.2	0.5	0.8	0.3	<0.1	<0.1	99.4
848-3A	699A	84.4	11.7	0.3	0.6	<0.1	0.5	1.4	0.5	<0.1	<0.1	99.3
838-01	709C	78.4	13.5	2.0	0.6	0.2	<0.1	2.8	0.6	<0.1	<0.1	98.1
854-3-3	709C	77.7	11.6	3.5	0.8	1.2	1.0	3.1	0.3			99.2
843-2	709C	77.4	11.4	3.6	0.8	1.0	1.1	3.3	0.4			98.9
843-1	709C	77.2	11.5	3.6	0.7	1.0	1.1	3.5	0.4			99.0
854-1-1	709C	76.9	18.8	1.3	0.4	0.0	0.0	1.8	0.7			99.9
854-3-2	709C	76.7	11.7	4.4	1.0	1.2	0.8	3.2	0.4	<0.1	<0.1	99.4
891-10	709C	76.1	12.8	5.6	0.4	0.5	0.4	3.2	0.4	<0.1	<0.1	99.4
891-11	709C	76.1	12.6	5.7	0.4	0.5	0.5	3.2	0.4	<0.1	<0.1	99.4
891-12	709C	76.0	13.0	3.2	1.0	1.8	1.6	2.9	0.4			99.8
891-8	709C	75.9	12.5	5.4	0.1	0.2	1.3	3.2	0.3			98.9
891-9	709C	75.7	12.4	5.4	0.1	0.4	1.3	3.2	0.3			98.8
892-1	709C	73.1	14.3	3.7	0.9	1.6	1.8	3.3	0.6			99.4
892-2	709C	73.1	14.7	3.8	0.9	1.6	1.8	3.0	0.6			99.5
892-3	709C	71.9	11.6	3.9	2.5	4.8	1.4	3.2	0.3			99.5
893-1	709C	68.4	17.3	7.8	1.2	0.3	0.9	2.8	0.7	<0.1	<0.1	99.4
893-2	709C	67.3	12.1	2.7	5.3	8.3	1.4	2.6	0.6			100.3
893-3	709C	67.0	11.0	6.6	7.3	4.1	1.7	1.7	0.3			99.6
894-1	709C	61.4	10.7	4.1	13.7	10.7	0.1	0.4	0.4			101.5
894-2	709C	56.3	8.5	7.4	14.4	11.9	0.5	0.7	0.2			99.9
894-3	709C	55.6	9.6	7.6	12.4	12.3	1.0	0.4	0.4	0.1	0.1	99.5
841-02	786A	64.3	5.7	2.6	11.5	15.3	<0.1	<0.1	<0.1	<0.1	<0.1	99.4
830-A1	1090B	86.6	11.0	0.3	0.4	0.3	<0.1	0.4	0.4	0.1	<0.1	99.5
830-B2	1090B	83.8	12.2	0.4	0.3	0.3	<0.1	1.9	0.5	<0.1	<0.1	99.5
891-3	1090B	82.9	13.4	0.3	0.1	<0.1	<0.1	0.5	0.5			97.7
891-5	1090B	79.3	13.8	0.5	0.1	0.1	0.2	2.6	0.6			97.2
891-2	1090B	78.8	8.6	7.0	0.2	0.4	0.5	2.3	0.3			98.1
830-B1	1090B	78.3	13.4	3.6	0.4	0.6	<0.1	2.6	0.4	0.1	<0.1	99.4
829-3	1090B	77.6	11.7	6.0	0.6	0.5	<0.1	2.5	0.4	<0.1	<0.1	99.4
891-1	1090B	77.5	16.0	0.8	0.2	0.1	0.1	2.4	0.7			97.8
891-4	1090B	77.4	11.4	3.9	0.6	0.7	0.5	3.1	0.4			97.9
891-7	1090B	77.1	10.8	6.9	0.4	0.3	0.4	2.3	0.3			98.6
833-1b	1090B	76.9	15.3	3.3	0.8	0.2	<0.1	2.3	0.7	<0.1	<0.1	99.4
833-1c	1090B	76.5	17.0	2.1	0.9	0.4	<0.1	2.0	0.7	<0.1	<0.1	99.6
829-2	1090B	73.3	10.0	5.2	5.2	2.0	1.0	2.1	0.4	<0.1	0.1	99.3
830B4	1090B	71.2	6.0	9.9	4.1	1.9	2.0	3.8	0.3	0.1	0.3	99.5
830-B3	1090B	70.5	11.5	5.7	3.6	2.5	1.6	2.9	0.5	0.2	0.4	99.5
891-6	1090B	68.8	12.1	2.3	5.0	7.3	0.8	2.4	0.8			99.5
890-3	1090B	61.4	8.7	7.0	8.7	6.9	2.0	4.1	0.3			99.1
840-4	1090B	59.6	10.0	10.2	9.9	4.7	1.5	3.2	<0.1			99.1
829-1	1090B	59.5	13.2	9.6	8.3	4.3	1.6	2.4	0.4	0.1	<0.1	99.4
890-2	1090B	57.5	8.6	8.1	11.3	11.3	1.0	1.6	0.3			99.7
890-1	1090B	57.1	6.2	8.0	14.6	11.7	1.2	1.6	<0.1			100.4

*Major oxide compositions are the averages of five spot analyses by energy-dispersive X-ray analysis.

†All iron is given as FeO.

§The analyses were normalized to 100 wt% and then standardized using a glass standard made by Corning Glass Company to have a composition similar to an australite tektite.

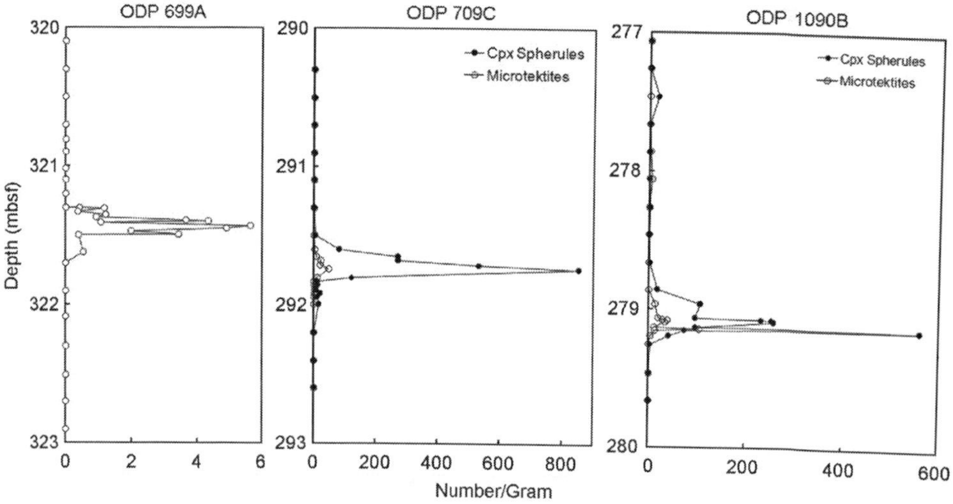

Figure 3. Distribution of cpx spherules and microtektites in cores from Ocean Drilling Program Holes 699A, 709C, and 1090B taken in the southern South Atlantic, northwestern Indian Ocean, and southeastern South Atlantic Ocean, respectively (mbsf—m below seafloor). Note the different horizontal scales. The number/g is for cpx spherules and/or microtektites >63 μm in size for Sites 699A and 1090B and for cpx spherules and microtektite >125 μm in size for Site 709C.

A large sample (~50 g), previously used for the shipboard study of interstitial water chemistry, was also studied. This sample was from Core 31X, Section 4, 145–150 cm, and it included the lower part of the spherule/microtektite layer. After removal of the carbonate fraction, using dilute HCl, and sieving, ~50 white opaque grains were recovered from the 125–250 μm size fraction. X-ray diffraction patterns were obtained for seven of these white opaque grains (Fig. 4A) using a Gandolfi X-ray camera. Six of the grains were found to be coesite with no trace of quartz and one was found to be quartz. Eighty translucent colorless grains from the 63–125 μm size fraction were studied using a petrographic microscope. Four of the grains exhibited one to two sets of planar features, which were probably planar deformation features (Figs. 4C and 4D).

ODP Hole 738B

Site 738 is on the southern part of the Kerguelen Plateau at a depth of 2252.5 mbsl (Shipboard Scientific Party, 1989). One of the authors (Bohaty) found glass spherules in Hole 738B-5H-2 at 127–144 cm. The glass spherules were found in

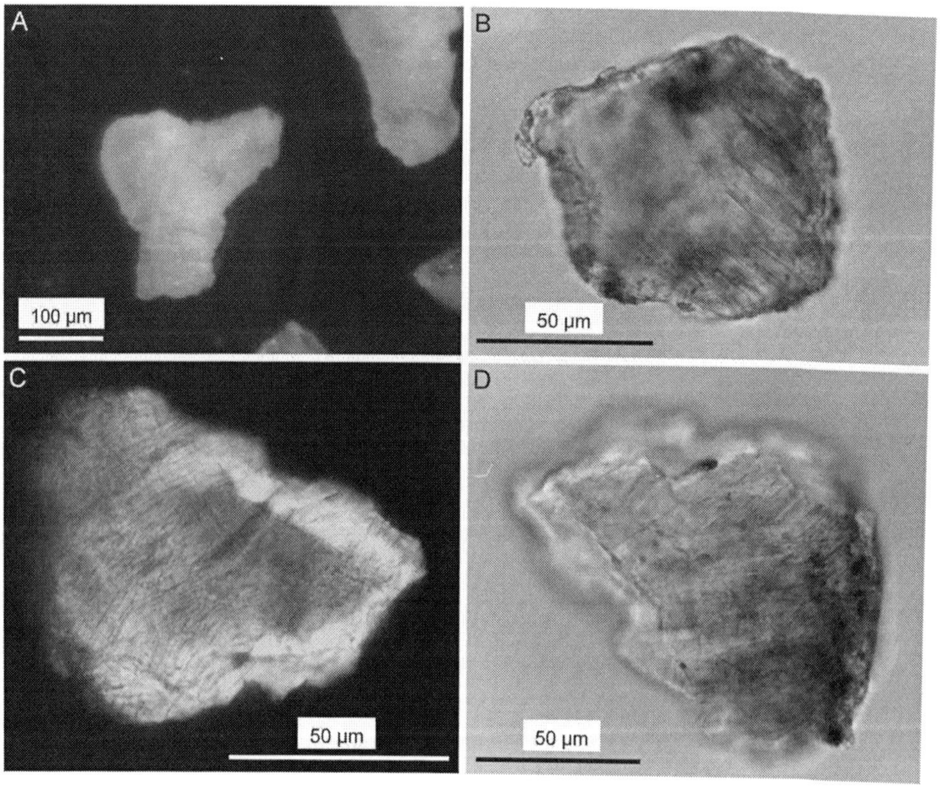

Figure 4. Optical microscope images of coesite and shocked quartz grains from the cpx spherule layer in Ocean Drilling Program Hole 709C in the northwestern Indian Ocean. (A) Coesite-rich grain taken in reflected light. (B) Quartz grain with one set of quasi-planar features taken in transmitted light. (C) Quartz grain with several sets of planar features taken in transmitted light under cross polarizers. (D) Quartz grain with at least two sets of planar features taken in transmitted light.

the >150 μm size fraction of six samples, with a peak abundance of ~2/g. No cpx spherules were identified. The glass spherules were mostly transparent colorless spheres, but oblate forms and teardrops were also found. Some of the spherules are a "smoky" color, and some are pale brown to pale yellow. The largest spherule has a diameter of ~350 μm. A few are composed of a cluster of smaller spheres fused together. Some of the transparent colorless spherules contain bubble cavities. Although no compositional data have been obtained for these spherules, we are reasonably confident that they belong to the cpx spherule layer based on biostratigraphic data and the association with an Ir anomaly. Additional studies will be required to confirm that the glass spherules are microtektites and that no cpx spherules are present. An Ir anomaly was found at this site, with a peak concentration within 6 cm of the peak abundance of the spherules. One of the authors (Bohaty) also determined oxygen isotope data for the sediments across the spherule layer. The Ir data and oxygen isotope data are discussed later.

ODP Hole 786A

Approximately 20 cpx spherules were recovered from ODP Hole 786A in the center of the Izu-Bonin forearc basin in the western Pacific Ocean at a water depth of 3070 mbsl (Shipboard Scientific Party, 1990). The spherules were recovered from the >125 μm size fraction of a sample taken at a depth of 40–41 cm in Section 2, Core 11. This section of the core was assigned to calcareous nannofossil CP15a subzone (Stabell et al., 1992) and to planktonic foraminifera zones P14–P16, but planktonic foraminifera are not very abundant, and their preservation is poor (Milner, 1992).

All the cpx spherules are opaque with a cream color. These spherules are unique in appearance compared to cpx spherules found at other sites. Some are broken or fragmented. Several broke into small pieces very easily. An X-ray diffraction pattern of one of the spherules shows that the crystalline phase is clinopyroxene (probably diopside). Scanning electron microscope (SEM) images indicate that the spherules are composed primarily of intergrown clinopyroxene crystals with little or no glass matrix (Fig. 2D). The major oxide compositions were determined for 10 of these spherules. Nine of the ten spherules have compositions similar to diopside: low silica (50–54 wt%) and Al_2O_3 (3–7 wt%) contents, and high MgO (~12–17 wt%) and CaO (~17–27 wt%) contents (Table 2). In addition, some spherules composed of feldspar were also found. The one microtektite recovered from this site has a major oxide composition similar to those of the cpx spherules, but with higher SiO_2, and slightly lower MgO, CaO, and especially Na_2O contents (Table 2). Ni-rich spinels were not observed in any of the spherules from this site. Unfortunately, due to the abundance of volcanic ash in the section of core where the cpx spherules were found, it was difficult to recover the spherules, and the layer could not be defined very well. No shock-metamorphosed grains were identified from this site.

ODP Hole 1090B

Site 1090 is located on the southern flank of the Agulhas Ridge in the South Atlantic at a depth of 3716 mbsl (Shipboard Scientific Party, 1999). Cpx spherules and microtektites were found over an interval of ~1.8 m, with a peak abundance of 562/g and 106/g, respectively, at 30X-5, 114–115 cm (~279.14 mbsf) (Fig. 3). The layer occurs in zone CP15b (Marino and Flores, 2002) and in magnetochron C16n.1n (Channell et al., 2003). Approximately 400 microtektites (>63 μm) and ~2500 cpx spherules (>63 μm) were recovered from this site. The largest microtektite was ~960 × 1140 μm in size. Approximately 55% are spherical; the rest are discs, cylinders, dumbbells, teardrops, or fragments. Most are transparent and colorless, but a few are transparent pale-brown or green. Approximately half of the cpx spherules are spherical; the rest are fragments. They range from translucent yellow to dark opaque. A few agglutinated foraminiferal tests were observed with cpx spherules in them. No shock-metamorphosed grains were observed at this site.

Sites Where Late Eocene Spherules and Impact Ejecta Were Not Found

We searched 14 additional DSDP and ODP sites for late Eocene cpx spherules and/or associated microtektites without success (Table 1). Two of the sites (ODP Holes 744A and 763B) have missing sections (uncored or a gap in the recovered section at a core break), and even though we found no microtektites or spherules, we cannot be sure that the layer is not present at these sites. We are reasonably certain that the cpx spherule layer is not present at four of the sites (Holes 111A, 522A, 543, 757B). An ~60-cm-thick layer of sediment containing rock fragments and mineral grains, including white opaque grains, was found between 61 and 120 cm depth in Section 4 of Core 27 (659.21–659.8 mbsf) at ODP Hole 1073A. XRD data indicate that at least some of the white opaque grains contain coesite. The ejecta layer at this site occurs in the CP15 calcareous nannofossil zone, at the same stratigraphic level as the North American microtektite layer at Site 612 (Wei, 1998). Thus, the coesite found at Hole 1073A probably was produced during the North American microtektite event (i.e., the Chesapeake Bay impact). If present, the cpx spherule layer should have been just below the ejecta layer; however, no cpx spherules were found. At the seven other sites, no cpx spherules/microtektites were found in the section where the layer was predicted to be based on biostratigraphy, but additional searches would be required to make sure that the spherules are not above or below the sampled intervals.

Iridium Data

We measured Ir concentrations through the late Eocene cpx spherule layer at five ODP holes: 689D, 699A, 709C, 738B, and 1090B (Tables 3–5). Table 6 includes peak Ir concentrations, Ir fluences, dry bulk density values, and the intervals integrated to arrive at the fluences, for each of the five holes in this study, as

TABLE 3. IRIDIUM CONCENTRATIONS IN SEDIMENTS FROM OCEAN DRILLING PROGRAM (ODP) HOLE 1090B

Hole-core-section	Depth (cm)	Depth (mbsf)	Ir (pg/g)	Hole-core-section	Depth (cm)	Depth (mbsf)	Ir (pg/g)
1090B-30X-1	6 to 7	272.06	26	1090B-30X-4	105–106	277.55	54
1090B-30X-1	13–14	272.13	20	1090B-30X-4	115–116	277.65	57
1090B-30X-1	18–19	272.18	26	1090B-30X-4	125–126	277.75	63
1090B-30X-1	23–24	272.23	11	1090B-30X-4	135–136	277.85	73
1090B-30X-1	30–31	272.3	12	1090B-30X-4	139–140	277.89	68
1090B-30X-1	35–36	272.35	16	1090B-30X-4	144–145	277.94	87
1090B-30X-1	42–43	272.42	17	1090B-30X-5	5–6	278.05	117
1090B-30X-1	47–48	272.47	13	1090B-30X-5	15–16	278.15	60
1090B-30X-1	52–53	272.52	19	1090B-30X-5	25–26	278.25	84
1090B-30X-1	62–63	272.62	11	1090B-30X-5	35–36	278.35	87
1090B-30X-1	72–73	272.72	11	1090B-30X-5	45–46	278.45	93
1090B-30X-1	77–78	272.77	22	1090B-30X-5	55–56	278.55	124
1090B-30X-1	83–84	272.83	33	1090B-30X-5	65–66	278.65	142
1090B-30X-1	88–89	272.88	18	1090B-30X-5	75–76	278.75	247
1090B-30X-1	92–93	272.92	13	1090B-30X-5	85–86	278.85	261
1090B-30X-1	97–98	272.97	11	1090B-30X-5	87–89	278.87	266
1090B-30X-1	101–102	273.01	12	1090B-30X-5	92–94	278.92	452
1090B-30X-1	107–108	273.07	16	1090B-30X-5	95–96	278.95	518
1090B-30X-1	112–113	273.12	14	1090B-30X-5	96–98	278.96	557
1090B-30X-1	117–118	273.17	13	1090B-30X-5	103–105.5	279.00	808
1090B-30X-1	124–125	273.24	28	1090B-30X-5	105–107	279.05	908
1090B-30X-1	134–135	273.34	21	1090B-30X-5	105–106	279.05	954
1090B-30X-1	144–145	273.44	19	1090B-30X-5	107–109	279.07	1046
1090B-30X-2	5–6	273.55	16	1090B-30X-5	113–115	279.13	873
1090B-30X-2	15–16	273.65	25	1090B-30X-5	115–116	279.15	752
1090B-30X-2	24–25	273.74	17	1090B-30X-5	117–119	279.17	596
1090B-30X-2	35–36	273.85	15	1090B-30X-5	122–124	279.22	174
1090B-30X-2	45–46	273.95	20	1090B-30X-5	125–126	279.25	250
1090B-30X-2	56–57	274.06	17	1090B-30X-5	126–127	279.26	483
1090B-30X-2	65–66	274.15	19	1090B-30X-5	126.5–129	279.27	156
1090B-30X-2	75–76	274.25	35	1090B-30X-5	129–134	279.29	155
1090B-30X-3	135–136	276.35	36	1090B-30X-5	135–136	279.35	105
1090B-30X-3	140–141	276.40	38	1090B-30X-5	145–146	279.45	80
1090B-30X-3	145–146	276.45	45	1090B-30X-6	5–6	279.55	115
1090B-30X-3	149–150	276.49	33	1090B-30X-6	15–16	279.65	76
1090B-30X-4	6–7	276.56	37	1090B-30X-6	25–26	279.75	55
1090B-30X-4	15–16	276.65	42	1090B-30X-6	35–36	279.85	63
1090B-30X-4	25–26	276.75	36	1090B-30X-6	45–46	279.95	64
1090B-30X-4	33–34	276.83	41	1090B-30X-6	55–56	280.05	83
1090B-30X-4	35–36	276.85	80	1090B-30X-6	65–66	280.15	59
1090B-30X-4	37–38	276.87	52	1090B-30X-6	75–76	280.25	52
1090B-30X-4	45–46	276.95	42	1090B-30X-6	85–86	280.35	63
1090B-30X-4	56–57	277.06	42	1090B-30X-6	95–96	280.45	51
1090B-30X-4	65–66	277.15	48	1090B-30X-6	105–106	280.55	41
1090B-30X-4	75–76	277.25	82	1090B-30X-6	115–116	280.65	41
1090B-30X-4	86–87	277.36	50	1090B-30X-6	124–125	280.74	24
1090B-30X-4	95–96	277.45	71	1090B-30X-6	135–136	280.85	38
				1090B-30X-6	145–146	280.95	46

Note: mbsf—meters below seafloor.

well as similar data from nine other localities, including ODP Hole 689B (Montanari et al., 1993). Each of the five cores has anomalous Ir concentrations, and, in each case, the Ir anomalies are essentially coincident with the peaks in spherule concentrations (Figs. 5 and 6). The peak concentrations in Holes 1090B, 699A, 709C, 689D, and 738B are 1046, 666, 309, 188, and 157 pg/g, respectively (Table 6). The net fluence for Ir in these holes ranges from 1.2 (738B) to 15 ng/cm^2 (1090B). The sites with the highest peak Ir concentrations also have the highest net fluence. There is also a correlation between the peak Ir concentration (or fluence) and the thickness of the Ir anomaly. The anomaly is only 20–30 cm thick at Sites 689 and 738, but at least 200 cm thick at Site 1090. These results are generally within the range of Ir concentrations from other localities reported in earlier studies.

The results for Ir concentrations in fractions extracted from our large-volume sample at 709C-31X-4, 145–150 cm, provide insight into the distribution of Ir within the sediment. This sample was an entire 5 cm section of the core that had been squeezed by the Shipboard Scientific Party for interstitial water studies. Our sieved samples were from half of this piece, which had a dry weight of 50.35 g. This sample was sieved at 63, 125, 250, and 500 μm. Most of the <63μm fraction was saved and freeze-dried.

TABLE 4. IRIDIUM CONCENTRATIONS IN SEDIMENTS FROM OCEAN DRILLING PROGRAM (ODP) HOLES 709C AND 699A

Hole-core-section	Depth (cm)	Depth (mbsf)	Ir (pg/g)	Hole-core-section	Depth (cm)	Depth (mbsf)	Ir (pg/g)
709C-31X-4	20–21	290.50	20	699A-35X-3	85–86	319.95	82
709C-31X-4	30–31	290.60	18	699A-35X-3	95–96	320.05	76
709C-31X-4	40–41	290.70	58	699A-35X-3	105–106	320.15	42
709C-31X-4	50–51	290.80	40	699A-35X-3	125–126	320.35	67
709C-31X-4	60–61	290.90	24	699A-35X-3	135–136	320.45	56
709C-31X-4	70–71	291.00	22	699A-35X-3	145–146	320.55	49
709C-31X-4	80–81	291.10	20	699A-35X-4	10–11	320.70	87
709C-31X-4	90–91	291.20	33	699A-35X-4	20–21	320.80	80
709C-31X-4	100–101	291.30	36	699A-35X-4	30–31	320.90	63
709C-31X-4	110–111	291.40	56	699A-35X-4	40–41	321.00	78
709C-31X-4	118–119	291.48	179	699A-35X-4	50–51	321.10	104
709C-31X-4	123–124	291.53	276	699A-35X-4	60–61	321.20	115
709C-31X-4	128–129	291.58	309	699A-35X-4	65–66	321.25	185
709C-31X-4	133–134	291.63	99	699A-35X-4	70–71	321.30	225
709C-31X-4	138–139	291.68	293	699A-35X-4	75–76	321.35	337
709C-31X-4	143–144	291.73	156	699A-35X-4	80–81	321.40	666
709C-31X-4	145–150	291.78	273	699A-35X-4	85–86	321.45	528
709C-31X-5	3–4	291.83	131	699A-35X-4	90–91	321.50	357
709C-31X-5	8–9	291.88	66	699A-35X-4	100–101	321.60	356
709C-31X-5	13–14	291.93	103	699A-35X-4	110–111	321.70	90
709C-31X-5	18–19	291.98	87	699A-35X-4	120–121	321.80	67
709C-31X-5	23–24	292.03	53	699A-35X-4	130–131	321.90	60
709C-31X-5	28–29	292.08	69	699A-35X-4	140–141	322.00	84
709C-31X-5	38–39	292.18	40	699A-35X-5	5–6	322.15	58
709C-31X-5	48–49	292.28	106	699A-35X-5	15–16	322.25	52
709C-31X-5	58–59	292.38	24	699A-35X-5	25–26	322.35	54
709C-31X-5	68–69	292.48	24				
709C-31X-5	78–79	292.58	16				
709C-31X-5	88–89	292.68	19				
709C-31X-5	98–99	292.78	17				

Note: mbsf—meters below seafloor.

The 63–125 and 125–250 μm fractions were acidified to remove carbonates and put through heavy liquid separation to remove the biogenic silica fraction. The remaining 63–125 μm silicate fraction was split magnetically, with the magnetic fraction containing mainly cpx spherules and microtektites, and the nonmagnetic fraction containing microtektites and mineral grains. Some of the latter were probably nonimpact grains. The 125–250, 250–500, and >500 μm fractions were handpicked to separate microtektites and cpx spherules, as well as biogenic carbonate and opal in the coarser sizes. It was in these fractions that coesite was initially discovered. Our Ir analyses were on four samples: the <63 μm fraction, the 63–125 μm magnetic split (cpx spherules and microtektites), a sample of cpx spherules 125–250 μm, and one of microtektites 250–500 μm. The microtektite sample was composed of nine individual spherules. The cpx spherules were not counted, but the 125–250 μm sample contained hundreds of cpx spherules, and the 63–125 μm magnetic concentrate likely contained thousands of cpx spherules and microtektites.

The fine fraction (<63 μm) of sediment had a measured Ir concentration of 161 pg/g, significantly less than the "bulk" sediment sample, which has 273 pg/g. This difference could be due to sample heterogeneity, as the bulk sample was a small piece taken from the larger sample, which spanned 5 cm of core. The lower concentration (161 pg/g) is consistent with samples from immediately above and below this sample in the core, which have values of 156 and 131 pg/g, respectively. The highest Ir concentrations are found in the 125–250 μm cpx spherules, which, at 1620 pg/g, are a factor of 10 higher than in the fine fraction. The 63–125 μm magnetic split is much less, only 401 pg/g. This value is close to that measured in the microtektites, which have 584 pg/g Ir; this is only 36% of that measured in the 125–250 μm cpx spherules, but it is higher than in any bulk sediment sample from this core. The Ir peak is irregular and spread out in 709C, but the highest concentration measured in sediment was 309 pg/g, ~20 cm higher in the core.

When we calculate where the Ir is actually found in these sediments, based on results from the fractions taken from the large sample of 709C-31X-4, 145–150 cm (Table 7), we find that the great majority is in the fine (<63 μm) fraction. The <63 μm fraction is 96.86% by weight of the sediment. Of the >63 μm fraction, biogenic opal and carbonate make up 2.59 wt% of the total sample mass, and the nonbiogenic fraction, composed mainly of microtektites and cpx spherules, is only 0.55 wt%. Using Ir concentrations of 161 pg/g for the <63 μm sediment, 0 pg/g for the coarse biogenic, and 1620 pg/g for the cpx spherules (125–250 μm in size), we estimate that 95% of the Ir is contained in the silt- and clay-sized portion of the sediment, whereas only 5% is in the actual impact spherules. The 5% is a generous estimate, as the actual Ir content of the nonbiogenic fraction is probably less than the 1620 pg/g we measured in the cpx spherule split and intermediate between this and the 584 pg/g measured in the microtektites.

TABLE 5. IRIDIUM CONCENTRATIONS IN SEDIMENTS FROM OCEAN DRILLING PROGRAM (ODP) HOLES 689D AND 738B

Hole-core-section	Depth (cm)	Depth (mbsf)	Ir (pg/g)	Hole-core-section	Depth (cm)	Depth (mbsf)	Ir (pg/g)
689D-11H-5	10–11	120.50	22	738B-5H-2	50–51	34.50	7
689D-11H-5	30–31	120.70	13	738B-5H-2	65–66	34.65	7
689D-11H-5	50–51	120.90	14	738B-5H-2	81–82	34.81	9
689D-11H-5	70–71	121.10	13	738B-5H-2	98–99	34.98	10
689D-11H-5	95–71	121.35	14	738B-5H-2	110–111	35.10	15
689D-11H-5	97.5–98.5	121.38	11	738B-5H-2	117–118	35.17	25
689D-11H-5	105–106	121.45	15	738B-5H-2	120–121	35.20	26
689D-11H-5	110–111	121.50	14	738B-5H-2	125–126	35.25	49
689D-11H-5	113–114	121.53	9	738B-5H-2	130–131	35.30	72
689D-11H-5	122–123	121.62	13	738B-5H-2	135–136	35.35	157
689D-11H-5	125–126	121.65	19	738B-5H-2	140–141	35.40	147
689D-11H-5	130–131	121.70	12	738B-5H-2	143–144	35.43	86
689D-11H-5	135–136	121.75	13	738B-5H-2	149–150	35.49	39
689D-11H-5	140–141	121.80	17	738B-5H-3	5–6	35.55	27
689D-11H-5	146–147	121.86	11	738B-5H-3	10–11	35.60	33
689D-11H-5	149–150	121.89	15	738B-5H-3	25–26	35.75	34
689D-11H-6	11–12	122.01	14	738B-5H-3	41–42	35.91	10
689D-11H-6	30–31	122.20	10	738B-5H-3	56–57	36.06	10
689D-11H-6	50–51	122.40	<7	738B-5H-3	70–71	36.20	10
689D-11H-6	70–71	122.60	12	738B-5H-3	90–91	36.40	14
689D-12H-2	25–26	125.85	7	738B-5H-3	120–121	36.70	7
689D-12H-2	50–51	126.10	12	738B-5H-3	145–146	36.95	<30
689D-12H-2	75–76	126.35	13	738B-5H-4	20–21	37.20	<16
689D-12H-2	85–86	126.45	19	738B-5H-4	50–51	37.50	10
689D-12H-2	95–96	126.55	25	738B-5H-4	80–81	37.80	<14
689D-12H-2	105–106	126.65	33	738B-5H-4	112.5–113.5	38.13	9
689D-12H-2	115–116	126.75	75	738B-5H-4	140–141	38.40	<16
689D-12H-2	125–126	126.85	188				
689D-12H-2	135–136	126.95	154				
689D-12H-2	145–146	127.05	47				
689D-12H-3	5–6	127.15	24				
689D-12H-3	25–26	127.35	18				
689D-12H-3	50–51	127.60	14				

Note: mbsf—meters below seafloor.

DISCUSSION

Biostratigraphic Position of the Cpx Spherule Layer

Due to the near-global distribution of the cpx spherule strewn field and the greater resistance to dissolution than microfossils, the cpx spherule layer provides a powerful tool for chronostratigraphic correlation of late Eocene strata. The last occurrence of *T. tetracantha* is used to define the *Calocyclas bandyca–Cryptopora ornata* zonal boundary, and earlier studies concluded that the cpx spherule layer was closely associated with this zonal boundary. ODP Hole 709C has a good late Eocene radiolarian assemblage, and, in agreement with previous sites, the cpx spherule layer is closely associated with the sharp drop in abundance of many of the aforementioned radiolarian species. However, in more recently used radiolarian zonations, the *C. bandyca* zone is referred to as the RP18 zone, and the *C. ornata* (or *ornatum*) zone is referred to as the RP19 zone. The RP18/RP19 zonal boundary is placed in magnetochron C15r and assigned an age of 35.16 Ma (Nigrini et al., 2005). In fact, all the last occurrences of the radiolarian species listed here are assigned ages that are younger than the cpx spherule layer (Nigrini et al., 2005), as is the case for Site 612 in the northwest Atlantic (Keller et al., 1987). Thus, although the cpx spherule layer is associated with a sharp drop in abundance of many of these radiolarian species, it appears that it is not associated with the extinction of these radiolarian species as previously thought.

All of the sites in the present study have good nannofossil zonations, and the cpx spherule layer in each of them lies within the CP15 zone. In previous studies (e.g., Glass et al., 1985), the cpx layer at three sites (Sites 167, 216, 292) was reported as being in the *Discoaster barbadiensis* zone, which is equivalent to the CP15 zone. Zone CP15 has been subdivided into an older CP15a subzone and a younger CP15b subzone based on the first occurrence of *Isthmolithus recurvus* (Okada and Bukry, 1980). At two of the sites in this study with well-defined cpx spherule layers (ODP Sites 699A and 1090B), the spherule layer is in a section of the core assigned to the CP15b subzone. At Site 709C, also with a well-defined cpx spherule layer, the layer may also be in the CP15b subzone, but the resolution is not good enough to be sure (Okada, 1990). At Site 786A, the cpx spherule layer appears to be within zone CP15a. At the other sites, zone CP15 was not divided into CP15a and CP15b. More recent calcareous nannofossil zonations do not divide the CP15 zone into the CP15a and CP15b subzones (e.g., Shipboard Scientific Party, 2004).

TABLE 6. IRIDIUM DATA FOR CLINOPYROXENE SPHERULE–BEARING SITES

Site/Hole	Location	Dry bulk density (g/cm^3)	References	Interval integrated (mbsf or thickness)*	Peak Ir (pg/g)	Ir fluence (ng/cm^2)	References
ODP Hole 1090B	Agulhas Ridge, S. Atlantic	0.46	1	278.1–280.1	1046	15	7, 8
ODP Hole 699A	Georgia Rise, S. Atlantic	0.79	2	320.7–321.8	666	13	8
ODP Hole 689B	Maud Rise, S. Atlantic	0.95	3	128.5–128.85	156	2.2	9
ODP Hole 689D	Maud Rise, S. Atlantic	0.95	3	126.7–127.0	188	2.5	8
ODP Hole 709C	Mascarene Plateau, W. Indian Ocean	1.2	4	291.4–292.0	309	8.8	8
ODP Hole 738B	Kerguelen Plateau, S. Indian Ocean	0.98	5	35.28–35.46	157	1.2	8
Massignano	Italy	2.5	6	0.29 m	200	4.7	9
Gay's Cove	Barbados, Caribbean Sea	2.5	6	0.30 m	200	12	10
RC9-58	Venezuela Basin, Caribbean Sea	1	6	2.58–2.90	4100	42	11
DSDP Site 149	Caribbean Sea	1	6	270.0–270.26	>410	>4	12
DSDP Site 292	Equatorial Pacific	1.3	14	350.15–350.75	40	1.5	13
DSDP Site 315	Equatorial Pacific	1.05	15	738.1–738.8	250	7	13
DSDP Site 462	Equatorial Pacific	1.04	16	343.5–344.55	220	15	13
DSDP Site 612	Equatorial Pacific	0.86	17	181.1–181.36	104	0.6	13

Note: References: 1—Shipboard Scientific Party (1999), 2—Shipboard Scientific Party (1988b), 3—Shipboard Scientific Party (1988a), 4—Shipboard Scientific Party (1988d), 5—Shipboard Scientific Party (1989), 6—dry bulk density assumed (see text), 7—Kyte (2001), 8—this work, 9—Montanari et al. (1993), 10—Sanfilippo et al. (1985), 11—Ganapathy (1982), 12—Alvarez et al. (1982), 13—Keller et al. (1987), 14—Bouma and Moore (1975), 15—Shipboard Scientific Party (1976), 16—Shipboard Scientific Party (1981), 17—Shipboard Scientific Party (1987).

*mbsf—meters below seafloor.

There is a problem with the placement of the cpx spherule layer in planktonic foraminiferal zonation. At most sites, the cpx spherule layer occurs in what originally was identified as the P16 or *Globigerinatheka cerroazulensis* zone (Glass et al., 1985; Keller et al., 1987). The zonations given for most of these sites were based on widely spaced samples, and the boundary between zone P16 and the underlying zone P15 was not well defined. Nevertheless, the large number of sites where the cpx spherule layer has been identified as being in the P16 zone is not easily dismissed. The most recent attempts to correlate between the various biostratigraphic zones and the geomagnetic polarity time scale indicate that the base of zone P16 has an age of 35.2 Ma (e.g., Shipboard Scientific Party, 2004). Thus, the cpx spherule layer should occur in the underlying P15 or *Globigerinatheka semiinvoluta* zone. It appears that either the age given for the P15/P16 zonal boundary is too young, or the cpx spherule layer occurs in zone P15. Unfortunately, none of the new spherule or ejecta-bearing sites has good planktonic foraminifera zonations for the late Eocene, and thus the new sites cannot help to determine whether the cpx spherule layer is in zone P15 or P16. Berggren and Pearson (2005) revised the Eocene planktonic foraminiferal zonation. In this zonation, the cpx spherule layer should occur in the E15 or *Globigerinatheka index* zone, which ranges from 34.3 to 35.8 Ma.

Wei (1995) found that the last occurrence (LO) of *Discoaster saipanensis* occurs above the North American microtektite/microkrystite (cpx spherule) layer couplet and that the LO of *Chiasmolithus grandis* occurs below the North American microtektite/microkrystite (cpx spherule) layer couplet. Both of these calcareous nannofossils taxa are found only in three of the new sites: Holes 699A (Crux, 1991), 709C (Okada, 1990), and 786A (Xu and Wise, 1992). Based on published data for these three sites, we find the same relationship that Wei (1995) observed. Three other sites (592, 1073A, 1090B) have *Discoaster saipanensis* present, and at all these sites, the LO is above the spherule layer, in agreement with Wei's (1995) findings. Some additional biostratigraphic datums have been identified that should bracket the cpx spherule layer (Table 8) (Shipboard Scientific Party, 2004).

Magnetostratigraphy of the Cpx Spherule Layer

At most sites, the cpx spherule–bearing late Eocene sections of the cores do not have good magnetostratigraphy. The Ir anomaly and spherule layer at Massignano was originally thought to be in chron 16n.2n (Bice and Montanari, 1988, as reported in Montanari et al., 1993), but in a later publication, the part of the outcrop containing the spherule layer was interpreted as the upper part of chron 16n.1n (Jovane et al., 2006). An Ir anomaly and accompanying cpx spherule layer was found in Core 14H, Section 6, from ODP Hole 689B (between 128.59 and 128.77 mbsf) (Montanari et al., 1993; Vonhof and Smit, 1999; Glass and Koeberl, 1999). The section in which the spherule layer was found has been assigned to chron 15N by

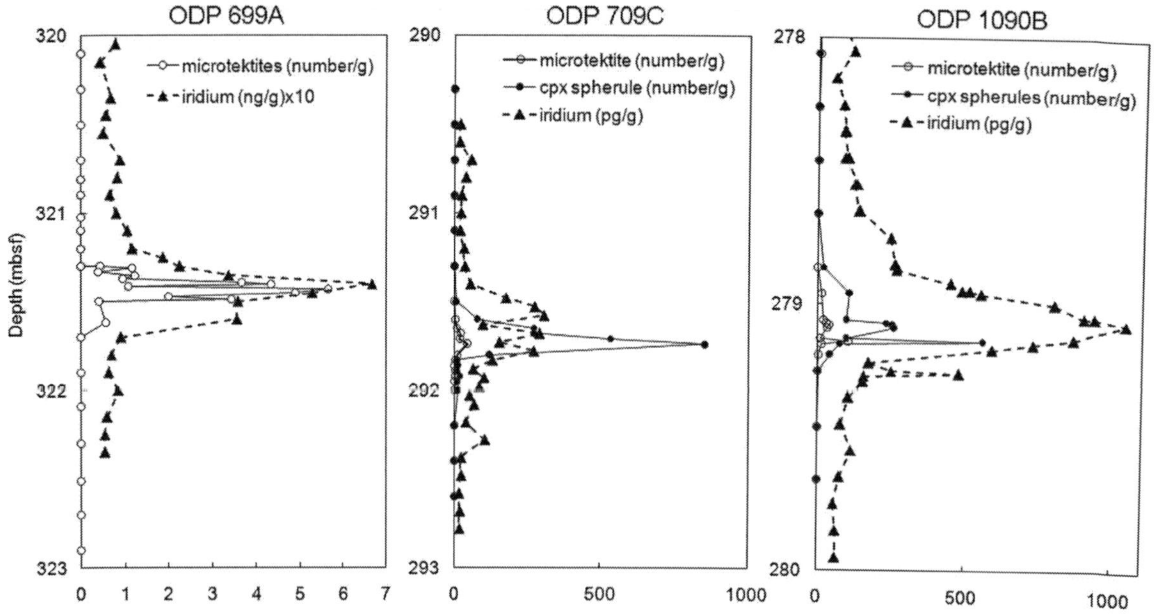

Figure 5. Vertical distribution of Ir, microtektites, and cpx spherules in cores from Ocean Drilling Program Holes 699A, 709C, and 1090B in the southern South Atlantic, northwestern Indian, and southeastern South Atlantic Oceans, respectively (mbsf—m below seafloor). Note different horizontal and vertical scales and different units for Ir concentrations. The number/g is for cpx spherules and/or microtektites >63 µm in size for Sites 699A and 1090B and for cpx spherules and microtektite >125 µm in size for Site 709C.

Spieß (1990) and to chron 16N by Stott and Kennett (1990). In a more recent study, Florindo and Roberts (2005) revised the magnetostratigraphy of Hole 689B and concluded that Core 14H, Section 6 (which contains the cpx spherule layer), occurs within chron 16n.1n.

At Site 1090 (Hole 1090B), the spherule layer appears to be within chron 16n, perhaps chron 16n.1n (Channell et al., 2003). Thus, the presently available data indicate that the cpx spherule layer is in chron 16n, probably c16n.1n. According Cande and Kent (1995), chron 16n.1n ranges between 35.343 and 35.526 Ma. This is consistent with an age of ca. 35.4 ± 0.3 Ma for the age of the cpx spherule layer as discussed in the "Introduction" section, and if the cpx spherule layer is within chron 16n.1n, this would confine the age of the spherule layer to between ca. 35.3 and 35.5 Ma. Thus, the age of the cpx spherule layer appears to be consistent with the age of the Popigai impact structure (35.7 ± 0.2 Ma; Bottomley et al., 1997), which is believed to be the source crater of the cpx spherule layer (Langenhorst, 1996; Whitehead et al., 2000; Liu et al., 2006).

Our conclusions regarding the biostratigraphic and magnetostratigraphic occurrence of the cpx spherule layer are summarized in Table 8 and Figure 7. The cpx spherule layer lies between the last occurrence (LO) of *Discoaster saipanensis* above the layer and the LO of *Chiasmolithus grandis* below the layer. The published ages for the LO of *Discoaster saipanensis* and the LO of *Chiasmolithus grandis* are 34.2 and 37.1 Ma, respectively (Shipboard Scientific Party, 2004), and thus they bracket the proposed age of the cpx spherule layer (i.e., 35.4 Ma).

Source of Silica-Rich Microtektites at ODP Sites 689, 699, and 1090

In previous studies, silica-rich (>75 wt% silica) microtektites were found in the cpx spherule layer in Core 14 from ODP Hole 689B in the South Atlantic Ocean (Vonhof and Smit, 1999; Glass and Koeberl, 1999). These microtektites have major oxide compositions similar to those of the North American microtektites, and Vonhof and Smit (1999) suggested that they may belong to the North American microtektite strewn field. Glass and Koeberl (1999) concurred with that conclusion based on the major oxide composition of the microtektites, the vertical distribution of microtektites and cpx spherules, and the estimated abundance of microtektites (>125 µm) at that site (~52/cm^2). The estimated abundance is close to the abundance predicted by extrapolation of the trend of abundance of North American microtektites versus distance from the proposed source crater (the Chesapeake Bay impact structure). At Hole 689B, the peak abundance of the microtektites appears to be ~2 cm above the peak abundance of the cpx spherules.

Part of the reason for looking for the cpx spherule layer at Hole 689D was to determine if the peak abundance of microtektites occurred above the peak abundance of cpx spherules as appears to be the case at Hole 689B. We found instead that, at Hole 689D, the peak abundance of the microtektites coincides with the peak abundance of the cpx spherules. In addition, microtektites from Hole 689D were sent to Dimitri Papanastassiou at CalTech for Sr and Nd isotopic study. These studies showed that

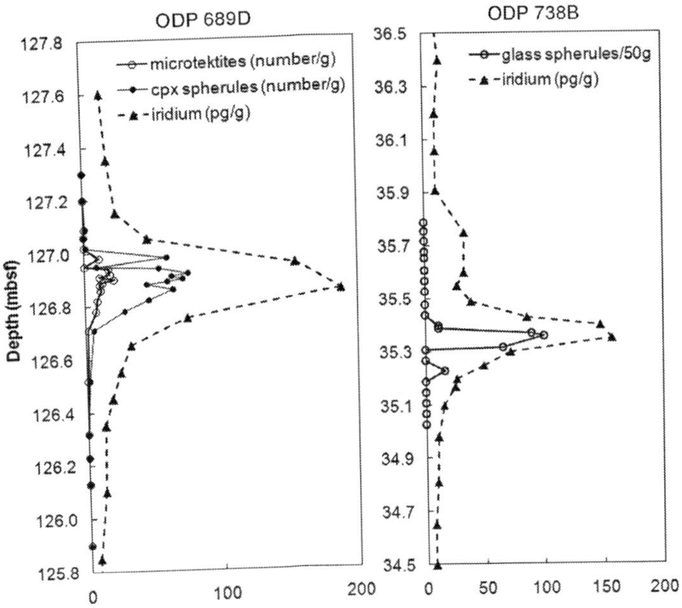

Figure 6. Vertical distribution of Ir and cpx spherules and/or microtektites in cores from Ocean Drilling Program Holes 689D and 738B in the southern South Atlantic and southern Indian Ocean, respectively (mbsf—m below seafloor). Note the difference in horizontal scales; for Site 738B, the number of microtektites is giver per 50 g of sediment. The number/g is for cpx spherules and microtektites >63 µm in size for Site 689D and for microtektite >150 µm in size for Site 709C.

the microtektites from ODP Hole 689D have Sr and Nd isotopic ratios similar to the associated cpx spherules but not to the North American tektites/microtektites (Fig. 8)(Liu et al., 2006). Thus, at ODP Site 689, the microtektites found in the cpx spherule layer do not appear to belong to the North American tektite/microtektite strewn field as previously proposed by Vonhof and Smit (1999) and Glass and Koeberl (1999).

Microtektites similar in appearance and composition to the transparent colorless to pale brown microtektite found at Hole 689B are also found at Hole 1090B. This site has an unusually high sediment accumulation rate for late Eocene sediments (~30 m/m.y.). Assuming that the North American microtektites are ~15 k.y. younger than the cpx spherules, and if the transparent colorless to pale brown microtektites at Hole 1090B belong to the North American microtektite layer, they should be ~45 cm above the cpx spherule layer. Since the microtektites at Hole 1090B are found at the same depth as the cpx spherules, it is unlikely that they belong to the North American event.

TABLE 7. IRIDIUM CONCENTRATION IN SEDIMENT FRACTIONS OF A SAMPLE FROM HOLE 709C-31X-4, 145–150 CM

Sample description	Sample mass (mg)	Iridium (pg/g)
Bulk sediment sample	92.6	273
<63 µm	80.4	161
63–125 µm magnetic fraction	21.0	401
125–250 µm cpx spherules	17.0	1620
250–500 µm microtektites	1.41	584

At ODP Hole 699A, also in the South Atlantic, we expected to find the cpx spherule layer, but only found microtektites, which we were tempted to interpret as North American microtektites. Like the North American microtektites, they are silica-rich with low MgO and CaO (Table 2). However, these microtektites are more similar in appearance and composition to the microtektites found in the cpx spherule layer at Sites 689 and 1090, and they are associated with an Ir anomaly (Fig. 5). Thus, although only microtektites were found at Hole 699A, we interpret them as belonging to the cpx strewn field. However, these microtektites have high SiO_2 content, and all have low Fe content, in contrast to the microtektites at Site 689, which have a broader range in SiO_2 content and a mixture of microtektites with low- and high-Fe content.

Hole 699A is the first site where only microtektites were found in what we interpret to be the cpx spherule layer. It is even more unusual, since none of the microtektites at this site has compositions similar to the compositions of the cpx spherules. One possible explanation is that the cpx spherules and low-

TABLE 8. BIOSTRATIGRAPHIC DATUMS NEAR THE CPX SPHERULE LAYER

Type*	Datum	Fossil group	Age (Ma)†
LO	*Discoaster saipanensis*	Calcareous nannofossil	34.2
LO	*Discoaster barbadiensis*	Calcareous nannofossil	34.3
LO	*Cribrocentrum reticulatum*	Calcareous nannofossil	34.9
LO	*Thyrsocyrtis tetracantha*	Radiolaria	34.9
FO	*Turborotalia cunialensis*	Planktonic foraminifera	35.2
LO	*Turborotalia pomeroli*	Planktonic foraminifera	35.3
LO	*Globigerinatheka semiinvoluta*	Planktonic foraminifera	35.3
	Cpx Spherule Layer		**35.4**
FO	*Cribrohantkenina inflata*	Planktonic foraminifera	35.5
FO	*Isthmolithus recurvus*	Calcareous nannofossil	36.0
FO	*Calocyclas bandyca*	Radiolaria	36.4
FO	*Chiasmolithus oamaruensis*	Calcareous nannofossil	37.0
LO	*Chiasmolithus grandis*	Calcareous nannofossil	37.1

*LO—last occurrence; FO—first occurrence.
†Ages of biostratigraphic datums from Shipboard Scientific Party (2004).

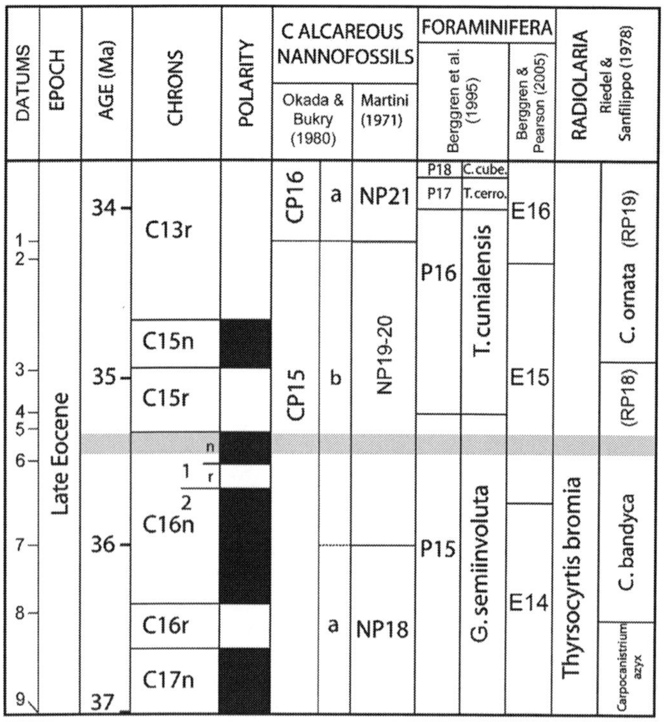

Figure 7. Late Eocene magnetostratigraphy and biostratigraphy. Horizontal gray band between ca. 35.3 and 35.4 Ma indicates the probable position of the cpx spherule layer. Ages for the various magnetostratigraphic and biostratigraphic zonations are taken primarily from Shipboard Scientific Party (2004). The datums in the left column are: (1) Last occurrence (LO) of *Discoaster saipanensis*, (2) LO of *Discoaster barbadiensis*, (3) LO of *Cribrocentrum reticulatum* and *Thyrsocyrtis tetracantha*, (4) First occurrence (FO) of *Turborotalia cunialensis*, (5) LO of *Turborotalia pomeroli* and *Globigerinatheka semiinvoluta*, (6) FO of *Cribrohantkenina inflata*, (7) FO of *Isthmolithus recurvus*, (8) FO of *Calocyclas bandyca*, (9) LO of *Chiasmolithus grandis*. (See Table 8 for more details.)

silica microtektites, with compositions similar to those of cpx spherules, were dissolved by bottom water that was undersaturated in silica. However, the water paleodepth was deep enough at this site to have precluded dissolution. In addition, the sediment contains ~20% biogenic silica, and biogenic silica is more soluble than the cpx spherules and silica-poor microtektites. Furthermore, the high percent of microtektites at this site (100%) is consistent with the trend of increasing percent of microtektites in the cpx spherule layer with increasing distance from Popigai impact crater. This is discussed in the section "Ray-Like Nature of the Cpx Strewn Field."

Major Oxide Compositions of Cpx Spherules and Microtektites

The major oxide compositions of the cpx spherules from ODP Holes 689D, 709C, 786A, and 1090B are similar to previously published major oxide compositions of cpx spherules from sites in the Gulf of Mexico, equatorial Pacific, and eastern equatorial Indian Ocean (Fig. 9). There are two light-colored cpx spherules from Hole 1090B that have high SiO_2 contents and low FeO, K_2O, and TiO_2 contents, but it is their low Al_2O_3 and high Na_2O contents that distinguish them from previously analyzed cpx spherules (Table 2; Fig. 9).

Previously published compositional data for microtektites found in the cpx spherule layer indicate that they can be divided into two groups (Vonhof and Smit, 1999; Glass and Koeberl, 1999; Glass et al., 2004a). One group includes microtektites with low SiO_2 contents (generally <75 wt%) and high MgO and CaO contents (generally >2 wt%), and the other group includes microtektites with high SiO_2 contents (generally >74 wt%) and low MgO and CaO contents (generally <1 and 2 wt%, respectively) (Fig. 10). Microtektites in the low-SiO_2 group have compositions similar to the cpx spherules (Glass et al., 2004a). The microtektites in the high-SiO_2 group have compositions similar to the composition of the North American microtektites, except for their low Na_2O contents and their iron contents, which have a much greater range for a given silica content. The high-SiO_2 microtektites are most abundant in the western equatorial Pacific (DSDP Site 292), Indian Ocean (DSDP Site 216), and South Atlantic Ocean (ODP Hole 689B), but they occur in the equatorial Pacific sites as well. They have not been reported from the Gulf of Mexico or Caribbean Sea sites, but the North American microtektite layer overlaps the cpx spherule layer at these sites, and, therefore, it would be difficult to distinguish between the high-SiO_2 microtektites and the North American microtektites. As far as we know, no one has searched for high-SiO_2 microtektites associated with the cpx spherules in the Gulf of Mexico and/or Caribbean Sea sites. The high-SiO_2 microtektite group can be divided into a high-FeO group (>2.5 wt%) and a low-FeO group (<2.5 wt%) (Fig. 10) (Glass and Koeberl, 1999). The microtektites in the low-FeO group are transparent colorless and generally have very low Na_2O contents (<0.1 wt%), whereas the microtektites in the high FeO-group are generally transparent pale brown and may have higher Na_2O contents (up to 1 wt%).

None of the DSDP Site 292 microtektites and only two of the microtektites from the equatorial Pacific sites fall in the high-SiO_2 group; however, numerous microtektites from Site 216 in the eastern equatorial Indian Ocean belong to the high-SiO_2 group (Fig. 10). Thus, microtektites from the various cpx spherule layer sites can be tied together compositionally, but there are geographic variations in composition within the strewn field. These compositional variations may be related to the stratigraphy of the target rocks in the source crater and will be discussed further in the section "Ray-Like Nature of the Cpx Strewn Field."

Microtektites from the cpx spherule layer found at the sites/holes reported in this paper have compositions similar to previously reported compositions of microtektites found in the cpx spherule layer at other sites (Fig. 11). Microtektites from the cpx spherule layer at ODP Hole 689D have compositions similar to the microtektites found at Hole 689B, and all but one belong to the high-SiO_2 group: four are high-FeO microtektites, and four are low-FeO microtektites (Fig. 11). We determined the

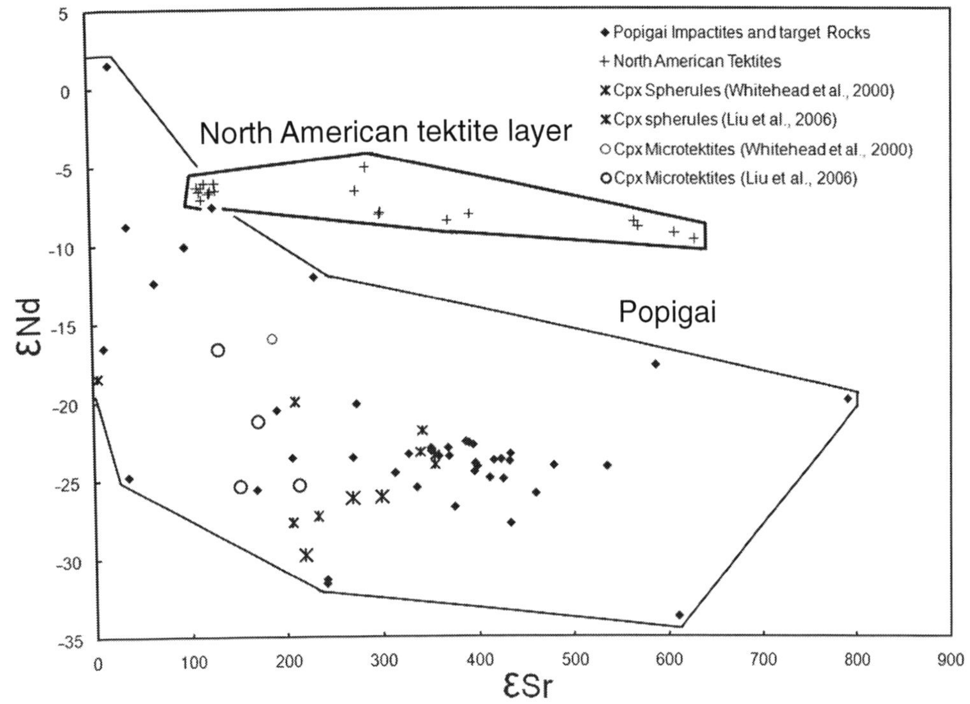

Figure 8. Plot of ε_{Nd} versus ε_{Sr} for cpx spherules, microtektites in the cpx spherule layer, North American tektites and microtektites, and Popigai target rocks and impactites. Data for the North American tektites and microtektites are from Shaw and Wasserburg (1982), Ngo et al. (1985), Stecher et al. (1989), Glass et al. (1998), and Whitehead et al. (2000). Data for cpx spherules and associated microtektites are from Whitehead et al. (2000) and Liu et al. (2006). Data for the Popigai crater target rocks and impactites are from Kettrup et al. (2003) and Hölker et al. (1997).

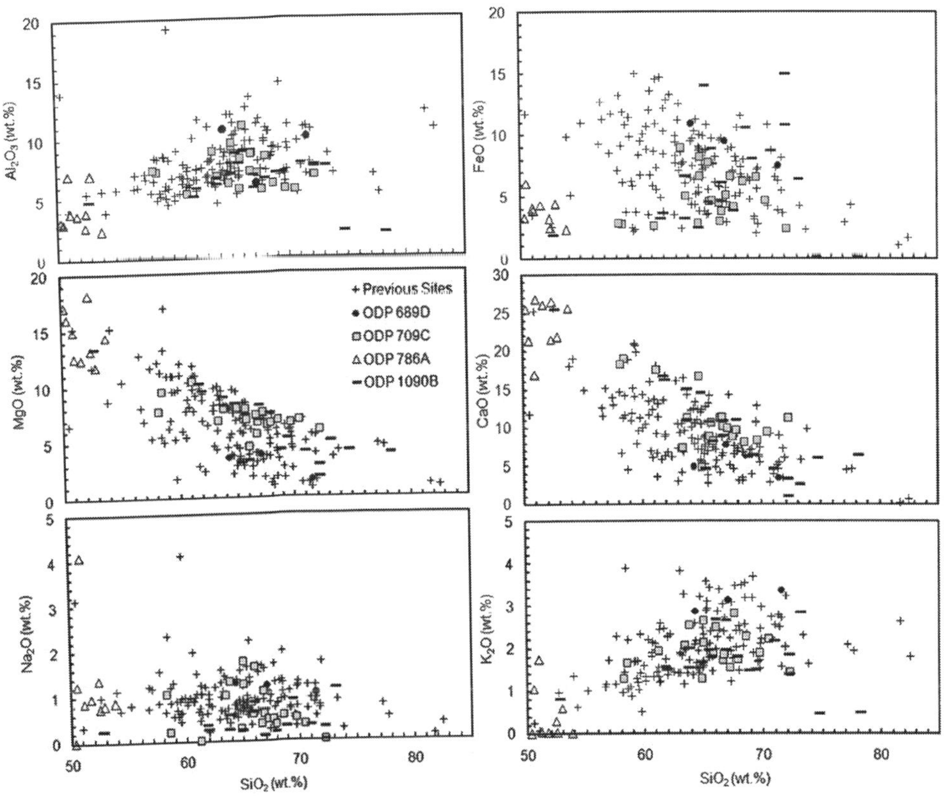

Figure 9. Plots of major oxide contents versus SiO_2 contents for cpx spherules from previous studies and Ocean Drilling Program (ODP) sites in this study (689D, 709C, 786A, 1090B). Previous sites include: Core RC9-58 and Deep Sea Drilling Project (DSDP) Site 149 in the Caribbean Sea; DSDP Site 94 in the Gulf of Mexico; DSDP Holes/Sites 69A, 70A, 162, 166, and 167 in the equatorial Pacific; DSDP Site 292 in the Philippine Sea; DSDP Site 216 in the northeastern Indian Ocean; and Ocean Drilling Program Hole (ODP) 689B in the southern South Atlantic. See Table 9 and Figure 1 for locations of these sites. Major oxide compositions for the spherules from these sites were obtained in the same manner and under the same conditions as the spherules in this study. The data were also standardized using the same glass standard as used in this study. The Hole 689B cpx spherule analyses were published in Glass and Koeberl (1999).

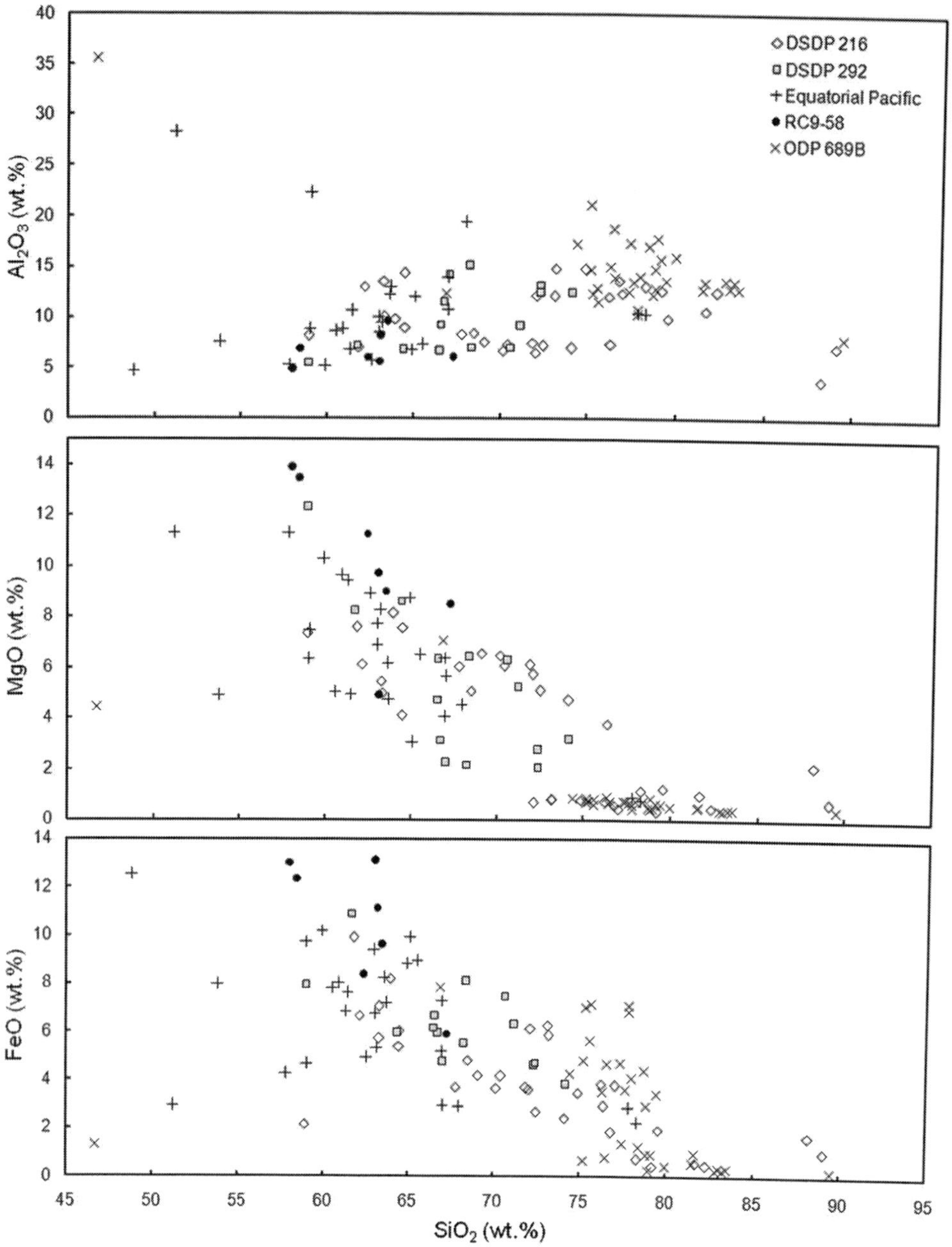

Figure 10. Plots of Al_2O_3, MgO, and FeO contents versus SiO_2 contents for microtektites in the cpx spherule layer at previously known sites: Core RC9-58 in the Caribbean Sea; Deep Sea Drilling Project (DSDP) Holes/Sites 69A, 70A, 166, and 167 in the equatorial Pacific; DSDP Site 292 in the Philippine Sea; DSDP Site 216 in the northeastern Indian Ocean; and Ocean Drilling Program (ODP) Hole 689B in the southern South Atlantic. See Table 9 and Figure 1 for location of these sites. Compositional data are from DuBois (1983), Glass et al. (1985), D'Hondt et al. (1987), Glass and Koeberl (1999), Glass et al. (2004a), and this study.

major oxide composition of 10 microtektites from Hole 699A. The Hole 699A microtektites all belong to the high-SiO_2, low-FeO subgroup (Table 2). They have the highest average SiO_2 content (86.7 ± 2.5 wt%) of any group of microtektites found in the cpx spherule layer, and two have the highest SiO_2 contents recorded for microtektites in the cpx spherule layer (83.8 and 86.6 wt%) (Table 2). Although similar to other high-SiO_2 microtektites in most respects, they apparently have higher Na_2O contents for their SiO_2 contents.

Twenty microtektites from Hole 709C were analyzed. Thirteen of the microtektites have compositions similar to the high-SiO_2 microtektites (Table 2; Fig. 11): two are in the low-FeO

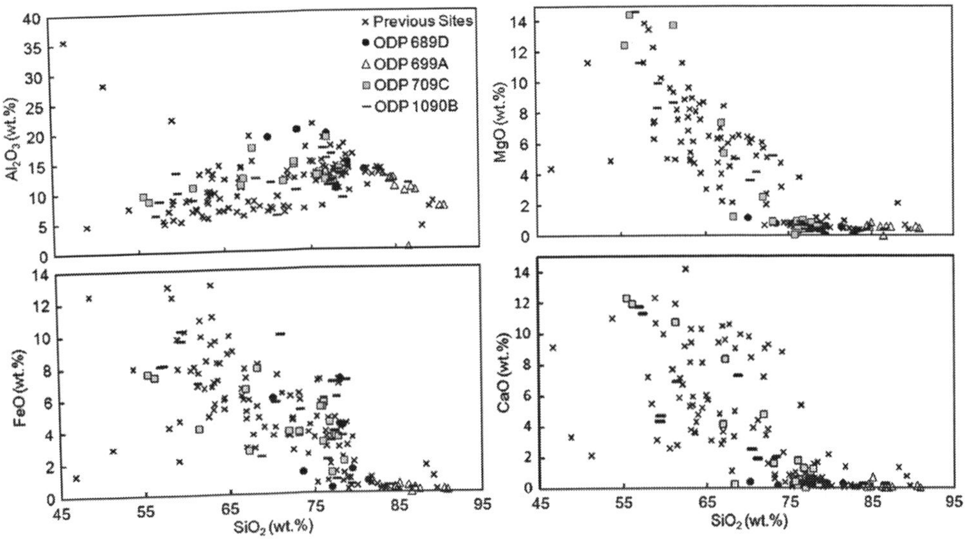

Figure 11. Plots of Al_2O_3, MgO, FeO, and CaO contents versus SiO_2 contents for microtektites in the cpx spherule layer at previously known sites and sites in this study. The previously known sites are: Core RC9-58 in the Caribbean Sea; Deep Sea Drilling Project (DSDP) Holes/Sites 69A, 70A, 166, and 167 in the equatorial Pacific; DSDP Site 292 in the Philippine Sea; DSDP Site 216 in the northeastern Indian Ocean; and Ocean Drilling Program (ODP) Hole 689B in the southern South Atlantic. See Tables 1 and 9 and Figure 1 for location of these sites and the sites in this study. Compositional data for microtektites from previous sites are from DuBois (1983), Glass et al. (1985), D'Hondt et al. (1987), Glass and Koeberl (1999), Glass et al. (2004a), and this study.

group, and 11 in the high-FeO group. Four are transparent green and have some of the highest MgO and CaO contents observed in microtektites from the cpx spherule layer. The major oxide contents were determined for 21 microtektites from the cpx spherule layer at ODP Hole 1090B; 12 are in the high-SiO_2 group, and nine are in the low-SiO_2 group. Six of the low-SiO_2 microtektites are transparent green. Five of them have high MgO contents and are similar in composition to the green microtektites at Hole 709C (Table 2). The high-SiO_2 microtektites from Hole 1090B are approximately equally divided between the high-FeO and low-FeO subgroups (Table 2). Two of the low-FeO microtektites are reddish in color, and one of those has a high FeO content (~9.9 wt%) for its SiO_2 content (~71 wt%).

Formation of Microkrystites versus Microtektites

Glass and Burns (1988) proposed the term "microkrystites" for impact spherules containing primary crystallites in order to distinguish them from microtektites, which, by definition, do not contain primary crystallites. These authors pointed out that the late Eocene crystal-bearing spherules (microkrystites) were more mafic in composition than most of the Australasian and North American microtektites and suggested that partial crystallization of microkrystites was due to their more mafic compositions. However, since the low-SiO_2 microtektites found in the cpx spherule layer have compositions indistinguishable from the compositions of the cpx spherules, composition, by itself, cannot have been the reason why the cpx spherules partly crystallized, while the associated microtektites did not. It appears, therefore, that the crystallinity of the more mafic impact spherules depends primarily on the cooling history. However, the cooling history may have been related to the size of the impact. We suggest that, compared with a small impact, in a large impact, the environment (impact plume and atmosphere) is heated to a higher temperature and stays hot longer, resulting in a lower average cooling rate for the spherules. For example, the high-Mg or bottle-green Australasian and Ivory Coast microtektites are always holohyaline, and not a single microkrystite has been found in either of these strewn fields. On the other hand, the more felsic impact spherules (i.e., those with SiO_2 > 78 wt%) apparently always have cooled too quickly to crystallize, regardless of the size of the impact.

Implications of Shocked Quartz and Coesite in the Cpx Spherule Layer

The presence of shocked quartz and coesite in the cpx layer at Site 709 in the northwest Indian Ocean and coesite in the cpx layer at Site 216 in the north-central Indian Ocean provides further evidence that the cpx spherule layer has an impact origin. In addition, it supports the previous conclusion (Glass et al., 2004b) that the pancake-shaped clay spherules associated with shock-metamorphosed quartz grains with PDFs at Massignano, Italy, are altered cpx spherules.

It is significant that the three sites (Massignano and Sites 216 and 709) where shocked quartz and/or coesite coexist in the cpx spherule layer are three of the sites closest to Popigai crater. Sites 292 and 786 (in the northwest Pacific) are closer to Popigai (Fig. 1), but, as far as we know, shocked quartz and coesite have not been searched for at Site 292. We did not search for shocked quartz and coesite at Site 786 because the spherule layer is not well-defined at this site, and volcanic ejecta have diluted any shocked quartz and coesite that might be present. At the time of the impact, Site 216 was the farthest site from Popigai that accumulated shocked quartz and coesite. Site 216 was ~9960 km (about ~100 crater diameters) from Popigai. Shocked quartz appears to have been distributed globally during the Chicxulub impact at the end of the Cretaceous. This means shocked quartz from Chicxulub was thrown as far as 20,000 km or 110 crater diameters. In the case of the Ivory Coast strewn field, no coesite or shocked quartz has been found in any of the

Ivory Coast microtektite-bearing cores. The closest Ivory Coast microtektite-bearing core site is ~1250 km (119 crater diameters) from the source crater (Bosumtwi). Thus, it appears that on Earth, unmelted impact ejecta (shocked quartz, etc.) can be thrown up to ~110 crater diameters from the source crater, but only melted ejecta (microtektites/microkrystites) occurs beyond ~110 crater diameters. We realize that the distance, in crater diameters, beyond which there only will be melted ejecta probably varies depending on the nature and velocity of the projectile, the angle of impact, and the nature of the target rock.

The source crater for the Australasian tektite/microtektite strewn field has not been found, but most authors have concluded that it is somewhere in the Indochina region (e.g., Stauffer, 1978; Ford, 1988; Schnetzler, 1992; Hartung and Koeberl, 1994; Lee and Wei, 2000; Ma et al., 2004; Glass and Koeberl, 2006). Since the Australasian source crater is only ~0.8 m.y. old and must be fairly large (>30 km; see Glass and Koeberl, 2006; Prasad et al. 2007), but has not yet been found, it may be offshore, perhaps in the Gulf of Tonkin area (Ma et al., 2004; Glass and Koeberl, 2006).

Australasian microtektites have been found at more than 60 sites, but coesite and shocked quartz have been found in only ~15 of the sites, all of which are within ~3000 km of the central Indochina region. Assuming the source crater is in the Gulf of Tonkin area (19°N latitude, 106°E longitude), the farthest site from the source crater that contains coesite and shocked quartz is ~3100 km from there. The closest site without coesite or shocked quartz is ~3200 km from there. If unmelted ejecta is found up to 110 crater diameters from the source crater and the Australasian tektite/microtektite source crater is near 19°N latitude, 106°E longitude, then the source crater for the Australasian strewn field would be ~29 km in diameter (i.e., 3200 km/110 = 28.6 km). This is smaller than the size of 43 ± 9 km estimated by Glass and Koeberl (2006) using an equation by Stöffler et al. (1975), which relates thickness of an ejecta layer to size of and distance from the source crater. Our estimate is closer to the size of 33 ± 8 km calculated by Prasad et al. (2007).

Nature of the Target Rock and Size of the Projectile

In addition to Massignano, flattened spherules are found in Cretaceous-Tertiary (K-T) boundary sediments in Umbria, Italy (Bohor and Glass, 1995). This flat shape may reflect deformation during sediment compaction. Shocked quartz grains containing planar deformation features (PDFs) have been recovered from the spherule layer at Massignano (Clymer et al., 1996; Langenhorst, 1996). Based on the nature of the PDFs, and the lack of coesite and stishovite in the spherule layer, Langenhorst (1996) concluded that the source rock was a quartz-rich, nonporous rock (e.g., quartzite) from the upper part of the target. He favored the Popigai impact structure as the source crater because of the presence of compact quartzites in its pre-impact cover. As discussed already, we have found coesite at Site 709 in the northwest Indian Ocean and Site 216 in the north-central Indian Ocean. The abundance of coesite in comparison to the abundance of shocked quartz with PDFs suggests that the target rock was porous (Kieffer et al., 1976; Grieve et al., 1996). This is also consistent with the source crater being Popigai, because the near-surface rocks at Popigai at the time of the impact included over a kilometer of Upper Proterozoic to Cretaceous clastic rocks (shales, siltstones, sandstones, and conglomerates) and carbonates (limestones and dolomites) (Vishnevsky and Montanari, 1999; Kettrup et al., 2003). The near-surface siltstones and sandstones of Permian and Cretaceous age could be the source of the coesite.

Sr-Nd isotopic data for the cpx spherules and associated microtektites support the hypothesis that the cpx spherules and microtektites were derived from the Popigai impact crater (Whitehead et al., 2000; Liu et al., 2006). Unfortunately, no compositional data have been published for the near-surface clastic and carbonate rocks from the crater area. Glass et al. (2004a) suggested that the Ca- and Mg-rich cpx spherules and associated microtektites may have had carbonates as a major component. However, the cpx spherules have high Cr, Ni, and Co contents, suggesting a meteoritic component (Glass et al., 2004a). This means that a portion of the MgO may be from meteoritic contamination. Glass et al. (2004a) also pointed out that the low Al_2O_3 contents of the cpx spherules indicate that shales and graywackes could not have been a major component and that the low silica content of the cpx spherules suggests that quartz arenites were not a major component. However, the high-SiO_2 microtektites found in the cpx spherule layer indicate that a SiO_2-rich rock, like a quartz arenite, must have been present in the target rock.

The Ir abundances are also consistent with Popigai as the source crater. The average Ir fluence for 12 sites (Table 6; excluding Site 149 which is incomplete, and using only one value for Site 689) is 9.8 ng/cm^2, and the geometric mean is 5.7 ng/cm^2. This is 10% of the geometric mean for the K-T boundary (Chicxulub) Ir fluence, which Donaldson and Hildebrand (2001) estimate at 55 ng/cm^2. Assuming similar projectile characteristics (density, composition, velocity), the projectile for the 100 km Popigai structure would have had ~17% of the Ir content of the projectile that created the ~180-km-diameter Chicxulub impact structure (Kring, 1995; Hildebrand et al., 1998). In a more recent paper, Gulick et al., (2008) estimated the diameter of the Chicxulub crater at ~200 km. If correct, the projectile for the Popigai structure would have had <10% of the Ir content of the projectile that formed the Chicxulub structure. Given the uncertainties of these estimates, we find the relative amounts of Ir in the cpx spherule layer and the K-T boundary to be consistent with the relative sizes of these craters.

Distribution of Iridium in the Sediments

An Ir anomaly appears to be a normal characteristic of the late Eocene cpx spherule layer. The anomaly has been identified in every known cpx spherule layer in which Ir has been measured (with the possible exception of DSDP Site 292; Keller et al., 1987). In that study, the Ir analyses were sporadic, with up to 30 cm gaps between measurements, so it is possible the anomaly was missed. There is only one locality where an Ir anomaly has been found without cpx spherules or associated microtektites. That locality is Barbados (Sanfilippo et al., 1985). The presence of the Ir anomaly without cpx spherules or associated microtektites could indicate that the Ir is largely contained in an ejecta component other than the cpx spherules. This possibility is consistent with our observation that 95% of the Ir is in the <63 μm fraction of the sediment. It also offers an explanation for the relatively high Ir fluence at Site 699, despite the absence of cpx spherules and the relatively low abundance of microtektites. An alternate possibility is that diagenesis may have destroyed the spherules. In Massignano, many of the spherules are coated with fine-grained spinels, but that is not typical of cpx spherules at other sites. Perhaps the spinel is a major carrier of the Ir that separates from the spherule surfaces at most localities and becomes concentrated in the fine sediments. Dissolution of the cpx spherules might be the reason they are absent at Barbados and Site 699. We presently have no data to constrain either of these possibilities, but dissolution may be the simplest explanation for this enigma.

Decameter Homogeneity within the Cpx Strewn Field

At Hole 689B, cpx spherules and microtektites are spread through a section of core ~60 cm thick (Glass and Koeberl, 1999), whereas at Hole 689D, they are spread over an ~70-cm-thick layer of sediment. At both holes, however, the spherules are concentrated in an ~20-cm-thick layer of sediment (Fig. 12).

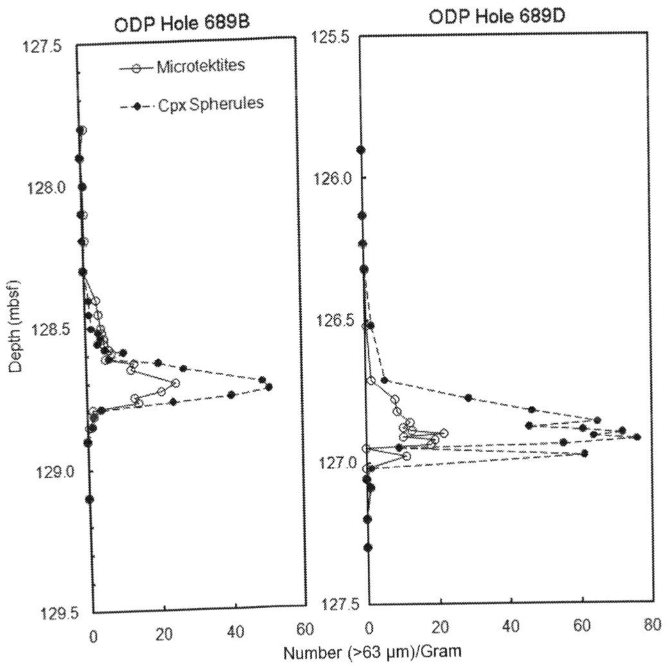

Figure 12. Vertical distribution of cpx spherules and associated microtektites in cores from Ocean Drilling Program (ODP) Holes 689B and 689D (mbsf—m below seafloor). Data for Hole 689B were modified after Glass and Koeberl (1999). Data for Hole 689D are from this study.

At the peak abundance, the number of cpx spherules (>63 μm in diameter) is ~51 g^{-1} and 76 g^{-1} for Holes 689B and 689D, respectively, and the number of microtektites (>63 μm in diameter) is ~25 g^{-1} and 22 g^{-1} for Holes 689B and 689D, respectively. The number of cpx spherules plus microtektites (>125 μm) per square centimeter at Holes 689B and 689D is estimated to be ~58 and 57, respectively. Similarly, the peak Ir concentrations, anomaly thickness, and Ir fluence at these two sites are nearly identical (Table 6). The appearance and major oxide compositions of the cpx spherules and microtektites from these two holes are also similar. The microtektites at both holes range from transparent colorless to transparent pale brown. At both holes, the microtektites generally have silica contents >70 wt% and MgO and CaO contents <1 wt% (Table 2). Furthermore, at both holes, the colorless microtektites have FeO contents less than 2 wt%, and the pale brown ones range in FeO contents from >2 wt% up to 7 (or more) wt%. The cpx spherules from Hole 689B can be divided into two groups based on their CaO contents: a high-Ca group with CaO contents between 11 and 18 wt% and a low-Ca group with CaO contents less than 9 wt% (Table 2). Three cpx spherules from Hole 689D were analyzed, and they have compositions similar to the composition of the cpx spherules in the low-Ca group. In summary, the cpx spherules and microtektites at Hole 689D are similar in abundance, appearance, and composition to those found at Hole 689B.

Ray-Like Distribution of the Cpx Strewn Field

In previous studies, cpx spherules have been reported at 20 sites around the globe (Table 9), although at four of the sites, the spherules have been altered to pyrite or clay. During the present study, cpx spherules and/or associated microtektites have been found at seven additional sites (Site/Hole 592, 699A, 703A, 709C, 786A, 738B, 1090B; Table 1). Thus, the cpx spherule layer has now been recognized at a total of 27 sites. For at least one of these sites (ODP 699A, and perhaps 738B), only microtektites were found in association with an Ir anomaly, but we believe that these microtektites belong to the cpx event for the reasons just discussed. At five of the sites (592, 612, 903, 904, and Massignano), the cpx spherules have been partly or completely replaced by pyrite or clay. Unaltered cpx spherules and associated microtektites have been found at the remaining 21 sites.

In order to describe the geographic distribution of cpx spherules and define the outline of the strewn field, we need to know where cpx spherules do not occur as well as where they do occur. Demonstrating that the cpx spherule layer does not occur at a given site is much more difficult than showing that it does. The failure to find cpx spherules at a site can be the result of several different factors other than the cpx spherules not having been deposited at that site. Possible reasons include poor core recovery, hiatus/unconformity, and poor or ambiguous biostratigraphy. In addition, it is possible that cpx spherules and/or microtektites may have fallen at one or more of these sites and were later removed by solution. It has been shown that silica-poor glass is more sus-

TABLE 9. SITES WHERE THE CPX SPHERULE LAYER HAS PREVIOUSLY BEEN FOUND

Site/Hole	Latitude	Longitude	Iridium anomaly	Comments	References
DSDP 65	4°21.21′N	176°59.16′E	n.d.	Not abundant, scattered	1
DSDP 69A	6°00.0′N	152°51.93′W	Yes		1,2,3
DSDP 70A	6°20.08′N	140°21.72′W	n.d.		1,2
DSDP 94	24°31.64′N	88°28.16′W	Yes	Ir anomaly not well defined	1–4
DSDP 149	15°06.25′N	69°21.85′W	Yes		1–5
DSDP 161A	10°14.25′N	139°57.21′W	n.d.	A few scattered cpx spherules	1
DSDP 162	14°52.19′N	140°02.61′W	n.d.	Scattered	1
DSDP 166	3°45.7′N	175°04.8′W	Yes		1,2
DSDP 167	7°0.4.1′N	176°49.5′W	Yes	Ir anomaly not well defined	1,2
DSDP 216	1°27.73′N	90°12.48′E	Yes	Abundant spherules	1,2,3
DSDP 291	12°48.43′N	127°49.85′E	n.d.	A few cpx spherules recovered	2
DSDP 292	15°49.11′N	124°39.05′E	Yes		1,2
DSDP 315A	4°10.26′N	159°31.54′W	n.d.		2
DSDP 462	7°14.25′N	165°01.83′E	n.d.	Abundant spherules	2
DSDP 612	38°49.21′N	72°46.73′W	Yes	Cpx spherules were pyritized	2,6
ODP 689B	64°31.01′S	3°06.00′E	Yes		7,8,9
ODP 903	38°56.30′N	72°49.03′W	n.d.	Cpx spherules were pyritized	6
ODP 904	38°51.81′N	72°46.08′W	n.d.	Cpx spherules were pyritized	6
RC9-58	14°33.4′N	70°48.6′W	Yes		10,11
Barbados	13°13′N	59°31′W	Yes	Ir anomaly, no cpx spherules	12
Massignano	43°32.22′N	13°35.60′E	Yes	Cpx spherules altered to clay	7,13

Note: 1—Glass et al. (1985), 2—Keller et al. (1987), 3—F. Asaro and H.V. Michel (1983, personal commun.) in DuBois (1983), 4—Glass and Zwart (1979), 5—Alvarez et al. (1982), 6—Glass et al. (1998), 7—Montanari et al. (1993), 8—Vonhof and Smit (1999), 9—Glass and Koeberl (1999), 10—Glass et al. (1973), 11—Glass et al. (1982), 12—Sanfilippo et al. (1985), 13—Glass et al. (2004b); DSDP—Deep Sea Drilling Project; ODP—Ocean Drilling Program; n.d.—no data.

ceptible to dissolution than is silica-rich glass and that the rate of dissolution increases with decreasing water depth (Glass, 1984).

In the present study, 13 cores were unsuccessfully searched for microtektites and/or cpx spherules. In previous studies by one of the authors (Glass) and numerous students over the past 35 yr, 16 Lamont Geological Observatory (now Lamont-Doherty Earth Observatory) cores, and 28 DSDP/ODP cores, involving a total of ~2000 samples, were unsuccessfully searched for cpx spherules and microtektites. Out of the total of 57 cores unsuccessfully searched for late Eocene microtektites and cpx spherules, 12 have good core recovery, no obvious evidence of a hiatus or unconformity, and reasonably good biostratigraphy (Table 10). We have used these twelve core sites to help define the geographic boundaries of the cpx strewn field.

We used the Plate Tracker Version 2.0 software by J. Eldridge, D. Walsh, and C.R. Scotese (PALEOMAP Software, Department of Geology, University of Texas, http://www.scotese.com/software.htm) to determine the position of the continents and core sites at ca. 35 Ma. The outline of the cpx spherule strewn field (Fig. 13) was influenced by an abstract

TABLE 10. SITES WHERE THE CPX SPHERULE LAYER IS APPARENTLY NOT PRESENT

Site/Hole*	Latitude	Longitude	Water depth (mbsl)†	Sediment type§	Interval searched (core-section-cm)#	Sample spacing (cm)
19	28°32.08′S	23°40.63′W	4685	Nanno ooze	5-4-17 to 6-6-142	20
73	01°54.58′S	23°40.63′W	4387	Rad nanno ooze	19-1-21 to 19-6-150	20
111A	50°25.572′N	46°22.050′W	1797	Clay	6R-3	10
214	11°20.21′S	88°43.08′E	1671	Nanno ooze	28-1-1 to 28-6-142	20
253	24°52.65′S	87°21.97′E	1962	Nanno ooze	13-2-16 to 14-6-143	20
317B	11°00.09′S	162°15.78′W	2598	Nanno ooze	35-5-9 to 37-6-149	20
366	05°40.7′N	19°51.1′W	2853	Nanno chalk	12-1-30 to 12-3-150	20
522A	26°6.840′S	5°07.782′W	4441	Nanno ooze	29H-CC to 3H-CC	20
543	15°42.738′N	58°39.222′W	5633	Pelagic clay	27R-4-1 to 27R-5-120	2–20
757B	17°01.458′S	88°10.889′E	1652	Nanno ooze	13H-1-1 to 15H-6-140	20
1073A	39°13.522′N	72°16.546′W	639	Nanno chalk	72X-1-1 to 72X-6-1	20
GPC3**	30°19.9′N	157°49.4′W	5705	Pelagic clay	1250–1450††	5

*Deep Sea Drilling Project (DSDP)/Ocean Drilling Program (ODP) sites/holes searched during the present research are indicated in bold.

†mbsl—meters below sea level.

§Nanno ooze—calcareous nannofossil ooze; nanno chalk—indurated nannofossil ooze; Rad—radiolarian.

#Core, section, and depth in section in centimeters.

**Giant piston core taken during Woods Hole Oceanographic Institution's Long Line Cruise 44.

††Depth in centimeters.

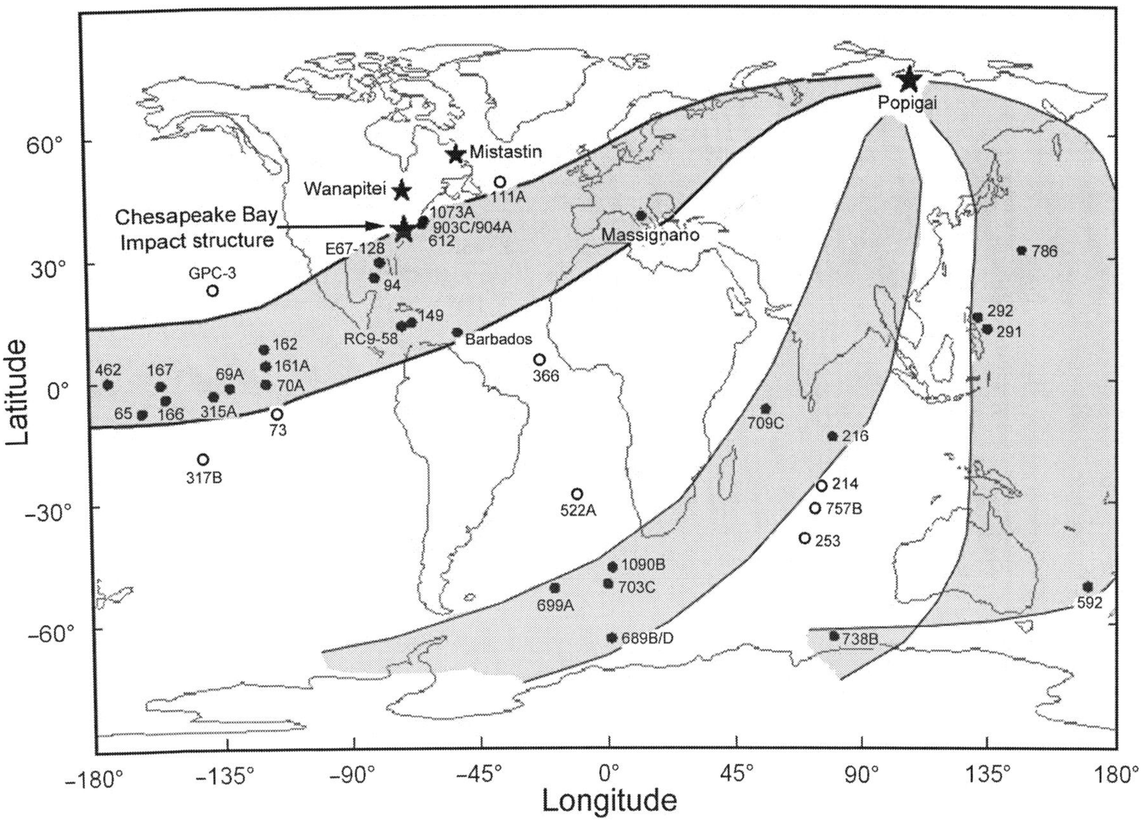

Figure 13. The cpx strewn field, as it was 35 Ma, showing the proposed ray pattern. Sites where the cpx spherule layer has been found are shown with solid circles; sites where the cpx spherule layer was not found are shown with open circles. For information about these sites, see Tables 1, 9, and 11. The ray that extends across Europe, the North Atlantic, the Caribbean Sea, and Gulf of Mexico, and across most of the equatorial Pacific follows high-angle ejecta trajectories from Popigai as proposed by Wrobel and Schultz (2003). The ray that extends across the Indian Ocean and into the South Atlantic follows nominal (45° angle) trajectories from Popigai as proposed by Wrobel and Schultz (2003). The third ray, also a ray that follows the nominal trajectories, extends down the western Pacific and adjacent seas, across the Tasman Sea, and into the Southern Ocean south of Australia, where it turns to the west.

published by Wrobel and Schultz (2003), in which they modeled ejecta trajectories from the Popigai impact taking into account Earth's rotation. We show three rays or ejecta lobes from Popigai impact crater that follow ejecta trajectories in Wrobel and Schultz's (2003) model. One ray follows high-angle trajectories across northern Europe and the North Atlantic Ocean through the Gulf of Mexico–Caribbean region, and then across much of the equatorial Pacific. A second ray follows the nominal (45° angle) ejection trajectories down across Asia, through the western Indian Ocean, and into the South Atlantic. The third ray follows nominal ejection trajectories down through the western Pacific and Tasman Sea and into the Antarctic region. The proposed rays are consistent with the present data set in terms of where spherules are present and where they appear to be absent. The abundance of spherules and percent microtektites appear to change systematically along the ray path that goes through the western Indian Ocean into the South Atlantic Ocean.

Along the Indian Ocean–South Atlantic ray (Fig. 13), the highest abundance of microtektites and cpx spherules, 9574 (>125 μm) cm^{-2}, is found at Hole 709C in the northwest Indian Ocean, and the abundance decreases to 8 cm^{-2} at Hole 699A in the South Atlantic (Table 11). The percent microtektites in the cpx spherule layer increases with distance from Popigai along the ray from ~4% by number at Hole 709C to 100% at Hole 699A (Table 11; Fig. 14). On the other hand, the percent Ca-rich, light-colored cpx spherules plus cryptocrystalline cpx spherules decrease distance from Popigai along this ray (Table 11). The Ir fluence (Table 11) shows no obvious trend with distance from Popigai along this ray. As discussed previously, coesite and shocked quartz grains with planar features have been found at Hole 709C. We have also recovered some white opaque grains from Site 216, which, according to their X-ray diffraction patterns, contain abundant coesite. Thus, the two sites in this ray that are closest to Popigai and have the highest abundance of spherules also contain shocked quartz grains and/or coesite. Except for Massignano, Italy, these are the only two sites where coesite and shocked quartz grains have been recovered from the cpx spherule layer.

TABLE 11. DATA FOR THE SITES IN EACH RAY

Site/Hole*	References	Distance from Popigai (km)	Abundance (No./cm²)†	Percent microtektites (>125 μm)	Percent light-colored cpx spherules§	Ir fluence (ng/cm²)
North Atlantic–Caribbean Sea–Gulf of Mexico–equatorial Pacific ray						
Massignano	1	6170	(46)	n.d.	n.d.	4.7
ODP 904A	2	9270	(206)	n.d.	n.d.	n.d.
ODP 903C	2	9270	n.d.	n.d.	n.d.	n.d.
DSDP 612	2,3	9270	123	n.d.	n.d.	0.6
Barbados#	4	13,070	n.a.	n.a.	n.a	12
DSDP 149	5	16,210	n.d.	n.d.	19	>4
DSDP 94	5	10,990	(84)	n.d.	20	n.d.
RC9-58	6	16,210	891	2	37	42
DSDP 162	7	18,490	244	9	n.d.	n.d.
DSDP 161A	7	18,810	NWDL	n.d.	n.d.	n.d.
DSDP 70A	7	19,050	5	(25)**	n.d.	n.d.
DSDP 315A	8	21,060	n.a.	n.d.	n.d.	7
DSDP 69A	7	23,420	926	4	51	86.6
DSDP 166	7	25,890	854	10	67	13.4
DSDP 167	7	26,100	40	<8	71	1.3
DSDP 65	7	27,510	NWDL	n.d.	n.d.	n.d.
DSDP 462	7	27,770	n.d.	n.d.	n.d.	15
Indian Ocean–South Atlantic ray						
ODP 709C	9	9930	9574	4	70	8.8
DSDP 216	7	10,650	2274	9	65	4.3
ODP 1090B	9	16,650	45	60	61	15
ODP 703A	9	17,030	<1	n.d.	n.d.	n.d.
ODP 699A	9	18,310	8	100	n/a	13
DSDP 689B & **D**	9,10	18,930	57	81	n.d.	2.5
Western Pacific–Tasman Sea ray						
ODP 786A	9	5270	(8)	5	95	n.d.
DSDP 292	7,8	6720	213	17	80	1.5
DSDP 291	8	7250	n.d.	n.d.	n.d.	n.d.
DSDP 592	8	15,320	(0)	n.d.	n.d.	n.d.
ODP 738B	8	20,640	(11)	(100)††	n.d.	1.2.

Note: n.d.—not determined; n.a. —not applicable due to diagenesis, low numbers, or no data. References: 1—Glass et al. (2004b), 2—Glass et al. (1998), 3—Glass (1989), 4—Sanfilippo et al. (1985), 5—Glass and Zwart (1979), 6—Glass et al. (1982), 7—Glass et al. (1985), 8—Keller et al. (1987), 9—This study, 10—Glass and Koeberl (1999).

*Bold indicates sites described in this study. Italics indicate sites mentioned in Keller et al. (1987). For each ray, the sites are listed in order of distance from Popigai along the ray.

†Number of spherules (>125 μm) per centimeter squared. The values in parentheses are not well known. NWDL means no well-defined layer, just scattered spherules.

§Includes cryptocrystalline cpx spherules.

#An Ir anomaly, but no cpx spherules, was found below the North American microtektite layer at this site.

**The percent microtektites at this site is based on only four spherules.

††Preliminary data. Further studies are needed.

We cannot document systematic changes in the western Pacific–Tasman Sea ray very well. The cpx layer at Site 786 is masked by abundant volcanic ejecta, the layer at Site 292 may have been affected by erosion and redeposition (Glass and Burns, 1987), the layer at Site 291 is not described in any detail (Keller et al., 1987), the spherules at Site 592 have been replaced with pyrite, and the spherule layer at Hole 738B needs additional study. The spherules recovered from Site 786 are more completely crystallized than in any other known site, and they have the highest average CaO contents (Table 2; Fig. 9). The cpx layer at Site 292 has a high percentage of light-colored cpx spherules (white opaque plus translucent cryptocrystalline) with high CaO contents (Table 11).

Only glass spherules (microtektites?), no cpx spherules, were recovered from ODP Hole 738B on the Kerguelen Plateau in the southern Indian Ocean. However, only the >150 μm size fraction was searched, and it is possible that cpx spherules may be present in the <150 μm size fractions at this site. Additional studies are required to confirm that the glass spherules are microtektites and that cpx spherules are absent. If absent, their absence may be due to their not having been deposited at this site, or it may be due to dissolution, as the water is relatively shallow (~2252 mbsl). However, if only microtektites were deposited at this site, this would be consistent with our findings: because of the nominal ejection angle, the proportion of microtektites increases with distance from Popigai.

The ray that crosses the North Atlantic and goes into the equatorial Pacific follows the high-angle ejecta trajectories proposed by Wrobel and Schultz (2003). Unfortunately, there are also problems with documenting changes in this ray. The cpx

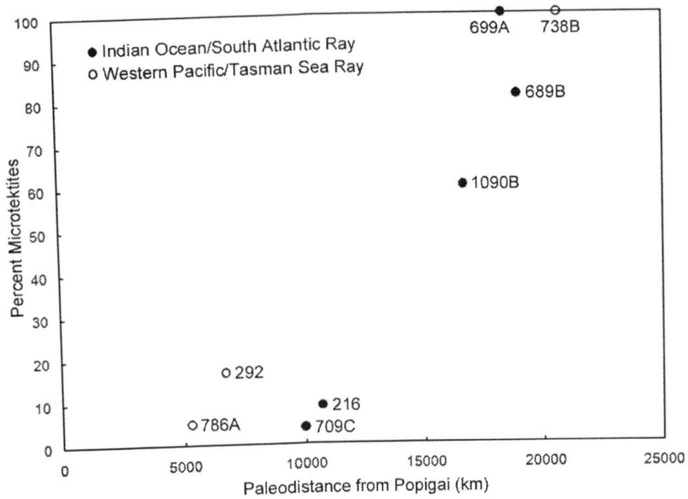

Figure 14. Plot of percent microtektites in the cpx spherule layer versus estimated paleodistance (for 35 Ma) from Popigai along the Indian Ocean–South Atlantic and western Pacific Ocean–Tasman Sea rays (see Table 11; Fig. 13).

spherules in the sites closest to Popigai (Sites 612, 903, and 904) have all been partly to completely replaced by pyrite, and these sites, as well as the sites in the Caribbean Sea and Gulf of Mexico, are in close proximity to the overlying North American microtektite layer, which makes determining the percent microtektites in the cpx layer difficult. In addition, the top of the layer at Site 149 in the Gulf of Mexico is truncated by a coring gap. The spherule layers at Sites 315 and 462 were mentioned in Keller et al. (1987) but were not described in any detail. Only a few scattered spherules were recovered from Sites 65 and 161A. Based on the sites for which we do have data (Core RC9-58 and DSDP Sites 69A, 166, and 167), there does not appear to be a consistent change in abundance of cpx spherules or percent microtektites in the cpx layer with distance from Popigai crater (Table 11). There does appear to be an increase in percent CaO-rich (cryptocrystalline plus light-colored) cpx spherules with increasing distance along this ray. The lowest percentage is at the site for Core RC9-58, which is closer to Popigai, and the highest percentage is at Site 167, which is the farthest site from Popigai for which we have data (Table 11).

According to the model by Wrobel and Schultz (2003), all of the high-angle ejecta thrown out from Popigai in an easterly direction looped around as it traveled to lower latitudes, where it then traveled westerly parallel to the equator. None of the high-angle ejecta was deposited in the Southern Hemisphere, but it was concentrated in a band ~30° wide just north of the equator. Thus, not all of the ejecta found at low latitude above the equator necessarily followed trajectories along the proposed high-angle ray path shown in Figure 13. An unresolved problem regarding the possible North Atlantic–Caribbean Sea–Gulf of Mexico–equatorial Pacific ray is that nearly half the sites in the equatorial Pacific are below the paleoequator in the 35 Ma reconstruction. According Wroble and Schultz' (2003) model, the high-angle ejecta from Popigai should not have been deposited below the equator. Thus, either the reconstruction is a little off, the model is incorrect, or the spherules found below the paleoequator are not from the high-angle trajectories.

As of the late 1990s, when the bulk of this research was being done, there were very few DSDP/ODP cores from the South Pacific, middle to high latitudes in the North Pacific, and equatorial to mid-latitudes in the South Atlantic containing late Eocene sections that could be searched for the cpx spherule layer (see Table 1; Fig. 1). Thus, there are large gaps in our knowledge regarding geographic distribution of the cpx spherules. Increased sampling density is necessary to test the proposed ray-like nature of the cpx spherule strewn field.

Evidence for Possible Biological Effects of the Cpx Spherule Impact

As of 2007, the Chesapeake Bay impact structure and the Popigai impact crater are the largest impact structures of Cenozoic age, and Popigai is one of the five largest terrestrial impact structures known to date. Their one-two punch within ~20 k.y. of each other should have had some global environmental consequences (Stöffler and Claeys, 1997). However, there is no evidence that the impacts that produced these impact structures were associated with a mass extinction event. Funakawa et al. (2006) concluded that a major radiolarian turnover and paleoceanographic change took place during magnetochron 16n.1n (ca. 35.5–35.7 Ma), or about the time that the North American and cpx strewn fields were formed. Several researchers have suggested that the impacts responsible for these strewn fields may have caused immediate climatic change and associated changes in ocean bioproductivity (e.g., Poag, 1997; Vonhof et al., 2000; Coccioni et al., 2000; Spezzaferri et al., 2002; Poag et al., 2004; Bodiselitsch et al., 2004). We have $\delta^{18}O$ data indicating a drop in temperature and/or increase in ice volume and some indirect evidence for a change in surface productivity coincident with the impact event that produced the cpx spherule layer.

In Hole 738B, $\delta^{18}O$ values for the bulk carbonate increased by ~0.25‰ directly above the spherule layer and Ir anomaly (Fig. 15). The change is abrupt, taking place over an interval of ~10 cm. The increase in $\delta^{18}O$ values indicates that the surface waters became cooler, and/or the ice volume increased immediately after the impact that produced the cpx spherule layer. This $\delta^{18}O$ increase has been documented at all the sites with apparently complete late Eocene sections on the Maud Rise and Kerguelen Plateau (i.e., Sites 689, 744, 748, as well as 738) (Vonhof et al., 2000; Bohaty and Zachos, 2003). Our data agree well with the $\delta^{18}O$ record published for Hole 689B by Vonhof et al. (2000). It is not clear if the change in $\delta^{18}O$ at this time was due to change in water temperature and/or ice volume, but changes in foraminifera assemblages at Holes 738 and 748 at this time suggest that there was at least some cooling component in the $\delta^{18}O$ signal. At Site 738, there is a major change in planktonic

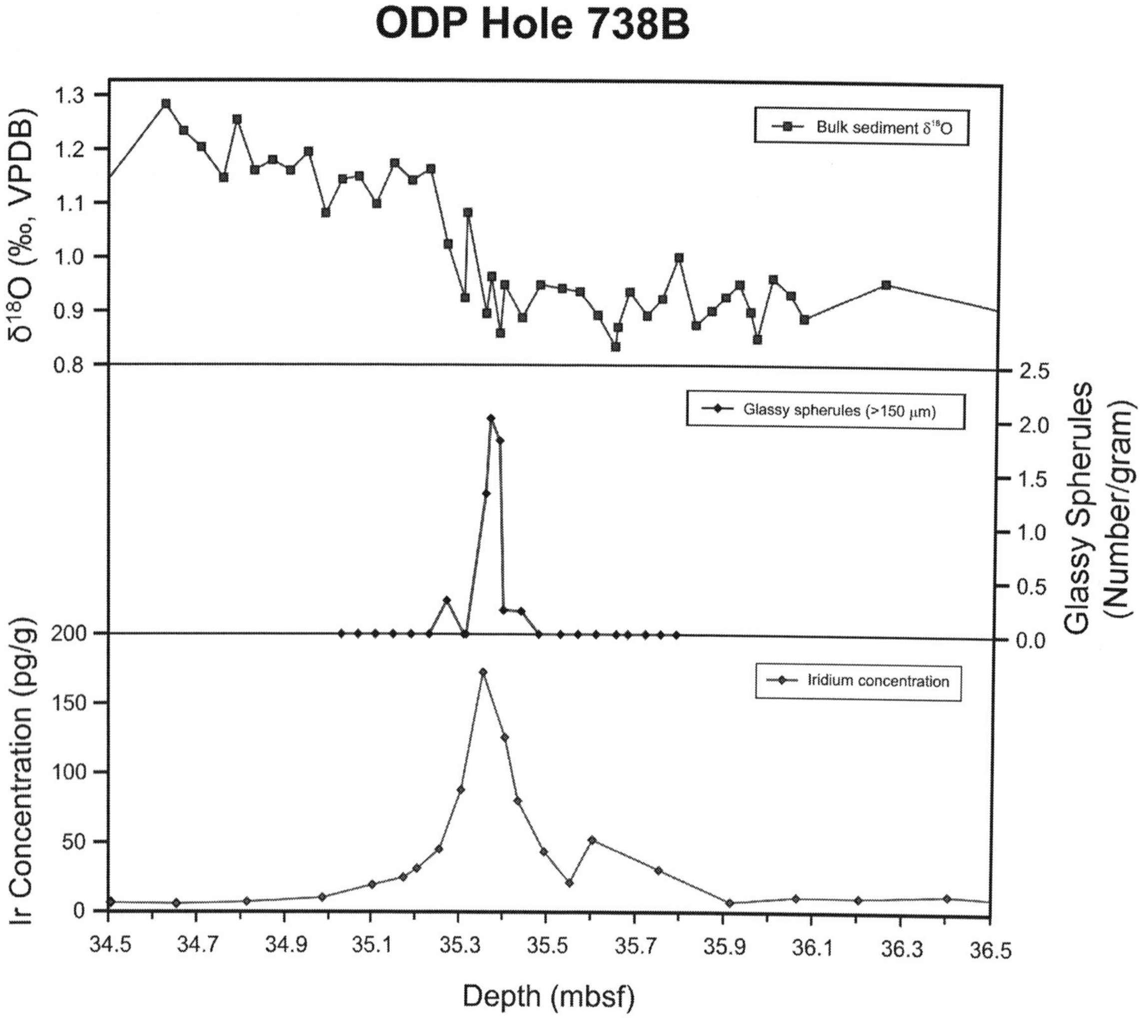

Figure 15. Vertical distribution of $\delta^{18}O$, glassy spherules, and Ir concentration in Sections 2–3 of Core 5H in Hole 738B on the southern Kerguelen Plateau in the southern Indian Ocean (mbsf—m below seafloor; VPDB—Vienna Peedee belemnite).

foraminifera assemblage across the Ir anomaly. The most obvious change is a decrease in the abundance of the surface-dwelling taxon *Globigerinatheka index*. Below the Ir anomaly, *G. index* is a major component of the assemblage, and, coincident with the Ir anomaly, the abundance of *G. index* decreases dramatically. Above the Ir anomaly, *G. index* is very rare to absent. In terms of relative abundance, *G. index* is replaced by deep-dwelling taxa, such as *Subottina* spp., suggesting cooling and/or a change in water-column thermal structure occurring at the same time as the impact event. As previously mentioned, the core samples studied in this project were disaggregated in water using ultrasonics and sieved into three size fractions: <63 μm, 63–125 μm, and >125 μm. At the three sites in this study with the best-defined spherule layer (Sites 689, 709, and 1090), the weight percent of the coarse fractions (63–125 μm and >125 μm) decreases just above the spherule layer (e.g., Fig. 16). Since the sediments at these sites are primarily biogenic (calcareous nannofossils, radiolarians, planktonic foraminifera, diatoms), the drop in percent coarse fraction above the spherule layer suggests that there was a change in surface productivity after the Popigai impact (assuming that Popigai is the source crater for the cpx spherules). However, the decrease in coarse fraction (primarily radiolarians and/or planktonic foraminifera) could be caused either by a decrease in microfossils or an increase in nannofossils. Since the microfossils make up only a small fraction of the sediment, a decrease in the coarse fraction due to a decrease in microfossils would not noticeably change the sediment accumulation rate, but an increase in nannofossils, enough to cause the observed decrease in coarse fraction, would be significant. Unfortunately, the biostratigraphic resolution is not good enough to determine whether or not there was an increase in sedimentation rate at these sites after the impact.

At Site 689, this low-percent coarse fraction continues for several meters above the spherule layer, which, based on the average sediment accumulation rate for the late Eocene section

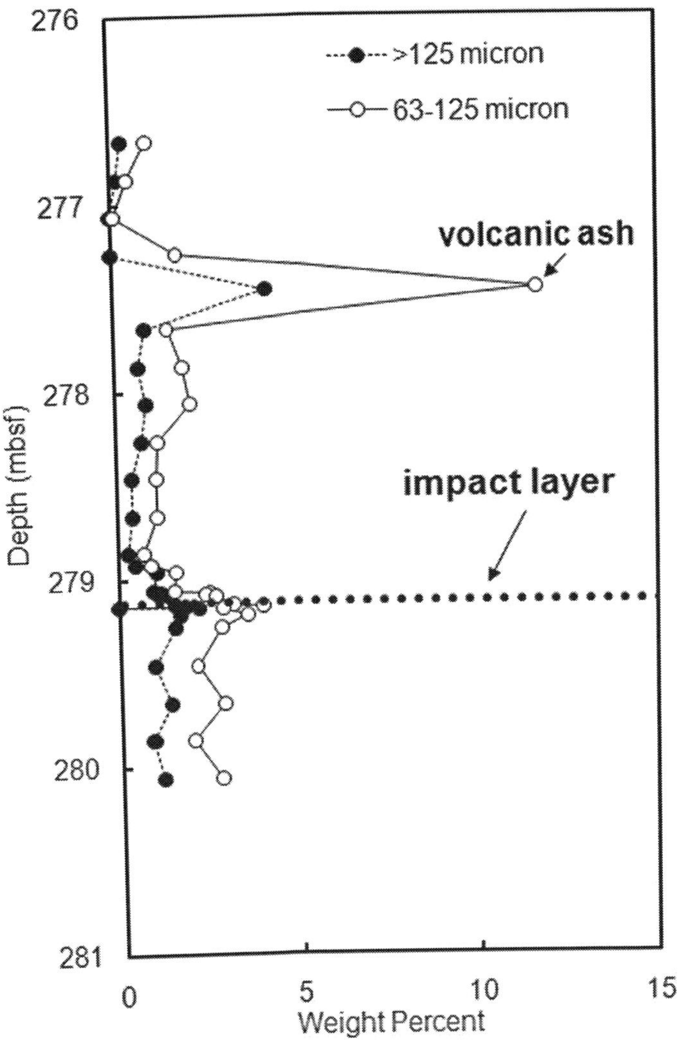

Figure 16. Weight percent coarse fractions (63–125 μm and >125 μm) versus depth in Core 30X from Hole 1090B in the South Atlantic Ocean.

at this site (4.45 m/m.y.), represents approximately several hundred thousand years. However, if the sedimentation rate was higher than average, then the period of reduced coarse fraction would have been proportionately shorter. Our size data do not go up high enough in the cores at the other two sites (709 and 1090) to define the upper limit of the low-percent coarse fraction, but the low-percent coarse fraction appears to have lasted for at least 180–190 k.y.

SUMMARY AND CONCLUSIONS

We report here the first documentation of cpx spherules and/or associated microtektites at seven DSDP/ODP sites/holes (592, 689D, 699A, 703C, 709C, 786A, 1090B) (Table 1). Glass spherules and a coincident Ir anomaly at Hole 738B probably also belong to the cpx spherule layer, but additional studies are needed. The spherules at DSDP Site 592, in the Tasman Sea, have been replaced with pyrite (Fig. 2A). Transparent, colorless, SiO_2-rich (average ~87 wt%) microtektites, but no cpx spherules, were recovered from Hole 699A in the South Atlantic. The microtektites are associated with an Ir anomaly at this site (Fig. 5). Although the microtektites at Hole 699A have major oxide compositions similar to those of the North American microtektites, we conclude that these microtektites belong to the cpx strewn field.

The greatest abundance of spherules (~9570 cm^{-2}) occurs at Hole 709C in the western Indian Ocean. Both cpx spherules and microtektites are found at this site (Figs. 2B and 2C). Coesite and shocked quartz with planar features (Fig. 4) were found in the cpx spherule layer at this site. We recovered ~20 cpx spherules, but no microtektites, at Hole 786A in the northwestern Pacific. However, searching for spherules at this site was difficult because of the abundance of volcanic ejecta. The cpx spherules at Site 786A are generally more crystalline (i.e., less glass matrix; Fig. 2D) than at other sites, and they have low SiO_2 and Al_2O_3 and high MgO contents compared with most other cpx spherules. Some of the spherules have the highest CaO (up to ~27 wt%) contents ever determined for cpx spherules (Fig. 9). Hole 1090B contains a well-defined cpx spherule layer. The layer at this site contains ~60% microtektites (>125 μm) by number.

Presently available data indicate that the cpx spherule layer occurs near the top of the *Calocyclas bandyca* zone (radiolarians), within the NP19/20 and CP15 (*Discoaster barbadiensis*) zones (calcareous nannofossils), and in magnetochron 16n.1n (Fig. 7). The presently accepted ages for the upper and lower boundaries of this subchron are 35.343 and 35.526 Ma, respectively. Thus, the age of the cpx spherule layer estimated from magnetochronology is consistent with the age extrapolated for the cpx spherule layer from the radiometric age of the North America tektites/microtektites, which are estimated (based on sediment accumulation rates) to be 10–15 k.y. younger than the cpx spherule layer.

Microtektites in the cpx spherule layer can be divided into two compositional groups: a high-SiO_2 group and a low-SiO_2 group with lower Al_2O_3 for a given SiO_2 content (Fig. 10). Microtektites in the high-SiO_2 group have compositions similar to the North American microtektites, but their Sr and Nd isotopic ratios are more similar to the cpx spherules than to the North American tektites/microtektites (Fig. 8) (Whitehead et al., 2000; Liu et al., 2006). Microtektites in the low-SiO_2 group have compositions indistinguishable from the cpx spherules (Table 2) (Glass et al., 2004a). This indicates that formation of microkrystites depends on temperature history as well as composition. Since there are low-silica microtektites in the Australasian and Ivory Coast microtektite layers, but no microkrystites, perhaps only larger impacts (impacts that produce craters 100 km in diameter or larger) produce microkrystites.

We found that an Ir anomaly is associated with the cpx layer at Holes 689D, 699A, 709C, 738B, and 1090B. The peak concentrations of Ir at these sites range between 156 and 1046 pg/g, and the calculated net fluence ranges from 1.2 (Hole 738B) to

15 ng/cm^2 (Hole 1090B). We calculate that the bulk of the Ir, ~95 wt%, is in the <63 μm size fraction, and 5% in impact spherules (>63 μm in size). Based on Ir data from the five sites in this study, plus data from seven additional sites, we estimate that the geometric mean of the Ir fluence for the cpx spherule layer is ~5.7 ng/cm^2. Assuming similar projectile characteristics and velocity, we determined that the projectile for the Popigai structure had ~10% of the Ir content of the projectile associated with the 180-km-diameter Chicxulub impact structure.

We found the cpx spherule layer at Hole 689D, which is only 24 m from Hole 689B, where the cpx layer had previously been found. We found that the abundance of spherules, thickness of the spherule-bearing layer, ratio between microtektites and cpx spherules, and associated Ir anomaly are very similar between these two holes. This indicates that the cpx spherule layer is fairly homogeneous over a distance of tens of meters, even in the most distal part of the strewn field.

The discovery of coesite and shocked quartz grains with planar features supports the impact origin for the cpx spherule layer and supports the conclusion that the pancake-shaped clay spherules (containing Ni-rich spinel crystallites), associated with shock-metamorphosed quartz grains exhibiting PDFs at Massignano, Italy, are diagenetically altered cpx spherules. Furthermore, the occurrence of coesite in the cpx spherule layer indicates that at least part of the target rock was porous, and the high-SiO_2 microtektites, with silica contents as high as ~91 wt%, indicate that part of the target rock was rich in SiO_2 (e.g., quartz arenite). These constraints on the target rock are consistent with Popigai being the source crater, since sandstones are present in the near-surface rocks at that location (Kettrup et al., 2003).

With the addition of the six (seven if Hole 738B is included) sites reported here, there are now at least 27 sites where the cpx spherule layer has been found (Tables 1, 9, and 11). In addition, at least 12 sites with good core recovery and good biostratigraphy have been sampled at fairly close intervals (generally 20 cm or less) through the biostratigraphic interval of interest but have failed to yield any cpx spherules and/or microtektites. The present data suggest to us that the cpx spherule strewn field may have a ray-like distribution pattern, in which the rays fall along ejecta trajectories from Popigai crater as proposed by Wrobel and Schultz (2003). We have proposed three rays (Fig. 13): One ray follows Wrobel and Schultz' (2003) high-angle trajectories across northern Europe and the North Atlantic, through the Gulf of Mexico–Caribbean region, and then across much of the Pacific just above the equator. A second ray follows nominal (45° angle) ejecta trajectories down through the Indian Ocean and into the South Atlantic. The third ray follows 45° angle ejecta trajectories down the western Pacific and into the Tasman Sea.

Along the North Atlantic–Caribbean Sea–Gulf of Mexico–equatorial Pacific ray, there appears to be an increase in the percent of high-Ca cryptocrystalline and light-colored cpx spherules, and there may be a slight increase in percent microtektites with distance from Popigai impact structure, but there is no apparent systematic change in abundance of cpx spherules with distance from Popigai (Table 11). This may be because, according to the model by Wrobel and Schultz (2003), all the high-angle ejecta eventually ended up traveling in a westerly direction parallel to the equator and was concentrated in a band ~30° wide. Thus, ejecta at any given site within ~30° north of the equator may have traveled along a multitude of trajectories.

The best-defined ray is the Indian Ocean–South Atlantic ray (Fig. 13). There are systematic changes in abundance and petrography (composition) with distance from Popigai along this ray (Table 11). The percent microtektites in the cpx spherule layer increases with distance from Popigai along this ray (Table 11; Fig. 14), whereas the abundance of spherules and the percent Ca-rich, light-colored cpx spherules plus cryptocrystalline cpx spherules decrease with distance from Popigai along this ray (Table 11). Shocked quartz and/or coesite have only been found associated with the cpx spherule layer at Site/Hole 216 and 709C, which are the two sites in this ray that are closest to Popigai.

Along the Western Pacific–Tasman Sea ray, the percent Ca-rich light-colored plus cryptocrystalline cpx spherules decreases from Hole 786A (~95% by number), the closest to Popigai, to Site 292 (~80%), which is farther from Popigai. In addition, the percent microtektites appears to increase from ~5% at Hole 786A, closest to Popigai, to perhaps 100% at Hole 738B, which is the farthest site from Popigai in this ray (Table 11). However, the present data for this proposed ray are of poor quality because of dilution by volcanic ash (Hole 786A), reworking (Site 292), poor recovery (Site 291), diagenetic alteration (Site 592), and incomplete documentation (Hole 738B).

The finding of an increase in $\delta^{18}O$ values immediately above the spherule layer at Hole 738B on the Kerguelen Plateau (Fig. 15) is similar to the results reported by Vonhof et al. (2000), who reported an increase in $\delta^{18}O$ values immediately above the spherule layer at Hole 689B on the Maud Rise. This abrupt increase in $\delta^{18}O$ values immediately above the cpx spherule layer suggests that the impact that produced the spherule layer may have caused a cooling event and/or an increase in glacial ice. An increase in the weight percent of the coarse fraction, which is primarily biotic, at some sites (Sites 689, 709, 1090) at this time may indicate a change in surface productivity that was triggered by the cooler climate.

ACKNOWLEDGMENTS

We thank the Ocean Drilling Program (ODP) for the samples used in this study. Paul Steineck and Detlef Warnke forwarded a bulk ODP sample from Hole 689D and five samples from Hole 1090B for use in this study. We thank Wylie Poag and Philippe Claeys for detailed critical and constructive reviews of the manuscript. This work was supported by National Science Foundation (NSF) grant EAR-9903811 to Glass and by National Aeronautics and Space Administration (NASA) grant NNG066F07G to Kyte. Bohaty was funded by NSF polar programs grant OPP-0338337.

REFERENCES CITED

Albin, E., and Wampler, J.M., 1996, New potassium-argon ages for georgiaites and the upper Eocene Dry Branch Formation (Twiggs Clay Member): Inferences about tektite stratigraphic occurrence [abs.]: Houston, Texas, Lunar and Planetary Institute, Lunar and Planetary Science, v. 27, p. 5–6.

Alvarez, W., Asaro, F., Michel, H.V., and Alvarez, L.W., 1982, Iridium anomaly approximately synchronous with terminal Eocene extinction: Science, v. 216, p. 886–888, doi: 10.1126/science.216.4548.886.

Berggren, W.A., and Pearson, P.N., 2005, A revised tropical to subtropical Paleogene planktonic foraminiferal zonation: Journal of Foraminiferal Research, v. 35, p. 279–298, doi: 10.2113/35.4.279.

Berggren, W.A., Kent, D.V., Swisher, C.C., III, and Aubry, M.-P., 1995, A revised Cenozoic geochronology and chronostratigraphy, *in* Berggren, W.A., Kent, D.V., Swisher, C.C., III, Aubry, M.-P., and Hardenbol, J., eds., Geochronology, Time Scales and Global Stratigraphic Correlation: SEPM (Society for Sedimentary Geology) Special Publication 54, p. 129–212.

Berry, L.G., 1974, Selected Powder Diffraction Data for Minerals—Search Manual: Joint Swarthmore, Pennsylvania, Committee on Powder Diffraction Standards, Publication DBM-1-123, 832 p.

Bice, D., and Montanari, A., 1988, Magnetic stratigraphy of the Massignano section across the Eocene-Oligocene boundary, *in* Premoli Silva, I, Coccioni, R., and Montanari, A., eds., The Eocene-Oligocene Boundary in the Marche-Umbria Basin (Italy): International Union of Geological Sciences (IUGS) Special Publication: Ancona, F. Ili Aniballi Publishers, p. 111–117.

Bodiselitsch, B., Montanari, A., Koeberl, C., and Coccioni, R., 2004, Delayed climate cooling in the late Eocene caused by multiple impacts: High-resolution geochemical studies at Massignano, Italy: Earth and Planetary Science Letters, v. 223, p. 283–302, doi: 10.1016/j.epsl.2004.04.028.

Bohaty, S.M., and Zachos, J.C., 2003, Significant Southern Ocean warming event in the later middle Eocene: Geology, v. 31, p. 1017–1020, doi: 10.1130/G19800.1.

Bohor, B.F., and Glass, B.P., 1995, Origin and diagenesis of K/T impact spherules—From Haiti to Wyoming and beyond: Meteoritics, v. 30, p. 182–198.

Bottomley, R.J., Grieve, R.A.F., York, D., and Masaitis, V., 1997, The age of the Popigai impact event and its relations to events at the Eocene/Oligocene boundary: Nature, v. 388, p. 365–368, doi: 10.1038/41073.

Bouma, A.H., and Moore, C.J., 1975, Physical properties of deep-sea sediments from the Philippine Sea and Sea of Japan, *in* Initial Reports of the Deep Sea Drilling Project, Volume 31: Washington, D.C., U.S. Government Printing Office, p. 535–568.

Cande, S.C., and Kent, D.V., 1995, Revised calibration of the geomagnetic polarity time scale for the late Cretaceous and Cenozoic: Journal of Geophysical Research, v. 100, p. 6093–6095, doi: 10.1029/94JB03098.

Channell, J.E.T., Galeotti, S., Martin, E.E., Billups, K., Scher, H.D., and Stoner, J.S., 2003, Eocene to Miocene magnetostratigraphy, biostratigraphy, and chemostratigraphy at ODP Site 1090 (sub-Antarctic South Atlantic): Geological Society of America Bulletin, v. 115, p. 607–623, doi: 10.1130/0016-7606(2003)115<0607:ETMMBA>2.0.CO;2.

Clymer, A.K., Bice, D.M., and Montanari, A., 1996, Shocked quartz from the late Eocene: Impact evidence from Massignano, Italy: Geology, v. 24, p. 483–486, doi: 10.1130/0091-7613(1996)024<0483:SQFTLE>2.3.CO;2.

Coccioni, R., Basso, D., Brinkhuis, H., Galeotti, S., Gardin, S., Monechi, S., and Spezzaferri, S., 2000, Marine biotic signals across a late Eocene impact layer at Massignano, Italy: Evidence for long-term environmental perturbations?: Terra Nova, v. 12, p. 258–263, doi: 10.1046/j.1365-3121.2000.00305.x.

Crux, J.A., 1991, Calcareous nannofossils recovered by Leg 114 in the sub-Antarctic South Atlantic Ocean, *in* Ciesielski, P.F., Kristoffersen, Y., Clement, B.M., et al., Proceedings of the Ocean Drilling Program, Scientific Results, Volume 114: College Station, Texas, U.S. Government Printing Office, p. 155–177.

Deutsch, A., and Koeberl, C., 2006, Establishing the link between the Chesapeake Bay impact structure and the North American tektite strewn field: The Sr-Nd isotopic evidence: Meteoritics & Planetary Science, v. 41, p. 689–703.

D'Hondt, S.L., Keller, G., and Stallard, R.F., 1987, Major element compositional variation within and between different late Eocene microtektite strewn fields: Meteoritics, v. 22, p. 61–79.

Donaldson, S., and Hildebrand, A.R., 2001, The global fluence of iridium at the Cretaceous-Tertiary boundary: Meteoritics & Planetary Science, v. 36, no. 9, supplement, p. A50.

DuBois, D.L., 1983, The Spatial and Temporal Relationships of the North American Microtektite Layer [M.S. thesis]: Newark, University of Delaware, 170 p.

Florindo, F., and Roberts, A.P., 2005, Eocene-Oligocene magnetobiochronology of ODP Sites 689 and 690, Maud Rise, Weddell Sea, Antarctica: Geological Society of America Bulletin, v. 117, p. 46–66, doi: 10.1130/B25541.1.

Ford, R.J., 1988, An empirical model for the Australasian tektite field: Australian Journal of Earth Sciences, v. 35, p. 483–490, doi: 10.1080/08120098808729464.

Funakawa, S., and Nishi, H., 2005, Late middle Eocene to late Oligocene radiolarian biostratigraphy in the Southern Ocean (Maud Rise, ODP Leg 113, Site 689): Marine Micropaleontology, v. 54, p. 213–247, doi: 10.1016/j.marmicro.2004.12.002.

Funakawa, S., Nishi, H., Moore, T.C., and Nigrini, C.A., 2006, Radiolarian faunal turnover and paleoceanographic change around the Eocene/Oligocene boundary in the central Pacific, ODP Leg 199, Holes 1218A, 1210A, and 1220A: Palaeogeography, Palaeoclimatology, Palaeoecology, v. 230, p. 183–203.

Ganapathy, R., 1982, Evidence for a major meteorite impact on the Earth 34 million years ago: Implication for Eocene extinctions: Science, v. 216, p. 885–886, doi: 10.1126/science.216.4548.885.

Gary, M., McAfee, R., Jr., and Wolf, C.L., 1972, Glossary of Geology: Washington, D.C., American Geological Institute, 805 p.

Glass, B.P., 1984, Solution of naturally-occurring glasses in the geological environment: Journal of Non-Crystalline Solids, v. 67, p. 265–286, doi: 10.1016/0022-3093(84)90154-6.

Glass, B.P., 1986, Late Eocene microtektites and clinopyroxene-bearing spherules, *in* Pomerol, Ch., and Premoli Silva, I. eds., Terminal Eocene Events: Amsterdam, Elsevier, p. 395–401.

Glass, B.P., 1989, North American tektite debris and impact ejecta from DSDP Site 612: Meteoritics, v. 24, p. 209–218.

Glass, B.P., 2002, Upper Eocene impact ejecta/spherule layers in marine sediments: Chemie der Erde, v. 62, p. 173–196, doi: 10.1078/0009-2819-00017.

Glass, B.P., and Burns, C.A., 1987, Late Eocene crystal-bearing spherules: Two layers or one?: Meteoritics, v. 22, p. 265–279.

Glass, B.P., and Burns, C.A., 1988, Microkrystites: A new term for impact-produced glassy spherules containing primary crystallites, *in* Ryder, G., ed., Proceedings of the Eighteenth Lunar and Planetary Science Conference: Houston, Texas, Lunar and Planetary Institute, p. 455–458.

Glass, B.P., and Koeberl, C., 1999, Ocean Drilling Project Hole 689B spherules and upper Eocene microtektite and clinopyroxene-bearing spherule strewn fields: Meteoritics & Planetary Science, v. 34, p. 197–208.

Glass, B.P., and Koeberl, C., 2006, Australasian microtektites and associated impact ejecta in the South China Sea and the Middle Pleistocene supereruption of Toba: Meteoritics & Planetary Science, v. 41, p. 305–326.

Glass, B.P., and Liu, S., 2001, Discovery of high-pressure $ZrSiO_4$ polymorph in naturally occurring shock-metamorphosed zircons: Geology, v. 29, p. 371–373, doi: 10.1130/0091-7613(2001)029<0371.DOHPZP>2.0.CO,2.

Glass, B.P., and Wu, J., 1993, Coesite and shocked quartz discovered in the Australasian and North American microtektite layers: Geology, v. 21, p. 435–438, doi: 10.1130/0091-7613(1993)021<0435:CASQDI>2.3.CO;2.

Glass, B.P., and Zwart, M.J., 1979, North American microtektites in Deep Sea Drilling Project cores from the Caribbean Sea and Gulf of Mexico: Geological Society of America Bulletin, v. 90, p. 595–602, doi: 10.1130/0016-7606(1979)90<595:NAMTID>2.0.CO;2.

Glass, B.P., Baker, R.N., Storzer, D., and Wagner, G.A., 1973, North American microtektites from the Caribbean Sea and their fission track age: Earth and Planetary Science Letters, v. 19, p. 184–192, doi: 10.1016/0012-821X(73)90113-1.

Glass, B.P., Swinki, M.B., and Zwart, P.A., 1979, Australasian, Ivory Coast and North American tektite strewn fields: Size, mass and correlation with geomagnetic reversals and other Earth events, *in* Merrill, R.B., Bogard, D.D., Hörz, F., McKay, D.S., and Robertson, P.C., eds., Proceedings of the Tenth Lunar and Planetary Science Conference: Houston, Texas, Lunar and Planetary Institute, p. 2535–2545.

Glass, B.P., DuBois, D.L., and Ganapathy, R., 1982, Relationship between an iridium anomaly and the North American microtektite layer in Core RC9-58 from the Caribbean Sea, *in* Proceedings of the Thirteenth Lunar and Planetary Science Conference: Journal of Geophysical Research, v. 87, supplement, p. A425–A428, doi: 10.1029/JB087iS01p0A425.

Glass, B.P., Burns, C.A., Crosbie, J.R., and DuBois, D.L., 1985, Late Eocene North American microtektites and clinopyroxene-bearing spherules, *in* Ryder, G., and Schubert, G., eds., Proceedings of the Sixteenth Lunar and Planetary Science Conference: Journal of Geophysical Research, v. 90, Supplement No. B, p. D175–D196.

Glass, B.P., Hall, C.M., and York, D., 1986, $^{40}Ar/^{39}Ar$ laser-probe dating of North American tektite fragments from Barbados and the age of the Eocene-Oligocene boundary: Chemical Geology–Isotope Geoscience Section, v. 59, p. 181–186.

Glass, B.P., Koeberl, C., Blum, J.D., and McHugh, C.M.G., 1998, Upper Eocene tektite and impact ejecta layer on the continental slope off New Jersey: Meteoritics & Planetary Science, v. 33, p. 229–241.

Glass, B.P., Huber, H., and Koeberl, C., 2004a, Geochemistry of Cenozoic microtektites and clinopyroxene-bearing spherules: Geochimica et Cosmochimica Acta, v. 68, p. 3971–4006, doi: 10.1016/j.gca.2004.02.026.

Glass, B.P., Liu, S., and Montanari, A., 2004b, Impact ejecta in upper Eocene deposits at Massignano, Italy: Meteoritics & Planetary Science, v. 39, p. 589–597.

Grieve, R.A.F., Langenhorst, F., and Stöffler, D., 1996, Shock metamorphism of quartz in nature and experiment: II. Significance in geoscience: Meteoritics, v. 31, p. 6–35.

Gulick, S.S., Barton, P.J., Christeson, G.L., Morgan, J.V., McDonald, M., Mendoza-Cervantes, K., Pearson, Z.F., Surendra, A., Urrutia-Fucugauchi, J., Vermeesch, P.M., and Warner, M.R., 2008, Importance of pre-impact crustal structure for the asymmetry of the Chicxulub impact crater: Nature Geoscience, v. 1, p. 131–135. Available from: www.nature.com/naturegeoscience; doi:10.1038/ngeo103.

Hailwood, E.A., and Clement, B.M., 1991, Magnetostratigraphy of Sites 703 and 704, Meteor Rise, southeastern South Atlantic, *in* Ciesielski, P.F., Kristoffersen, Y., et al., Proceedings of the Ocean Drilling Program, Scientific Results, Volume 114: Washington, D.C., U.S. Government Printing Office, p. 367–386.

Hartung, J.B., and Koeberl, C., 1994, In search of the Australasian tektite source crater: The Tonle Sap hypothesis: Meteoritics, v. 29, p. 411–416.

Hazel, J.E., 1989, Chronostratigraphy of upper Eocene microspherules: Palaios, v. 4, p. 318–329, doi: 10.2307/3514556.

Hildebrand, A.R., Pilkington, M., Ortiz-Aleman, C., Chavez, R.E., Urrutis-Fucugauchi, J., Conners, M., Graniel-Castro, E., Camara-Zi, A., Halpenny, J.A., and Niehaus, D., 1998, Mapping Chicxulub crater structure with gravity and seismic reflection data, *in* Grady, M.M., Hutchison, R., McCall, G.J.H., and Rothery, D.A., eds., Meteorites: Flux with Time and Impact Effects: Geological Society of London Special Publication 140, p. 153–173.

Hölker, T., Deutsch, A., and Masaitis, V.L., 1997, Nd-Sr isotope signatures of impactites from the Popigai impact crater (Russia): Houston, Texas, Lunar and Planetary Institute, Lunar and Planetary Science Conference, v. 28, abstract 1356 [CD-ROM].

Horton, J.W., Jr., and Izett, G.A., 2005, Crystalline ejecta and shocked minerals of the Chesapeake Bay impact structure, USGS-NASA Langley core, Hampton, Virginia, with supplemental constraints on the age of impact, *in* Horton, J.W., Jr., et al., eds., Studies of the Chesapeake Bay Impact Structure—The USGS-NASA Langley Corehole, Hampton, Virginia, and Related Coreholes and Geophysical Surveys: U.S. Geological Survey Professional Paper 1688, p. E1–E30.

John, C., and Glass, B.P., 1974, Clinopyroxene-bearing spherules associated with North American microtektites: Geology, v. 2, p. 599–602, doi: 10.1130/0091-7613(1974)2<599:CGSAWN>2.0.CO;2.

Johnson, D.A., 1990, Radiolarian biostratigraphy in the central Indian Ocean, Leg 115, *in* Duncan, R.A., Backman, J., Peterson, L.C., et al., Proceedings of the Ocean Drilling Program, Scientific Results, Volume 115: Washington, D.C., U.S. Government Printing Office, p. 395–409.

Jovane, L., Florindo, F., Sprovieri, M., and Pälike, H., 2006, Astronomic calibration of the late Eocene/early Oligocene Massignano section (central Italy): Geochemistry, Geophysics, Geosystems, v. 7, no. 7, p. Q07012, doi: 10.1029/2005GC001195.

Keller, G., D'Hondt, S.L., and Vallier, T.L., 1983, Multiple microtektite horizons in upper Eocene marine sediments: No evidence for mass extinctions: Science, v. 221, p. 150–152, doi: 10.1126/science.221.4606.150.

Keller, G., D'Hondt, S.L., Orth, C.J., Gilmore, J.S., Oliver, P.Q., Shoemaker, E.M., and Molina, E., 1987, Late Eocene impact microspherules: Stratigraphy, age and geochemistry: Meteoritics, v. 22, p. 25–60.

Kettrup, B., Deutsch, A., and Masaitis, V.L., 2003, Homogeneous impact melts produced by a heterogeneous target? Sr-Nd isotopic evidence from the Popigai Crater, Russia: Geochimica et Cosmochimica Acta, v. 67, p. 733–750, doi: 10.1016/S0016-7037(02)01143-2.

Kieffer, S.W., Phakey, P.P., and Christie, J.M., 1976, Shock processes in porous quartzite: Transmission electron microscope observation and theory: Contributions to Mineralogy and Petrology, v. 59, p. 41–93, doi: 10.1007/BF00375110.

Koeberl, C., Poag, C.W., Reimold, W.U., and Brandt, D., 1996, Origin of the Chesapeake Bay structure and the source of the North American tektites: Science, v. 271, p. 1263–1266, doi: 10.1126/science.271.5253.1263.

Kring, D.A., 1995, The dimensions of Chicxulub impact crater and the impact melt sheet: Journal of Geophysical Research, v. 100, p. 16,979–16,986, doi: 10.1029/95JE01768.

Kyte, F.T., 2001, Identification of late Eocene impact deposits at ODP Site 1090, *in* Gersonde, R., Hodell, D.A., and Blum, P., et al., Proceedings of the Ocean Drilling Program, Scientific Results, Volume 177: Washington, D.C., U.S. Government Printing Office, p. 1–9 [online]. Available from: http://www-odp.tamu.edu/publications/177_SR/VOLUME/CHAPTERS/SR177_04.PDF (accessed 14 February 2008).

Kyte, F.T., and Liu, S., 2002, Iridium and spherules in late Eocene impact deposits: Houston, Texas, Lunar and Planetary Institute, Lunar and Planetary Science Conference, v. 33, abstract 1981 [CD-ROM].

Langenhorst, F., 1996, Characteristics of shocked quartz in late Eocene impact ejecta from Massignano (Ancona, Italy): Clues to shock conditions and source crater: Geology, v. 24, p. 487–490, doi: 10.1130/0091-7613(1996)024<0487:COSQIL>2.3.CO;2.

Lee, M.-Y., and Wei, K.-Y., 2000, Australasian microtektites in the South China Sea and the West Philippine Sea: Implications for age, size, and location of the impact crater: Meteoritics & Planetary Science, v. 35, p. 1151–1155.

Liu, S., 2006, Characterization, Geographic Distribution, and Number of Upper-Eocene Impact Ejecta Layers and Their Correlations with Source Craters [Ph.D. dissertation]: Newark, Delaware, University of Delaware, 284 p.

Liu, S., Papanastassiou, D.A., Ngo, H.H., and Glass, B.P., 2006, Sr and Nd analyses of upper Eocene spherules and their implications for target rocks: Meteoritics & Planetary Science, v. 41, p. 705–714.

Ma, P., Aggrey, K., Tonzola, C., Schnabel, C., de Nicloa, P., Herzog, G.F., Wasson, J.T., Glass, B.P., Brown, L., Tera, F., Middleton, R., and Klein, J., 2004, Beryllium-10 in Australasian tektites: Constraints on the location of the source crater: Geochimica et Cosmochimica Acta, v. 68, p. 3883–3896, doi: 10.1016/j.gca.2004.03.026.

Madile, M., and Monechi, S., 1991, Late Eocene to early Oligocene calcareous nannofossil assemblages from Sites 699 and 703, sub-Antarctic South Atlantic Ocean, *in* Ciesielski, P.F., Kristoffersen, Y., Clement, B.M., et al., Proceedings of the Ocean Drilling Program, Scientific Results, Volume 114: Washington, D.C., U.S. Government Printing Office, p. 179–192.

Marchand, E., and Whitehead, J., 2002, Statistical evaluation of compositional differences between upper Eocene impact ejecta layers: Mathematical Geology, v. 34, p. 555–572, doi: 10.1023/A:1016094928524.

Marino, M., and Flores, J.A., 2002, Data report: Calcareous nannofossil data from the Eocene to Oligocene, *in* Gersonde, R., Hodell, D.A., and Blum, P., et al., Proceedings of the Ocean Drilling Program, Scientific Results, Volume 177 [Online]: http://www.odp.tamu.edu/publications/177_SR/VOLUME/CHAPTERS/SR177_08.PDF (accessed 14 February 2008).

Martini, E., 1971, Standard Tertiary and Quaternary calcareous nannoplankton zonation, *in* Farinacci, A., ed., Proceedings of the 2nd Planktonic Conference, 1970: Rome, Tecnoscienza, v. 2, p. 739–785.

Maurrasse, F.J.M.R., and Glass, B.P., 1976, Radiolarian stratigraphy and North American microtektites in Caribbean RC9–58: Implications concerning late Eocene radiolarian chronology and the age of the Eocene-Oligocene boundary, *in* Caüsse, R., ed., Proceedings of the 7th Caribbean Geological Conference, Guadeloupe, France, June 30–July 12, 1974: Bureau de Recherches Geologiques et Minieres, p. 205–212.

Miller, K.G., Berggren, W.A., Zhang, J., and Palmer-Julson, A.A., 1991, Biostratigraphy and isotope stratigraphy of upper Eocene microtektites at Site 612; how many impacts?: Palaios, v. 6, p. 17–38, doi: 10.2307/3514951.

Milner, G.J., 1992, Middle Eocene to early Oligocene foraminifers from the Izu-Bonin forearc, *in* Fryer, P., Pearce, J.A., Stokking, L.B., et al., Proceedings of the Ocean Drilling Program, Scientific Results, Volume 125: Washington, D.C., U.S. Government Printing Office, p. 71–90.

Montanari, A., Asaro, F., Michel, H.V., and Kennett, J.P., 1993, Iridium anomalies of late Eocene age at Massignano (Italy), and ODP Site 689B (Maud Rise, Antarctic): Palaios, v. 8, p. 420–437, doi: 10.2307/3515017.

Ngo, H., Wasserburg, G.J., and Glass, B.P., 1985, Nd and Sr isotopic compositions of tektite material from Barbados and their relationship to North American tektites: Geochimica et Cosmochimica Acta, v. 49, p. 1479–1485, doi: 10.1016/0016-7037(85)90297-2.

Nigrini, C., Sanfilippo, A., and Moore, T.J., Jr., 2005, Cenozoic radiolarian biostratigraphy: A magnetostratigraphic chronology of Cenozoic sequences from ODP Sites 1218, 1219, and 1220, equatorial Pacific, *in* Wilson, P.A., Lyle, M., Firth, J.V., eds., Scientific results of the Ocean Drilling Program, v. 199: College Station, Texas, U.S. Government Printing Office, p. 1–6, http://www-odp.tamu.edu/publications/199_SR/VOLUME/CHAPTERS/225.PDF.

Nocchi, M., Amici, E., and Premoli Silva, I., 1991, Planktonic foramininferal biostratigraphy and paleoenvironmental interpretation of Paleogene faunas from the sub-Antarctic transect, Leg 114, *in* Ciesielski, P.F., Krisstofferson, Y., et al., Proceedings of the Ocean Drilling Program, Scientific Results, Volume 125: Washington, D.C., U.S. Government Printing Office, p. 233–279.

Obradovich, J.D., Snee, L., and Izett, G.A., 1989, Is there more than one glassy impact layer in the late Eocene?: Geological Society of America Abstracts with Programs, v. 21, no. 7, p. A134.

Okada, H., 1990, Quaternary and Paleogene calcareous nannofossils, Leg 115, *in* Duncan, R.A., Backman, J., Peterson, L.C., et al., Proceedings of the Ocean Drilling Program, Scientific Results, Volume 115: Washington, D.C., U.S. Government Printing Office, p. 129–174.

Okada, H., and Bukry, D., 1980, Supplementary modifications and introduction of code numbers to the low-latitude coccolith biostratigraphic zonation (Bukry, 1973, 1975): Marine Micropaleontology, v. 5, p. 321–325.

Pierrard, O., Robin, E., Rocchia, R., and Montanari, A., 1998, Extraterrestrial Ni-rich spinel in upper Eocene sediments from Massignano, Italy: Geology, v. 26, p. 307–310.

Poag, C.W., 1997, Roadblocks on the kill curve: Testing the Raup hypothesis: Palaios, v. 12, p. 582–590, doi: 10.2307/3515413.

Poag, C.W., Powars, D.S., Poppe, L.J., and Mixon, R.B., 1994, Meteoroid mayhem in Ole Virginny: Source of the North American tektite strewn field: Geology, v. 22, p. 691–694, doi: 10.1130/0091-7613(1994)022<0691:MMIOVS>2.3.CO;2.

Poag, C.W., Koeberl, C., and Reimold, W.U., 2004, The Chesapeake Bay Crater—Geology and Geophysics of a Late Eocene Submarine Impact Structure: Heidelberg, Germany, Springer, 522 p.

Prasad, M.S., Mahale, V.P., and Kodagali, V.N., 2007, New sites of Australasian microtektites in the central Indian Ocean: Implications for the location and size of source crater: Journal of Geophysical Research, v. 112, p. E06007, doi: 10.1029/2006JE002857.

Premoli Silva, I., and Spezzaferri, S., 1990, Paleogene planktonic foraminifer biostratigraphy and paleoenvironmental remarks on Paleogene sediments from Indian Ocean Sites, Leg 115, *in* Duncan, R.A., Backman, J., Peterson, L.C., et al., Proceedings of the Ocean Drilling Program, Scientific Results, Volume 115: Washington, D.C., U.S. Government Printing Office, p. 277–314.

Prothero, D.R., Ivany, L., and Nesbitt, E., eds., 2003, From Greenhouse to Icehouse: The Marine Eocene–Oligocene Transition: New York, Columbia University Press, 560 p.

Riedel, W.R., and Sanfilippo, A., 1978, Stratigraphy and evolution of tropical Cenozoic radiolarians: Micropaleontology, v. 24, p. 61–96, doi: 10.2307/1485420.

Sanfilippo, A., Riedel, W.R., Glass, B.P., and Kyte, F.T., 1985, Late Eocene microtektites and radiolarian extinctions on Barbados: Nature, v. 314, p. 613–615, doi: 10.1038/314613a0.

Schellenberg, S.A., Brinkhuis, H., Stickley, C.E., Fuller, M., Kyte, F.T., and Williams, G.L., 2004, The Cretaceous/Paleogene transition on the East Tasman Plateau, southwestern Pacific, *in* Exon, N.F., Kennett, J.P., and Malone, M.J., eds., Climate Evolution in the Southern Ocean and Australia's Cenozoic Flight Northward from Antarctica: American Geophysical Union Geophysical Monograph 151, p. 93–112.

Schnetzler, C.C., 1992, Mechanism of Muong Nong–type tektite formation and speculation on the source of Australasian tektites: Meteoritics, v. 27, p. 154–165.

Shaw, H.F., and Wasserburg, G.J., 1982, Age and provenance of the target materials for tektites and possible impactites as inferred from Sm-Nd and Rb-Sr systematics: Earth and Planetary Science Letters, v. 60, p. 155–177, doi: 10.1016/0012-821X(82)90001-2.

Shipboard Scientific Party, 1976, Site 315, *in* Schlanger, S.O., Jackson, E.D., et al., Initial Reports of the Ocean Drilling Program, Volume 33: Washington, D.C., U.S. Government Printing Office, p. 37–104.

Shipboard Scientific Party, 1981, Site 462, *in* Larson, R.L., Schlanger, S.O., et al., Initial Reports of the Ocean Drilling Program, Volume 61: Washington, D.C., U.S. Government Printing Office, p. 19–394.

Shipboard Scientific Party, 1986, Site 592: Lord Howe Rise, 36°S, *in* Kennett, J.P., von der Borch, C.C., et al., Initial Reports of the Deep Sea Drilling Project, Volume 90: Washington, D.C., U.S. Government Printing Office, p. 487–550.

Shipboard Scientific Party, 1987, Site 612, *in* Poag, C.W., Watts, A.B., et al., Initial Reports of the Ocean Drilling Program, Volume 95: Washington, D.C., U.S. Government Printing Office, p. 31–153.

Shipboard Scientific Party, 1988a, Site 689, *in* Barker, P.F., Kennett, J.P., et al., Initial Reports of the Ocean Drilling Program, Volume 113: Washington, D.C., U.S. Government Printing Office, p. 89–181.

Shipboard Scientific Party, 1988b, Site 699, *in* Ciesielski, P.F., Krisstofferson, Y., et al., Initial Reports of the Ocean Drilling Program, Volume 110: Washington, D.C., U.S. Government Printing Office, p. 389–485.

Shipboard Scientific Party, 1988c, Site 703, *in* Ciesielski, P.F., Krisstofferson, Y., et al., Initial Reports of the Ocean Drilling Program, Volume 110: Washington, D.C., U.S. Government Printing Office, p. 549–619.

Shipboard Scientific Party, 1988d, Site 709, *in* Brockman, J., Duncan, R.A., et al., Initial Reports of the Ocean Drilling Program, Volume 121: Washington, D.C., U.S. Government Printing Office, p. 305–358.

Shipboard Scientific Party, 1989, Site 738, *in* Barron, J., Larsen, B., et al., Initial Reports of the Ocean Drilling Program, Volume 119: Washington, D.C., U.S. Government Printing Office, p. 229–288.

Shipboard Scientific Party, 1990, Site 786, *in* Fryer, P., Pearce, J.A., Stokking, L.B., et al., Initial Reports of the Ocean Drilling Program, Volume 125: Washington, D.C., U.S. Government Printing Office, p. 313–363.

Shipboard Scientific Party, 1999, Site 1090, *in* Gersonde, R., Hodell, D.A., Blum, P., et al., Initial Reports of the Ocean Drilling Program, Volume 177: Washington, D.C., U.S. Government Printing Office, p. 1–101 [Online]. Available from: http://www.odp.tamu.edu/publications/177_IR/VOLUME/CHAPTERS/CHAP_105.PDF (accessed 14 February 2008).

Shipboard Scientific Party, 2004, Explanatory notes, *in* Tucholke, B.E., Sibuet, J.-C., Klaus, A., et al., Initial Reports of the Ocean Drilling Program, Volume 210: Washington, D.C., U.S. Government Printing Office, p. 1–69, doi: 10.2973/odp.proc.ir.210.102.2004.

Spezzaferri, S., Basso, D., and Coccioni, R., 2002, Late Eocene planktonic foraminiferal response to an extraterrestrial impact at Massignano GSSP (northeastern Apennines, Italy): Journal of Foraminiferal Research, v. 32, p. 188–199, doi: 10.2113/0320188.

Spieß, V., 1990, Cenozoic magnetostratigraphy of Leg 113 drill sites, Maud Rise, Weddell Sea, Antarctica, *in* Barker, P.F., Kennett, J.P., et al., Proceedings of the Ocean Drilling Program, Scientific Results, Volume 113: Washington, D.C., U.S. Government Printing Office, p. 261–315.

Stabell, B., Ali, J., Ciampo, G., Milner, G., Wang, Y.-J., and Xu, Y., 1992, Biostratigraphic summary, Leg 125, *in* Fryer, P., Pearce, J.A., Stokking, L.B., et al., Proceedings of the Ocean Drilling Program, Scientific Results, Volume 125: Washington, D.C., U.S. Government Printing Office, p. 615–622.

Stauffer, P.H., 1978, Anatomy of the Australasian tektite strewn field and the probable size of its source crater, *in* Nutalaya, P., ed., Proceedings of the Third Regional Conference on the Geology and Mineral Resources of Southeast Asia: Bangkok, Thailand, Asian Institute of Technology, p. 285–289.

Stecher, O., Ngo, H.H., Papanastassiou, D.A., and Wasserburg, G.J., 1989, Nd and Sr isotopic evidence for the origin of tektite material from DSDP Site 612 off the New Jersey coast: Meteoritics, v. 24, p. 89–98.

Stöffler, D., and Claeys, P., 1997, Earth rocked by combination punch: Nature, v. 388, p. 331–332, doi: 10.1038/40981.

Stöffler, D., Gault, D.E., Wedekind, J., and Polkowski, G., 1975, Experimental hypervelocity impact into quartz sand: Distribution and shock metamorphism of ejecta: Journal of Geophysical Research, v. 80, p. 4062–4077, doi: 10.1029/JB080i029p04062.

Stott, L.D., and Kennett, J.P., 1990, Antarctic Paleogene planktonic foraminifer biostratigraphy: ODP Leg 113, Sites 689 and 690, *in* Barker, P.F.,

Kennett, J.P., et al., Proceedings of the Ocean Drilling Program, Scientific Results, Volume 113: Washington, D.C., U.S. Government Printing Office, p. 549–569.

Thein, J.E., 1987, A tektite layer in upper Eocene sediments of the New Jersey continental slope (Site 612, Leg 95), *in* Poag, C.W., Watts, A.B., Cousin, M., et al., Initial Reports of the Deep Sea Drilling Project, Volume 95: Washington, D.C., U.S. Government Printing Office, p. 565–579.

Vishnevsky, S., and Montanari, A., 1999, Popigai impact structure (Arctic Siberia, Russia): Geology, petrology, geochemistry, and geochronology of glass-bearing impactites, *in* Dressler, B.O., and Sharpton, V.L. eds., Large Meteorite Impacts and Planetary Evolution II: Geological Society of America Special Paper 339, p. 19–59.

Vonhof, H.B., and Smit, J., 1999, Late Eocene microkrystites and microtektites at Maud Rise (Ocean Drilling Project Hole 689B; Southern Ocean) suggest a global extension of the approximately 35.5 Ma Pacific impact ejecta strewn field: Meteoritics & Planetary Science, v. 34, p. 747–755.

Vonhof, H.B., Smit, J., Brinkhuis, H., Montanari, A., and Nederbragt, A.J., 2000, Global cooling accelerated by early late Eocene impacts?: Geology, v. 28, p. 687–690, doi: 10.1130/0091-7613(2000)28<687:GCABEL>2.0.CO;2.

Wei, W., 1995, How many impact-generated microspherule layers in the upper Eocene?: Palaeogeography, Palaeoclimatology, Palaeoecology, v. 114, p. 101–110, doi: 10.1016/0031-0182(95)00075-W.

Wei, W., 1998, Calcareous nannofossils from the New Jersey continental margin, *in* Christie-Blick, N., Austin, J.A., Jr., Malone, M.J., et al., Proceedings of the Ocean Drilling Program, Scientific Results, Volume 174A: Washington, D.C., U.S. Government Printing Office, p. 1–16. Available from: http://www-odp.tamu.edu/publications/174A_SR/VOLUME/CHAPTERS/SR174a05.PDF (accessed 14 February 2008).

Wei, W., and Wise, S.W., Jr., 1990, Middle Eocene to Pleistocene calcareous nannofossils recovered by Ocean Drilling Program Leg 113 in the Weddell Sea, *in* Barker, P.F., Kennett, J.P., et al., Proceedings of the Ocean Drilling Program, Scientific Results, Volume 113: Washington, D.C., U.S. Government Printing Office, p. 639–666.

Whitehead, J., Papanastassiou, D.A., Spray, J.G., Grieve, R.A.F., and Wasserburg, G.J., 2000, Late Eocene impact ejecta: Geochemical and isotopic connections with the Popigai impact structure: Earth and Planetary Science Letters, v. 181, p. 473–487, doi: 10.1016/S0012-821X(00)00225-9.

Wrobel, K.E., and Schultz, P.H., 2003, The effect of rotation on the deposition of terrestrial impact ejecta: Houston, Texas, Lunar and Planetary Institute, Lunar and Planetary Science Conference, v. 34, abstract 1190 [CD-ROM].

Xu, Y., and Wise, S.W., Jr., 1992, Middle Eocene to Miocene calcareous nannofossils of Leg 125 from the western Pacific Ocean, *in* Fryer, P., Pearce, J.A., Stokking, L.B., et al., Proceedings of the Ocean Drilling Program, Scientific Results, Volume 125: Washington, D.C., U.S. Government Printing Office, p. 43–70.

Manuscript Accepted by the Society 16 September 2008

Printed in the USA

The Geological Society of America
Special Paper 452
2009

A search for extraterrestrial chromite in the late Eocene Massignano section, central Italy

Birger Schmitz*
Anders Cronholm
Department of Geology, Lund University, Sölvegatan 12, SE-22362 Lund, Sweden

Alessandro Montanari
Osservatorio Geologico di Coldigioco, 62020 Frontale di Apiro, Italy

ABSTRACT

The late Eocene may have been a period with an enhanced flux of extraterrestrial matter to Earth related either to a comet or an asteroid shower. The evidence comes from two very large and several medium-sized impact craters, at least two microtektite-microkrystite layers, and a stratigraphic interval with enhanced extraterrestrial ^{3}He, all within the period ca. 36.3–34.3 Ma. Here, we show that the distribution of sediment-dispersed extraterrestrial (ordinary chondritic) chromite (EC) grains in the Massignano section, central Italy, can be used to test whether the flux of ordinary chondritic matter to Earth was enhanced in the late Eocene. In twelve limestone samples, each weighing ~12–15 kg, from 1.25 m to 10.25 m above the base of the section, only 1 EC grain was found. Based on the total amount of limestone analyzed, 167 kg, this corresponds to 0.006 EC grain kg^{-1} limestone. This is a factor of five lower than the 0.029 EC grain kg^{-1} recovered in 210 kg of latest Cretaceous–Paleocene limestone from the Bottaccione Gorge section at Gubbio, central Italy. The difference can readily be explained by an approximately threefold higher sedimentation rate in the late Eocene at Massignano. In essence, our results speak against a late Eocene asteroid shower. Apparently, there was no significant increase in the flux of extraterrestrial chromite at this time, such as that after the disruption of the L-chondrite parent body in the mid-Ordovician, when the EC flux was enhanced by two orders of magnitude. We also discuss the potential to search for lunar minerals in the Massignano section in order to test the recent hypothesis that late Eocene ^{3}He enrichments originated from impact-ejected lunar regolith.

Keywords: extraterrestrial chromite, asteroid shower, micrometeorite, impact ejecta, late Eocene.

*birger.schmitz@geol.lu.se

Schmitz, B., Cronholm, A., and Montanari, A., 2009, A search for extraterrestrial chromite in the late Eocene Massignano section, central Italy, *in* Koeberl, C., and Montanari, A., eds., The Late Eocene Earth—Hothouse, Icehouse, and Impacts: Geological Society of America Special Paper 452, p. 71–82, doi: 10.1130/2009.2452(05). For permission to copy, contact editing@geosociety.org.

INTRODUCTION

Global changes such as cooling and stepwise floral and faunal turnovers in the late Eocene have been suggested to be at least partially related to either a comet or an asteroid shower (Hut et al., 1987; Bodiselitsch et al., 2004; Tagle and Claeys, 2004). Supporting evidence includes an overrepresentation of impact craters of this age, enhanced extraterrestrial ^{3}He concentrations in sediments over an ~2.2 m.y. period, and, in the same interval, three iridium anomalies, of which one has associated shocked quartz, and two microtektite-microkrystite layers (Montanari et al., 1993, 1998; Clymer et al., 1996; Farley et al., 1998; Bodiselitsch et al., 2004). The two largest impact craters of the Cenozoic, Popigai, 100 km in diameter, and Chesapeake Bay, 85 km in diameter, formed at 36–35 Ma, and a number of smaller craters with comparable ages may be part of the same event (Montanari and Koeberl, 2000). The sedimentary ^{3}He enrichments have been considered to more likely represent a comet shower (Farley et al., 1998; Bodiselitsch et al., 2004), whereas platinum group element patterns for ejecta melts from the Popigai crater have been related to a stony chondritic projectile, possibly an L chondrite (Tagle and Claeys, 2004, 2005). Recently, Fritz et al. (2007) suggested that an asteroid (L chondrite) shower in the late Eocene led to an increased impact rate on both Earth and the Moon, and that impact-ejected lunar regolith was transported to Earth, resulting in the ^{3}He enrichments in marine sediments.

The question of comet or asteroid shower may be resolved by comparisons with the middle Ordovician, a time when the L-chondrite parent body disrupted, leading to a dramatically enhanced flux of extraterrestrial matter to Earth (Schmitz et al., 2001, 2003, 2008). This is the largest documented asteroid breakup event during the Phanerozoic, and a major fraction of the L-chondritic meteorites that fall onto Earth have argon gas retention ages around 470 Ma attesting to the event (Korochantseva et al., 2007). No other asteroid breakup has left any similar prominent imprint on present-day meteorite argon ages. In marine limestone formed at 470 Ma in the mid-Ordovician, more than 80 fossil L-chondritic meteorites (1–20 cm in diameter) and a two-orders-of-magnitude enrichment in sediment-dispersed L-chondritic chromite grains, mainly from decomposed unmelted micrometeorites, attest to the parent-body breakup (Schmitz et al., 2001; Schmitz, 2008, personal commun.). According to modeling studies, a breakup event of this magnitude should be accompanied by an asteroid shower to Earth (Zappalà et al., 1998). There is some evidence of an enhanced number of impact craters on Earth at this time, and an asteroid shower could have played a role in the coeval rapid diversification of invertebrate life (Schmitz et al., 2001, 2008).

Analogous to the mid-Ordovician, a potential asteroid shower in the late Eocene may have resulted in excess amounts of extraterrestrial (ordinary chondritic) chromite (EC) grains (>63 μm) in sediments. Here, we present results of searches for EC grains in twelve limestone samples, each ~12–15 kg in size, from the late Eocene ^{3}He enriched interval of the Massignano section near Ancona in central Italy (Fig. 1). Nine of the samples represent "background" samples chosen at arbitrary levels throughout the section, and three samples correspond to three levels with enhanced iridium concentrations (see Montanari et al., 1993; Bodiselitsch et al., 2004). One of the Ir-enriched levels, at 5.6 m, is a confirmed impact ejecta layer with abundant shocked quartz, Ni-rich spinels, 200 ppt Ir, and common pancake spherules, representing relics of clinopyroxene-bearing microkrystites

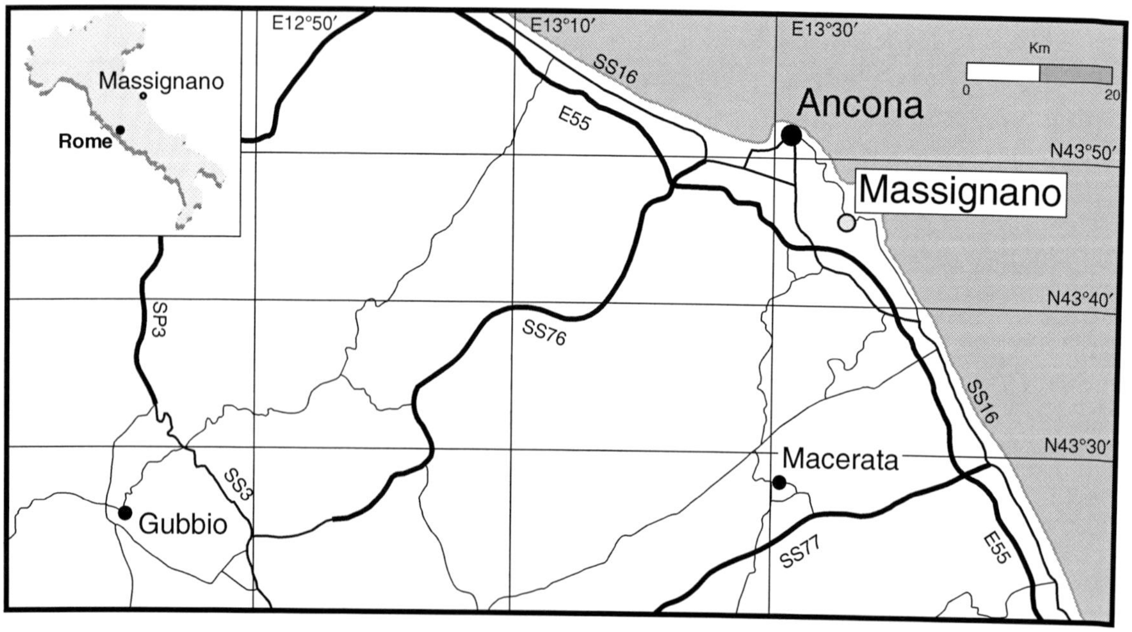

Figure 1. Map of area around Ancona, eastern central Italy, and the position of the Massignano section. Note that the town of Gubbio, where the Bottaccione section is exposed, is also indicated.

(e.g., Montanari and Koeberl, 2000, and references therein). The layer may represent the distal ejecta from the Popigai crater, but this link is mainly based on the approximate coincidence in time. The two other Ir-enriched levels have not yielded any shocked quartz or other compelling impact ejecta constituents, and their possible impact association remains enigmatic (Montanari et al., 1993; Clymer et al., 1996; Bodiselitsch et al., 2004).

We note that the chromite grains discussed here have a quite different origin and composition than the Ni- and Cr-rich spinels in the Massignano impact ejecta layer, dealt with in studies by, for example, Pierrard et al. (1998) and Glass et al. (2004). The latter spinels occur as tiny crystals (a few microns) in fragile stringers on pancake spherules, and formed in the heat during the impact event, whereas the coarse chromite studied here represents an original component (on average 0.25 wt%) in equilibrated ordinary chondrites (Keil, 1962; Bridges et al., 2007).

MATERIALS AND METHODS

Dry weights of the limestone samples studied ranged from 11.6 to 15.1 kg, and the samples were collected from 1.25 m to 10.25 m above the base of the section (Table 1), all within the 13-m-thick ^{3}He enriched interval of Farley et al. (1998). In the Massignano section, meter levels relative to the base of the section are marked by persistent metal plates, and the sample depths in our study refer to this meter scale, like in most previous studies of the section (see, e.g., Montanari et al., 1993). The three samples from Ir-enriched levels were collected at 5.60, 6.15, and 10.25 m. All samples were decalcified in 6 *M* hydrochloric acid at room temperature and then sieved at 32 µm. The samples were readily dissolvable, and there was no need to break them into smaller pieces prior to HCl leaching. The remaining HCl residues were leached with 18 *M* hydrofluoric acid at room temperature for about 2 d, with occasional stirring. The HF residues were recovered with a 32 µm sieve. The acid-insoluble fraction in the size range 63–355 µm was searched under the binocular microscope, and all suspected chromite (i.e., black, opaque) grains were picked with a fine brush for element analyses.

All suspected chromite grains were first mounted on carbon tape and analyzed for elements in an unpolished state. Performed in the right way, these analyses are mostly sufficient to tell whether a grain is extraterrestrial chromite or not. All the confirmed chrome spinel (including chromite) grains were mounted in epoxy resin, polished flat using a 1 µm diamond slurry, and again analyzed for major and trace elements, but with the aim of producing high-quality data. It is unavoidable for a minor fraction of the chrome spinel grains to be lost during the polishing, but by first analyzing all the grains in an unpolished state, we were able to control whether any EC grains were lost. The quantitative element analyses were performed with an Oxford Inca X-sight energy-dispersive spectrometer with a Si detector, mounted on a Hitachi S-3400 scanning electron microscope. Cobalt was used to monitor drift of the instrument. An acceleration voltage of 15 kV, a sample current ~1 nA, and a counting live-time of 80 s were used. Precision of analyses was typically better than 1–4 rel%. Analytical accuracy was controlled by repeated analyses of the USNM 117075 (Smithsonian) chromite reference standard (Jarosewich et al., 1980).

RESULTS

We follow the practice of Schmitz and Häggström (2006) and divide sediment-dispersed Cr-rich spinels into two groups: extraterrestrial (ordinary chondritic) chromite (EC) and "other" Cr-rich spinels (OC). For a grain to be classified as an EC grain, it has to have a composition within narrowly defined ranges for Cr_2O_3 (~55–60 wt%), FeO (~25–30 wt%), Al_2O_3 (~5–8 wt%), TiO_2 (~2.0–3.5 wt%), V_2O_3 (~0.6–0.9 wt%), and MgO (~1.5–4 wt%) (Schmitz and Häggström, 2006). Only grains with compositions that lie within the ranges of all elements listed are considered to be EC grains. The OC grains, on the other hand, have a wide compositional range, and the vast majority has a terrestrial origin, but, in rare cases, they may represent unusual types of extraterrestrial spinels.

In total, 374 chrome spinel grains were found in the 167 kg of limestone analyzed, and of these, only 1 grain can be classified as EC; the remaining are OC grains (Table 1). The single EC grain was found in the sample at 6.15 m (grain no. Cr 12), i.e., from one of the two Ir-enriched levels where supporting evidence for an impact relation is lacking. As shown in Figure 2, the grain is not so well preserved, having some rounded edges and a general abraded appearance, and compared with a typical mid-Ordovician EC grain, the latter has a more fresh appearance. From the analyses of the unpolished chrome spinel grains, it could be ascertained that of the five grains lost during polishing, all were OC grains.

DISCUSSION

The method of extracting EC grains from bulk sediment was originally developed as a complement to the search for mid-Ordovician fossil meteorites (Schmitz et al., 2003; Schmitz and Häggström, 2006). In fossil meteorites, chromite is the only common relict mineral, but much information about the precursor meteorites can be gained from the mineral. For example, with chromite analyses, it is possible to classify the fossil meteorites to group and petrographic type, and to determine their cosmic-ray exposure histories (e.g., Schmitz et al., 2001; Heck et al., 2004; Bridges et al., 2007; Greenwood et al., 2007). The mid-Ordovician sediment-dispersed EC grains contain abundant solar-wind implanted neon, showing that most grains originate from micrometeorites that have undergone decomposition on the seafloor (Heck et al., 2008). In mid-Ordovician condensed strata in southern Sweden, the beds that formed after the L-chondrite disruption event contain one to three EC grains kg^{-1} of limestone, whereas in beds that formed at the same rate during the ~5 m.y. immediately prior to this event, 379 kg of limestone only yielded five EC grains (Schmitz and Häggström, 2006).

TABLE 1. ELEMENT CONCENTRATIONS (WT%) IN CR-RICH SPINELS EXTRACTED FROM LATE EOCENE SEDIMENTS IN THE MASSIGNANO QUARRY, CENTRAL ITALY*

Grains	Cr_2O_3	Al_2O_3	MgO	TiO_2	V_2O_3	FeO	MnO	ZnO	Total†	Fe#§	Cr##
MAS 1.25–1.35 m (13.9 kg)**											
Cr 1	45.11	26.79	10.96	B.D.††	B.D.	16.27	B.D.	B.D.	99.1	45.4	53.0
Cr 2	46.73	21.20	9.44	B.D.	B.D.	21.61	B.D.	0.36	99.3	54.9	59.5
Cr 3	50.05	15.38	9.95	B.D.	B.D.	23.05	B.D.	B.D.	98.4	51.5	68.6
Cr 4	44.04	24.91	12.87	B.D.	B.D.	17.31	B.D.	B.D.	99.1	41.2	54.2
Cr 5	45.37	23.14	12.21	B.D.	B.D.	18.15	B.D.	B.D.	98.9	43.4	56.8
Cr 6	47.55	21.35	12.22	B.D.	B.D.	16.20	B.D.	B.D.	97.3	42.2	59.9
Cr 7	50.12	19.16	11.30	B.D.	B.D.	18.66	B.D.	B.D.	99.2	46.6	63.7
Cr 8	41.97	19.57	9.26	0.72	B.D.	27.27	0.48	B.D.	99.3	56.1	59.0
Cr 9	42.54	27.58	13.95	B.D.	0.34	15.26	B.D.	B.D.	99.7	37.9	50.8
Cr 10	44.54	24.21	12.64	B.D.	B.D.	16.69	B.D.	B.D.	98.1	41.5	55.2
Cr 11	42.54	26.12	13.90	B.D.	B.D.	15.73	B.D.	B.D.	98.3	36.8	52.2
Cr 12	46.21	22.72	12.74	B.D.	0.37	16.70	B.D.	B.D.	98.7	41.4	57.7
Cr 13	50.99	5.61	4.20	B.D.	0.42	35.72	0.66	B.D.	97.6	77.4	85.9
Cr 14	44.74	23.33	12.56	B.D.	B.D.	17.31	B.D.	B.D.	97.9	41.5	56.3
Cr 15	38.29	29.76	13.76	B.D.	B.D.	16.12	B.D.	B.D.	97.9	38.1	46.3
Cr 16	41.93	16.53	8.02	2.53	0.52	27.56	B.D.	B.D.	97.1	62.8	63.0
Cr 17	43.74	25.84	12.27	B.D.	B.D.	18.06	B.D.	B.D.	99.9	44.3	53.2
Cr 18	47.81	23.17	13.59	B.D.	B.D.	16.57	B.D.	B.D.	101.1	38.8	58.0
Cr 19	56.30	12.49	10.90	B.D.	B.D.	18.88	B.D.	B.D.	98.6	46.6	75.1
Cr 20	44.73	24.35	11.39	0.36	B.D.	19.79	B.D.	B.D.	100.6	48.4	55.2
Cr 21	38.27	30.27	12.91	B.D.	B.D.	17.53	0.44	B.D.	99.4	41.8	45.6
Cr 22	50.64	17.79	10.41	B.D.	0.30	19.62	B.D.	B.D.	98.8	50.3	65.6
Cr 23	54.75	14.50	10.33	B.D.	0.40	19.30	B.D.	B.D.	99.3	50.3	71.7
Cr 24	54.86	15.80	8.73	B.D.	B.D.	20.80	B.D.	B.D.	100.2	57.2	70.0
Cr 25	59.92	10.34	8.42	B.D.	B.D.	21.56	0.62	B.D.	100.9	57.7	79.0
Cr 26	50.89	18.68	11.52	B.D.	B.D.	18.08	B.D.	B.D.	99.2	45.5	64.6
Cr 27	29.66	34.89	15.77	0.48	B.D.	16.39	B.D.	B.D.	97.2	31.7	36.3
Cr 28	44.04	23.50	12.21	B.D.	B.D.	17.94	B.D.	B.D.	97.7	43.0	55.7
Cr 29	55.87	11.54	7.84	B.D.	B.D.	23.15	B.D.	B.D.	98.4	60.5	76.5
Cr 30	51.83	17.32	10.47	B.D.	B.D.	19.48	B.D.	B.D.	99.1	49.8	66.7
Cr 31	49.54	16.80	11.22	B.D.	B.D.	21.64	B.D.	B.D.	99.2	46.5	66.4
Cr 32	46.16	23.17	13.10	B.D.	B.D.	16.52	B.D.	B.D.	99.0	39.7	57.2
MAS 1.95–2.20 m (13.9 kg)											
Cr 1	34.41	30.32	11.37	B.D.	0.35	23.44	0.39	B.D.	100.3	49.0	43.2
Cr 2	48.08	22.97	9.61	B.D.	B.D.	18.66	B.D.	0.40	99.7	52.1	58.4
Cr 3	49.31	20.45	12.18	B.D.	B.D.	18.83	B.D.	B.D.	100.8	43.9	61.8
Cr 4	52.41	18.40	10.91	B.D.	B.D.	17.52	B.D.	B.D.	99.2	47.4	65.6
Cr 5	43.74	19.15	11.07	0.37	B.D.	26.01	B.D.	B.D.	100.3	48.7	60.5
Cr 6	44.85	25.99	11.16	B.D.	B.D.	16.16	B.D.	B.D.	98.2	44.8	53.6
Cr 7	56.76	12.72	9.63	B.D.	B.D.	19.28	B.D.	B.D.	98.4	52.3	75.0
Cr 8	50.66	18.79	10.68	B.D.	0.30	19.64	B.D.	B.D.	100.1	49.9	64.4
Cr 9	52.75	16.68	11.33	B.D.	B.D.	19.68	B.D.	B.D.	100.4	46.5	68.8
Cr 10	49.46	21.06	12.35	B.D.	B.D.	16.82	B.D.	B.D.	99.7	42.7	61.2
Cr 11	42.67	27.33	12.53	B.D.	B.D.	18.07	B.D.	B.D.	100.6	43.8	51.1
Cr 12	36.69	26.99	11.43	B.D.	0.31	25.29	B.D.	B.D.	100.7	48.8	47.7
Cr 13	27.09	41.23	16.71	B.D.	B.D.	14.45	B.D.	B.D.	99.5	30.2	30.6
Cr 14	50.59	19.41	11.65	B.D.	B.D.	17.53	B.D.	B.D.	99.2	45.1	63.6
Cr 15	53.06	2.71	1.17	0.47	B.D.	33.74	0.84	0.62	92.6	92.9	92.9
Cr 16	36.14	29.99	13.25	B.D.	B.D.	19.58	B.D.	B.D.	99.0	40.8	44.7
Cr 17	48.57	20.64	11.42	B.D.	B.D.	18.38	B.D.	B.D.	99.0	46.3	61.2
Cr 18	44.75	24.92	14.19	B.D.	B.D.	15.87	B.D.	B.D.	99.7	36.0	54.6
Cr 19	42.20	19.38	8.68	0.46	B.D.	27.19	B.D.	0.32	98.2	58.2	59.4
Cr 20	49.68	20.53	12.34	B.D.	B.D.	17.61	B.D.	B.D.	100.2	42.9	61.9
Cr 21	42.25	22.85	10.11	B.D.	B.D.	23.60	B.D.	B.D.	98.8	52.5	55.4
Cr 22	47.82	21.11	11.74	B.D.	B.D.	17.50	B.D.	B.D.	98.2	44.6	60.3
Cr 23	43.63	26.29	13.01	B.D.	B.D.	17.65	B.D.	B.D.	100.6	41.6	52.7
Cr 24	52.72	15.74	10.78	B.D.	0.33	18.62	B.D.	B.D.	98.2	48.0	69.2
Cr 25	42.40	27.11	12.73	B.D.	B.D.	18.37	B.D.	B.D.	100.6	43.0	51.2
Cr 26	47.42	23.48	12.46	B.D.	B.D.	17.57	B.D.	B.D.	100.9	43.4	57.5
Cr 27	47.23	20.68	12.30	0.29	B.D.	18.87	B.D.	B.D.	99.4	43.1	60.5
Cr 28	46.06	21.72	12.59	B.D.	B.D.	18.84	0.36	B.D.	99.6	41.3	58.7
Cr 29	41.68	26.24	13.30	B.D.	B.D.	17.11	B.D.	B.D.	98.3	39.3	51.6

(*continued*)

TABLE 1. ELEMENT CONCENTRATIONS (WT%) IN CR-RICH SPINELS EXTRACTED FROM LATE EOCENE SEDIMENTS IN THE MASSIGNANO QUARRY, CENTRAL ITALY* (*continued*)

Grains	Cr_2O_3	Al_2O_3	MgO	TiO_2	V_2O_3	FeO	MnO	ZnO	Total[†]	Fe#[§]	Cr#[#]
MAS 3.05–3.35 m (15.0 kg)											
Cr 1	51.63	16.96	11.24	B.D.	0.37	20.66	B.D.	B.D.	100.9	47.5	67.1
Cr 2	43.50	21.56	10.28	B.D.	0.39	24.02	0.48	B.D.	100.5	51.8	57.5
Cr 3	40.41	18.96	10.43	2.12	0.36	28.17	B.D.	B.D.	100.5	53.7	58.8
Cr 4	48.81	21.08	11.80	B.D.	B.D.	18.11	B.D.	B.D.	99.8	45.1	60.8
Cr 5	49.12	16.93	10.33	B.D.	0.48	23.07	B.D.	B.D.	99.9	51.2	66.1
Cr 6	42.82	15.12	7.54	0.82	0.66	33.01	B.D.	B.D.	100.0	64.3	65.5
Cr 7	59.86	12.11	11.84	B.D.	B.D.	15.88	B.D.	B.D.	99.7	42.8	76.8
Cr 8	51.46	18.19	11.07	B.D.	B.D.	19.22	B.D.	B.D.	99.9	47.7	65.5
Cr 9	27.93	40.75	15.07	B.D.	B.D.	16.82	B.D.	B.D.	100.6	36.8	31.5
Cr 10	45.48	25.07	13.23	B.D.	B.D.	15.96	B.D.	B.D.	99.7	40.0	54.9
Cr 11	56.92	13.38	10.75	B.D.	B.D.	18.15	B.D.	B.D.	99.2	47.7	74.0
Cr 12	51.27	17.55	10.86	B.D.	B.D.	19.10	B.D.	B.D.	98.8	48.0	66.2
Cr 13	50.38	18.77	11.85	B.D.	B.D.	18.78	B.D.	B.D.	99.8	44.4	54.3
Cr 14	47.74	20.09	9.81	B.D.	B.D.	21.64	B.D.	B.D.	99.3	53.4	61.4
Cr 15	44.25	25.16	12.55	B.D.	0.30	17.54	B.D.	B.D.	99.8	43.2	54.1
Cr 16	20.93	47.48	16.92	B.D.	B.D.	15.24	B.D.	B.D.	100.6	31.5	22.8
Cr 17	45.72	23.96	11.40	B.D.	0.50	17.87	0.42	B.D.	99.9	46.8	56.1
Cr 18	48.82	22.15	13.20	B.D.	B.D.	16.22	B.D.	B.D.	100.4	39.7	59.6
Cr 19	45.65	16.12	7.79	B.D.	B.D.	29.29	B.D.	0.56	99.4	61.5	65.5
Cr 20	39.96	21.62	6.43	B.D.	B.D.	30.02	B.D.	0.46	98.5	68.4	55.3
Cr 21	52.25	14.90	7.82	B.D.	B.D.	23.94	B.D.	B.D.	98.9	61.4	70.2
Cr 22	50.05	17.67	10.91	B.D.	B.D.	19.66	B.D.	B.D.	98.3	47.6	65.5
MAS 4.80–5.10 m (14.6 kg)											
Cr 1	38.61	29.99	14.18	B.D.	B.D.	17.65	0.00	B.D.	100.4	37.7	46.3
Cr 2	54.27	14.11	9.31	B.D.	B.D.	20.69	0.00	B.D.	98.4	54.1	72.1
Cr 3	45.70	19.97	9.58	B.D.	0.37	23.21	0.51	B.D.	99.3	54.1	60.5
Cr 4	47.16	12.26	5.28	0.34	B.D.	32.25	0.74	B.D.	98.0	72.7	72.1
Cr 5	41.21	27.40	13.68	B.D.	B.D.	15.76	B.D.	B.D.	98.0	37.9	50.2
Cr 6	34.31	9.01	2.40	1.75	B.D.	49.83	0.67	0.33	98.3	87.6	71.9
Cr 7	39.70	16.67	8.70	1.54	0.59	31.41	B.D.	B.D.	98.6	59.6	61.5
Cr 8	43.43	25.42	12.61	B.D.	B.D.	17.00	B.D.	B.D.	98.5	42.1	53.4
Cr 9	45.33	22.62	11.74	B.D.	B.D.	18.87	B.D.	B.D.	98.6	45.2	57.3
Cr 10	32.03	10.20	7.26	6.19	B.D.	43.95	B.D.	B.D.	99.6	68.7	67.8
Cr 11	59.39	11.01	10.41	B.D.	B.D.	18.99	B.D.	B.D.	99.8	49.0	78.3
Cr 12	46.84	18.67	9.33	B.D.	B.D.	24.22	0.69	B.D.	99.8	54.7	62.7
Cr 13	46.86	19.64	11.67	B.D.	B.D.	20.47	B.D.	B.D.	98.6	44.9	61.5
Cr 14	52.22	16.45	10.55	B.D.	B.D.	21.25	B.D.	B.D.	100.5	49.9	68.0
Cr 15	52.80	18.05	12.43	B.D.	B.D.	16.59	0.56	B.D.	100.4	41.2	66.2
Cr 16	51.86	17.20	11.56	B.D.	B.D.	19.12	B.D.	B.D.	99.8	45.3	66.9
Cr 17	56.68	13.34	10.32	B.D.	B.D.	18.13	B.D.	B.D.	98.5	49.3	74.0
Cr 18	45.99	23.45	12.61	B.D.	B.D.	16.70	B.D.	B.D.	98.8	41.7	56.8
Cr 19	56.07	13.97	9.60	B.D.	B.D.	19.05	B.D.	B.D.	98.7	52.7	72.9
MAS 5.60–5.65 m (13.9 kg)											
Cr 1	41.88	26.70	13.33	B.D.	B.D.	17.51	B.D.	B.D.	99.1	39.9	51.3
Cr 2	21.56	39.83	15.73	1.00	B.D.	20.99	B.D.	B.D.	99.1	34.9	26.6
Cr 3	31.30	35.83	15.01	B.D.	B.D.	16.47	B.D.	B.D.	98.6	34.9	36.9
Cr 4	59.22	10.45	11.15	B.D.	B.D.	17.93	0.51	B.D.	99.3	44.5	79.2
Cr 5	44.79	19.35	10.72	0.81	B.D.	25.51	B.D.	B.D.	101.2	51.1	60.8
Cr 6	59.44	4.14	3.17	B.D.	0.40	31.04	B.D.	0.19	98.4	83.0	90.6
Cr 7	45.56	23.75	12.33	B.D.	B.D.	18.32	B.D.	B.D.	100.0	43.6	56.3
Cr 8	52.18	18.34	9.63	B.D.	B.D.	18.45	B.D.	B.D.	98.6	51.8	65.6
Cr 9	37.17	11.90	5.20	2.86	0.70	40.92	B.D.	B.D.	98.8	75.7	67.7
Cr 10	46.91	23.00	12.35	B.D.	B.D.	17.93	B.D.	B.D.	100.2	43.4	57.8
Cr 11	31.42	24.04	9.14	B.D.	B.D.	33.77	B.D.	B.D.	98.4	57.0	46.7
Cr 12	35.36	32.76	14.76	B.D.	B.D.	17.00	B.D.	B.D.	99.9	35.8	42.0
Cr 13	46.58	22.16	13.31	B.D.	B.D.	17.10	B.D.	B.D.	99.2	38.7	58.5
Cr 14	59.72	8.43	6.82	B.D.	B.D.	23.39	B.D.	0.70	99.1	64.5	82.6
Cr 15	52.29	15.80	11.63	B.D.	B.D.	18.62	B.D.	B.D.	98.3	44.0	68.9
Cr 16	45.47	22.47	10.08	B.D.	B.D.	21.03	B.D.	0.38	99.4	52.3	57.6
Cr 17	43.94	23.11	13.05	B.D.	B.D.	18.24	B.D.	B.D.	98.3	39.7	56.0
Cr 18	21.05	46.82	17.39	B.D.	B.D.	14.36	B.D.	B.D.	99.6	29.2	23.2
Cr 19	42.29	26.34	11.90	B.D.	B.D.	18.75	B.D.	B.D.	99.3	45.7	51.8
Cr 20	47.43	21.75	10.64	B.D.	B.D.	18.37	B.D.	B.D.	98.2	49.2	59.4
Cr 21	57.36	7.64	0.62	B.D.	B.D.	32.87	1.53	0.54	100.6	96.5	83.4

(*continued*)

TABLE 1. ELEMENT CONCENTRATIONS (WT%) IN CR-RICH SPINELS EXTRACTED FROM LATE EOCENE SEDIMENTS IN THE MASSIGNANO QUARRY, CENTRAL ITALY* (*continued*)

Grains	Cr_2O_3	Al_2O_3	MgO	TiO_2	V_2O_3	FeO	MnO	ZnO	Total[†]	Fe#[§]	Cr#[#]
MAS 5.60–5.65 m (13.9 kg) (*continued*)											
Cr 22	36.13	11.30	6.38	2.71	0.57	41.59	B.D.	B.D.	98.7	70.2	68.2
Cr 23	46.94	23.58	12.73	B.D.	B.D.	16.84	B.D.	B.D.	100.1	41.9	57.2
Cr 24	37.16	27.90	13.20	0.34	B.D.	20.05	B.D.	B.D.	98.7	40.8	47.2
Cr 25	45.35	21.28	10.01	B.D.	B.D.	21.56	0.69	B.D.	98.9	51.7	58.8
Cr 26	63.69	7.27	11.10	B.D.	B.D.	16.72	B.D.	B.D.	98.8	44.5	85.5
Cr 27	45.75	22.84	12.65	B.D.	B.D.	17.18	B.D.	B.D.	98.4	41.3	57.3
Cr 28	48.01	12.68	6.24	0.24	B.D.	28.93	2.09	B.D.	98.2	66.6	71.7
Cr 29	45.48	24.25	14.19	B.D.	B.D.	15.21	B.D.	B.D.	99.1	35.5	55.7
Cr 30	31.21	34.06	13.01	B.D.	B.D.	21.42	B.D.	B.D.	99.7	43.0	38.1
Cr 31	44.40	24.96	12.98	B.D.	B.D.	17.26	B.D.	B.D.	99.6	41.0	54.4
Cr 32	38.09	29.70	13.70	B.D.	B.D.	17.47	B.D.	B.D.	99.0	38.8	46.2
Cr 33	39.97	29.58	15.58	B.D.	B.D.	14.06	B.D.	B.D.	99.2	31.3	47.5
Cr 34	46.89	22.29	14.80	B.D.	B.D.	14.51	0.73	1.00	100.2	30.2	58.5
Cr 35	46.90	22.17	12.65	B.D.	B.D.	17.35	B.D.	B.D.	99.1	41.4	58.8
MAS 5.70–5.80 m (13.8 kg)											
Cr 1	43.06	17.03	9.27	0.48	B.D.	28.47	B.D.	B.D.	98.3	55.5	62.9
Cr 2	47.31	20.13	11.35	B.D.	B.D.	19.40	B.D.	B.D.	98.2	46.2	61.2
Cr 3	58.15	7.78	5.14	B.D.	B.D.	26.25	0.68	B.D.	98.0	72.5	83.4
Cr 4	58.48	12.51	12.64	B.D.	B.D.	16.40	B.D.	B.D.	100.0	39.6	75.8
Cr 5	29.81	38.46	15.22	B.D.	B.D.	15.34	B.D.	B.D.	98.8	34.8	34.2
Cr 6	37.28	4.78	0.98	0.16	B.D.	37.10	1.74	0.97	83.0	93.1	83.9
Cr 7	47.81	21.31	12.15	B.D.	B.D.	18.97	B.D.	B.D.	100.2	43.9	60.1
Cr 8	21.56	4.94	1.48	4.69	0.59	64.48	0.60	B.D.	98.3	92.9	74.5
Cr 9	39.64	29.06	12.75	B.D.	B.D.	17.13	B.D.	B.D.	98.6	42.4	47.8
Cr 10	51.61	13.13	9.91	B.D.	B.D.	23.87	B.D.	B.D.	98.5	51.3	72.5
Cr 11	29.50	37.01	17.33	0.34	B.D.	14.05	B.D.	B.D.	98.2	26.6	34.8
Cr 12	42.63	19.00	10.28	0.81	0.47	27.04	B.D.	B.D.	100.2	52.9	60.1
Cr 13	61.42	8.86	5.93	B.D.	B.D.	21.76	0.65	B.D.	98.6	67.3	82.3
Cr 14	57.49	13.67	10.65	B.D.	B.D.	18.84	B.D.	B.D.	100.7	48.9	73.8
Cr 15	46.05	19.31	10.31	0.45	B.D.	23.61	B.D.	B.D.	99.7	51.8	61.5
Cr 16	43.06	17.03	9.27	0.48	B.D.	28.47	B.D.	B.D.	98.3	55.5	62.9
MAS 6.15–6.25 m (14.0 kg)											
Cr 1	48.52	18.39	10.96	B.D.	B.D.	20.66	B.D.	B.D.	98.5	47.7	63.9
Cr 2	45.92	23.97	13.25	B.D.	B.D.	16.12	B.D.	B.D.	99.3	39.4	56.2
Cr 3	35.75	30.22	12.04	B.D.	B.D.	20.90	0.16	0.57	99.6	45.2	44.2
Cr 4	44.86	19.08	9.22	B.D.	B.D.	27.59	B.D.	B.D.	100.8	56.5	61.2
Cr 5	44.26	26.13	13.76	B.D.	B.D.	16.21	B.D.	B.D.	100.4	38.4	53.2
Cr 6	40.17	18.62	6.52	1.63	0.56	29.58	2.01	0.96	100.1	67.4	59.1
Cr 7	43.55	24.20	12.69	B.D.	B.D.	19.85	B.D.	B.D.	100.3	42.4	54.7
Cr 8	44.10	22.70	10.98	B.D.	B.D.	23.01	B.D.	B.D.	100.8	49.5	56.6
Cr 9	53.86	13.03	8.07	B.D.	B.D.	23.59	B.D.	B.D.	98.6	59.8	73.5
Cr 10	37.33	21.88	6.16	0.28	0.35	32.86	0.72	0.26	99.8	70.1	53.4
Cr 11	48.92	21.71	11.16	B.D.	B.D.	18.04	B.D.	B.D.	99.8	47.6	60.2
Cr 12	58.87	5.16	2.42	1.72	0.71	28.85	1.29	0.36	99.4	87.0	88.4
Cr 13	35.65	28.40	12.64	0.53	B.D.	21.42	B.D.	B.D.	98.6	43.5	45.7
Cr 14	30.11	34.25	14.17	0.48	B.D.	19.65	B.D.	B.D.	98.7	38.6	37.1
Cr 15	43.92	25.46	11.97	B.D.	B.D.	19.14	B.D.	B.D.	100.5	45.7	53.6
Cr 16	44.05	25.83	12.38	B.D.	B.D.	18.16	B.D.	B.D.	100.4	44.0	53.3
Cr 17	47.14	19.23	10.74	B.D.	B.D.	22.87	B.D.	B.D.	100.0	49.5	62.2
Cr 18	60.06	6.79	4.91	B.D.	B.D.	26.89	B.D.	B.D.	98.7	74.3	85.6
Cr 19	34.42	33.13	14.70	B.D.	B.D.	15.84	B.D.	B.D.	98.1	35.2	41.1
Cr 20	32.58	34.21	14.23	B.D.	B.D.	17.17	B.D.	B.D.	98.2	37.4	39.0
Cr 21	33.48	36.18	15.68	B.D.	B.D.	15.11	B.D.	B.D.	100.5	33.3	38.3
Cr 22	30.39	37.64	15.28	0.45	B.D.	16.60	B.D.	B.D.	100.4	35.8	35.1
Cr 23	48.15	19.41	11.82	B.D.	B.D.	20.48	0.75	B.D.	100.6	44.0	62.5
MAS 6.70–6.90 m (14.1 kg)											
Cr 1	57.45	15.84	9.19	B.D.	B.D.	16.66	B.D.	B.D.	99.1	50.4	70.9
Cr 2	32.07	37.46	14.83	B.D.	B.D.	15.21	B.D.	B.D.	99.6	36.5	36.5
Cr 3	47.68	19.74	10.77	B.D.	B.D.	21.65	B.D.	B.D.	99.8	49.4	61.8
Cr 4	50.70	19.19	12.01	B.D.	B.D.	18.15	B.D.	B.D.	100.1	43.9	63.9
Cr 5	43.88	25.59	13.54	B.D.	B.D.	16.58	B.D.	B.D.	99.6	38.8	53.5
Cr 6	34.97	33.77	13.76	B.D.	B.D.	17.67	B.D.	B.D.	100.2	40.1	41.0
Cr 7	55.60	22.50	6.98	B.D.	B.D.	15.18	B.D.	B.D.	100.3	55.0	62.4
Cr 8	46.76	22.40	11.94	B.D.	B.D.	18.65	B.D.	B.D.	99.8	44.8	58.3

(*continued*)

TABLE 1. ELEMENT CONCENTRATIONS (WT%) IN CR-RICH SPINELS EXTRACTED FROM LATE EOCENE SEDIMENTS IN THE MASSIGNANO QUARRY, CENTRAL ITALY* (*continued*)

Grains	Cr_2O_3	Al_2O_3	MgO	TiO_2	V_2O_3	FeO	MnO	ZnO	Total[†]	Fe#[§]	Cr#[#]
MAS 6.70–6.90 m (14.1 kg) (*continued*)											
Cr 9	33.15	31.18	14.97	0.66	B.D.	19.63	B.D.	B.D.	99.6	35.4	41.6
Cr 10	55.85	14.78	5.52	B.D.	0.28	23.25	B.D.	B.D.	99.7	70.3	71.7
Cr 11	41.63	24.05	11.57	B.D.	B.D.	21.83	B.D.	B.D.	99.1	46.5	53.7
Cr 12	43.92	26.94	13.07	B.D.	B.D.	15.87	B.D.	B.D.	99.8	40.5	52.2
Cr 13	39.53	30.32	13.72	B.D.	B.D.	16.23	B.D.	B.D.	99.8	39.3	46.6
Cr 14	54.37	20.92	9.89	B.D.	B.D.	14.19	B.D.	B.D.	99.4	44.6	63.5
Cr 15	51.13	26.83	8.31	B.D.	0.30	12.79	B.D.	B.D.	99.4	46.3	56.1
Cr 16	48.98	29.21	7.61	B.D.	B.D.	14.53	B.D.	B.D.	100.3	51.7	52.9
Cr 17	45.03	27.52	8.66	B.D.	B.D.	17.03	B.D.	B.D.	98.3	52.5	52.3
Cr 18	46.63	23.82	12.58	B.D.	B.D.	17.67	B.D.	B.D.	100.7	42.9	56.8
Cr 19	41.28	15.41	8.41	0.97	0.39	33.05	B.D.	B.D.	99.5	60.3	64.2
Cr 20	45.81	20.37	12.00	0.62	B.D.	21.89	B.D.	B.D.	100.7	45.4	60.1
Cr 21	55.84	5.77	2.49	B.D.	B.D.	33.01	1.33	0.33	98.8	86.0	86.6
Cr 22	43.20	26.64	13.99	B.D.	B.D.	16.65	B.D.	B.D.	100.5	37.7	52.1
Cr 23	36.54	30.37	12.93	B.D.	B.D.	18.95	B.D.	B.D.	98.8	42.1	44.7
Cr 24	56.81	12.20	8.28	B.D.	B.D.	23.52	B.D.	B.D.	100.8	59.5	75.7
MAS 8.20–8.40 m (14.2 kg)											
Cr 1	61.25	10.64	9.48	B.D.	0.49	16.31	1.77	B.D.	99.9	49.1	79.4
Cr 2	42.08	24.80	12.66	0.74	B.D.	19.93	B.D.	B.D.	100.2	43.5	53.2
Cr 3	30.89	8.69	1.01	2.02	B.D.	56.68	1.09	B.D.	100.4	94.9	70.4
Cr 4	47.22	22.61	12.41	B.D.	B.D.	17.59	B.D.	B.D.	99.8	42.9	58.3
Cr 5	50.31	18.58	10.18	B.D.	B.D.	20.44	B.D.	B.D.	99.5	51.6	64.5
Cr 6	49.93	18.04	10.95	0.27	B.D.	20.15	B.D.	B.D.	99.3	48.3	65.0
Cr 7	46.57	24.38	10.88	B.D.	B.D.	17.59	B.D.	B.D.	99.4	47.6	56.2
Cr 8	46.22	24.67	12.72	B.D.	B.D.	16.03	B.D.	B.D.	99.6	41.4	55.7
Cr 9	46.97	23.07	10.42	B.D.	B.D.	19.59	B.D.	B.D.	100.1	51.3	57.7
Cr 10	45.98	25.80	13.07	B.D.	B.D.	15.96	B.D.	B.D.	100.8	40.7	54.4
Cr 11	45.62	26.18	12.06	B.D.	B.D.	15.84	B.D.	B.D.	99.7	42.4	53.9
Cr 12	51.68	9.42	11.06	0.86	B.D.	27.14	B.D.	B.D.	100.2	47.2	78.6
Cr 13	51.59	12.40	6.95	1.22	0.80	25.88	0.53	B.D.	99.4	66.2	73.6
Cr 14	34.81	28.91	14.20	1.18	B.D.	20.70	B.D.	B.D.	99.8	38.7	44.7
Cr 15	60.68	7.48	12.53	B.D.	B.D.	19.03	B.D.	B.D.	99.7	38.7	84.5
Cr 16	37.42	25.22	9.25	0.37	B.D.	27.28	B.D.	B.D.	99.6	57.4	49.9
Cr 17	41.78	18.23	9.24	0.90	B.D.	28.53	B.D.	B.D.	98.7	56.4	60.6
Cr 18	44.24	23.16	12.51	B.D.	0.19	18.00	0.57	B.D.	98.7	41.4	56.2
Cr 19	48.69	20.61	11.67	B.D.	B.D.	18.05	0.74	B.D.	99.8	44.5	61.3
Cr 20	44.38	17.02	8.90	0.42	B.D.	28.87	B.D.	B.D.	99.6	57.6	63.6
Cr 21	37.06	23.31	9.47	B.D.	0.36	29.22	B.D.	B.D.	99.4	56.0	51.6
Cr 22	50.12	15.23	10.40	0.73	B.D.	22.98	B.D.	B.D.	99.5	50.8	68.8
Cr 23	41.88	26.50	12.65	B.D.	B.D.	18.62	B.D.	B.D.	99.7	42.7	51.5
Cr 24	44.99	22.62	11.52	B.D.	B.D.	19.50	B.D.	B.D.	98.6	46.2	57.2
Cr 25	38.94	22.33	11.67	0.72	0.40	25.03	0.20	B.D.	99.3	46.9	53.9
Cr 26	43.31	26.20	11.99	B.D.	B.D.	18.63	B.D.	B.D.	100.1	45.6	52.6
Cr 27	51.82	11.80	9.06	1.01	0.47	25.13	B.D.	B.D.	99.3	56.7	74.7
Cr 28	44.20	27.67	11.70	B.D.	B.D.	14.43	B.D.	B.D.	98.0	40.9	51.7
Cr 29	40.21	24.94	11.80	B.D.	B.D.	22.37	B.D.	B.D.	99.3	45.9	51.9
Cr 30	42.67	17.24	7.24	0.98	B.D.	31.30	B.D.	B.D.	99.4	65.6	62.4
Cr 31	29.75	12.24	3.88	1.83	0.54	50.72	B.D.	B.D.	99.0	81.3	62.0
Cr 32	27.76	40.72	16.47	B.D.	B.D.	14.32	B.D.	B.D.	99.3	30.8	31.4
Cr 33	56.62	12.29	10.89	B.D.	B.D.	19.51	0.53	B.D.	99.8	46.4	75.5
Cr 34	55.57	11.86	6.99	B.D.	B.D.	24.70	0.74	B.D.	99.9	64.4	75.9
Cr 35	45.29	23.77	12.40	B.D.	B.D.	18.70	B.D.	B.D.	100.2	43.4	56.1
Cr 36	35.76	25.02	13.00	1.67	B.D.	25.15	B.D.	B.D.	100.6	43.7	48.9
MAS 9.00–9.15 m (14.1 kg)											
Cr 1	60.14	4.00	2.28	0.40	0.41	32.20	B.D.	B.D.	99.4	88.0	91.0
Cr 2	51.90	18.19	11.42	B.D.	B.D.	18.25	B.D.	B.D.	99.8	46.1	65.7
Cr 3	46.81	20.81	11.12	B.D.	B.D.	20.73	B.D.	B.D.	99.5	47.9	60.1
Cr 4	30.13	38.11	15.19	B.D.	B.D.	15.54	B.D.	B.D.	99.0	34.9	34.6
Cr 5	49.08	20.97	13.01	B.D.	B.D.	15.97	B.D.	B.D.	99.0	39.6	61.1
Cr 6	39.93	16.76	6.06	1.88	0.58	34.45	B.D.	B.D.	99.7	71.8	61.5
Cr 7	57.07	10.84	7.02	B.D.	B.D.	25.48	B.D.	B.D.	100.4	65.0	77.9
Cr 8	51.90	16.74	11.54	B.D.	B.D.	19.53	B.D.	B.D.	99.7	45.3	67.5
Cr 9	48.58	19.97	11.18	B.D.	B.D.	19.67	B.D.	B.D.	99.4	47.4	62.0
Cr 10	52.76	16.66	12.05	B.D.	B.D.	18.49	B.D.	B.D.	100.0	43.2	68.0

(*continued*)

TABLE 1. ELEMENT CONCENTRATIONS (WT%) IN CR-RICH SPINELS EXTRACTED FROM LATE EOCENE SEDIMENTS IN THE MASSIGNANO QUARRY, CENTRAL ITALY* (*continued*)

Grains	Cr_2O_3	Al_2O_3	MgO	TiO_2	V_2O_3	FeO	MnO	ZnO	Total†	Fe#§	Cr##
MAS 9.00–9.15 m (14.1 kg) (*continued*)											
Cr 11	50.72	18.83	12.15	B.D.	0.42	17.77	B.D.	B.D.	99.9	43.6	64.4
Cr 12	47.60	22.94	12.66	B.D.	B.D.	17.29	B.D.	B.D.	100.5	42.2	58.2
Cr 13	53.83	14.30	9.36	B.D.	B.D.	21.65	B.D.	B.D.	99.1	54.2	71.6
Cr 14	36.82	32.19	14.01	B.D.	B.D.	16.09	B.D.	B.D.	99.1	38.2	43.4
Cr 15	51.57	16.28	11.63	B.D.	0.35	18.71	B.D.	B.D.	98.5	44.5	68.0
Cr 16	40.58	29.88	14.12	B.D.	B.D.	15.75	B.D.	B.D.	100.3	37.8	47.7
Cr 17	49.29	22.06	11.81	B.D.	B.D.	17.09	B.D.	B.D.	100.3	44.8	60.0
Cr 18	47.02	21.64	11.92	B.D.	0.41	18.12	B.D.	B.D.	99.1	44.8	59.3
Cr 19	59.25	11.68	11.03	B.D.	B.D.	18.14	B.D.	B.D.	100.1	46.5	77.3
Cr 20	44.56	23.24	12.36	B.D.	B.D.	18.98	B.D.	B.D.	99.1	43.0	56.3
Cr 21	57.49	11.33	7.74	B.D.	B.D.	22.80	B.D.	B.D.	99.4	61.3	77.3
Cr 22	34.01	33.12	13.60	B.D.	B.D.	19.55	B.D.	B.D.	100.3	40.7	40.8
Cr 23	38.04	15.28	7.84	1.45	0.58	35.56	B.D.	B.D.	98.8	63.2	62.5
Cr 24	37.47	27.29	14.52	0.71	B.D.	18.97	B.D.	B.D.	99.0	36.0	47.9
Cr 25	33.47	33.61	13.90	B.D.	B.D.	17.63	B.D.	B.D.	98.6	38.8	40.0
Cr 26	39.37	10.90	5.21	B.D.	B.D.	43.40	B.D.	0.11	99.0	73.5	70.8
Cr 27	50.29	20.31	11.23	B.D.	B.D.	18.07	B.D.	B.D.	99.9	47.5	62.4
Cr 28	46.60	22.14	11.83	B.D.	B.D.	17.62	B.D.	B.D.	98.2	44.5	58.5
Cr 29	58.00	12.68	8.80	B.D.	B.D.	19.54	B.D.	B.D.	99.0	55.5	75.4
Cr 30	42.91	21.66	6.45	B.D.	B.D.	27.02	0.58	0.90	99.5	67.7	57.1
Cr 31	59.29	7.67	3.30	B.D.	B.D.	27.92	0.83	0.39	99.4	82.1	83.8
Cr 32	46.03	20.66	10.19	0.36	B.D.	22.93	B.D.	B.D.	100.2	52.6	59.9
Cr 33	42.33	25.19	9.91	B.D.	B.D.	21.19	B.D.	B.D.	98.6	53.7	53.0
Cr 34	49.76	20.40	10.27	B.D.	0.34	18.47	B.D.	B.D.	99.2	50.2	62.1
Cr 35	47.05	21.18	11.37	B.D.	0.37	19.46	B.D.	B.D.	99.4	47.2	59.8
Cr 36	48.88	22.47	12.48	B.D.	B.D.	16.85	B.D.	B.D.	100.7	43.0	59.3
Cr 37	37.29	3.52	2.76	B.D.	0.42	55.38	B.D.	B.D.	99.4	85.5	87.7
Cr 38	55.80	2.13	2.92	B.D.	0.57	35.87	0.80	0.42	98.5	83.7	94.6
Cr 39	48.21	20.06	11.12	B.D.	B.D.	19.32	B.D.	B.D.	98.7	47.4	61.7
Cr 40	34.44	34.65	15.22	B.D.	B.D.	15.42	B.D.	B.D.	99.7	34.4	40.0
Cr 41	48.02	21.48	11.81	B.D.	B.D.	17.64	B.D.	B.D.	98.9	44.8	60.0
Cr 42	40.70	28.68	13.15	B.D.	B.D.	17.77	B.D.	B.D.	100.3	41.5	48.8
Cr 43	32.76	33.44	15.14	B.D.	B.D.	18.82	B.D.	B.D.	100.2	34.7	39.6
Cr 44	41.45	27.76	14.16	B.D.	B.D.	16.62	B.D.	B.D.	100.0	37.0	50.0
Cr 45	48.98	19.70	11.14	B.D.	B.D.	18.97	B.D.	B.D.	98.8	47.2	62.5
Cr 46	54.62	14.30	9.80	B.D.	B.D.	21.37	B.D.	B.D.	100.1	52.6	71.9
Cr 47	42.89	25.51	12.10	0.37	B.D.	18.98	B.D.	B.D.	99.8	45.4	53.0
Cr 48	30.70	31.35	8.61	B.D.	0.58	28.97	B.D.	B.D.	100.2	61.4	39.6
Cr 49	43.85	25.43	13.59	B.D.	B.D.	17.27	B.D.	B.D.	100.1	38.8	53.6
Cr 50	47.20	22.57	13.71	B.D.	B.D.	16.97	B.D.	B.D.	100.4	37.8	58.4
Cr 51	40.49	29.26	13.45	B.D.	B.D.	16.63	B.D.	B.D.	99.8	40.1	48.1
Cr 52	53.46	14.62	8.71	B.D.	B.D.	22.07	B.D.	B.D.	98.9	57.2	71.0
Cr 53	44.68	21.71	10.89	B.D.	B.D.	21.70	B.D.	B.D.	99.0	48.9	58.0
Cr 54	64.12	9.62	5.99	B.D.	B.D.	19.12	B.D.	0.47	99.3	64.1	81.7
Cr 55	21.65	48.00	15.82	B.D.	B.D.	14.59	B.D.	B.D.	100.1	34.1	23.2
Cr 56	31.77	36.66	15.52	B.D.	B.D.	15.52	B.D.	B.D.	99.5	33.6	36.8
Cr 57	47.13	18.06	11.76	1.43	B.D.	22.21	B.D.	B.D.	100.6	46.9	63.6
Cr 58	37.19	31.75	15.37	B.D.	B.D.	15.20	B.D.	B.D.	99.5	32.9	44.0
Cr 59	45.47	23.41	12.00	B.D.	B.D.	18.94	B.D.	B.D.	99.8	44.8	56.6
Cr 60	47.85	21.30	10.81	B.D.	B.D.	19.94	B.D.	B.D.	99.9	49.5	60.1
Cr 61	51.34	18.49	9.72	B.D.	0.32	18.55	B.D.	B.D.	98.4	51.7	65.1
Cr 62	45.89	23.22	12.37	B.D.	B.D.	18.36	B.D.	B.D.	99.8	43.2	57.0
Cr 63	46.96	13.69	1.31	0.41	B.D.	35.48	B.D.	0.52	98.4	93.1	69.7
Cr 64	44.70	25.42	10.62	B.D.	B.D.	18.60	B.D.	B.D.	99.4	49.6	54.1
Cr 65	42.61	25.73	12.93	B.D.	B.D.	17.37	B.D.	B.D.	98.6	40.9	52.6
Cr 66	46.62	22.01	11.26	B.D.	B.D.	19.10	B.D.	B.D.	99.0	47.3	58.7
Cr 67	42.23	27.86	13.54	B.D.	B.D.	16.52	B.D.	B.D.	100.2	39.6	50.4
Cr 68	42.22	19.93	7.44	B.D.	B.D.	28.26	B.D.	0.51	98.4	63.4	58.7
Cr 69	43.49	25.13	12.69	B.D.	B.D.	17.44	B.D.	B.D.	98.8	41.8	53.7
Cr 70	46.54	23.04	12.65	B.D.	B.D.	17.05	B.D.	B.D.	99.3	41.7	57.5
Cr 71	48.08	21.61	12.76	B.D.	B.D.	16.44	B.D.	B.D.	98.9	40.7	59.9
MAS 10.00–10.15 m (14.0 kg)											
Cr 1	47.82	18.50	10.20	B.D.	0.32	21.73	B.D.	B.D.	98.6	51.3	63.4
Cr 2	56.02	12.94	9.35	B.D.	0.32	20.19	B.D.	B.D.	98.8	54.1	74.4
Cr 3	55.19	11.90	8.13	B.D.	B.D.	25.22	B.D.	B.D.	100.4	60.0	75.7

(*continued*)

TABLE 1. ELEMENT CONCENTRATIONS (WT%) IN CR-RICH SPINELS EXTRACTED FROM LATE EOCENE SEDIMENTS IN THE MASSIGNANO QUARRY, CENTRAL ITALY* (*continued*)

Grains	Cr_2O_3	Al_2O_3	MgO	TiO_2	V_2O_3	FeO	MnO	ZnO	Total†	Fe#§	Cr##
MAS 10.00–10.15 m (14.0 kg) (*continued*)											
Cr 4	45.63	25.68	14.13	B.D.	B.D.	14.47	B.D.	B.D.	99.9	36.5	54.4
Cr 5	41.88	27.12	12.49	B.D.	0.34	18.42	B.D.	B.D.	100.2	44.1	50.9
Cr 6	45.22	24.20	13.05	B.D.	B.D.	17.10	B.D.	B.D.	99.6	40.5	55.6
Cr 7	40.94	27.20	12.03	B.D.	B.D.	19.46	B.D.	0.25	99.9	45.2	50.2
Cr 8	44.31	25.40	12.89	B.D.	B.D.	18.00	B.D.	B.D.	100.6	41.9	53.9
Cr 9	32.05	34.21	13.81	B.D.	B.D.	18.72	B.D.	B.D.	98.8	38.4	38.6
Cr 10	48.06	22.47	10.84	B.D.	B.D.	17.28	B.D.	B.D.	98.7	47.2	58.9
Cr 11	38.18	30.77	14.07	B.D.	B.D.	16.67	B.D.	B.D.	99.7	37.9	45.4
Cr 12	46.35	15.51	9.93	1.37	B.D.	25.97	B.D.	B.D.	99.1	53.5	66.7
Cr 13	24.99	43.54	17.31	B.D.	B.D.	13.42	B.D.	B.D.	99.3	28.4	27.8
Cr 14	57.20	11.83	8.98	B.D.	B.D.	21.19	B.D.	B.D.	99.2	55.5	76.4
Cr 15	59.69	10.74	11.45	B.D.	B.D.	17.95	B.D.	B.D.	99.8	44.3	78.8
Cr 16	41.85	26.81	12.62	0.25	0.22	18.05	0.48	B.D.	100.3	43.0	51.1
Cr 17	39.07	28.20	13.14	B.D.	B.D.	18.13	B.D.	B.D.	98.5	40.6	48.2
Cr 18	40.88	28.55	14.79	B.D.	B.D.	14.86	B.D.	B.D.	99.1	34.2	49.0
Cr 19	47.14	21.39	11.74	B.D.	B.D.	18.68	B.D.	B.D.	99.0	45.1	59.6
Cr 20	52.55	14.60	9.22	B.D.	0.38	21.60	0.85	B.D.	99.2	54.2	70.7
Cr 21	49.51	19.87	11.80	B.D.	B.D.	18.02	B.D.	B.D.	99.2	44.6	62.6
Cr 22	48.39	21.46	12.68	B.D.	B.D.	16.44	B.D.	B.D.	99.0	41.1	60.2
Cr 23	45.60	23.39	12.76	B.D.	B.D.	16.85	B.D.	B.D.	98.6	41.0	56.7
Cr 24	40.40	26.48	11.30	B.D.	0.26	19.57	B.D.	B.D.	98.0	47.9	50.6
Cr 25	50.47	16.23	9.81	B.D.	B.D.	22.83	B.D.	B.D.	99.3	52.7	67.6
Cr 26	54.62	15.93	10.75	B.D.	B.D.	18.58	B.D.	B.D.	99.9	48.6	69.7
Cr 27	47.88	19.39	11.05	B.D.	B.D.	22.10	B.D.	B.D.	100.4	48.4	62.3
Cr 28	42.77	22.90	9.60	0.25	B.D.	24.09	0.45	B.D.	100.1	54.9	55.6
MAS 10.25–10.35 m (11.6 kg)											
Cr 1	37.32	36.42	13.10	B.D.	B.D.	11.86	B.D.	B.D.	98.7	33.7	40.7
Cr 2	39.84	28.79	13.80	B.D.	B.D.	16.22	B.D.	B.D.	98.7	38.0	48.1
Cr 3	43.38	23.30	11.28	B.D.	B.D.	21.68	B.D.	B.D.	99.6	47.8	55.5
Cr 4	40.09	28.38	12.92	B.D.	B.D.	18.00	B.D.	B.D.	99.4	41.9	48.6
Cr 5	45.93	23.77	15.13	B.D.	0.30	15.20	B.D.	B.D.	100.3	32.5	56.4
Cr 6	40.69	29.64	12.02	B.D.	B.D.	17.88	B.D.	B.D.	100.2	45.5	47.9
Cr 7	39.60	29.86	13.51	B.D.	B.D.	16.95	B.D.	B.D.	99.9	40.1	47.1
Cr 8	47.59	22.19	13.93	B.D.	B.D.	16.36	B.D.	B.D.	100.1	36.6	59.0
Cr 9	40.24	22.24	12.88	0.69	B.D.	22.73	B.D.	B.D.	98.8	41.4	54.8
Cr 10	42.66	26.68	13.89	B.D.	B.D.	16.69	0.57	B.D.	100.5	37.2	51.7
Cr 11	46.75	23.71	11.74	B.D.	0.25	16.41	B.D.	B.D.	98.9	44.0	56.9
Cr 12	41.52	32.36	12.21	B.D.	B.D.	13.98	B.D.	B.D.	100.1	39.1	46.2
Cr 13	41.93	21.59	12.40	0.91	B.D.	22.43	0.57	B.D.	99.8	43.2	56.6
Cr 14	37.97	31.95	14.94	B.D.	B.D.	14.52	B.D.	B.D.	99.4	34.6	44.3
Cr 15	59.32	7.04	3.87	B.D.	0.33	29.18	B.D.	0.60	100.4	79.7	85.0
Cr 16	46.63	16.81	10.82	0.57	B.D.	23.82	B.D.	B.D.	98.7	48.8	65.0
Cr 17	48.57	20.56	11.47	B.D.	B.D.	18.92	B.D.	B.D.	99.5	46.3	61.3
Cr 18	58.93	8.93	6.51	B.D.	0.50	24.16	B.D.	B.D.	99.0	67.0	81.6
Cr 19	42.07	26.71	13.17	B.D.	B.D.	16.11	B.D.	B.D.	98.1	39.8	51.4
Cr 20	45.61	24.16	11.84	B.D.	B.D.	16.87	B.D.	B.D.	98.5	44.4	55.9
Cr 21	50.29	20.53	10.01	B.D.	0.36	19.33	0.41	B.D.	100.9	52.0	62.2
Cr 22	37.79	15.82	6.13	1.33	B.D.	39.36	B.D.	B.D.	100.4	71.0	61.6
Cr 23	50.47	20.32	10.84	B.D.	0.32	18.41	B.D.	B.D.	100.4	48.8	62.5
Cr 24	33.29	36.26	15.62	B.D.	B.D.	14.53	B.D.	B.D.	99.7	33.2	38.1
Cr 25	53.60	18.79	8.12	B.D.	B.D.	18.37	B.D.	B.D.	98.9	55.9	65.7
Cr 26	56.74	6.31	3.44	B.D.	0.42	32.00	0.52	0.50	99.9	81.6	85.8
Cr 27	48.69	15.86	9.99	1.43	0.39	23.15	B.D.	B.D.	99.5	53.8	67.3
Cr 28	36.75	7.16	5.49	5.34	0.77	44.44	B.D.	B.D.	99.9	75.6	77.5
Cr 29	40.57	19.10	9.09	0.36	B.D.	29.99	B.D.	B.D.	99.1	56.9	58.8
Cr 30	47.23	20.93	11.02	B.D.	0.35	20.58	B.D.	B.D.	100.1	48.9	60.2
Cr 31	51.41	17.84	10.96	B.D.	0.36	19.14	B.D.	B.D.	99.7	48.3	65.9
Cr 32	48.81	21.74	10.72	B.D.	B.D.	17.73	B.D.	B.D.	99.0	48.1	60.1
Cr 33	36.64	32.86	14.66	B.D.	B.D.	15.59	B.D.	B.D.	99.8	36.1	42.8
Cr 34	47.51	23.18	12.39	B.D.	B.D.	17.20	B.D.	B.D.	100.3	43.3	57.9

*All analyses were on polished grains. Five grains were lost during polishing.

†The total values represent a mean of a minimum of two analyses.

§Fe#: mole% Fe/(Fe + Mg).

#Cr#: mole% Cr/(Cr + Al).

**Relative to the base of the section; see main text.

††B.D.—below detection limit.

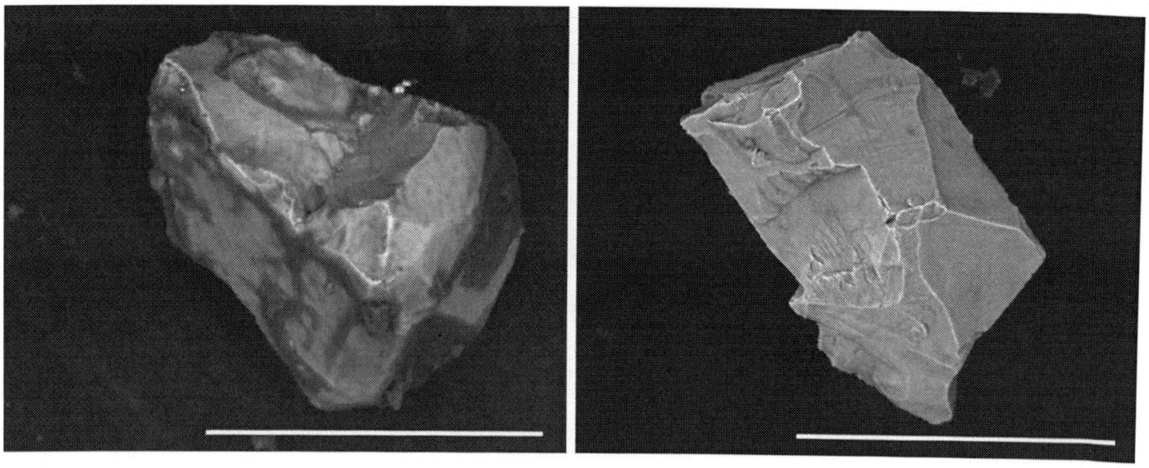

Figure 2. Left: The only EC grain found in this study. It was found in the sample from 6.15 m above the base of the Massignano section. Note the slightly rounded appearance, indicating possibly a reworked origin. Right: For comparison, a mid-Ordovician extraterrestrial chromite grain from the upper part of the Puxi River section in China (Schmitz et al., 2008). Many or most of the extraterrestrial chromite grains recovered from limestone that formed directly after the mid-Ordovician asteroid breakup event are characterized by their sharp edges and often excellent preservation. Scale bar in both images is 100 µm.

Similar trends have been confirmed at several sites in Sweden and now also in the Puxi River section in the Hubei Province in China (Schmitz et al., 2003, 2008; Schmitz and Häggström, 2006). This is strong support for a dramatic increase in the flux of extraterrestrial matter in the submeter size fraction at ca. 470 Ma.

Based on the experience from the mid-Ordovician limestone, it should be possible to test, using quantitative searches for EC grains in late Eocene sediments, whether an asteroid shower, with accompanying increase in the flux of EC-carrying meteorites and micrometeorites, occurred also at that time. Our finding in this study of only one EC grain in 167 kg of limestone from the ^{3}He-enriched part of the Massignano section is a very strong indication that the EC flux was not significantly enhanced at this time. In a previous study, we also searched for extraterrestrial chromite in 210 kg of condensed limestone from the latest Maastrichtian and Paleocene section in the Bottaccione Gorge at Gubbio (Cronholm and Schmitz, 2007). The limestone studied there represents eight levels across 24 m of section. Only six EC grains were found, which is in accord with the predisruption values for similarly condensed limestone in the mid-Ordovician. Using well-established sedimentation rates for the Gubbio section (e.g., Mukhopadhyay et al., 2001), the results indicate an EC flux of ~0.26 grain m^{-2} $k.y.^{-1}$. The sedimentation rate in the late Eocene part of the Massignano section is about three times higher than in the latest Cretaceous–Paleocene at Gubbio (see, e.g., Montanari et al., 1993; Cronholm and Schmitz, 2007). An estimate of the late Eocene EC flux based on the finding of only 1 grain in 167 kg of rock gives 0.16 grain m^{-2} $k.y.^{-1}$, similar to the value for the Gubbio section. This is similar also to the EC flux of 0.09 EC grain m^{-2} $k.y.^{-1}$ for the mid-Ordovician before the asteroid breakup event (based on 5 grains in 379 kg of limestone), but it is two orders of magnitude lower than the 13 EC grains m^{-2} $k.y.^{-1}$ after the event (based on 276 EC grains in 148 kg of limestone) (Schmitz and Häggström, 2006; Cronholm and Schmitz, 2007). We emphasize that flux rates based on a single or a few grains have large statistical uncertainties and are significant for comparison mainly at an order-of-magnitude scale.

Our data thus give no support of a chondritic asteroid shower in the late Eocene as envisioned by Tagle and Claeys (2004, 2005). Their claim was based mainly on studies of Popigai crater impact melts, for which platinum group element profiles indicated an L-chondritic impactor. A chondritic impactor for the Popigai crater agrees with Cr-isotopic measurements of the late Eocene layers with (altered) clinopyroxene-bearing spherules at Massignano and Ocean Drilling Program (ODP) Site 709c (Kyte et al., 2004), but the Cr isotopes cannot distinguish between different groups of ordinary chondrites. The clinopyroxene-spherule layers at the two sites are generally considered to represent the same ejecta event, probably associated with the formation of the Popigai crater (Glass et al., 2004). Similar to Farley (this volume), however, we are skeptical about the use of element signatures to determine impactor types. For example, recently Tagle et al. (2008) suggested, based on Ir and Cr measurements of the mid-Ordovician Lockne crater resurge deposits, that the Lockne impactor was a nonmagmatic iron body, similar to a group of meteorites that is uncommon among recent meteorite falls. This is contradictory to the findings of Alwmark and Schmitz (2007), who showed that the same resurge deposits are extremely rich in L-chondritic chromite grains (>75 grains kg^{-1}), representing relict fragments of the impactor. The approach by Tagle et al. (2008) assumes that Ir is immobile in sediments, which is inconsistent with our experience of Ir behavior in mid-Ordovician fossil meteorites (Schmitz et al., 1996). Other studies have shown that Ir is highly mobile in ejecta layers in marine sediments (e.g.,

Wallace et al., 1990; Colodner et al., 1992). Our recent oxygen isotopic measurements of chromite grains from the Lockne crater resurge deposits confirm an L-chondritic origin (Schmitz et al., 2008, personal commun.).

The fact that the single EC grain found in the present study at 6.15 m is associated with a small Ir-enrichment (~250 ppt) could reflect that it is a preserved fragment of an impactor, perhaps a small body that broke up in the water column, but the grain may as well reflect the normal rain of micrometeorites to Earth. Only studies of additional very large samples can show if the bed at 6.15 m is richer in extraterrestrial chromite than other parts of the section. The slightly rounded edges of the grain may indicate that it was transported on the seafloor for some time before being trapped in the sediment (Fig. 2).

The detailed origin of the many OC grains in the Massignano section is beyond the scope of this study, but there is no doubt that almost all or all of them are terrestrial. Most of the grains are similar in composition to OC grains found by Cronholm and Schmitz (2007) in middle Paleocene sediments in the Bottaccione Gorge section at Gubbio, supporting the existence of a long-term local source for this type of chrome spinel grains. However, OC grains are much more common in the Massignano section, with on average 2.24 grains kg^{-1}, compared to Gubbio limestone, with 0.07 OC grain kg^{-1}. The difference may reflect a more pelagic situation in the Paleocene and development to more hemipelagic conditions in the late Eocene, as indicated also by increasing sedimentation rates.

The hypothesis of Fritz et al. (2007) of a lunar origin of ^{3}He in late Eocene sediments is testable, for example, by searches for relict minerals from lunar micrometeorites and meteorites in the Massignano sediments. The minerals with greatest potential are ilmenites and chrome spinels because they are common in lunar volcanic rocks, very durable in sedimentary environments on Earth, and can be readily retrieved by different acid-leaching processes. Ilmenites are abundant in the Massignano sediments, but they were mostly destroyed in the strong HF acid used in the present study; otherwise we could have searched for Mg-rich ilmenites with a possible lunar origin (Papike et al., 1998). Any lunar ilmenites, however, would be diluted with large amounts of terrestrial ilmenites, and the compositional differences between the two types of ilmenites are not straightforward; hence, this would be a very tedious and difficult project. With regard to chrome spinels, it cannot be entirely ruled out that one or a few of the OC grains in Table 1 actually originated from the Moon. The problem is that, whereas the dominant type of chromite in equilibrated chondrites has a very narrow compositional range, which also is very unique and different from terrestrial chromite, lunar chrome spinels have a wide range of compositions, and complex overlaps exist with terrestrial chrome spinels (Papike et al., 1998). With the present level of understanding, it is questionable whether a lunar origin of a chrome spinel grain can be ascertained only by its element composition.

The L-chondrite disruption event in the mid-Ordovician is the largest documented asteroid breakup event during the past 3 b.y. (Keil et al., 1994), and it is highly plausible that it led to a dramatic increase in the flux of extraterrestrial matter to Earth. There is no indication in argon gas retention ages from recent meteorites for any major asteroid disruption event during the last 100 m.y., but cosmic-ray exposure ages show that secondary collisional events involving smaller bodies have occurred. In fact, cosmic-ray exposure ages show that a major fraction of the L chondrites falling on Earth today originated from such a meteorite dispersal event at ca. 40 Ma (Wieler and Graf, 2001). The close temporal relation of this event to the Popigai and Chesapeake Bay impact events could be consistent with a slight increase in the flux of chondritic matter to Earth, but such an increase was most likely not larger than a factor of a few, judging from the results in the present study.

CONCLUSIONS

Only one ordinary chondritic chromite grain was found in large limestone samples, totaling at 167 kg and collected at representative levels throughout the late Eocene part of the Massignano section. Considering the sedimentation rates at Massignano, and by comparison with similar studies for other periods in Earth history, this is the approximate number of grains to be expected if the flux rate of chondritic matter to Earth was normal or close to normal during the late Eocene. We thus find no evidence of any chondritic meteorite or micrometeorite shower in the late Eocene, and by implication, suggest that neither was there any "asteroid shower" at this time, although a slight increase (a factor of a few) in the asteroid flux cannot be ruled out.

ACKNOWLEDGMENTS

We thank C. Koeberl for his patience and encouragement in the editorial processing of this paper, L. Ferrière and F.T. Kyte for helpful reviews of a previous, quite different version of the paper, M. Nolvi and G. Thorelli for heroic assistance with the acid leaching, and C. Alwmark for valuable comments and suggestions. The study was supported by the Swedish Research Council.

REFERENCES CITED

Alwmark, C., and Schmitz, B., 2007, Extraterrestrial chromite in the resurge deposits of the early Late Ordovician Lockne Crater, central Sweden: Earth and Planetary Science Letters, v. 253, p. 291–303, doi: 10.1016/j.epsl.2006.10.034.

Bodiselitsch, B., Montanari, A., Koeberl, C., and Coccioni, R., 2004, Delayed climate cooling in the late Eocene caused by multiple impacts: High-resolution geochemical studies at Massignano, Italy: Earth and Planetary Science Letters, v. 223, p. 283–302, doi: 10.1016/j.epsl.2004.04.028.

Bridges, J.C., Schmitz, B., Hutchison, R., Greenwood, R.C., Tassinari, M., and Franchi, I.A., 2007, Petrographic classification of Middle Ordovician fossil meteorites from Sweden: Meteoritics & Planetary Science, v. 42, p. 1781–1789.

Clymer, A.K., Bice, D.M., and Montanari, A., 1996, Shocked quartz from the late Eocene: Impact evidence from Massignano, Italy: Geology, v. 24, p. 483–486, doi: 10.1130/0091-7613(1996)024<0483:SQFTLE>2.3.CO;2.

Colodner, D.C., Boyle, E.A., Edmond, J.M., and Thomson, J., 1992, Post-depositional mobility of platinum, iridium and rhenium in marine sediments: Nature, v. 358, p. 402–404, doi: 10.1038/358402a0.

Cronholm, A., and Schmitz, B., 2007, Extraterrestrial chromite in latest Maastrichtian and Paleocene pelagic limestone at Gubbio, Italy: The flux of unmelted ordinary chondrites: Meteoritics & Planetary Science, v. 42, p. 2099–2109.

Farley, K.A., Montanari, A., Shoemaker, E.M., and Shoemaker, C.S., 1998, Geochemical evidence for a comet shower in the late Eocene: Science, v. 280, p. 1250–1253, doi: 10.1126/science.280.5367.1250.

Farley, K.A., 2008, this volume, Late Eocene and late Miocene cosmic dust events: Comet showers, asteroid collisions, or lunar impacts?, *in* Koeberl, C., and Montanari, A., eds., The Late Eocene Earth—Hothouse, Icehouse, and Impacts: Geological Society of America Special Paper 452, doi: 10.1130/2008.2452(03).

Fritz, J., Tagle, R., and Artemieva, N., 2007, Lunar helium-3 in marine sediments: Implications for a late Eocene asteroid shower: Icarus, v. 189, p. 591–594, doi: 10.1016/j.icarus.2007.03.026.

Glass, B.P., Liu, S., and Montanari, A., 2004, Impact ejecta in upper Eocene deposits at Massignano, Italy: Meteoritics & Planetary Science, v. 39, p. 589–597.

Greenwood, R.C., Schmitz, B., Bridges, J.B., Hutchison, R., and Franchi, I.A., 2007, Disruption of the L chondrite parent body: New oxygen isotope evidence from Ordovician relict chromite grains: Earth and Planetary Science Letters, v. 262, p. 204–213, doi: 10.1016/j.epsl.2007.07.048.

Heck, P.R., Schmitz, B., Baur, H., Halliday, A.N., and Wieler, R., 2004, Fast delivery of meteorites to Earth after a major asteroid collision: Nature, v. 430, p. 323–325, doi: 10.1038/nature02736.

Heck, P.R., Schmitz, B., Baur, H., and Wieler, R., 2008, Noble gases in fossil micrometeorites and meteorites from 470 Myr old sediments from southern Sweden and new evidence for the L-chondrite parent body breakup event: Meteoritics & Planetary Science, v. 43, p. 517–528.

Hut, P., Alvarez, W., Elder, W.P., Hansen, T., Kauffman, E.G., Keller, G., Shoemaker, E.M., and Weissman, P.R., 1987, Comet showers as a cause of mass extinctions: Nature, v. 329, p. 118–126, doi: 10.1038/329118a0.

Jarosewich, E., Nelen, J.A., and Norberg, J.A., 1980, Reference samples for electron microprobe analysis: Geostandards Newsletter, v. 4, p. 43–47, doi: 10.1111/j.1751-908X.1980.tb00273.x.

Keil, K., 1962, On the phase composition of meteorites: Journal of Geophysical Research, v. 67, p. 4055–4061, doi: 10.1029/JZ067i010p04055.

Keil, K., Haack, H., and Scott, E.R.D., 1994, Catastrophic fragmentation of asteroids: Evidence from meteorites: Planetary and Space Science, v. 42, p. 1109–1122, doi: 10.1016/0032-0633(94)90011-6.

Korochantseva, E.V., Trieloff, M., Lorenz, C.A., Buykin, A.I., Ivanova, M.A., Schwarz, W.H., Hopp, J., and Jessberger, E.K., 2007, L-chondrite asteroid breakup tied to Ordovician meteorite shower by multiple isochron ^{40}Ar-^{39}Ar dating: Meteoritics & Planetary Science, v. 42, p. 113–130.

Kyte, F.T., Shukolyukov, A., Hildebrand, A.R., Lugmair, G.W., and Hanova, J., 2004, Initial Cr-isotopic and iridium measurements of concentrates from late-Eocene cpx-spherule deposits: Lunar and Planetary Science Conference XXXV, abstract 1824, 2 p.

Montanari, A., and Koeberl, C., 2000, Impact Stratigraphy—The Italian Record: Lecture Notes in Earth Sciences, Volume 93: Heidelberg-Berlin, Springer Verlag, p. 1–364.

Montanari, A., Asaro, F., Michel, H.V., and Kennett, J.P., 1993, Iridium anomalies of late Eocene age at Massignano (Italy), and ODP Site 689B (Maud Rise, Antarctic): Palaios, v. 8, p. 420–437, doi: 10.2307/3515017.

Montanari, A., Campo Bagatin, A., and Farinella, P., 1998, Earth cratering record and impact energy flux in the last 150 Ma: Planetary and Space Science, v. 46, p. 271–281, doi: 10.1016/S0032-0633(97)00130-X.

Mukhopadhyay, S., Farley, K.A., and Montanari, A., 2001, A 35 Myr record of helium in pelagic limestone from Italy: Implications for interplanetary dust accretion from the early Maastrichtian to the middle Eocene: Geochimica et Cosmochimica Acta, v. 65, p. 653–669, doi: 10.1016/S0016-7037(00)00555-X.

Papike, J.J., Ryder, G., and Shearer, C.K., 1998, Lunar samples, *in* Papike, J.J., ed., Planetary Materials, Mineralogical Society of America Reviews in Mineralogy, v. 36, ch. 5, 234 p.

Pierrard, O., Robin, E., Rocchia, R., and Montanari, A., 1998, Extraterrestrial Ni-rich spinel in Upper Eocene sediments from Massignano, Italy: Geology, v. 26, p. 307–310, doi: 10.1130/0091-7613(1998)026<0307:ENRSIU>2.3.CO;2.

Schmitz, B., and Häggström, T., 2006, Extraterrestrial chromite in Middle Ordovician marine limestone at Kinnekulle, southern Sweden—Traces of a major asteroid breakup event: Meteoritics & Planetary Science, v. 41, p. 455–466.

Schmitz, B., Lindström, M., Asaro, F., and Tassinari, M., 1996, Geochemistry of meteorite-rich marine limestone strata and fossil meteorites from the Lower Ordovician at Kinnekulle, Sweden: Earth and Planetary Science Letters, v. 145, p. 31–48, doi: 10.1016/S0012-821X(96)00205-1.

Schmitz, B., Tassinari, M., and Peucker-Ehrenbrink, B., 2001, A rain of ordinary chondritic meteorites in the early Ordovician: Earth and Planetary Science Letters, v. 194, p. 1–15, doi: 10.1016/S0012-821X(01)00559-3.

Schmitz, B., Häggström, T., and Tassinari, M., 2003, Sediment-dispersed extraterrestrial chromite traces a major asteroid disruption event: Science, v. 300, p. 961–964, doi: 10.1126/science.1082182.

Schmitz, B., Harper, D.A.T., Peucker-Ehrenbrink, B., Stouge, S., Alwmark, C., Cronholm, A., Bergström, S.M., Tassinari, M., and Wang, X., 2008, Asteroid breakup linked to the Great Ordovician Biodiversification Event: Nature Geoscience, v. 1, p. 49–53, doi: 10.1038/ngeo.2007.37.

Tagle, R., and Claeys, P., 2004, Comet or asteroid shower in the late Eocene?: Science, v. 305, p. 492, doi: 10.1126/science.1098481.

Tagle, R., and Claeys, P., 2005, An ordinary chondrite impactor for the Popigai crater, Siberia: Geochimica et Cosmochimica Acta, v. 69, p. 2877–2889, doi: 10.1016/j.gca.2004.11.024.

Tagle, R., Schmitt, R.T., and Erzinger, J., 2008, The Lockne impact is not related to the Ordovician L-chondrite shower: Lunar and Planetary Science Conference XXXIX, abstract 1418, 2 p.

Wallace, M.W., Gostin, V.A., and Keays, R.R., 1990, Acraman impact ejecta and host shales: Evidence for low-temperature mobilization of iridium and other platinoids: Geology, v. 18, p. 132–135, doi: 10.1130/0091-7613(1990)018<0132:AIEAHS>2.3.CO;2.

Wieler, R., and Graf, T., 2001, Cosmic ray exposure history of meteorites, *in* Peucker-Ehrenbrink, B., and Schmitz, B., eds., Accretion of Extraterrestrial Matter Throughout Earth's History: New York, Kluwer Academic/Plenum Publishers, p. 221–240.

Zappalà, V., Cellino, A., Gladman, B.J., Manley, S., and Migliorini, F., 1998, Asteroid showers on Earth after family breakup events: Icarus, v. 134, p. 176–179, doi: 10.1006/icar.1998.5946.

Manuscript Accepted by the Society 16 September 2008

Printed in the USA

The Geological Society of America
Special Paper 452
2009

Stable isotopic response to late Eocene extraterrestrial impacts

Aimee E. Pusz*
Kenneth G. Miller
James D. Wright
Department of Earth and Planetary Sciences, Rutgers University, Piscataway, New Jersey 08854, USA

Miriam E. Katz
Department of Earth and Environmental Sciences, Rensselaer Polytechnic Institute, Troy, New York 12180, USA

Benjamin S. Cramer
Department of Geological Sciences, University of Oregon, Eugene, Oregon 97403, USA

Dennis V. Kent
Department of Earth and Planetary Sciences, Rutgers University, Piscataway, New Jersey 08854, USA, and
Lamont-Doherty Earth Observatory, Palisades, New York 10964, USA

ABSTRACT

We evaluated the age of two Upper Eocene impact ejecta layers (North American microtektites linked to the Chesapeake Bay impact structure and clinopyroxene [cpx] spherules from the Popigai crater) and the global effects of the associated impact events. The reported occurrence of cpx spherules from the Popigai impact structure at South Atlantic ODP Site 1090 within the middle of magnetochron C16n.1n yields a magnetochronologic age of 35.4 Ma. We generated high-resolution stable isotope records at Sites 1090, 612 (New Jersey slope), and Caribbean core RC9-58 that show: (1) a 0.5‰ $\delta^{13}C$ decrease in bulk-carbonate at Site 1090 coincident with the Popigai cpx spherule layer, and (2) a 0.4‰–0.5‰ decrease in deep-water benthic foraminiferal $\delta^{13}C$ values across the Popigai impact ejecta layer at Site 612 and core RC9-58. We conclude that the $\delta^{13}C$ excursion associated with Popigai was a global event throughout the marine realm that can be correlated to magnetochron C16n.1n. The amplitude of this excursion (~0.5‰) is within the limits of natural variability, suggesting it was caused by a decrease in carbon export productivity, potentially triggered by the impact event(s). North American microtektites associated with the Chesapeake Bay impact occur stratigraphically above the Popigai cpx spherules at Site 612 and core RC9-58. We found no definite evidence of a $\delta^{13}C$ anomaly associated with the North American microtektite layer, though further studies are warranted. High-resolution bulk-carbonate and benthic foraminiferal $\delta^{18}O$ records show no global temperature change associated with the cpx spherule or North American microtektite layers.

Keywords: impacts, climate, foraminifera, stable isotopes.

*Present address: Department of Geological Sciences, University of South Carolina, Columbia, South Carolina 29208, USA; apusz@geol.sc.edu.

Pusz, A.E., Miller, K.G., Wright, J.D., Katz, M.E., Cramer, B.S., and Kent, D.V., 2009, Stable isotopic response to late Eocene extraterrestrial impacts, *in* Koeberl, C., and Montanari, A., eds., The Late Eocene Earth—Hothouse, Icehouse, and Impacts: Geological Society of America Special Paper 452, p. 83–95, doi: 10.1130/2009.2452(06). For permission to copy, contact editing@geosociety.org.

INTRODUCTION

Two of the three largest impact craters found on Earth younger than 200 Ma (Popigai and the Chesapeake Bay) are well preserved, yet the environmental effects of these events are poorly understood. The 100-km-diameter Popigai crater, located in northern Siberia, generated clinopyroxene-bearing (cpx) spherules (Whitehead et al., 2000; Glass et al., 2004a; Liu et al., 2006) that are stratigraphically slightly older than the North American microtektite layer (Glass et al., 1998). The North American microtektite layer is thought to be associated with the 85-km-diameter Chesapeake Bay impact structure, located on the eastern shore of North America (Glass et al., 1985, 1998; Glass and Burns, 1987; Glass and Koeberl 1999; Poag et al., 1994; Koeberl et al., 1996; Deutsch and Koeberl, 2006). In situ fragments of tektites that belong to the North American microtektite layer have been radiometrically dated to ca. 35.4 ± 0.6 Ma (2σ) (Glass et al., 1986) and 35.5 ± 0.3 Ma (Obradovich et al., 1989). Recent radiometric dates on four North American tektites produced an age of ca. 35.3 ± 0.2 Ma (2σ) (Horton and Izett, 2005). Thus, the age of the Chesapeake Bay impact structure ranges from 35.3 Ma to 35.5 Ma, assuming the North American tektites/microtektites were formed by the Chesapeake Bay impact event. Cpx spherules have not been radiometrically dated; however, impact melt rocks occurring as irregular melt bodies and coherent sheets from the Popigai crater yielded an age of 35.7 ± 0.2 Ma (2σ) (Bottomley et al., 1997). Cpx spherules associated with the cpx spherule strewn field have been found in the western North Atlantic, South Atlantic, Gulf of Mexico, Caribbean Sea, equatorial Pacific, and eastern equatorial Indian Ocean (Fig. 1) (Glass, 2000, 2002; Glass and Burns, 1987; Glass et al., 1985, 1998; Keller et al., 1987; Glass and Koeberl, 1999; Liu et al., this volume). The Upper Eocene cpx spherule layer also contains clear microtektites, leucocratic and melanocratic microkrystites, and shocked minerals (Clymer et al., 1996; Langenhorst, 1996), as well as a positive iridium anomaly (200 parts per trillion) and Ni-rich spinel crystals (e.g., Pierrard et al., 1998; Glass et al., 2004b). North American microtektites found in sediments from the western North Atlantic, Barbados, Gulf of Mexico, and Caribbean Sea (Fig. 1) have been correlated to the North American tektite/microtektite strewn field based on geographic proximity, age, and geochemical evidence (Donnelly and Chao, 1973; Glass et al., 1982, 1985, 1998; Saunders et al., 1984; Sanfilippo et al., 1985; Keller, 1986; Thein, 1987; Glass and Burns, 1987; Glass and Koeberl, 1999; Deutsch and Koeberl, 2006). The North American microtektite layer also consists of tektite fragments, shocked mineral and rock fragments, and reidite, a high-pressure polymorph of zircon (e.g., Glass, 1989; Glass et al., 2002; Harris et al., 2004).

Both impact structures are well preserved, yet the global environmental response to these large and near-synchronous impacts remains poorly understood. No major extinctions are associated with the late Eocene impact events in benthic foraminifera (Miller et al., 1992), calcareous nannoplankton (Aubry, 1993), or planktonic foraminifera (Pearson et al., 2006), and late Eocene terrestrial biota was largely unaffected (Prothero, 1985). Evidence of a postimpact biological dead zone that is confined to a localized region at the Chesapeake Bay impact site has been reported in the Chickahominy Formation (Poag et al., 2004). One possible exception to the lack of global biotic effects is the reported abrupt decline in abundance of four (or five) low-latitude radiolarian species associated with the cpx spherule layer (Glass and Zwart, 1977; Glass et al., 1985; Sanfilippo et al., 1985; Glass and Burns, 1987; Glass, 2000). Detailed knowledge of environmental and paleoceanographic effects of these extraterrestrial impacts is lacking, in part, because most late Eocene stable isotopic studies have lacked sufficient resolution to decipher any response of Earth's climate system.

Previous studies noted that an extraterrestrial impact of an ice/volatile object could potentially cause a large (>0.5‰), rapid (<10 k.y.), negative excursion in global $\delta^{13}C$ records (Wilde and Quinby-Hunt, 1997). Such an explanation of a comet impact has been postulated for the carbon isotope excursion at the Paleocene-Eocene boundary (Kent et al., 2003; Cramer and Kent, 2005; see Schmitz et al. [2004] for an alternative view). If large comet impacts can cause large and rapid $\delta^{13}C$ excursions, then a large anomaly should be associated with the late Eocene impact events, if they were indeed part of a comet shower (Farley et al., 1998). However, published stable isotopic records lack the resolution to investigate this prediction adequately: (1) a closely sampled stable isotopic record from Ocean Drilling Program (ODP) Site 689 (Vonhof et al., 2000) lacks precise magnetochronologic age control; (2) isotopic records from Massignano, Italy, suffer from diagenesis (Bodiselitsch et al., 2004); and (3) other deep-sea records lack critical coverage of this time interval (Zachos et al., 2001) or precise calibration to the late Eocene impact record (e.g., ODP Site 1218; Coxall et al., 2005).

Previous studies (Vonhof et al., 2000) from Southern Ocean Site 689 found an ~0.5‰ $\delta^{13}C$ decrease in benthic foraminifera associated with an Ir anomaly (Montanari et al., 1993) and cpx spherule layer (Glass and Koeberl, 1999) at 128.7 m below seafloor (mbsf). Bodiselitsch et al. (2004) performed bulk-carbonate analyses on samples from Massignano, Italy, and reported that the lowest $\delta^{13}C$ and $\delta^{18}O$ values occur in the same stratigraphic interval as the upper Ir anomaly (6.17 m). However, foraminifera from Massignano are described as being altered considerably by diagenesis (Vonhof et al., 2000) and, therefore, may record postdepositional changes to the carbonate's isotopic composition.

We chose South Atlantic Site 1090 (Fig. 1) for this investigation because benthic foraminifera are well preserved, the identified cpx spherule layer is marked by an Ir anomaly (~950 pg/g) (Kyte and Liu, 2002), and the published magnetostratigraphic age control is excellent (Channell et al., 2003). The Upper Eocene impact ejecta layer at Site 1090 was first interpreted to consist of a cpx spherule layer containing microtektites (Kyte and Liu, 2002). Subsequent work at nearby Site 689 (Vonhof and Smit, 1999; Glass and Koeberl, 1999) concluded that microtektites from the cpx spherule layer belong to the North American microtektite

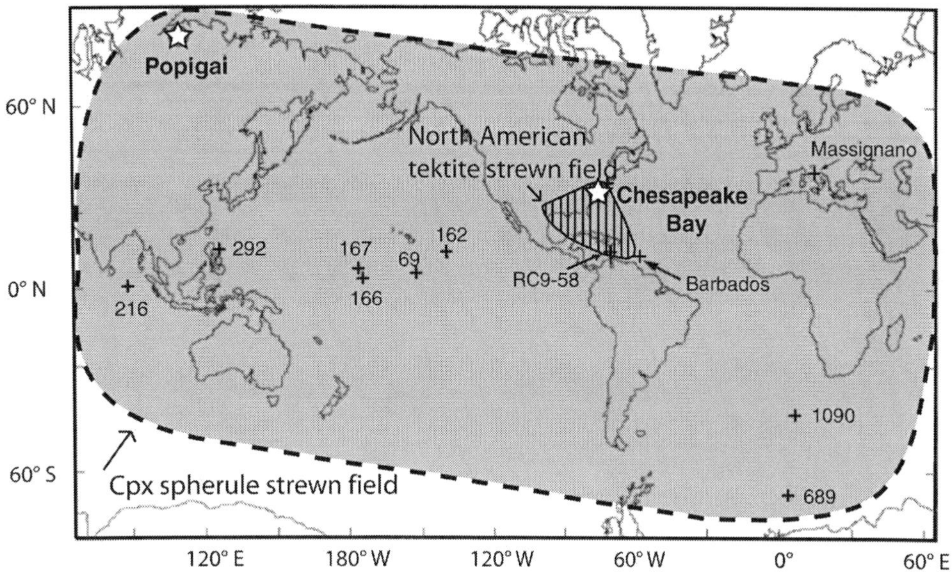

Figure 1. Map of the North American tektite/microtektite strewn field (black lines) and Popigai's clinopyroxene (cpx) spherule strewn field (gray shading). Popigai and Chesapeake Bay craters are shown by the white stars. Map also displays the geographic distribution of sites with identified Upper Eocene cpx spherules (+ signs) from Popigai event (modified from Glass et al., 2004a).

strewn field; this would be the first report of the North American strewn field outside of the North American region (Fig. 1). In contrast, Liu et al. (2006) used radiogenic isotopic studies to interpret the microtektites at Sites 1090 and 689 as exclusively related to the cpx spherule layer and the Popigai impact. We follow their interpretation, which is consistent with the mapped distributions of both the cpx spherule layer and the North American tektite/microtektite strewn field (Fig. 1).

We also analyzed New Jersey slope Deep Sea Drilling Project (DSDP) Site 612 and eastern Caribbean Sea core RC9-58 (Fig. 1) to generate benthic foraminiferal $\delta^{13}C$ records across Upper Eocene impact ejecta layers identified at these locations (Glass et al., 1982; Thein, 1987; Glass, 1989). Multiple high-resolution benthic foraminiferal $\delta^{13}C$ records from this study are compared with previous work (Vonhof et al., 2000) in order to evaluate the possibility of a $\delta^{13}C$ anomaly associated with the late Eocene extraterrestrial impact events.

MATERIALS AND METHODS

ODP Site 1090 (42°54′S, 08°53.98′E) is located in the South Atlantic on the southern flank of the Agulhas Ridge in 3702 m water depth (Shipboard Scientific Party ODP, 1999). Site 1090 paleowater depths of ~3200 m at 34 Ma were estimated using the equation paleodepth = initial depth + $300t^{1/2}$, with a crustal age of 83 Ma, and initial depth of 2700 m (i.e., average ocean crust), and accounting for the effects of sediment loading (Miller et al., 1986; using updated constant of Stein and Stein, 1992). The Upper Eocene cpx spherule layer (Fig. 2; 279 mbsf) is located in sediments described as radiolarian and nannofossil-diatom ooze (Shipboard Scientific Party ODP, 1999; Kyte, 2001). We generated a benthic foraminiferal $\delta^{13}C$ record at Site 1090 across the cpx spherule layer (279 mbsf) from 270 to 291 mbsf with a 20 cm (8 k.y.) sampling interval (Table 1). From 266 to 269 mbsf, we increased the sampling interval to every 40 cm (16 k.y.). A high-resolution bulk-carbonate isotopic record was constructed across the cpx spherule layer using a sampling interval every ~5 cm (2 k.y.) (Fig. 2; Table 2).

We produced two additional late Eocene benthic foraminiferal $\delta^{13}C$ records, DSDP Site 612 (Table 3) and Lamont-Doherty Earth Observatory (LDEO) piston core RC9-58 (Table 4), which we compare to published data from ODP Site 689 (Fig. 3). DSDP Site 612 (38°49.21′N, 72°46.43′W) is located in 1404 m water depth, with an Eocene paleowater depth of ~1000 m (Miller and Katz, 1987). Site 612 sediments are composed of chalk and biosiliceous calcareous ooze in core 21X section 5, between 110 and 119 cm (181.3 mbsf) (Fig. 3) (Thein, 1987; Glass, 1989). Site 612 shows 3–4 cm of separation between the highest abundance of North American microtektites and that of the cpx spherules (Glass, 1989), equivalent to 1.2–4 k.y., assuming continuous sedimentation (Fig. 3) (Miller et al., 1991). The peak concentration of North American microtektites occurs from 112 to 113 cm, and the highest occurrence of cpx spherules is positioned between 115 and 117 cm (Fig. 3) (Glass, 1989).

Caribbean Sea core RC9-58 (14°33.4′N, 70°48.6′W) is located in 3548 m water depth (Saito et al., 1974) with a late Eocene paleowater depth >1000 m (Baker and Glass, 1974). Core RC9-58 displays 25 cm of separation between the North American microtektite layer and the cpx spherule layer, with its coupled Ir anomaly (Fig. 3) (Ganapathy, 1982; Glass et al., 1982). The equivalent age separation equals ~25 k.y. (Maurrasse and Glass, 1976).

An average of four to five benthic foraminiferal specimens (of the epifaunal genus *Cibicidoides*) per sample were selected for stable isotopic analysis on a Micromass Optima mass spectrometer at the Stable Isotope Laboratory in the Department of Earth and Planetary Sciences at Rutgers University. Duplicate samples were intermittently run to ensure reproducibility of analyses. Samples were reacted with phosphoric acid for 15 min at 90 °C in a multiprep peripheral device and reported against

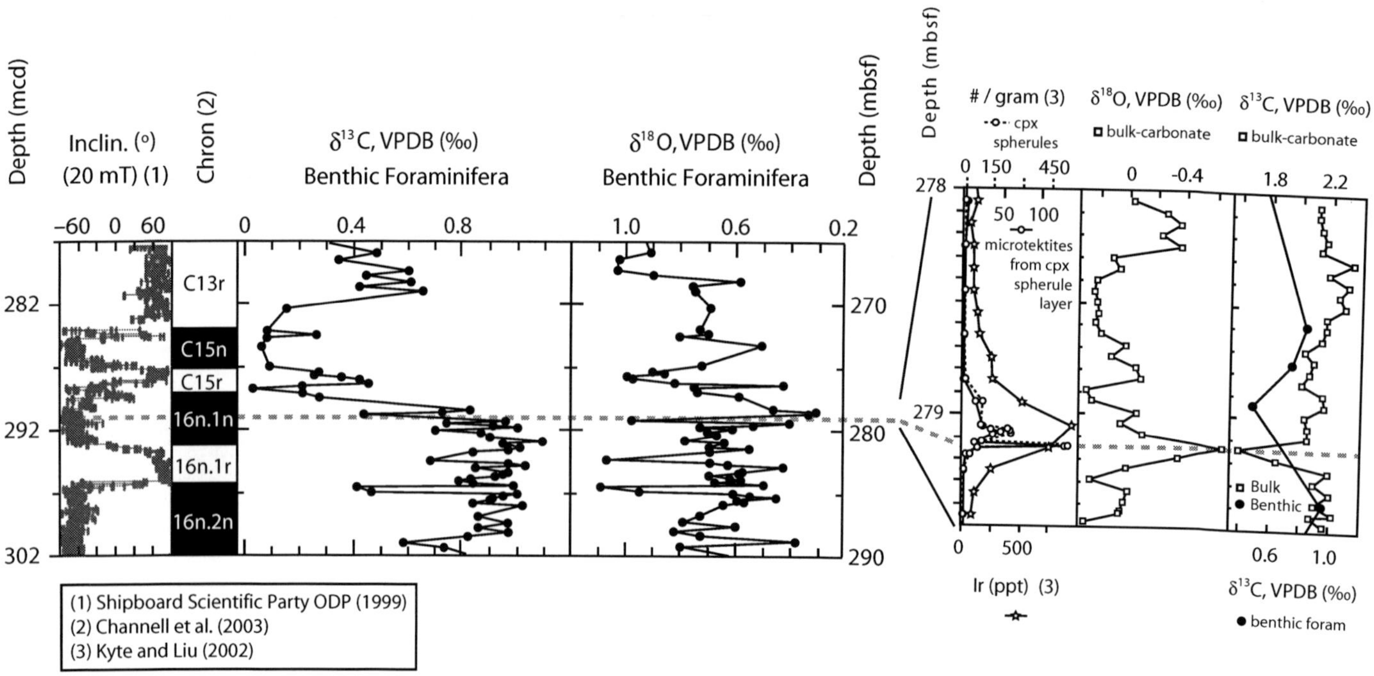

Figure 2. South Atlantic Ocean Drilling Program (ODP) Site 1090 shipboard magnetic inclination (Shipboard Scientific Party, 1999; Channell et al., 2003) plotted with benthic foraminiferal $\delta^{13}C$ and $\delta^{18}O$ data across the clinopyroxene (cpx) spherule layer at ~279 m below seafloor (mbsf) in chron C16n.1n, with a corresponding magnetochronologic age of ca. 35.4 Ma. Meters below seafloor (mbsf) and meters composite depth (mcd) are both displayed to match the work for the magnetic inclination (Channell et al., 2003) and cpx spherule layer (Kyte and Liu, 2002) references. Enlarged plots show Popigai cpx spherule layer, microtektites belonging to the cpx spherule layer, and iridium abundances at Site 1090 plotted next to high-resolution bulk-carbonate $\delta^{18}O$ and $\delta^{13}C$ data. VPDB—Vienna PeeDee belemnite.

the V-PDB (Vienna PeeDee belemnite) standard ($\delta^{18}O$ = –2.20‰ and $\delta^{13}C$ = 1.95‰; following Coplen et al., 1983). The laboratory standard was checked regularly against NBS-19, which has a $\delta^{18}O$ offset equal to 0.04‰ and a $\delta^{13}C$ offset of 0.10‰. The laboratory standard error (1σ) is equal to 0.08‰ for $\delta^{18}O$ and 0.05‰ for $\delta^{13}C$.

RESULTS

ODP Site 1090 provides a first-order correlation of the Upper Eocene cpx spherule layer to the geomagnetic polarity time scale (GPTS) (Cande and Kent, 1995; Berggren et al., 1995) in chron C16n.1n (279 mbsf; Fig. 2; Channell et al., 2003). Chron C16n.1n has an estimated magnetochronologic age of 35.526–35.343 Ma, and the cpx spherule layer occurs near the middle of the chronozone. Based on interpolations of sedimentation rates between the chron C16n.1n boundaries, the cpx spherule layer has a corresponding magnetochronologic age of 35.43 Ma, which is consistent with published radiometric ages for North American tektites/microtektites ranging from ca. 35.3 to 35.5 Ma (Glass et al., 1986; Obradovich et al., 1989; Horton and Izett, 2005), and slightly offset from Popigai crater melt rocks dated at 35.7 ± 0.2 Ma (2σ) (Bottomley et al., 1997).

Distinct $\delta^{13}C$ decreases are associated with the Upper Eocene cpx spherule layer at ODP Site 1090 (Fig. 2). An ~0.5‰ transient $\delta^{13}C$ anomaly in bulk carbonate occurs precisely at the cpx spherule layer (279.15 mbsf) at this site: the $\delta^{13}C$ decrease occurs within 5 cm (~2 k.y.) of the peak cpx spherule concentration and recovers to near pre-excursion values within 5 cm (~2 k.y.) (Fig. 2). Benthic foraminiferal $\delta^{13}C$ values decrease across the cpx spherule layer (Fig. 2) and return to pre-excursion values ~20 cm (~8 k.y.) above the cpx spherule layer, though our resolution is not sufficient to provide more detail. The transient $\delta^{13}C$ decrease is superimposed on a longer-term decrease (occurring within 1.5 m, equivalent to ~60 k.y.) and numerous 0.2‰–0.5‰ $\delta^{13}C$ changes that reflect variability on a scale of 10^4–10^5 yr (Fig. 2). Site 1090 bulk-carbonate $\delta^{18}O$ values show a distinct 0.5‰ decrease at the cpx spherule layer; however, there is no significant change in benthic foraminiferal $\delta^{18}O$ at the layer (Fig. 2).

DSDP Site 612 displays the cpx spherule layer and North American microtektite layer (Glass, 1989; Miller et al., 1991). We show the onset of a negative (~0.5‰) $\delta^{13}C$ anomaly in benthic foraminifera associated with the cpx spherule layer (Fig. 3). Site 612 benthic foraminiferal $\delta^{13}C$ values are ~0.7‰ just below the cpx spherule layer and drop to ~0.2‰ at 1 cm above the layer (Fig. 3). Insufficient carbonate content prevented us from obtaining benthic and bulk-carbonate data across the peak abundance of North American microtektites at Site 612 in core 21X, section 5, 110–112 cm (Fig. 3).

Core RC9-58 shows that minimum benthic foraminiferal $\delta^{13}C$ values occur at 2.9 m, immediately below the cpx spherule layer (2.8 m) (Fig. 3). The 10 cm offset between the $\delta^{13}C$ minimum and Ir/cpx spherule peak is within the range for biological mixing depths of deep-sea sediments (Guinasso and Schink, 1975) and is most likely the result of bioturbation of rare benthic foraminiferal specimens versus more common cpx spherules. RC9-58 benthic foraminiferal $\delta^{13}C$ values show no significant change across the stratigraphically higher North American microtektite layer (Fig. 3). Bulk-carbonate $\delta^{13}C$ values at RC9-58 show minima at 2.8 m, coincident with the cpx spherule layer, at 2.4 m, ~10 cm above the North American microtektite layer, and at 2.1 m (Fig. 3). Large variations in the bulk $\delta^{13}C$ at RC9-58 contrast with minimal to no variation in the bulk $\delta^{13}C$ record at Site 1090; the only communality is a minimum at the cpx spherule layer at both sites, which, in concert with benthic foraminiferal values, is interpreted as the only interval of whole-ocean $\delta^{13}C$ change associated with the cpx spherule layer.

ODP Site 1090 benthic foraminiferal $\delta^{18}O$ values show minimal variation across the cpx spherule layer (Figs. 2 and 4). A transient (~0.4‰) $\delta^{18}O$ increase in benthic foraminifera occurs ~25 cm below the cpx spherule layer, which is one of several 0.3‰–0.4‰ $\delta^{18}O$ changes that reflect variability on a scale of 10^4–10^5 yr (Fig. 2). However, no significant variation in benthic foraminiferal $\delta^{18}O$ values occurs across the cpx spherule layer at Site 1090 (Figs. 2 and 4). Site 1090 bulk-carbonate $\delta^{18}O$ values display a transient ~0.5‰ decrease at the cpx spherule layer (Figs. 2 and 4). Site 1090 bulk-carbonate $\delta^{18}O$ values equal ~0.11‰ just 10 cm below the cpx spherule layer and drop to ~–0.55‰ right at the layer (Figs. 2 and 4).

DSDP Site 612 benthic foraminiferal $\delta^{18}O$ values show an overall decrease of ~0.4‰ across the cpx spherule layer (Fig. 4). Site 612 benthic foraminiferal $\delta^{18}O$ ratios center on 0.35‰–0.4‰ at 2 cm below the cpx spherule layer and decrease to 0‰ at 1–2 cm above the layer (Fig. 4). Core RC9-58 benthic foraminiferal $\delta^{18}O$ values display no change across the cpx spherule layer, but they show a positive shift equal to ~0.4‰ occurring 20 cm above the layer (Fig. 4). RC9-58 benthic foraminiferal $\delta^{18}O$ ratios remain steady across the North American microtektite layer around ~0.8‰ (Fig. 4). Core RC9-58 bulk-carbonate $\delta^{18}O$ values show an increase of ~0.4‰ at 10 cm above the cpx spherule layer, but they decrease by ~0.3‰ just 10 cm above the North American microtektite layer (Fig. 4). RC9-58 bulk-carbonate $\delta^{18}O$ values remain steady around ~–0.6‰ throughout the youngest part of the core from 2.4 to 2.0 m (Fig. 4).

DISCUSSION

Timing and Correlation

The amount of offset in timing between the Popigai and Chesapeake impact events remains unclear. The duration between the two impact events is estimated at less than 25 k.y. from Caribbean core RC9-58, based on constant sedimentation rates across this interval (Fig. 3) (Maurrasse and Glass, 1976; Miller et al., 1991). Although the peak abundance of cpx spherules occurs ~4 cm below that of North American microtektites at DSDP Site 612 (Fig. 3), they occur within the North American microtektite/tektite layer (also consisting of tektite fragments from ~116–117 cm) (Glass, 1989) due to bioturbation. The peak abundance of cpx spherules and North American microtektites at Site 612 also occurs just above a hiatus, which may reflect unusually low sedimentation rates. As noted already, isotopic (Liu et al., 2006), vertical distributional, and global distributional data (Fig. 1) argue that Southern Ocean ODP Sites 1090 and 689 preserve only a single impact ejecta layer (cpx spherule layer associated with Popigai). At Barbados, the North American microtektite layer and a stratigraphically lower Ir anomaly (Sanfilippo et al.., 1985) are separated by 26 cm (~13 k.y.). The inference from stratigraphic constraints of <25 k.y. separation between the two impacts is significantly better than the constraint from radiometric dating (0–400 k.y.).

We show that the Popigai cpx spherule layer is associated with a carbon isotopic excursion (~0.5‰) in bulk carbonate (ODP Site 1090 and core RC9-58) and benthic foraminifera (DSDP Site 612, ODP Site 689, and core RC9-58) (Fig. 3). The timing of this anomaly (ca. 35.43 Ma) is coeval with an ~0.4‰ benthic foraminiferal $\delta^{13}C$ decrease at equatorial Pacific ODP Site 1218 (Fig. 3; Coxall et al., 2005; Lear et al., 2004). Of the sites with impact ejecta layer(s) (core RC9-58; Sites 612, 1090, and 689; Fig. 3), two have reasonably complete magnetostratigraphic records (Sites 1090 and 689). The most recent magnetostratigraphic interpretation at ODP Site 689 (Florindo and Roberts, 2005) suggests that chronozone C16n.1r is absent from Hole 689B, making it difficult to obtain a first-order correlation to the geomagnetic polarity time scale. However, the cpx spherule layer and Ir anomaly found at Hole 689B are also identified at nearby Hole 689D (Liu et al., this volume) in chronozone C16n.1n, which is consistent with the first-order correlation at ODP Site 1090 (Fig. 2). We show that the ~0.5‰ $\delta^{13}C$ excursion is associated with the Upper Eocene cpx spherule layer found at Sites 612, 689, 1090, and core RC9-58 (Figs. 2 and 3). The wide distribution of the carbon isotopic anomaly, both geographically (Atlantic and Pacific Oceans) and oceanographically (bulk-carbonates representing surface waters, Pacific and Southern Ocean deep water, and Atlantic intermediate water), suggests that this is a global marine event that should also be recognizable in the terrestrial realm. We predict that this global event can be correlated consistently to magnetochron C16n.1n.

Marine Oxygen Isotope Response to Popigai and Chesapeake Bay Impacts

Benthic foraminiferal $\delta^{18}O$ values show no consistent trend or variation at Sites 1090, 612, and core RC9-58 (Fig. 4). At low-latitude Site 612, a 0.4‰ decrease (Fig. 4) indicates a maximum warming of 2 °C in intermediate waters.

TABLE 1. BENTHIC FORAMINIFERA (*CIBICIDOIDES* SPP.) STABLE ISOTOPE DATA FOR OCEAN DRILLING PROGRAM SITE 1090

Sample (core-section-depth [in cm])	Depth (mbsf)*	Age (Ma)†	$\delta^{13}C$ (‰, VPDB)§	$\delta^{18}O$ (‰, VPDB)§
29X-3-84 (cm)	266.14	34.499	0.48	0.91
29X-3-136	266.66	34.513	0.34	1.03
29X-4-65	267.45	34.534	0.60	1.03
29X-4-107	267.87	34.545	0.44	0.90
29X-5-5	268.35	34.558	0.61	0.59
29X-5-43.5	268.73	34.568	0.42	0.76
29X-5-86	269.16	34.579	0.65	0.75
29X-6-64	270.44	34.613	0.15	0.69
29X-7-18.5	272.18	34.671	0.08	0.73
30X-1-58.5	272.58	34.709	0.26	0.70
30X-1-76	272.76	34.726	0.08	0.80
30X-1-142	273.42	34.788	0.06	0.51
30X-2-148.5	274.98	34.937	0.09	0.73
30X-3-58.5	275.58	35.055	0.27	0.90
30X-3-76	275.76	35.091	0.25	0.85
30X-3-96	275.96	35.131	0.35	0.99
30X-3-116	276.16	35.171	0.42	0.97
30X-3-137.5	276.37	35.214	0.45	0.82
30X-4-6	276.56	35.252	0.21	0.43
30X-4-26	276.76	35.292	0.03	0.75
30X-4-66	277.16	35.349	0.21	0.74
30X-4-104	277.54	35.364	0.27	0.59
30X-5-58.5	278.58	35.407	0.83	0.47
30X-5-75.5	278.75	35.414	0.72	0.31
30X-5-94	278.94	35.421	0.43	0.34
30X-5-105 (Ir peak)#	279.05	35.426	N.D.**	N.D.**
30X-5-114 (cpx spherule)††	279.14	35.430	N.D.**	N.D.**
30X-5-138	279.38	35.439	0.96	0.98
30X-6-12	279.62	35.449	0.74	0.40
30X-6-33	279.83	35.457	0.91	0.54
30X-6-48.5	279.98	35.464	1.00	0.73
30X-6-67	280.17	35.471	0.70	0.61
30X-6-87	280.37	35.479	0.87	0.70
30X-6-107.5	280.57	35.488	0.90	0.67
30X-7-5.5	281.05	35.507	1.09	0.78
30X-7-25.5	281.25	35.515	0.95	0.65
30X-7-44	281.44	35.523	1.01	0.70
31X-1-6.5	281.76	35.542	0.97	0.56

(*continued*)

Low-latitude core RC9-58 shows no significant variation in benthic foraminiferal $\delta^{18}O$ values across the cpx spherule layer, but an increase of ~0.4‰ occurs 20 cm above the layer (Fig. 4). This suggests a delayed (~20 k.y.) cooling with a maximum temperature change of 2 °C associated with the Popigai impact event. However, no significant change in benthic foraminiferal $\delta^{18}O$ values occurs across the North American microtektite layer in core RC9-58. Higher-latitude Site 1090 also displays no major change in benthic foraminiferal $\delta^{18}O$ values across the cpx spherule layer, indicating that no temperature variation is associated with the Popigai impact event at this site. Previous work (Vonhof et al., 2000) at Site 689 found an ~0.5‰ increase in benthic foraminiferal $\delta^{18}O$ values (Fig. 4), indicating a 2.5 °C cooling. Overall, benthic foraminiferal $\delta^{18}O$ values at Site 612 indicate a 2 °C warming across the cpx spherule layer, core RC9-58 shows a delayed (20 k.y.) 2 °C cooling, Site 1090 shows no significant temperature variation, and Site 689 displays a 2.5 °C cooling (Vonhof et al., 2000) associated with the Popigai impact event (Fig. 4). Benthic foraminiferal values from core RC9-58 show no temperature change associated with the North American microtektite layer, and consequently the Chesapeake Bay impact event.

Bulk-carbonate $\delta^{18}O$ values from Site 1090 indicate a 2.5 °C warming in surface waters associated with the Popigai impact event, but core RC9-58 shows a maximum 2 °C cooling of surface waters within 10 k.y. of the event (Fig. 4). Core RC9-58 displays a maximum warming of 1.5 °C in surface waters occurring within 10 k.y. of the Chesapeake Bay impact event (Fig. 4). Bulk-carbonate $\delta^{18}O$ data from ODP Site 738 (southern Kerguelan Plateau; Liu et al., this volume) show an increase of 0.3‰ across the cpx spherule layer, suggesting a 1.5 °C cooling of surface waters associated with Popigai. This increase is consistent with our core RC9-58 data, but it differs from our South Atlantic Site 1090 data. Coccioni et al. (this volume) found planktonic foraminiferal assemblages that indicate a cooling associated with the Popigai event and a warming linked to the Chesapeake Bay impact. A comparison of these records suggests that any cooling is regional in extent.

TABLE 1. BENTHIC FORAMINIFERA (*CIBICIDOIDES* SPP.) STABLE ISOTOPE DATA FOR OCEAN DRILLING PROGRAM SITE 1090 (*continued*)

Sample (core-section-depth [in cm])	Depth (mbsf)*	Age (Ma)†	$\delta^{13}C$ (‰, VPDB)§	$\delta^{18}O$ (‰, VPDB)§
31X-1-25.5	281.95	35.554	0.84	0.69
31X-1-85	282.55	35.592	0.68	1.07
31X-1-105	282.75	35.604	0.97	0.70
31X-1-125	282.95	35.617	1.03	0.63
31X-1-145.5	283.15	35.630	0.85	0.43
31X-2-18	283.38	35.645	0.97	0.57
31X-2-37	283.57	35.657	0.95	0.59
31X-2-57	283.77	35.669	0.92	0.69
31X-2-75.5	283.95	35.681	0.83	0.62
31X-2-95.5	284.15	35.691	0.79	0.58
31X-2-113.5	284.33	35.698	0.84	0.67
31X-2-135	284.55	35.708	0.99	0.50
31X-3-6	284.76	35.717	0.41	1.09
31X-3-25	284.95	35.725	0.46	0.95
31X-3-45	285.15	35.733	1.00	0.61
31X-3-63	285.33	35.741	0.95	0.55
31X-3-83	285.53	35.749	0.91	0.46
31X-3-105	285.75	35.759	0.90	0.60
31X-3-123	285.93	35.766	0.84	0.58
31X-3-144	286.14	35.775	1.02	0.64
31X-4-82.5	287.02	35.813	0.86	0.73
31X-4-122	287.42	35.830	0.97	0.79
31X-5-12	287.82	35.847	0.86	0.61
31X-5-48.5	288.18	35.863	0.97	0.82
31X-5-88	288.58	35.880	0.82	0.73
31X-5-134	289.04	35.899	0.58	0.39
31X-6-25	289.44	35.916	0.73	0.79
31X-6-104.5	290.24	35.951	0.88	0.59
31X-6-144	290.64	35.968	0.88	0.84
31X-7-30	291.00	35.983	0.50	0.56
31X-CC-18	291.41	36.000	0.81	1.12

*mbsf—meters below seafloor.

†Ages were calculated by interpolating sedimentation rates between the chron boundaries from Channell et al. (2003).

§VPDB—Vienna PeeDee belemnite standard.

#Ir peak—marks depth of the Ir peak abundance, not complete extent of layer.

**N.D.—no data.

††cpx spherule—marks depth of the cpx spherule layer's highest abundance, not complete extent of layer.

Marine Carbon Isotope Response to Popigai and Chesapeake Bay Impacts

Bulk-carbonate isotopic data coincident with the Popigai cpx spherule layer from ODP Site 1090 (Figs. 2 and 3) and previous work at Site 689 (Vonhof et al., 2000) suggest a rapid (<< 10 k.y.) and short-lived (<10 yr) $\delta^{13}C$ response in surface waters. The bulk-carbonate $\delta^{13}C$ anomaly records a negative excursion of ~0.5‰ that is similar to the benthic foraminiferal $\delta^{13}C$ change across the Popigai cpx spherule layer from Site 612, core RC9-58, and previous work at Site 689 (Vonhof et al., 2000) (Fig. 3). ODP Site 1090 benthic foraminiferal $\delta^{13}C$ data show a $\delta^{13}C$ decrease no larger than 0.5‰ (Fig. 2), comparable to the amplitude of background variability. Benthic foraminiferal $\delta^{13}C$ data from the sites with the Upper Eocene cpx spherule layer (Sites 1090, 612, and 689, and core RC9-58) and Site 1218 all show a maximum response of 0.5‰ throughout the ocean carbon system (Fig. 3). Additional $\delta^{13}C$ data are needed to demonstrate that this anomaly also occurred in the terrestrial carbon system in order to evaluate the global response of the carbon cycle system.

We also lack sufficient $\delta^{13}C$ data to fully evaluate the marine carbon isotope response to the Chesapeake Bay impact event. Only two locations examined (Site 612 and core RC9-58) definitely contain microtektites belonging to the North American tektite/microtektite strewn field (Glass, 1989; Glass et al., 1982). Initial benthic foraminiferal data from RC9-58 suggest that a $\delta^{13}C$ anomaly is not associated with the North American microtektite layer, but further analyses at a higher resolution are needed in core RC9-58 and at Site 612 to help separate the global effects between the Popigai and Chesapeake Bay impact events.

Causes and Implications of Marine Carbon Isotope Response

The $\delta^{13}C$ decrease (~0.5‰) we found in bulk-carbonate and benthic foraminiferal values correlates with the cpx spherule layer but not the North American microtektite layer. The ~0.5‰

TABLE 2. BULK-CARBONATE STABLE ISOTOPE DATA FOR OCEAN DRILLING PROGRAM SITE 1090, HOLE B, CORE 30X, SECTION 5

Depth (cm)	Depth (mbsf)*	Age (Ma)†	$\delta^{13}C$ (‰, VPDB)§	$\delta^{18}O$ (‰, VPDB)§
5	278.05	35.3853	2.12	0.07
10	278.10	35.3873	2.12	–0.16
15	278.15	35.3894	2.14	–0.26
20	278.20	35.3914	2.17	–0.13
25	278.25	35.3934	2.14	–0.26
30	278.30	35.3955	2.35	0.21
35	278.35	35.3975	2.19	0.16
40	278.40	35.3995	2.32	0.32
45	278.45	35.4016	2.26	0.34
50	278.50	35.4036	2.30	0.32
55	278.55	35.4056	2.18	0.31
60	278.60	35.4077	2.18	0.33
65	278.65	35.4097	2.15	0.29
70	278.70	35.4117	2.04	0.11
75	278.75	35.4138	2.10	0.22
80	278.80	35.4158	2.07	0.05
85	278.85	35.4178	2.02	0.01
90	278.90	35.4199	2.16	0.39
95	278.95	35.4219	2.17	0.35
100	279.00	35.4239	2.04	0.04
105 Ir peak#	279.05	35.4260	2.06	0.15
110	279.10	35.4280	2.06	0.00
115 cpx spherule**	279.15	35.4300	1.60	–0.55
120	279.20	35.4321	1.85	–0.25
125	279.25	35.4341	2.20	0.11
130	279.30	35.4361	2.11	0.36
135	279.35	35.4382	2.21	0.10
140	279.40	35.4402	2.11	0.13
144	279.44	35.4418	2.23	0.15
145	279.45	35.4422	2.08	0.17
149	279.49	35.4439	2.17	0.41

*mbsf—meters below seafloor.

†Ages were calculated by interpolating sedimentation rates between the chron boundaries from Channell et al. (2003).

§VPDB—Vienna PeeDee belemnite.

#Ir peak—marks depth of the Ir peak abundance, not complete extent of layer.

**cpx spherule—marks depth of the cpx spherule layer's highest abundance, not complete extent of layer.

$\delta^{13}C$ anomaly was an abrupt (<10 k.y.), short-lived (< 10 yr), and apparently a global event. We considered, and then rejected, the possibility that the ~0.5‰ $\delta^{13}C$ anomaly resulted from the injection of chondritic or cometary carbon associated with the impactors that formed the Popigai and perhaps Chesapeake Bay craters. Reports of a late Eocene comet shower were based in part on extraterrestrial ^3He concentrations from Massignano, Italy (Farley et al., 1998). Impact ejecta deposits associated with the Chesapeake Bay impact structure lack typical extraterrestrial concentrations of certain siderophile, platinum group elements (PGEs), such as Ir (Tagle and Claeys, 2005), and may in fact be due to a cometary impact. However, impactites from the Popigai crater contain PGEs that indicate the impactor was an L-type ordinary chondrite (Tagle and Claeys, 2005). Chromium isotope data from the cpx spherules indicate that the Popigai impactor was an ordinary chondrite (Kyte et al., 2004), in addition to Cr, Ni, and Co interelement ratios, which are also consistent with an L- or LL-type chondrite (Glass et al., 2004a). Chondritic asteroids typically contain very low amounts of carbon (median 0.1 wt% carbon; Makjanic et al., 1993), compared to ~25 wt% carbon in comets (Jessberger and Kissel, 1991). Thus, if Popigai were formed by the impact of an asteroid (Kyte et al., 2004; Glass et al., 2004a; Tagle and Claeys, 2005), then it could not possibly explain the observed $\delta^{13}C$ anomaly. The 0.5‰ magnitude of the ocean $\delta^{13}C$ excursion is also too large to be explained by the size of a cometary impactor that possibly formed the Chesapeake crater. Typical cometary $\delta^{13}C$ values range from –110‰ to –45‰ (Messenger, 2000; Wyckoff et al., 2000; Arpigny et al., 2003); generation of a negative $\delta^{13}C$ anomaly of 0.5‰ by a cometary

TABLE 3. BENTHIC FORAMINIFERA (*CIBICIDOIDES* SPP.) STABLE ISOTOPE DATA FOR DEEP SEA DRILLING PROJECT SITE 612, CORE 21X, SECTION 5

Core	Section	Interval (cm)	$\delta^{13}C$ (‰, VPDB)*	$\delta^{18}O$ (‰, VPDB)*
21X	5	114	0.21	0.01
21X	5	115	0.16	0.20
21X	5	116	0.31	0.09
21X	5	117	0.64	0.31
21X	5	118	0.68	0.35

*VPDB—Vienna PeeDee belemnite.

TABLE 4. BENTHIC FORAMINIFERA (*CIBICIDOIDES* SPP.) AND BULK-CARBONATE STABLE ISOTOPE DATA FOR LAMONT-DOHERTY EARTH OBSERVATORY CARIBBEAN CORE RC9-58

Depth (m)	$\delta^{13}C$ (‰, VPDB)* Benthic Foram	$\delta^{18}O$ (‰, VPDB)* Benthic Foram	$\delta^{13}C$ (‰, VPDB)* Bulk CO_3	$\delta^{18}O$ (‰, VPDB)* Bulk CO_3
2.00	1.09	0.76	2.54	–0.64
2.09	1.19	0.70	2.51	–0.68
2.20	1.16	0.74	2.81	–0.52
2.30	1.06	0.75	2.68	–0.57
2.35	1.23	0.80	2.57	–0.49
2.46	1.11	0.84	2.74	–0.65
2.61	1.04	0.81	2.93	–0.37
2.69	0.83	0.47	2.79	–0.37
2.80	0.62	0.49	2.59	–0.81
2.90	0.55	0.60	2.76	–0.70
3.01	0.92	0.49	2.83	–0.63

*VPDB—Vienna PeeDee belemnite.

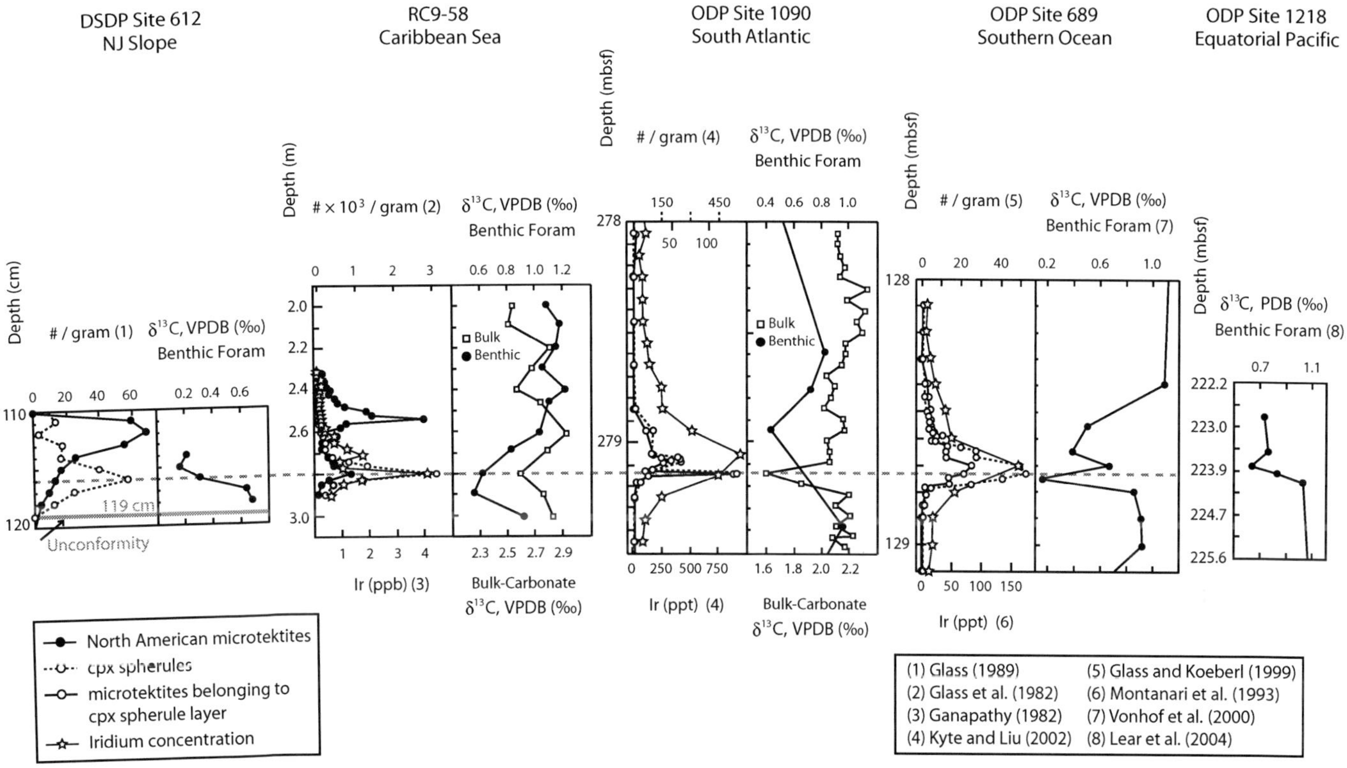

Figure 3. Synthesis of high-resolution late Eocene benthic foraminifera and bulk-carbonate $\delta^{13}C$ data from New Jersey (NJ) Slope Deep Sea Drilling Project (DSDP) Site 612, Caribbean core RC9-58, and South Atlantic Ocean Drilling Program (ODP) Site 1090 (all this study), and previous work at Southern Ocean ODP Site 689 (Vonhof et al., 2000), and equatorial Pacific ODP Site 1218 (Lear et al., 2004) across the identified Upper Eocene cpx spherule layer and the North American microtektite layer, except Site 1218. Data from Site 1218 are shown to illustrate that the carbon isotopic response was global according to our age model. VPDB—Vienna PeeDee belemnite; mbsf—m below seafloor.

impact requires 170 Gt (at –110‰) to 410 Gt (at –45‰) of extraterrestrial carbon (refer to Fig. 8 in Kent et al., 2003), which corresponds to 9.5- and 12.5-km-diameter comets, respectively. Modeling studies of the Chesapeake Bay impact structure (Collins and Wünnemann, 2005) indicate that the Chesapeake Bay impactor had a diameter of ~3.2 km. A 4-km-diameter comet (at –45‰) would only contribute ~13 Gt of carbon into the deep ocean reservoir, which can only explain 0.016‰ of the 0.5‰ observed $\delta^{13}C$ excursion. Even if Popigai was formed by the impact of a comet and not an L-type chondrite, the 100-km-diameter crater and associated 5-km-diameter impactor would only explain 25 Gt of carbon, well below the necessary 170–410 Gt.

We calculated the effects of other potential sources of low $\delta^{13}C$ carbon that may have contributed to the observed 0.5‰

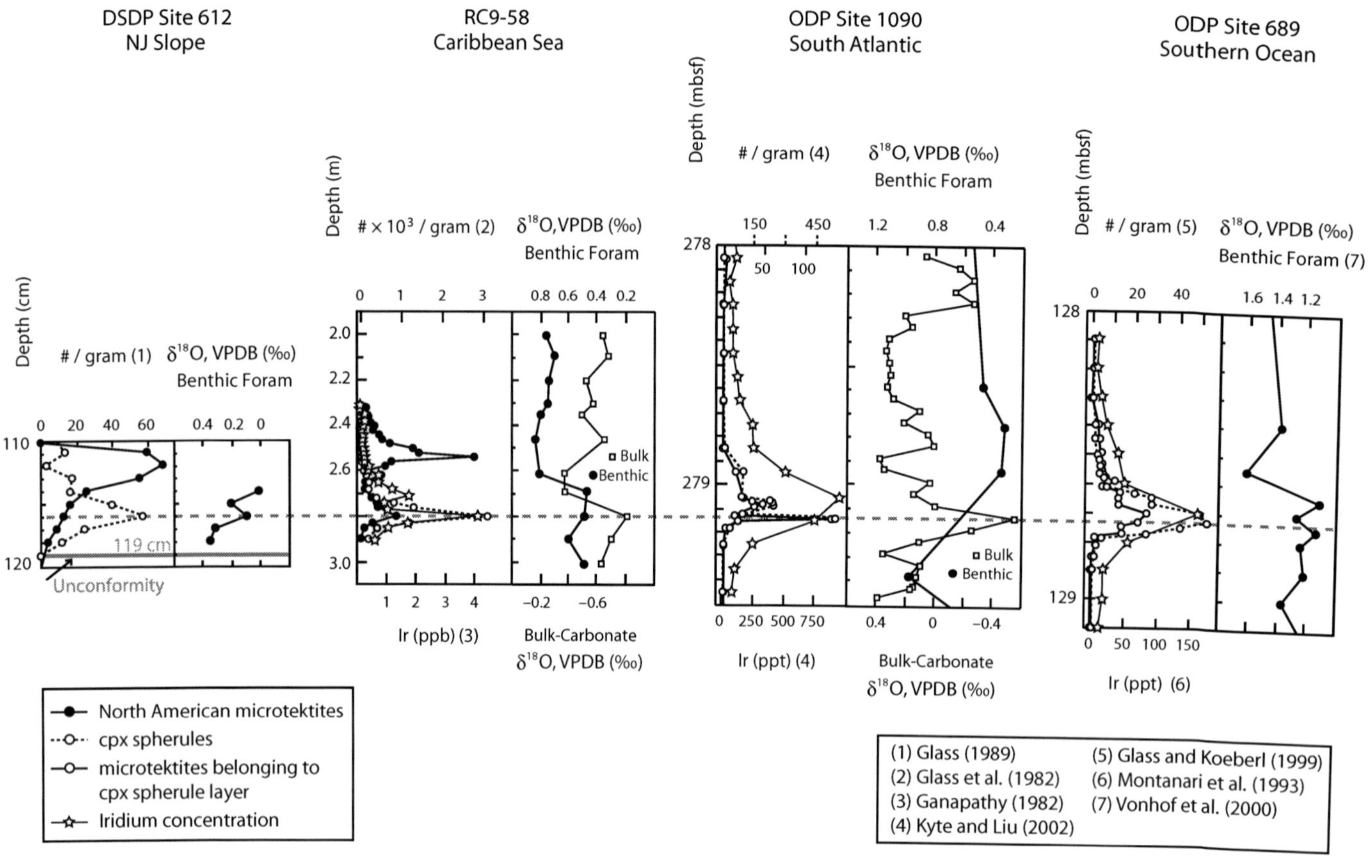

Figure 4. Compilation of high-resolution late Eocene benthic foraminifera and bulk-carbonate $\delta^{18}O$ data from New Jersey (NJ) Slope Deep Sea Drilling Project (DSDP) Site 612, Caribbean core RC9-58, and South Atlantic Ocean Drilling Program (ODP) Site 1090 (all this study), and previous work at Southern Ocean ODP Site 689 (Vonhof et al., 2000) across the identified Upper Eocene cpx spherule layer and the North American microtektite layer. VPDB—Vienna PeeDee belemnite; mbsf—m below seafloor.

negative $\delta^{13}C$ anomaly. One possible source is the dissociation of ~310 Gt of methane hydrates stored on continental margins, which are thought to be capable of abruptly releasing massive amounts of readily oxidized ^{12}C-rich carbon ($\delta^{13}C$ = −60‰) into the ocean-atmosphere system (e.g., Dickens et al., 1995, 1997; Katz et al., 1999, 2001), perhaps through slope failure triggered by an impact. We know of no data to support such a release, but we cannot exclude this hypothesis. Alternatively, a portion of the 0.5‰ oceanic $\delta^{13}C$ decrease could be attributed to the rapid burning of terrestrial organic matter in the form of peat deposits, as has been invoked for the Paleocene-Eocene thermal maximum (Kurtz et al., 2003), perhaps triggered by intense heat from an incoming comet or L-type chondrite. However, terrestrial organic reservoirs cannot readily explain the rapid 0.5‰ $\delta^{13}C$ decrease because this excursion requires 740 Gt of carbon from organic matter (−25‰), an amount comparable to the entire modern living terrestrial carbon reservoir (Falkowski, 2005). Furthermore, the lack of evidence for a mass extinction event argues against such a planetary holocaust. One final source that may have contributed to the observed $\delta^{13}C$ anomaly could be a drop in export carbon productivity following an impact. For example, impact-induced mass mortality in surface waters is hypothesized for the cause of the negative $\delta^{13}C$ anomaly (2‰) at the Cretaceous-Paleogene boundary (Hsu et al., 1982). Although a mass extinction in surface waters is apparently not associated with the late Eocene impact event(s), this hypothesis is readily testable because it predicts a carbonate preservational pulse versus the inverse prediction by the methane clathrate hypothesis.

From an empirical perspective, a $\delta^{13}C$ excursion with 0.5‰ amplitude is not atypical for the ODP Site 1090 benthic foraminiferal $\delta^{13}C$ record (Fig. 2). Eccentricity-controlled variability in Paleogene and Neogene deep-sea $\delta^{13}C$ records has maximum amplitudes of 1‰, with average values of ~0.5‰ (Woodruff and Savin, 1991; Zachos et al., 1996, 1997; Cramer et al., 2003). Changes of this magnitude on eccentricity time scales are explainable by orbitally forced variations in the ratio of carbonate to organic carbon burial (Cramer et al., 2003; Pälike et al., 2006a, 2006b). The 0.5‰ $\delta^{13}C$ decrease associated with the Popigai impact that we document here is apparently more rapid (<10 k.y.), but high-resolution (<4 k.y. sampling) benthic foraminiferal $\delta^{13}C$ records from the late Paleogene–Neogene (e.g., Pälike et al., 2006a, 2006b) show variability of similar magnitude and abruptness.

As noted herein, it is difficult to explain a whole-ocean $\delta^{13}C$ excursion of ~0.5‰ magnitude and <10 k.y. onset using standard carbon budgets. The observation that the $\delta^{13}C$ excursion coincident with the Popigai impact was a rapid (< 10 k.y.) whole-ocean event is therefore of interest for more general understanding of the carbon cycle. Variability of similar scale (~0.5‰ amplitude and <10 k.y.) is present in individual Oligocene to Miocene high-resolution benthic foraminiferal $\delta^{13}C$ records from the equatorial Atlantic and Pacific Oceans (Pälike et al., 2006a, 2006b), but, to our knowledge, it has not been demonstrated to be synchronous among geographically disparate sites. It is plausible that the Popigai impact triggered a carbon cycle response similar to that which occurs under the influence of natural climate variability. This hypothesis predicts that variability of similar scale in high-resolution $\delta^{13}C$ records is generally reflective of a global, rather than local, signal.

The large, rapid, and short-lived $\delta^{13}C$ change also implies either that the carbon cycle is less buffered than believed at present, or that our understanding of the modern to glacial carbon cycle is a poor framework for interpretation of past $\delta^{13}C$ variability. For example, the whole-ocean $\delta^{13}C$ change of 0.3‰ during the last glacial-interglacial cycle has perplexed paleoceanographers for decades (e.g., Shackleton, 1977; Broecker, 1982; Broecker and Peng, 1989), and we are reporting a larger, possibly more rapid change in whole-ocean $\delta^{13}C$. Reliance on the modern glacial change may lead to an exaggerated view of the severity of climate perturbations in the older geologic record, potentially implying a significantly smaller (~50%) ocean carbon reservoir, a larger (200%) terrestrial carbon reservoir, or the existence of some as-yet-unrecognized carbon reservoir.

CONCLUSIONS

ODP Site 1090 provides a first-order correlation of the identified Upper Eocene cpx spherule layer to the geomagnetic polarity time scale (Cande and Kent, 1995; Berggren et al., 1995) in chron C16n.1n (279 mbsf), with a corresponding magnetochronologic age of 35.43 Ma. Bulk-carbonate data from Site 1090 show large (~0.5‰), rapid, short-lived, and negative $\delta^{13}C$ anomalies at the Upper Eocene cpx spherule layer (279 mbsf) that are interpreted to reflect a transient response in surface waters. Three new benthic foraminiferal records from South Atlantic ODP Site 1090, New Jersey slope DSDP Site 612, and Caribbean core RC9-58 also show a coeval (~–0.5‰) $\delta^{13}C$ excursion associated with the cpx spherule layer and perhaps with the North American microtektite layer. The significant (~0.5‰), rapid (<<10 k.y.), short-lived (<10 yr), and negative $\delta^{13}C$ excursion found in benthic foraminifera associated with the cpx spherule layer in core RC9-58, at Sites 1090 and 612, and from previous work at Site 689 (Vonhof et al., 2000) is interpreted to be a global event. The $\delta^{13}C$ excursion may reflect a greater environmental perturbation associated with the impact(s) than has previously been recognized. Alternatively, it may imply a significantly different carbon budget than is generally assumed. We determined there was no global temperature change associated with the cpx spherule or North American microtektite layers, and consequently the Popigai and Chesapeake Bay impact events.

ACKNOWLEDGMENTS

We thank L.E. Edwards and B.P. Glass for reviews, the Ocean Drilling Program (ODP) for supplying samples, including Deep Sea Drilling Project (DSDP) Site 612 archival sections for bulk-carbonate across the Upper Eocene impact ejecta layers. The Lamont-Doherty Earth Observatory Deep-Sea Sample Repository provided samples from Caribbean core RC9-58. This work was supported by National Science Foundation (NSF) grants OPP-0424940 (Kent), EAR06-06693 (Miller), and OCE06-23256 (Katz, Miller, Wade, Wright).

REFERENCES CITED

Arpigny, C., Jehin, E., Manfroid, J., Hutsemekers, D., Schulz, R., Stuwe, J., Zucconi, J.M., and Ilyin, I., 2003, Anomalous nitrogen isotope ratios in comets: Science, v. 301, p. 1522–1524, doi: 10.1126/science.1086711.

Aubry, M.P., 1993, Late Paleogene Calcareous nannoplankton evolution: A tale of climatic deterioration, *in* Prothero, D.R., and Berggren, W.A., eds., Eocene/Oligocene Climatic and Biotic Evolution: Princeton, New Jersey, Princeton University Press, p. 272–309.

Baker, R.N., and Glass, B.P., 1974, Microtektites as test components of Caribbean arenaceous foraminifera: Micropaleontology, v. 20, no. 2, p. 231–235, doi: 10.2307/1485063.

Berggren, W.A., Kent, D.V., Swisher, C.C., III, and Aubry, M.-P., 1995, A revised Cenozoic geochronology and chronostratigraphy, *in* Berggren, W.A., Kent, D.V., Aubry, M.-P., and Hardenbol, J., eds., Geochronology, Time Scales and Global Stratigraphic Correlations: SEPM (Society for Sedimentary Geology) Special Publication 54, p. 129–212.

Bodiselitsch, B., Montanari, A., Koeberl, C., and Coccioni, R., 2004, Delayed climate cooling in the late Eocene caused by multiple impacts: High-resolution geochemical studies at Massignano, Italy: Earth and Planetary Science Letters, v. 223, p. 283–302, doi: 10.1016/j.epsl.2004.04.028.

Bottomley, R., Grieve, R., York, D., and Masaitis, V., 1997, The age of the Popigai impact event and its relation to events at the Eocene/Oligocene boundary: Nature, v. 388, p. 365–368, doi: 10.1038/41073.

Broecker, W.S., 1982, Ocean chemistry during glacial time: Geochimica et Cosmochimica Acta, v. 46, p. 1689–1705, doi: 10.1016/0016-7037(82)90110-7.

Broecker, W.S., and Peng, T.H., 1989, The cause of the glacial and interglacial atmospheric CO_2 change: A polar alkalinity hypothesis: Global Biogeochemical Cycles, v. 3, p. 215–240, doi: 10.1029/GB003i003p00215.

Cande, S.C., and Kent, D.V., 1995, Revised calibration of the geomagnetic polarity time scale for the late Cretaceous and Cenozoic: Journal of Geophysical Research, v. 100, p. 6093–6095, doi: 10.1029/94JB03098.

Channell, J.E.T., Galeotti, S., Martin, E.E., Billups, K., Scher, H.D., and Stoner, J.S., 2003, Eocene to Miocene magnetostratigraphy, biostratigraphy, and chemostratigraphy at ODP Site 1090 (sub-Antarctic South Atlantic): Geological Society of America Bulletin, v. 115, p. 607–623, doi: 10.1130/0016-7606(2003)115<0607:ETMMBA>2.0.CO;2.

Clymer, A.K., Bice, D.M., and Montanari, A., 1996, Shocked quartz in the late Eocene: Impact evidence from Massignano, Italy: Geology, v. 24, p. 483–486, doi: 10.1130/0091-7613(1996)024<0483:SQFTLE>2.3.CO;2.

Coccioni, R., Frontalini, F., and Spezzaferri, S., 2009, this volume, Late Eocene impact-induced climate and hydrological changes: Evidences from the Massignano GSSP (central Italy), *in* Koeberl, C., and Montanari, A., eds., The Late Eocene Earth—Hothouse, Icehouse, and Impacts: Geological Society of America Special Paper 452, doi: 10.1130/2009.2452(07).

Collins, G.S., and Wünnemann, K., 2005, How big was the Chesapeake Bay impact? Insight from numerical modeling: Geology, v. 33, p. 925–928, doi: 10.1130/G21854.1.

Coplen, T.B., Kendall, C., and Hopple, J., 1983, Comparison of stable isotope reference samples: Nature, v. 302, p. 236–238, doi: 10.1038/302236a0.

Coxall, H.K., Wilson, P.A., Pälike, H., Lear, C.H., and Backman, J., 2005, Rapid step-wise onset of Antarctic glaciation and deeper calcite compensation in the Pacific Ocean: Nature, v. 433, p. 53–57, doi:10/1038/nature03135.

Cramer, B.S., and Kent, D.V., 2005, Bolide summer: The Paleocene/Eocene thermal maximum as a response to an extraterrestrial trigger: Palaeogeography, Palaeoclimatology, Palaeoecology, v. 224, p. 144–166, doi: 10.1016/j.palaeo.2005.03.040.

Cramer, B.S., Wright, J.D., Kent, D.V., and Aubry, M.-P., 2003, Orbital climate forcing of $\delta^{13}C$ excursions in the late Paleocene–early Eocene (chrons C24n-C25n): Paleoceanography, v. 18, p. 1097, doi: 10.1029/2003PA000909.

Deutsch, A., and Koeberl, C., 2006, Establishing the link between the Chesapeake Bay impact structure and the North American tektite strewn field: The Sr-Nd isotopic evidence: Meteoritics & Planetary Science, v. 41, p. 689–703.

Dickens, G.R., O'Neil, J.R., Rea, D.K., and Owen, R.M., 1995, Dissociation of oceanic methane hydrate as a cause of the carbon isotope excursion at the end of the Paleocene: Paleoceanography, v. 10, p. 965–971, doi: 10.1029/95PA02087.

Dickens, G.R., Castillo, M.M., and Walker, J.C., 1997, A blast of gas in the latest Paleocene: Simulating first-order effects of massive dissociation of oceanic methane hydrate: Geology, v. 25, p. 259–262, doi: 10.1130/0091-7613(1997)025<0259:ABOGIT>2.3.CO;2.

Donnelly, T.W., and Chao, E.C.T., 1973, Microtektites of late Eocene age from the eastern Caribbean Sea, *in* Edgar, N.T., Saunders, J.B., et al., Initial Reports of the Deep Sea Drilling Project, Volume 15: Washington, D.C., U.S. Government Printing Office, p. 1031–1037.

Falkowski, P.G., 2005, Biogeochemistry of primary production in the sea, *in* Schlesinger, W.H., ed., Biogeochemistry: Treatise on Geochemistry, v. 8, p. 185–214.

Farley, K.A., Montanari, A., Shoemaker, E.M., and Shoemaker, C.S., 1998, Geochemical evidence for a comet shower in the late Eocene: Science, v. 280, p. 1250–1253, doi: 10.1126/science.280.5367.1250.

Florindo, F., and Roberts, A.P., 2005, Eocene-Oligocene magnetobiochronology of ODP Sites 689 and 690, Maud Rise, Weddell Sea, Antarctica: Geological Society of America Bulletin, v. 117, p. 46–66, doi: 10.1130/B25541.1.

Ganapathy, R., 1982, Evidence for a major meteorite impact on the Earth 34 million years ago: Implication for Eocene extinctions: Science, v. 216, p. 885–886, doi: 10.1126/science.216.4548.885.

Glass, B.P., 1989, North American tektite debris and impact ejecta from DSDP Site 612: Meteoritics, v. 24, p. 209–218.

Glass, B.P., 2000, Cenozoic microtektites and clinopyroxene-bearing spherule layers marine sediments, *in* Detre, C.H., ed., Terrestrial and Cosmic Spherules: Budapest, Akademiai Kiado, p. 55–71.

Glass, B.P., 2002, Upper Eocene impact ejecta/spherule layers in marine sediments: Chemie der Erde, v. 62, p. 173–196, doi: 10.1078/0009-2819-00017.

Glass, B.P., and Burns, C.A., 1987, Late Eocene crystal-bearing spherules: Two layers or one?: Meteoritics, v. 22, p. 265–279.

Glass, B.P., and Koeberl, C., 1999, ODP Project Hole 689B spherules and Upper Eocene microtektite and clinopyroxene-bearing spherule strewn fields: Meteoritics & Planetary Science, v. 34, p. 197–208.

Glass, B.P., and Zwart, M.J., 1977, North American microtektites, radiolarian extinctions and the age of the Eocene-Oligocene boundary, *in* Swain, F.M., ed., Stratigraphic Micropaleontology of the Atlantic Basin and Borderlands: Amsterdam, Elsevier, p. 553–568.

Glass, B.P., DuBois, D.L., and Ganapathy, R., 1982, Relationship between an Ir anomaly and the North American microtektite layer in core RC9-58 from the Caribbean Sea: Journal of Geophysical Research, v. 87, p. A425–A428, doi: 10.1029/JB087iS01p0A425.

Glass, B.P., Burns, C.A., Crosbie, J.R., and DuBois, D.L., 1985, Late Eocene North American microtektites and clinopyroxene-bearing spherules, *in* Proceedings of the Sixteenth Lunar and Planetary Science Conference: Houston, Texas, Lunar and Planetary Institute, p. D175–D196.

Glass, B.P., Hall, C.M., and York, D., 1986, $^{40}Ar/^{39}Ar$ laser probe dating of North American tektite fragments from Barbados and the age of the Eocene-Oligocene boundary: Chemical Geology, v. 59, p. 181–186.

Glass, B.P., Koeberl, C., Blum, J.D., and McHugh, C.M.G., 1998, Upper Eocene tektite ejecta layer on the continental slope off New Jersey: Meteoritics & Planetary Science, v. 33, p. 229–241.

Glass, B.P., Liu, S., and Leavens, P.B., 2002, Reidite: An impact-produced high-pressure polymorph of zircon found in marine sediments: The American Mineralogist, v. 87, p. 562–565.

Glass, B.P., Huber, H., and Koeberl, C., 2004a, Geochemistry of Cenozoic microtektites and clinopyroxene-bearing spherules: Geochimica et Cosmochimica Acta, v. 68, p. 3971–4006, doi: 10.1016/j.gca.2004.02.026.

Glass, B.P., Liu, S., and Montanari, A., 2004b, Impact ejecta in Upper Eocene deposits at Massignano, Italy: Meteoritics & Planetary Science, v. 39, p. 589–597.

Guinasso, N.L., and Schink, D.R., 1975, Quantitative estimates of biological mixing rates in abyssal sediments: Journal of Geophysical Research, v. 80, p. 3032–3043, doi: 10.1029/JC080i021p03032.

Harris, R.S., Roden, M.S., Schroeder, P.A., Holland, S.M., Duncan, M.S., and Albin, E.F., 2004, Upper Eocene impact horizon in east-central Georgia: Geology, v. 32, p. 717–720, doi: 10.1130/G20562.1.

Horton, J.W., and Izett, G.A., 2005, Crystalline rock ejecta and shocked minerals of the CBIS, USGS-NASA Langley core, Hampton, Virginia, *in* Horton, J.R., Powars, D.S., and Gohn, G.S., eds., Studies of the Chesapeake Bay Impact Structure: U.S. Geological Survey Professional Paper 1688, p. E1–E30.

Hsu, K.J., He, Q., McKenzie, J.A., Weissert, H., Perch-Nielsen, K., Oberhänsli, H., Kelts, K., LaBrecque, J., Tauxe, L., Krähenbühl, U., Percival, S.F., Jr., Wright, R., Karpoff, A.M., Petersen, N., Tucker, P., Poore, R.Z., Gombos, A.M., Pisciotto, K., Carman, M.F., Jr., and Schreiber, E., 1982, Mass mortality and its environmental and evolutionary consequences: Science, v. 216, p. 249–256, doi: 10.1126/science.216.4543.249.

Jessberger, E.K., and Kissel, J., 1991, Chemical properties of cometary dust and a note on carbon isotopes, *in* Newburn, R.L., Neugebauer, M., and Rahe, J.H., eds., Comets in the Post-Halley Era: Amsterdam, Kluwer Academic, p. 1075–1092.

Katz, M.E., Pak, D.K., Dickens, G.R., and Miller, K.G., 1999, The source and fate of massive carbon input during the latest Paleocene thermal maximum: Science, v. 286, p. 1531–1533, doi: 10.1126/science.286.5444.1531.

Katz, M.E., Cramer, B.S., Mountain, G., Katz, S., and Miller, K.G., 2001, Uncorking the bottle: What triggered the Paleocene/Eocene thermal maximum methane release?: Paleoceanography, v. 16, p. 549–562, doi: 10.1029/2000PA000615.

Keller, G., 1986, Late Eocene impact events and stepwise mass extinctions, *in* Pomerol, C.H., and Premoli-Silva, I., eds., Terminal Eocene Events: Amsterdam, Elsevier, p. 403–412.

Keller, G., D'Hondt, S.L., Orth, C.J., Gilmore, J.S., Oliver, P.Q., Shoemaker, E.M., and Molina, E., 1987, Late Eocene microspherules: Stratigraphy, age, and geochemistry: Meteoritics, v. 22, p. 25–60.

Kent, D.V., Cramer, B.S., Lanci, L., Wang, D., Wright, J.D., and Van der Voo, R., 2003, A case for a comet impact trigger for the Paleocene/Eocene thermal maximum and carbon isotope excursion: Earth and Planetary Science Letters, v. 211, p. 13–26, doi: 10.1016/S0012-821X(03)00188-2.

Koeberl, C., Poag, C.W., Reimold, W.U., and Brandy, D., 1996, Impact origin of the Chesapeake Bay structure, source of the North American tektites: Science, v. 271, p. 1263–1266, doi: 10.1126/science.271.5253.1263.

Kurtz, A.C., Kump, L.R., Arthur, M.A., Zachos, J.C., and Paytan, A., 2003, Early Cenozoic decoupling of the global carbon and sulfur cycles: Paleoceanography, v. 18, no. 4, doi: 10.1029/2003PA000908.

Kyte, F.T., 2001, Identification of late Eocene impact deposits at ODP Site 1090, *in* Gersonde, R., Hodell, D.A., and Blum, P., et al., Scientific Results, Ocean Drilling Program, Volume 177: Washington, D.C., U.S. Government Printing Office, p. 1–9 [Online]. Available from: http://www-odp.tamu.edu/publications/177_SR/VOLUME/CHAPTERS/SR177_04.PDF.

Kyte, F.T., and Liu, S., 2002, Iridium and spherules in late Eocene impact deposits: Lunar and Planetary Institute, Houston, Texas, U.S.A., Lunar and Planetary Science, v. 33, Abstract A1981 (CD-ROM).

Kyte, F.T., Shukolyukov, A., Hildebrand, A.R., Lugmair, G.W., and Hanova, J., 2004. Initial Cr isotopic and iridium measurements of concentrates from Late-Eocene cpx-spherule deposits: Lunar and Planetary Institute, Houston, Texas, Lunar and Planetary Science, v. 35, abstract A1824 (CD-ROM).

Langenhorst, F., 1996, Characteristics of shocked quartz in late Eocene impact ejecta from Massignano (Ancona, Italy): Clues to shock conditions and source crater: Geology, v. 24, p. 487–490, doi: 10.1130/0091-7613(1996)024<0487:COSQIL>2.3.CO;2.

Lear, C.H., Rosenthal, Y., Coxall, H.K., and Wilson, P.A., 2004, Late Eocene to early Miocene ice sheet dynamics and the global carbon cycle: Paleoceanography, v. 19, p. PA4015, doi: 10.1029/2004PA001039.

Liu, S., Papanastassiou, D.A., Ngo, H.H., and Glass, B.P., 2006, Sr and Nd analyses of Upper Eocene spherules and their implications for target rocks: Meteoritics & Planetary Science, v. 41, p. 705–714.

Liu, S., Glass, B.P., Kyte, F.T., and Bohaty, S.M., 2009, this volume, The late Eocene clinopyroxene-bearing spherule layer: New sites, nature of the strewn field, Ir data, and discovery of coesite and shocked quartz, *in* Koeberl, C., and Montanari, A., eds., The Late Eocene Earth—Hothouse,

Icehouse, and Impacts: Geological Society of America Special Paper 452, doi: 10.1130/2009.2452(04).

Makjanic, J., Vis, R.D., Hovenier, J.W., and Heymann, D., 1993, Carbon in the matrices of ordinary chondrites: Meteoritics, v. 28, p. 63–70.

Maurrasse, F.J.M.R., and Glass, B.P., 1976, Radiolarian stratigraphy and North American microtektites in Caribbean RC9-58; implications concerning late Eocene radiolarian chronology and the age of the Eocene-Oligocene boundary, *in* Proceedings of the 7th Caribbean Geological Conference, July 1974, Guadeloupe, p. 205–212.

Messenger, S., 2000, Identification of molecular-cloud material in interplanetary dust particles: Nature, v. 404, p. 968–971, doi: 10.1038/35010053.

Miller, K.G., and Katz, M.E., 1987, Eocene benthic foraminiferal biofacies of the New Jersey Transect, *in* Poag, C.W., Watts, A., et al., Initial Reports of the Deep Sea Drilling Project, Volume 95: Washington, D.C., U.S. Government Printing Office, p. 253–265.

Miller, K.G., Fairbanks, R.G., and Thomas, E., 1986, Benthic foraminiferal carbon isotope records and the development of abyssal circulation in the eastern North Atlantic, *in* Kidd, R., Ruddiman, W.F., et al., Initial Reports of the Deep Sea Drilling Project, Volume 94: Washington, D.C., U.S. Government Printing Office, p. 981–996.

Miller, K.G., Berggren, W.A., Zhang, J., and Palmer-Julson, A., 1991, Biostratigraphy and isotope stratigraphy of Upper Eocene microtektites at Site 612: How many impacts?: Palaios, v. 6, p. 17–38, doi: 10.2307/3514951.

Miller, K.G., Katz, M.E., and Berggren, W.A., 1992, Cenozoic deep-sea benthic foraminifera: A tale of three turnovers, *in* Takayanagi, Y., and Saito, T., eds., Studies of Benthic Foraminifera; Proceedings of the 4th International Symposium on Benthic Foraminifera: Sendai, Japan, Tokai University Press, p. 67–75.

Montanari, A., Asaro, F., Michel, H.V., and Kennett, J.P., 1993, Iridium anomalies of late Eocene age at Massignano (Italy) and ODP Site 689B, Maud Rise, Antarctica: Palaios, v. 8, p. 420–437, doi: 10.2307/3515017.

Obradovich, J.D., Snee, L.W., and Izett, G.A., 1989, Is there more than one glassy impact layer in the late Eocene?: Geological Society of America Abstracts with Programs, v. 21, no. 7, p. A134.

Pälike, H., Frazier, J., and Zachos, J.C., 2006a, Extended orbitally forced palaeoclimatic records from the equatorial Atlantic Ceara Rise: Quaternary Science Reviews, v. 25, p. 3138–3149, doi: 10.1016/j.quascirev.2006.02.011.

Pälike, H., Norris, R.D., Herrle, J.O., Wilson, P.A., Coxall, H.K., Lear, C.H., Shackleton, N.J., Tripati, A.K., and Wade, B.S., 2006b, The heartbeat of the Oligocene climate system: Science, v. 314, p. 1894–1898, doi: 10.1126/science.1133822.

Pearson, P.N., Olsson, R.K., Huber, B.T., Hemleben, C., and Berggren, W.A., 2006, Atlas of Eocene Planktonic Foraminifera: Cushman Foundation of Foraminiferal Research Special Publication 41, 514 p.

Pierrard, O., Robin, E., Rocchia, R., and Montanari, A., 1998, Extraterrestrial Ni-rich spinel in Upper Eocene sediments from Massignano, Italy: Geology, v. 26, p. 307–310, doi: 10.1130/0091-7613(1998)026<0307:ENRSIU>2.3.CO;2.

Poag, W.C., Powars, D.S., Poppe, L.J., and Mixon, R.B., 1994, Meteoroid mayhem in Ole Virginny: Source of the North American tektite strewn field: Geology, v. 22, p. 691–694, doi: 10.1130/0091-7613(1994)022<0691:MMIOVS>2.3.CO;2.

Poag, W.C., Koeberl, C., and Reimold, W.U., 2004, The Chesapeake Bay Crater: New York, Springer, 522 p.

Prothero, D.R., 1985, North American mammalian diversity and Eocene-Oligocene extinctions: Paleobiology, v. 11, p. 389–405.

Saito, T., Burckle, L.H., and Hays, J.D., 1974, Implications of some pre-Quaternary sediment cores and dredgings, *in* Hay, W.W., ed., Studies in Palaeo-oceanography: SEPM (Society for Sedimentary Geology) Special Publication 20, p. 6–36.

Sanfilippo, A., Riedel, W.R., Glass, B.P., and Kyte, F.T., 1985, Late Eocene microtektites and radiolarian extinctions on Barbados: Nature, v. 314, p. 613–615, doi: 10.1038/314613a0.

Saunders, J.B., Bernoulli, D., Muller-Merz, E., Oberhansli, H., Perch-Nielsen, K., Riedel, W.R., Sanfilippo, A., and Torrini, R., Jr., 1984, Stratigraphy of the late middle Eocene to early Oligocene in the Bath Cliff section, Barbados, West Indies: Micropaleontology, v. 30, p. 390–425, doi: 10.2307/1485710.

Schmitz, B., Peucker-Ehrenbrink, B., Heilmann-Clausen, C., Åberg, G., Asaro, F., and Lee, C.A., 2004, Basaltic explosive volcanism, but no comet impact, at the Paleocene-Eocene boundary: High-resolution chemical and isotopic records from Egypt, Spain and Denmark: Earth and Planetary Science Letters, v. 225, p. 1–17, doi: 10.1016/j.epsl.2004.06.017.

Shackleton, N.J., 1977, Carbon-13 in *Uvigerina*: Tropical rainforest history and the equatorial Pacific carbonate dissolution cycles, *in* Anderson, N.R., and Malahoff, A., eds., The Fate of Fossil Fuel CO_2 in the Oceans, New York, Plenum, p. 401–427.

Shipboard Scientific Party ODP, 1999, Site 1090, *in* Gersonde, R., Hodell, D.A., Blum, P., et al., Proceedings of the Ocean Drilling Program, Initial Reports Leg 177: College Station, Texas, Ocean Drilling Program, doi: 10.2973/odp.proc.ir.177.105.1999.

Stein, C.A., and Stein, S., 1992, A model for the global variation in oceanic depth and heat flow with lithospheric age: Nature, v. 359, p. 123–129, doi: 10.1038/359123a0.

Tagle, R., and Claeys, P., 2005, An ordinary chondrite impactor for the Popigai crater, Siberia: Geochimica et Cosmochimica Acta, v. 69, p. 2877–2889, doi: 10.1016/j.gca.2004.11.024.

Thein, J., 1987, A tektite layer in Upper Eocene sediments of the New Jersey continental slope (Site 612, Leg 95), *in* Poag, C.W., Watts, A., et al., Initial Reports of the Deep Sea Drilling Project, Volume 95: Washington, D.C., U.S. Government Printing Office, p. 565–579.

Vonhof, H.B., and Smit, J., 1999, Late Eocene microkrystites and microtektites at Maud Rise (Ocean Drilling Program Hole 698B: Southern Ocean) suggest a global extension of the approximately 35.5 Ma Pacific impact ejecta strewn field: Meteoritics & Planetary Science, v. 34, no. 5, p. 747–755.

Vonhof, H.B., Smit, J., Brinkhuis, H., Montanari, A., and Nederbragt, A.J., 2000, Global cooling accelerated by early late Eocene impacts?: Geology, v. 28, p. 687–690, doi: 10.1130/0091-7613(2000)28<687:GCABEL>2.0.CO;2.

Whitehead, J., Papanastassiou, D.A., Spray, J.G., Grieve, R.A.F., and Wasserburg, G.J., 2000, Late Eocene impact ejecta: Geochemical and isotopic connections with the Popigai impact structure: Earth and Planetary Science Letters, v. 181, p. 473–487, doi: 10.1016/S0012-821X(00)00225-9.

Wilde, P., and Quinby-Hunt, M.S., 1997, Collisions with ice/volatile objects; geological implications—A qualitative treatment: Palaeogeography, Palaeoclimatology, Palaeoecology, v. 132, p. 47–63, doi: 10.1016/S0031-0182(97)00050-3.

Woodruff, F., and Savin, S.M., 1991, Mid-Miocene isotope stratigraphy in the deep sea: High-resolution correlations, paleoclimatic cycles, and sediment preservation: Paleoceanography, v. 6, p. 755–806, doi: 10.1029/91PA02561.

Wyckoff, S., Kleine, M., Peterson, B., Wehinger, P., and Ziurys, L., 2000, Carbon isotope abundances in comets: The Astrophysical Journal, v. 535, p. 991–999, doi: 10.1086/308863.

Zachos, J.C., Quinn, T.M., and Salamy, K.A., 1996, High-resolution (10^4 years) deep-sea foraminiferal stable isotope records of the Eocene-Oligocene climate transition: Paleoceanography, v. 11, p. 251–266, doi: 10.1029/96PA00571.

Zachos, J.C., Flower, B.P., and Paul, H., 1997, Orbitally paced climate oscillations across the Oligocene/Miocene boundary: Nature, v. 388, p. 567–570, doi: 10.1038/41528.

Zachos, J.C., Pagani, M., Sloan, L., Thomas, E., and Billups, K., 2001, Trends, rhythms, and aberrations in global climate 65 Ma to present: Science, v. 292, p. 686–693, doi: 10.1126/science.1059412.

Manuscript Accepted by the Society 16 September 2008

The Geological Society of America
Special Paper 452
2009

Late Eocene impact-induced climate and hydrological changes: Evidence from the Massignano global stratotype section and point (central Italy)

Rodolfo Coccioni*
Fabrizio Frontalini
Dipartimento di Scienze dell'Uomo, dell'Ambiente e della Natura, Università degli Studi di Urbino "Carlo Bo," Campus Scientifico, Località Crocicchia, 61029 Urbino, Italy

Silvia Spezzaferri
Department of Geosciences, University of Fribourg, Chemin du Musée 6, Pérolles CH-1700, Fribourg, Switzerland

ABSTRACT

The Eocene-Oligocene transition marks the passage from "greenhouse" conditions to an "icehouse state" with progressive global cooling starting in the early middle Eocene. The late Eocene is also characterized by a high concentration of extraterrestrial impacts, the effects of which, on living organisms and climatic changes, are still not understood. We carried out a high-resolution investigation on planktonic foraminiferal assemblages in an 8-m-thick segment of the Massignano global stratotype section and point for the Eocene-Oligocene boundary with the aim of assessing the effects that the impacts may have had on the environment and this group of organisms. The studied interval is punctuated by three late Eocene iridium-rich layers, several cosmic signatures, and enhanced levels of ^{3}He. The two lower closely spaced iridium anomalies are possibly linked to the Popigai and Chesapeake Bay impact events, respectively, whereas no particular impact event can be assigned to the third anomaly, even if it might be correlated with some large craters. Interpretation of data suggests that all the impacts had no abrupt, dramatic effects on planktonic foraminifera. However, acting as forcing factors, they induced some environmental perturbations and may have contributed to remarkable climate changes superimposed on the general late Eocene cooling trend. The Popigai and Chesapeake Bay impact events triggered significant changes in the water mass structure, in terms of stratification and trophic resources, associated with some climatic excursions that took place within chron C16n.1n and chron C15r and at the transition between planktonic foraminiferal zones P15 and P16.

The short-term warming pulse recognized after the Popigai impact might have been due to greenhouse effects produced by injection of CO_2 into the atmosphere and/or the release of methane hydrate after the impact itself. The dynamic between

*cron@info-net.it.

Coccioni, R., Frontalini, F., and Spezzaferri, S., 2009, Late Eocene impact-induced climate and hydrological changes: Evidence from the Massignano global stratotype section and point (central Italy), *in* Koeberl, C., and Montanari, A., eds., The Late Eocene Earth—Hothouse, Icehouse, and Impacts: Geological Society of America Special Paper 452, p. 97–118, doi: 10.1130/2009.2452(07). For permission to copy, contact editing@geosociety.org.

hydrological and climate changes across the impactoclastic layers as observed at Massignano displays different features at each impact event, probably due to the context in which each occurred in terms of impactor size, location, and target rocks. The relatively long duration of the enhanced cooling following the Chesapeake Bay impact suggests that this event induced a progressive cooling and triggered a feedback mechanism that sustained the initial impact-induced changes. Similar patterns of climatic excursions reported worldwide across the equivalent impact-ejecta horizons indicate that the impact-induced climate changes recorded at Massignano appear to be global in extent.

Keywords: planktonic foraminifera, impacts, hydrological changes, climate changes, late Eocene, Massignano, central Italy.

INTRODUCTION

The transition from Eocene to Oligocene marks one of the most important steps in Earth's climate history during the Cenozoic, when the "greenhouse" world of the early Eocene changed into the "icehouse" conditions of the early Oligocene (Miller et al., 1991; Zachos et al., 2001, for context). This transition was characterized by global cooling, changes in oceanic circulation, and stepwise faunal and floral turnovers and extinctions, together with a major tectonic reorganization and intensive volcanism (e.g., Boersma and Premoli Silva, 1991; Prothero, 1994; Coxall et al., 2000, 2005; Zachos et al., 2001; Prothero et al., 2003; Wade and Pal̆ike, 2004; Tripati et al., 2005; Coxall and Pearson, 2007). The late Eocene (37.2–33.7 Ma) was also marked by a high concentration of impact craters, ejecta layers, and an elevated flux of interplanetary dust particles (e.g., Farley et al., 1998; Montanari and Koeberl, 2000). The two largest craters of the Cenozoic Era, Popigai (northern Siberia), 100 km in diameter, and Chesapeake Bay (offshore southeastern Virginia), 85 km in diameter, formed almost simultaneously at ca. 35 Ma and were accompanied by several others essentially isochronous small craters (e.g., Wanapitei and Mistastin, Canada, and Logoisk, Belarus). These craters may have been part of the same event (e.g., Koeberl et al., 1996; Bottomley et al., 1997; Montanari et al., 1998; Poag et al., 2004; Koeberl, 2007). Enhanced levels of ^{3}He clearly indicate that these impacts were related to an ~2.5-m.y.-long comet or asteroid shower, which produced impact debris (Farley et al., 1998; Tagle and Claeys, 2004; Claeys, 2007; Farley, 2007).

This multiple impact event is recorded in the Massignano section (northeastern Apennines, central Italy), the currently accepted global stratotype section and point (GSSP) for the Eocene-Oligocene boundary (Premoli Silva and Jenkins, 1993). Three Ir-rich layers occur, respectively, at 5.61 m stratigraphic level from the base (msl), from 6.15 to 6.18 msl, with a maximum value at 6.17 msl, and from 10.24 to 10.30 msl, with a maximum value at 10.28 msl (Montanari et al., 1993; Bodiselitsch et al., 2004), and they are associated with major spikes of extraterrestrial ^{3}He (Farley et al., 1998). The lowest Ir-rich layer contains high concentrations of shocked quartz grains (Clymer et al., 1995, 1996; Langenhorst and Clymer, 1996) and Ni-rich spinel crystals (Pierrard et al., 1998), and it has been interpreted as an impactoclastic horizon dated as old as 35.7 ± 0.4 Ma (Montanari et al., 1993). According to Bodiselitsch et al. (2004), the Ir anomaly at 5.61 msl is possibly linked to the Popigai impact, dated as old as 35.7 ± 0.2 Ma (Bottomley et al., 1997), and the Ir anomaly at 6.17 msl might be derived from the Chesapeake Bay impact, which has an inferred age of 35.2 ± 0.3 Ma to 35.5 ± 0.3 Ma (Poag et al., 2004). No particular impact event can be assigned to the highest Ir anomaly layer, even if it might be correlated with some large craters (e.g., Wanapitei, Mistastin, and Logoisk). The late Eocene multiple impact event may have played an important role related to the deterioration of the global climate at the end of the Eocene Epoch (Coccioni et al., 2000; Vonhof et al., 2000; Spezzaferri et al., 2002), even if the effects on the climate and on the biosphere are still not understood. The Massignano GSSP, which has well-established integrated stratigraphy and geochronology available for the entire section, provides a unique opportunity to explore whether the late Eocene impact events had any effect on the biosphere and climate with regard to short- and long-term changes. In this study, we focus on the late Eocene 4–12 msl segment of the Massignano section and investigate the response of planktonic foraminiferal assemblages to impact-induced environmental perturbations based on quantitative analyses and statistical treatment of data.

THE MASSIGNANO GSSP AND THE 4–12 MSL SEGMENT

The Massignano GSSP is exposed at 43°32′13″N, 13°35′36″E, in an abandoned quarry located along the provincial road of the Conero Regional Natural Park. The section is named after the nearby village of Massignano and is located at ~4 km northwest of the town of Sirolo, along the Adriatic coast of central Italy (Fig. 1). The marine sedimentary sequence at Massignano section spans the interval from the latest Eocene to the earliest Oligocene and belongs to the Umbria-Marche sedimentary succession, which represents the northwestern sector of the Neotethys Ocean. During the Eocene-Oligocene transition, in this region, the sediments were deposited in an upper to lower bathyal setting at a paleodepth of 1000–1500 m (Coccioni and Galeotti, 2003).

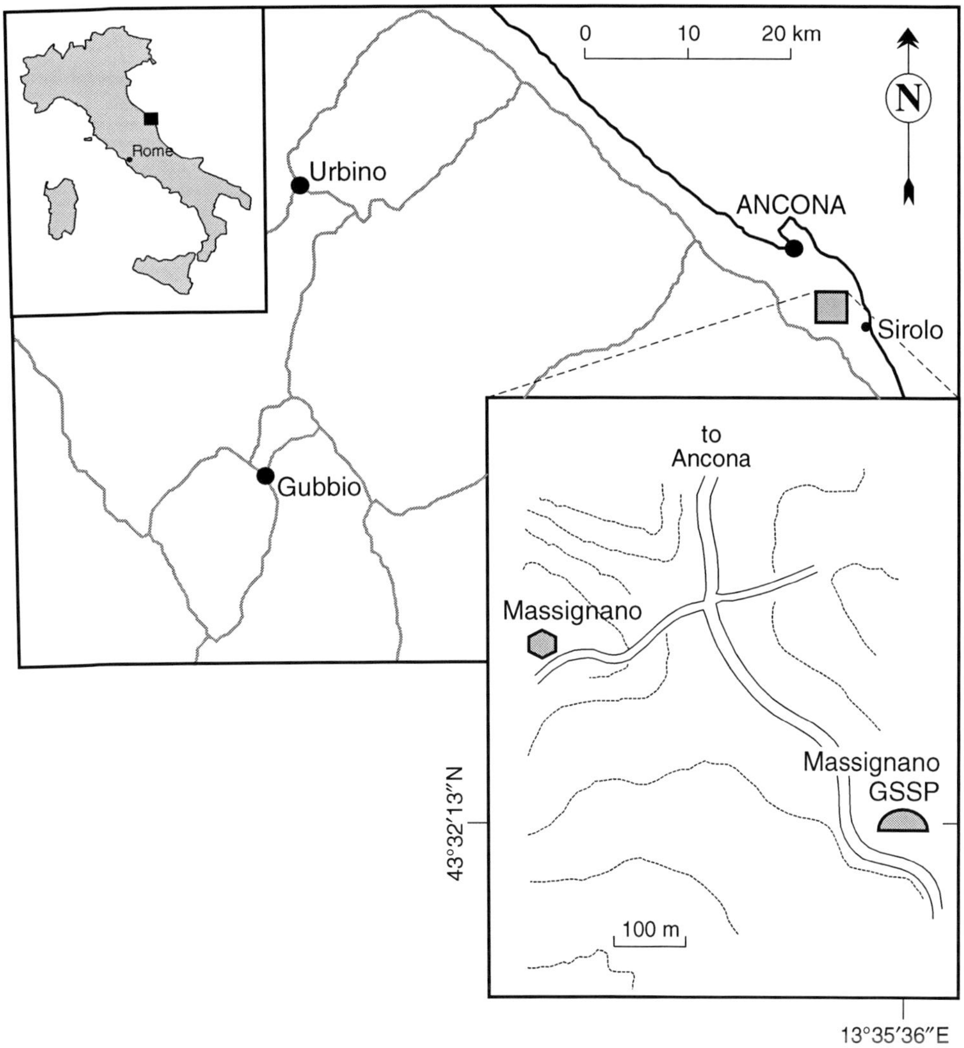

Figure 1. Location map of the Massignano global stratotype section and point (GSSP) for the Eocene-Oligocene boundary.

The exposed continuous and complete sequence is 23 m thick and consists of reddish and greenish-gray hemipelagic marls, calcareous marls, and marly limestones, rich in well-preserved benthic and planktonic microfossils. The sediments span the upper part of the Scaglia Variegata Formation (12 m thick) and the lower part of the Scaglia Cinerea Formation (11 m thick). In the lower to middle part of the section, they are intercalated with several biotite-rich volcano-sedimentary layers (Coccioni et al., 1988) (Figs. 2 and 3). In particular, the 4–12 msl segment contains five biotite-rich volcaniclastic layers at 5.25, 5.8, 6.5, 7.25, and 7.75 msl, respectively (Figs. 2, 4, and 5). Two reddish intervals interrupted by greenish-gray intervals occur in the studied segment. The sediments show evidence of only minor amounts of bioturbation (Coccioni et al., 1988), which appears limited to ~15–20 cm of vertical dispersion (Coccioni et al., 2000; Huber et al., 2001).

The Massignano GSSP spans the calcareous nannofossil zones NP18 to NP21 of Martini (1971) and CP15a to CP16 of Okada and Bukry (1980), the planktonic foraminiferal zones P15 to P18 of Blow (1969, 1979) and Berggren et al. (1995) and zones E14 to O1 of Berggren and Pearson (2006), the benthic foraminiferal zones BB4 to BB5 of Berggren and Miller (1989), and the dinocyst zones *Melitasphaeridium pseudorecurvatum* (Mps) to *Areosphaeridium diktyoplokus* (Adi) of Brinkhuis and Biffi (1993) (Coccioni et al., 1988, 2000; Brinkhuis and Biffi, 1993; Spezzaferri et al., 2002; Monechi et al., 2000) (Fig. 2). The Eocene-Oligocene boundary occurs at 19 msl where the planktonic foraminiferal *Hantkenina* and *Cribrohantkenina*, Eocene genera of the Hantkeninidae, both become extinct (Coccioni et al., 1988; Nocchi et al., 1988a; Premoli Silva and Jenkins, 1993).

The Massignano GSSP spans the interval from chron C16n.2n to chron C13n (Bice and Montanari, 1988; Lowrie and Lanci, 1994; Jovane et al., 2004, 2006, 2007, and this volume) (Fig. 2). K-Ar and Rb-Sr dating of biotite-rich layers in central Italian magnetostratigraphic sections (Montanari et al., 1988)

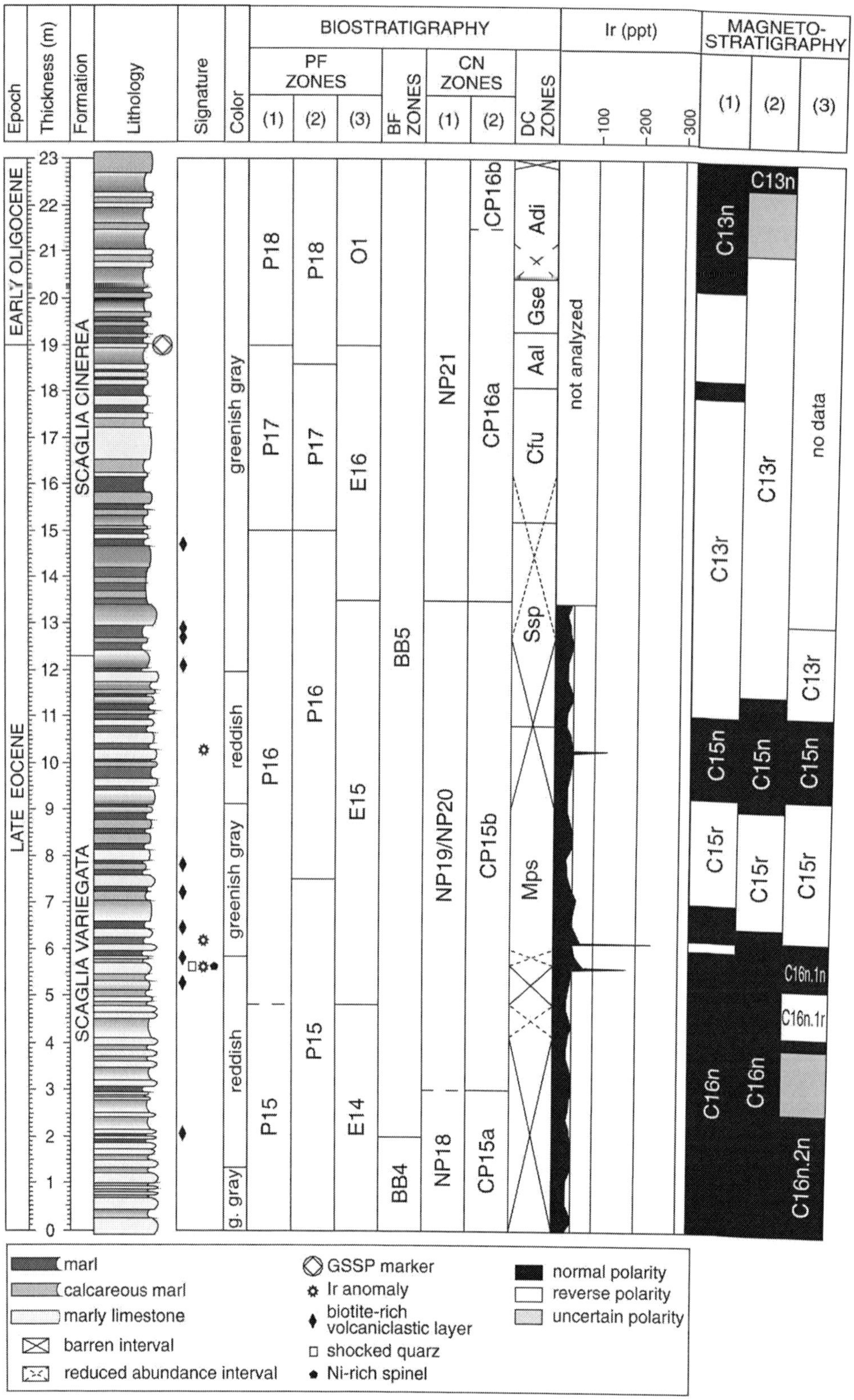

Figure 2. Synoptic stratigraphic scheme of the Massignano global stratotype section and point (GSSP). Lithostratigraphy is after Coccioni et al. (1988) and Jovane et al. (2004, this volume). Biostratigraphy is after Coccioni et al. (1988), Brinkhuis and Biffi (1993), Monechi et al. (2000), Spezzaferri et al. (2002), Coccioni and Galeotti (2003), and Van Mourik and Brinkhuis (2005). Different biozonal schemes are reported according to their authors. Planktonic foraminifera (PF): (1) Blow (1969, 1979), (2) Berggren et al. (1995), and (3) Berggren and Pearson (2006). Benthic foraminifera (BF): Berggren and Miller (1989). Calcareous nannoplankton (CN): (1) Martini (1971) and (2) Okada and Bukry (1980). Dinoflagellate cysts (DC): Brinkhuis and Biffi (1993). Signatures are after Coccioni et al. (1988), Montanari et al. (1988, 1993), Clymer et al. (1995, 1996), Langenhorst and Clymer (1996), Pierrard et al. (1998), and Jovane et al. (2004). Iridium profile is after Montanari et al. (1993) and Bodiselitsch et al. (2004). Magnetostratigraphy is after (1) Bice and Montanari (1988), (2) Lowrie and Lanci (1994), and (3) Jovane et al. (2004, 2007, this volume).

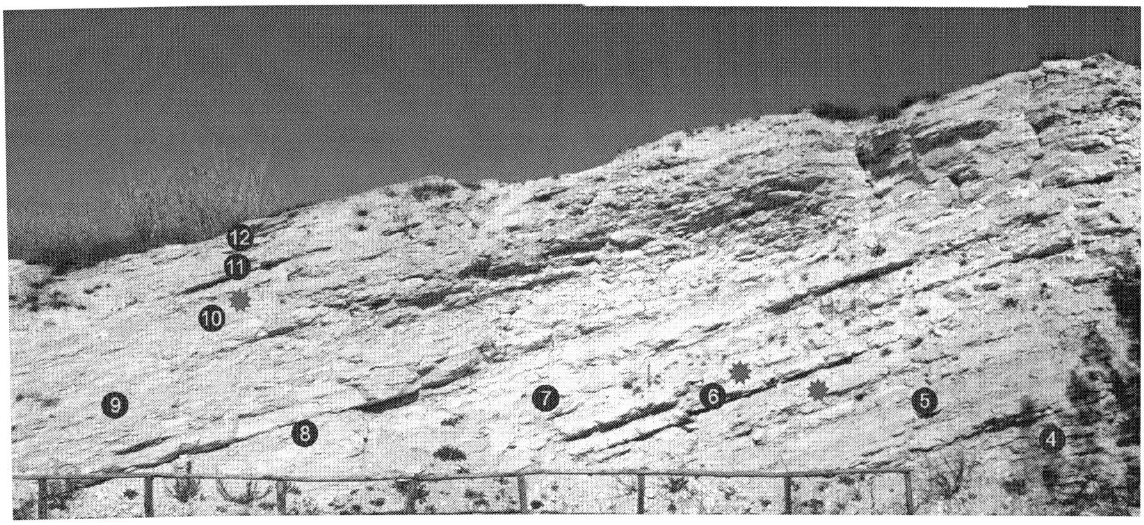

Figure 3. Outcrop photo of the 4–12 m stratigraphic level from the base (msl) segment at Massignano. Numbers correspond to the meters of the measured section. Asterisks mark locations of the impactoclastic layers at 5.61, 6.17, and 10.28 msl.

gave an estimated age of 33.7 ± 0.5 Ma for the Eocene-Oligocene boundary, which has been incorporated as a calibration point in the latest polarity time scales (Cande and Kent, 1992, 1995; Berggren et al., 1995). Additional results from $^{40}Ar/^{39}Ar$ dating of biotites (Odin et al., 1991) and U/Pb dating of zircons and monazites (Oberli and Meier, 1991) were combined with the earlier data to yield revised mean ages of 35.4 ± 0.3 Ma, 34.5 ± 0.3 Ma, and 34.2 ± 0.2 Ma for the biotite horizons at 7.25, 12.9, and 14.7 msl, respectively. A new magnetostratigraphic data set, combined with the results of the astronomical tuning of the Massignano section (Jovane et al., 2006), allowed researchers to refine the dates for the bio- and chronostratigraphic events previously recorded throughout the succession and to place the Eocene-Oligocene boundary at 33.7 Ma, 0.2 m.y. younger than in the time scale of Gradstein et al. (2004) (Jovane et al., 2007). Based on the magnetostratigraphy provided by Jovane et al. (2007) and the numerical ages from Cande and Kent (1995), the mean sedimentation rate of the 4–12 msl segment at Massignano is 0.73 cm/k.y. Accordingly, this segment would span ~1.1 m.y. (Figs. 4 and 5).

MATERIALS AND METHODS

High-resolution sampling of the 4–12 msl segment at Massignano GSSP was conducted at intervals of 5–10 cm, collecting a total of 87 samples. Samples were prepared using standard techniques for preparation of washed residues from marls, calcareous marls, and marly limestones. Samples were disaggregated with hydrogen peroxide and rinsed through a 40 µm sieve. The washed residues were additionally soaked in Desogen for several hours and then immersed in an ultrasonic bath for a few minutes. The final washed residues were dry-sieved into three different size fractions: 40–150 µm, 150–250 µm, and >250 µm. The residues were studied with a binocular microscope for their planktonic foraminiferal content. The 40 µm mesh sieve was used to retain very small species, such as small acarininids, chiloguembelinids, tenuitellids, and pseudohastigerinids, which have important paleoceanographic implications (e.g., Premoli Silva and Boersma, 1988; Spezzaferri, 1995; Pearson et al., 2006).

Quantitative analyses of planktonic foraminiferal assemblages were performed on the same samples previously used for biostratigraphy, following the standard method of Haq et al. (1977), and applied by Premoli Silva and Boersma (1989), Spezzaferri and Premoli Silva (1991), and Spezzaferri (1995). Generic and specific concepts adopted here are those of Premoli Silva and Boersma (1989), Spezzaferri (1994), and Pearson et al. (2006). Planktonic foraminiferal species were then assembled, whenever possible, in groups displaying phylogenetic affinities and environmental significance as reported in Premoli Silva and Boersma (1988) and Spezzaferri (1995). Grouping was necessary because the scarcity of specimens of some species sometimes prevented a correct paleoclimatic interpretation. Species and group abundances were then transformed to percentages of the total planktonic foraminiferal fauna, and percentage curves were constructed. Abundance data of planktonic foraminifers were double-square-root transformed (no standardization, no further species reduction) in order to limit the contribution of the most abundant, ubiquitous species with respect to the less abundant species and to simplify the interpretation of the data structure (Field et al., 1982).

Multivariate statistical analysis was performed with the software PAST version 1.67. The data were used in hierarchical agglomerative clustering based on Bray-Curtis similarity (Clifford and Stephenson, 1975). On the basis of the same similarity matrix, samples were ordered by nMDS (non-metric multidimensional scaling; Kruskal, 1977). The nMDS plot has no dimensions and no axes and can be arbitrarily scaled, located, rotated, or inverted; it gives the relationship of samples relative to each other.

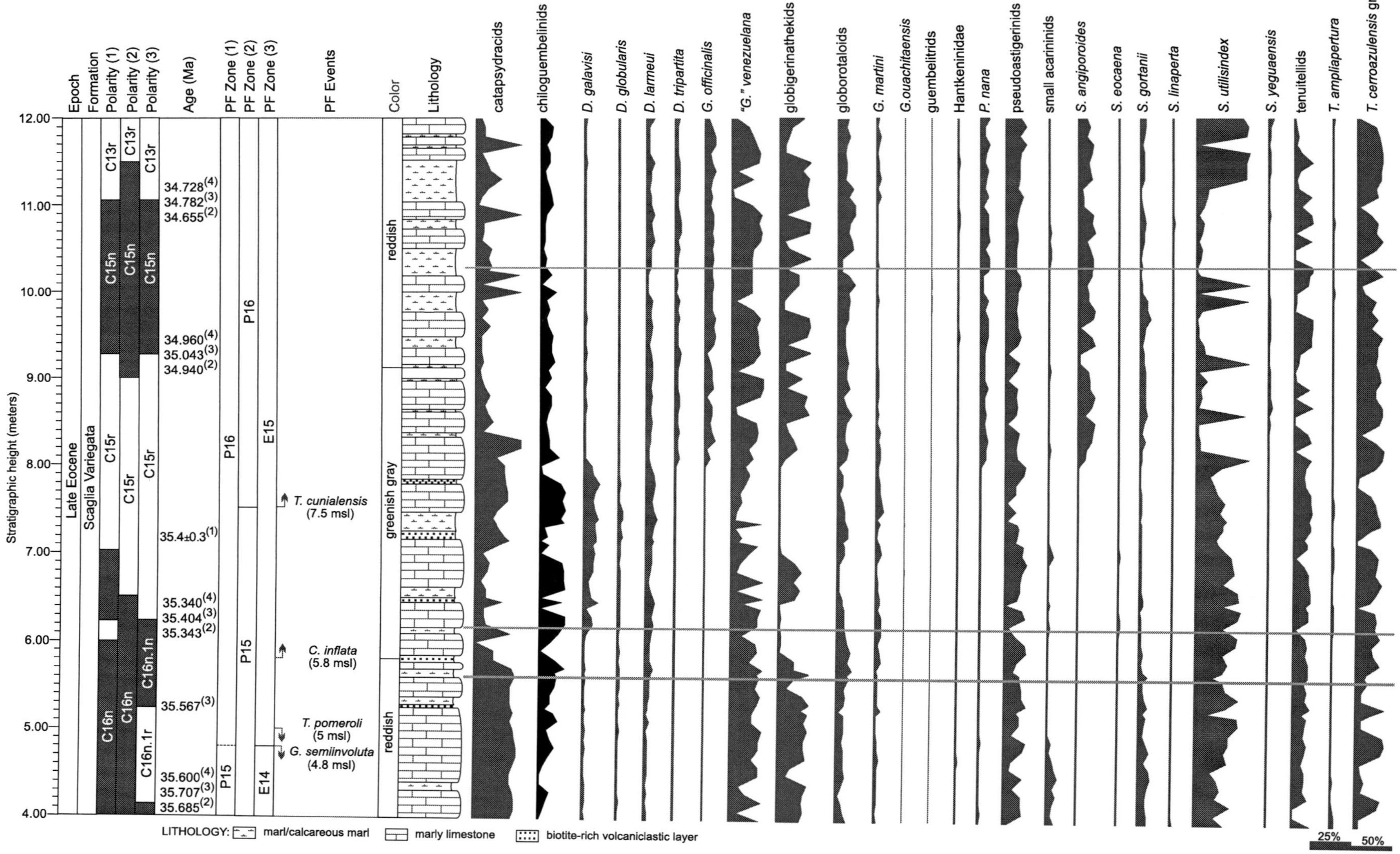

Figure 4. Stratigraphy of the 4–12 m stratigraphic level from the base (msl) segment at Massignano plotted against the relative abundances of planktonic foraminiferal species and groups. Magnetic polarity and planktonic foraminiferal (PF) zones are as in Figure 2. Numerical ages are after (1) Montanari et al. (1988), (2) Cande and Kent (1995), (3) Gradstein et al. (2004), and (4) Jovane et al. (2006). Thin gray lines mark the impactoclastic layers.

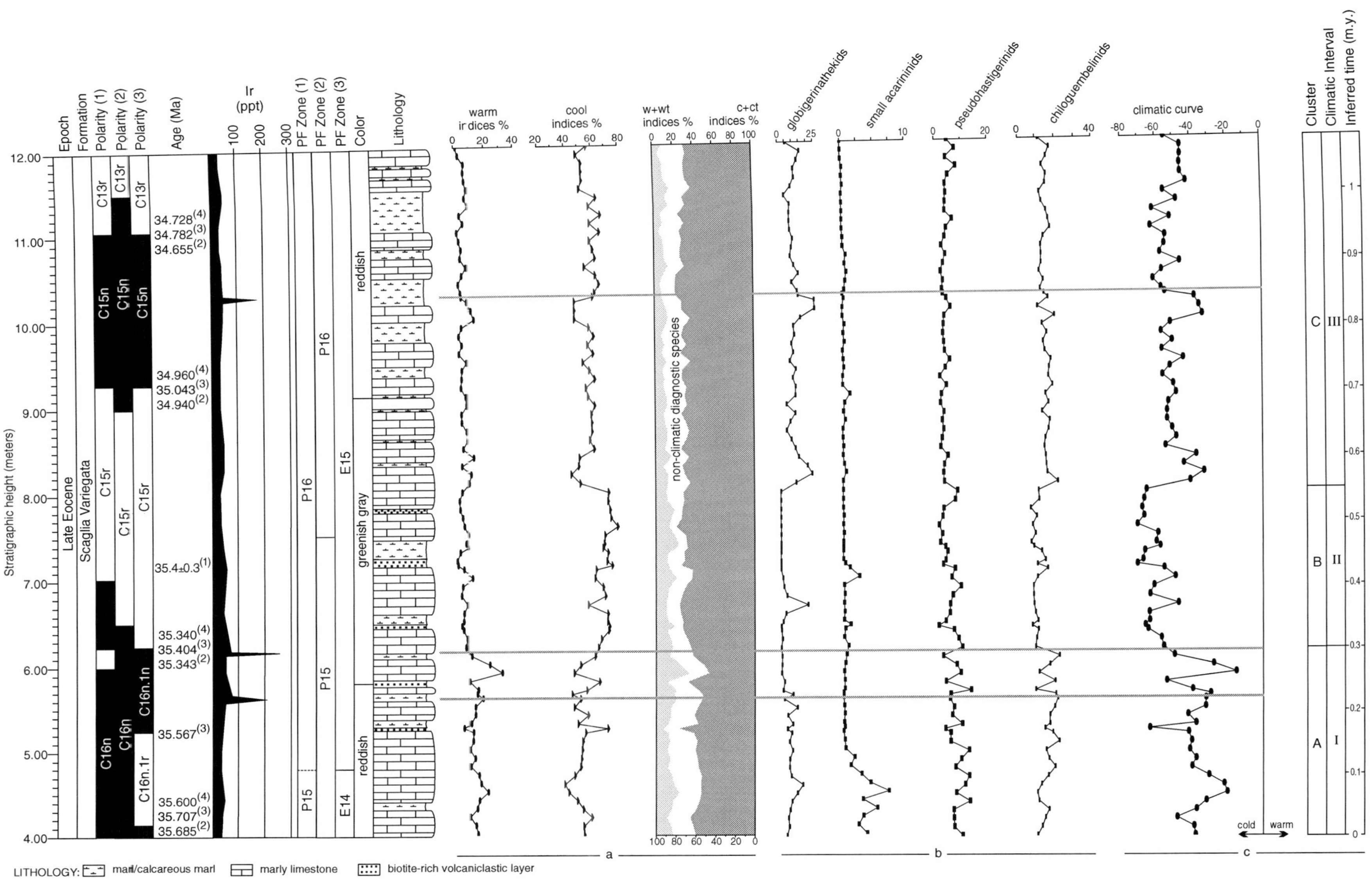

Figure 5. Stratigraphy of the 4–12 m stratigraphic level from the base (msl) segment at Massignano plotted against: (A) the percentage curves of warm and cool indices, warm + warm-temperate, and cool + cool-temperate indices; (B) abundance percentage curves of globigerinathekids, small acarininids, pseudohastigerinids, and chiloguembelinids; (C) climatic curve, clusters, and intervals with inferred time. The cool–cool-temperate index curve is reversed to emphasize the fluctuations in abundance of the non-climatic diagnostic species. Magnetic polarity, iridium profile, and planktonic foraminiferal (PF) zones are as in Figure 2. Numerical ages are as in Figure 4. Thin gray lines mark the impactoclastic layers.

RESULTS

Planktonic Foraminiferal Assemblages and Biostratigraphy

Planktonic foraminiferal assemblages are relatively well preserved, rich, and well diversified throughout the 4–12 msl segment (Fig. 4; Table 1). Specimens often display a slight calcitic overgrowth or infilling, which prevented reliable oxygen and carbon isotopes analyses; consequently, our observations are only based on assemblages.

The 4–12 msl segment is dominated by catapsydracids (*Catapsydrax dissimilis*, *C. unicavus*), chiloguembelinids, "*Globigerina*" *venezuelana*, globigerinathekids (*Globigerinateka index*, *G. luterbacheri*, *G. semiinvoluta*, *G. subconglobata*, *G. tropicalis*), globorotaloids (*Globorotaloides euvariabilis*, *Globorotaloides* sp. 1), pseudohastigerinids, *Subbotina utilizindex*, tenuitellids, and *Turborotalia cerroazulensis* group (*T. cerroazulensis*, *T. cocoaensis*, *T. cunialensis*, *T. pomeroli*), which occur throughout (Fig. 4; Table 1). *Globoturborotalita ouachitaensis*, Hantkeninidae (*Cribrohantkenina inflata*, *Hantkenina alabamensis*), *Subbotina eocaena*, *S. linaperta*, *S. yeguaensis*, *Turborotalia ampliapertura*, and guembelitrids occur discontinuously and in very low numbers. *Subbotina utilizindex* and *T. ampliapertura*, which according to Pearson et al. (2006) have their first occurrence (FO) in zone E15, are already present within the uppermost part of zone E14. *Dentoglobigerina tripartita*, *Globigerina officinalis*, *Paragloborotalia nana*, and *Subbotina angiporoides* are continuously present only from the upper half of the studied segment.

The bioevents identified in the studied segment are (from bottom to top): (1) last occurrence (LO) of *Globigerinatheka semiinvoluta* (4.8 msl), (2) LO of *Turborotalia pomeroli* (5 msl), (3) FO of *Cribrohantkenina inflata* (5.8 msl), and (4) FO of *Turborotalia cunialensis* (7.5 msl) (Coccioni et al., 1988, 2000; Spezzaferri et al., 2002) (Fig. 4).

The 4–12 msl segment spans the uppermost part of zone P15 to the upper part of zone P16 of Blow (1969, 1979), the upper part of zone P15 to the middle part of zone P16 of Berggren et al. (1995), and the uppermost part of zone E14 to the lowermost part of zone O1 of Berggren and Pearson (2006) (Figs. 2, 4, and 5).

Paleoenvironmental Remarks

The paleoclimatic significance and the paleoecology of Eocene planktonic foraminifera are generally deduced from their latitudinal abundance patterns, stable isotopic paleobiology, and from comparison of these parameters to those of modern counterparts.

The paleoecology (preferred biogeographical distribution and life position in the water column) and the paleoclimatic significance of the recognized species and groups are based on the abundant literature, largely following Douglas and Savin (1978), Boersma et al. (1979, 1987), Keller (1983), Poore and Matthews (1984), Nocchi et al. (1988b), Premoli Silva and Boersma (1988, 1989), Stott et al. (1990), Barrera and Huber (1991, 1993), Boersma and Premoli Silva (1991), Spezzaferri and Premoli Silva (1991), Pearson et al. (1993, 1997, 2001, 2006), Spezzaferri (1995), Van Eijden and Ganssen (1995), Pearson and Palmer (1999), Coxall et al. (2000), and Wade and Kroon (2002).

According to the aforementioned literature, we consider the following to be warm-water indices: the small acarininids, *Globoturborotalita martini*, Hantkeninidae, and the *T. cerroazulensis* group. As cool-water indices, we consider catapsydracids, *Globigerina officinalis*, globorotaloids, low-spired subbotininds (*S. angiporoides*, *S. eocaena*, *S. linaperta*, *S. utilizindex*), "*Globigerina*" *venezuelana*, and tenuitellids. We regard globigerinathekids, *P. nana*, and *T. ampliapertura* as cool temperate-water indices, and as warm-temperate water indices, we consider *Dentoglobigerina galavisi*, *D. larmeui*, and *D. tripartita*. Preferred geographic distribution and habitat and paleoclimatic significance of planktonic foraminiferal species and groups identified at Massignano are summarized in Table 2, and Figure 5 shows the abundance curves of the paleoclimatic indicators listed here.

The curve of the warm-water indices shows values ranging from 0.7% to 30.1%, decreasing upward. Positive peaks of 20.9%, 30.1%, and 12.1% occur at 4.5, 5.9, and 10 msl, respectively. The curve of the cool-water indices can be subdivided into three intervals: the first one from 4 to 6.1 msl, with average values around 52% and positive peaks of 70% and 64%, respectively, at 5.25 and 5.8 msl; the second one from 6.2 to 8 msl, with average values around 68%; the third one from 8 msl to the top of the studied segment, with average values around 56% (Fig. 5). The abundance curves of the warm and warm-temperate indices and of the cool and cool-temperate indices show a pattern similar to that obtained for the warm- and cool-water indices (Fig. 5). The abundance curve of the globigerinathekids shows higher numbers in the upper half of the studied segment, with average values increasing from 4% to 11%. Values range from 0% to ~25% at 10.1 msl (Fig. 5). The curve of the small acarininids shows values ranging up to 7.1%. Relatively higher abundances are observed from the bottom of the studied segment up to 4.9 msl, with a peak at 4.5 msl. From here upward, the small acarininids are absent or occur with very low numbers (Fig. 5). The curves of pseudohastigerinids and chiloguembelinids show marked fluctuations from the bottom of the studied segment up to 6.2 msl and an evidently opposite trend in the interval between the impactoclastic layers related to the Popigai and Chesapeake Bay impact events (Fig. 5).

Climatic Curve

The climatic curve is derived from the algebraic sum of percent abundance of warm (positive) and cool (negative) indices as proposed by Cita et al. (1977) and applied by Spezzaferri and Premoli Silva (1991) and Spezzaferri (1995). In the biotite-rich volcaniclastic layers, only minor abundances of warm-water indices are observed (Fig. 5). Therefore, we consider the planktonic foraminifera–based curve to be reliable.

The climatic curve throughout the 4–12 msl segment allows us to recognize three discrete climatic intervals (CI-I to CI-III, from bottom to top) separated by two major shifts: the relatively warmest CI-I interval extends from 4 to 6.1 msl, with an average value of −38.1%, and shows two significant warmer pulses in the intervals from 4.4 to 4.7 msl and from 5.8 to 6 msl and two major cooler pulses at 5.25 and 5.8 msl; the relatively cooler CI-II extends from 6.2 to 8 msl, with an average value of −63.4%; the relatively warmer CI-III extends from 8.1 msl up to the top of the studied segment, with an average value of −50.6%, and shows two significant warmer pulses in the intervals from 8.1 to 8.4 msl and from 10 to 10.2 msl (Fig. 5).

The first two closely spaced impact events fall within the CI-I, and the last impact lies within the CI-III interval. Approximately 5.5 k.y. after the first impact (Popigai), some warmer conditions than those directly below the impact itself are recorded by the sample at 5.65 msl. Subsequently, climatic episodes alternate between, from bottom to top: (1) a cooling episode, up to 5.8 msl, therefore lasting for ~20 k.y., (2) a marked warming pulse, lasting for ~27 k.y., between 5.8 and 6 msl, and (3) a further cooling episode, lasting for ~23 k.y. from 6.0 msl up to the second impact (Chesapeake Bay). The relatively cooler CI-II interval that followed the Chesapeake Bay impact lasted for ~247 k.y. The last impact falls within a cooling episode occurring from 10 to 10.4 msl and lasting for ~55 k.y. About ~16 k.y. after this impact, a subsequent return to less cool conditions occurs that extends up to the top of the studied segment (Fig. 5).

Statistical Treatment of Data

Since patterns of community structures are often not readily apparent (Clarke and Warwick, 1994), we conducted statistical analyses of our data to better identify and characterize changes in planktonic foraminiferal assemblages, and to relate them to changing environmental conditions following the impact events.

At the 84% level of Bray-Curtis similarity, three clusters (A, B, and C in Figs. 6 and 7) were recognized. At the same level of dissimilarity, the differences among clusters were identified by the average abundance of species and/or groups. Cluster A differs from cluster B by higher percentages of chiloguembelinids, globigerinathekids, small acarininids, *Subbotina gortanii*, pseudohastigerinids and *T. cerroazulensis* group and lower percentages of catapsydracids, *D. galavisi*, and *Dentoglobigerina larmeui*. Cluster B differs from cluster C by higher percentages of catapsydracids, *D. galavisi*, *D. larmeui*, "*G.*" *venezuelana*, *S. utilizindex*, and lower percentages of chiloguembelinids, *G. officinalis*, globigerinathekids, *P. nana*, *S. angiporoides*, tenuitellids, and *T. cerroazulensis* group. Cluster C differs from cluster A by higher percentages of catapsydracids, globigerinathekids, globorotaloids, tenuitellids, *S. angiporoides*, and *P. nana*, and lower percentages of small acarininids, *S. utilizindex*, and *T. cerroazulensis* group.

The planktonic foraminiferal assemblages of cluster C can be considered relatively cooler than those of cluster A but relatively warmer than those of cluster B (Figs. 6 and 7). In fact, cluster A shows the lowest abundance of cool species and concurrently the highest abundance of warm species, such as small acarininids and *T. cerroazulensis* group. Moreover, cluster B has the highest abundances of catapsydracids, "*G.*" *venezuelana*, and *S. utilizindex* and the lowest abundance of *T. cerroazulensis* group and can be considered to be the coolest cluster of the studied segment.

The clusters recognized through the statistical treatment of our data fit well with the climatic intervals documented by means of the climatic curve (Figs. 5, 6 ,and 7). Remarkably, the line IMPACT 1, which corresponds to the Ir anomaly at 5.61 msl (Popigai impact), allows us to subdivide cluster A into subclusters A1 (pre-impact 1) and A2 (postimpact 1 to pre-impact 2), with only one exception, the sample found at 5.65 msl (Fig. 7). Subcluster A2 differs from subcluster A1 by lower abundances of catapsydracids, globigerinathekids, and small acarininids, and higher abundances of *D. larmeui*, globorotaloids, and *T. cerroazulensis* group. The most significant feature recognized through the multivariate statistical analysis in the nMDS plot is the line IMPACT 2, which corresponds to the Ir anomaly at 6.17 msl (Chesapeake Bay impact) (Fig. 7). This line clearly separates cluster A from cluster B, which correspond to CI-I and CI-II, respectively. The multivariate statistical analysis shows that significant changes in planktonic foraminiferal assemblages did not take place across the stratigraphical interval, including the Ir anomaly at 10.28 msl (line IMPACT 3) (Fig. 7).

DISCUSSION

High-Resolution Impact Stratigraphy, Geochronology, and Magnetochronology at Massignano: Implications for Global Correlation

On the basis of the available data from the Massignano GSSP, we provide here an updated magnetobiostratigraphic, geochronologic, and magnetochronologic constraint for the late Eocene impactoclastic layers at Massignano that will be useful for global correlation of equivalent horizons.

In terms of biostratigraphy, both the Ir anomalies at 5.61 and 6.17 msl, linked respectively to the Popigai and Chesapeake Bay impact events, fall in the uppermost part of planktonic foraminiferal zone P15 of Berggren et al. (1995), which is in the lowermost part of E15 of Berggren and Pearson (2006) and in the middle part of calcareous nannofossil zone NP19/20 of Martini (1971) (Figs. 2, 4, and 5). The Ir anomaly at 10.28 msl lies in the middle part of planktonic foraminiferal zone P16 of Berggren et al. (1995), which is the upper part of zone E15 of Berggren and Pearson (2006) and calcareous nannofossil zone NP19/20 of Martini (1971) (Figs. 2, 4, and 5).

As for magnetostratigraphy, different data are available. Following Bice and Montanari (1988) and Lowrie and Lanci (1994), the Popigai and Chesapeake Bay impacts fall within the uppermost part of chron C16n, whereas following Jovane et al. (2007), they are more precisely constrained within the upper half of chron C16n.1n. The last impact consistently lies within chron C15n (Figs. 2, 4, and 5).

TABLE 1. ABSOLUTE ABUNDANCES OF PLANKTONIC FORAMINIFERAL SPECIES AND GROUPS IN SAMPLES FROM THE 4–12 MSL SEGMENT AT MASSIGNANO

Stratigraphic level from base (m)	Chiloguembelinids	Catapsydracids	*D. galavisi*	*D. globularis*	*D. larmeui*	*D. tripartita*	*G. officinalis*	*"G." venezuelana*	Globigerinathekids	Globorotaloids	*Gt. martini*	*Gt. ouachitaensis*	Guembelitrids	Hantkeninidae	*P. nana*	Pseudohastigerinids	Small acarininids	*S. angiporoides*	*S. eocaena*	*S. gortanii*	*S. linaperta*	*S. utilizindex*	*S. yeguaensis*	Tenuitellids	*T. ampliapertura*	*T. cerroazulensis* gr.	Total
12	112	112	10	3	35	20	91	103	52	50	5	0	0	0	46	48	0	42	4	30	2	42	26	92	11	12	948
11.9	162	127	1	0	7	5	29	22	144	39	3	0	0	4	17	72	0	56	0	24	0	72	10	106	0	26	926
11.8	143	123	0	0	8	9	43	39	121	52	15	0	0	4	22	39	0	40	0	18	2	64	5	112	0	46	905
11.7	127	85	1	0	13	8	43	51	104	35	5	0	0	4	45	78	2	80	0	10	2	102	0	101	0	54	950
11.6	146	113	3	0	13	13	16	67	106	25	1	0	0	0	28	48	1	62	0	19	0	85	6	128	1	62	943
11.5	144	142	10	2	41	17	31	61	83	38	7	0	1	10	23	37	2	52	0	23	3	71	9	91	0	68	966
11.4	112	166	0	0	22	16	45	101	34	39	6	0	0	4	39	39	0	61	0	19	0	63	5	120	0	58	949
11.3	130	238	5	0	22	25	27	36	77	28	23	0	0	1	32	37	0	68	0	21	0	71	3	117	0	64	1025
11.2	142	150	2	0	8	15	16	3	67	72	1	0	0	2	17	32	0	99	0	3	0	164	1	115	1	29	939
11.1	151	109	0	0	19	6	30	5	66	45	16	0	0	3	38	60	0	126	0	0	1	151	0	73	1	34	934
11	121	74	0	0	12	6	11	22	53	67	2	0	0	1	9	29	0	89	0	4	1	98	2	97	0	8	706
10.9	126	84	0	1	37	16	38	77	89	82	1	0	0	7	38	31	1	79	0	18	2	103	8	74	1	22	935
10.8	117	121	0	0	21	24	40	61	82	62	2	0	0	7	40	19	1	67	0	27	8	132	3	82	5	36	957
10.7	117	127	1	0	24	26	29	74	73	52	2	0	0	1	35	33	6	83	0	20	3	120	0	95	1	24	946
10.6	119	118	2	0	43	14	38	63	98	72	1	0	0	3	49	26	4	43	0	23	4	104	0	76	0	63	963
10.5	107	169	0	0	18	20	28	42	127	72	4	0	1	1	20	16	7	56	0	9	2	118	0	96	0	31	944
10.4	124	152	0	0	20	7	33	32	81	102	3	0	1	3	47	23	5	75	0	4	0	112	0	91	0	11	926
10.3	110	168	4	0	19	9	24	30	119	101	6	0	0	0	44	20	3	72	0	8	0	93	0	88	1	23	942
10.25	130	156	2	0	14	9	27	25	124	109	2	0	0	4	38	23	0	63	0	4	0	97	0	90	1	35	953
10.2	145	76	1	0	0	4	28	31	226	44	0	0	0	1	13	36	2	62	0	2	4	111	0	67	0	66	919
10.1	96	111	2	0	0	9	41	27	235	85	1	0	2	2	7	51	0	34	0	5	2	80	0	53	2	95	940
10	186	84	2	0	0	4	30	28	142	52	1	0	1	3	13	29	0	45	0	14	4	96	0	106	10	116	966
9.9	119	116	3	1	30	18	32	97	92	54	5	0	2	3	35	25	3	50	0	36	0	79	9	107	0	30	946
9.8	142	155	0	0	23	10	31	35	97	46	1	0	0	0	19	23	1	83	0	35	0	124	6	94	0	19	944
9.7	125	117	10	1	17	20	44	65	78	50	0	0	0	1	36	27	3	76	0	54	0	94	14	86	0	37	955
9.6	129	162	0	0	25	13	36	49	87	24	2	0	1	0	37	29	0	95	0	24	1	105	7	83	0	19	928
9.5	162	113	4	0	20	12	46	53	66	33	3	0	1	5	40	47	1	68	0	22	0	101	7	78	0	63	945
9.4	156	120	2	0	14	10	42	58	109	28	2	0	1	0	33	32	2	89	0	13	1	137	4	73	0	36	962
9.3	146	113	6	0	15	24	6	62	84	37	3	0	0	2	19	10	0	93	0	16	1	137	3	117	0	34	928
9.2	171	125	0	0	27	17	20	35	106	49	2	0	2	0	20	35	0	92	0	20	0	86	3	110	0	31	951
9.1	136	109	0	0	28	12	28	109	105	37	5	0	1	2	22	16	11	51	0	25	1	104	10	97	2	47	958
9	151	114	8	0	27	10	25	89	42	36	9	0	1	4	19	13	3	58	0	27	0	157	7	108	0	56	964
8.9	117	137	3	1	23	14	39	83	102	58	6	0	0	4	35	26	1	57	0	29	1	110	8	70	2	28	954
8.8	154	158	2	0	26	20	12	63	68	23	6	0	0	1	22	19	4	82	1	32	0	119	7	94	0	28	941
8.7	150	129	8	0	17	17	28	99	43	38	1	0	0	3	18	24	0	69	0	30	0	107	15	90	0	67	953
8.6	137	157	4	0	19	12	31	104	77	67	7	0	0	1	31	19	0	34	0	38	2	63	4	99	2	68	976
8.5	124	176	2	0	24	12	20	45	99	33	1	0	0	2	20	12	1	84	0	20	0	103	6	95	0	42	921
8.4	134	123	0	0	16	18	28	53	123	24	16	0	0	3	24	40	0	74	0	8	3	102	1	60	0	94	944
8.3	136	87	0	3	20	10	45	48	182	28	10	0	0	0	33	26	2	63	0	29	1	113	1	73	1	25	936
8.2	143	86	1	0	6	10	10	18	217	51	15	0	0	0	32	23	5	43	0	12	2	114	1	91	0	76	956
8.1	199	195	3	0	11	15	20	34	107	41	19	0	1	0	30	25	0	28	0	9	3	84	1	77	0	64	966
8	82	203	34	1	15	1	0	127	1	2	7	0	0	0	0	62	0	0	0	3	0	196	0	38	3	25	800
7.9	88	205	43	3	30	1	0	169	0	20	9	0	0	0	1	57	0	0	1	4	0	151	0	36	0	9	827

(*continued*)

TABLE 1. ABSOLUTE ABUNDANCES OF PLANKTONIC FORAMINIFERAL SPECIES AND GROUPS IN SAMPLES FROM THE 4–12 MSL SEGMENT AT MASSIGNANO (*continued*)

Stratigraphic level from base (m)	Chiloguembelinids	Catapsydracids	*D. galavisi*	*D. globularis*	*D. larmeui*	*D. tripartita*	*G. officinalis*	*"G." venezuelana*	Globigerinathekids	Globorotaloids	*Gt. martini*	*Gt. ouachitaensis*	Guembelitrids	Hantkeninidae	*P. nana*	Pseudohastigerinids	Small acarininids	*S. angiporoides*	*S. eocaena*	*S. gortanii*	*S. linaperta*	*S. utilizindex*	*S. yeguaensis*	Tenuitellids	*T. ampliapertura*	*T. cerroazulensis* gr.	Total
7.8	49	212	76	5	48	0	0	131	0	20	19	0	0	0	1	19	0	0	1	4	0	170	0	44	3	2	804
7.7	80	237	51	2	38	1	0	112	0	7	15	1	0	0	2	17	1	0	1	5	0	200	0	68	3	21	862
7.6	58	218	37	3	23	1	0	111	0	15	21	0	0	0	1	5	0	0	2	8	0	226	0	65	5	23	822
7.5	79	188	54	15	34	3	1	100	2	25	39	1	0	0	0	16	1	0	1	7	0	221	0	79	5	32	903
7.4	53	155	68	20	59	1	1	79	1	25	1	1	0	0	0	8	0	0	3	1	0	235	0	62	2	53	828
7.35	68	175	56	7	50	2	2	71	2	38	16	2	0	0	0	27	0	0	3	5	0	225	0	69	4	52	874
7.3	108	191	51	5	33	0	0	94	0	42	9	1	0	0	0	36	0	0	3	9	0	232	0	76	1	15	906
7.2	132	234	62	15	32	0	0	132	0	46	4	0	0	0	0	30	0	0	2	9	0	207	0	40	2	3	950
7.15	86	254	65	10	34	0	0	109	0	20	7	0	0	1	0	19	3	0	2	9	0	207	0	56	0	1	883
7.1	138	203	49	7	29	1	0	51	0	25	22	0	0	2	0	62	9	0	3	12	1	228	0	53	5	15	915
7	84	114	48	10	30	2	0	14	12	29	29	1	0	0	0	48	22	0	6	8	3	269	0	92	7	41	869
6.9	69	154	48	4	28	1	0	121	18	10	2	0	0	1	0	84	1	0	3	8	0	291	0	37	7	36	923
6.8	69	145	49	10	18	1	1	103	38	11	2	2	0	0	2	53	2	0	2	10	0	296	0	60	4	27	905
6.7	70	112	17	5	8	0	1	81	175	9	6	0	0	0	0	44	1	0	2	12	1	230	0	66	2	59	901
6.6	79	137	24	9	22	0	0	100	32	11	9	1	0	4	1	44	2	0	3	9	1	324	0	71	4	35	922
6.5	93	136	33	10	33	1	2	83	15	16	20	1	0	1	1	30	1	0	4	17	1	333	0	72	0	23	926
6.45	59	201	69	7	55	0	0	127	0	14	19	0	0	0	1	3	9	0	3	13	2	220	0	39	7	3	851
6.4	90	124	12	8	39	1	2	71	3	28	8	1	0	0	0	57	3	0	4	2	3	333	0	65	5	31	890
6.3	81	121	32	5	31	1	2	55	8	23	28	0	0	0	2	72	1	0	5	5	2	312	0	63	10	34	893
6.2	78	134	49	8	39	0	2	55	1	29	27	1	0	0	1	86	7	0	5	7	2	239	0	103	6	23	902
6.1	184	25	7	3	22	0	0	123	6	35	29	1	0	1	4	18	4	0	0	7	0	300	0	41	0	54	864
6	143	11	0	4	19	1	0	45	5	42	11	0	0	0	2	67	3	0	0	8	0	299	0	49	0	181	890
5.9	97	22	2	8	19	0	0	51	4	67	14	0	0	0	0	83	1	0	0	15	0	237	0	42	1	264	927
5.8	154	19	0	1	13	0	0	162	20	42	22	0	0	0	0	26	2	0	0	5	0	233	0	60	0	47	806
5.7	74	37	10	6	34	0	0	78	13	70	4	0	0	1	3	111	2	0	0	18	0	225	0	60	0	117	863
5.65	165	42	4	5	29	0	0	54	73	45	9	0	0	0	1	43	0	0	0	43	0	169	0	60	0	107	849
5.6	178	52	7	7	14	0	0	65	19	55	8	0	0	0	0	42	0	0	0	18	0	229	0	39	1	145	879
5.5	162	24	4	7	14	0	0	41	100	30	12	0	0	0	0	55	1	0	0	20	0	243	0	51	3	94	861
5.4	140	26	2	3	14	0	0	67	43	34	10	0	0	0	1	50	2	0	0	25	0	310	0	46	0	97	870
5.3	140	25	3	9	10	0	2	53	74	27	9	0	0	0	0	82	1	0	0	47	0	244	0	66	0	75	867
5.25	175	53	8	0	11	0	0	580	58	12	37	0	0	0	0	39	2	0	0	29	0	225	0	40	3	27	1299
5.2	143	43	3	1	2	0	0	33	61	22	3	0	0	0	0	41	0	0	0	30	0	281	0	52	1	87	803
5.1	179	25	4	0	15	0	0	33	51	26	1	0	0	0	0	42	2	0	0	26	0	306	0	46	1	89	846
5	122	47	9	7	3	0	1	78	71	26	4	0	0	4	1	106	3	0	0	20	0	258	0	32	4	71	867
4.9	136	52	3	2	9	0	2	60	54	18	10	0	0	0	5	80	15	0	0	34	0	258	0	51	2	82	873
4.8	159	58	7	4	8	0	0	62	51	28	12	0	0	1	0	57	9	0	0	48	0	227	0	42	1	58	832
4.7	122	57	1	0	5	0	0	33	58	25	16	0	0	4	0	98	22	0	0	23	0	228	0	20	2	80	794
4.6	92	37	1	0	8	0	0	53	123	21	11	0	0	0	0	84	33	0	0	48	0	147	0	43	1	83	785
4.5	81	25	2	3	9	0	0	57	99	30	6	0	0	0	1	61	60	0	0	60	0	195	0	43	5	111	848
4.4	87	28	6	0	3	0	0	109	65	13	10	0	0	0	1	105	25	0	0	44	0	212	0	35	0	91	834
4.3	124	40	5	0	4	0	0	134	51	8	10	0	0	0	0	50	42	0	0	34	0	174	0	54	2	57	789
4.2	120	54	6	6	4	0	0	76	46	17	11	0	0	0	0	56	27	0	0	35	2	269	0	91	10	42	872
4.1	104	49	4	2	8	0	0	104	50	23	4	0	0	0	7	56	20	0	1	42	1	180	0	89	20	87	851
4	83	33	8	0	10	0	0	66	37	41	5	0	0	0	1	82	31	0	0	39	5	226	0	75	18	84	844

TABLE 2. PALEOECOLOGY* AND PALEOCLIMATIC SIGNIFICANCE OF THE RECOGNIZED SPECIES AND GROUPS FROM THE 4–12 MSL SEGMENT AT MASSIGNANO[†]

Taxa/groups	Biogeographic distribution	Habitat	Paleoclimatic significance
Catapsydrax dissimilis	Cosmopolitan	Deep water	Cool
Catapsydrax unicavus	Cosmopolitan	Deep water	Cool
Chiloguembelinids	Mid-latitudes	Thermocline/oxygen minimum layer	Non–climate diagnostic
Dentoglobigerina galavisi	Low to mid-latitudes	Deep water	Warm temperate
Dentoglobigerina globularis	No data	No data	No data
Dentoglobigerina larmeui	Mid-latitudes	High thermocline	Probably warm-temperate
Dentoglobigerina tripartita	Low to mid-latitudes	High thermocline	Warm temperate
Globigerina officinalis	Low to mid-latitudes	Mixed layer	Cool
"*Globigerina*" *venezuelana*	Mid-latitudes	Deep water	Probably cool
Globigerinatheka index	Cosmopolitan, more abundant in high latitudes	Mixed layer	Cool temperate
Globigerinatheka luterbacheri	Common in mid-latitudes, very rare or absent in tropical regions	No data	Cool temperate
Globigerinatheka semiinvoluta	Low to mid-latitudes	Mixed layer	Cool temperate
Globigerinatheka subconglobata	Cosmopolitan	Mixed layer	Cool temperate
Globigerinatheka tropicalis	Mid- to low latitudes	No data	Cool temperate
Globorotaloides eovariabilis	High latitudes	No data	Cool
Globorotaloides sp. 1	Cosmopolitan	Probably deep water	Probably cool
Globoturborotalita martini	Mid- to low latitudes	Mixed layer	Warm
Globoturborotalita ouachitaensis	Mid- to low latitudes	No data	Non–climate diagnostic
Guembelitrids	Cosmopolitan	Probably upwelling	Probably cool
Hantkeninidae	Mid- to low latitudes	Mixed layer	Warm
Paragloborotalia nana	Low to mid-latitudes	Deep water	Cool temperate
Pseudohastigerinids	Low to mid-latitudes	Mixed layer to thermocline/oxygen minimum layer	Non–climate diagnostic
Small acarininids	Cosmopolitan, most common in low latitudes	Mixed layer	Warm
Subbotina angiporoides	Cosmopolitan, most common at high latitudes	Thermocline	Cool
Subbotina eocaena	Mid- to high latitudes	Deep water	Cool
Subbotina gortanii	Low to mid-latitudes	Deep water	Non–climate diagnostic
Subbotina linaperta	Cosmopolitan, most common at southern, high latitudes	Deep water	Cool
Subbotina utilisindex	Cosmopolitan, most common at high latitudes	Deep water	Cool
Subbotina yeguaensis	Low to high latitudes	No data	No data
Tenuitellids	Cosmopolitan, most common at high latitudes	Mixed layer	Cool
Turborotalia ampliapertura	Cosmopolitan	Mixed layer to deep water	Cool temperate
Turborotalia cerroazulensis	Cosmopolitan	Mixed layer to thermocline	Warm
Turborotalia cocoaensis	Cosmopolitan, more restricted to tropical latitudes than *T. cerroazulensis*	Mixed layer to thermocline	Warm
Turborotalia cunialensis	Cosmopolitan, more restricted to tropical latitudes than *T. cerroazulensis*	No data	Warm
Turborotalia pomeroli	Cosmopolitan, most common in northern mid-latitudes	Mixed layer	Warm

*Preferred biogeographical distribution and life position in the water column.

[†]Largely following Douglas and Savin (1978), Boersma et al. (1979, 1987), Keller (1983), Poore and Matthews (1984), Nocchi et al. (1988b), Premoli Silva and Boersma (1988, 1989), Stott et al. (1990), Barrera and Huber (1991, 1993), Boersma and Premoli Silva (1991), Spezzaferri and Premoli Silva (1991), Pearson et al. (1993, 1997, 2001, 2006), Spezzaferri (1995), Van Eijden and Ganssen (1995), Pearson and Palmer (1999), Coxall et al. (2000), and Wade and Kroon (2002).

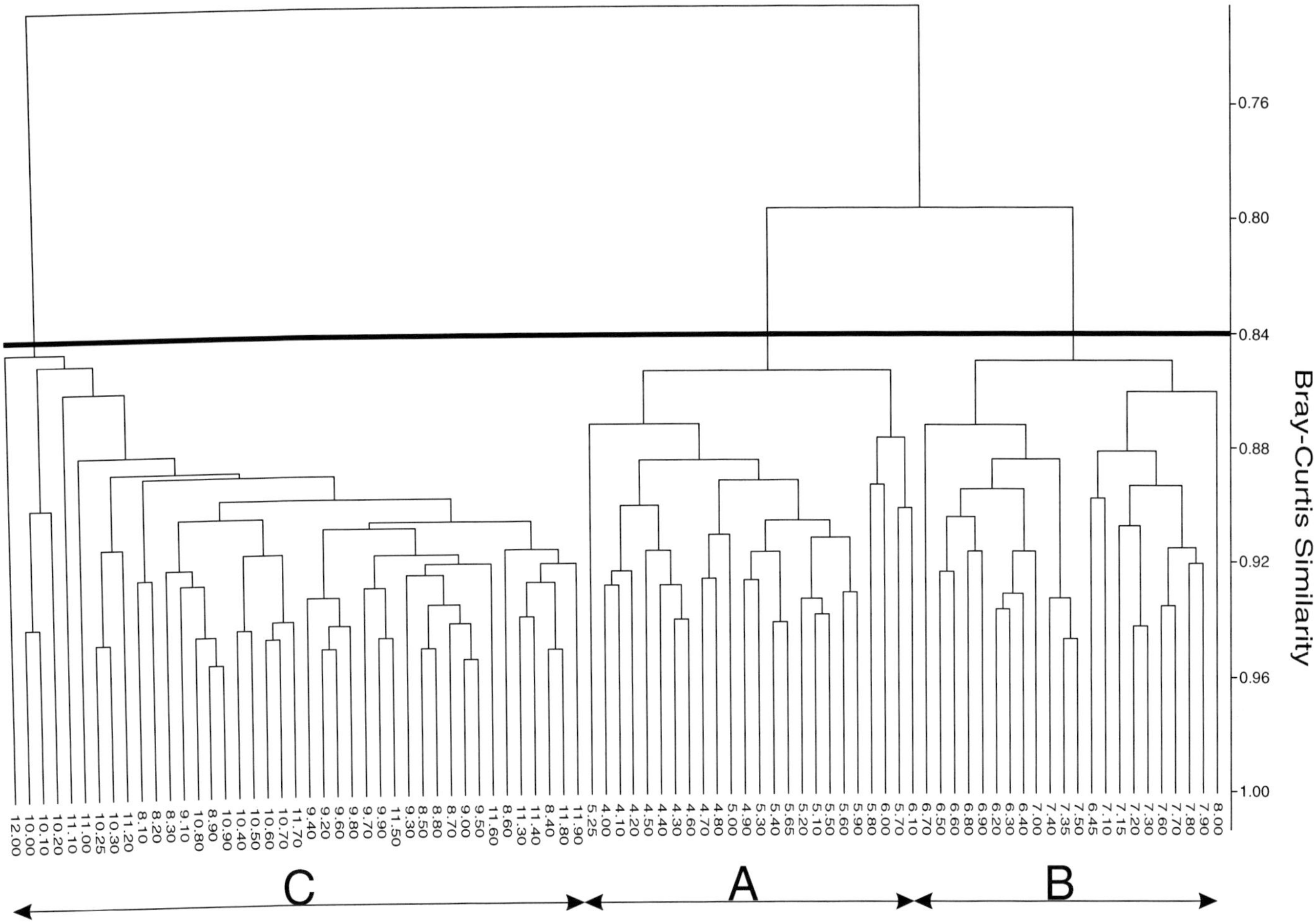

Figure 6. Classification of samples of the 4–12 m stratigraphic level from the base (msl) segment at Massignano using hierarchical agglomerative cluster analysis based on the Bray-Curtis similarity matrix. Three clusters (A, B, and C) were identified at the 84% level of Bray-Curtis similarity.

In terms of geochronology, an interpolated age of 35.7 ± 0.4 Ma for the impactoclastic layer at 5.61 msl was derived from radioisotopic dating performed in the volcaniclastic layers at Massignano (Montanari et al., 1993) (Figs. 4 and 5). On the basis of the magnetochronologic ages provided by Cande and Kent (1995), Gradstein et al. (2004), and Jovane et al. (2006) for chrons C16n.2r, C16.1n, and C15n (Figs. 4 and 5), the numerical ages for the Popigai, the Chesapeake Bay, and the unidentified impact events at Massignano would be 35.417–35.493 Ma, 35.346–35.411 Ma, and 34.778–34.894 Ma, respectively. On the basis of the mean sedimentation rate of 0.73 cm/k.y. (Jovane et al., 2007) and of the magnetochronologic ages, the time interval between the Popigai and Chesapeake Bay impacts would span ~77 k.y. on average.

A comparison of the high-resolution magnetobiostratigraphic and geochronologic data from the Massignano GSSP with those from Ocean Drilling Program (ODP) Holes 689B (Maud Rise, Southern Ocean) and 1090B (South Atlantic sector of the Southern Ocean), and the Chickahominy Formation in the Kiptopeke core hole (offshore southeastern Virginia) provides a valuable time constraint and global correlation for the late Eocene impact-ejecta horizons.

At ODP Hole 689B, the Antarctic impact-ejecta horizon that is interpreted to represent the clinopyroxene-bearing spherule-strewn field of the North America tektite strewn field, providing an $^{40}Ar/^{39}Ar$ date of 35.5 Ma (Glass et al., 1982; Vonhof et al., 2000), lies within chron C16n.1n (Florindo and Roberts, 2005). This horizon falls in the uppermost part of planktonic foraminiferal *Globorotaloides suteri* (AP12) zone (Stott and Kennett, 1990), which would correspond to the uppermost part of zone P15 of Berggren et al. (1995) and to the calcareous nannofossil NP19/20 equivalent zone (Wei and Wise, 1992; see also Berggren et al., 1995).

According to Pusz et al. (this volume), no distinct stratigraphic separation is found between the microtektite and microkrystite layers at Site 1090, recognized as its own layer (Liu et al., 2000; Kyte and Liu, 2002). Following the magnetostratigraphic age control provided by Channell et al. (2003), the Ir anomaly located 10 cm above the two stratigraphically coeval, but geochemically distinct, Chesapeake Bay impact structure microtektite

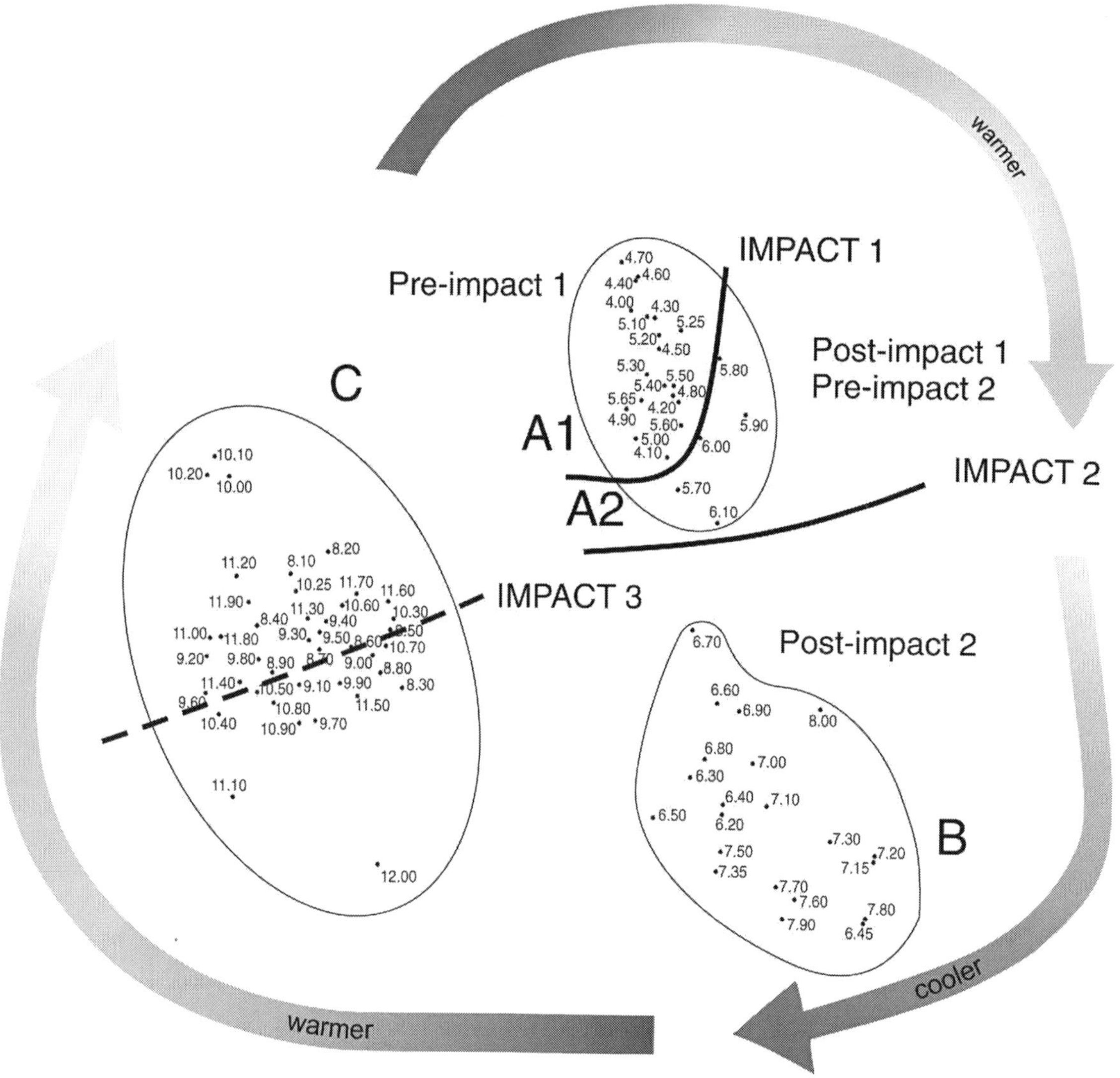

Figure 7. Interpretation of clusters identified in the dendrogram (Fig. 6) using non-metric multidimensional scaling (nMDS). The ordination by nMDS is an iterative procedure to represent "distance" of samples from a multidimensional space on the base of rank dissimilarities. The plot has no defined scale. The stress value (0.13) represents the distortion involved in compressing the data to a small number of dimensions; stress < 0.2 in 2D means a useful representation of data (Kruskal, 1977; Field et al., 1982). In this study, the line IMPACT 1, which corresponds to the Ir anomaly at 5.61 msl, allows us to subdivide cluster A into subcluster A1 (pre-impact 1) and subcluster A2 (postimpact 1 to pre-impact 2). The most significant feature recognized in the nMDS plot is the line IMPACT 2, which corresponds to the Ir anomaly at 6.17 msl. This line clearly separates cluster A from cluster B, corresponding to CI-I and CI-II, respectively. According to the multivariate statistical results, any significant change in planktonic foraminiferal assemblages took place across the Ir anomaly at 10.28 msl (dotted line IMPACT 3). The shaded bands with arrows indicate the temperature water gradient based on climate curve.

and Popigai microkrystite layers (both at 35.43 Ma) falls within chron C16n.1n, with an interpolated magnetochronologic age of 35.426 Ma (Miller et al., 2007; Pusz et al., this volume). This is consistent with published radiometric ages ranging from 35.3 to 35.7 Ma (Glass et al., 1982; Obradovich et al., 1989; Bottomley et al., 1997; Horton and Izett, 2005). According to the planktonic foraminiferal biostratigraphy provided at the ODP Site 1090 by Galeotti et al. (2002), the microtektite and microkrystite layers fall in the upper part of the *Globigerinatheka index–Globorotaloides suteri* (AP11–AP12) zones of Stott and Kennett (1990), which would correspond to the upper part of zones P15–P16 of Berggren et al. (1995) and in the middle part of zones NP19/20 of Martini (1971), following the nannofossil calcareous biostratigraphy established by Marino and Flores (2002).

According to the magnetobiochronology provided by Poag et al. (2003, 2004) for the continuous interval of the Chickahominy

Formation in the Kiptopeke core hole, which lies conformably above the Chesapeake Bay impact crater, the Chesapeake Bay impact would have taken place in chron C16n.2n and within planktonic foraminiferal zone P15 of Berggren et al. (1995), with an extrapolated age of 35.78 Ma. However, the new magnetostratigraphic data from the Massignano GSSP together with those from ODP Holes 689B and 1090B cast reliable doubt on the interpretation of the magnetostratigraphic record of the Kiptopeke core hole, which, as a consequence, needs to be refined.

Planktonic Foraminiferal Overall Trends during the Late Eocene

The end of the Eocene was a critical interval within the transition from the Eocene "greenhouse" world to the Oligocene "icehouse" world marked by significant evolutionary turnovers of marine biota and changes of oceanographic and atmospheric circulation patterns (Benson, 1975, 1990; Kennett, 1977; Miller et al., 1987, 1991, 2005; Lear et al., 2000, 2008; Zachos et al., 2001; Coxall et al., 2005; Pagani et al., 2005; Tripati et al., 2005; Coxall and Pearson, 2007). Through a study based on planktonic foraminifera assemblages from some Deep Sea Drilling Program (DSDP) sites in the low-latitude Atlantic (Site 363), Pacific (Site 292), and Indian (Site 219) Oceans and a high-latitude South Pacific site (Site 277), Keller (1983) observed that the Eocene to early Oligocene cooling proceeded by pulses separated by relatively stable climate conditions. These pulses were characterized by rapid cooling indicated by the oxygen isotope record, major faunal turnovers, and hiatuses. She identified one of the major pulses in the interval across the transition between planktonic foraminiferal zone P15 and P16 of Berggren and Miller (1988) and Berggren et al. (1995) and observed that this pulse was accompanied by successive replacement of warm middle Eocene surface planktonic foraminiferal species by cooler late Eocene intermediate- and deep-water species. According to Keller (1983), the disappearance of the surface dwellers suggests that in high- and low-latitude oceans, during the planktonic foraminiferal zone P15–P16 interval, significantly cooler temperatures prevailed in the upper 400–500 m of the water column. Boersma and Premoli Silva (1986) and Keller et al. (1992) pointed out that long-term trend of planktonic foraminifera evolutionary turnover from the middle Eocene into the Oligocene largely involved extinction of warm-water, tropical, surface-dwelling species.

Several authors have speculated about the causes that produced the late Eocene cooling and foraminiferal faunal reorganization. Keller (1983) and Keller et al. (1992) related the late Eocene cooling and water mass reorganization to significant ice-cap development and increased thermal gradients between the equator and the poles, and Berger et al. (1981) observed that cooling phases may have been enhanced by increased albedo. Boersma et al. (1987) explained the reorganization of the planktonic foraminiferal assemblage structures in zone P16 by a shoaling of the nutricline and oxygen minimum layer. Boersma and Premoli Silva (1991) and Hallock et al. (1991) attributed the variations in the structure of the planktonic foraminiferal communities from the late middle Eocene through the middle Oligocene to reduced temperatures, shoaling of the thermocline, and increased delivery of nutrients to oceanic surface waters. These changes produced a contraction of the trophic resource regional gradient and reduction of oligotrophic and warm surface-water habitats (Hallock, 1987). According to Coxall and Pearson (2007), the extinction at and near the Eocene-Oligocene boundary might therefore have been due to rapid environmental change and cooling, and it may also have been influenced by changing water mass stratification and patterns of biological productivity.

The discovery of some bolide-impact craters and associated breccias, ejecta horizons, and cosmic signatures (e.g., Montanari et al., 1993, 1998; Clymer et al., 1995, 1996; Langenhorst and Clymer, 1996; Poag and Aubry, 1995; Koeberl et al., 1996; Bottomley et al., 1997; Farley et al., 1998; Pierrard et al., 1998; Liu et al., 2000; Kyte, 2001; Kyte and Liu, 2002; Poag et al., 2002, 2003, 2004; Bodiselitsch et al., 2004) in the late Eocene has led to speculation that the late Eocene impact events may have played an important role related to the biotic turnover and the deterioration of the global climate at the end of the Eocene Epoch, even if their effects on the climate and on the biosphere are still not completely clear (e.g., Keller, 1986; Brinkhuis and Coccioni, 1995; Coccioni et al., 2000; Vonhof et al., 2000; Poag et al., 2002, 2003, 2004; Spezzaferri et al., 2002; Bodiselitsch et al., 2004).

Minor extinction episodes and assemblage shifts among planktonic (e.g., Keller, 1986; Coccioni et al., 2000; Spezzaferri et al., 2002) and benthic (Poag et al., 2004) foraminifera, dinocysts (Brinkhuis and Coccioni, 1995; Vonhof et al., 2000), and stable oxygen and carbon isotope excursions associated with the impact-ejecta horizons at several sites (Vonhof et al., 2000; Poag et al., 2002, 2003, 2004; Bodiselitsch et al., 2004; Pusz et al., 2006, 2007, this volume; Miller et al., 2007) have been taken as evidence for global-scale short- and long-term environmental disturbance related to the impacts. According to Coxall and Pearson (2007), it seems unlikely that the impacts had a major influence on Earth's biosphere at the Eocene-Oligocene boundary, and they have to be ruled out as a possible cause of the climatic transition. It is possible, however, that climatic disruption caused by atmospheric dust loading associated with the late Eocene impacts, coupled with an ice-albedo feedback mechanism that amplified impact-induced climatic cooling, may have contributed to global cooling in the run-up to the Eocene-Oligocene transition.

Late Eocene Impact-Induced Environmental Perturbations, Including Climate and Water Mass Structure Changes?: Evidence from the Massignano GSSP

Several lines of evidence from nannofossil and foraminiferal assemblages (e.g., Keller, 1983; Quilty, 2001; Persico and Villa, 2004; Coccioni et al., 2000; Spezzaferri et al., 2002), dinoflagellate cyst palynology (e.g., Brinkhuis, 1992; Coccioni et al., 2000), clay mineralogy (e.g., Diester-Haass et al., 1996; Ehrmann, 1997), oxygen isotopes (e.g., Kennett and Stott, 1990; Miller et al., 1992;

Vonhof et al., 2000; Bohaty and Zachos, 2003; Poag et al., 2003, 2004; Bodiselitsch et al., 2004; Pusz et al., 2007, this volume), and environmental magnetism (Sagnotti et al., 1998; Jovane et al., 2004) indicate that the major cooling event at the Eocene-Oligocene boundary was predated by distinct late Eocene cooling pulses, separated by relatively stable warming phases.

Our data show that the late Eocene impact events recorded in the 4–12 msl segment of the Massignano GSSP had no dramatic effects on planktonic foraminifera. However, acting as forcing factors, they induced some environmental perturbations and may have contributed to noticeable climate changes superimposed on the general late Eocene cooling trend.

Based on field evidence (Coccioni et al., 1988) and on the redistribution of extraterrestrial iridium (Huber et al., 2001) at Massignano, we exclude the effect of bioturbation and the resultant vertical displacement of biostratigraphic and paleoclimatic events. Accordingly, the paleoclimatic trend at Massignano can be considered uncorrupted.

The thermocline is the one of the main features responsible for partitioning of the water column and therefore of the vertical distribution of planktonic organisms, especially foraminifera. Either generally warmer surface water or the presence of a deeper thermocline quantitatively favors shallow and warm-water dwellers, while a shallower thermocline quantitatively favors deeper-dwelling and cool-water forms (e.g., Kennett and Barker, 1990).

The generally high abundance of cool- and deep-water species, the low abundances of warm and temperate forms, and the climatic curve from the 4–12 msl segment indicate that the studied interval was characterized by generally cool conditions (Fig. 5), a relatively shallow thermocline, as deduced by the generally high abundances of cool- and deep-water dwellers, and consequently a reduced surface-layer thickness.

Within a time of global cooling, however, abundance variations of the recognized paleoclimatic species and groups document different climate conditions throughout the studied segment at Massignano. Relatively warmer conditions (CI-I) occurred from the base of the studied segment up to the Chesapeake Bay impact event, then relatively coolest conditions (CI-II) persisted for a few hundreds of thousands of years, and finally relatively less cool conditions (CI-III) developed up to the top of the segment (Fig. 5). The statistical treatment of our data corroborates the paleoclimatic trend detected by the fluctuations of the planktonic foraminiferal climatic indices (Fig. 7). Based on our observations, we interpret lines IMPACT 1 and IMPACT 2 in the nMDS plot to represent the temperature gradient (Fig. 7), and we relate the gaps between subcluster A1 and subcluster A2 and between subcluster A2 and cluster B to a prevailing warming pulse and a cooling episode after the Popigai and Chesapeake Bay impact events, respectively.

Both the Popigai and Chesapeake Bay impacts liberated energy, probably equivalent to some 10 million megatons of TNT (trinitrotoluene) (Montanari and Koeberl, 2000). These impacts may have had some short-term effects on the environment by injecting dust into the atmosphere, inducing acid rain, cooling, and inhibiting photosynthesis (e.g., Gerstl and Zardecki, 1982; Toon et al., 1994; Kring, 2000).

A slight decrease in abundance of globigerinathekids, which have a symbiotic life strategy (Boersma et al., 1987; Pearson et al., 1993), took place just after the Popigai impact and might have been related to increased water turbidity and/or inhibition of photosynthesis following the impact event itself, which prevented the proliferation of algal symbionts (Figs. 4 and 5). Acarininids are also known as symbiont-bearing species (Coxall et al., 2000), but they are very rare throughout the studied interval (Figs. 4 and 5), and it was not possible to verify their decrease in abundance after the impact events. The recognized distinct opposite trends of the low-oxygen-tolerant chiloguembelinids and pseudohastigerinids may indicate enhanced competition between them during times of environmental stress. There was a slight reduction in the number of chiloguembelinids after the second impact event, which may reflect a very moderate enrichment of O_2 in the water masses (Figs. 4 and 5).

At Massignano, an increase in productivity in surface waters across the Popigai and Chesapeake Bay impact intervals persisting for ~137 k.y. might be inferred from a strong abundance increase of the cosmopolitan, calcareous dinoflagellate species *Thalassiphora pelagica*, which is interpreted to be a cold-water or productivity indicator (Coccioni et al., 2000). In fact, this species is known to be associated with anoxic sediments (e.g., Köthe, 1990) and with the major cooling event recorded worldwide in the early Oligocene (Brinkhuis and Biffi, 1993; Zachos et al., 1996).

Changes in ventilation and flushing at the seafloor, reflecting significant reorganization of the water mass structure after the Popigai and Chesapeake Bay impact events, might be indicated by sudden changes in sediment color (e.g., Coccioni et al., 2000; Spezzaferri et al., 2002) (Figs. 2–5).

Quantitative changes in the planktonic foraminiferal assemblages indicate a slight warming pulse ~5 k.y. after the deposition of the impactoclastic layer related to the Popigai impact. This discrete pulse is followed by alternating cooler and warmer intervals. However, as highlighted by subcluster A2, the interval between the Popigai and Chesapeake Bay impacts, lasting for ~77 k.y., was characterized by prevailing warmer condition compared to the conditions encountered prior to the first impact (Figs. 5 and 7).

The Chesapeake Bay impact took place within a following cooling episode. Bodiselitsch et al. (2004) documented a negative shift in bulk carbonate $\delta^{18}O$ and $\delta^{13}C$ close to the Ir-rich level at 6.17 msl compared to the continuously decreasing trend over the entire Massignano section. According to these authors, despite the fact that oxygen isotope values at Massignano are clearly diagenetically influenced, the negative excursion of $\delta^{18}O$ might provide evidence of a short warming pulse. After the Chesapeake Bay impact, an enhanced cooling (CI-II) interval started (Fig. 5; cluster B in Figs. 6 and 7), as testified by the increase in abundance of cool and cool-temperature water species and groups. This change in abundance occurred along with a concurrent decrease in abundance of warm and warm-temperature water species and groups, which reveals a contraction of the mixed and surface layers, and consequent decrease of niches in the warmer surface water. This relatively long-term cooling episode lasted for ~247 k.y. and was accompanied by a shallowing of the thermocline, which occurred in the interval spanning the transition between planktonic foraminiferal zones P15 and P16 of Berggren et al. (1995), which is in the lower part of zone E15 of Berggren and Pearson (2006), and it may have been related to the cooling pulse identified by Keller (1983) in the same biostratigraphic interval.

Oxygen isotope records at ODP Hole 689B indicate accelerated climatic cooling, possibly combined with increased surface-water productivity deduced by $\delta^{13}C$ data and calcareous dinocyst abundances, coinciding with the two nearly contemporaneous impact events (Vonhof et al., 2000). The relatively long duration of the cooling pulse that followed the impact events led Vonhof et al. (2000) to suggest that a feedback mechanism, such as a global albedo increase due to extended snow and ice cover, sustained the initial impact-induced changes. We agree with this hypothesis to explain the enhanced cooling following the Chesapeake Bay impact and corresponding to interval CI-II, where a rapid increase of $\delta^{13}C$ and $\delta^{18}O$ values occurs (Bodiselitsch et al., 2004) together with an increase in abundance of *Thalassiphora pelagica* (Coccioni et al., 2000).

The enhanced cooling terminated within the middle part of chron C15r and the lowermost part of planktonic foraminiferal zone P16 of Berggren et al. (1995), which is in the lower part of zone E15 of Berggren and Pearson (2006) (Fig. 5). From there on, climate conditions intermediate (CI-III) between the warmest conditions (CI-I) and the coolest conditions (CI-II), as recorded through the studied segment, developed and persisted (Fig. 5), as also evidenced by cluster C (Figs. 6 and 7).

The third impact, which lies within the CI-III interval, led to an enhanced cooling that was followed after ~15 k.y. by a subsequent return to less cool conditions, which extend up to the top of the studied segment (Fig. 5).

However, this impact does not appear to have produced significant environmental perturbations, as evidenced by the unimportant fluctuations of the planktonic foraminiferal assemblages (Figs. 5 and 7). According to Bodiselitsch et al. (2004), negative excursions in bulk carbonate $\delta^{18}O$ and $\delta^{13}C$ occur close to the Ir-rich level at 10.28 msl, which might indicate a sudden warming pulse caused by this impact event. Again, on the basis of carbon and oxygen isotope ratio data, it seems that seawater temperature did not remarkably change in between the two last Ir-enhanced intervals at Massignano (Bodiselitsch et al., 2004). However, the average $\delta^{13}C$ values differ somewhat between the two regions, which could indicate an incursion of colder, more vigorous bottom waters and an increase of biomass and productivity during this time span as seen in the deep-water benthic foraminifera faunal turnover recognized in the same interval by Coccioni and Galeotti (2003).

Based on this line of evidence, the climatic effects of the late Eocene impact events as documented at Massignano therefore appear to be not univocal. In fact, it seems that the Popigai impact was followed by a prevailing warming pulse, whereas the other two impact events were followed by an enhanced cooling.

It is well known that the environmental and the biological effects of an impact event must be evaluated in terms of impactor size as well as in the context in which it occurs in terms of location and target rocks (e.g., Kring, 2000). The Popigai and Chesapeake Bay impactors are estimated to have been 5 km (Tagle and Claeys, 2004) and 3.2 km (Collins and Wünnemann, 2005) in diameter, respectively. The Popigai impact occurred on the continent, whereas the Chesapeake Bay impact took place on the continental shelf. The target rocks of the Popigai structure are generally granitic gneisses overlain by sandstone and carbonates (Bottomley et al., 1997) produced by the impact of an ordinary chondrite body (Masaitis and Raikhlin, 1986). The Chesapeake Bay impact structure is developed in a mixed-target substrate composed of granitoids and metasedimentary rocks overlain by dominantly siliciclastic, sedimentary rocks (Koeberl et al., 1996; Poag et al., 2002). According to Bodiselitsch et al. (2004), the Ir anomaly at 10.28 msl may be also derived from an impact into a continental shelf, similar to the Chesapeake Bay impact. Kent et al. (2003) suggested that release of methane hydrates from mechanical disruption of sediments as a result of an impact could cause a greenhouse effect, which is revealed by negative excursions in the carbon and oxygen isotope records. In view of this, Bodiselitsch et al. (2004) speculated that the warm pulses at 6.17 and 10.28 msl could be due to the release of large amounts of seafloor methane hydrate during and after the last two impact events that induced the negative excursions in $\delta^{13}C$ in both Ir-enhanced regions. A similar hypothesis has been evaluated by Pusz et al. (2007, this volume) on the basis of the significant $\delta^{13}C$ negative excursion in both bulk and benthic foraminiferal carbonate coincident with the impact-ejecta horizons at ODP Site 1090 and DSDP Site 612. According to our data, a release of methane large enough to induce a greenhouse scenario can be evaluated only for the Popigai impact, which was followed by a prevailing warming pulse.

On the basis of the relationship among the Chesapeake Bay and Montagnais impacts and the oxygen isotope (inferred paleotemperature) and extraterrestrial ^3He curves, Poag et al. (2002, 2003, 2004) stated that the general cooling trend from the middle Eocene to the early Oligocene was interrupted by long- and short-term warming episodes. These episodes, possibly triggered by a multiple impact event, would have released greenhouse gases into the atmosphere during the late Eocene comet or asteroid shower and would have been superimposed on an overall cooling trend. Poag et al. (2002, 2003) proposed that this impact-generated greenhouse warming may have postponed a pending late Eocene mass extinction until the early Oligocene. More specifically, Poag et al. (2003, 2004) recognized three distinct subpulses of warm climate, W-1 to W-3, that followed the Chesapeake Bay impact and are indicated by lower oxygen and carbon isotope ratio values. According to these authors, the warm pulses correlate with an interval of increased ^3He at Massignano. In particular, subpulse W-1 correlates with chron C16n.2n and the lower part of chron C16n.1r and falls within zone P15 of Berggren et al.

(1995), and subpulse W-2 coincides with chron C16n.1n and the lower two-thirds of chron C15r and lies in the transition between planktonic foraminiferal zones P15 and P16 of Berggren et al. (1995). However, as reported in "High-Resolution Impact Stratigraphy, Geochronology, and Magnetochronology at Massignano: Implications for Global Correlation," the correlation of these subpulses with the magnetostratigraphy at Kiptopeke core hole has to be refined.

At Massignano, Bodiselitsch et al. (2004) correlated the Ir anomalies at 6.17 and 10.28 msl, respectively, with warm subpulses W-1 and W-2 of Poag et al. (2003, 2004). Following this and according to the refined magnetobiostratigraphy provided by Jovane et al. (2007, this volume), the warm subpulse W-1 would fall in the uppermost part of chron C16n.1n and in the upper part of planktonic foraminiferal zone P15 Berggren et al. (1995), which is in the lower part of zone E15 Berggren and Pearson (2006). Warm subpulse W-2 would lie in the middle part of chron C15n and in the lower part of planktonic foraminiferal zone P16 Berggren et al. (1995), which is in the middle part of zone E15 Berggren and Pearson (2006) (Figs. 4 and 5). However, according to the magnetobiostratigraphy currently available for the Massignano GSSP and our climatic curve, correlation of the warm subpulse W-2 of Poag et al. (2003, 2004) with the warm pulse recognized at Massignano in the interval between 8.1 and 8.4 msl and well constrained in the middle part of chron C15r and in the lowermost part of planktonic foraminiferal zone P16 of Berggren et al. (1995) (Fig. 5) would be more consistent. Moreover, it fits well with the prominent negative value of $\delta^{18}O$ recorded at 8.2 msl by Bodiselitsch et al. (2004).

The main climatic episodes documented at Massignano by the abundance variations of the planktonic foraminiferal paleoclimatic indices are largely mirrored by the alternating relatively warm and humid intervals with cold and dry intervals recognized in the late Eocene on a global scale through high-resolution environmental magnetic studies carried out at CIROS-1 drill core from McMurdo Sound (Antarctica) (Sagnotti et al., 1998) and at Massignano (Jovane et al., 2004). The existence of alternating warm and cool intervals at the Eocene-Oligocene transition has also been revealed through high-resolution paleoclimatic analyses of calcareous nannofossils at the Southern Ocean ODP Holes 689D (Maud Rise, Weddell Sea, Antarctica) and 744A (Kerguelen Plateau, Antarctica) (Persico and Villa, 2004). In particular, from the uppermost part of chron C16n.2n up to the lowermost part of chron C13r, three warm intervals alternate with two cool intervals. These intervals are well constrained by the magnetostratigraphy provided by Roberts et al. (2003) and Florindo and Roberts (2005). We can point out that in terms of magnetobiostratigraphy, these intervals fit very well with the climatic intervals we recognize at Massignano, including the cool intervals detected at the transition between chron C16n.1r and chron C16n.1n, which would be correlated with the cooling recorded between 5.20 and 5.30 msl.

The consequences of the late Eocene impacts seem to be much smaller than those seen following the Chicxulub impact at the Cretaceous-Paleogene boundary (e.g., Alvarez et al., 1980; Kring, 2007), which led to the well-known mass extinction event (e.g., Sepkoski, 1986).

Further high-resolution studies, possibly on millennial time scales, from high-quality sediment archives should provide reliable data sets to document and to improve understanding of impact-induced biotic and climatic changes and to more accurately correlate them with marine and continental records.

CONCLUSIONS

The three late Eocene impact events recorded in the Massignano GSSP had no abrupt, dramatic effects on planktonic foraminifera. However, the high-resolution quantitative study and the multivariate statistical analysis revealed noteworthy changes in the assemblages occurring after the impacts. Acting as forcing factors, these impacts induced some environmental perturbations and may have contributed to significant climate changes superimposed on the general late Eocene cooling trend. These climate changes displayed different features at each impact event, probably due to the context in which each occurred (e.g., target rocks).

In particular, the first two closely spaced impacts related to the Popigai and Chesapeake Bay craters revealed significant changes in the water mass structure, in terms of stratification and trophic resources, associated with some warming and cooling episodes that took place within chron C16n.1n and chron C15r and at the transition between planktonic foraminiferal zones P15 and P16 of Berggren et al. (1995).

The prevailing warming pulse recognized after the Popigai impact might have been due to greenhouse effects produced by injection of CO_2 into the atmosphere and/or the release of methane hydrate after the impact itself. The relatively long duration of the enhanced cooling following the Chesapeake Bay impact suggests that this impact induced a progressive cooling and triggered a feedback mechanism that sustained the initial impact-induced changes. Similar patterns of climatic excursions reported across the equivalent impact-ejecta horizons in the Pacific, Indian, Atlantic-Caribbean, and Southern Oceans, as well as in Antarctica, indicate that climate changes recorded at Massignano seem to have had a global, rather than local or regional, significance.

ACKNOWLEDGMENTS

The authors warmly thank F. Hilgen, K.G. Miller, C. Koeberl, and A.E. Pusz for their criticism and helpful suggestions, which greatly improved the manuscript. We also thank Carla Bucci for technical support. This is publication 24 of the Centro di Geobiologia of Urbino University.

REFERENCES CITED

Alvarez, L.W., Alvarez, W., Asaro, F., and Michel, H.V., 1980, Extraterrestrial cause for the Cretaceous-Tertiary extinction: Science, v. 208, p. 1095–1108, doi: 10.1126/science.208.4448.1095.

Barrera, E., and Huber, B.T., 1991, Paleogene and early Neogene oceanography of the southern Indian Ocean; Leg 119 foraminifer stable isotope results, *in* Larsen, B., and Baldauf, J.G., et al., Proceedings of the Ocean Drilling Program, Scientific Results, Volume 119: College Station, Texas, Ocean Drilling Program, p. 736–774.

Barrera, E., and Huber, B.T., 1993, Eocene to Oligocene oceanography and temperatures in the Antarctic Indian Ocean, *in* Kennett, J.P., and Warnke, D.A., eds., The Antarctic Paleoenvironment: A Perspective on Global Change, Part II. Antarctic Research Series: Washington, D.C., American Geophysical Union, v. 60, p. 49–65.

Benson, R., 1975, The origin of the psycrosphere as recorded in changes of deep-sea ostracods assemblages: Lethaia, v. 8, p. 69–83, doi: 10.1111/j.1502-3931.1975.tb00919.x.

Benson, R.H., 1990, Ostracoda and the discovery of global Cainozoic palaeoceanographical events, *in* Whatley, R., and Maybury, C., eds., Ostracoda and Global Events: London, Chapman and Hall, p. 41–58.

Berger, W.H., Vincent, E., and Thierstein, H.R., 1981, The deep sea record: Major steps in Cenozoic ocean evolution: Society of Economic Paleontologists and Mineralogists (SEPM) Special Publication 32, p. 489–504.

Berggren, W.A., and Miller, K.G., 1988, Paleogene tropical planktonic foraminiferal biostratigraphy and magnetobiochronology: Micropaleontology, v. 34, p. 362–380, doi: 10.2307/1485604.

Berggren, W.A., and Miller, K.G., 1989, Cenozoic bathyal and abyssal calcareous benthic foraminiferal zonation: Micropaleontology, v. 35, p. 308–320, doi: 10.2307/1485674.

Berggren, W.A., and Pearson, P.N., 2006, Tropical to subtropical planktonic foraminiferal zonation of the Eocene and Oligocene, *in* Pearson, P.N., Olsson, R.K., Huber, B.T., Hemleben, Ch., and Berggren, W.A., eds., Atlas of Eocene Planktonic Foraminifera: Cushman Foundation for Foraminiferal Research, Special Publication 41, Greenville, North Carolina, p. 29–40.

Berggren, W.A., Kent, D.V., Swisher, C.C., III, and Aubry, M.-P., 1995, A revised Cenozoic geochronology and chronostratigraphy, *in* Berggren, W.A., Kent, D.V., Swisher, C.C., III, Aubry, M.-P., and Hardenbol, J., eds., Geochronology, Time Scales, and Global Stratigraphic Correlation: SEPM (Society for Sedimentary Geology) Special Publication 54, p. 129–212.

Bice, D., and Montanari, A., 1988, Magnetic stratigraphy of the Massignano section across the Eocene/Oligocene boundary, *in* Premoli Silva, I., Coccioni, R., and Montanari, A., eds., The Eocene-Oligocene Boundary in the Marche-Umbria Basin (Italy): Aniballi, Ancona, International Subcommission on Paleogene Stratigraphy, Special Publication, p. 111–117.

Blow, W.H., 1969, Late middle Eocene to Recent planktonic foraminiferal biostratigraphy, *in* Bronnimann, P., and Renz, H.H., eds., Proceedings of the First International Conference on Planktonic Microfossils (Geneva, 1967): Leiden, E.J. Brill, p. 199–421.

Blow, W.H., 1979, The Cenozoic Globigerinida: A study of the morphology, taxonomy, evolutionary relationships and the stratigraphical distribution of some Globigerinida (mainly Globigerinacea), Volume 3: Leiden, E.J. Brill, 1413 p.

Bodiselitsch, B., Montanari, A., Koeberl, C., and Coccioni, R., 2004, Delayed climate cooling in the late Eocene caused by multiple impacts: High-resolution geochemical studies at Massignano, Italy: Earth and Planetary Science Letters, v. 223, p. 283–302, doi: 10.1016/j.epsl.2004.04.028.

Boersma, A., and Premoli Silva, I., 1986, Terminal Eocene events—Planktonic foraminifera and isotopic evidence, *in* Pomerol, C., and Premoli Silva, I., eds., Terminal Eocene Events: Development in Paleontology and Stratigraphy, Elsevier, Amsterdam, p. 213–224.

Boersma, A., and Premoli Silva, I., 1991, Distribution of Paleogene planktonic foraminifera: Analogies with the Recent?: Palaeogeography, Palaeoclimatology, Palaeoecology, v. 83, p. 29–48, doi: 10.1016/0031-0182(91)90074-2.

Boersma, A., Shackleton, N.J., Hall, M.A., and Given, Q., 1979, Carbon and oxygen isotope records at DSDP Site 384 (North Atlantic) and some Paleocene paleotemperature and carbon isotope variations in the Atlantic Ocean, *in* Tucholke, B.E., Vogt, P.R., et al., Initial Reports of the Deep Sea Drilling Project, Volume 43: Washington, D.C., U.S. Government Printing Office, p. 695–715.

Boersma, A., Premoli Silva, I., and Shackleton, N.J., 1987, Atlantic Eocene planktonic foraminiferal paleohydrographic indicators and stable isotope paleoceanography: Paleoceanography, v. 3, p. 287–331.

Bohaty, S.M., and Zachos, J.C., 2003, Significant Southern Ocean warming event in the late middle Eocene: Geology, v. 31, p. 1017–1020, doi: 10.1130/G19800.1.

Bottomley, R., Grieve, R., York, D., and Masaitis, V., 1997, The age of the Popigai impact event and its relation to events at the Eocene/Oligocene boundary: Nature, v. 388, p. 365–368, doi: 10.1038/41073.

Brinkhuis, H., 1992, Late Paleogene dinoflagellate cysts with special reference to the Eocene/Oligocene boundary, *in* Prothero, D.R., and Berggren, W.A., eds., Eocene-Oligocene Climatic and Biotic Evolution: Princeton, New Jersey, Princeton University Press, p. 327–340.

Brinkhuis, H., and Biffi, U., 1993, Dinoflagellate cyst stratigraphy of the Eocene/Oligocene transition in central Italy: Marine Micropaleontology, v. 22, p. 131–183, doi: 10.1016/0377-8398(93)90007-K.

Brinkhuis, H., and Coccioni, R., 1995, Is there a relation between dinocysts changes and iridium after all? Preliminary results from the Massignano section, *in* Montanari, A., and Coccioni, R., eds., The Effects of Impacts on the Atmosphere and Biosphere with Regard to Short- and Long-Term Changes: Abstracts and Field Trips: Ancona, Italy, European Science Foundation, p. 40.

Cande, S.C., and Kent, D.V., 1992, A new geomagnetic polarity time scale for the Late Cretaceous and Cenozoic: Journal of Geophysical Research, v. 97, p. 13,917–13,951, doi: 10.1029/92JB01202.

Cande, S.C., and Kent, D.V., 1995, Revised calibration of the geomagnetic polarity time scale for the Late Cretaceous and Cenozoic: Journal of Geophysical Research, v. 100, p. 6093–6095, doi: 10.1029/94JB03098.

Channell, J.E.T., Galeotti, S., Martin, E., Billups, K., Scher, H.D., and Stoner, J.S., 2003, Eocene to Miocene magnetostratigraphy, biostratigraphy, and chemostratigraphy at ODP Site 1090 (sub-Antarctic South Atlantic): Geological Society of America Bulletin, v. 115, p. 607–623, doi: 10.1130/0016-7606(2003)115<0607:ETMMBA>2.0.CO;2.

Cita, M.B., Vergnaud-Grazzini, C., Robert, C., Chamley, H., Ciaranfi, N., and D'Onofrio, S., 1977, Paleoclimatic record of a long deep sea core from the Eastern Mediterranean: Quaternary Research, v. 8, p. 205–235, doi: 10.1016/0033-5894(77)90046-1.

Claeys, P., 2007, Cause(s) of the elevated bombardment in the late Eocene: Eos (Transactions, American Geophysical Union), v. 88, no. 52, Fall Meeting supplement, abstract U21E-01.

Clarke, K.R., and Warwick, R.M., 1994, Change in Marine Communities: An Approach to Statistical Analysis and Interpretation: Plymouth, Plymouth Marine Laboratory, 144 p.

Clifford, H.T., and Stephenson, W., 1975, An Introduction to Numerical Classification: New York, Academic Press, 229 p.

Clymer, A.K., Bice, D.M., and Montanari, A., 1995, Shocked quartz in the late Eocene: Bolide impact evidence from Massignano, Italy, *in* Montanari, A., and Coccioni, R., eds., The Effects of Impacts on the Atmosphere and Biosphere with Regard to Short- and Long-Term Changes: Abstracts and Field Trips: Ancona, Italy, European Science Foundation, p. 60–61.

Clymer, A.K., Bice, D.M., and Montanari, A., 1996, Shocked quartz from the late Eocene: Impact evidence from Massignano, Italy: Geology, v. 24, p. 483–486, doi: 10.1130/0091-7613(1996)024<0483:SQFTLE>2.3.CO;2.

Coccioni, R., and Galeotti, S., 2003, Deep-water benthic foraminiferal events from the Massignano Eocene/Oligocene boundary stratotype section and point (central Italy): Biostratigraphic, paleoecologic, and paleoceanographic implications, *in* Prothero, D.R., Ivant, L.C., and Nesbitt, E., eds., From Greenhouse to Icehouse: The Marine Eocene-Oligocene Transition: New York, Columbia University Press, p. 438–452.

Coccioni, R., Monaco, P., Monechi, S., Nocchi, M., and Parisi, G., 1988, Biostratigraphy of the Eocene/Oligocene boundary at Massignano (Ancona, Italy), *in* Premoli Silva, I., Coccioni, R., and Montanari, A., eds., The Eocene-Oligocene Boundary in the Marche-Umbria Basin (Italy): Aniballi, Ancona, International Subcommission on Paleogene Stratigraphy, Special Publication, p. 59–80.

Coccioni, R., Basso, D., Brinkhuis, H., Galeotti, S., Gardin, S., Monechi, S., and Spezzaferri, S., 2000, Marine biotic signals across a late Eocene impact layer at Massignano, Italy: Evidence for long-term environmental perturbations?: Terra Nova, v. 12, p. 258–263, doi: 10.1046/j.1365-3121.2000.00305.x.

Collins, G.S., and Wünnemann, K., 2005, How big was the Chesapeake Bay impact? Insight from numerical modelling: Geology, v. 33, p. 925–928, doi: 10.1130/G21854.1.

Coxall, H.K., and Pearson, P.N., 2007, The Eocene-Oligocene transition, *in* Williams, M., Hayward, A., Gregory, J., and Schmidt, D., eds., Deep Time Perspectives on Climate Change: Marrying the Signal from Computer Models and Biological Proxies: The Micropalaeontological Society Special Publication 2, p. 351–387.

Coxall, H.K., Pearson, P.N., and Shackleton, N.J., 2000, Hantkeninids depth adaptation: An evolving life strategy in a changing ocean: Geology, v. 28, p. 87–90, doi: 10.1130/0091-7613(2000)28<87:HDAAEL>2.0.CO;2.

Coxall, H.K., Wilson, P.A., Pälike, H., Lear, C.H., and Backman, J., 2005, Rapid stepwise onset of Antarctic glaciation and deeper calcite compensation in the Pacific Ocean: Nature, v. 433, p. 53–57, doi: 10.1038/nature03135.

Diester-Haass, L., Robert, C., and Chamley, H., 1996, The Eocene-Oligocene preglacial–glacial transition in the Atlantic sector of the Southern Ocean (ODP Site 690): Marine Geology, v. 131, p. 123–149, doi: 10.1016/0025-3227(95)00174-3.

Douglas, W., and Savin, S.M., 1978, Oxygen isotopic evidence for the depth stratification of Tertiary and Cretaceous planktic foraminifera: Marine Micropaleontology, v. 3, p. 175–196, doi: 10.1016/0377-8398(78)90004-X.

Ehrmann, W.U., 1997, Smectite concentrations and crystallinities: Indications for Eocene age of glaciomarine sediments in the CIROS-1 drill hole, McMurdo Sound, Antarctica, *in* Ricci, C.A., ed., The Antarctic Region, Geological Evolution and Processes: Siena, Museo Nazionale dell'Antartide, p. 771–780.

Farley, K.A., 2007, The late Eocene extraterrestrial dust event: Comet or asteroid shower?, *in* Montanari, A., Koeberl, C., Coccioni, R., and Hilgen, F., eds., The Late Eocene Earth: Hot House, Ice House, and Impacts—Abstract with Program and Field Trip Guide: Ancona, Italy, Osservatorio Geologico di Coldigioco Frontale di Apiro, Geological Society of America Penrose Conference, p. 22–23.

Farley, K.A., Montanari, A., Shoemaker, E.M., and Shoemaker, C.S., 1998, Geochemical evidence for a comet shower in the late Eocene: Science, v. 280, p. 1250–1253, doi: 10.1126/science.280.5367.1250.

Field, J.G., Clarke, K.R., and Warwick, R.M., 1982, A practical strategy for analysing multispecies distribution patterns: Marine Ecology Progress Series, v. 8, p. 37–52, doi: 10.3354/meps008037.

Florindo, F., and Roberts, A.P., 2005, Eocene-Oligocene magnetobiochronology of ODP Sites 689 and 690, Maud Rise, Weddell Sea, Antarctica: Geological Society of America Bulletin, v. 117, p. 46–66, doi: 10.1130/B25541.1.

Galeotti, S., Coccioni, R., and Gersonde, R., 2002, Middle Eocene–early Pliocene subantarctic planktic foraminiferal biostratigraphy of Site 1090, Agulhas Ridge: Marine Micropaleontology, v. 45, p. 357–381, doi: 10.1016/S0377-8398(02)00035-X.

Gerstl, S.A.W., and Zardecki, A., 1982, Reduction of photosynthetically active radiation under extreme stratospheric aerosol loads, *in* Silver, L.T., and Schultz, P.H., eds., Geological Implications of Impacts of Large Asteroids and Comets on the Earth: Geological Society of America Special Paper 190, p. 201–210.

Glass, B.P., Dubois, D.L., and Ganapathy, R., 1982, Relationship between an iridium anomaly and the North American microtektite layer in core RC9-58 from the Caribbean Sea, *in* Boynton, W.V., and Ahrens, T.J., eds., Proceedings of the 13th Lunar and Planetary Science Conference: Journal of Geophysical Research, v. 90, part I, p. D175–D196.

Gradstein, F.M., Ogg, J.G., and Smith, A.G., 2004, A Geologic Time Scale: Cambridge, Cambridge University Press, 589 p.

Hallock, P., 1987, Fluctuations in the trophic resource continuum: A factor in global diversity cycles?: Paleoceanography, v. 2, p. 457–471, doi: 10.1029/PA002i005p00457.

Hallock, P., Premoli Silva, I., and Boersma, A., 1991, Similarities between planktonic and larger foraminiferal evolutionary trends through Paleogene paleoceanographic changes: Palaeogeography, Palaeoclimatology, Palaeoecology, v. 83, p. 49–64, doi: 10.1016/0031-0182(91)90075-3.

Haq, B.U., Premoli Silva, I., and Lohman, G.P., 1977, Calcareous plankton paleobiogeographic evidence for major climatic fluctuations in the early Cenozoic Atlantic Ocean: Journal of Geophysical Research, v. 82, p. 3861–3876, doi: 10.1029/JC082i027p03861.

Horton, J.W., and Izett, G.A., 2005, Crystalline-rock ejecta and shocked minerals of the Chesapeake Bay impact structure: The USGS-NASA Langley corehole, Hampton, Virginia, with supplement constraints on the age of the impact, *in* Horton, J.W., Powars, D.S., and Gohn, G., eds., Studies of the Chesapeake Bay Impact Structure: Reston, Virginia, U.S. Geological Survey, p. E1–E29.

Huber, H., Koeberl, C., King, D.T., Petruny, L.W., and Montanari, A., 2001, Effects of bioturbation through the late Eocene impactoclastic layer near Massignano, Italy, *in* Buffetaut, E., and Koeberl, C., eds., Geological and Biological Effects of Impact Events: Berlin, Springer-Verlag, p. 197–216.

Jovane, L., Florindo, F., and Dinarès-Turell, J., 2004, Environmental magnetic record of paleoclimate change from the Eocene-Oligocene stratotype section, Massignano, Italy: Geophysical Research Letters, v. 31, p. L15601, doi: 10.1029/2004GL020554.

Jovane, L., Florindo, F., Sprovieri, M., and Pälike, H., 2006, Astronomic calibration of the late Eocene/early Oligocene Massignano section (central Italy): Geochemistry, Geophysics, Geosystems, v. 7, p. Q07012, doi: 10.1029/2005GC001195.

Jovane, L., Sprovieri, M., Florindo, F., Acton, G., Coccioni, R., Dall'Antonia, B., and Dinarès-Turrell, J., 2007, Eocene-Oligocene paleoceanographic changes in the stratotype section, Massignano, Italy: Clues from rock magnetism and stable isotopes: Journal of Geophysical Research, v. 112, p. B11101, doi: 10.1029/2007JB004963.

Jovane, L., Coccioni, R., Marsili, A., and Acton, G., 2009, this volume, The late Eocene greenhouse-icehouse transition: Observations from the Massignano global stratotoype section and point (GSSP), *in* Koeberl, C., and Montanari, A., eds., The Late Eocene Earth—Hothouse, Icehouse, and Impacts: Geological Society of America Special Paper 452, doi: 10.1130/2009.2452(10).

Keller, G., 1983, Paleoclimatic analyses of middle Eocene through Oligocene planktic foraminiferal faunas: Palaeogeography, Palaeoclimatology, Palaeoecology, v. 43, p. 73–94, doi: 10.1016/0031-0182(83)90049-4.

Keller, G., 1986, Stepwise mass extinctions and impact events late Eocene to early Oligocene: Marine Micropaleontology, v. 10, p. 267–293, doi: 10.1016/0377-8398(86)90032-0.

Keller, G., Macleod, N., and Barrera, E., 1992, Eocene-Oligocene faunal turnover in planktic foraminifera, and Antarctic glaciation, *in* Prothero, D.R., and Berggren, W.A., eds., Eocene-Oligocene Climatic and Biotic Evolution: Princeton, New Jersey, Princeton University Press, p. 218–244.

Kennett, J.P., 1977, Cenozoic evolution of Antarctic glaciation, the circum–Antarctic Ocean, and their impact on global palaeoceanography: Journal of Geophysical Research, v. 82, p. 3843–3860, doi: 10.1029/JC082i027p03843.

Kennett, J.P., and Barker, P.F., 1990, Latest Cretaceous to Cenozoic climate and oceanographic developments in the Weddell Sea, Antarctica: An ocean-drilling perspective, *in* Barker, P.F., Kennett, J.P., O'Connell, S., and Pisias, N.G., et al., Proceedings of the Ocean Drilling Program, Scientific Results, Volume 113: College Station, Texas, Ocean Drilling Program, p. 937–960.

Kennett, J.P., and Stott, L.D., 1990, Proteus and proto-Oceanus: Ancestral Paleogene oceans as revealed from Antarctic stable isotopic results; ODP Leg 113, *in* Barker, P.F., Kennett, J.P., O'Connell, S., and Pisias, N.G., et al., Proceedings of the Ocean Drilling Program, Scientific Results, Volume 113: College Station, Texas, Ocean Drilling Program, p. 865–880.

Kent, D.V., Cramer, B.S., Lanci, L., Wang, D., Wright, J.D., and Van der Voo, R., 2003, A case for a comet impact trigger for the Paleocene/Eocene thermal maximum and carbon isotope excursion: Earth and Planetary Science Letters, v. 211, p. 13–26, doi: 10.1016/S0012-821X(03)00188-2.

Koeberl, C., 2007, Late Eocene craters and impactoclastic layers—An overview, *in* Montanari, A., Koeberl, C., Coccioni, R., and Hilgen, F., eds., The Late Eocene Earth: Hot House, Ice House, and Impacts—Abstract with Program and Field Trip Guide: Ancona, Italy, Osservatorio Geologico di Coldigioco Frontale di Apiro, Geological Society of America Penrose Conference, p. 43–45.

Koeberl, C., Poag, C.W., Reimold, W.H., and Brandt, D., 1996, Impact origin of the Chesapeake Bay structure and the source of the North American tektites: Science, v. 271, p. 1263–1266, doi: 10.1126/science.271.5253.1263.

Köthe, A., 1990, Paleogene dinoflagellates from north West Germany: Geologisches Jahrbuch A, v. 118, 111 p.

Kring, D.A., 2000, Impact events and their effect on the origin, evolution and distribution of life: GSA Today, v. 10, no. 8, p. 1–7.

Kring, D.A., 2007, The Chicxulub impact event and its environmental consequences at the Cretaceous–Tertiary boundary, *in* Coccioni, R., Monechi, S., and Rampino, M.R., eds., Cretaceous-Paleogene Boundary Events: Palaeogeography, Palaeoclimatology, Palaeoecology, v. 255, p. 4–21.

Kruskal, J.B., 1977, Multidimensional scaling and other methods for discovering structure, *in* Enslein, K., Ralston, A., and Wilf, H.S., eds., Statistical Methods for Digital Computers: New York, John Wiley, p. 296–339.

Kyte, F.T., 2001, Identification of late Eocene impact deposits at ODP Site 1090, *in* Gersonde, R., Hodell, D.A., and Blum, P., et al., Proceedings of the Ocean Drilling Program, Scientific Results, Volume 177: College Station, Texas, Ocean Drilling Program, p. 1–9.

Kyte, F.T., and Liu, S., 2002, Iridium and spherules in late Eocene impact deposits: Houston, Texas, Lunar and Planetary Institute, v. 33, abstract A1981 (CD-ROM).

Langenhorst, F., and Clymer, A.K., 1996, Characteristics of shocked quartz in late Eocene impact ejecta from Massignano (Ancona, Italy): Clues to shock conditions and source crater: Geology, v. 24, p. 487–490, doi: 10.1130/0091-7613(1996)024<0487:COSQIL>2.3.CO;2.

Lear, C.H., Elderfield, H., and Wilson, P.A., 2000, Cenozoic deep-sea temperatures and global ice volumes from Mg/Ca in benthic foraminiferal calcite: Science, v. 287, p. 269–272, doi: 10.1126/science.287.5451.269.

Lear, C.H., Bailey, T.R., Pearson, P.N., Coxall, H.K., and Rosenthal, Y., 2008, Cooling and ice growth across the Eocene-Oligocene transition: Geology, v. 36, p. 251–254, doi: 10.1130/G24584A.1.

Liu, S., Kyte, F.T., Glass, B.P., and Gersonde, R., 2000, Upper Eocene spherules at ODP Site 1090B: Meteoritics & Planetary Science, v. 35, abstract A98–99.

Lowrie, W., and Lanci, L., 1994, Magnetostratigraphy of Eocene/Oligocene boundary sections in Italy: No evidence for short subchrons within chrons 12R and 13R: Earth and Planetary Science Letters, v. 126, p. 247–258, doi: 10.1016/0012-821X(94)90110-4.

Marino, M., and Flores, J.A., 2002, Middle Eocene to early Oligocene calcareous nannofossil stratigraphy at Leg 117 Site 1090: Marine Micropaleontology, v. 45, p. 383–398, doi: 10.1016/S0377-8398(02)00036-1.

Martini, E., 1971, Standard Tertiary and Quaternary calcareous nannoplankton zonation, *in* Farinacci, A., ed., Proceedings of the Second Planktonic Conference, Rome 1970: Rome, Tecnoscienza, v. 2, p. 739–785.

Masaitis, V.L., and Raikhlin, A.I., 1986, The Popigai crater formed by the impact of common chondrite: Doklady Akademii Nauk SSSR, v. 286, p. 1476–1478.

Miller, K., Fairbanks, R., and Mountain, G., 1987, Tertiary oxygen isotope synthesis, sea level history, and continental margin erosion: Paleoceanography, v. 2, p. 1–19, doi: 10.1029/PA002i001p00001.

Miller, K.G., Wright, J.D., and Fairbanks, R.G., 1991, Unlocking the Ice House: Oligocene-Miocene oxygen isotopes, eustasy, and margin erosion: Journal of Geophysical Research, v. 96, p. 6829–6848, doi: 10.1029/90JB02015.

Miller, K.G., Katz, M.E., and Berggren, W.A., 1992, Cenozoic deep-sea benthic foraminifera: A tale of three turnovers, *in* Takayanagi, Y., and Saito, T., eds., Studies in Benthic Foraminifera; Proceedings of the Fourth International Symposium on Benthic Foraminifera (Benthos '90): Sendai, Japan, Tokai University Press, p. 67–75.

Miller, K.G., Wright, J.D., and Browning, J.V., 2005, Visions of ice sheets in a greenhouse world: Marine Geology, v. 217, p. 215–231, doi: 10.1016/j.margeo.2005.02.007.

Miller, K.G., Wright, J.D., Katz, M.E., Wade, B.S., Browning, J.V., and Rosenthal, Y., 2007, Passing a climate threshold at the Eocene-Oligocene transition: Antarctic ice takes the driver's seat, *in* Montanari, A., Koeberl, C., Coccioni, R., and Hilgen, F., eds., The Late Eocene Earth: Hot House, Ice House, and Impacts—Abstract with Program and Field Trip Guide: Ancona, Italy, Osservatorio Geologico di Coldigioco Frontale di Apiro, Geological Society of America Penrose Conference, p. 49–50.

Monechi, S., Buccianti, A., and Gardin, S., 2000, Biotic signals from nannoflora across the iridium anomaly in the Upper Eocene of the Massignano section: Evidence from statistical analysis: Marine Micropaleontology, v. 39, p. 219–237, doi: 10.1016/S0377-8398(00)00022-0.

Montanari, A., and Koeberl, C., 2000, Impact Stratigraphy—The Italian Record: Lecture Notes in Earth Sciences, v. 93, 364 p.

Montanari, A., Deino, A.L., Drake, R.E., Turrin, B.D., Depaolo, D.J., Odin, G.S., Curtis, G.H., Alvarez, W., and Bice, D.M., 1988, Radioisotopic dating of the Eocene-Oligocene boundary in the pelagic sequence of the Northern Apennines, *in* Premoli Silva, I., Coccioni, R., and A. Montanari, eds., The Eocene-Oligocene Boundary in the Marche-Umbria Basin (Italy): Aniballi, Ancona, International Subcommission on Paleogene Stratigraphy, Special Publication, p. 195–208.

Montanari, A., Asaro, F., Michel, H.V., and Kennett, J.P., 1993, Iridium anomalies of late Eocene age at Massignano (Italy), and ODP Site 689B (Maud Rise, Antarctic): Palaios, v. 8, p. 420–437, doi: 10.2307/3515017.

Montanari, A., Campo, A., and Farinella, P., 1998, Earth cratering record and impact energy flux in the last 150 Ma: Planetary and Space Science, v. 46, p. 271–281, doi: 10.1016/S0032-0633(97)00130-X.

Nocchi, M., Monechi, S., Coccioni, R., Madile, M., Monaco, P., Orlando, M., Parisi, G., and Premoli Silva, I., 1988a, The extinction of Hantkeninidae as a marker for recognizing the Eocene-Oligocene boundary: A proposal, *in* Premoli Silva, I., Coccioni, R., and Montanari, A., eds., The Eocene/Oligocene Boundary in the Marche-Umbria Basin (Italy): Ancona, Italy, International Subcommission on Paleogene Stratigraphy, Special Publication, p. 249–252.

Nocchi, M., Parisi, G., Monaco, P., Monechi, S., and Madile, M., 1988b, Eocene and early Oligocene micropaleontology and paleoenvironments in SE Umbria, Italy: Palaeogeography, Palaeoclimatology, Palaeoecology, v. 67, p. 181–244, doi: 10.1016/0031-0182(88)90154-X.

Oberli, F., and Meier, M., 1991, Age of Eocene-Oligocene boundary in the Marche-Umbria basin, Italy, by high resolution U/Th-Pb dating: Terra Abstracts, v. 3, p. 286.

Obradovich, J.D., Snee, L., and Izett, G.A., 1989, Is there more than one glassy impact layer in the late Eocene?: Geological Society of America Abstracts with Programs, v. 21, no. 6, p. A134.

Odin, G.S., Montanari, A., Deino, A., Drake, R., Gruise, P., Kreuzer, H., and Rex, D.C., 1991, Reliability of volcano-sedimentary biotite ages around the Eocene/Oligocene boundary: Chemical Geology, v. 86, p. 203–224.

Okada, H., and Bukry, D., 1980, Supplementary modification and introduction of code numbers to the low-latitude coccolith biostratigraphic zonation (Bukry, 1973; 1975): Marine Micropaleontology, v. 5, p. 321–325, doi: 10.1016/0377-8398(80)90016-X.

Pagani, M., Zachos, J., Freeman, K.H., Bohaty, S., and Tipple, B., 2005, Marked change in atmospheric carbon dioxide concentrations during the Oligocene: Science, v. 309, p. 600–603, doi: 10.1126/science.1110063.

Pearson, P.N., and Palmer, M.R., 1999, Middle Eocene seawater pH and atmospheric carbon dioxide concentrations: Science, v. 284, p. 1824–1826, doi: 10.1126/science.284.5421.1824.

Pearson, P.N., Shackleton, N.J., and Hall, M.A., 1993, Stable isotope paleoecology of middle Eocene planktonic foraminifera and multi-species isotope stratigraphy, DSDP Site 523, South Atlantic: Journal of Foraminiferal Research, v. 23, p. 123–140.

Pearson, P.N., Shackleton, N.J., Weedon, G.P., and Hall, M.A., 1997, Multispecies planktonic foraminifer stable isotope stratigraphy through Oligocene/Miocene boundary climatic cycles, Site 926, *in* Shackleton, N.J., Curry, W.B., Richter, C., and Bralower, T.J., et al., Proceedings of the Ocean Drilling Program, Scientific Results, Volume 154: College Station, Texas, Ocean Drilling Program, p. 441–449.

Pearson, P.N., Ditchfield, P.W., Singano, S., Harcourt-Brown, K.G., Nicholas, C.J., Olsson, R.K., Shackleton, N.J., and Hall, M.A., 2001, Warm tropical sea surface temperatures in the Late Cretaceous and Eocene Epochs: Nature, v. 413, p. 481–487, doi: 10.1038/35097000.

Pearson, P.N., Olsson, R.K., Hemleben, C., Huber, B.T., and Berggren, W.A., 2006, Atlas of Eocene Planktonic Foraminifera: Cushman Foundation of Foraminiferal Research Special Publication 41, 514 p.

Persico, D., and Villa, G., 2004, Eocene-Oligocene calcareous nannofossils from Maud Rise and Kerguelen Plateau (Antarctica): Palaeoecological and palaeoceanographic implications: Marine Micropaleontology, v. 52, p. 153–179, doi: 10.1016/j.marmicro.2004.05.002.

Pierrard, O., Robin, E., Rocchia, R., and Montanari, A., 1998, Extraterrestrial Ni-rich spinel in Upper Eocene sediments from Massignano, Italy: Geology, v. 26, p. 307–310, doi: 10.1130/0091-7613(1998)026<0307:ENRSIU>2.3.CO;2.

Poag, C.W., and Aubry, M.P., 1995, Upper Eocene impactites of the U.S. East Coast: Depositional origins, biostratigraphic framework and correlation: Palaios, v. 10, p. 16–43, doi: 10.2307/3515005.

Poag, C.W., Plescia, J.B., and Molzer, P.C., 2002, Ancient impact structures on modern continental shelves: The Chesapeake Bay, Montagnais, and Toms Canyon craters, Atlantic margin of North America: Deep-Sea Research, Part II, v. 49, p. 1081–1102, doi: 10.1016/S0967-0645(01)00144-8.

Poag, C.W., Mankinene, E., and Norris, R.D., 2003, Late Eocene impacts: Geologic record, correlation, and paleoenvironmental consequences, *in* Prothero, D.R., Yvany, L.C., and Nesbitt, E., eds., From Greenhouse to Icehouse: The Marine Eocene-Oligocene Transition: New York, Columbia University Press, p. 495–510.

Poag, C.W., Koeberl, C., and Reimold, W.U., 2004, The Chesapeake Bay crater—Geology and geophysics of a late Eocene submarine impact structure: Heidelberg, Germany, Springer, 522 p.

Poore, R.Z., and Matthews, R.K., 1984, Late Eocene–Oligocene oxygen and carbon isotope record from South Atlantic Ocean, Deep Sea Drilling Project Site 522, *in* Hsü, K.J., LaBrecque, J.L., et al., Initial Reports of the Deep Sea Drilling Project, Volume 73: Washington, D.C., U.S. Government Printing Office, p. 725–735.

Premoli Silva, I., and Boersma, A., 1988, Atlantic Eocene planktonic foraminiferal biogeography and paleohydrographic indicators: Palaeogeography, Palaeoclimatology, Palaeoecology, v. 67, p. 315–356, doi: 10.1016/0031-0182(88)90159-9.

Premoli Silva, I., and Boersma, A., 1989, Atlantic Paleogene planktonic foraminiferal bioprovincial indices: Marine Micropaleontology, v. 14, p. 357–371, doi: 10.1016/0377-8398(89)90019-4.

Premoli Silva, I., and Jenkins, D.G., 1993, Decision on the Eocene-Oligocene boundary stratotype: Episodes, v. 16, p. 379–382.

Prothero, D.R., 1994, The Eocene-Oligocene Transition—Paradise Lost: New York, Columbia University Press, 291 p.

Prothero, D.R., Ivany, L., and Nesbitt, E., 2003, From Greenhouse to Icehouse: The Marine Eocene-Oligocene Transition: New York, Columbia University Press, 541 p.

Pusz, A.E., Miller, K.G., Wright, J.D., Browning, J.V., Wade, B.S., and Kent, D.V., 2006, Upper Eocene microtektites discovered in Alabama: A first-order correlation to the GPTS: Geological Society of America Abstracts with Programs, v. 38, no. 7, p. 120.

Pusz, A.E., Miller, K.G., and Wright, J.D., 2007, Late Eocene impacts associated with a global carbon cycle perturbation, *in* Montanari, A., Koeberl, C., Coccioni, R., and Hilgen, F., eds., The Late Eocene Earth: Hot House, Ice House, and Impacts—Abstract with Program and Field Trip Guide: Ancona, Italy, Osservatorio Geologico di Coldigioco Frontale di Apiro, Geological Society of America Penrose Conference, p. 62–63.

Pusz, A.E., Miller, K.G., Wright, J.D., Katz, M.E., Cramer, B.S., and Kent, D.V., 2009, this volume, Stable isotopic response to late Eocene extraterrestrial impacts, *in* Koeberl, C., and Montanari, A., eds., The Late Eocene Earth—Hothouse, Icehouse, and Impacts: Geological Society of America Special Paper 452, doi: 10.1130/2009.2452(06).

Quilty, P.G., 2001, Late Eocene foraminifers and palaeoenvironment, Cascade Seamount, southwest Pacific Ocean: Implications for seamount subsidence and Australia-Antarctica Eocene correlation: Australian Journal of Earth Sciences, v. 48, p. 633–641.

Roberts, A.P., Bicknell, S.J., Byatt, J., Bohaty, S.M., Florindo, F., and Harwood, D.M., 2003, Magnetostratigraphic calibration of Southern Ocean diatom datums from the Eocene–Oligocene of Kerguelen Plateau (Ocean Drilling Program Sites 744 and 748), *in* Florindo, F., Cooper, A.K., and O'Brien, P.A., eds., Antarctic Cenozoic Palaeoenvironments: Geologic Record and Models: Palaeogeography, Palaeoclimatology, Palaeoecology, v. 198, p. 145–168.

Sagnotti, L., Florindo, F., Verosub, K.L., Wilson, G.S., and Roberts, A.P., 1998, Environmental magnetic record of Antarctic palaeoclimate from Eocene-Oligocene glaciomarine sediments, Victoria Land margin: Geophysical Journal International, v. 134, p. 653–662, doi: 10.1046/j.1365-246x.1998.00559.x.

Sepkoski, J.J., Jr., 1986, Phanerozoic overview of mass extinction, *in* Raup, D.M., and Jablonski, D., eds., Patterns and Processes in the History of Life: Berlin, Springer Verlag, p. 277–295.

Spezzaferri, S., 1994, Planktonic foraminiferal biostratigraphy and taxonomy of the Oligocene and Lower Miocene in the oceanic record: An overview: Paleontographia Italica, v. 81, 187 p.

Spezzaferri, S., 1995, Planktonic foraminiferal paleoclimatic implications across the Oligocene-Miocene transition in the oceanic record (Atlantic, Indian and South Pacific): Palaeogeography, Palaeoclimatology, Palaeoecology, v. 114, p. 43–74, doi: 10.1016/0031-0182(95)00076-X.

Spezzaferri, S., and Premoli Silva, I., 1991, Oligocene planktonic foraminiferal biostratigraphy and paleoclimatic interpretation from Hole 538A, DSDP Leg 77, Gulf of Mexico: Palaeogeography, Palaeoclimatology, Palaeoecology, v. 83, p. 217–263, doi: 10.1016/0031-0182(91)90080-B.

Spezzaferri, S., Basso, D., and Coccioni, R., 2002, Late Eocene planktonic foraminiferal response to an extraterrestrial impact at Massignano GSSP (northeastern Apennines, Italy): Journal of Foraminiferal Research, v. 32, p. 188–199, doi: 10.2113/0320188.

Stott, L.D., and Kennett, J.P., 1990, Antarctic Paleogene planktonic foraminifer biostratigraphy: ODP Leg 113, Sites 689 and 690, *in* Barker, P.F., Kennett, J.P., O'Connell, S., and Pisias, N.G., et al., Proceedings of the Ocean Drilling Program, Scientific Results, Volume 113: College Station, Texas, Ocean Drilling Program, p. 549–569.

Stott, L.D., Kennett, J.P., Shackleton, N.J., and Corfield, R.M., 1990, The evolution of Antarctic surface waters during the Paleogene: Inferences from the stable isotopic composition of planktonic foraminifers, ODP Leg 113, *in* Barker, P.F., Kennett, J.P., O'Connell, S., and Pisias, N.G., et al., Proceedings of the Ocean Drilling Program, Scientific Results, Volume 113: College Station, Texas, Ocean Drilling Program, p. 849–863.

Tagle, R., and Claeys, Ph., 2004, Comet or asteroid shower in the late Eocene?: Science, v. 305, p. 492, doi: 10.1126/science.1098481.

Toon, O.B., Zahnle, K., Turco, R.P., and Covey, C., 1994, Environmental perturbations caused by asteroid impacts, *in* Gehrels, T., ed., Hazards due to Comets and Asteroids: Tucson, University of Arizona Press, p. 791–826.

Tripati, A., Backman, J., Elderfield, H., and Ferretti, P., 2005, Eocene bipolar glaciation associated with global carbon cycle changes: Nature, v. 436, p. 341–346, doi: 10.1038/nature03874.

Van Eijden, A.J.M., and Ganssen, G.M., 1995, An Oligocene multi-species foraminiferal oxygen and carbon isotope record from ODP Hole 758A (Indian Ocean): Paleoceanographic and paleoecologic implications: Marine Micropaleontology, v. 25, p. 47–65, doi: 10.1016/0377-8398(94)00028-L.

Van Mourik, C., and Brinkhuis, H., 2005, The Massignano Eocene-Oligocene golden spike section revisited: Stratigraphy, v. 2, p. 13–30.

Vonhof, H.B., Smit, J., Brinkhuis, H., Montanari, A., and Nederbragt, A.J., 2000, Global cooling accelerated by early late Eocene impacts?: Geology, v. 28, p. 687–690, doi: 10.1130/0091-7613(2000)28<687:GCABEL>2.0.CO;2.

Wade, B.S., and Kroon, D., 2002, Middle Eocene regional climate instability: Evidence from the western North Atlantic: Geology, v. 30, p. 1011–1014, doi: 10.1130/0091-7613(2002)030<1011:MERCIE>2.0.CO;2.

Wade, B.S., and Pälike, H., 2004, Oligocene climate dynamics: Paleoceanography, v. 19, p. PA4019, doi: 10.1029/2004PA001042.

Wei, W., and Wise, S.W., Jr., 1992, Eocene–Oligocene calcareous nannofossil magnetobiochronology of the Southern Ocean: Newsletters on Stratigraphy, v. 26, p. 119–132.

Zachos, J.C., Quinn, T.M., and Salamy, K.A., 1996, High resolution (104 years) deep sea foraminiferal stable isotope records of the Eocene-Oligocene climate transitions: Paleoceanography, v. 11, p. 251–266, doi: 10.1029/96PA00571.

Zachos, J.C., Pagani, M., Sloan, L., Thomas, E., and Billups, K., 2001, Trends, rhythms, and aberrations in global climate 65 Ma to Present: Science, v. 292, p. 686–693, doi: 10.1126/science.1059412.

Manuscript Accepted by the Society 16 September 2008

Printed in the USA

The Geological Society of America
Special Paper 452
2009

Evidence for a change in Milankovitch forcing caused by extraterrestrial events at Massignano, Italy, Eocene-Oligocene boundary GSSP

Rachel E. Brown*†
Carleton College, One North College Street, Northfield, Minnesota 55057, USA

Christian Koeberl†
Center of Earth Sciences, University of Vienna, Althanstrasse 14, A-1090 Vienna, Austria

Alessandro Montanari†
Osservatorio Geologico di Coldigioco, I-62020 Frontale di Apiro (MC), Italy,

David M. Bice†
The Pennsylvania State University, Department of Geosciences, 503 Deike Building, University Park, Pennsylvania 16802, USA

ABSTRACT

High-resolution spectral analyses of four climate proxies from Massignano, Italy (Eocene-Oligocene boundary global stratotype section and point [GSSP]), indicate that the deposition of this rhythmically bedded sedimentary sequence was controlled by Milankovitch orbital cycles. An inverse relationship between the magnetic susceptibility record and the co-varied calcium carbonate, $\delta^{18}O$, and $\delta^{13}C$ records is indicative of a climate model in which limestones represent dry/cold periods, while marly limestones represent warm/wet periods. Through pattern matching of band-pass filtered signals with the La2004 eccentricity curve, we propose an astrochronological calibration for this important time period. Constrained by three radioisotopically dated volcanic ashes and based on a band-pass version of eccentricity that exhibits expected amplitude modulations, our astrochronology yields a refined age for the Eocene-Oligocene boundary of 33.91 ± 0.05 Ma. Orbital forcing is less pronounced in the lower portion of the Massignano section (meter levels 0–15), which contains evidence of several impact events and a 2.2-m.y.-long comet/asteroid shower. We propose that substantial, nonperiodic climate alterations caused by this period of enhanced extraterrestrial activity mask the Milankovitch climate cycles. Possible mechanisms for the exaggeration of impact-related climatic changes include the ice-albedo feedback or the combined effect of impact-related atmospheric alterations with ongoing dust-particle loading associated with the comet/asteroid shower.

Keywords: Eocene/Oligocene boundary, Milankovitch theory, impacts, spectral analysis, stable isotopes.

*Corresponding author. Present address: University of California, Santa Cruz, 1156 High Street, Santa Cruz, California 95064, USA.
†E-mails: rbrown@pmc.ucsc.edu; christian.koeberl@univie.ac.at; sandro.ogc@fastnet.it; dbice@geosc.psu.edu.

Brown, R.E., Koeberl, C., Montanari, A., and Bice, D.M., 2009, Evidence for a change in Milankovitch forcing caused by extraterrestrial events at Massignano, Italy, Eocene-Oligocene boundary GSSP, *in* Koeberl, C., and Montanari, A., eds., The Late Eocene Earth—Hothouse, Icehouse, and Impacts: Geological Society of America Special Paper 452, p. 119–137, doi: 10.1130/2009.2452(08). For permission to copy, contact editing@geosociety.org.

INTRODUCTION

The mid- to late Eocene is characterized by accelerated global cooling coincident with the appearance of Antarctic polar ice sheets (Fig. 1) (e.g., Prothero, 1994; Zachos et al., 1994, 2001; Zanazzi et al., 2007). This long-term climate cooling trend is reflected by an increase in marine oxygen isotope values (e.g., Mackensen and Ehrmann, 1992; Zachos et al., 1994), an ~55 m sea-level fall (Miller et al., 2008), and the occurrence of significant biotic turnovers (Berggren and Prothero, 1992). The cooling trend, however, is further complicated by the occurrence of multiple bolide impact events (e.g., Glass and Koeberl, 1999; Montanari and Koeberl, 2000; Deutsch and Koeberl, 2006), including the nearly contemporaneous Chesapeake Bay and Popigai impacts. These and other impact events, which are possibly related to an extensive late Eocene comet shower (Farley et al., 1998) or asteroid shower (Tagle and Claeys, 2004), may also have had a significant impact on global climate (Vonhof et al., 2000; Bodiselitsch et al., 2004).

The pelagic sediments of the Umbria-Marche Basin, Italy, which were deposited during the period of Eocene-Oligocene cooling, provide insight into the manner in which local sedimentation and regional climate patterns were affected by both ice-sheet development and impact events. The 23-m-thick sequence of pelagic limestone and marly limestone exposed at the Massignano Quarry (global stratotype section and point [GSSP] for the Eocene-Oligocene boundary), located in the easternmost part of the basin (Fig. 2), is ideal for cyclostratigraphic analysis, as it is both continuous and undisturbed (Cotillion, 1995). Moreover, the sequence contains three radioisotopically dated volcanic ashes (Montanari et al., 1988, 1993; Montanari and Koeberl, 2000, and references therein), which provide time constraints for the section and allow for correlation between the stratigraphic climate record and mathematically predicted orbital variations (e.g., Laskar et al., 2004).

In this study, we analyzed four different high-resolution climate proxies in the Massignano section, including carbonate content, magnetic susceptibility, and bulk-rock oxygen and carbon stable isotopes, in order to provide insight into orbital forcing on the Eocene-Oligocene climate system and determine precise astrochronological ages for stratigraphic horizons.

GEOLOGIC SETTING

The Massignano section (43°32′13″N; 13°35′36″E; Jovane et al., 2007), which became the GSSP for the Eocene-Oligocene boundary in 1993 (Premoli-Silva and Jenkins, 1993), consists of 23 m of alternating pale green- and pink-colored pelagic limestones and marly limestones (Fig. 3). As part of

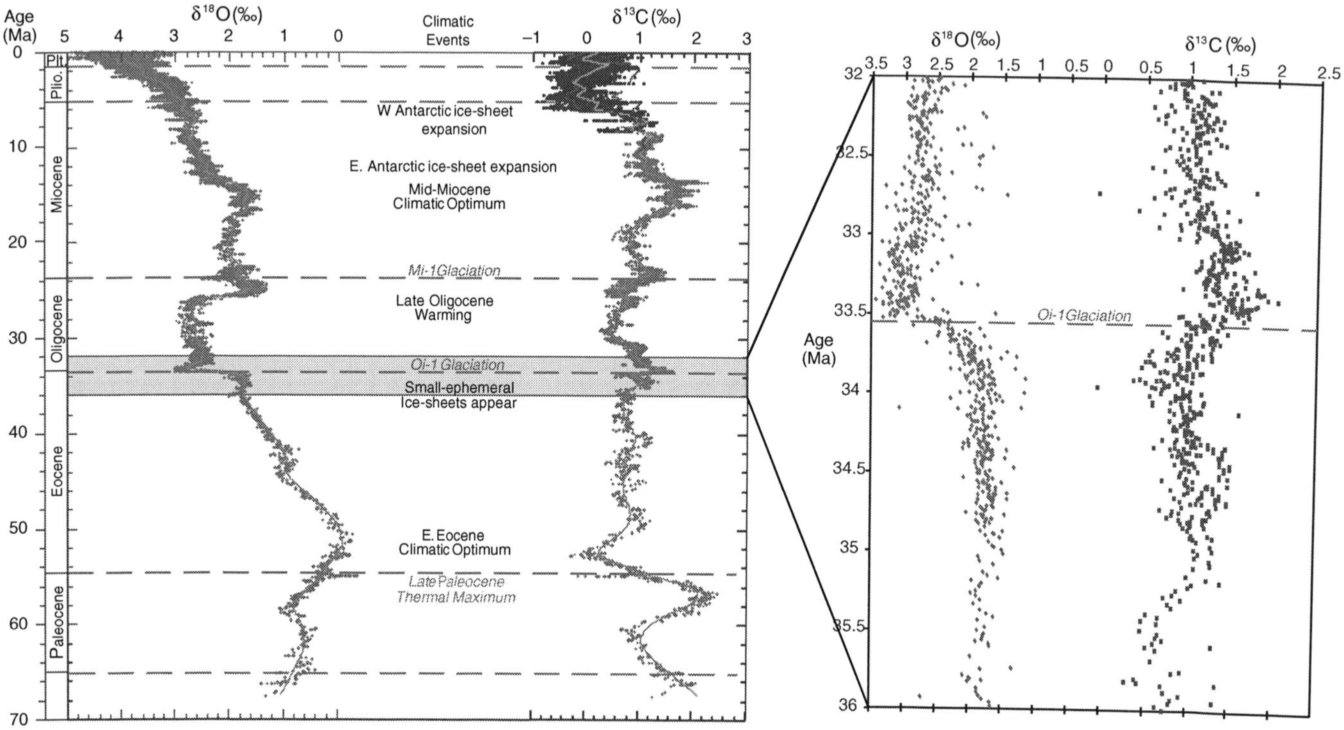

Figure 1. Global deep-sea carbon and oxygen isotope records compiled from pelagic sediments at over 40 Deep Sea Drilling Project (DSDP) and Ocean Drilling Program (ODP) sites. The ^{18}O trends reflect changes in global ice volume and temperature, while the ^{13}C trends are primarily indicative of changes in productivity. The section of interest between 32 and 36 Ma, outlined by the gray box at the left, is isolated on the right. Note the change in scale in the inset (after Zachos et al., 2001). The abbreviations in the upper left hand corner are for Pliocene and Pleistocene.

the Umbria-Marche sedimentary basin (paleolatitude ~34°N; Dercourt et al., 1993), the sequence was deposited in the Neotethys Ocean at a paleodepth of ~1000–1500 m (Coccioni and Galeotti, 2003). The section contains two distinctive formations, the Scaglia Variegata Formation and Scaglia Cinerea Formation, which span an interval from the latest Eocene to early Oligocene (e.g., Lowrie and Lanci, 1994).

As the type section for the Eocene-Oligocene boundary, Massignano has been the subject of a number of detailed studies integrating litho-, bio-, magneto-, and chemostratigraphy (see, e.g., Premoli-Silva et al., 1988; Montanari and Koeberl, 2000; Jovane at el., 2004, and references therein). Three radioisotopically dated biotite-rich volcano-sedimentary layers are found at meter levels 7.2, 12.7, and 14.7, and they provide independent age constraints necessary for compelling astronomical dating (Montanari et al., 1988; Montanari et al., 1993; Montanari and Koeberl, 2000, and references therein). Montanari et al. (1993) reported revised mean ages of 35.4 ± 0.3 Ma, 34.5 ± 0.3 Ma, and 34.2 ± 0.2 Ma for these three layers, respectively. In addition, the section contains several other biotite-rich volcano-sedimentary layers as well as impactoclastic layers marked by Ir anomalies (Montanari et al., 1988, 1993), shocked quartz grains (Clymer et al., 1996; Langenhorst, 1996), and finally, a broad ^{3}He anomaly (Farley et al., 1998), which has been interpreted as evidence for a comet shower lasting ~2.2 m.y.

FIELD AND LABORATORY METHODS

Sample Collection

Following the work of Coccioni et al. (1988), we relogged the Massignano section and sampled between meter levels 0.5 and 23 at a 5 cm sampling resolution (Fig. 3). The section is heavily bioturbated (Huber et al., 2001), making a higher sampling resolution inadvisable. Samples weighing a minimum of 20 g were primarily collected using a Bosch PbH 200 RE masonry drill with a 12 mm drill bit. Marly layers not conducive to drilling were collected as hand samples. Samples were dried, powdered (using a brass mortar and pestle), and sieved to a grain size of 1–2 mm to ensure homogeneity.

Figure 2. Location map of the Massignano section, Marche region, Italy. Landsat image in the upper right is courtesy of the Global Land Cover Facility.

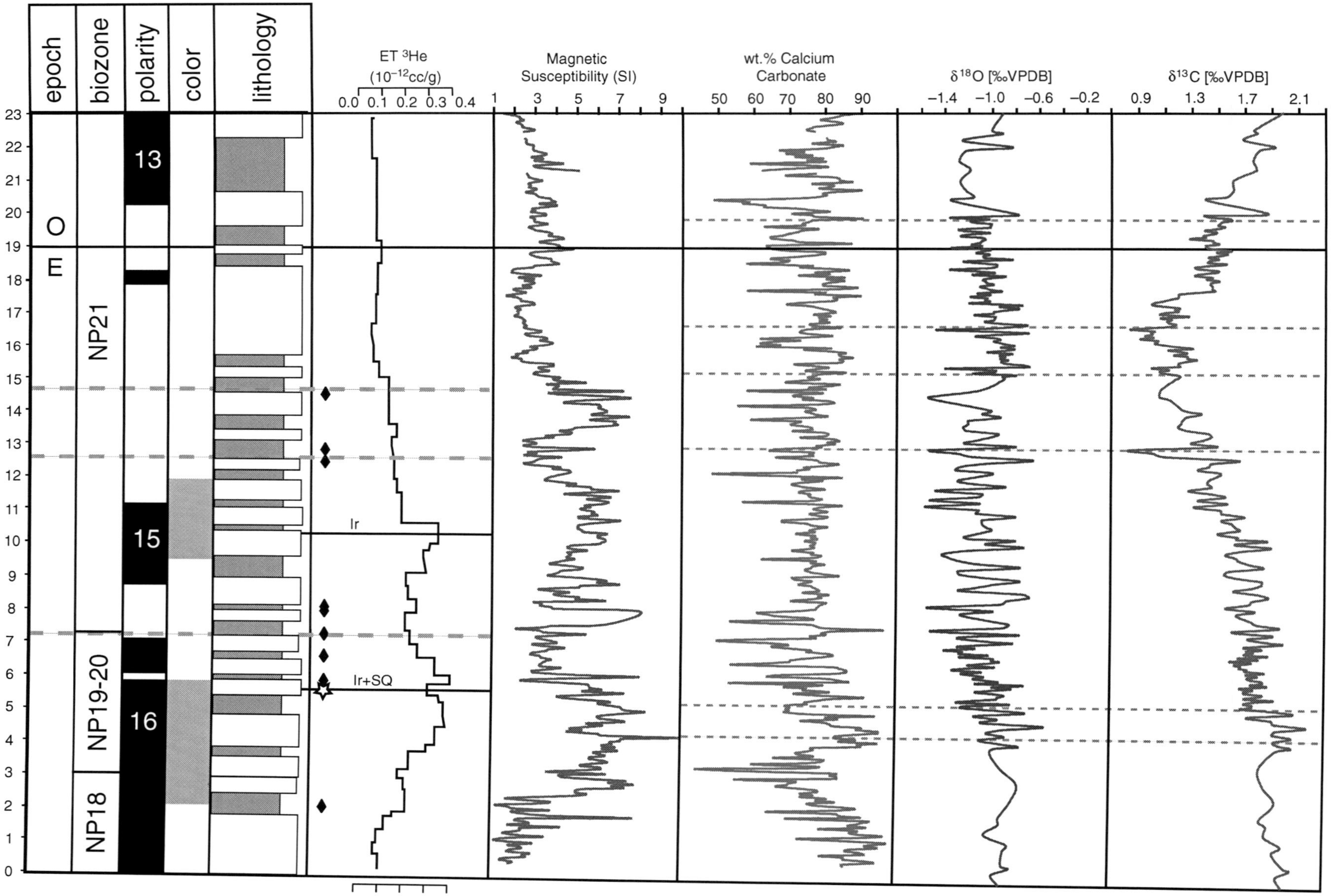

Figure 3. Lithostratigraphy, ^{3}He content (Farley et al., 1998), magnetic susceptibility, carbonate content, and oxygen and carbon stable isotope records (this study; Bodiselitsch et al., 2004) from the Massignano section. The Eocene-Oligocene boundary is located at meter level 19. The dark boxes under "color" represent reddish intervals, horizontal dashed lines denote dated volcanic ashes (Montanari et al., 1988), diamonds signify biotite-rich layers, and the star is indicative of shocked quartz (Clymer et al., 1996). Note that the carbonate content co-varies with the isotope data but exhibits a primarily inverse relationship with magnetic susceptibility. Ashes have been removed from the magnetic susceptibility record to accentuate variability. See Premoli-Silva et al. (1988) and references therein for further information on the biozones and magnetic polarity. E—Eocene; O—Oligocene; VPDB—Vienna Peedee belemnite.

Percent Carbonate and Magnetic Susceptibility

The carbonate content determinations and magnetic susceptibility analyses were completed at the Osservatorio Geologico di Coldigioco. We measured the calcium carbonate content using a Dietrich-Fruling water calcimeter with ±2 rel% precision. Samples weighing between 300 and 320 mg (and powdered to a grain size < 250 μm) were reacted in excess with a 10 vol% HCl in water solution for 2 min. Standards (Carrara marble) were run after every 20 samples to ensure proper calibration.

Mass specific magnetic susceptibility measurements were carried out on a Bartington MS2 and a Bartington MS2B dual-frequency sensor on low frequency (0.465 kHz) and ×0.1 sensitivity. Samples were measured at a constant volume, and their masses were noted. Air measurements were made between samples to correct for thermally induced drift.

Bulk-Rock Stable Isotopes

Oxygen and carbon stable isotopic compositions were obtained from bulk-rock carbonate at 5 cm intervals in two short intervals of the Massignano section: meter levels 5.25–7.1 and 15–20 (Fig. 3). The first of these intervals contains an impactoclastic layer at meter level 5.6, and a small iridium anomaly at meter level 6.2 (Montanari et al., 1993; Bodiselitsch et al., 2004), while the second contains the Eocene-Oligocene boundary at meter level 19. Stable isotope analyses were carried out at the Pennsylvania State University on a Finnigan MAT 252 isotope-ratio mass spectrometer with common acid bath. Samples were reacted in 100% anhydrous phosphoric acid at 90 °C. The isotope data are reported in per mil deviations from the international Vienna Peedee belemnite (V-PDB) carbonate standard, to which the data have been calibrated with NBS-19. Analytical precision for $\delta^{18}O$ and $\delta^{13}C$ values is 0.1‰ and 0.07‰, respectively.

The use of bulk carbonates in stable isotope analysis is convenient because it requires very small amounts of sample material and also allows for the rapid analysis of numerous samples. One concern with this method, however, is that because the bulk samples represent a mixture of carbonates from different sources, the $\delta^{18}O$ of seawater will not be accurately represented (Stoll and Schrag, 2000). While this concern is valid, comparisons of single species foraminiferal $\delta^{18}O$ records with those of bulk carbonates show that bulk carbonates do in fact accurately represent changes in both sea-surface temperature and the $\delta^{18}O$ of seawater when environmental changes are universal, affecting species throughout the water column (Shackleton et al., 1993; Schrag et al., 1995). Another potential problem with using bulk carbonates is that weathering and diagenesis may alter the original carbon and oxygen isotopic compositions (Banner and Hanson, 1990). Weathering at Massignano is minimized, since the excavation of the quarry front was discontinued some 30 yr ago, and the outcrop is now maintained by the Monte Cònero Regional Park. Furthermore, Odin et al. (1988) found no evidence of carbonate recrystallization at Massignano, though scanning electron microscope (SEM) analysis has revealed that foraminifers from the section contain secondary, blocky calcite (Vonhof et al., 2000). Still, as long as sediments in the same section experience recrystallization at the same rate through time, they will contain about the same amount of secondary calcite. The mean $\delta^{18}O$ values may be shifted in either direction, but the important primary variations are preserved (Stoll and Schrag, 2000).

Spectral Analysis

We conducted spectral analyses of the four proxy data sets in Matlab 5.2 using algorithms modified from Muller and MacDonald (2000). Outliers, namely ash deposits with anomalously high magnetic susceptibility values, were removed from the data set, requiring linear interpolation between points. We then performed fast Fourier transforms (FFTs) of the data (detrended and padded with zeros) in the stratigraphic domain, which yielded a set of spectral peaks representing cycles per stratigraphic centimeter. The spectral power used here is the complex conjugate of the Fourier coefficients normalized to unit mean power (Muller and MacDonald, 2000). We evaluated the statistical significance of the spectral peaks by generating a 95% confidence level from a Monte Carlo noise simulation (Muller and MacDonald, 2000). In this approach, 1000 random number sets with the same dimensions as the data are generated and then run through a FFT to find spectral power at each frequency. The 95% confidence level was estimated from these 1000 spectral results by calculating the mean value for each frequency three times. Peaks rising above this 95% confidence level were considered to be statistically significant. We assume that prominent, statistically significant peaks in the magnetic susceptibility data that are related to one another by the same ratios found between Milankovitch cycles represent eccentricity, obliquity, and precession, making it possible to correlate meter level to age through the calculation of an average sedimentation rate. For the Massignano section, we calculated an average sedimentation rate of 10.6 m/m.y., which is consistent with the possible range of sedimentation rates suggested by the independently dated volcanic ashes (between 4.4 m/m.y. and 10.8 m/m.y. using the revised ash dates reported by Montanari et al., 1993), and it is in agreement with sedimentation rates suggested by magnetostratigraphy (Lowrie and Lanci, 1994). The 23 m section, therefore, spans a time period of 2.17 m.y., with an average time interval of 4.72 k.y. between samples (every 5 cm). Spectral peaks with periods greater than ~800 k.y. are therefore not significant in the Massignano section, as they were repeated fewer than three times during deposition.

Amplitude modulations were evaluated for both the mathematically predicted orbital cycles as well as for the cycles in our $CaCO_3$ content data following the approach of Grippo et al. (2004) and Mitchell et al. (2008). This analysis provides a means for testing the hypothesis that the Milankovitch-band cycles are indeed orbital cycles—if they are, then we expect them to have the predicted amplitude modulations. The first step of this process involved isolating the presumed orbital signals from the raw data using Gaussian band-pass filters (eccentricity, 145–75 k.y.; obliquity, 50–35 k.y.; and precession, 25–17 k.y.). We then applied a Hilbert transform to each of the presumed orbital signals, which resulted in an enveloping curve that graphically illustrates the amplitude modulation. We then applied a FFT to this enveloping curve in order to identify the spectral character of the amplitude modulation. A similar amplitude modulation analysis of the Laskar 2004 (La2004) orbital parameters allowed us to compare our presumed orbital signals with the mathematically predicted cycles.

To examine the stratigraphic changes in spectral power, the result of likely variations in environmental conditions and sediment accumulation rates, we employed a sliding-window technique, or evolutionary spectral analysis, in which FFTs were performed on a specified portion or "window" of the data that shifted incrementally through the time series. By looking at the continuity of peaks through time, we can establish whether they represent long-term cyclicity or whether they are the result of random variations and climatic noise (Muller and MacDonald, 2000). Because the choice of window size can affect which cycles appear important, we used a variety of window sizes for the analysis and noted which frequencies were present, at least in part, in multiple windows. Frequencies observed in a range of window sizes and that were continuous through time are considered to reflect actual cyclic trends in the data. Prior to the sliding-window analyses, we applied broad band-pass filters to smooth the data, thereby removing long-period, low-frequency contributions that may be swamping the higher-frequency signals.

Astronomical dating was similarly accomplished through the use of broad band-pass filtering. Variations in the carbonate and magnetic susceptibility data between 140 and 85 k.y. were isolated so as to emphasize the variance in the short eccentricity band. These filtered data were then used for correlation via constrained pattern matching with the La2004 theoretical eccentricity curve of Laskar et al. (2004). The uncertainties in astronomical dating depend on (1) the astronomical solution applied; (2) the assumption of a constant sedimentation rate; and (3) the assumption that there is no lag between orbital forcing and its sedimentary expression (Kuiper et al., 2008). Laskar et al. (2004) estimated a maximum error in their calculations of 30 k.y. over the last 50 m.y. If we assume our tuning to be correct, then the error in the astronomical dates will be about half of the cycle period to which we are matching (short eccentricity), so for this study, about ±50 k.y.

RESULTS

Percent Carbonate and Magnetic Susceptibility Data

The carbonate content (% by weight $CaCO_3$) of the limestones and marly limestones of the Massignano section varies from 47.4 wt% to 96.1 wt%, with a mean of 76.9 wt% (Fig. 3). There is no trend through the section. Conversely, the magnetic susceptibility data exhibit an upsection trend of decreasing variability when the anomalously high ash-layer values are disregarded (Fig. 3). The magnetic susceptibility values fluctuate between 1.42 and 123 (SI units), averaging 4.72 SI. The highest values fall within the ash layers. Both proxy curves correspond well with lithology. The two data sets are inversely related, such that high carbonate percentages correspond with low magnetic susceptibility values.

Bulk-Rock Stable Isotope Data

The oxygen and carbon isotopic data (Fig. 3) from this study, combined with those of Bodiselitsch et al. (2004), correlate well with lithology and appear to co-vary, such that $\delta^{18}O$ highs correspond with $\delta^{13}C$ highs, though this relationship is less clear in the lower section. A decreasing trend is visible in the $\delta^{13}C$ data through the majority of the section, followed by an increasing trend from about meter-level 17 to the top. The $\delta^{13}C$ values from the lower high-resolution interval (5.35–7.10 m) fluctuate between 1.53‰ and 1.83‰ (averaging 1.7‰), while values in the upper section (m 15–20) range from 0.8‰ to 1.6‰ (with an average of 1.28‰) and exhibit a marked upsection increase.

Trends in the $\delta^{18}O$ data are less pronounced, but there appears to be a general decrease in values from meter level 3 to 8, followed by an increase and a subsequent decrease from about meter level 15 to 21. For meters 15–20, the $\delta^{18}O$ values range from –0.73‰ to –1.51‰ (averaging –1.05‰). Values are similar between meters 5.35 and 7.10, varying from –0.84‰ and –1.43‰ (averaging –1.16‰).

In comparison with the global benthic foraminiferal $\delta^{18}O$ values listed in Figure 1, we can see that the isotope values are on the heavier side. The $\delta^{18}O$ values reported in this study average about –1‰ as opposed to an average of about –2‰ in the global benthic compilation for this time period. Similarly, the $\delta^{13}C$ values from this study average ~1.5‰, rather than 1‰. These differences likely result from two sources: (1) in this study, we used bulk-rock stable isotopes rather than isotopes from benthic foraminifera, and (2) the paleolatitude of the study site is ~34°N (Dercourt et al., 1993), and so it is a lower-latitude site where temperatures and rates of productivity are generally higher.

Notably, the significant positive isotope excursion that marks the Eocene-Oligocene boundary in the global benthic record (Fig. 1) is lacking at Massignano. Both the $\delta^{18}O$ and $\delta^{13}C$ isotope records do, however, begin to show a slight increase in values near the top of the record. The $\delta^{13}C$ record begins to increase slightly before the boundary at about meter level 17, while the $\delta^{18}O$ record does not begin to increase until after the boundary at about meter level 21. It may be that only the beginning of the first step of the two-step Eocene-Oligocene transition proposed by Coxall et al. (2005) is preserved at Massignano, as the section ends just a few meters above the Eocene-Oligocene boundary, and the more dramatic second step is not preserved here.

Spectral Analysis

Fast Fourier Transform

Spectral analysis results for the percent carbonate and magnetic susceptibility data (Fig. 4) were obtained using a linear sedimentation rate of 10.6 m/m.y., which was calculated by matching peaks in the meter-level domain to known astronomical cycles. The plots are quite noisy, particularly that for the $CaCO_3$ content, and while orbital peaks are present, they do not all emerge significantly above the noise level. Prominent spectral power peaks for the $CaCO_3$ content occur at 315, 60, 41, 30, and 24 k.y., and for magnetic susceptibility, peaks occur at 254, 116, 43, and 24 k.y., all of which exceed the 95% confidence limit.

Knowledge of the comet shower and multiple volcanic ashes in the lower and central portion of the section prompted a split analysis of the outcrop. The results for the split section are shown in Figures 5 and 6. The plots for the upper portion (meters 15–23) are considerably less noisy. Common elements to all four proxies include a peak or cluster of peaks at 36–43 k.y., and pronounced peaks at 95 or 97, as well as at 23–25 k.y. Both the magnetic susceptibility and $CaCO_3$ content plots exhibit strong peaks at ~125–127 k.y. and ~95–97 k.y. This second peak is also notable in the plot for $\delta^{13}C$, while the peaks with the greatest spectral power in $\delta^{18}O$ plot are those at 23 and 19 k.y. Spectral peaks with periods greater than 250 k.y. are not considered significant, as the section under scrutiny spans just 755 k.y. Although neither rises above the 95% confidence interval, a peak at 55 k.y. is prominent in three of the records, as is a peak at 30 or 31 k.y. A 55 k.y. peak may reflect either the 54 k.y. obliquity component or the first harmonic of the 100 k.y. eccentricity (F. Hilgen, 2008, personal commun.). According to Mix et al. (1995), a 29–31 k.y. peak may be caused by nonlinear coupling of eccentricity and obliquity; however, in this case, a strong 41 k.y. obliquity component would also be expected. Alternatively, a 29–31 k.y. peak could be explained by the combination of precession components, or of precession and obliquity (von Dobeneck and Schmieder, 1999).

The spectral plots for the lower portion are reminiscent of the whole-section plots, exhibiting an elevated level of noise (Fig. 6). The most prominent peaks in the magnetic susceptibility and $CaCO_3$ data that are possibly orbital in origin are at 44 and 24 k.y. A peak seen in the $CaCO_3$ data at 58 k.y. is not echoed in the magnetic susceptibility record. In the $\delta^{13}C$ record, peaks at 245 and 150 k.y. are the only ones to rise above the 95% confidence interval, though peaks at 123, 86, and 56 k.y. all seem to stand out from the rest. Finally, in the $\delta^{18}O$ record, prominent peaks at 144, 77, 22, and 18 k.y. appear significant.

Amplitude Modulations

Amplitude modulations, which reflect a beat between fundamental astronomical frequencies, are not easily introduced into a data set via the tuning process (Pälike et al., 2001) and can, therefore, provide insight into how strongly an astronomical forcing is represented in the data (Laskar, 1999). Amplitude modulations often cited in the literature include the 400 k.y. modulation of climatic precession by eccentricity, as well as the 1.2 m.y. obliquity modulation (e.g., Lourens and Hilgen, 1997; Turco et al., 2001; Grippo et al., 2004; Pälike et al., 2004). All three of the spectral plots displaying amplitude modulations of the Massignano percent carbonate display amplitude modulation peaks that are in agreement with the peaks expected from the mathematically predicted Laskar et al. (2004) curve (Fig. 7). The amplitude modulations exhibited by the Massignano percent carbonate short

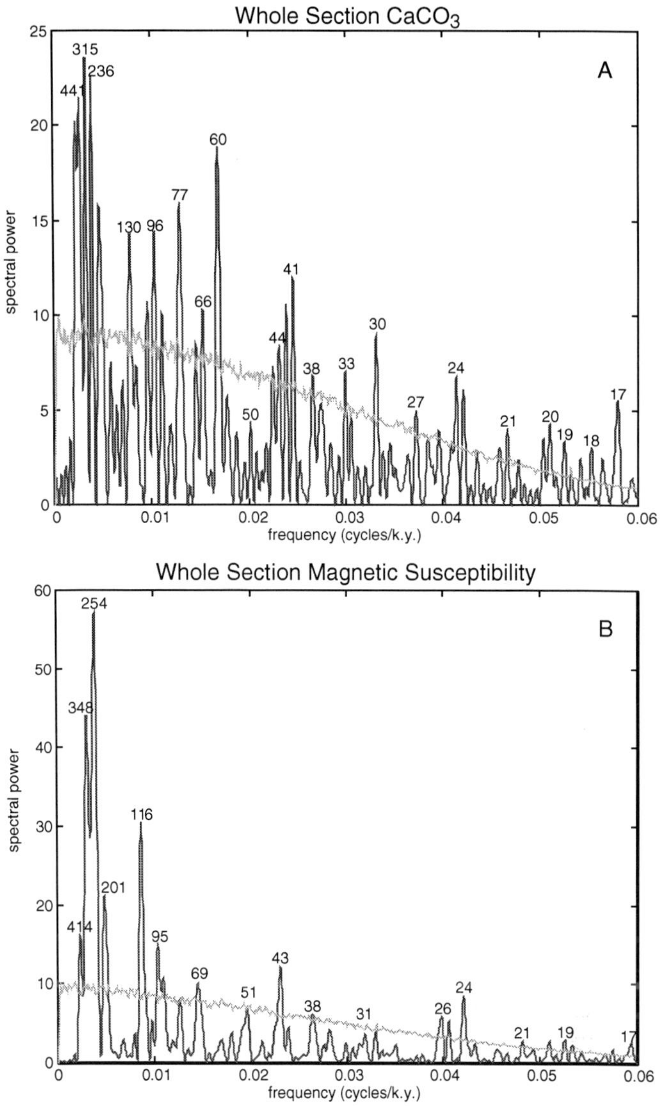

Figure 4. Spectral analysis results for (A) $CaCO_3$ content and (B) magnetic susceptibility for the entire Massignano section using a sedimentation rate of 10.6 m/m.y. and 10–700 k.y. band-pass filter to remove long-period, low-frequency contributions. The black curve (with peaks labeled in k.y.) represents the various cycles present in the data. The light-gray line is the 95% confidence limit; peaks rising above this curve can be attributed to true cyclicity with 95% confidence. While orbital peaks are present, other peaks appear equally or even more pronounced.

Figure 5. Spectral analysis results for (A) carbon isotopes, (B) oxygen isotopes, (C) magnetic susceptibility (MS), and (D) calcium carbonate content for the upper portion (meters 15–23) at Massignano, Italy. Data from Bodiselitsch et al. (2004) were combined with data from this study to produce the carbon and oxygen stable isotope plots. Again, the black curve with peaks labeled in k.y. represents the cycles present in the data, while the light-gray curve is the 95% confidence limit. These plots are considerably less noisy than those for the whole section. Common elements in the four plots include a peak or peak cluster at 36–43 k.y., and pronounced peaks at 95 or 97 k.y., as well as at 23–25 k.y. Both the eccentricity and precession peaks consistently clear the light-gray 95% confidence line; however, the obliquity peak does not.

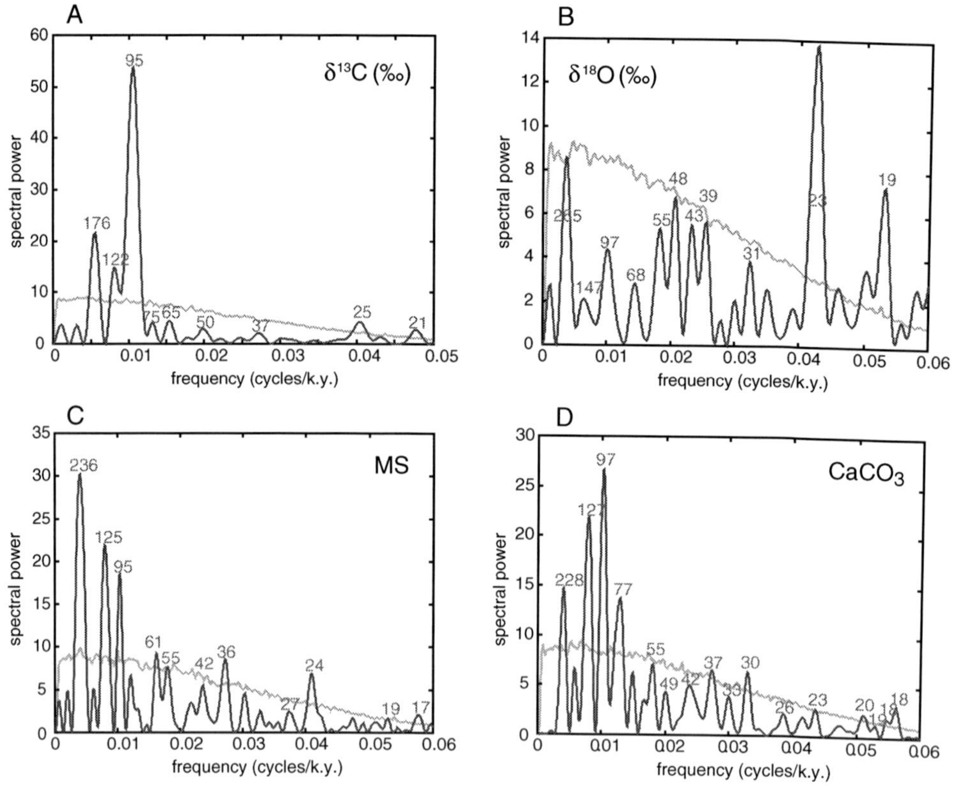

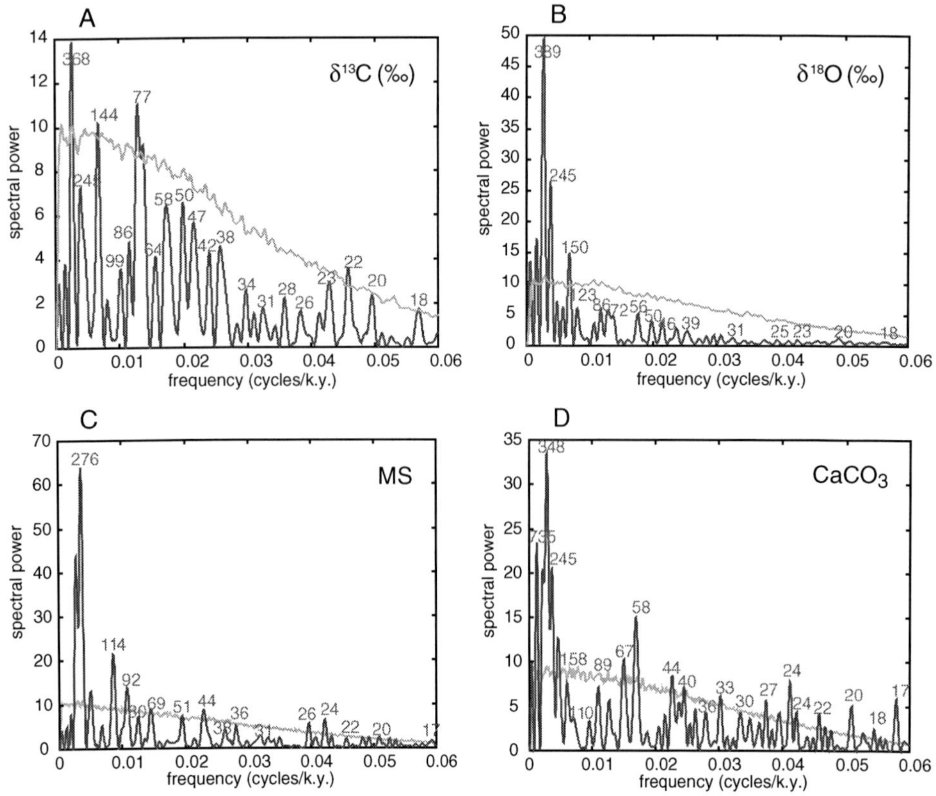

Figure 6. Spectral analysis results for all four proxies in the lower portion (meters 0.5–15) of the Massignano section: (A) carbon isotopes, (B) oxygen isotopes, (C) magnetic susceptibility (MS), and (D) calcium carbonate content. Data from Bodiselitsch et al. (2004) were combined with data from this study to produce the carbon and oxygen stable isotope plots. The black curve represents the various cycles present in the data, and the light-gray line is the 95% confidence limit. Cycles are labeled in k.y. Similar to the spectral plots for the entire section, these plots exhibit a significant amount of noise.

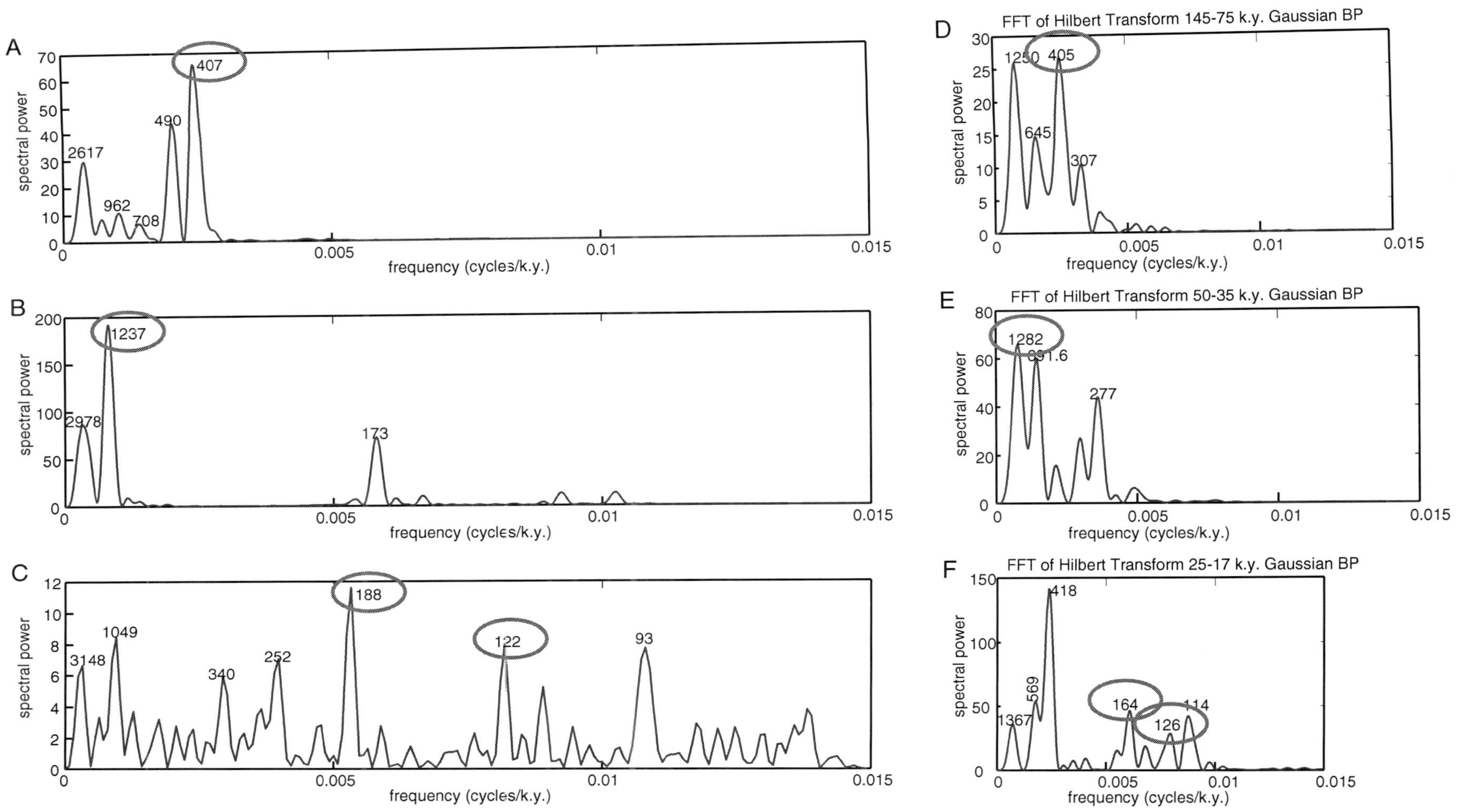

Figure 7. Amplitude modulations of Laskar et al. (2004) for (A) short eccentricity, (B) obliquity, and (C) precession, as well as of the Massignano section calcium carbonate (D) short eccentricity, (E) obliquity, and (F) precession signals. These plots were obtained following the approach of Grippo et al. (2004). We employed a Hilbert transform to find the enveloping curve of a Gaussian band-passed (BP) signal followed by a fast Fourier transform (FFT) of the enveloping curve. Notable and consistent amplitude modulations are denoted by the ovals.

eccentricity signal include a 405 k.y. peak, as well as peaks at 1250, 645, and 307 k.y. For the obliquity signal, the most prominent peak has a period of ~1.2 m.y., with two other peaks at ~692 and 277 k.y. Finally, amplitude modulations of the precessional signal include a conspicuous peak with a period of ~418 k.y., as well as peaks at ~569, 164, 114, 126, and 1367 k.y.

Evolutionary Spectral Analyses

In the upper section, by comparing results from sliding-window analyses using both medium and small window sizes, it is apparent that long-term cyclic variations are most prevalent in the $CaCO_3$, magnetic susceptibility, and $\delta^{13}C$ data. While all four proxies exhibit coherent spectral peaks in the medium-sized windows (200–400 k.y.) (Fig. 8), continuous, low-frequency, ~95–118 k.y. peaks are also evident in the $CaCO_3$, magnetic susceptibility, and $\delta^{13}C$ data in the smaller windows (100–226 k.y.) (Fig. 9). Furthermore, faintly continuous peaks are also visible at ~38–42 k.y. and 23–26 k.y. The peaks within the $\delta^{18}O$ data appear primarily discontinuous in the smaller window. Despite this, because three of the proxy records display coherent spectral peaks in multiple window sizes, it is reasonable to assert that the cyclicity apparent from the initial spectral analyses is not a result of random depositional variations, but of repeated climatic or depositional cycles.

DISCUSSION

Magnetic Susceptibility

Variability within the magnetic susceptibility record of a particular rock sequence can be explained in a number of ways. For one, hematite records a much lower magnetic susceptibil-

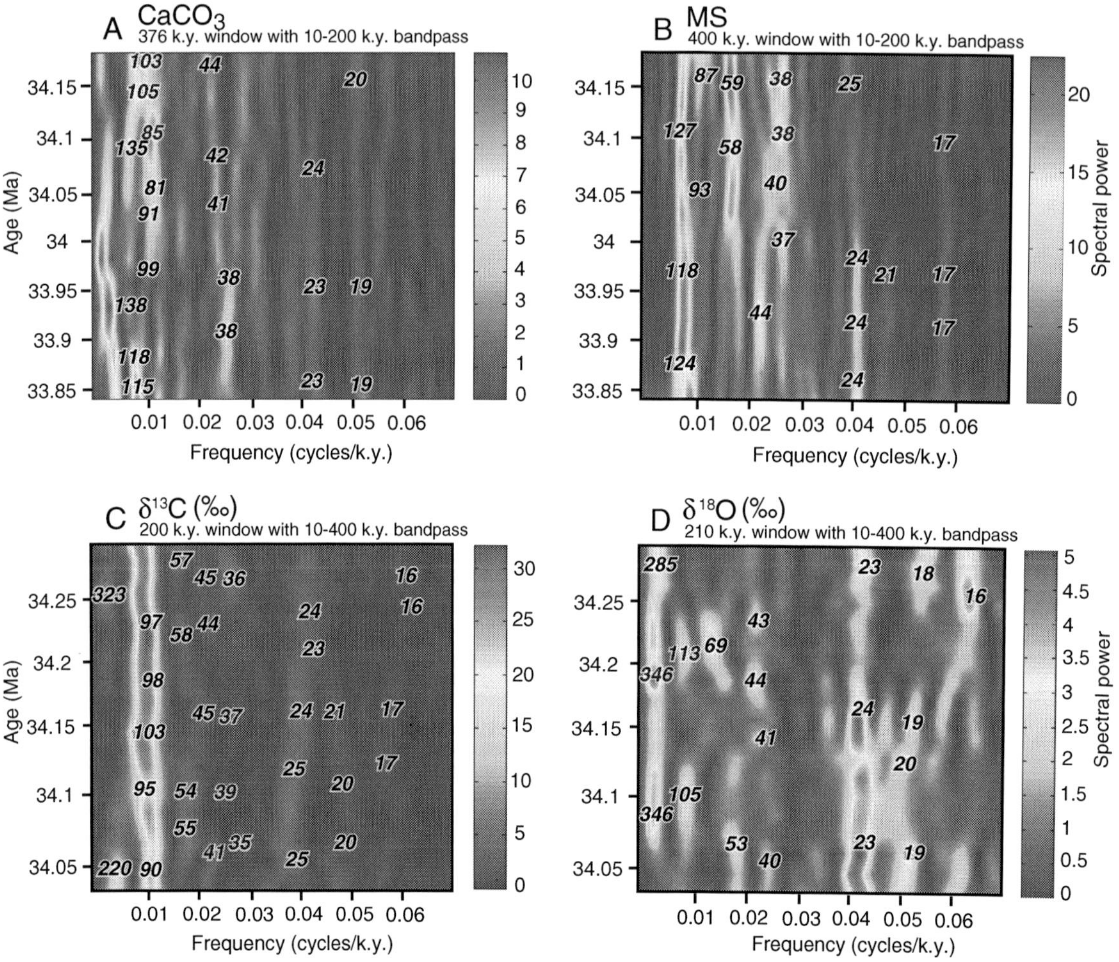

Figure 8. Results for the upper section from the medium-sized sliding-window spectral analyses, which are used to reveal changes in the spectral power of different frequencies over time from: (A) carbonate content (m 15–23), (B) magnetic susceptibility (m 15–23), (C) $\delta^{13}C$ (m 15–20), and (D) $\delta^{18}O$ (m 15–20). A medium-sized window is between two-fifths to one-half of the time scale in question, in this case, ~750 k.y. for plots A and B and ~470 k.y. for plots C and D. Spectral power scales are displayed to the right of each plot. The *y*-axis represents age in millions of years of the sliding-window midpoints.

ity value than magnetite, so a change in magnetic susceptibility could be illustrating a change in mineralogy. While the Massignano section does exhibit packages of reddish-tinted sediments, possibly indicative of elevated hematite content, these sections do not correspond with relatively lower magnetic susceptibility values. Contrarily, the reddish interval between meter levels 2 and 5 exhibits magnetic susceptibility values that are elevated in comparison with the rest of the sequence (Fig. 3). An alternate explanation hinges on terrigenous input, the primary source of iron-bearing minerals in pelagic sediments. Magnetic susceptibility variations are also thought to reflect variations in weathering and erosion caused by sea-level fluctuation or alterations in wet/dry climate cycling (Ellwood et al., 2000; Mader et al., 2004). Changes in sea level affect the supply of detrital material to the marine environment, as drops in base level associated with regressive periods enhance continental erosion (Crick et al., 1997). The variability in the magnetic susceptibility signal of the Massignano section could be linked to either climatically controlled continental erosion or, conversely, to variations in biogenic carbonate production, which would dilute an otherwise constant sediment supply.

Percent Carbonate and Stable Isotopes

Percent carbonate highs represent periods of increased carbonate production, decreased carbonate dissolution, or periods in which a steady supply of carbonate is periodically less diluted by fluxes in terrigenous input (Einsele and Ricken, 1991). Insight into which interpretation of calcium carbonate highs might best fit the Massignano section can be found by comparing the calcium carbonate content with the stable isotope data. At Massignano, there is a generally persistent covariance in the $CaCO_3$, $\delta^{13}C$, and $\delta^{18}O$ records (Fig. 3), which implies that the fluctuation in these parameters might be linked to a common paleoenvironmental cause. High carbonate accumulation corresponds to high planktonic productivity, evidenced by heightened $\delta^{13}C$ values

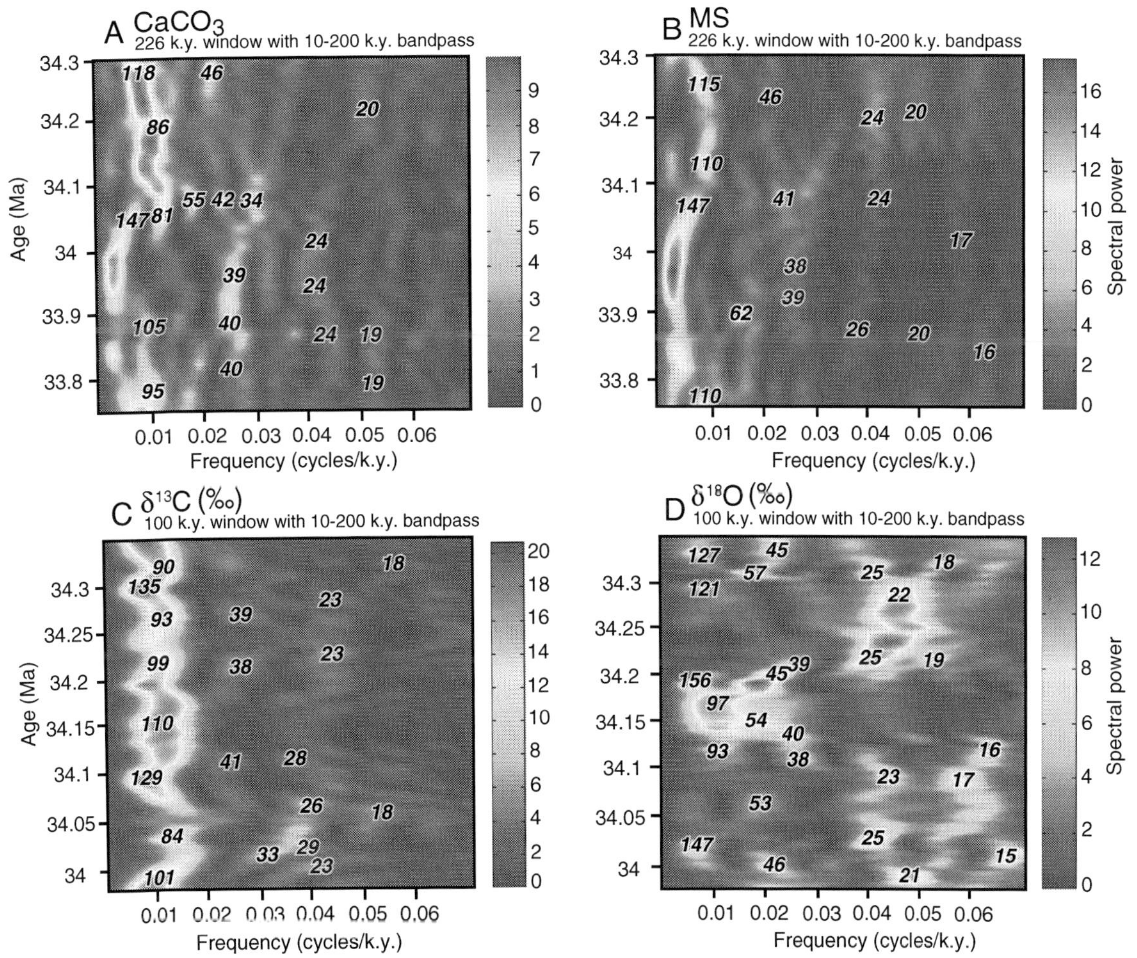

Figure 9. Results for the upper section from the small-sized sliding-window spectral analyses, which are used to reveal changes in the spectral power of different frequencies over time. Small-sized windows range between one-third and one-sixth of the time scale in question (100–226 k.y.). Spectral power scales are displayed to the right of each graph. Relatively continuous, low-frequency ~95–118 k.y. peaks are evident in the CaCO3, magnetic susceptibility, and $\delta^{13}C$ data.

resulting from the preferential fractionation of oceanic ^{12}C by marine photosynthesizers (Bickert, 2000). High carbonate accumulation is simultaneously matched with increased $\delta^{18}O$ values, implying high salinity conditions or low water temperatures at the time of limestone deposition. Following Mader et al. (2004) and in agreement with Jovane et al. (2006), we interpret the correspondence of stable isotope highs with high carbonate contents as an indication that Eocene-Oligocene Mediterranean limestone formation occurred during dry, cold periods marked by higher productivity. It is likely that increased ice volume and the appearance of small ephemeral ice sheets in the late Eocene caused a sea-level drop and a corresponding slight increase in continental erosion, providing the nutrients necessary for high planktonic productivity. Also, an increase in deep water formation tied to ice-sheet expansion would be balanced by increased upwelling of nutrient-rich waters in other parts of the oceans. In light of the inverse relationship between $CaCO_3$ content and magnetic susceptibility, it is reasonable to imagine that marly layers were deposited during relatively wet (and possibly warm) periods with a greater fluvial influx of detrital matter. It is important to note, however, that while ice-sheet development is a global phenomenon, productivity changes in a site like the Mediterranean probably occurred on a more local scale. Cold periods, therefore, may not be equated with high productivity at other latitudes.

Stable Isotopes and Orbital Cycle Strength

The spectral power of the orbital cycles of eccentricity, obliquity, and precession varies significantly from proxy to proxy. Most notably, the $\delta^{18}O$ spectral plot for meter levels 15–20 exhibits its statistically significant high-frequency precessional peaks that dwarf the other possible orbital signals. This is in stark contrast with the $\delta^{13}C$ spectral plot for the same interval, in which the low-frequency eccentricity cycle greatly exceeds the 95% confidence limit (Fig. 5). One possible explanation for these differences lies in the response times of the environmental cycles to which the proxies are tied. The $\delta^{13}C$ values of marine carbonates are linked to the global carbon cycle, which, with many reservoirs, has a slow response time (relative to the hydrologic cycle). The eccentricity cycle, then, may be amplified by the long oceanic residence time of carbon, which was similarly noted by Pälike et al. (2006) for long-period eccentricity (405 k.y.). Contrarily, the hydrologic cycle, to which the $\delta^{18}O$ values of marine carbonates are tied, is characterized by much faster relative response times and therefore may amplify precessional cycles. For example, it is believed that the African monsoon undergoes large changes in tune with the precessional cycle (e.g., Pokras and Mix, 1987; Tuenter et al., 2003).

Correlation with the Astronomical Time Scale

The spectral analysis results indicate that the Massignano sequence's rhythmic bedding reflects orbitally paced climate variations. Even though eccentricity weakly affects insulation, the prominent ~95 k.y. peak common to all four proxies (particularly in the upper section) implies that the late Eocene and early Oligocene climate experienced eccentricity forcing. Others have similarly found the eccentricity signal to be disproportionately significant as far back as the Miocene, with strength far greater than its theoretical contribution to insolation would suggest (e.g., Fischer, 1991; Clemens and Tiedemann, 1997; van Vugt et al., 2001; Cleaveland et al., 2002). This disproportional relationship of eccentricity to climate could perhaps be explained by an amplifying mechanism, such as the ice-albedo feedback (Cleaveland et al., 2002; see also DeConto et al. 2007). With Antarctic ice sheets emerging in the late Eocene (e.g., Miller et al., 1991; Zachos et al., 1992), it is conceivable that this feedback mechanism would have come into play during the deposition of the Massignano section. Alternatively, the ~95 k.y. cycle may not represent the direct effect of eccentricity on insolation, but the modulation of the precessional amplitude by eccentricity.

We achieved an astronomical correlation between Laskar's eccentricity curve and Massignano's smoothed magnetic susceptibility and calcium carbonate data (Fig. 10) through pattern matching constrained by the dated volcanic ashes at meter levels 7.2, 12.7, and 14.7. Following Cleaveland et al. (2002), eccentricity highs were matched with magnetic susceptibility highs and carbonate content lows. For pattern matching, we used the smoothed magnetic susceptibility data from the entire Massignano section (meters 0.5–23). Of all four proxies, the magnetic susceptibility data exhibit the least amount of noise, and Milankovitch peaks rise above the 95% confidence level in both the whole-section and lower-section spectral plots (Figs. 4 and 6). In order to confirm the validity of the match, we similarly matched the smoothed $CaCO_3$ content data from the upper portion of the sequence to Laskar eccentricity and found the two correlations to be in agreement (Fig. 11). The correlation yields a refined date for the Eocene-Oligocene boundary of 33.91 ± 0.05 Ma, which is just 0.2 m.y. older than the previously proposed date of 33.7 ± 0.4 Ma (Montanari et al., 1988). Precise ages of 35.13 ± 0.05 Ma, 34.55 ± 0.05 Ma, and 34.32 ± 0.05 Ma are also obtained for the three radioisotopically dated volcanic ashes of at meter levels 7.2, 12.7, and 14.7, respectively. All three precise ages fall within the error bars of the previous radiometric dates of 35.4 ± 0.3 Ma, 34.5 ± 0.3 Ma, and 34.2 ± 0.2 Ma (Montanari et al., 1993).

These refined dates are not in agreement with the age of 33.714 Ma for the Eocene-Oligocene boundary proposed by Jovane et al. (2006). Differences may arise for the following reasons: (1) Using the Lowrie and Lanci (1994) magnetostratigraphic study at Massignano combined with numerical ages from Cande and Kent (1995), Jovane et al. (2004) established a linear sedimentation rate of ~6.9 m/m.y. for the Scaglia Variegata Formation, which they subsequently applied to the whole of the Massignano section (both the Scaglia Variegata and Scaglia Cinerea Formations); and (2) Jovane et al. (2006) started their tuning with a visual match for the 400 k.y. eccentricity cycle and subsequently tuned their records to obliquity. While the La2004 numerical solution is accurate for eccentricity over the last

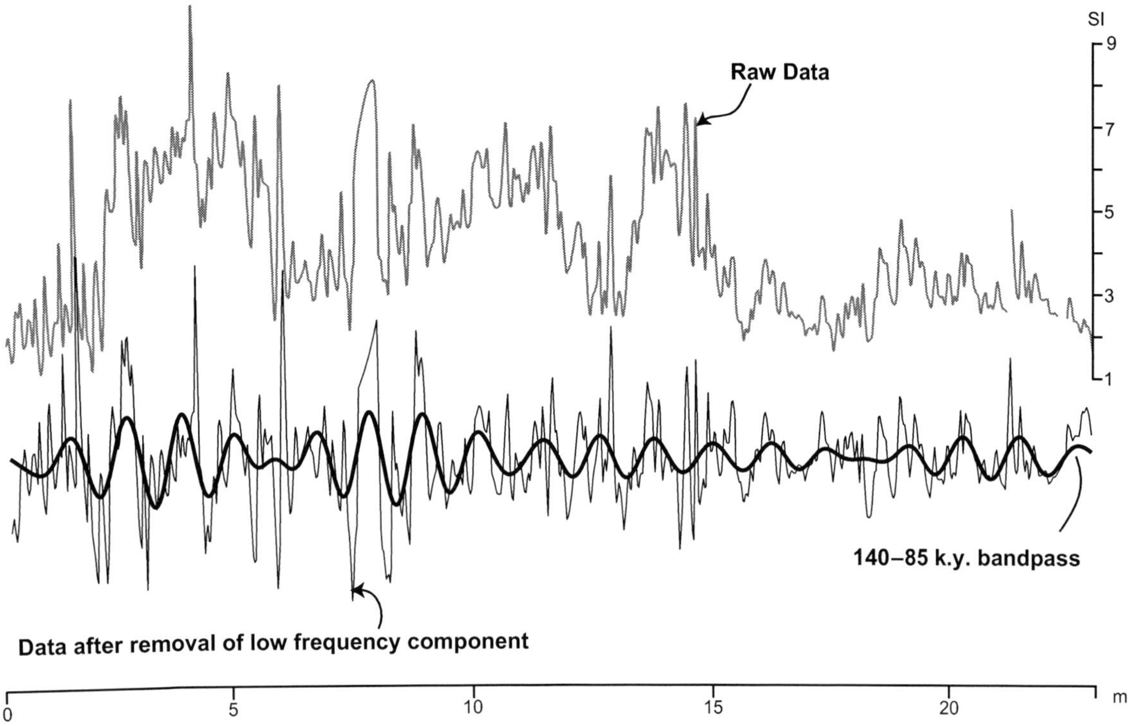

Figure 10. Massignano magnetic susceptibility proxy data shown adjacent to a version of the data with the low-frequency component removed and overlain by a smoothed version of the raw data (45–140 k.y. band-pass filter).

40 m.y., this is not necessarily the case for precession and obliquity, as they are dependent on the Earth-Moon system in addition to the orbital part of the solution (Hilgen et al., 2006; see also Pälike et al., 2004).

Impacts, Comet/Asteroid Showers, and Climate Cycles

While we and others (e.g., Jovane et al., 2006) have shown that some portions of the Massignano section preserve a clear record of Milankovitch orbital cycles, it is also true that the signatures of impact events (and a comet/asteroid shower) are prevalent in parts of the section, allowing us to explore the question of whether or not extraterrestrial events can alter the preservation of "normal" Milankovitch cycles. The lower portion of the outcrop contains evidence of at least two, possibly three, extraterrestrial impacts, including the giant Popigai and Chesapeake Bay impact events (Montanari and Koeberl, 2000; Bodiselitsch et al., 2004; Glass et al., 2004). The Popigai impact crater, a 100-km-diameter structure located in Siberia, Russia, has been dated at 35.7 ± 0.8 Ma, and the 85-km-diameter Chesapeake Bay impact crater is indistinguishable in age (e.g., Bottomley et al., 1997; Koeberl et al., 1996). Evidence of these events has been reported in numerous other early late Eocene sedimentary successions (e.g., Glass et al., 1985; Koeberl and Glass, 1988; see also review in Koeberl, this volume). At Massignano, an impactoclastic layer at ~5.6 m is marked by an iridium anomaly (Montanari et al., 1993), shocked quartz (Clymer et al., 1996; Langenhorst, 1996), extraterrestrial Ni-rich spinel and altered microkrystites (Pierrard et al., 1998), and, finally, a ^{3}He anomaly (Farley et al., 1998). Impactoclastic layers at meter levels 6.19 and 10.25 are similarly distinguished by ^{3}He anomalies, as well as increased iridium abundances (Bodiselitsch et al., 2004). The ^{3}He anomaly at Massignano is in fact quite broad, spanning from about meter level 2 to 15, with its maximum value coincident with the Ir anomaly at 5.6 m (Farley et al., 1998). Farley et al. (1998) argued that this ^{3}He enhancement is attributable to an early late Eocene comet shower lasting ~2.2 m.y. In contrast, Tagle and Claeys (2004) suggested that the increase in delivery of interplanetary dust particles (IDPs) implied by the ^{3}He anomaly could instead be explained by an asteroid shower.

Both individual impact events and comet/asteroid showers have the potential to affect global climate. An impact occurring on a continental shelf, such as the Chesapeake Bay impact (e.g., Koeberl et al., 1996; Poag et al., 2004), might be coupled with the release of methane hydrates, and certainly would deliver a significant amount of water vapor into the atmosphere, both of which contribute to the greenhouse effect and induce global warming, besides changing the Earth's albedo. A terrestrial impact, such as the Popigai impact event (e.g., Bottomley et al., 1997; Tagle and Claeys, 2005), on the other hand, can be linked to global cooling and a corresponding decrease in bioproductivity (Toon et al., 1997). Global cooling could also be attributed to the dust-particle loading of the atmosphere associated with the increased levels of planetary dust inherent to a comet shower. At Massignano,

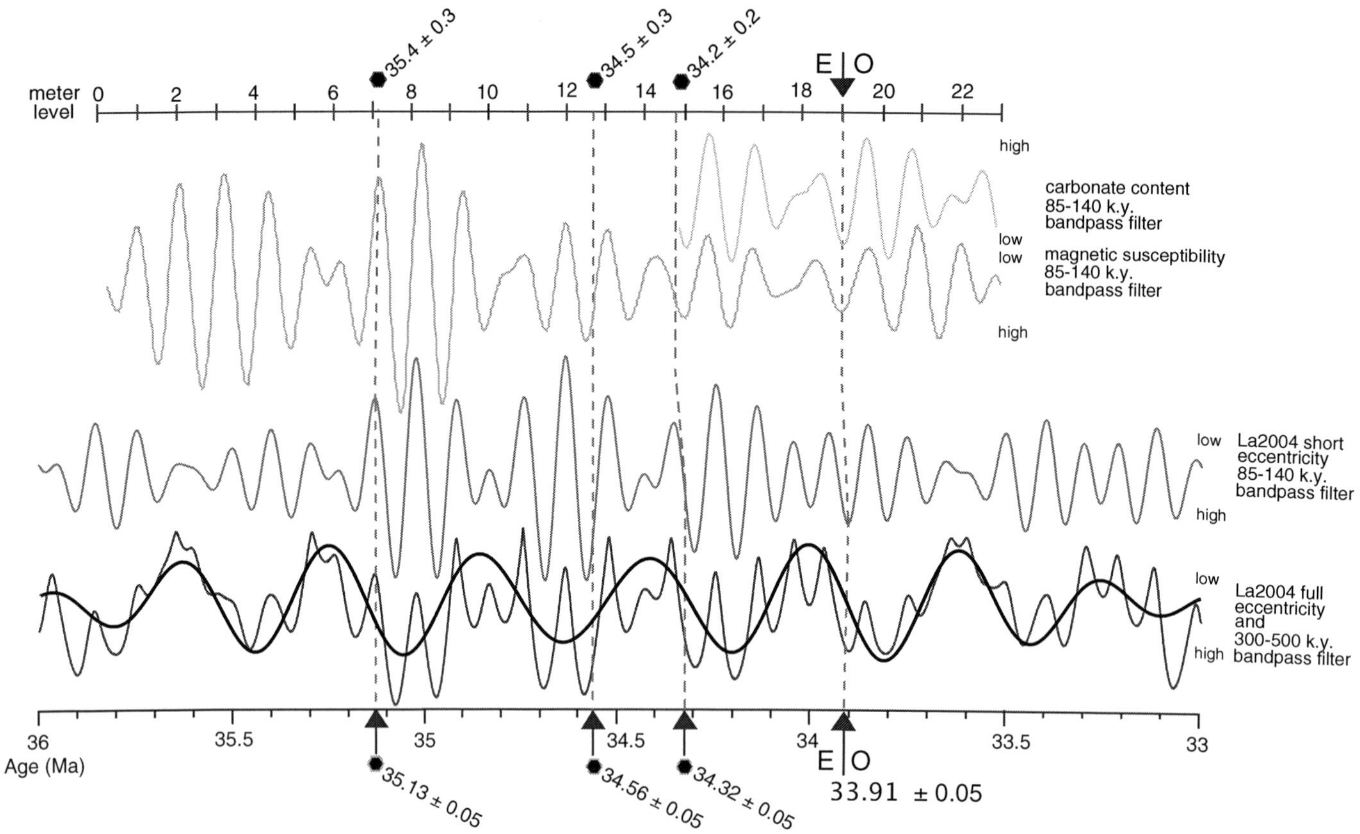

Figure 11. Correlations between Laskar full eccentricity, Laskar short eccentricity (85–140 k.y. band-pass), Massignano magnetic susceptibility, and Massignano carbonate content. Both the magnetic susceptibility data and the $CaCO_3$ content have been smoothed by a 85–140 k.y. band-pass filter to isolate the short eccentricity signal. The filtered curves overlie the raw magnetic susceptibility and $CaCO_3$ data, shown in light gray. Pattern matching constrained by the three volcanic ashes yields a date for the Eocene-Oligocene boundary of 33.91 ± 0.05 Ma, somewhat older than the previously determined date of 33.7 ± 0.4 (Montanari et al., 1988). The smoothed $CaCO_3$ data confirm the pattern match made with the smoothed magnetic susceptibility curve. E—Eocene; O—Oligocene.

based on stable isotope variations, Bodiselitsch et al. (2004) proposed that impact-induced warm pulses punctuate the general trend of continuous Eocene-Oligocene cooling. Vonhof et al. (2000) similarly used stable isotope records from Maud Rise to suggest that late Eocene impact events (ca. 35.5 Ma) accelerated the long-term cooling trend. Because the biotic response to the impacts was greater than would be expected, Vonhof et al. (2000) invoked the possibility of a feedback mechanism, such as the ice-albedo feedback, which could have prolonged the duration of impact-induced climate cooling. The ice-albedo feedback has also been shown to be modulated by Milankovitch periodicities (see Muller and MacDonald, 2000, and references therein); however, it is possible that the climate alterations caused by the impact events could overprint or swamp the longer-term climate cycling events paced by orbital cycles.

The $CaCO_3$ content and magnetic susceptibility sliding-window plots from the Massignano section (Figs. 12 and 13) indicate that the record of Milankovitch cyclicity, while persistent throughout the section, changes during and immediately following extraterrestrial events. In the $CaCO_3$ plot with a 50–10 k.y. band-pass filter, beginning at a window midpoint of ~5.6 m and extending to ~13 m, the relative strength of the obliquity signal is masked by stronger power in some discontinuous 26–30 k.y. cycles. A similar, though less pronounced pattern of discontinuous, stronger-power 30–36 k.y. cycles is visible in the 50–10 k.y. band-pass filter magnetic susceptibility plot, again beginning at a window midpoint of ~5.6 m and extending to ~12–13 m. Most remarkably, the onset of this pattern is coincident with the Ir anomaly and shocked-quartz layer at meter level 5.6. The disruption is also probably at least partially coupled to the ongoing comet/asteroid shower extending from about meter level 2 to 15.

Possible causes for the masked record of Milankovitch cycles include both global and regional mechanisms. Global mechanisms for impact-related climatic changes include the ice-albedo feedback or impact-related atmospheric alterations combined with ongoing interplanetary dust particle input associated with the comet or asteroid shower. On a more regional scale, the record of Milankovitch cycles could be altered by changes in the rate of sedimentation on a time scale comparable to the Milankovitch cycles.

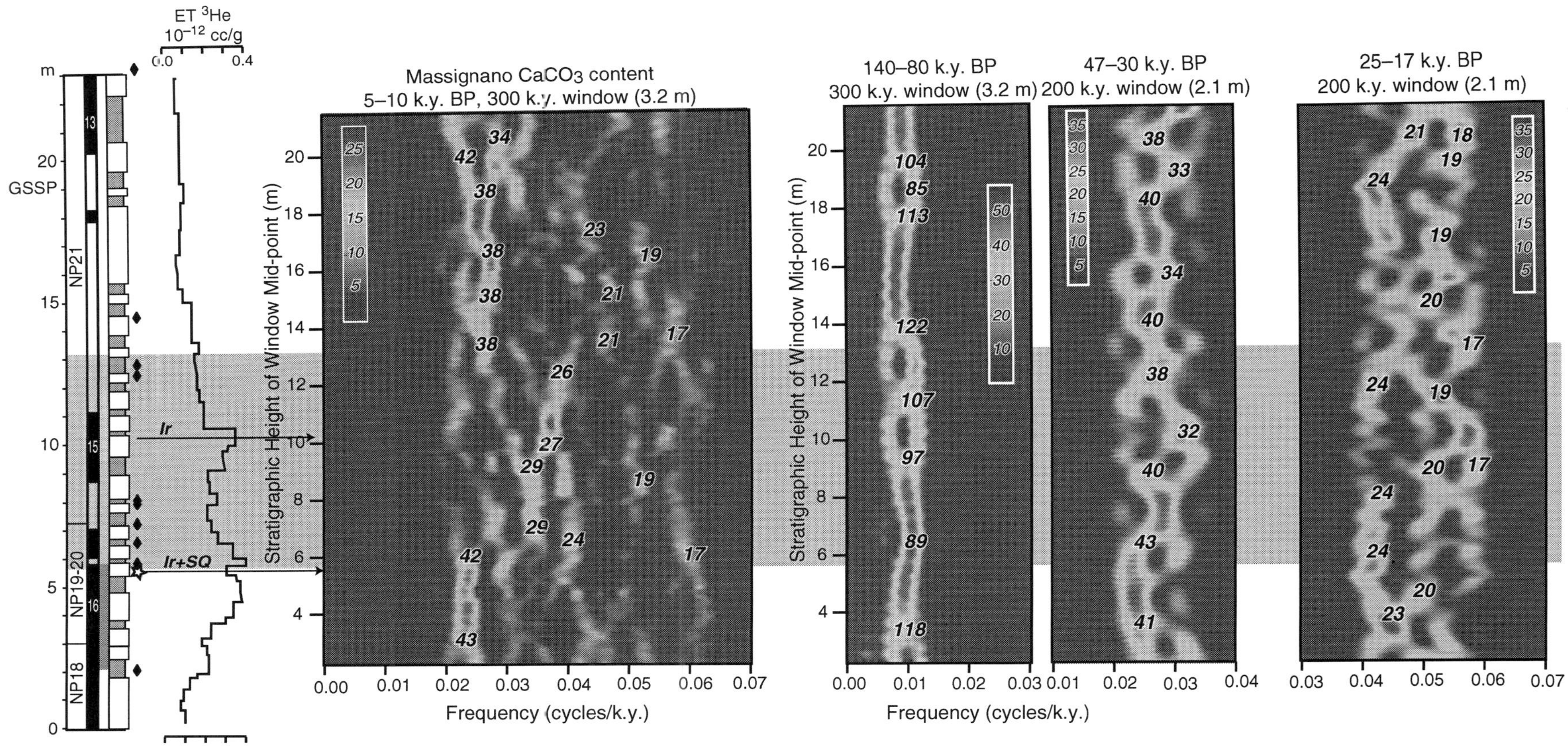

Figure 12. Evolutionary spectral analysis results from the whole-section $CaCO_3$ data. BP stands for band-pass filter. Milankovitch cycles persist through the zone of the comet shower; however, the relative strength of the obliquity signal is masked by stronger power in some discontinuous 26–30 k.y. cycles. The onset of this zone is the Ir anomaly and shocked-quartz layer at meter level 5.6. GSSP—global stratotype section and point.

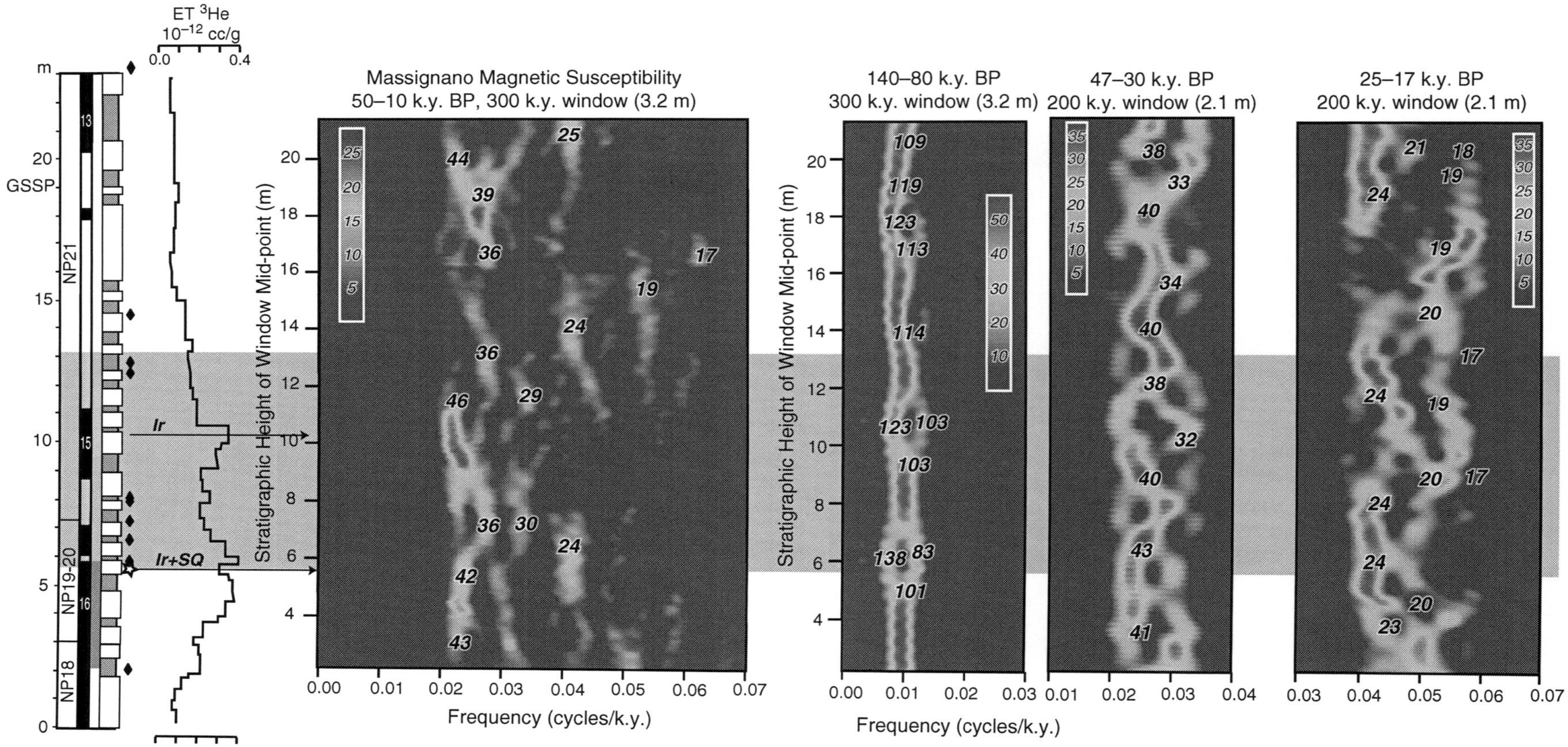

Figure 13. Evolutionary spectral analysis results from the whole-section magnetic susceptibility data. BP stands for band-pass filter. Milankovitch cycles similarly seem to persist through the zone of the comet/asteroid shower; however, the relative strength of the obliquity signal is once again masked by stronger power in some discontinuous 26–30 k.y. cycles, beginning at the 5.6 m Ir anomaly. GSSP—global stratotype section and point.

Certainly, disruptions in the sliding-window plots can in part be attributable to changes in sedimentation rate; some of the lateral oscillations of Milankovitch peaks in the plots are indicative of this. If we were to apply a different linear sedimentation rate to the section, the relative positions and powers of the peaks would not change; rather only the frequencies correlated to the peaks. Trying to vary sedimentation rate over time in a more complex fashion is difficult to justify, as sedimentation rates are difficult to constrain at such a fine resolution. Furthermore, changes in sedimentation rate do not preclude extraterrestrial disruptions, as changes could and likely would be caused by these events.

While it is intriguing that extraterrestrial events appear to have affected global climate on longer time scales, questions remain. Milankovitch forcing is certainly obscured in the $CaCO_3$, $\delta^{18}O$, and $\delta^{13}C$ proxy records of the lower Massignano section, but the magnetic susceptibility record emerges largely unscathed. It is curious that this particular proxy would be less susceptible to impact-related climatic overprinting than the other proxies. Perhaps the usual fluvial or wind-deposited detrital influx is augmented by the slowly accreting extraterrestrial dust, preserving some level of cyclicity. Despite lingering questions, it is reasonable to assert that impact-related climatic changes interacted with the continuous record of longer-term Milankovitch cycles in such a way that they were disrupted or obscured during the deposition of Eocene sediments at Massignano.

CONCLUSIONS

Spectral analyses of four high-resolution climate proxies indicate that deposition at Massignano was orbitally controlled. The relationship between these proxies, where stable isotope highs correspond to calcium carbonate highs and magnetic susceptibility lows, helps to elucidate the paleoclimatic conditions imposed by orbital forcing. Marly limestones are inferred to represent wet/warm periods, while limestones represent dry/cold periods characterized by enhanced productivity. By means of pattern matching, we are able correlate the presumed eccentricity signal in the Massignano magnetic susceptibility and calcium carbonate data to La2004, providing astronomical ages for the entire section. This correlation yields a refined date for the Eocene-Oligocene boundary of 33.91 ± 0.05 Ma.

While Milankovitch cyclicity is persistent throughout the section, the record of this cyclicity changes in the lower portion of the outcrop, which contains three possible impact events and a ^3He anomaly indicative of a comet/asteroid shower ~2.2 m.y. in duration. Climate alterations caused by these extraterrestrial events were most likely exaggerated such that Milankovitch forcing was overprinted and obscured. Possible mechanisms for the exaggeration of impact-related climatic changes include: (1) on a more regional scale, changes in sedimentation rate on a time scale comparable to Milankovitch cycles; or (2) more globally, the ice-albedo feedback or a combination of ongoing atmospheric dust-particle loading with impact-related atmospheric alterations. We encourage analyses of other climate records of this age in order to establish the extent of this phenomenon.

ACKNOWLEDGMENTS

We thank Laura Cleaveland and Alessandro Grippo for their thoughtful reviews as well as Frits Hilgen and Rodolfo Coccioni for additional helpful comments, all of which helped to improve this paper. This work was supported by the Coldigioco Research Fund (Montanari and Bice), the Carleton College Duncan Stewart Fellowship (Brown), and the Austrian Science Foundation FWF, project P18862-N10 (Koeberl).

REFERENCES CITED

Banner, J.L., and Hanson, G.N., 1990, Calculation of simultaneous isotopic and trace element variations during water-rock interaction with applications to carbonate diagenesis: Geochimica et Cosmochimica Acta, v. 54, p. 3123–3137, doi: 10.1016/0016-7037(90)90128-8.

Berggren, W.A., and Prothero, D.R., 1992, Eocene-Oligocene Climatic and Biotic Evolution: Princeton, New Jersey, Princeton University Press, 568 p.

Bickert, T., 2000, Influence of geochemical processes on stable isotope distribution in marine sediments, *in* Schulz, H.D., and Zabel, M., eds., Marine Geochemistry: Berlin, Springer, p. 309–333.

Bodiselitsch, B., Montanari, A., Koeberl, C., and Coccioni, R., 2004, Delayed climate cooling in the late Eocene caused by multiple impacts: High-resolution geochemical studies at Massignano, Italy: Earth and Planetary Science Letters, v. 223, p. 283–302, doi: 10.1016/j.epsl.2004.04.028.

Bottomley, R., Grieve, R., York, D., and Masaitis, V., 1997, The age of the Popigai impact event and its relation to events at the Eocene/Oligocene boundary: Nature, v. 388, p. 365–368, doi: 10.1038/41073.

Cande, S.C., and Kent, D.V., 1995, Revised calibration of the geomagnetic polarity timescale for the Late Cretaceous and Cenozoic: Journal of Geophysical Research, v. 100, p. 6093–6095, doi: 10.1029/94JB03098.

Cleaveland, L.C., Jensen, J., Goese, S., Bice, D.M., and Montanari, A., 2002, Cyclostratigraphic analysis of pelagic carbonates at Monte dei Corvi (Ancona, Italy) and astronomical correlation of the Serravallian-Tortonian boundary: Geology, v. 30, p. 931–934, doi: 10.1130/0091-7613(2002)030 <0931:CAOPCA>2.0.CO;2.

Clemens, S.C., and Tiedemann, R., 1997, Eccentricity forcing of Pliocene–early Pleistocene climate revealed in a marine oxygen-isotope record: Nature, v. 385, p. 801–804, doi: 10.1038/385801a0.

Clymer, A.K., Bice, D.M., and Montanari, A., 1996, Shocked quartz in the late Eocene: Impact evidence from Massignano, Italy: Geology, v. 24, p. 483–486, doi: 10.1130/0091-7613(1996)024<0483:SQFTLE>2.3.CO;2.

Coccioni, R., and Galeotti, S., 2003, Deep-water foraminiferal events from the Massignano Eocene/Oligocene boundary stratotype, central Italy, *in* Prothero, D.R., Ivany, L.C., and Nesbitt, E.A., eds., From Greenhouse to Icehouse: The Marine Eocene-Oligocene Transition: New York, Columbia University Press, p. 438–452.

Coccioni, R., Monaco, P., Monechi, S., Nocchi, M., and Parisi, G., 1988, Biostratigraphy of the Eocene-Oligocene boundary at Massignano (Ancona, Italy), *in* Premoli-Silva, I., Coccioni, R., and Montanari, A., eds., The Eocene-Oligocene Boundary in the Marche-Umbria Basin (Italy): Ancona, Italy, International Union of Geological Sciences, p. 59–80.

Cotillion, P., 1995, Constraints for using high-frequency sedimentary cycles in cyclostratigraphy, *in* House, M.R., and Gale, A.S., eds., Orbital Forcing Timescales and Cyclostratigraphy: Geological Society of London Special Publication 85, p. 133–141.

Coxall, H.K., Wilson, P.A., Pälike, H., Lear, C.H., and Backman, J., 2005, Rapid stepwise onset of Antarctic glaciation and deeper calcite compensation in the Pacific Ocean: Nature, v. 433, p. 53–57, doi: 10.1038/nature03135.

Crick, R.E., Ellwood, B.B., Hassani, A.E., Feist, R., and Hladil, J., 1997, Magnetosusceptibility event and cyclostratigraphy (MSEC) of the Eifelian-Givetian GSSP and associated boundary sequences in North Africa and Europe: Episodes, v. 20, p. 167–174.

DeConto, R., Pollard, D., and Harwood, D., 2007, Sea ice feedback and Cenozoic evolution of Antarctic climate and ice sheets: Paleoceanography, v. 22, PA3214, doi: 10.1029/2006PA001350, 18 p.

Dercourt, J., Ricou, L.E., and Vrielynck, B., eds., 1993, Atlas Tethys Palaeoenvironmental Maps: Paris, Gauthier Villars, 307 p.

Deutsch, A., and Koeberl, C., 2006, Establishing the link between the Chesapeake Bay impact structure and the North American tektite strewn field: The Sr-Nd isotopic evidence: Meteoritics & Planetary Science, v. 41, p. 689–703.

Einsele, G., and Ricken, W., 1991, Limestone-marl alteration—An overview, *in* Einsele, G., Ricken, W., and Seilacher, A., eds., Cycles and Events in Stratigraphy: Berlin, Springer-Verlag, p. 23–47.

Ellwood, B.B., Crick, R.E., El Hassani, A., Benoist, S.L., and Young, R.H., 2000, Magnetosusceptibility event and cyclostratigraphy method applied to marine rocks: Detrital input versus carbonate productivity: Geology, v. 28, p. 1135–1138, doi: 10.1130/0091-7613(2000)28<1135:MEACMA>2.0.CO;2.

Farley, K.A., Montanari, A., Shoemaker, E.M., and Shoemaker, C.S., 1998, Geochemical evidence for a comet shower in the late Eocene: Science, v. 280, p. 1250–1253, doi: 10.1126/science.280.5367.1250.

Fischer, A.G., 1991, Orbital cyclicity in Mesozoic strata, *in* Einsele, G., Ricken, W., and Seilacher, A., eds., Cycles and Events in Stratigraphy: Berlin, Springer-Verlag, p. 48–62.

Glass, B.P., and Koeberl, C., 1999, Ocean Drilling Project Hole 689B spherules and upper Eocene microtektites and clinopyroxene-bearing spherule strewn fields: Meteoritics & Planetary Science, v. 34, p. 185–196.

Glass, B.P., Burns, C.A., Crosbie, J.R., and DuBois, D.L., 1985, Late Eocene North American microtektites and clinopyroxene-bearing spherules: Journal of Geophysical Research, v. 90, p. D175–D196.

Glass, B.P., Liu, S.B., and Montanari, A., 2004, Impact ejecta in upper Eocene deposits at Massignano, Italy: Meteoritics & Planetary Science, v. 39, p. 589–597.

Grippo, A., Fischer, A.G., Hinnov, L.A., Herbert, T.D., and Premoli-Silva, I., 2004, Cyclostratigraphy and chronology of the Albian stage (Italy), *in* D'Argenio, B., Fischer, A.G., Premoli-Silva, I., Weissert, H., and Ferreri, V., eds., Cyclostratigraphy: Approaches and Case Histories: Society for Sedimentary Geology Special Publication 81, p. 57–81.

Hilgen, F., Brinkhuis, H., and Zachariasse, W.J., 2006, Unit stratotypes for global stages: The Neogene perspective: Earth-Science Reviews, v. 74, p. 113–125.

Huber, H., Koeberl, C., King, D.T., Petruni, L.W., and Montanari, A., 2001, Effects of bioturbation through the late Eocene impactoclastic layer near Massignano, Italy, *in* Buffetaut, E., and Koeberl, C., eds., Geological and Biological Effects of Impact Events, Impact Studies: Berlin, Springer Verlag, p. 197–216.

Jovane, L., Florindo, F., Dinares-Turell, J., and Turrell, S., 2004, Environmental magnetic record of paleoclimate change from the Eocene-Oligocene stratotype section, Massignano, Italy: Geophysical Research Letters, v. 31, p. L15601, doi: 10.1029/2004GL020554, 4 p.

Jovane, L., Florindo, F., Sprovieri, M., and Pälike, H., 2006, Astronomic calibration of the late Eocene/early Oligocene Massignano section (central Italy): Geochemistry, Geophysics, Geosystems, v. 7, doi: 10.1029/2005GC001195, 10 p.

Jovane, L., Sprovieri, M., Florindo, F., Acton, G., Coccioni, R., Dall'Antonia, B., and Dinarès-Turrell, J., 2007, Eocene-Oligocene paleoceanographic changes in the stratotype section, Massignano, Italy: Clues from rock magnetism and stable isotopes: Journal of Geophysical Research, v. 112, B11101, doi: 10.1029/2007JB004963, 16 p.

Koeberl, C., and Glass, B.P., 1988, Chemical composition of North American microtektites and tektite fragments from Barbados and DSDP Site 612 on the continental slope off New Jersey: Earth and Planetary Science Letters, v. 87, p. 286–292, doi: 10.1016/0012-821X(88)90016-7.

Koeberl, C., Poag, C.W., Reimold, W.U., and Brandt, D., 1996, Impact origin of Chesapeake Bay structure and the source of North American tektites: Science, v. 271, p. 1263–1266, doi: 10.1126/science.271.5253.1263.

Kuiper, K.F., Deino, A., Hilgen, F.J., Krijgsman, W., Renne, P.R., and Wijbrans, J.R., 2008, Synchronizing rock clocks of Earth history: Science, v. 320, p. 500–504, doi: 10.1126/science.1154339.

Langenhorst, F., 1996, Characteristics of shocked quartz in late Eocene impact ejecta from Massignano (Ancona, Italy): Clues to shock conditions and source crater: Geology, v. 24, p. 487–490, doi: 10.1130/0091-7613(1996)024<0487:COSQIL>2.3.CO;2.

Laskar, J., 1999, The limits of Earth orbital calculations for geological time scale use: Philosophical Transactions of the Royal Society of London, v. 357, p. 1735–1759.

Laskar, J., Robutel, P., Joutel, F., Gastineau, M., Correia, A.C.M., and Levrard, B., 2004, A long-term numerical solution for the insolation quantities of the Earth: Astronomy & Astrophysics, v. 428, p. 261–285, doi: 10.1051/0004-6361:20041335.

Lourens, L.J., and Hilgen, F.J., 1997, Long-periodic variations in the Earth's obliquity and their relation to third-order eustatic cycles and late Neogene glaciations: Quaternary International, v. 40, p. 43–52, doi: 10.1016/S1040-6182(96)00060-2.

Lowrie, W., and Lanci, L., 1994, Magnetostratigraphy of Eocene-Oligocene boundary sections in Italy: No evidence for short subchrons within chrons 12R and 13R: Earth and Planetary Science Letters, v. 126, p. 247–258, doi: 10.1016/0012-821X(94)90110-4.

Mackensen, A., and Ehrmann, W.U., 1992, Middle Eocene through early Oligocene climate history and paleoceanography in the Southern Ocean: Stable oxygen and carbon isotopes from ODP Sites on Maud Rise and Kerguelen Plateau: Marine Geology, v. 108, p. 1–27, doi: 10.1016/0025-3227(92)90210-9.

Mader, D., Cleaveland, L.C., Bice, D.M., Montanari, A., and Koeberl, C., 2004, High-resolution cyclostratigraphic analysis of multiple climate proxies from a short Langhian pelagic succession in the Conero Riviera, Ancona (Italy): Palaeogeography, Palaeoclimatology, Palaeoecology, v. 211, p. 325–344, doi: 10.1016/j.palaeo.2004.06.001.

Miller, K.G., Wright, J.D., and Fairbanks, R.G., 1991, Unlocking the ice house: Oligocene-Miocene oxygen isotopes, eustasy, and margin erosion: Journal of Geophysical Research, v. 96, p. 6829–6848, doi: 10.1029/90JB02015.

Miller, K.G., Browning, J.V., Aubry, M.P., Wade, B.S., Katz, M.E., Kulpecz, A.A., and Wright, J.D., 2008, Eocene-Oligocene global climate and sea-level changes: St. Stephens Quarry, Alabama: Geological Society of America Bulletin, v. 120, p. 34–53.

Mitchell, R.N., Bice, D.M., Montanari, A., Cleaveland, L.C., Christianson, K.T., Coccioni, R., and Hinnov, L.A., 2008, Oceanic anoxic cycles? Orbital prelude to the Bonarelli Level (OAE 2): Earth and Planetary Science Letters, v. 267, p. 1–16, doi: 10.1016/j.epsl.2007.11.026.

Mix, A.C., Le, J., and Shackleton, N.J., 1995, Benthic foraminiferal stable isotope stratigraphy of Site 846: 0–1.8 Ma, *in* Pisias, N.G., Mayer, L.A., Janecek, T.R., Palmer-Julson, A., and van Andel, T. H., et al., Proceedings of the Ocean Drilling Program, Scientific Results, Volume 138: College Station, Texas, Ocean Drilling Program, p. 839–856.

Montanari, A., and Koeberl, C., 2000, Impact Stratigraphy: The Italian Record, Lecture Notes in Earth Sciences: Berlin, Springer-Verlag, 364 p.

Montanari, A., Deino, A.L., Drake, R.E., Turrin, B.D., DePaolo, D.J., Odin, G.S., Curtis, G.H., Alvarez, W., and Bice, D.M., 1988, Radioisotopic dating of the Eocene-Oligocene boundary in the pelagic sequence of the northeastern Apennines, *in* Premoli-Silva, I., Coccioni, R., and Montanari, A., eds., The Eocene-Oligocene Boundary in the Marche-Umbria Basin (Italy): Ancona, Italy, International Union of Geological Sciences, p. 195–208.

Montanari, A., Asaro, F., Kennett, J.P., and Michel, E., 1993, Iridium anomalies of late Eocene age at Massignano (Italy), and in ODP Site 689B (Maud Rise, Antarctica): Palaios, v. 8, p. 420–437, doi: 10.2307/3515017.

Muller, R.A., and MacDonald, G.J., 2000, Ice Ages and Astronomical Causes: Data, Spectral Analysis and Mechanisms: Chichester, Praxis Publishing Ltd., 318 p.

Odin, G.S., Clauser, S., and Renard, M., 1988, Sedimentological and geochemical data on the Eocene-Oligocene boundary at Massignano (Apennines, Italy), *in* Premoli-Silva, I., Coccioni, R., and Montanari, A., eds., The Eocene-Oligocene Boundary in the Marche-Umbria Basin (Italy): Ancona, Italy, International Union of Geological Sciences, p. 175–186.

Pälike, H., Shackleton, N.J., and Röhl, U., 2001, Astronomical forcing in late Eocene marine sediments: Earth and Planetary Science Letters, v. 193, p. 589–602, doi: 10.1016/S0012-821X(01)00501-5.

Pälike, H., Laskar, J., and Shackleton, N.J., 2004, Geologic constraints on the chaotic diffusion of the solar system: Geology, v. 32, p. 929–932, doi: 10.1130/G20750.1.

Pälike, H., Norris, R.D., Herrle, J.O., Wilson, P.A., Coxall, H.K., Lear, C.H., Shackleton, N.J., Tripati, A.K., and Wade, B.S., 2006, The heartbeat of the Oligocene climate system: Science, v. 314, p. 1894–1898, doi: 10.1126/science.1133822.

Pierrard, O., Robin, E., Rocchia, R., and Montanari, A., 1998, Extraterrestrial Ni-rich spinel in Upper Eocene sediments from Massignano, Italy: Geology,

v. 26, p. 307–310, doi: 10.1130/0091-7613(1998)026<0307:ENRSIU>2.3.CO;2.

Poag, C.W., Koeberl, C., and Reimold, W.U., 2004, The Chesapeake Bay Crater: Geology and Geophysics of a Late Eocene Submarine Impact Structure: Berlin, Springer, 522 p.

Pokras, E.M., and Mix, A.C., 1987, Earth's precession cycle and Quaternary climatic change in tropical Africa: Nature, v. 326, p. 486–487, doi: 10.1038/326486a0.

Premoli-Silva, I., and Jenkins, D.J., 1993, Decision on the Eocene-Oligocene boundary stratotype: Episodes, v. 16, p. 379–381.

Premoli-Silva, I., Coccioni, R., and Montanari, A., 1988, The Eocene-Oligocene Boundary in the Marche-Umbria Basin (Italy): Ancona, Italy, International Union of Geological Sciences, 268 p.

Prothero, D.R., 1994, The Eocene-Oligocene Transition: Paradise Lost: New York, Columbia University Press, 291 p.

Schrag, D.P., DePaolo, D.J., and Richter, F.M., 1995, Reconstructing past sea-surface temperatures—correcting for diagenesis of bulk marine carbonate: Geochimica et Cosmochimica Acta, v. 59, p. 2265–2278, doi: 10.1016/0016-7037(95)00105-9.

Shackleton, N.J., Hall, M.A., and Pate, D., 1993, High-resolution stable isotope stratigraphy from bulk sediment: Paleoceanography, v. 8, p. 141–148, doi: 10.1029/93PA00123.

Stoll, H.M., and Schrag, D.P., 2000, High-resolution stable isotope records from the Upper Cretaceous rocks of Italy and Spain: Glacial episodes in a greenhouse planet?: Geological Society of America Bulletin, v. 112, p. 308–319, doi: 10.1130/0016-7606(2000)112<0308:HRSIRF>2.3.CO;2.

Tagle, R., and Claeys, P., 2004, Comet or asteroid shower in the late Eocene?: Science, v. 305, p. 492, doi: 10.1126/science.1098481.

Tagle, R., and Claeys, P., 2005, An ordinary chondrite impactor for the Popigai crater, Siberia: Geochimica et Cosmochimica Acta, v. 69, p. 2877–2889, doi: 10.1016/j.gca.2004.11.024.

Toon, O.B., Zahnle, K., Morrison, D., Turco, R.P., and Covey, C., 1997, Environmental perturbations caused by the impacts of asteroids and comets: Reviews of Geophysics, v. 35, p. 41–78, doi: 10.1029/96RG03038.

Tuenter, E., Weber, S.L., Hilgen, F.J., and Lourens, L.J., 2003, The response of the African summer monsoon to remote and local forcing due to precession and obliquity: Global and Planetary Change, v. 36, p. 219–235, doi: 10.1016/S0921-8181(02)00196-0.

Turco, E., Hilgen, F.J., Lourens, L.J., Shackleton, N.J., and Zachariasse, W.J., 2001, Punctuated evolution of global climate cooling during the late middle to early late Miocene: High-resolution planktonic foraminiferal and oxygen isotope records from the Mediterranean: Paleoceanography, v. 16, p. 405–423, doi: 10.1029/2000PA000509.

van Vugt, N., Langereis, C.G., and Hilgen, F.J., 2001, Orbital forcing in Pliocene-Pleistocene Mediterranean lacustrine deposits: Dominant expression of eccentricity versus precession: Palaeogeography, Palaeoclimatology, Palaeoecology, v. 172, p. 193–205, doi: 10.1016/S0031-0182(01)00270-X.

von Dobeneck, T., and Schmieder, F., 1999, Using rock magnetic proxy records for orbital tuning and extended time series analyses into the super- and sub-Milankovitch bands, *in* Fischer, G., and Wefer, G., eds., Use of Proxies in Paleoceanography: Examples from the South Atlantic: Berlin, Springer, p. 601–633.

Vonhof, H.B., Smit, J., Brinkhuis, H., Montanari, A., and Nederbragt, A.J., 2000, Global cooling accelerated by early late Eocene impacts?: Geology, v. 28, p. 687–690, doi: 10.1130/0091-7613(2000)28<687:GCABEL>2.0.CO;2.

Zachos, J.C., Breza, J., and Wise, S.W., 1992, Earliest Oligocene ice-sheet expansion on East Antarctica: Stable isotope and sedimentological data from Kerguelen Plateau: Geology, v. 20, p. 569–573, doi: 10.1130/0091-7613(1992)020<0569:EOISEO>2.3.CO;2.

Zachos, J.C., Stott, L.D., and Lohmann, K.C., 1994, Evolution of early Cenozoic marine temperatures: Paleoceanography, v. 9, p. 353–387, doi: 10.1029/93PA03266.

Zachos, J.C., Pagani, M., Sloan, L., Thomas, E., and Billups, K., 2001, Trends, rhythms, and aberrations in global climate 65 Ma to present: Science, v. 292, p. 686–693, doi: 10.1126/science.1059412.

Zanazzi, A., Kohn, M.J., MacFadden, B.J., and Terry, D.O., 2007, Large temperature drop across the Eocene-Oligocene transition in central North America: Nature, v. 445, p. 639–642, doi: 10.1038/nature05551.

Manuscript Accepted by the Society 16 September 2008

Printed in the USA

The Geological Society of America
Special Paper 452
2009

A critical evaluation of the numerical age of the Eocene-Oligocene boundary

Frits J. Hilgen
Faculty of Geosciences, Department of Earth Sciences, Utrecht University, Budapestlaan 4, 3584 CD Utrecht, Netherlands

Klaudia F. Kuiper
Faculty of Geosciences, Department of Earth Sciences, Utrecht University, Budapestlaan 4, 3584 CD Utrecht, Netherlands, and *Faculty of Earth and Life Sciences, Institute of Earth Sciences, Vrije Universiteit Amsterdam, De Boelelaan 1085, 1081 HV Amsterdam, Netherlands*

ABSTRACT

Published radioisotopic (K/Ar, $^{40}Ar/^{39}Ar$, and Rb/Sr) and astronomical ages for the Eocene-Oligocene boundary are essentially consistent at ca. 33.8 ± 0.1 Ma, but the $^{40}Ar/^{39}Ar$ ages have been calculated relative to an outdated age of 27.83–27.84 Ma for the Fish Canyon Tuff sanidine dating standard. Application of a revised age of 28.02 Ma, or the new astronomically calibrated age of 28.201 Ma, leads to significant discrepancies, while others are eliminated. In particular, the astronomically tuned ages of ca. 33.79 Ma at Ocean Drilling Program (ODP) Site 1218 and of 33.90–33.95 Ma at Massignano–Monte Cagnero are now in good agreement with recalculated (alternative) $^{40}Ar/^{39}Ar$ sanidine ages for the boundary as derived from the volcanic ignimbrite complex in New Mexico and for the Persistent White Layer (PWL) ash bed in North America, which is supposed to closely correspond to the boundary. This mutual consistency suggests that the tuning is correct at the scale of the 400 k.y. eccentricity cycle.

Evidently, additional single-crystal $^{40}Ar/^{39}Ar$ sanidine dates from the tuffs in North America and independent checks on the astronomical tuning and the intercalibration between the astronomical and $^{40}Ar/^{39}Ar$ dating methods are needed to definitively solve the problem of the numerical age of the Eocene-Oligocene boundary. It is anticipated that such analyses and tests will be carried in the coming years as part of the international Earthtime initiative and associated projects to significantly improve the geological time scale. Clearly, an accurate and precise dating of the Eocene-Oligocene boundary is crucial if we are to unravel the underlying cause of the major climate transition associated with it.

Keywords: numerical age, Eocene-Oligocene boundary, astronomical dating, radioisotopic dating, Greenhouse-Icehouse tradition.

Hilgen, F.J., and Kuiper, K.F., 2009, A critical evaluation of the numerical age of the Eocene-Oligocene boundary, *in* Koeberl, C., and Montanari, A., eds., The Late Eocene Earth—Hothouse, Icehouse, and Impacts: Geological Society of America Special Paper 452, p. 139–148, doi: 10.1130/2009.2452(09). For permission to copy, contact editing@geosociety.org.

INTRODUCTION

The Eocene-Oligocene boundary, formally defined by the last occurrence of the planktonic foraminiferal taxon of the hantkeninids in the Massignano section in northern Italy (Premoli Silva and Jenkins, 1993), is of special interest because it marks the transition from the Paleocene-Eocene greenhouse to the Oligocene to Holocene icehouse world. The (biostratigraphic) definition of the boundary was questioned by Brinkhuis and Visscher (1995). They argued that the boundary did not correspond to the top of the (uppermost) Eocene stage of the Priabonian in its type section and, hence, that it would be preferable to redefine it so as to coincide with oxygen isotope stage Oi-1, the slightly younger major shift in $\delta^{18}O$. However, the present definition clearly marks an important faunal event firmly positioned in between two major steps in $\delta^{18}O$ that delineate the boundary interval (Pearson et al., 2008). Evidently, a highly accurate and precise dating of the boundary is crucial to unraveling the origin of the climate transition associated with it. Such an excellent time control is especially important in view of the potential control of long-period astronomical forcing (Coxall et al., 2005).

In the 1950s and beginning of the 1960s, the then-upcoming K/Ar method resulted in the rapid and significant growth in the number of available radioisotopic dates for the Cenozoic. However, this growth also resulted in serious discrepancies in the numerical age estimate of the Eocene-Oligocene boundary, one of the age tie-points used to construct the geomagnetic polarity time scale, which forms the backbone of our geological time scale. The discrepancies in boundary ages, ranging between 32.5 and 38.5 Ma (e.g., Evernden et al., 1964; LaBrecque et al., 1977; Glass and Crosbie, 1982; Montanari et al., 1985; Berggren et al., 1985), were essentially solved by much improved age dating, including single-crystal $^{40}Ar/^{39}Ar$ ages, and a revised magnetostratigraphic calibration of the continental record in North America to the geomagnetic polarity time scale. The revised age estimate for the boundary in North America was ca. 33.91 Ma ± 0.06 (Swisher and Prothero, 1990) (note that all uncertainties are reported at 68% [1σ] confidence level; uncertainties originally reported at 95% confidence level are halved for consistency). This age is consistent, within error, with a K/Ar and Rb/Sr biotite-derived age estimate of 33.7 ± 0.2 Ma for the boundary in the pelagic record of the Northern Apennines (Montanari et al., 1988; Odin et al., 1991).

The same years also witnessed the extension of astronomical tuning from the last 800,000 yr (SPECMAP: Imbrie et al., 1984) to the Miocene-Pliocene boundary, dated astronomically at 5.32 Ma (Shackleton et al., 1990; Hilgen, 1991). The tuning was soon extended further back in time, covering the Miocene and, more recently, the entire Oligocene (Hilgen et al., 1995; Shackleton et al., 1999; Pälike et al., 2006). In particular, Ocean Drilling Program (ODP) Site 1218 was instrumental to extend the tuning into the latest Eocene, providing an astronomical age of 33.79 Ma for the Eocene-Oligocene boundary (Pälike et al., 2006). In addition, different tunings were published based on high-resolution proxy records from the Massignano section and core itself, with ages ranging from 33.714 to 34.1 Ma (Jovane et al., 2006; van Mourik et al., 2006; Brown et al., this volume; Hyland et al., this volume). Most of these ages proved consistent with the K/Ar and $^{40}Ar/^{39}Ar$ ages obtained more than 15 yr earlier. However, the $^{40}Ar/^{39}Ar$ (not K/Ar) ages had been calculated relative to an age of 27.84 Ma for the Fish Canyon Tuff (FCT) sanidine dating standard. This age has subsequently been revised to 28.02 ± 0.28 Ma (Renne et al., 1998). This age, which is consistent within error with the previous age of 27.84 Ma, was adopted in the 2004 geologic time scale (Gradstein et al., 2004).

Recently, the FCT sanidine standard was astronomically calibrated using single-crystal sanidine ages of ash layers intercalated in tuned marine sections in the Mediterranean in a rigorous attempt to reduce the unacceptably large full error (≥2.5%) in $^{40}Ar/^{39}Ar$ dating (Kuiper et al., 2008). This calibration resulted in an age of 28.201 ± 0.022 Ma for the FCT sanidine. In this paper, we will briefly review the numerical dating history of the Eocene-Oligocene boundary and discuss the implications of both the revised and new, astronomically calibrated ages of the FCT sanidine dating standard for the recently published astronomical ages of the Eocene-Oligocene boundary. In particular, we aim to test independently whether the tunings are correct on the scale of the ~400 k.y. eccentricity cycle.

HISTORICAL OVERVIEW

We first provide a short historical overview of the (most important) radioisotopic and astronomical age estimates for the Eocene-Oligocene boundary that have been published over the years. We will start with the radioisotopic ages before the more recently obtained astronomical ages and the intercalibration between the two independent dating methods will be dealt with.

Radioisotopic Dating

The initial crude U/Pb dating of the Cenozoic by Holmes (Holmes, 1960) was rapidly expanded with the development of the K/Ar dating method in the 1950s and the publication of numerous K/Ar ages that came out of it in the 1960s. Part of the K/Ar work focused on the Eocene-Oligocene boundary interval, in particular, as exposed in the continental record of North America and in the marine pelagic succession of the Northern Apennines in Italy. However, other records played an important role as well.

North America

Relevant K/Ar ages for the NALMAs (North American Land Mammal Ages) of the Cenozoic continental succession in North America were first published by Evernden et al. (1964). They showed that the Chadronian mammal age traditionally considered to be Oligocene in age ranged from ca. 37 to 32 Ma. These ages, in combination with the magnetostratigraphy (e.g., Prothero et al., 1982), were used to arrive at an age of 36.5 Ma or the Eocene-Oligocene boundary, which was incorporated s an age calibration point in the Cenozoic time scale of, for xample, Berggren et al. (1985). However, this age was much lder than the age estimate of 32.3 ± 0.9 Ma for the boundary ased on extrapolation of sedimentation rates from a microtekite layer in Deep Sea Drilling Project (DSDP) cores dated at 34.2 ± 0.3 Ma by K/Ar and fission-track dating (Glass and Crosbie, 1982). This younger boundary age was subsequently revised to 34.4 ± 0.3 Ma, using $^{40}Ar/^{39}Ar$ laser-probe dating of seven tektite fragments recovered ~26 m below the Eocene-Oligocene boundary in a deep-sea succession exposed at Bath Cliffs on Barbados (Glass et al., 1986).

A similar discrepancy was also found when the ages from continental North America were compared with the (final) boundary age based on K/Ar and Rb/Sr biotite ages of volcanic ash beds intercalated in the pelagic succession in northern Italy (see following). These discrepancies were solved by combining single-crystal $^{40}Ar/^{39}Ar$ anorthoclase and biotite dates with a revised calibration of the magnetostratigraphy of the continental sections to the geomagnetic polarity time scale (Swisher and Prothero, 1990; Prothero and Swisher, 1992). As a consequence, the Eocene-Oligocene boundary was shown to coincide with the Chadronian-Orellian mammal age boundary (rather than with the Duchesnian-Chadronian boundary), and the revised radioisotope-derived age estimate was ca. 33.9 Ma (Swisher and Prothero, 1990; Table 1). This age is above all based on the dating of the Persistent White Layer (PWL) ash at Dilts Ranch (Fig. 1). Note, however, that Swisher and Prothero (1990) used an age of 27.84 Ma for the FCT sanidine mineral dating standard to obtain their $^{40}Ar/^{39}Ar$ ages.

The revised age for the Eocene-Oligocene boundary proved consistent with stepwise heating and single-crystal $^{40}Ar/^{39}Ar$ sanidine ages of 34.32 ± 0.05 (incremental heating) and 34.31 ± 0.05 Ma (single-crystal laser fusion), respectively, and a $^{40}Ar/^{39}Ar$ biotite age 34.28 ± 0.04 Ma for bentonites intercalated in the upper part of the Yazoo Clay Formation of the Gulf Coast (Obradovich et al., 1993; see also Berggren et al., 1992). This interval is biostratigraphically assigned to (upper part) zone NP20 and P17.

An alternative attempt to radioisotopically constrain the age of the Eocene-Oligocene boundary in North America came from a detailed study of ignimbrites in New Mexico and adjacent areas (McIntosh et al., 1992). These authors combined total fusion $^{40}Ar/^{39}Ar$ sanidine dates and magnetostratigraphy of—a discontinuous record of—ignimbrites to arrive at a most plausible age of 33.1–34.4 Ma for C13R, and 33.4 Ma for the Eocene-Oligocene boundary (Table 1). Their ages, which were calculated relative to an age of 27.83 Ma for the FCT sanidine, are significantly younger than the ages that came out of the continental succession in North America (see also Fig. 2).

Northern Apennines

The pelagic succession of the Northern Apennines and its excellent integrated magnetobiochemostratigraphy (Lowrie et al., 1982) offered a unique opportunity to date the Eocene-Oligocene boundary due to the intercalation of syndepositional volcanic ash layers (Montanari et al., 1985). An initial age estimate of 35.7 ± 0.2 Ma was derived from radioisotopic K/Ar and Rb/Sr biotite ages of five stratigraphic levels in the classic Contessa section, including an age of 35.4 ± 0.2 Ma for the younger end of chron C13n (Montanari et al., 1985). This age was consistent with other time scales published around that time (e.g., Berggren et al., 1985). Already existing, younger K/Ar and Rb/Sr biotite dates from the Massignano section were not included because of potential miscorrelations with the much better constrained magnetobiostratigraphy in the Contessa section. However, ongoing studies of the Massignano section (Nocchi et al., 1986; Coccioni et al., 1988) showed that the age discrepancies could not be attributed to uncertainties in the position of biostratigraphic events resulting from lack of sample resolution. In fact, the age of 35.4 Ma for the top of chron C13n at Contessa was shown to be unreliable due to geochemical inhomogeneity of the biotite sample that had produced the discrepantly old age (Montanari, 1988). This was confirmed by the fact that the suspect biotite sample yielded $^{40}Ar/^{39}Ar$ ages varying between 33 and 49 Ma (Deino et al., 1988). Using all the reliable K/Ar and Rb/Sr dates from the

TABLE 1. K/Ar AND $^{40}Ar/^{39}Ar$ AGES FOR C13n(o) AND THE E-O BOUNDARY

Authors	FCT age	Location	C13n(o)	E-O boundary
M 88/O 91		Massignano		33.7 ± 0.2
SP 1990	27.84	N. America		33.91 ± 0.06
SP 1990	28.02	N. America		34.13 ± 0.06
SP 1990	28.201	N. America		34.35 ± 0.06
O 1995	27.84	N. America		33.59 ± 0.02
O 1995	28.02	N. America		33.81 ± 0.02
O 1995	28.201	N. America		34.03 ± 0.02
McI 1992	27.84	New Mexico	ca. 33.1	ca. 33.4
McI 1992	28.02	New Mexico	ca. 33.31	ca. 33.62
McI 1992	28.201	New Mexico	ca. 33.54	ca. 33.85

Note: Published $^{40}Ar/^{39}Ar$ ages are also calculated to ages of 28.02 and 28.201 Ma for the Fish Canyon tuff (FCT) sanidine dating standard. References: M 88 Montanari et al. (1988); O 91–Odin et al. (1991); SP 1990–Swisher and Prothero (1990); O 1995–Obradovich et al. (1995); McI 1992–McIntosh et al. (1992).

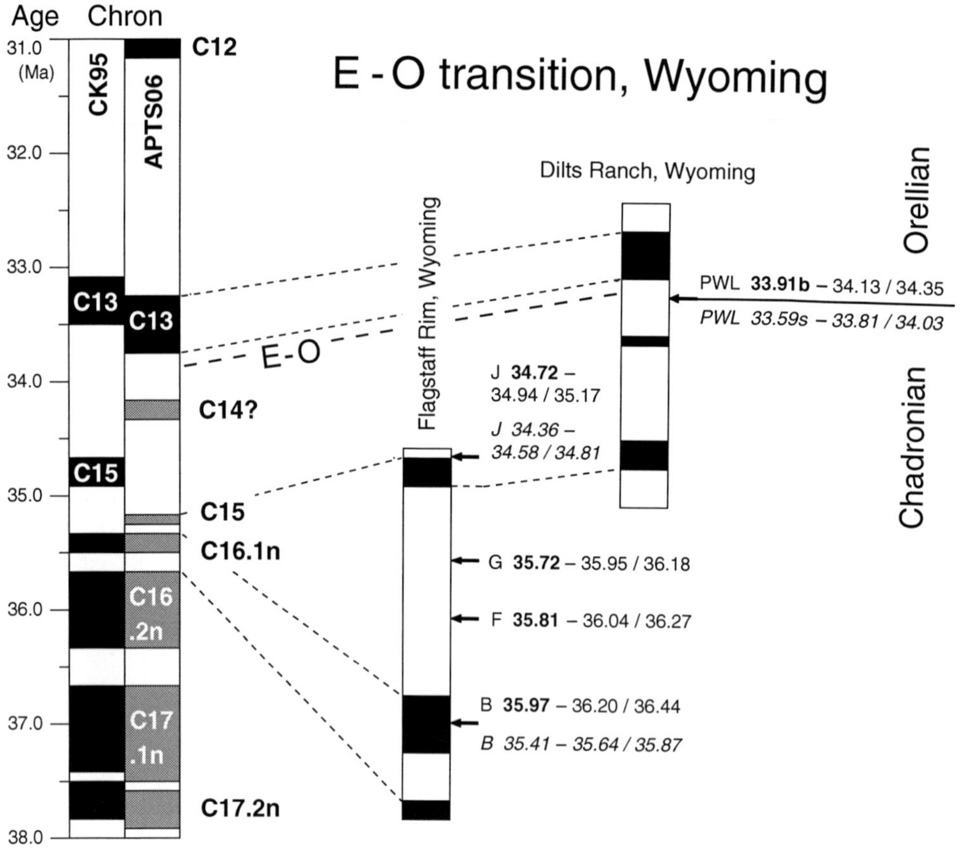

Figure 1. Comparison of combined single-crystal $^{40}Ar/^{39}Ar$ anorthoclase and biotite ages of ash layers and magnetostratigraphic records of the Flagstaff Rim and Dilts Ranch sections with published polarity time scales of Cande and Kent (1995) and the astronomical polarity time scale (APTS06) of Pälike et al. (2006). The radioisotopic ages both show the original age published in Swisher and Prothero (1990) relative to Fish Canyon tuff (FCT) sanidine at 27.84 Ma in bold as well as recalculated older ages relative to FCT sanidine at 28.02 and 28.201 Ma. For the Persistent White Layer (PWL) ash bed, only a single crystal biotite (b) age is available. The original and similarly recalculated $^{40}Ar/^{39}Ar$ ages of Obradovich et al. (1995) for the same ash layers are shown in italics. B-J denotes Lone Tree ash layers at Flagstaff Rim, while PWL denotes the Persistent White Layer at Dilts Range. E-O stands for Eocene-Oligocene.

pelagic succession, a revised age of 33.7 ± 0.2 Ma was calculated for the Eocene-Oligocene boundary (Montanari et al., 1988; Odin et al., 1991; Table 1).

Stepwise-heating, multigrain $^{40}Ar/^{39}Ar$ biotite ages showing spectra with reliable plateaus were reported by Odin et al. (1991) from two ash layers in the Massignano section, namely an age of 34.1 ± 0.1 Ma for the ash bed at 14.7 m (approximately middle part of chron C13r) and an age of 34.9 ± 0.1 Ma for the ash bed at 12.7 m. These ages are consistent with the age of 33.7 Ma for the Eocene-Oligocene boundary.

Astronomical Dating

Astronomical tuned ages have only recently been obtained for the Eocene-Oligocene boundary. These ages come from ODP Site 1218 in the equatorial Pacific and from the Eocene-Oligocene boundary sections at Massignano and Monte Cagnero in Italy.

ODP Site 1218

The astronomical tuning of the Eocene-Oligocene boundary interval was only recently achieved following the tuning of the Miocene at ODP Site 925 (Shackleton et al., 1999). An important foundation for this accomplishment came from ODP Site 1218 with its excellent magnetostratigraphy and calcareous plankton biostratigraphy. Starting from the shipboard stratigraphy, an initial astronomical age model was generated from physical property proxy records reflecting $CaCO_3$ percentage. The original age model was then refined by fine-scale adjustment of the time scale, first by tuning the prominent ~405 k.y. cycle in the carbon isotope time series, and followed by the tuning of shorter astronomical frequencies observed in the data. The last step consisted of a further refinement of this age model, obtained by manually matching the data with the selected astronomical target curve, with as few tie-points as possible, by algorithmic means. This final tuning resulted in an age of ca. 33.65 Ma for oxygen isotope stage Oi-1 and of ca. 33.79 Ma for the Eocene-Oligocene boundary (Pälike et al., 2006; Table 2).

Massignano–Monte Cagnero

Other tuning attempts focused their attention on the Eocene-Oligocene boundary stratotype section of Massignano itself. Jovane et al. (2006) used a high-resolution magnetic susceptibility record for the tuning. After confirming a deterministic external control on the susceptibility fluctuations, they used the magnetostratigraphic calibration to the geomagnetic polarity time scale of CK95 (Cande and Kent, 1995), and the resultant age for the reversal boundaries, as a starting point for their tuning. The first-order tuning was based on visually correlating the 2.84 m

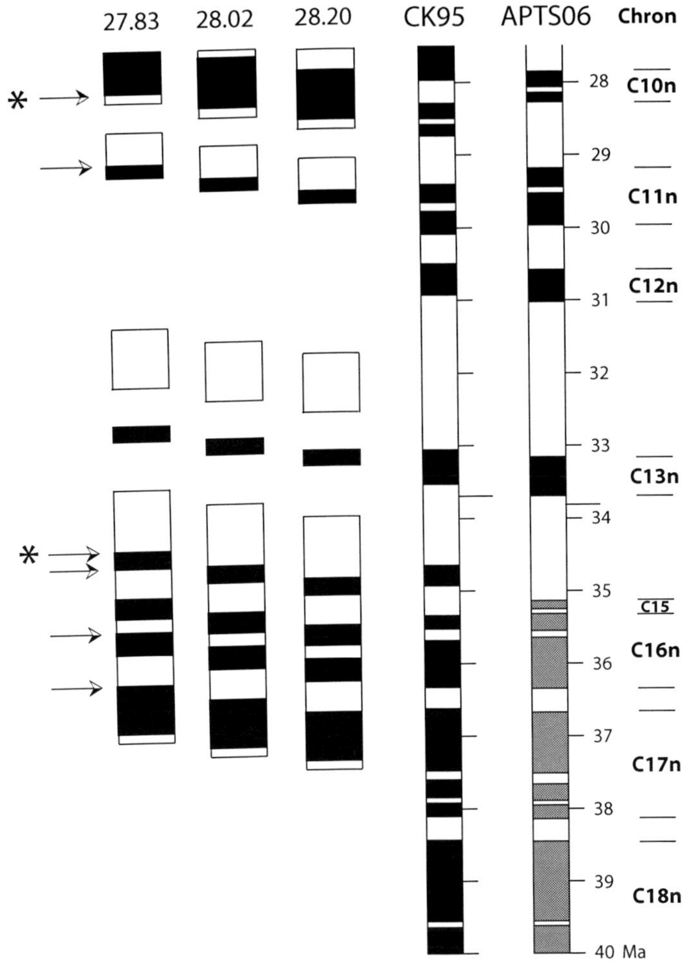

Figure 2. Interpreted polarity model for the ignimbrites in New Mexico (McIntosh et al., 1992) recalculated for different ages of the Fish Canyon Tuff (FCT) sanidine dating standard and comparison with the geomagnetic polarity time scale of Cande and Kent (1995) and the astronomical polarity time scale (APTS06) of Pälike et al. (2006). Arrows mark the better constrained polarity reversals. Asterisks mark the reversals that were used as tie points by McIntosh et al. (1992).

(~405 k.y.) cycle in susceptibility to the eccentricity time series, assuming that susceptibility minima correspond to minimum eccentricity. This first-order tuning was refined by correlating basic limestone-marl cycles with an average thickness of 0.3 m to the 41 k.y. cycle of obliquity. For this purpose, they tuned consecutive obliquity maxima to the midpoint of successive marl beds and the midpoint of the limestone beds to obliquity minima, thus achieving an orbital tuning of the section with a finer resolution than the eccentricity-based correlation. The tuning was verified by the 1.2 m.y. cycle in the amplitude modulation of obliquity. The tuning resulted in an astronomical age of 33.714 Ma for the Eocene-Oligocene boundary (Table 2), i.e., in agreement with the 33.7 Ma reported in Berggren et al. (1995).

Van Mourik et al. (2006) started their tuning from a dinoflagellate cyst–derived sea-surface temperature (SST) record largely derived from samples of the Massicore drilled ~100 m south of the Massignano section with the aim of, among others, establishing a more reliable magnetostratigraphy (Lanci et al., 1996). The SST record reveals a clear cooling trend, on top of which two distinct cold pulses can be distinguished; these pulses were interpreted to reflect the Oi-1a and Oi-1b glacial oxygen isotope events. In addition, two prominent cycles can be discerned in the depth domain with thicknesses of 8.2 and 2 m. Given the magnetostratigraphic age constraints, these SST cycles may reflect ~400 and ~100 k.y. eccentricity.

Two options exist for the tuning of the eccentricity-related dinoflagellate cyclicity to the Laskar et al. (2004) solution. The first option, following the Coxall et al. (2005) age model developed for ODP Site 1218, results in an age of ca. 33.65 Ma for the Oi-1 dinoflagellate cold pulse and an age of ca. 33.75 Ma for the Eocene-Oligocene boundary. The other option starts from the assumption that the Oi-1 event was related to the next older 400 k.y. eccentricity minimum and the 1.2 m.y. minimum in obliquity amplitude (= node) with the same age; similar phase relations have also been found for younger glaciation events in Oligocene and Miocene times. The latter tuning would imply ages older by ~400 k.y., giving an age of ca. 34.0 Ma for the Oi-1 and ca. 34.1 Ma for the Eocene-Oligocene boundary (Table 2). Ages according to both options differ from the age of 33.9 Ma adopted in the 2004 geologic time scale for the Eocene-Oligocene boundary (Gradstein et al., 2004).

Brown et al. (this volume) used band-pass–filtered magnetic susceptibility and $CaCO_3$ data to establish an astronomical tuning of the Massignano section to the eccentricity time series of La2004. The tuning, which is constrained by radioisotopically dated volcanic ash layers and by amplitude variations in the filtered ~100 k.y. (106 cm) $CaCO_3$ component that reflect 400 k.y. eccentricity, resulted in an age of 33.90 ± 0.01 Ma for the Eocene-Oligocene boundary (Table 2). It is important to note

TABLE 2. PUBLISHED ASTRONOMICAL AGES FOR Oi-1, C13n (o), AND THE EOCENE-OLIGOCENE (E-O) BOUNDARY IN Ma

Authors	Coxall et al., 2005	Pälike et al., 2006	Jovane et al., 2006	van Mourik et al., 2006 option 1	van Mourik et al., 2006 option 2	Brown et al., this volume	Hyland et al., this volume
Event/Section	Site 1218	Site 1218	Massignano	Massicore	Massicore	Massignano	Cagnero
Oi-1 (Ma)	33.65			33.65	34.0		
C13n (o) (Ma)	33.71	33.705				33.75	
E-O boundary (Ma)	33.90	33.79	33.714	33.75	34.1	33.90 ± 0.01	33.95

that the radioisotopic constraints come from K/Ar dating of biotites, although limited numbers of $^{40}Ar/^{39}Ar$ biotite ages are also available for these ash layers (Odin et al., 1991).

Hyland et al. (this volume) used the astronomical age of 33.90 of Brown et al. (this volume) and an $^{40}Ar/^{39}Ar$ biotite age for a younger ash layer (see Coccioni et al., 2008) to constrain their tuning of the filtered 5.65 m $CaCO_3$ component in the Monte Cagnero section (northern Italy) to the ~400 k.y. eccentricity cycle of La2004. The tuning, which resulted in an age of 33.95 Ma for the Eocene-Oligocene boundary (Table 2), was verified by detailed correlations between the filtered ~100 k.y. $CaCO_3$ cycle components in the Massignano and Monte Cagnero sections. These components reveal a remarkable similarity in phase and amplitude and correspond well with the La2004 eccentricity time series.

The different cyclostratigraphic interpretation (i.e., inferred cycle thickness of the 100 k.y. cycle) of van Mourik et al. (2006) as compared with Jovane et al. (2006), Brown et al. (this volume), and Hyland et al. (this volume) may partly be explained by an increase in sedimentation rate starting around the Eocene-Oligocene boundary at Massignano. However, integrated stratigraphic correlations between various sections in the area (see Figure 7 in Hyland et al., this volume) do not support this explanation and lend credit to a thickness of ~1 m for the 100 k.y. cycle.

$^{40}Ar/^{39}Ar$—Astronomical Intercalibration

The astronomical ages for the Eocene-Oligocene boundary at ODP Site 1218 and Massignano (except for the older tuning option of van Mourik et al., 2006) and the K/Ar and $^{40}Ar/^{39}Ar$ ages for the same boundary in the marine record of northern Italy and the continental record of North America are in good apparent agreement. However, it should be noted that the uncertainty in the North American single-crystal $^{40}Ar/^{39}Ar$ dates only represents the analytical error, and that the full error is approximately one order of magnitude larger, i.e., in the order of 2.5% as compared with analytical errors of 0.2%–0.3% reported for the anorthoclase and biotite dates (e.g., Kuiper et al., 2008). Besides, these ages differ significantly from the age estimate based on $^{40}Ar/^{39}Ar$ dating of the volcanic ignimbrite complex.

Moreover, all $^{40}Ar/^{39}Ar$ ages were calculated relative to an age of 27.84 (or 27.83) Ma for the Fish Canyon Tuff dating standard, an age that has subsequently been revised to 28.02 ± 0.28 Ma (Renne et al., 1998). This revised age still has a large uncertainty and is within error consistent with the earlier age. For a more accurate and precise comparison of the $^{40}Ar/^{39}Ar$ and astronomical ages, it is important that the intercalibration between the two dating methods be into account. Following earlier attempts, the $^{40}Ar/^{39}Ar$ method was calibrated to the astronomical method by fitting single-crystal $^{40}Ar/^{39}Ar$ sanidine ages of volcanic ash beds intercalated in astronomically tuned Miocene marine sections in the Mediterranean to their astronomical age (Kuiper et al., 2008). This exercise resulted in a, in principle, much more accurate and precise astronomically calibrated age of 28.201 ± 0.022 Ma for the FCT sanidine, and in concomitantly older $^{40}Ar/^{39}Ar$ ages for all samples dated earlier by the $^{40}Ar/^{39}Ar$ dating method.

IMPLICATIONS AND DISCUSSION

The revised age of 28.02 Ma and the astronomically calibrated age of 28.201 ± 0.022 Ma for the FCT sanidine dating standard have almost no consequences for the age of the Eocene-Oligocene boundary in the Northern Apennines, since this age was based on conventional K/Ar dating, the ages of which remain unaffected. Only some $^{40}Ar/^{39}Ar$ ages are presently available, but they were not employed in calculating the age of 33.7 ± 0.2 Ma for the boundary, and their reliability is not always guaranteed (Deino et al., 1988). The more reliable $^{40}Ar/^{39}Ar$ biotite ages for the two ash layers intercalated in the middle and lower part of chron C13r at Massignano were originally calculated using LP-6 biotite as neutron flux monitor (Odin et al., 1991), a standard which is now known to be heterogeneous in age. Nevertheless, we recalculated these data assuming that Odin et al. (1991) used an age of 127.7 Ma for LP-6 (Odin et al., 1982). Combined with intercalibration factors of 1.305 ± 0.003 for LP-6 and GA1550 (Spell and McDougall, 2003) and 3.5957 ± 0.0038 for GA1550 and FCT (Renne et al., 1998), recalculated ages arrive at 33.9–34.1–34.4 Ma and 34.7–34.9–35.2 Ma for FCT at 27.84–28.02–28.201 Ma, respectively. However, both conventional K/Ar biotite ages and $^{40}Ar/^{39}Ar$ biotite ages should preferably not be taken into account when the numerical ages of chronostratigraphic horizons such as the Eocene-Oligocene boundary are established (Kuiper et al., 2008).

Application of the revised FCT sanidine age results in significantly older ages for the ash layers in the continental record of North America, which were previously in reasonably good agreement with the astronomical ages. The age of the Persistent White Layer (PWL) ash at Dilts Ranch, which coincides approximately with the Chadronian-Orellian (NALMA) boundary and the Eocene-Oligocene boundary, increases from 33.91 to 34.12 and 34.35 Ma, respectively (Table 1). Clearly, these ages are too old for the Eocene-Oligocene boundary if we assume that the astronomical age of ca. 33.79 Ma at ODP Site 1218 or the tuned ages of 33.90–33.95 Ma at Massignano–Monte Cagnero are about correct. In fact, this offset may suggest that the tuning of the Eocene-Oligocene boundary is off by one 400 k.y. eccentricity cycle, as suggested by van Mourik et al. (2006) (Table 2). This 400 k.y. cycle, which is particularly strong in benthic $\delta^{13}C$, was used for first-order tuning of the ODP Site 1218 record. This implies that a possible 400 k.y. tuning error, if it exists, may not originate from the Oligocene tuning but may still arise from the Miocene tuning at Ceara Rise (Shackleton et al., 2000). However, the latter tuning is consistent with the assumption of (an almost) constant seafloor spreading rate of the Australian-Antarctic plate pair (Lourens et al., 2004) and with the independent tuning of other ODP sites (Billups et al., 2004). Moreover, significant coherencies were found in the very long astronomical frequency band of the spectrum. Nevertheless, the astronomical age of the Oligocene-Miocene boundary remains to be tested independently preferably by means of (astronomically calibrated) $^{40}Ar/^{39}Ar$ dating. This is not only important for the tuning of the entire Oligocene but also for the tuning of the lower to middle Miocene, including the Oligocene-Miocene boundary itself. Even so, alleged errors in the tuning of the Oligocene-Miocene boundary interval by as much as 1.2 m.y. (Wilson et al., 2002) are considered rather implausible.

However, the astronomically calibrated $^{40}Ar/^{39}Ar$ age of 34.35 Ma for the Persistent White Layer (Table 1) is still too old, even if we start from the older option of 34.1 Ma for the astronomical age of the boundary (van Mourik et al., 2006), but this relatively old age also opens the possibility that the normal polarity interval above the Persistent White Layer corresponds to one of the short normal polarity intervals identified within chron C13r in magnetostratigraphic records from deep-sea cores and land-based marine sections (Tauxe et al., 1983; Premoli Silva et al., 1988). Unfortunately, these older paleomagnetic records lack the resolution necessary for a more detailed analysis of the events, while a thorough rock magnetic study has not been carried out as a check on the reliability of the data. Nevertheless, the normal polarity interval above the Persistent White Layer at Dilts Ranch may correspond with the well-defined longer normal interval provisionally labeled C13r.1n at Site 1218; this normal interval has been assigned tentative astronomical ages of 34.151 and 34.285 Ma (Pälike et al., 2006), which would fit excellently with the revised Persistent White Layer age. Short normal polarity intervals in C13r may well correspond to four tiny wiggles observed in seafloor anomaly profiles (e.g., Cande and Kent, 1992). However, the detailed analysis of seafloor anomaly profiles makes it unlikely that a normal polarity subchron in C13r exists with a prolonged duration (i.e., longer than 20 k.y.). Moreover, the high-quality and high-resolution magnetostratigraphic record from the Massicore drilled close to the Massignano section did not reveal any evidence for short normal events in C13r, but core recovery was admittedly somewhat less than 100% (Lanci et al., 1996).

Such an alternative interpretation of the upper normal interval at Dilts Ranch is also inconsistent with the supposed magneto-biostratigraphic correlations between sections in the Douglas area (e.g., Dilts Ranch) and Toadstool Park in Nebraska (e.g., Swisher and Prothero, 1990; Terry, 2001). At Toadstool Park, the identification of C13n is convincing in view of the upward extension of the magnetostratigraphy to C10r, which is well constrained by $^{40}Ar/^{39}Ar$ dating (Swisher and Prothero, 1990). Unfortunately, the detailed correlation of individual ash layers between Toadstool Park and Dilts Ranch is still not completely solved (Larson and Evanoff, 1998). The identification of the upper normal interval at Dilts Ranch as C13n is further corroborated by an independent single-crystal $^{40}Ar/^{39}Ar$ sanidine date of 33.59 ± 0.02 Ma for the Persistent White Layer ash layer in the Douglas area (Obradovich et al., 1995; Table 1), which deviates significantly from the biotite age of Swisher and Prothero (1990). Recalculated ages for this sanidine age arrive at 33.81 and 34.03 Ma for FCT sanidine at 28.02 and 28.201 Ma, respectively (Table 1). Especially one age of 33.81 Ma is in excellent agreement with the astronomical age of 33.79 Ma for the Eocene-Oligocene boundary at Site 1218 (Pälike et al., 2006), suggesting that the age of 28.02 Ma for the FCT sanidine is actually more accurate than the astronomically calibrated age of 28.201 Ma. Conversely, the age of 34.03 Ma is somewhat closer to the tuned ages of 33.90–33.95 Ma for the Eocene-Oligocene boundary at Massignano–Monte Cagnero (Brown et al., this volume; Hyland et al., this volume; Tables 1 and 2).

However, the ages of the older ash beds at Flagstaff Rim given in Swisher and Prothero (1990) increase by almost 0.5 m.y., becoming older than 36 Ma (Fig. 1). Evidently, the combined magnetostratigraphy and recalculated $^{40}Ar/^{39}Ar$ ages of Swisher and Prothero (1990) do not fit when they are compared with the most recent geomagnetic polarity time scales of 2004 (Gradstein et al., 2004) and the Astronomical Polarity Time Scale (APTS) (Pälike et al., 2006; Fig. 1). A possible explanation for this lack of consistency might still be an error in the magnetostratigraphic record of the Flagstaff Rim section, although other options such as hiatuses or condensed intervals in the continental succession cannot be excluded. However, also in this case, the single-crystal $^{40}Ar/^{39}Ar$ sanidine ages of Obradovich et al. (1995) are significantly younger than the ages published by Swisher and Prothero (1990) for the same ash layers (Fig. 1). In fact, the observed discrepancies largely disappear if we start from these younger ages recalculated against 28.02 and 28.201 Ma for FCT sanidine. The rather systematic offset between the $^{40}Ar/^{39}Ar$ ages of Swisher and Prothero (1990) and Obradovich et al. (1995) suggests that the aberrantly old ages of Swisher and Prothero (1990) may be responsible for the discrepancies (Fig. 1). Around the same time, Swisher et al. (1993) published single-crystal $^{40}Ar/^{39}Ar$ ages of ca. 66.0 Ma for the Cretaceous-Paleogene boundary that were later revised to ca. 65.0 Ma (against FCT sanidine at 27.84 Ma). The initial aberrantly old ages were related to the chemical preparation technique applied to ash samples intertwined in coal, but this cannot explain the older ages of Swisher and Prothero (1990) as compared with Obradovich et al. (1995) in the Eocene-Oligocene boundary interval (Swisher, 2007, personal commun.). Nevertheless, it is intriguing that the ratio difference is remarkably the same.

Application of revised FCT sanidine ages to the $^{40}Ar/^{39}Ar$ ages of the magnetostratigraphically constrained New Mexico ignimbrite succession results in significantly older ages of 33.3–33.5 Ma for C13n(o) and of 33.6–33.9 Ma for the Eocene-Oligocene boundary when recalculated against an age of 28.02–28.201 Ma for the FCT sanidine (Table 1; Fig. 2). Clearly, these ages, which are not very well constrained, are now in much better agreement with the recently published astronomical ages if we exclude the older tuning option presented in van Mourik et al. (2006). However, recalculated $^{40}Ar/^{39}Ar$ ages for the much better constrained R-N reversal between their polarity intervals D and C (ca. 28.19–28.37 Ma) and the N-R reversal between F and E (ca. 29.20–29.39 Ma) are also now in relatively good agreement with the astronomical ages for reversals at Site ODP

1218, which correspond to the same C10n.2n(o) and C11n.1n(y) chron boundaries (Fig. 2).

A very well-dated older reversal in the ignimbrite succession is the N to R polarity transition that supposedly corresponds to the top of C15n (McIntosh et al., 1992). The age of 34.36 Ma is exactly identical to the single-crystal $^{40}Ar/^{39}Ar$ sanidine age for ash bed J in the Flagstaff Rim section in North America (of Obradovich et al., 1995). This ash is found directly on top of a normal polarity interval that corresponds to C15n as well. This precise age may already provide tight constraints on the astronomical tuning of the Upper Eocene, if the ages of Obradovich et al. (1995) are correct (see previous discussion), and if the calibration of the Flagstaff Rim magnetostratigraphy and the ignimbrite polarity sequence to the geomagnetic polarity time scale are both correct. The recalculated astronomically calibrated $^{40}Ar/^{39}Ar$ age for this reversal arrives at 34.81–34.82 Ma. This age is at odds with the preliminary astronomical age of 35.126 Ma for the same reversal at ODP Site 1218 (Pälike et al., 2006), but it is within error consistent with the astronomical ages of 34.74 Ma (Brown et al., this volume) for the top of C15n at Massignano. It is also in good agreement with the tuned age of 34.728 Ma for this reversal in the same section (Jovane et al., 2006, 2007), but in this case, the age of 33.714 Ma for the Eocene-Oligocene boundary is inconsistent with the recalculated age of 34.03 Ma for the Persistent White Layer ash bed at Dilts Ranch (see Tables 1 and 2). It should be mentioned here that we followed the preferred calibration of the ignimbrite polarity sequence to the geomagnetic polarity time scale of McIntosh et al. (1992).

The $^{40}Ar/^{39}Ar$ sanidine ages of the bentonite in the upper part of the Yazoo Formation increase from 34.31–34.32 to 34.53–34.54 or 34.76–34.77 ± 0.05 Ma, while the biotite age increases from 34.28 to 34.50 or 34.73 ± 0.04 Ma, depending whether we use the age of 28.02 or 28.201 Ma for FCT sanidine. Starting from astronomical ages of 33.79 Ma or 33.90–33.95 Ma for the Eocene-Oligocene boundary, these ages appear to be too old in view of the biostratigraphic assignment of the bentonites to the upper part of zone NP20 and P17. However, biostratigraphic control in the western part of the outcrop belt is poor as age diagnostic calcareous plankton taxa become less common due to the shallow(er) marine conditions (Obradovich et al., 1993).

The $^{40}Ar/^{39}Ar$ age of the tektite fragments found at Bath Cliffs and supposedly coming from the Chesapeake impact increases from 35.4 to 35.6 or 35.9 ± 0.3 Ma. Unfortunately, the analytical uncertainty is too large to be very helpful in solving the problem of the numerical age of the Eocene-Oligocene boundary.

It should be emphasized that all recalculated $^{40}Ar/^{39}Ar$ ages using the astronomically calibrated age of the FCT sanidine dating standard critically depend on the accuracy of the intercalibration (Kuiper et al., 2008). The intercalibration still needs to be independently tested using a different set of ash layers than the Melilla ashes on which it was based. These ash beds should allow single-crystal sanidine dating and be intercalated in well-tuned marine sections, preferably of older pre-Messinian age. In this respect results of an interlaboratory $^{40}Ar/^{39}Ar$ comparison project carried out within the framework of the Earthtime initiative that became recently available should be taken into consideration (Heizler et al., 2008; Heizler, personal commun.). The comparison revealed that age differences of up to 1–1.5% are still present between 20 laboratories even when standards (Fish Canyon) and unknowns (Alder Creek) are irradiated in one irradiation batch and consequently distributed between laboratories and analyzed. The cause of this scatter is not completely clear at this stage, although variable data reduction procedures might explain some of it. Evidently, the results of this intercalibration experiment have important consequences for some of the issues that we tried to solve in the present paper, and the argon community should solve these issues, before continuing high precision time scale work.

CONCLUSIONS

Despite attempts to date the Eocene-Oligocene boundary more accurately and precisely, the problem of its numerical age remains at present unsolved because existing astronomical ages and $^{40}Ar/^{39}Ar$ ages recalculated to a revised and astronomically calibrated age for the FCT sanidine dating standard are partly inconsistent with each other. This inconsistency may stem from an error—of one 400 k.y. eccentricity cycle—in the tuning of the boundary interval at ODP Site 1218 and at Massignano–Monte Cagnero (of Pälike et al., 2006; Brown et al., this volume; Hyland et al., this volume), an error in the intercalibration of the astronomical and $^{40}Ar/^{39}Ar$ dating methods (of Kuiper et al., 2008), an error in the magnetostratigraphy of the continental sections in North America and/or the calibration of this magnetostratigraphy to the geomagnetic polarity time scale, or a problem encountered in $^{40}Ar/^{39}Ar$ dating (i.e., sample preparation, interlaboratory bias). Nevertheless, it is encouraging that the tuned ages for the Eocene-Oligocene boundary are now largely consistent with a recalculated alternative single-crystal $^{40}Ar/^{39}Ar$ sanidine age for the Persistent White Layer ash bed and a recalculated age estimate for the Eocene-Oligocene boundary based on the combined $^{40}Ar/^{39}Ar$ and magnetostratigraphic dating of ignimbrites in New Mexico.

Clearly, testing the astronomical age of the Eocene-Oligocene boundary by means of astronomically calibrated $^{40}Ar/^{39}Ar$ dating is possible and necessary, especially in view of the remaining inconsistencies. Evidently, additional single-crystal $^{40}Ar/^{39}Ar$ sanidine dates from the tuffs in North America and independent checks on the astronomical tuning and the intercalibration between the astronomical and $^{40}Ar/^{39}Ar$ dating methods are needed to definitively solve the problem of the numerical age of the Eocene-Oligocene boundary. It is anticipated that such analyses and tests will be carried in the coming years as part of the current international Earthtime initiative and associated projects (http://www.earth-time.org/; http://www.earthtime-eu.eu/Earthtime_Europe/) to significantly improve the geological time scale. Clearly, an accurate and precise dating of the Eocene-Oligocene boundary is crucial if we are to unravel the cause(s) underlying the major climate transition associated with it.

REFERENCES CITED

Berggren, W.A., Kent, D.V., and van Couvering, J.A., 1985, The Neogene: Part 2. Neogene geochronology and chronostratigraphy, *in* Snelling, N.J., ed., The Chronology of the Geological Record: Geological Society of London Memoir 10, p. 211–260.

Berggren, W.A., Kent, D.V., Obradovich, J.D., and Swisher, C.C., III, 1992, Towards a revised Paleogene geochronology, *in* Prothero, D.R., and Berggren, W.A., eds., Eocene-Oligocene Climatic and Biotic Evolution: Princeton Series in Geology and Paleontology: Princeton, Princeton University Press, p. 29–45.

Berggren, W.A., Kent, D.V., Swisher, C.C., III, and Aubry, M.-P., 1995, A revised Cenozoic geochronology and chronostratigraphy, *in* Berggren, W.A., Kent, D.V., Aubry, M.-P., and Hardenbol, J., eds., Geochronology, Time Scales and Global Stratigraphic Correlation: Society of Economic Paleontologists and Mineralogists Special Publication 54, p. 129–212.

Billups, K., Pälike, H., Channell, J.E.T., Zachos, J.C., and Shackleton, N.J., 2004, Astronomic calibration of the late Oligocene through early Miocene geomagnetic polarity time scale: Earth and Planetary Science Letters, v. 224, p. 33–44, doi: 10.1016/j.epsl.2004.05.004.

Brinkhuis, H., and Visscher, H., 1995, The upper boundary of the Eocene Series: A reappraisal based on dinoflagellate cyst biostratigraphy and sequence stratigraphy, *in* Berggren, W.A., Kent, D.V., Aubry, M.-P., and Hardenbol, J., eds., Geochronology, Time Scales and Global Stratigraphic Correlation: Society of Economic Paleontologists and Mineralogists (SEPM) Special Publication 54, p. 295–304.

Brown, R.E., Koeberl, C., Montanari, A., and Bice, D.M., 2009, this volume, Evidence for a change in Milankovitch forcing caused by extraterrestrial events at Massignano, Italy, Eocene-Oligocene boundary GSSP, *in* Koeberl, C., and Montanari, A., eds., The Late Eocene Earth—Hothouse, Icehouse, and Impacts: Geological Society of America Special Paper 452, doi: 10.1130/2009.2452(08).

Cande, S.C., and Kent, D.V., 1992, A new geomagnetic polarity time scale for the late Cretaceous and Cenozoic: Journal of Geophysical Research, v. 97, p. 13,917–13,951.

Cande, S.C., and Kent, D.V., 1995, Revised calibration of the geomagnetic polarity time scale for the late Cretaceous and Cenozoic: Journal of Geophysical Research, v. 100, p. 6093–6095, doi: 10.1029/94JB03098.

Coccioni, R., Monaco, P., Monechi, S., Nocchi, M., and Parisi, G., 1988, Biostratigraphy of the Eocene-Oligocene boundary at Massignano (Ancona, Italy), *in* Premoli Silva, I., Coccioni, R., and Montanari, A., eds., The Eocene/Oligocene Boundary in the Marche-Umbria Basin (Italy): International Subcommission on Paleogene Stratigraphy Special Publication: Ancona, Italy, Industrie Grafiche Fratelli Aniballi, p. 59–80.

Coccioni, R., Marsili, A., Montanari, A., Bellanca, A., Neri, R., Bice, D.M., Brinkhuis, H., Church, N., Macalady, A., McDaniel, A., Deino, A., Lirer, F., Sprovieri, M., Maiorano, P., Monechi, S., Nini, C., Nocchi, M., Pross, J., Rochette, P., Sagnotti, L., Tateo, F., Touchard, Y., Van Simaeys, S., and Williams, G.L., 2008, Integrated stratigraphy of the Oligocene pelagic sequence in the Umbria-Marche Basin (northeastern Apennines, Italy): A potential global stratotype section and point (GSSP) for the Rupelian/Chattian boundary: Geological Society of America Bulletin, v. 120, p. 487–511, doi: 10.1130/B25988.1.

Coxall, H.K., Wilson, P.A., Pälike, H., Lear, C.H., and Backman, J., 2005, Rapid stepwise onset of Antarctic glaciation and deeper calcite compensation in the Pacific Ocean: Nature, v. 433, p. 53–57, doi: 10.1038/nature03135.

Deino, A.L., Drake, R.E., Curtis, G.H., and Montanari, A., 1988, Preliminary laser-fusion $^{40}Ar/^{39}Ar$ dating results from Oligocene biotites of the Contessa Quarry section (Gubbio, Italy), *in* Premoli Silva, I., Coccioni, R., and Montanari, A., eds., The Eocene/Oligocene Boundary in the Marche-Umbria Basin (Italy): International Subcommission on Paleogene Stratigraphy Special Publication: Ancona, Italy, Industrie Grafiche Fratelli Aniballi, p. 229–238.

Evernden, J.F., Savage, D.E., Curtis, G.H., and James, G.T., 1964, Potassium-argon dates and the Cenozoic mammalian chronology of North America: American Journal of Science, v. 262, p. 145–198.

Glass, B.P., and Crosbie, J.R., 1982, Age of Eocene/Oligocene boundary based on extrapolation from North American microtektite layer: The American Association of Petroleum Geologists Bulletin, v. 66, p. 471–476.

Glass, B.P., Hall, C.M., and York, D., 1986, $^{40}Ar/^{39}Ar$ laser-probe dating of North American tektite fragments from Barbados and the age of Eocene-Oligocene boundary: Chemical Geology–Isotope Geoscience Section, v. 59, p. 181–186.

Gradstein, F.M., Ogg, J.G., and Smith, A.G., eds., 2004, A Geologic Time Scale 2004: Cambridge, Cambridge University Press, 589 p.

Heizler, M., and EARTHTIME Ar working group, 2008, Argon laboratory intercomparison efforts for the EARTHTIME initiative: Geophysical Research Abstracts, v. 10, EGU2008-A-11478, 2008.

Hilgen, F.J., 1991, Extension of the astronomically calibrated (polarity) time scale to the Miocene-Pliocene boundary: Earth and Planetary Science Letters, v. 107, p. 349–368, doi: 10.1016/0012-821X(91)90082-S.

Hilgen, F.J., Krijgsman, W., Langereis, C.G., Lourens, L.J., Santarelli, A., and Zachariasse, W.J., 1995, Extending the astronomical (polarity) time scale into the Miocene: Earth and Planetary Science Letters, v. 136, p. 495–510, doi: 10.1016/0012-821X(95)00207-S.

Holmes, A., 1960, A revised geological time-scale: Transactions of the Geological Society of Edinburgh, v. 17, p. 183–216.

Hyland, E., Murphy, B., Varela, P., Marks, K., Colwell, L., Tori, F., Monechi, S., Cleaveland, L., Brinkhuis, H., van Mourik, C.A., Coccioni, C., Bice, D., and Montanari, A., 2009, this volume, Integrated stratigraphy and astrochronologic calibration of the Eocene-Oligocene transition in the Monte Cagnero section (northeastern Apennines, Italy): A potential parastratotype for the Massignano GSSP, *in* Koeberl, C., and Montanari, A., eds., The Late Eocene Earth—Hothouse, Icehouse, and Impacts: Geological Society of America Special Paper 452, doi: 10.1130/2009.2452(19).

Imbrie, J., Hays, J.D., Martinson, D.G., McIntyre, A., Mix, A.C., Morley, J.J., Pisias, N.G., Prell, W.L., and Shackleton, N.J., 1984, The orbital theory of Pleistocene climate: Support from a revised chronology of the marine $\delta^{18}O$ record, *in* Berger, A., Imbrie, J., Hays, J.J., Kukla, G., and Saltzman, B., eds., Milankovitch and Climate, Understanding the Response to Astronomical Forcing: Mathematical and Physical Sciences: Norwell, Massachusetts, D. Reidel Publishing Company, NATO ASI Series, Series C, v. 126, p. 269–305.

Jovane, L., Florindo, F., Sprovieri, M., and Pälike, H., 2006, Astronomic calibration of the late Eocene/early Oligocene Massignano section (central Italy): Geochemistry, Geophysics, Geosystems, v. 7, doi: 10.1029/2005GC001195.

Jovane, L., Sprovieri, M., Florindo, F., Acton, G., Coccioni, R., Dall'Antonia, B., and Dinarès-Turell, J., 2007, Eocene-Oligocene paleoceanographic changes in the stratotype section, Massignano, Italy: Clues from rock magnetism and stable isotopes: Journal of Geophysical Research, v. 112, B11101, doi:10.1029/2007JB004963.

Kuiper, K.F., Deino, A., Hilgen, F.J., Krijgsman, W., Renne, P.R., and Wijbrans, J.R., 2008, Synchronizing rock clocks of Earth history: Science, v. 320, p. 500–504, doi: 10.1126/science.1154339.

LaBrecque, J.L., Kent, D.V., and Cande, S.C., 1977, Revised magnetic polarity time-scale for the Late Cretaceous and Cenozoic time: Geology, v. 5, p. 330–335, doi: 10.1130/0091-7613(1977)5<330:RMPTSF>2.0.CO;2.

Lanci, L., Lowrie, W., and Montanari, A., 1996, Magnetostratigraphy of the Eocene/Oligocene boundary in a short drillcore: Earth and Planetary Science Letters, v. 143, p. 37–48, doi: 10.1016/0012-821X(96)00136-7.

Larson, E., and Evanoff, E.E., 1998, An overview of the tuffs of the White River Sequence, *in* Terry, D.O., Jr., LaGarry, H.E., and Hunt, R.M., eds., Depositional Environments, Lithostratigraphy, and Biostratigraphy of the White River and Arikaree Groups (Late Eocene to Early Miocene, North America): Geological Society of America Special Paper 325, p. 1–14.

Laskar, J., Robutel, P., Joutel, F., Gastineau, M., Correia, A.C.M., and Levrard, B., 2004, A long term numerical solution for the insolation quantities of the Earth: Astronomy & Astrophysics, v. 428, p. 261–285, doi: 10.1051/0004-6361:20041335.

Lourens, L.J., Hilgen, F.J., Laskar, J., Shackleton, N.J., and Wilson, D., 2004, The Neogene Period, *in* Gradstein, F., Ogg, J., and Smith, A., eds., A Geological Time Scale 2004: Cambridge, Cambridge University Press, p. 409–440.

Lowrie, W., Alvarez, W., Napoleone, G., Perch-Nielsen, K., Premoli-Silva, I., and Toumarkine, M., 1982, Paleogene magnetic stratigraphy in Umbrian pelagic carbonate rocks: Geological Society of America Bulletin, v. 93, p. 414–432, doi: 10.1130/0016-7606(1982)93<414:PMSIUP>2.0.CO;2.

McIntosh, W.C., Geissman, J.W., Chapin, C.E., Kunk, M.J., and Henry, C.D., 1992, Calibration of the latest Eocene-Oligocene geomagnetic polarity time scale using $^{40}Ar/^{39}Ar$ dated ignimbrites: Geology, v. 20, p. 459–463, doi: 10.1130/0091-7613(1992)020<0459:COTLEO>2.3.CO;2.

Montanari, A., 1988, Geochemical characterization of volcanic biotites from the Upper Eocene–Upper Miocene pelagic sequence of the northeastern Apennines, *in* Premoli Silva, I., Coccioni, R., and Montanari, A., eds., The Eocene/Oligocene Boundary in the Marche-Umbria Basin (Italy): International Subcommission on Paleogene Stratigraphy Special Publication: Ancona, Italy, Industrie Grafiche Fratelli Aniballi, p. 209–227.

Montanari, A., Drake, R., Bice, D.M., Alvarez, W., Curtis, G.H., Turrin, B.D., and DePaolo, D.J., 1985, Radiometric time scale for the upper Eocene and Oligocene based on K/Ar and Rb/Sr dating of volcanic biotites from the pelagic sequence of Gubbio, Italy: Geology, v. 13, p. 596–599, doi: 10.1130/0091-7613(1985)13<596:RTSFTU>2.0.CO;2.

Montanari, A., Deino, A.L., Drake, R.E., Turrin, B.D., DePaolo, D.J., Odin, G.S., Curtis, G.H., Alvarez, W., and Bice, D.M., 1988, Radioisotopic dating of the Eocene-Oligocene boundary in the pelagic sequence of the northeastern Apennines, *in* Premoli Silva, I., Coccioni, R., and Montanari, A., eds., The Eocene/Oligocene Boundary in the Marche-Umbria Basin (Italy): International Subcommission on Paleogene Stratigraphy Special Publication: Ancona, Italy, Industrie Grafiche Fratelli Aniballi, p. 195–208.

Nocchi, M., Parisi, G., Monaco, P., Monechi, S., Madile, M., Napoleone, G., Ripepe, M., Orlando, M., Premoli Silva, I., and Bice, D.M., 1986, The Eocene–Oligocene boundary in the Umbrian pelagic sequences, Italy, *in* Pomerol, C., and Premoli Silva, I., eds., Terminal Eocene Events: Developments in Palaeontology and Stratigraphy, Volume 9: Amsterdam, Elsevier, p. 24–40.

Obradovich, J.D., Dockery, D.T., and Swisher, C.C., III, 1993, ^{40}Ar/^{39}Ar ages of bentonites in the upper part of the Yazoo Formation (Upper Eocene), west-central Mississippi: Mississippi Geology, v. 14, p. 1–9.

Obradovich, J.D., Evanoff, E., and Larson, E.E., 1995, Revised single-crystal laser-fusion ^{40}Ar/^{39}Ar ages of Chadronian tuffs in the White River Formation of Wyoming: Geological Society of America Abstracts with Programs, v. 27, no. 3, p. A-77.

Odin, G.S., Montanari, A., Deino, A., Drake, R., Guise, P.G., Kreuzer, H., and Rex, D.C., 1991, Reliability of volcano-sedimentary biotite ages across the Eocene-Oligocene boundary (Apennines, Italy): Chemical Geology–Isotope Geoscience Section, v. 86, p. 203–224, doi: 10.1016/0168-9622(91)90050-7.

Odin, G.S., et al., 1982, Interlaboratory standards for dating purposes, *in* Odin, G.S., ed., Numerical Dating in Stratigraphy: New York, John Wiley & Sons, p. 123–158.

Pälike, H., Norris, R.D., Herrle, J.O., Wilson, P.A., Coxall, H.K., Lear, C.H., Shackleton, N.J., Tripati, A.K., and Wade, B.S., 2006, The heartbeat of the Oligocene climate system: Science, v. 314, p. 1894–1898, doi: 10.1126/science.1133822.

Pearson, P.N., McMillan, I.K., Wade, B.S., Jones, T.D., Coxall, H.K., Bown, P.R., and Lear, C.H., 2008, Extinction and environmental change across the Eocene-Oligocene boundary in Tanzania: Geology, v. 36, p. 179–182, doi: 10.1130/G24308A.1.

Premoli Silva, I., and Jenkins, D.J., 1993, Decision on the Eocene-Oligocene boundary stratotype: Episodes, v. 16, p. 379–381.

Premoli Silva, I., Orlando, M., Monechi, S., Madile, M., Napoleone, G., and Ripepe, M., 1988, Calcareous plankton biostratigraphy and magnetostratigraphy at the Eocene-Oligocene transition in the Gubbio area, *in* Premoli Silva, I., Coccioni, R., and Montanari, A., eds., The Eocene/Oligocene Boundary in the Marche-Umbria Basin (Italy): International Subcommission on Paleogene Stratigraphy Special Publication: Ancona, Italy, Industrie Grafiche Fratelli Aniballi, p. 137–161.

Prothero, D.R., and Swisher, C.C., III, 1992, Magnetostratigraphy and geochronology of the terrestrial Eocene-Oligocene transition in North America, *in* Prothero, D.R., and Berggren, W.A., eds., Eocene-Oligocene Climatic and Biotic Evolution: Princeton Series in Geology and Paleontology: Princeton, Princeton University Press, p. 46–73.

Prothero, D.R., Denham, C.R., and Farmer, H.G., 1982, Oligocene calibration of the magnetic polarity time scale: Geology, v. 10, p. 650–653, doi: 10.1130/0091-7613(1982)10<650:OCOTMP>2.0.CO;2.

Renne, P.R., Swisher, C.C., Deino, A.L., Karner, D.B., Owens, T.L., and DePaolo, D.J., 1998, Intercalibration of standards, absolute ages and uncertainties in ^{40}Ar/^{39}Ar dating: Chemical Geology, v. 145, p. 117–152, doi: 10.1016/S0009-2541(97)00159-9.

Shackleton, N.J., Berger, A., and Peltier, W.R., 1990, An alternative astronomical calibration of the lower Pleistocene timescale based on ODP Site 677: Transactions of the Royal Society of Edinburgh, v. 81, p. 251–261.

Shackleton, N.J., Crowhurst, S.J., Weedon, G.P., and Laskar, J., 1999, Astronomical calibration of Oligocene-Miocene time: Philosophical Transactions of the Royal Society of London, Series A–Mathematical and Physical Sciences, v. 357, p. 1907–1929, doi: 10.1098/rsta.1999.0407.

Shackleton, N.J., Hall, M.A., Raffi, I., Tauxe, L., and Zachos, J., 2000, Astronomical calibration age for the Oligocene-Miocene boundary: Geology, v. 28, p. 447–450, doi: 10.1130/0091-7613(2000)28<447:ACAFTO>2.0.CO;2.

Spell, T.L., and McDougall, I., 2003, Characterization and calibration of ^{40}Ar/^{39}Ar dating standards: Chemical Geology, v. 198, p. 189–211, doi: 10.1016/S0009-2541(03)00005-6.

Swisher, C.C., III, and Prothero, D.R., 1990, Single-crystal ^{40}Ar/^{39}Ar dating of the Eocene-Oligocene transition in North America: Science, v. 249, p. 760–762, doi: 10.1126/science.249.4970.760.

Swisher, C.C., III, Dingus, L., and Butler, R.F., 1993, ^{40}Ar/^{39}Ar dating and magnetostratigraphic correlation of the terrestrial Cretaceous-Paleogene boundary and Puercan mammal age, Hell Creek–Tullock Formations, Eastern Montana: Canadian Journal of Earth Sciences, v. 30, p. 1981–1996.

Tauxe, L., Tucker, P., Petersen, N.P., and LaBrecque, J.L., 1983, The magnetostratigraphy of Leg 73 sediments: Palaeogeography, Palaeoclimatology, Palaeoecology, v. 42, p. 65–90, doi: 10.1016/0031-0182(83)90039-1.

Terry, D.O., 2001, Paleopedology of the Chadron Formation of northwestern Nebraska: Implications for paleoclimate change in the North American Midcontinent across the Eocene-Oligocene boundary: Palaeogeography, Palaeoclimatology, Palaeoecology, v. 168, p. 1–38, doi: 10.1016/S0031-0182(00)00248-0.

Van Mourik, C.A., Lourens, L.J., Brinkhuis, H., Pälike, H., Hilgen, F.J., Montanari, A., and Coccioni, R., 2006, From greenhouse to icehouse at the Massignano Eocene-Oligocene GSSP: Implications for the cause, timing and effect of the Oi-1 glaciation, *in* van Mourik, C.A., ed., The Greenhouse-Icehouse Transition, A Dinoflagellate Perspective [Ph.D. thesis]: Stockholm University, Stockholm, Meddelanden från Stockholms universitets institution för geologi och geokemi No 327.

Wilson, G.S., Lavelle, M., McIntosh, W.C., Roberts, A.P., Harwood, D.M., Watkins, D.K., Villa, G., Bohaty, S.M., Fielding, C.R., Florindo, F., Sagnotti, L., Naish, T.R., Shere, R.P., and Verosub, K.L., 2002, Integrated chronostratigraphic calibration of the Oligocene-Miocene boundary at 24.0 Ma from the CRP-2A drill core, Ross Sea, Antarctica: Geology, v. 30, p. 1043–1046, doi: 10.1130/0091-7613(2002)030<1043:ICCOTO>2.0.CO;2.

Manuscript Accepted by the Society 16 September 2008

The Geological Society of America
Special Paper 452
2009

The late Eocene greenhouse-icehouse transition: Observations from the Massignano global stratotype section and point (GSSP)

Luigi Jovane*
Istituto Nazionale di Geofisica e Vulcanologia, Via di Vigna Murata 605, 00143 Rome, Italy, and
Geology Department, Western Washington University, 516 High Street, Bellingham, Washington 98225, USA

Rodolfo Coccioni
Andrea Marsili
Dipartimento di Scienze dell'Uomo, dell'Ambiente e della Natura, Università degli Studi di Urbino "Carlo Bo," Campus Scientifico, Località Crocicchia, 61029 Urbino, Italy

Gary Acton
Department of Geology, University of California–Davis, One Shields Avenue, Davis, California 95616, USA

ABSTRACT

The interval from the middle Eocene to early Oligocene represents one of the most significant transitions in Earth's climate, during which greenhouse conditions were supplanted by icehouse conditions of the present day. This global transition was preceded by a long-term cooling phase and short-term uncorrelated variations in several marine proxies, which indicate paleoceanographic instabilities prior to the key climatic transition. We integrate previous multidisciplinary studies with recent data from the Massignano section (Umbria-Marche Basin) and summarize interpretations of studies from the past 20 yr that have been based on the Eocene-Oligocene boundary global stratotype section and point (GSSP). Based on the many data sets from this section, with an emphasis on the rock magnetic data, we propose that the fluctuations and final cessation of a westward subtropical Eocene Neotethys (STENT) current were drivers in the climatic transition from greenhouse to icehouse conditions. Our hypothesis considers that the global variable climatic conditions were synchronous to large changes in circulation in the western Neotethys Ocean, which were primarily due to paleogeographic and sea-level changes. Specifically, the closing of the gateway between the Arabian and Eurasian plates, through coupling of sea-level changes, could represent the threshold or one of the triggers that caused the paleoceanographic variations in the Neotethys and in global ocean circulation patterns during the end of the Eocene.

Keywords: Eocene-Oligocene boundary global stratotype section and point, Massignano section, Neotethys paleoceanographic circulation, integrated stratigraphy, Italy.

*luigijovane@gmail.com.

Jovane, L., Coccioni, R., Marsili, A., and Acton, G., 2009, The late Eocene greenhouse-icehouse transition: Observations from the Massignano global stratotype section and point (GSSP), *in* Koeberl, C., and Montanari, A., eds., The Late Eocene Earth—Hothouse, Icehouse, and Impacts: Geological Society of America Special Paper 452, p. 149–168, doi: 10.1130/2009.2452(10). For permission to copy, contact editing@geosociety.org.

THE MASSIGNANO GLOBAL STRATOTYPE SECTION AND POINT (GSSP) FOR THE EOCENE-OLIGOCENE BOUNDARY: TWENTY YEARS OF SCIENTIFIC HISTORY

The first study of the stratigraphical setting of the Massignano section was carried out by Baumann in 1970. Several years later, numerous cooperative studies were undertaken to precisely locate the Eocene-Oligocene boundary in the continuous and well-exposed sequence at Massignano (Coccioni et al., 1986; Montanari and Bice, 1986; Odin, 1986; Odin et al., 1986).

The International Union of Geological Sciences Subcommission on Paleogene Stratigraphy (IUGS-SPS) organized a meeting led by Isabella Premoli Silva, Rodolfo Coccioni, and Alessandro Montanari at Monte Cònero, near Ancona, in October 1987. For this occasion, several new interdisciplinary stratigraphic and geochronologic data were presented about the Massignano section and other equivalent Eocene-Oligocene sections in the Umbria-Marche Basin. The Massignano section became a focus, integrating stratigraphic studies that combined litho-, bio-, magneto-, and chemostratigraphic records with direct radioisotopic dating. In fact, during the IUGS-SPS meeting, it became clear that the Massignano site was the most likely candidate for the global stratotype section and point (GSSP) for the Eocene-Oligocene boundary.

One year later, several interdisciplinary studies that had been discussed at the Monte Cònero meeting were published in an IUGS-SPS special volume (Premoli Silva et al., 1988). This publication, which covered biostratigraphy (Coccioni, 1988; Coccioni et al., 1988; Nocchi et al., 1988; Parisi and Coccioni, 1988), magnetostratigraphy (Bice and Montanari, 1988), sedimentary petrology and chemostratigraphy (Odin et al., 1988a; Capo and De Paolo, 1988), and geochronology (Montanari, 1988; Montanari et al., 1988; Odin et al., 1988b), showed the potential of this site to be elected as stratotype section for the Eocene-Oligocene boundary (Odin and Montanari, 1988).

In 1989, a formal proposal promoting the Massignano section as the global stratotype section and point (GSSP) for the Eocene-Oligocene boundary was presented at the 28th International Geological Congress (IGC) in Washington DC (Montanari and Odin, 1989; Premoli Silva et al., 1989). In 1992, at the 29th International Geological Congress in Kyoto, the Commission on Stratigraphy of the International Union of Geological Sciences formally established the Massignano section as the GSSP for the Eocene-Oligocene boundary (Premoli Silva and Jenkins, 1993; and discussion of Brinkhuis and Visscher, 1995).

In the following years, numerous interdisciplinary studies were carried out at Massignano by a number of international research teams. They aimed to further refine the lithostratigraphic and mineralogic (Mattias et al., 1992), biostratigraphic and paleontologic (Bellagamba and Coccioni, 1990; Gonzalvo and Molina, 1992; Molina et al., 1993), and chemostratigraphic and geochronologic (Odin and Montanari, 1989; Montanari, 1990; Oberli and Meier, 1991; Odin et al., 1991) records of the GSSP Eocene-Oligocene section.

In the mid-1990s, an important discovery was made in the sediments that occur near the 5.61 m stratigraphic level from the base (msl) of the Massignano section, where evidence of cosmic influence was recognized. An impactoclastic layer, possibly of worldwide occurrence, containing an iridium anomaly (Montanari et al., 1993), shocked quartz (Clymer et al., 1995, 1996; Langenhorst and Clymer, 1996), extraterrestrial Ni-rich spinel and altered microkrystites (Pierrard et al., 1998), and a prominent anomaly in ^{3}He (Farley et al., 1998), was interpreted by Farley and co-workers as evidence of a comet shower that lasted ~2.5 m.y. This finding prompted a number of high-resolution stratigraphic and detailed interdisciplinary studies on this section. These studies aimed to assess the effects that the inferred cosmic events (impacts and comet shower) may have had on the terminal Eocene marine biota and on regional or global environmental changes (Brinkhuis and Coccioni, 1995; Vonhof et al., 1995, 1998; Gardin et al., 1999; Coccioni et al., 2000; Monechi et al., 2000; Huber et al., 2002; Spezzaferri et al., 2002; Coccioni and Galeotti, 2003; Dall'Antonia et al., 2003; Van Mourik and Brinkhuis, 2005; Jovane et al., 2004, 2006, 2007a).

High-resolution investigations of planktonic and benthic foraminifera, calcareous nannofossils, and dinocysts across the impact event (5.61 msl) provided evidence of important changes in climate and water mass structure (Brinkhuis and Coccioni, 1995; Coccioni et al., 2000; Monechi et al., 2000; Vonhof et al., 2000; Spezzaferri et al., 2002). Researchers recognized a distinct cooling pulse and a shallowing of the thermocline just after the impact event. In addition, Coccioni et al. (2000) suggested that rapid variations in lithological color could be an expression of changes of seafloor ventilation and related flushing. Benthic foraminifera suggest that the relatively young, cold bottom waters forming at high latitudes reached the western Tethys lower bathyal settings only in the middle part of zone P16 of Berggren et al. (1995) (Coccioni and Galeotti, 2003). Dall'Antonia et al. (2003), studying the deep-sea ostracods at high resolution, also documented changes in the water mass properties with cooling pulses between 4.1 msl and 6.3 msl related to the influence of cold bottom waters and an intensification of the cooling trend and increase of the primary productivity across the Eocene-Oligocene boundary. Bodiselitsch et al. (2004) carried out a high-resolution geochemical investigation in the late Eocene sediments aimed at recognizing the influence of the impact events on seawater temperature. These authors provided evidence that the general late Eocene cooling trend was interrupted by warm pulses triggered by multiple impact events.

Also during the 1990s, the paleomagnetism of the Massignano section was studied by William Lowrie and Luca Lanci. They obtained new magnetostratigraphy from 32 oriented samples that spanned the section (Lowrie and Lanci, 1994). The same authors also studied a 39.4-m-long drill core, referred to as Massicore, from 110 m south of the Massignano quarry at 42°32′09.6″N, 13°35′34.3″E (Lanci et al., 1996). They compared the paleomagnetism of Massignano section with the Massicore paleomagnetic results, presented the magnetostratigraphies for the sections, and evaluated the presence of chron C14 and tiny wiggles in chron C13n (Lanci et al., 1996; Lanci and Lowrie, 1997; Lanci et al., 1998). In those works, they also completed the first environmental magnetism study on the section and illustrated the influence of astronomical forcing.

The expected early Oligocene paleomagnetic direction at the Massignano site has a declination of 6° and an inclination of 56° based on the "African" apparent polar wander reference path of Besse and Courtillot (1991). Lowrie and Lanci (1994) found an observed declination of 13°, from which they inferred a clockwise rotation of 7°, and an observed inclination of 45° (11° shallower than expected, possibly owing to deposition and compaction of magnetic grains). The mean paleomagnetic direction of Massicore (declination = 9.5°, inclination = 39°, α_{95} = 2.5, and κ = 12.67) was observed to be similar to that of Massignano (Lanci et al., 1996).

Jovane et al. (2004) conducted a revised study of the paleomagnetism and rock magnetism of the section from 4 msl to 20 msl. They constructed a new stratigraphic log and collected unoriented rock samples every 5 cm for a total 304 samples. The environmental magnetism data set showed intervals characterized by high concentrations of magnetic minerals and variations in magnetic grain size and composition of the magnetic minerals unrelated to lithological changes. This pattern, which was related to climatic changes, had a surprising correspondence with other magnetic records from Antarctica, CIROS-1 (Jovane et al., 2004). The new magnetic records from Massignano, particularly the magnetic susceptibility, and the calcium carbonate content record were analyzed to see if they contained periodicities related to orbital oscillations. The variation of obliquity of Earth's axis was observed to be a strong driver of the climatic variations (Jovane et al., 2006). This permitted them to tune the section to the target astronomical curve (Laskar et al., 2004) and to calibrate the geomagnetic inversions recognized along the section. New magnetostratigraphic studies became necessary to carefully calibrate the magnetic reversal boundaries and other climatic and biostratigraphic events. Therefore, they collected 82 oriented samples from 0 msl to 13 msl. Previous magnetostratigraphic boundaries had been confirmed, refined, and calibrated using the astronomical calibration by Jovane et al. (2006, 2007a). The same authors, using statistical analyses that assess clustering of variables, showed that the magnetic variations along the section are related to a climatic switching on-and-off process that correlates directly to paleoceanographic changes (Jovane et al., 2007a). These findings magnify the importance on Massignano section not only as the GSSP for the Eocene-Oligocene boundary, but also as a sensitive recorder of abrupt and bimodal paleoceanographic changes in the Neotethys realm at the end of the Eocene.

Astronomical calibration of the Massignano section (Jovane et al., 2006) aids in refining sedimentation rates, relative ages of polarity chrons and estimates of the age of the Eocene/Oligocene boundary observed within the section. From tuning of the 400-k.y.-filtered susceptibility record from Massignano to the 400-k.y. period part of the target curve of Laskar et al. (2004), they found that the ages for the polarity chrons and the Eocene/Oligocene boundary in the Cande and Kent (1995) time scale fit the Massignano section well. Current estimates of the absolute ages of polarity chrons and the Eocene/Oligocene boundary, however, depend highly on radioisotopic ages of 33.9 to 34.6 Ma from volcanic ash within the Massignano section (Montanari et al., 1988) and a radioisotopic age of 34.6 ± 0.6 Ma from a biotite ash layer that is 4.3 m below the Eocene/Oligocene boundary in the Massignano section (Premoli Silva and Jenkins, 1993). Even with these constraints, the exact age of this boundary is still debated, although the last two major time scales have provided statistically indistinguishable ages of 33.7 ± 0.4 Ma (Cande and Kent, 1992, 1995) and 33.9 ± 0.1 Ma (Gradstein et al., 2004). Both use similar methods for estimation ages of polarity chrons within this time interval, which are based on fitting cubic splines through age versus distance data for marine magnetic anomalies in the South Atlantic, and both rely on the summary of age data provided in Odin et al. (1991). Cande and Kent (1992) specifically use the age of the Eocene/Oligocene boundary provided by Odin et al. (1991) as a tie point in their time scale, whereas Gradstein et al. (2004) (GTS04) use other ages provided by Odin et al. (1991). Gradstein et al. (2004) also mention the radioisotopic date of Premoli Silva and Jenkins (1993), but do not appear to use it as a tie point. Premoli Silva and Jenkins, (1993) had previously used their own radioisotopic date and a sedimentation rate of 0.71 cm/k.y. to extrapolate that the age of the Eocene/Oligocene boundary was ~34 Ma. On the basis of astronomical calibration (Jovane et al., 2006) and the new magnetostratigraphy, Jovane et al. (2007a) determined that the sedimentation rate oscillates from 0.6 to 0.8 cm/k.y. along the Massignano section. Linear interpolation between consecutive reversal boundaries, gives a mean sedimentation rate of 0.73 cm/k.y. With this value along with the age constraints from previous works, Jovane et al. (2007a) estimated the age of the Eocene/Oligocene boundary to be ~33.7 Ma, with an uncertainty that depends highly on the few absolute ages and which is surely >0.1 m.y. We use this age as our preferred estimate, although we note that elsewhere in this volume ages of 33.90 to 33.95 are used based on GTS04 (Brown et al., 2009; Hyland et al., 2009). Ultimately, the somewhat small but sometimes contentious differences in age estimates for the Eocene/Oligocene boundary will need to be resolved with additional radioisotopic ages and astronomical tuning, as discussed by Hilgen and Kruiper (2009, this volume).

GEOLOGICAL SETTING AND STRATIGRAPHY

The Massignano Eocene-Oligocene GSSP is located at 43°32′13″N, 13°35′36″E in an abandoned quarry, along the provincial road of the Cònero Regional Natural Park. The section is named after the nearby village of Massignano and is ~4 km northwest of the town of Sirolo, on the Adriatic coast of central Italy (Figs. 1 and 2). The section resides within the Monte Cònero headland, which represents the eastern and active thrusts

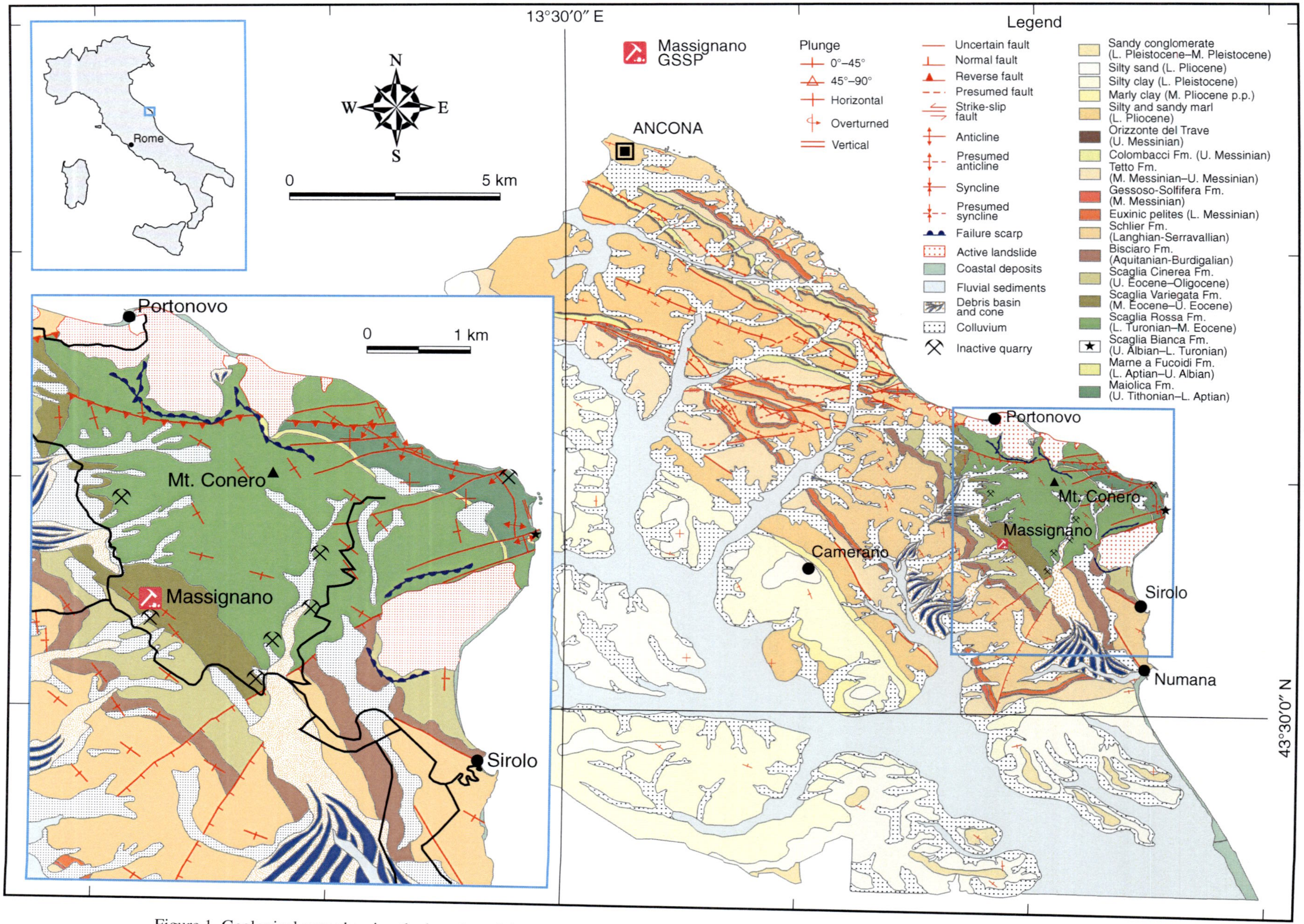

Figure 1. Geological map showing the location of the Massignano global stratotype section and point (GSSP) for the Eocene-Oligocene boundary on Mt. Conero structure (modified after Coccioni et al., 1997).

Figure 2. Outcrop photo of the Massignano Global Stratotype Section and Point (GSSP) taken in March 2008 with the meter stratigraphic levels (msl) from the base. The Eocene-Oligocene boundary occurs at 19 msl. For scale see also Figure 3.

of the northern Apennines of Italy (Chiarabba et al., 2005). As result of postdepositional tectonics, the Massignano beds are tilted 21° to the NW (strike N318°W).

The marine sediments of the Massignano section cover the uppermost Eocene to lowermost Oligocene and belong to the outer members of the Umbria-Marche sedimentary succession. This sedimentary succession was deposited in the Umbria-Marche Basin, which represents the northwestern sector of the Neotethys Ocean. This sector of the basin had been subsiding constantly from the Triassic to the Messinian (Marchegiani et al., 1999). During the end of the Eocene and the beginning of the Oligocene, the sediments of this zone of the Umbria-Marche Basin were deposited in the upper part of the lower bathyal setting, at a paleodepth of 1000–1500 m (Coccioni and Galeotti, 2003).

The Massignano section is composed of 23 m of sediments (Figs. 2 and 3) of the Umbria-Marche succession; the Scaglia Variegata Formation makes up the lower part of the section, and the Scaglia Cinerea Formation makes up the upper part. The Scaglia Variegata Formation is an alternation of reddish/greenish-gray marls and calcareous marls. This formation spans the top of the early Eocene to the end of the late Eocene. The Scaglia Variegata Formation is represented by three members: the lowest member consists of ~50 m of beds of alternated reddish violet-colored limestone layers, 15–100 cm thick, and violet marls layers, 20–200 cm thick, with intercalations of white marl layers; the middle member consists of ~20 m of marly limestones and yellow-green and gray marls layers that are 15–35 cm thick; the upper member consists of ~7 m of 15-cm-thick red calcareous marl layers. The Scaglia Variegata Formation turns into gray marls and 5–30-cm-thick marly limestones of the Scaglia Cinerea Formation at 12 msl (meters stratigraphic level from the base) (Coccioni et al., 1988). This formation, at Massignano, is deposited for the first 7 m in the Eocene and 4 m in the Oligocene.

The Massignano section spans the geological interval from the latest Eocene to the early Oligocene (e.g., Premoli Silva et al., 1988; Lowrie and Lanci, 1994; Lanci and Lowrie, 1997). The Eocene-Oligocene boundary occurs at a stratigraphic level of 19 msl, in the lower part of the Scaglia Cinerea Formation, (Premoli Silva et al., 1988; Premoli Silva and Jenkins, 1993). In the section, numerous biotite-rich layers related to volcanic events occur at 2.04, 5.25, 5.8, 6.5, 7.25, 7.75, 12.1, 12.7, 12.9, and 14.7 msl. Three iridium-rich layers occur at 5.61, 6.17, and 10.28 msl, and these have been interpreted as impactoclastic events (Montanari et al., 1988, 1993; Montanari and Koeberl, 2000; Bodiselitsch et al., 2004).

Biostratigraphically, the Massignano GSSP spans calcareous nannofossil zones NP18 to NP21 of Martini (1971) and CP15a to CP16 of Okada and Bukry (1980), planktonic foraminiferal zones P15 to P18 of Blow (1969, 1979) and Berggren et al. (1995) and zones E14 to O1 of Berggren and Pearson (2006), benthic foraminiferal zones BB4 to BB5 of Berggren and Miller (1989), and dinocyst zones *Melitasphaeridium pseudorecurvatum* (Mps) to *Areosphaeridium diktyoplokus* (Adi) of Brinkhuis and Biffi (1993) (Coccioni et al., 1988, 2000; Brinkhuis and Biffi, 1993; Spezzaferri

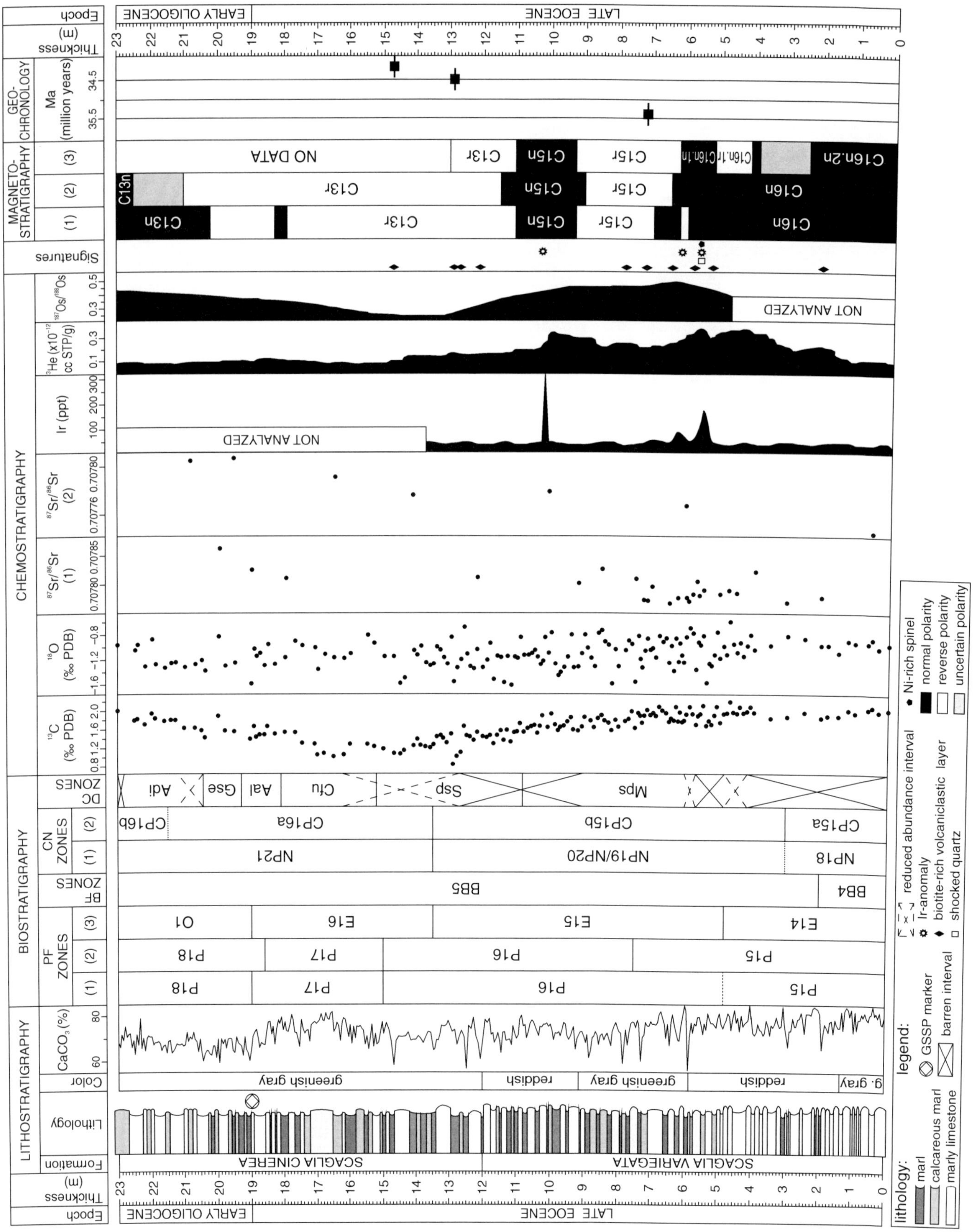
Epoch
Thickness (m)
GEO-CHRONOLOGY
Ma (million years)
MAGNETO-STRATIGRAPHY
Signatures
CHEMOSTRATIGRAPHY
$^{187}Os/^{188}Os$
^{3}He (x10^{-12} cc STP/g)
Ir (ppt)
$^{87}Sr/^{86}Sr$ (2)
$^{87}Sr/^{86}Sr$ (1)
$\delta^{18}O$ (‰ PDB)
$\delta^{13}C$ (‰ PDB)
BIOSTRATIGRAPHY
DC ZONES
CN ZONES
BF ZONES
PF ZONES
LITHOSTRATIGRAPHY
$CaCO_3$ (%)
Color
Lithology
Formation
EARLY OLIGOCENE
LATE EOCENE
NO DATA
NOT ANALYZED
C13n
C13r
C15n
C15r
C16n
C16n.1n
C16n.1r
C16n.2n
Adi
Gse
Aal
Cfu
Ssp
Mps
CP16b
CP16a
CP15b
CP15a
NP21
NP19/NP20
NP18
BB5
BB4
O1
E16
E15
E14
P18
P17
P16
P15
greenish gray
reddish
g. gray
SCAGLIA CINEREA
SCAGLIA VARIEGATA
lithology:
marl
calcareous marl
marly limestone
legend:
GSSP marker
barren interval
reduced abundance interval
Ir-anomaly
biotite-rich volcaniclastic layer
shocked quartz
Ni-rich spinel
normal polarity
reverse polarity
uncertain polarity

Figure 3. Synoptic scheme of the integrated litho-, bio-, chemo-, and magnetostratigraphy, and geochronology of the Massignano global stratotype section and point (GSSP) for the Eocene-Oligocene boundary. Lithostratigraphy is after Coccioni et al. (1988), Jovane et al. (2004), and this paper. Biostratigraphy is after Coccioni et al. (1988), Brinkhuis and Biffi (1993), Monechi et al. (2000), Spezzaferri et al. (2002), Coccioni and Galeotti (2003), and Van Mourik and Brinkhuis (2005). Different biozonal schemes according to their authors are reported. Planktonic foraminifera (PF): Blow (1969, 1979) (1), Berggren et al. (1995) (2), and Berggren and Pearson (2006) (3). Benthic foraminifera (BF): Berggren and Miller (1989). Calcareous nannoplankton (CN): Martini (1971) (1) and Okada and Bukry (1980) (2). Dinoflagellate cysts (DC): Brinkhuis and Biffi (1993). Chemostratigraphy: Carbon and oxygen isotope records are after Bodiselitsch et al. (2004); strontium isotope record is after Vonhof et al. (1998) (1) and Montanari et al. (1991) (2); iridium concentration record is after Montanari et al. (1993); extraterrestrial ^{3}He concentration record is after Farley et al. (1998); osmium isotope record is after Ravizza and Peucker-Ehrenbrink (2003). Signatures after Coccioni et al. (1988), Montanari et al. (1988, 1993), Clymer et al. (1995, 1996), Langenhorst and Clymer (1996), Pierrard et al. (1998), and Jovane et al. (2004, 2007a). Magnetostratigraphy is after (1) Bice and Montanari (1988), (2) Lowrie and Lanci (1994), and (3) Jovane et al. (2007a). Geochronology is after Montanari et al. (1988) and Odin et al. (1988b). PDB—Peedee belemnite.

et al., 2002; Monechi et al., 2000) (Fig. 3). The Eocene-Oligocene boundary occurs at 19 msl where both the planktonic foraminiferal Hantkenina and Cribrohantkenina, Eocene genera of the Hantkeninidae, become extinct (Coccioni et al., 1988; Nocchi et al., 1988; Premoli Silva and Jenkins, 1993).

EOCENE-OLIGOCENE GLOBAL CLIMATE CHANGES AND A NEW HYPOTHESIS

A combination of different factors is responsible for the variations in the amount and distribution of solar radiation, and heat transport over Earth's surface during the greenhouse-icehouse transition. Stable isotope records in the deep sea show that the polar regions and the global climate became colder over the past ~50 m.y. following the early Eocene climatic optimum (EECO) (e.g., Miller et al., 1991, 2005a, 2005b; Zachos et al., 2001; Lear et al., 2000; Coxall et al., 2005; Edgar et al., 2007). This cooling did not occur as a single gradual change, but was composed of a series of distinct cooling steps or intervals within a general cooling trend, which included a number of shorter term changes to relatively warm or cool conditions (e.g., Shackleton and Kennett, 1975; Kennett and Shackleton, 1976; Kennett and Stott, 1990; Miller et al., 1991; Brinkhuis, 1992; Zachos et al., 1994, 1996; Diester-Haass et al., 1996; Ehrmann, 1997; Vonhof et al., 2000; Quilty, 2001; Zachos et al., 2001; Persico and Villa, 2004; Scher and Martin, 2004; Jovane et al., 2004, 2007a, 2007b; Miller et al., 2008; Katz et al., 2008; Billups, 2008; Ivany et al., 2008; Merico et al., 2008). An important global event at ca. 40 Ma (Jovane et al., 2007b) was the middle Eocene climatic optimum (MECO), which caused a transient rise in pCO_2 levels (Bohaty and Zachos, 2003). During the Eocene, overall sea-surface temperatures were warm (e.g., Pearson and Palmer, 2000; Pearson et al., 2001, 2007), even though short and abrupt cooling events and phases were recorded from different climatic proxies at the end of this period, marking the initial stages toward modern thermohaline circulation (e.g., Thomas and Gooday, 1996; Sloan and Rea, 1996; Thomas et al., 2003; Via and Thomas, 2006).

Significant faunal and floral turnover has been recognized in both marine and terrestrial biota corresponding to global oceanographic circulation reorganization (Prothero and Berggren, 1992; Prothero, 1994; Prothero et al., 2003). The most significant climatic event in the long-term cooling trend of the Cenozoic era appears to be the oxygen isotope Oi-1 cooling event at ca. 33.5 Ma (e.g., Miller et al., 1991), which occurred shortly after the Eocene-Oligocene boundary. This event was part of an abrupt increase in $\delta^{18}O$, reflecting a combination of global cooling and ice growth in Antarctica (e.g., Shackleton and Kennett, 1975; Zachos et al., 2001; Coxall et al., 2005: Zachos and Kump, 2005; Lear et al., 2000, 2004, 2008), which have been correlated to changes in atmospheric CO_2 concentration (DeConto and Pollard, 2003; Pagani et al., 2005; Lyle et al, 2007) or, otherwise, to opening of Southern Ocean gateways (e.g., Huber and Sloan, 2001; Lawver and Gahagan, 2003; Barker et al., 2007; Livermore et al., 2005, 2007).

A corollary component to this global climatic transition has recently been proposed by Jovane et al. (2007a, 2007b) based on Eocene data from the Neotethys sector. The hypothesis, referred to as the STENT current hypothesis, suggests that the closing of the Neotethys gateway, which occurred as a result of local and regional tectonics, in combination with local, regional, or global sea-level changes, could represent the trigger or one of the causes that drove the global variable climatic condition.

MATERIALS AND METHODOLOGIES

A lithological log is presented for the Massignano section in Figure 3, which elaborates on the details of the section from 0 msl to 4 msl and from 20 msl to 23 msl, intervals which were not covered in Jovane et al. (2004, 2007a). In those segments, we took one sample every 5 cm (140 total samples) after removing weathered surface from the outcrop.

We followed the procedures outlined in detail in Jovane et al. (2007b) in order to have compatible data sets. Summarizing, each sample was weighed to allow mass normalization of studied properties. We measured the low-field magnetic susceptibility using a KLY-2 (AGICO) Kappabridge susceptibility-meter with an operating frequency of 920 Hz and a magnetic induction of 0.4 mT (noise level 2×10^{-10} m^3 kg^{-1}). Anhysteretic remanent magnetization (ARM) and isothermal remanent magnetization (IRM) were measured using an automated pass-through 2G Enterprises cryogenic magnetometer (noise level 3×10^{-9} A m^2 kg^{-1}). The ARM was imparted in a 100 mT peak alternating field with a 0.05 mT biasing field. The IRM was imparted in a 1000 mT field and then in backfield to 300 mT ($BIRM_{300}$).

Calcium carbonate content measurement was performed using the "Dietrich-Fruhling" calcimeter. The method is based on the measurement of CO_2 volume developed by chloridric acid reacting with the powdered rock. The total carbonate present is obtained using formulae that take into account the pressure, temperature, amount of previously weighed sediment, and the volume of CO_2 developed. Standards were measured every 20 samples to verify the appropriate calibration.

RESULTS

In the Massignano GSSP, the $CaCO_3$ content fluctuates widely from 56% to 92%, with a mean value of ~76%, which decreases slightly upward in the sequence. The greatest amplitude of fluctuations occurs close to the boundary between the Scaglia Variegata and Scaglia Cinerea Formations. The minima in calcium carbonate content are related to impactoclastic and volcaniclastic layers. No clear relationship exists between the calcium carbonate content and color of the sediments. There is a small long-term correlation between the magnetic mineral concentration zones and the $CaCO_3$ profile. Moreover, in the higher frequencies, there is also an opposite relation (anticorrelation) (see long-term eccentricity and obliquity in Figure 7 of Jovane et al., 2006).

Environmental magnetic parameters (concentration, composition, and dimension of magnetic grains) were defined studying ARM, IRMs, and indirect parameters from them. ARM magnetically excites only finer magnetic minerals, while IRM excites all of them (concentration). Consequently, the ratio between them represents magnetic grain size (dimension). Backfield magnetizations permit us to define HIRM ($HIRM_{300} = [IRM + BIRM_{300}]/2$) and S-ratio ($S_{300} = [BIRM_{300}/IRM]$), which gives information about the coercivities (composition) of the magnetic minerals.

The sediments of the Massignano GSSP contain abundant and generally well-preserved microfossils, including calcareous nannofossils, planktonic and benthic foraminifera, dinoflagellate cysts, prasinophyte algae, bisaccate pollen, and ostracods (Coccioni et al., 1988; Bellagamba and Coccioni, 1990; Brinkhuis and Biffi, 1993; Coccioni and Galeotti, 2003; Dall'Antonia et al., 2003). Reworking is limited to the intense bioturbation by *Chondrites* and *Planolites*, which, however, does not affect the biostratigraphic resolution. Calcareous nannofossils and planktonic foraminifera, with diverse genera and species typical of late Eocene–early Oligocene low latitude, are the main components of the microfossil assemblages. The nannofloral content is also typical of open-ocean assemblages. The palynological samples are generally rich and dominated by dinoflagellate cysts and prasinophyte algae. The palynological record is, however, discontinuous within the Scaglia Variegata Formation. The ostracod fauna shows fairly low abundance and diversity throughout the section, and assemblages display a close affinity with the coeval deep faunas of the North Atlantic. Several well-differentiated and sequential bioevents have been identified in the Massignano GSSP and are shown in Figures 3 and 4 and Table 1.

Rock-magnetism analyses for the intervals 0–4 msl and 20–23 msl are combined with those obtained in Jovane et al. (2004, 2007a) in Figures 4 and 5. Our new results confirm that in the Scaglia Variegata Formation, the main magnetic carrier is a complex mixture of magnetite, maghemite, hematite, and trace of goethite. On the contrary, the Scaglia Cinerea Formation is characterized by low concentrations of high coercivity minerals, apart from a lower interval where the hematite contribution is more abundant.

As noted by Jovane et al. (2007a) from a statistical analysis, the magnetic properties primarily alternate between two long-term modes throughout the Massignano section that are not directly related to lithostratigraphic variations. One mode is characterized by higher magnetic mineral concentration, finer magnetic grain size, and higher coercivities, which are associated with compositions that are mainly magnetite with some hematite. The other mode is characterized by lower magnetic mineral concentration, coarser grain sizes, and lower coercivities, which are associated with magnetic compositions that are completely dominated by pseudo–single-domain (PSD) to multidomain (MD) magnetite. We refer to the high-concentration, high-coercivity, and fine-grained magnetic mode as the HHFM mode and the low-concentration, low-coercivity, and coarse-grained magnetic mode as the LLCM mode (Figs. 4 and 5).

Based on the alternations between these modes, we divided the Massignano section into seven zones (Figs. 4 and 5). Six of these zones were already presented in previous works (Jovane et al., 2004, 2007a), with the upper zone now extending to the top of the section. We also added a new zone in this study for the stratigraphic section below 2.4 msl, which has properties consistent with the LLCM mode. The magnetic minerals, at finer scale, show a relation with short-term lithological alternation.

The presence of a new zone above the cooling event coinciding with the iridium-rich layer confirms that these magnetic variations are driven by climatic changes and are not directly related to any global impactoclastic abrupt change. The finding of another new zone at the bottom of the section reinforces the ~400 k.y. cycle in magnetic properties already proposed by Jovane et al. (2004).

We suggest that the bimodal trend in the climatic proxies is related to an on-and-off switching process in the paleoceanographic setting in the western Neotethys. The new data set from the top of the section shows that susceptibility decreases consistently after the Eocene-Oligocene boundary at 19 msl.

NEOTETHYS PALEOGEOGRAPHY

Our definition of the Neotethys Ocean follows Dercourt et al. (1986) as the ocean dividing Africa and Arabia to the south from Eurasia to the north. The suffix Neo (new) is used to differentiate the Neotethys of the Alpine-Himalayan orogeny from the Paleotethys of the Hercynian orogeny.

Before the Alpine-Himalayan orogeny and the opening of the Red Sea, the Neotethys Ocean was a wide ocean separat-

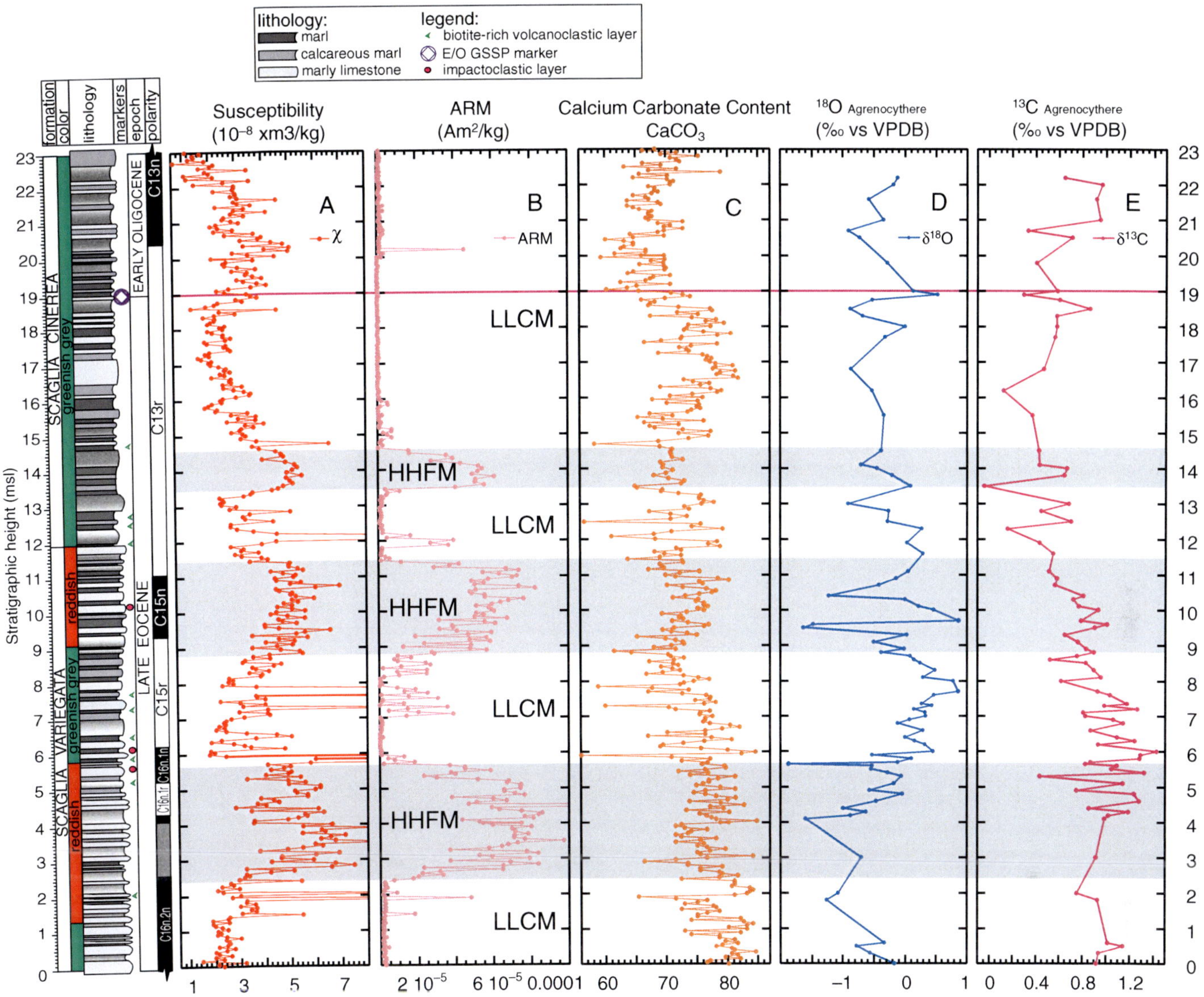

Figure 4. Lithostratigraphic log and impactoclastic and volcaniclastic layers shown with (A) low-field magnetic susceptibility (χ), (B) anhysteretic remanent magnetization (ARM), (C) calcium carbonate content ($CaCO_{3\%}$), (D) oxygen isotopes, and (E) carbon isotopes for the global stratotype section and point (GSSP) for the Eocene-Oligocene boundary. Numerical ages are from Cande and Kent (1995). HHFM—high concentration, high coercivity, and fine-grained magnetic; LLCM—low concentration, low coercivity, and coarse-grained magnetic (modified from Jovane et al., 2004). VPDB—Vienna Peedee belemnite.

ing the Eurasian plate and the African-Arabian plate (Dercourt et al., 1986, 1993, 2000; Popov and Stolyarov, 1996; Rögl, 1996, 1998; Gaetani et al., 2003; Popov et al., 1993, 2004). During this period, the Italian offshore platforms and basins were continually subsiding in a tropical realm, which permitted the deposition of hundreds of meters of Cenozoic carbonate sediments (Colacicchi and Passeri, 1981; Marchegiani et al., 1999). The Neotethys Ocean once extended from the Southern Europe of today to North African regions and from Turkey to the Himalayan Chain.

During the Cretaceous period, a Tethyan Circumglobal Current (TCC) flowed westward throughout the oceans at subtropical latitudes (Luyendyk et al., 1972, Oberhänsli and Hsü, 1986; Barron, 1987; Barron et al., 1993, 1995; Premoli Silva et al., 1995; Stanley, 1995; Johnson et al., 1996; Bush, 1997; Poulsen et al., 1998, 2001). In the same period, clockwise gyres with an eastward flow circulated in the northern margin of the Tethys (Barron and Peterson, 1989a, 1989b, 1990, 1991). This massive current was able to warm the polar regions and cool the tropical regions (e.g., Hotinski and Toggweiler, 2003). Subsequently, during the Paleogene and early Eocene periods, the Neotethys continued to be an important thermal connection between the Indian and the Atlantic Oceans (Orue-Extebarria and Lamolda, 1985; Oberhänsli and Hsü, 1986; Harzhauser et al., 2002; Popov et al., 2004). However, the paleoceanographic setting for the Eocene period is still

TABLE 1. STRATIGRAPHIC POSITION (METERS STRATIGRAPHIC LEVEL) OF THE PLANKTONIC AND BENTHIC FORAMINIFERA, CALCAREOUS NANNOFOSSIL, AND DINOFLAGELLATE CYST EVENTS IN THE MASSIGNANO GSSP

Bioevents	Height in section (msl)	Astronomical age (Ma)
Planktonic foraminifera		
LO *Globigerinatheka semiinvoluta* (Spezzaferri et al., 2002)	4.8	35.52
LO *Turborotalia pomeroli* (Coccioni et al., 1988)	5	35.495
FO *Cribrohantkenina inflata* (Coccioni et a., 1988)	5.8	35.39
FO *Turborotalia cunialensis* (Coccioni et al., 1988)	7.5	35.184
LO *Globigerinatheka luterbacheri* (Coccioni et a., 1988)	12.9	34.482
LO *Globigerinatheka index* (Coccioni et al., 1988)	13.5	34.4
LO *Cribrohantkenina inflata* (Coccioni et al., 1988)	15	34.22
LO *Turborotalia cocoaensis, Turborotalia cunialensis* (Coccioni et al., 1988)	18.6	33.757
LO *Hantkeninidae* (Coccioni et al., 1988)	19	33.703
Benthic foraminifera		
FO *Cibicidoides pseudoungerianus* (Coccioni and Galeotti, 2003)	0	
LO *Cibicidoides subhaidingerii* (Coccioni and Galeotti, 2003)	1.5	
LO *Cibicidoides truncanus* (Coccioni and Galeotti, 2003)	2	
LCO *Nuttallides truempyi* (Coccioni and Galeotti, 2003)	2.5	
LO *Nuttallides truempyi* (Coccioni and Galeotti, 2003)	9	34.994
FO *Cibicidoides cookei* (Coccioni and Galeotti, 2003)	23	
Calcareous nannofossils		
FCO *Isthmolithus recurvus* (Tori and Monechi, 2007)	3	
FO *Reticulofenestra oamaruensis* (Monechi et al., 2000)	5.5	35.431
LO *Cribrocentrum reticulatum* (Tori and Monechi, 2007)	9.2	34.968
LO *Cyclococcolithina kingii* (Coccioni et al., 1988)	9.95	34.871
LCO *Calcidiscus protoannulus* (Tori and Monechi, 2007)	11.6	34.652
LO *Chiasmolithus oamaruensis* (Coccioni et al., 1988)	12.7	34.509
LO *Discoaster barbadiensis, Discoaster saipanensis* (Coccioni et al., 1988)	13.6	34.388
Acme *Clausicoccus* (Tori and Monechi, 2007)	21.5	
Dinoflagellate cysts		
LO *Melitasphaeridium pseudorecurvatum* (Brinkhuis and Biffi, 1993)	8.7	35.035
LO *Schematophora speciosa* (Brinkhuis and Biffi, 1993)	14.7	34.253
FO *Achomosphaera alcicornu* (Brinkhuis and Biffi, 1993)	18.3	33.798
FO *Glaphyrocysta semitecta* (Brinkhuis and Biffi, 1993)	19.5	33.634
LO *Hemiplacophora seminulifera* (Brinkhuis and Biffi, 1993)	20.4	
Ostracods		
FO *Ambocythere* sp. 1 (Dall'Antonia et al., 2003)	1.8	
FO Genus et sp. indet. 1 (Dall'Antonia et al., 2003)	4.1	35.606
FO *Dutoitella eocenica* (Dall'Antonia et al., 2003)	4.3	35.583
FO *Legitimocythere presequenta* (Dall'Antonia et al., 2003)	6.3	35.329
LO *Ambocythere* sp. 1 (Dall'Antonia et al., 2003)	20.5	

Note: Astronomical ages are based on Jovane et al. (2006). Abbreviations: msl—meters stratigraphic level; FO—first occurrence; LO—last occurrence; FCO—first common occurrence; LCO—last common occurrence.

ambiguous (e.g., Barron, 1987; Dercourt et al., 2000; Huber and Sloan, 2001; Hotinski and Toggweiler, 2003; Scher and Martin, 2004). After the Eocene, this trans-Atlantic relationship ended, as evidenced by fossil assemblages that changed from cosmopolitan to endemic (Barron and Peterson, 1991; Premoli Silva et al., 1995; Harzhauser et al., 2002; Lawver and Gahagan, 2003) as the sea level dropped in the southeastern part of the Neotethys Sea (Dercourt et al., 2000). Middle and late Eocene stable isotopes from Italian outcrop (Jovane et al., 2007a, 2007b) show variation comparable to those once found in various deep-ocean cores (Bohaty and Zachos, 2003; Zachos et al., 1996; Diester-Haass and Zahn, 1996; Diester-Haass et al., 1996), signifying that the Neotethys was directly in communication with other oceans.

Major paleoceanographic changes have been recorded in biogeographic proxies from the Atlantic sector (Orue-Extebarria and Lamolda, 1985) to the Himalaya-Tibetan sector (Rögl, 1996, 1998; Popov and Stolyarov, 1996; Popov et al., 1993, 2004). These changes are synchronous to the Eocene-Oligocene boundary in the Eurasian configuration (Schulz et al., 2005).

India's rapid northward motion slowed abruptly around 57 Ma (Acton, 1999), which has been interpreted to have resulted from India colliding with an intraoceanic island arc, and subsequent continental-continental collision occurring at ca. 34 Ma (Aitchison et al., 2007). Arabia-Eurasia continental collision probably was under way before the Oligocene (23–25 Ma) and slightly after the late Eocene (35 Ma); associated precursors perhaps started with Campanian-Maastrichtian ophiolite obduction onto the Arabian passive continental margin (e.g., Yilmaz, 1993; Sharland et al., 2001; Blanc et al., 2003; Allen et al., 2004; Agard et al., 2005; Allen and Armstrong, 2008). Alternatively, Hessami

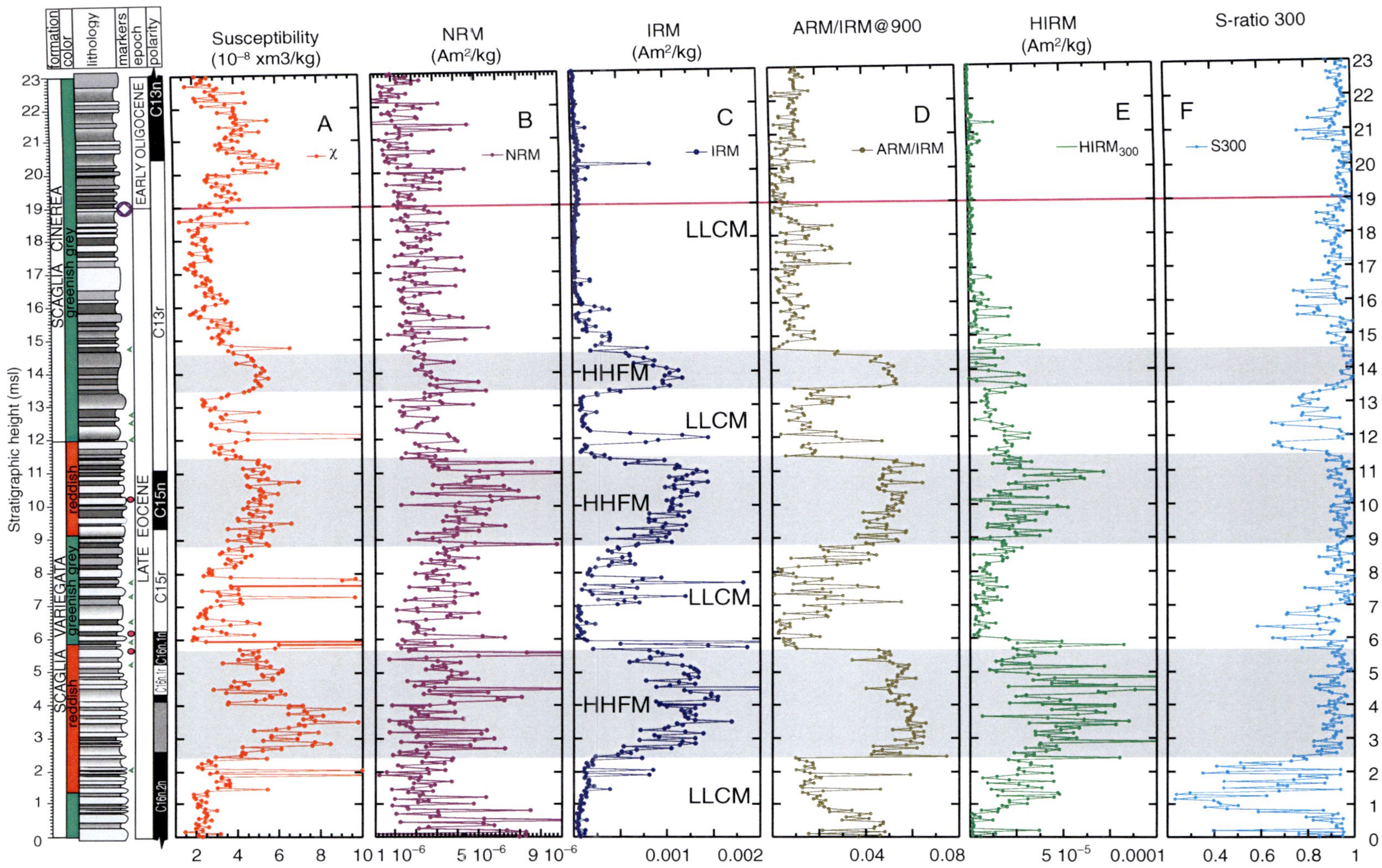

Figure 5. Lithostratigraphic log and impactoclastic and volcaniclastic layers shown with (A) low-field magnetic susceptibility, (B) natural remanent magnetization (NRM), (C) isothermal remanent magnetization (IRM), (D) ARM/IRM@900, (E) HIRM300, and (F) S-ratio$_{300}$. Numerical ages are from Cande and Kent (1995). HHFM—high concentration, high coercivity, and fine-grained magnetic; LLCM—low concentration, low coercivity, and coarse-grained magnetic (modified from Jovane et al., 2004).

et al. (2001) and Hempton (1987) have suggested that the first collision occurred in the late Eocene, while the definitive collision is dated to 14–10 Ma (Bernard et al., 2006). The Arabia-Eurasia collision extends from the Aegean Sea to Iran and between the northern side of the Caucasus–Kopet Dagh and the Persian Gulf, and it is mostly concentrated in present-day Turkey and Iran (Allen et al., 2004). The southern part of the eastern Neotethys, closed by the Arabian plate subduction under Eurasia, represents several microcontinents now preserved within the Zagros belt (Blanc et al., 2003; McClay et al., 2004; Agard et al., 2005). Hence, seaways existed between Eurasia and Africa-Arabia and between Eurasia and India that provided pathways for currents from the Pacific and Indian Oceans into the Atlantic Ocean, and their closure continued until about the Eocene-Oligocene boundary.

The Italian carbonate sections for the middle Eocene to early Oligocene time interval, in particular, the pelagic carbonates of the Scaglia Rossa, Scaglia Variegata, and Scaglia Cinerea Formations of the Umbria-Marche Basin, represent the northern margin of the western Neotethys. Due to the continuity, preservation, and correlation, these sections are considered type sections for this interval (Rio et al., 2003). As such, they are well situated geographically and temporally to provide evidence of changes in the Neotethys current as well as to record the climatic transition from greenhouse to icehouse conditions.

DISCUSSION—THE SUBTROPICAL EOCENE NEOTETHYS (STENT) CURRENT HYPOTHESIS

Variations in sea-bottom ventilation and productivity have been noted in various studies of the Umbria-Marche Basin (Coccioni et al., 2000; Dall'Antonia et al., 2003; Coccioni and Galeotti, 2003; Verducci and Nocchi, 2004; Jovane et al., 2004, 2007a, 2007b) from the middle Eocene to the early Oligocene time interval. These variations have been interpreted as arising from changes in the paleoceanographic conditions of the Neotethys during this period.

Evidence that the Umbria-Marche Basin represents global importance for understanding the climate changes during the Eocene was presented in a recent study on magnetic properties and stable isotopes. On the Contessa Highway in the middle Eocene section, a distinct $\delta^{13}C$ positive shift of ~0.6‰ with duration of ~600 k.y. was found extending from the top of chron C18r through to subchron C18n.2n (Jovane et al., 2007b), comparable to the middle Eocene climatic optimum stable isotope anomaly. The occurrence of the middle Eocene climatic optimum event in the Mediterranean-Tethys region deserves recognition because of its implications for global climate warming and its underlying causes.

At Massignano, the stable isotope and environmental magnetism data provide a complex history of the evolution of the bottom water ~3.0 m.y. preceding the global cooling event near the Eocene-Oligocene boundary (Jovane et al., 2007a). The $\delta^{13}C$ and $\delta^{18}O$ values of the Massignano sediments show alternations between relatively sluggish deep circulation and incursions of cooler oceanic water and/or increasing surface productivity (Jovane et al., 2007a). Jovane et al. (2004, 2007a, 2007b) argued that the presence, abundance, and composition of the magnetic minerals are related to paleoclimatic events that occurred in the deep-sea environment. The environmental magnetism record contains a statistically significant bimodal fluctuation (Jovane et al., 2007a) between the LLCM mode and the HHFM mode, which we propose was related to changes in the Neotethys current. Superimposed on a constant low concentration of medium- to coarse-grained magnetite are pulses of layers rich in finer-grained magnetite and hematite. An on-off switching of an environmental marine process controlled this particular pattern. We suggest the process was related to paleoclimatic changes that could arise from a climatic modulation of transport mechanisms of the sediments induced by (1) global sea-level changes and/or (2) tectonic closing of the Neotethyan Ocean. In addition, a further explanation may be that the long-term variations in magnetic properties are linked to variations of sea-bottom ventilation.

The magnetic material in a carbonate depositional environment is commonly influenced by detrital sources originating generally from nearby landmasses, although other contributions have been considered (Verosub and Roberts, 1995). More broadly, the magnetic minerals could originate in ocean sediments from biogenic sources as the results of (1) precipitation, (2) dehydration, (3) crystallization as authigenic oxides (e.g., magnetite) and as hydroxides (e.g., goethite and hematite), or from (4) chemical or microbial reduction (anoxic basin and fecal pellets) (Sunda, 2001). Thus, we suggest that the presence of fine hematite-rich sediments is an expression of high concentration of iron in the basin where the sediments were deposited. This incidence is also related to high biogenic productivity. In the Neotethys, during the Eocene, layers rich in fine hematite occur, and there are changes in deep-sea conditions, productivity, and freshening (Coccioni et al., 2000; Dall'Antonia et al., 2003; Harzhauser et al., 2002; Coccioni and Galeotti, 2003; Verducci and Nocchi, 2004; Jovane et al., 2004, 2007a, 2007b). Thus, due to the variety of potential sources, a better understanding of the mechanism that drives the presence of magnetic minerals is needed in order to explain the abundance of fine magnetic minerals in certain periods.

At present, iron dissolved in seawater is very low (0.05 nM near surface), yet it represents one of the most important controls of bioproductivity (e.g., de Jong et al., 1998; de Baar et al., 1995; de Baar and de Jong, 2001; de Baar et al., 2005; Blain et al., 2007; Cassar et al., 2007). The Antarctic Circumpolar Current (ACC) transports high quantities of dissolved iron for thousands of kilometers, reaching values of 9.7 nM (Löscher et al., 1997). The origin of this iron is dust coming from southern South America (Gaiero, 2007) that is captured by the sea surface in the Drake Passage area (de Baar and de Jong, 2001). The dissolved iron usually has a very short residence time in seawater. However, the ACC keeps it in suspension for months on the surface and for years in the bottom water. These bottom waters rich in iron are the main cause of bioproductivity wherever there is upwelling (de Baar and de Jong, 2001). Vivier et al. (2004) showed that there are important similarities between variations in the ACC model and a simulated satellite magnetic anomaly. Another idea is the one of Cather et al. (2008, personal commun.) who hypothesized that dissolved iron from ignimbrite eruptions might be linked to icehouse periods. However, we maintain that this present day iron-ACC association could be applied to historic global currents.

Changes in the circulation of the Neotethys were directly related to the tectonic evolution of the connection between the Indian and Atlantic Oceans and the Central American gateway. The General Circulation Model (GCM) simulation and Nd isotope record show that the location of deep-water formation in the middle Eocene occurred at a low latitude in the Neotethys and North Atlantic Oceans (as consequence of the high salinities and runoff) (e.g., Barron and Peterson, 1991; Bice et al., 1997; Huber and Sloan, 2001; Scher and Martin, 2004; Via and Thomas, 2006). The asymmetry of ocean heat transport and a meridional overturning throughout the Eocene testify that there was high southward heat transport that decreased during the progressive closure of the gateways (Bice et al., 2000). Neodymium isotopes in fish teeth show that another source of deep water in the middle Eocene was the Southern Ocean, which increased its importance around ca. 40.8 Ma (Scher and Martin, 2004; see discussion in Jovane et al., 2007b). Using the same technique, Via and Thomas (2006) showed that the beginning of the Atlantic thermohaline circulation and the onset of deep-water formation in the Norwegian-Greenland Sea (North Atlantic had just opened) occurred in the early Oligocene. Dercourt et al. (2000) maintained that during the Lutetian (46–40 Ma), the Tethyan shallow current was still flowing from west to east, though there was a moderate decrease in its flux. In contrast, Lawver and Gahagan (2003) inferred that a westward current through the Tethys presumably ended in the early Eocene. As stated already, recent tectonic reconstructions indicate that the Arabian-Eurasian gateway remained open during the Eocene, and it closed in the Oligocene.

Therefore, we propose that a warm subtropical Eocene Neotethys (STENT) current continued throughout all the oceans in a westward direction (Fig. 6). The STENT was still able to maintain temperate high latitudes, similar to the TCC in the Cretaceous, but it was weakening in stream while increased formation of deep-water began in the North Pacific (Thomas, 2004), Southern Ocean (Scher and Martin, 2004), and North Atlantic (Via and Thomas, 2006). The STENT was rich in dissolved iron from the presence of several sources of dust and runoff material. The closure of the Neotethys Ocean should have caused a dramatic change in the paleoceanography and in the bioproductivity processes, causing serious repercussion in the global CO_2 budget. However, this event did not happen abruptly; rather, it consisted of several alternations between periods when the STENT was able to pass through the Arabian-European gateway and phases when the current was obstructed. The intensity of this current would have been influenced by sea-level changes and climatic variation that were induced by astronomical oscillations. Paleontologic and geochemical evidence (Premoli Silva et al., 1995; Coccioni et al., 2000; Dall'Antonia et al., 2003; Harzhauser et al., 2002; Coccioni and Galeotti, 2003; Verducci and Nocchi, 2004; Jovane et al., 2004, 2007a, 2007b) confirms these alternations during the end of the Eocene. Fluctuations in primary productivity, several episodes of biotic turnovers, variations in magnetic properties and changes in other climatic proxies are the main evidences.

Additional sedimentological, biostratigraphic and magnetostratigraphic analyses are required in the Neotethys area to further elucidate whether these events were synchronous. Current studies have not yet confirmed the synchronous occurrence of fluctuations in magnetic properties (Jovane et al., 2004, 2007a, 2007b) with inferred paleoproductivity (Coccioni and Galeotti, 2003), paleoceanographic and paleoclimatic variations (Coccioni et al., 2000; Dall'Antonia et al., 2003; Verducci and Nocchi, 2004) and the deposition of red beds (Hu et al., 2005) in the Umbria-Marche Basin and Neotethys sector. Oceanic drilling in the Indian Ocean has not yet provided any high-resolution magnetic or paleoclimatic data sets, but an important rise in magnetic properties is recorded in Ocean Drilling Program (ODP) Holes 738B and 744A at 36.2 Ma (Jovane, 2006). Assuming the presence of the STENT current, this occurrence could be related to the southward contortion of the STENT in a wide counterclockwise Indian Ocean gyre during the closure of the Neotethys and/or to climatic variations occurring in Antarctica during establishment of the ACC (Barker et al., 2007). Gourlan et al. (2008) confirmed that at 30 Ma, a shallow-water, counterclockwise-current gyre had developed in the Indian Ocean.

Synthesizing the currently available data obtained from the Massignano section, we hypothesize that the paleoceanography of the Neotethys was characterized by alternating periods of low productivity, representing times when deposition was driven by local terrigenous material (expressed as coarser magnetite only), and periods of high productivity and freshening (expressed as abundant fine-grained magnetite plus hematite layers and coarser magnetite in the background). The high productivity and freshening periods represent times when the STENT was able to pass through the Arabia-Eurasian gateway, bringing large amounts of fresh and deep-sea waters rich in dissolved iron. This "gateway" hypothesis may provide a parallel explanation for the climatic fluctuations at the end of the Eocene and the global climatic Eocene-Oligocene instability before icehouse conditions prevailed. The STENT current hypothesis, in relation to the closure of the Neotethys Ocean, could have played an important or subordinate role for the reduction of atmospheric CO_2 following the mechanisms of Allen and Armstrong (2008). Actually, the STENT current occurred coeval with the following major paleogeographical events: (1) the coupling of the opening of the Australia-Antarctic and Drake Passage (Lawver and Gahagan, 2003) with the isolation of Antarctica, (2) the beginning of the Arabian-Eurasian continent-continent collision, expressed as arc magmatism, the (3) establishment of the Alpine-Himalayan chain and its subsequent aridification (Dupont-Nivet et al., 2007), and (4) low-latitude silicate weathering (Raymo and Ruddiman, 1992) in relation to previous collisional event.

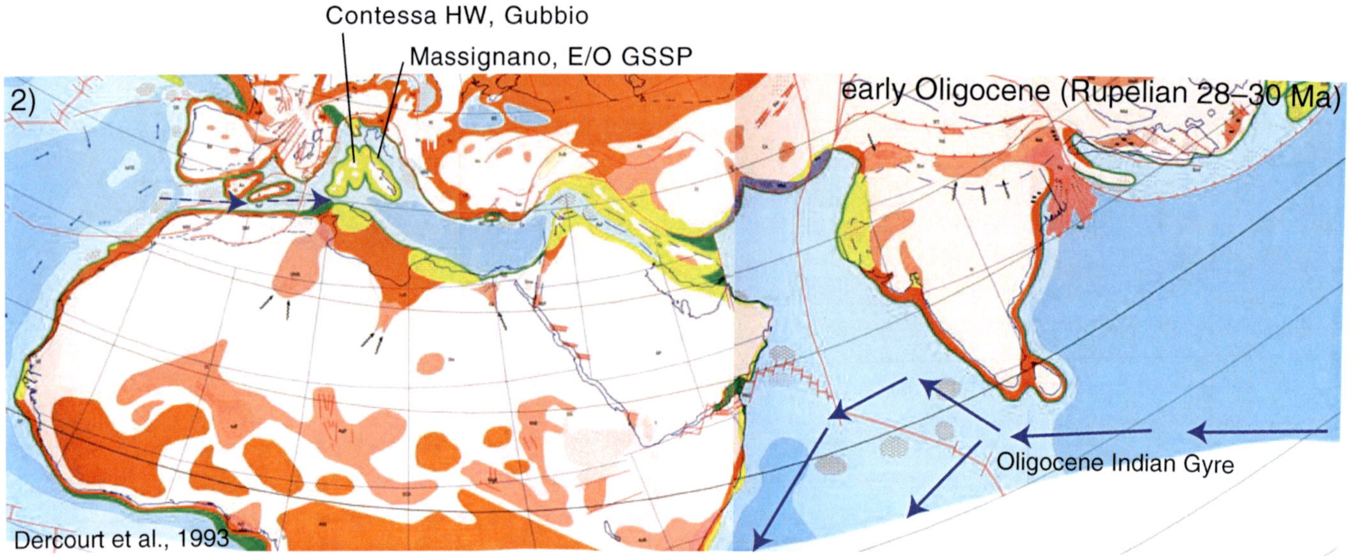

Figure 6. Eocene-Oligocene (Priabonian-Rupelian) paleogeography reconstruction of Dercourt et al. (1993) and blue arrows showing the circulation of the subtropical Eocene Neotethys (STENT) current (1) when the Arabian-Eurasia gateway was open and (2) when it was closed. The STENT current is interpreted to have brought high amounts of dissolved iron, which is supported by the occurrence of layers rich in fine-grained magnetite and hematite. GSSP—global stratotype section and point; E/O—Eocene-Oligocene transition.

CONCLUSIONS

The Massignano section has been studied extensively over the past 20 yr owing to its stratigraphic significance as GSSP for the Eocene-Oligocene boundary and because it contains a record of one of the most important climatic periods in Earth's history. Rock magnetic properties in conjunction with other paleoclimatic proxies from the Massignano section indicate the occurrence of environmental changes that alternated between two modes. We propose the STENT current hypothesis, which states that these bimodal magnetic properties resulted from a switching on and off of the current passing through the narrow Neotethys Ocean as this gateway was progressively closing. These events were the precursors to the main Oi-1 cooling event, implicating on the closing of the Neotethys Ocean as a factor in the transition from greenhouse to icehouse conditions. The hypothesis is consistent with recent tectonic studies that show that the collision between the Arabia and Eurasia plates occurred synchronous with the Eocene-Oligocene boundary. While the gateway was open, as it was during the Eocene, a current would have flowed through the oceans in a counterclockwise direction, funneling through the Neotethys (STENT). The flow of this current from the Indian Ocean to the Atlantic Ocean was dependent on bathymetry in the gateway, which, in the final stages of closure, is hypothesized to have been sensitive to local, regional, or global sea-level fluctuations and/or tectonic and sedimentary events in the gateway area. As the STENT fluctuated between being open or closed, the climate oscillated between warmer and cooler conditions. Finally, as the STENT closed permanently, the climate made a transition into icehouse conditions.

ACKNOWLEDGMENTS

We are grateful to Christian Koeberl and Sandro Montanari as editors, and David Bice and Christopher Poulsen for their reviews and comments. Eric Barron, Henk Brinkhuis, Helen Coxall, Fabio Florindo, Bernie Housen, Pierpaolo Mattias, Isabel Montanez, Dave Osleger, Isabella Premoli Silva, Andy Roberts, Leonardo Sagnotti, Geraldine Sarthou, Fabio Speranza, Mario Sprovieri, Ken Verosub, and Paul Wilson provided interesting discussion related to the hypothesis. This is publication 25 of the Centro di Geobiologia of the Urbino University.

REFERENCES CITED

Acton, G.D., 1999, Apparent polar wander of India since the Cretaceous with implications for regional tectonics and true polar wander, *in* Radhakrishna, T., and Piper, J.D.A., eds., The Indian Subcontinent and Gondwana: A Palaeomagnetic and Rock Magnetic Perspective: Memoir Geological Society of India 44, p. 129–175.

Agard, P., Omrani, J., Jolivet, L., and Mouthereau, F., 2005, Convergence history across Zagros (Iran): Constraints from collision and earlier deformation: International Journal of Earth Sciences, v. 94, p. 401–419, doi: 10.1007/s00531-005-0481-4.

Aitchison, J.C., Ali, J.R., and Davis, A.M., 2007, When and where did India and Asia collide?: Journal of Geophysical Research, v. 112, B05423, doi: 10.1029/2006JB004706.

Allen, M.B., and Armstrong, H.A., 2008, Arabia-Eurasia collision and the forcing of mid-Cenozoic global cooling: Palaeogeography, Palaeoclimatology, Palaeoecology, v. 265, p. 52–58, doi: 10.1016/j.palaeo.2008.04.021.

Allen, M.B., Jackson, J., and Walker, R., 2004, Late Cenozoic reorganization of the Arabia-Eurasia collision and the comparison of short-term and long-term deformation rates: Tectonics, v. 23, TC2008, doi: 10.1029/2003TC001530.

Barker, P.F., Filippelli, G.M., Florindo, F., Martin, E.E., and Scher, H.D., 2007, Onset and role of the Antarctic Circumpolar Current: Deep-Sea Research, Part II–Topical Studies in Oceanography, v. 54, p. 2388–2398, doi: 10.1016/j.dsr2.2007.07.028.

Barron, E.J., 1987, Eocene equator-to-pole surface ocean temperatures: A significant climate problem?: Paleoceanography, v. 2, p. 729–739, doi: 10.1029/PA002i006p00729.

Barron, E.J., and Peterson, W.H., 1989, Model simulation of the Cretaceous ocean circulation: Science, v. 244, p. 684–686, doi: 10.1126/science.244.4905.684.

Barron, E.J., and Peterson, W.H., 1990, Mid-Cretaceous ocean circulation: Results from model sensitivity studies: Paleoceanography, v. 5, p. 319–337, doi: 10.1029/PA005i003p00319.

Barron, E.J., and Peterson, W.H., 1991, The Cenozoic ocean circulation based on ocean General Circulation Model results: Palaeogeography, Palaeoclimatology, Palaeoecology, v. 83, p. 1–28, doi: 10.1016/0031-0182(91)90073-Z.

Barron, E.J., Peterson, W.H., Pollard, D., and Thompson, S., 1993, Past climate and the role of ocean heat transport: Model simulations for the Cretaceous: Paleoceanography, v. 8, p. 785–798, doi: 10.1029/93PA02227.

Barron, E.J., Fawcett, P.J., Peterson, W.H., Pollard, D., and Thompson, S., 1995, A "simulation" of mid-Cretaceous climate: Paleoceanography, v. 10, p. 953–962, doi: 10.1029/95PA01624.

Baumann, P., 1970, Micropaläontologische und stratigraphische Untersuchung der obereozänen-oligozänen Scaglia im Zentralen Apennin (Italien): Eclogae Geologicae Helvetiae, v. 63, p. 1133–1211.

Bellagamba, M., and Coccioni, R., 1990, Deep-water agglutinated foraminifera from the Massignano section (Ancona, Italy): A proposed stratotype for the Eocene-Oligocene boundary, *in* Hemleben, Ch., Kaminski, M.A., Khunt, W., and Scott D., eds., Paleoecology, Biostratigraphy, Paleoceanography and Taxonomy of Agglutinated Foraminifera: Dordrecht, Netherlands, Kluwer, NATO ASI Series, v. 327, p. 883–921.

Berggren, W.A., and Miller, K.G., 1989, Cenozoic bathyal and abyssal calcareous benthic foraminiferal zonation: Micropaleontology, v. 35, p. 308–320, doi: 10.2307/1485674.

Berggren, W.A., and Pearson, P.N., 2006, Tropical to subtropical planktonic foraminiferal zonation of the Eocene and Oligocene, *in* Pearson, P.N., Olsson, R.K., Huber, B.T. Hemleben Ch., and Berggren W.A., eds., Atlas of Eocene Planktonic Foraminifera: Cushman Foundation Special Publication 41, p. 29–40.

Berggren, W.A., Kent, D.V., Swisher, C.C., III, and Aubry, M.-P., 1995, A revised Cenozoic geochronology and chronostratigraphy, *in* Berggren, W.A., Kent, D.V., Swisher, C.C., III, Aubry, M.-P., and Hardenbol, J., eds., Geochronology, Time Scales, and Global Stratigraphic Correlation: SEPM (Society for Sedimentary Geology) Special Publication 54, p. 129–212.

Bernard, G., Stockli, D.F., Grove, M., Axen, G.J., Lam, P.S., and Hassanzadeh, J., 2006, Thermal histories from the central Alborz Mountains, northern Iran: Implications for the spatial and temporal distribution of deformation in northern Iran: Geological Society of America Bulletin, v. 118, p. 1507–1521.

Besse, J., and Courtillot, V., 1991, Revised and synthetic apparent polar wander paths of the African, Eurasian, North American and Indian plates, and true polar wander since 200 Ma: Journal of Geophysical Research, v. 96, p. 4029–4050, doi: 10.1029/90JB01916.

Bice, D.M., and Montanari, A., 1988, Magnetic stratigraphy of the Massignano section across the Eocene-Oligocene boundary, *in* Premoli Silva, I., Coccioni, R., and Montanari, A., eds., The Eocene/Oligocene Boundary in the Marche-Umbria Basin (Italy): International Subcommission on Paleogene Stratigraphy Special Publication: Ancona, Industrie Grafiche Fratelli Aniballi, p. 111–117.

Bice, K.L., Barron, E.J., and Peterson, W.H., 1997, Continental runoff and early Cenozoic bottom-water sources: Geology, v. 25, p. 951–954, doi: 10.1130/0091-7613(1997)025<0951:CRAECB>2.3.CO;2.

Bice, K.L., Scotese, C.R., Seidov, D., and Barron, E.J., 2000, Quantifying the role of geographic change in Cenozoic ocean heat transport using uncoupled atmosphere and ocean models: Palaeogeography, Palaeoclimatology, Palaeoecology, v. 161, p. 295–310, doi: 10.1016/S0031-0182(00)00072-9.

Billups, K., 2008, A tale of two climates: Nature Geoscience, v. 1, p. 294–295.

Blain, B., Quéguiner, B., Armand, L., Belviso, S., Bombled, B., Bopp, L., Bowie, A., Brunet, C., Brussaard, C., Carlotti, F., Christaki, U., Corbiere, A., Durand, I., Ebersbach, F., Fuda, J.-L, Garcia, N., Gerringa, L., Griffiths, B., Guigue, C., Guillerm, C., Jacquet, S., Jeandel, C., Laan, P., Lefevre, D., Lo Monaco, C., Malits, A., Mosseri, J., Obernosterer, I., Park, Y.-H., Picheral, M., Pondaven, P., Remenyi, T., Sandroni, V., Sarthou, G., Savoye, N., Scouarnec, L., Souhaut, M., Thuiller, D., Timmermans, K., Trull, T., Uitz, J., van Beek, P., Veldhuis, M., Vincent, D., Viollier, E., Vongl, L., and Wagener, T., 2007, Effect of natural iron fertilization sequestration in the Southern Ocean: Nature, v. 446, p. 1070–1074.

Blanc, E.J.-P., Allen, M.B., Inger, S., and Hassani, H., 2003, Structural styles in the Zagros simple folded zone, Iran: Journal of the Geological Society of London, v. 160, p. 401–412, doi: 10.1144/0016-764902-110.

Blow, W.H., 1969, Late middle Eocene to Recent planktonic foraminiferal biostratigraphy, *in* Bronnimann, P., and Renz, H.H., eds., Proceedings of the First International Conference on Planktonic Microfossils (Geneva, 1967): Leiden, E.J. Brill, p. 199–421.

Blow, W.H., 1979, The Cenozoic Globigerinida: A study of the morphology, taxonomy, evolutionary relationships and the stratigraphical distribution of some Globigerinida (mainly Globigerinacea), Volume 3: Leiden, E.J. Brill, 1413 p.

Bodiselitsch, B., Montanari, A., Koeberl, C., and Coccioni, R., 2004, Delayed climate cooling in the late Eocene caused by multiple impacts: High-resolution geochemical studies at Massignano, Italy: Earth and Planetary Science Letters, v. 223, p. 283–302, doi: 10.1016/j.epsl.2004.04.028.

Bohaty, S.M., and Zachos, J.C., 2003, Significant Southern Ocean warming event in the late middle Eocene: Geology, v. 31, p. 1017–1020, doi: 10.1130/G19800.1.

Brinkhuis, H., 1992, Late Paleogene dinoflagellate cysts with special reference to the Eocene/Oligocene boundary, *in* Prothero, D.R., and Berggren, W.A., eds., Eocene-Oligocene Climatic and Biotic Evolution: Princeton, New Jersey, Princeton University Press, p. 327–340.

Brinkhuis, H., and Biffi, U., 1993, Dinoflagellate cyst stratigraphy of the Eocene/Oligocene transition in central Italy: Marine Micropaleontology, v. 22, p. 131–183, doi: 10.1016/0377-8398(93)90007-K.

Brinkhuis, H., and Coccioni, R., 1995, Is there a relation between dinocyst changes and iridium after all? Preliminary results from the Massignano section, *in* Montanari, A., and Coccioni, R., eds., The Effects of Impacts on the Atmosphere and Biosphere with Regard to Short- and Long-Term Changes: Abstracts and Field Trips: Ancona, Italy, European Science Foundation, p. 40.

Brinkhuis, H., and Visscher, H., 1995, The upper boundary of the Eocene Series; a reappraisal based on dinoflagellate cyst biostratigraphy and sequence stratigraphy, *in* Berggren, W.A., Kent, D.V., Swisher, C.C., III, Aubry, M.P., and Hardenbol, J., eds., Geochronology, Time Scales and Global Stratigraphic Correlation: SEPM (Society for Sedimentary Geology) Special Publication 54, p. 129–212.

Brown, R.E., Koeberl, C., Montanari, A., and Bice D.M., 2009, this volume, Evidence for a change in Milankovitch forcing caused by extraterrestrial events at Massignano, Italy, Eocene-Oligocene Boundary GSSP, *in* Koeberl, C., and Montanari, A., eds., The Late Eocene Earth—Hothouse, Icehouse, and Impacts: Geological Society of America Special Paper 452, doi: 10.1130/2009.2452(08).

Bush, A.B., 1997, Numerical simulation of the Cretaceous Tethys Circumglobal Current: Science, v. 275, p. 807–810, doi: 10.1126/science.275.5301.807.

Cande, S.C., and Kent, D.V., 1995, Revised calibration of the geomagnetic polarity time scale for the Late Cretaceous and Cenozoic: Journal of Geophysical Research, v. 100, p. 6093–6095, doi: 10.1029/94JB03098.

Capo, R.C., and De Paolo, D.J., 1988, Sr isotopic analysis of marine carbonates from the Massignano section across the Eocene/Oligocene boundary, *in* Premoli Silva, I., Coccioni, R., and Montanari, A., eds., The Eocene/Oligocene Boundary in the Marche-Umbria Basin (Italy): International Subcommission on Paleogene Stratigraphy Special Publication: Ancona, Industrie Grafiche Fratelli Aniballi, p. 189–192.

Cassar, N., Bender, M.L., Barnett, B.A., Fan, S., Moxim, W.J., Levy, H., and Tilbrook, B., 2007, The Southern Ocean biological response to aeolian iron deposition: Science, v. 317, p. 1067–1070.

Chiarabba, C., Jovane, L., and Di Stefano, R., 2005, A new view of Italian seismicity using 20 years of instrumental recordings: Tectonophysics, v. 395, p. 251–268, doi: 10.1016/j.tecto.2004.09.013.

Clymer, A.K., Bice, D.M., and Montanari, A., 1995, Shocked quartz in the late Eocene: Bolide impact evidence from Massignano, Italy, *in* Montanari, A., and Coccioni, R., eds., The Effects of Impacts on the Atmosphere and Biosphere with Regard to Short- and Long-Term Changes: Abstracts and Field Trips: Ancona, Italy, European Science Foundation, p. 60–61.

Clymer, A.K., Bice, D.M., and Montanari, A., 1996, Shocked quartz from the late Eocene: Impact evidence from Massignano, Italy: Geology, v. 24, p. 483–486, doi: 10.1130/0091-7613(1996)024<0483:SQFTLE>2.3.CO;2.

Coccioni, R., 1988, The genera Hantkenina and Cribrohantkenina (Foraminifera) in the Massignano section, *in* Premoli Silva, I., Coccioni, R., and Montanari, A., eds., The Eocene/Oligocene Boundary in the Marche-Umbria Basin (Italy): International Subcommission on Paleogene Stratigraphy Special Publication: Ancona, Industrie Grafiche Fratelli Aniballi, p. 81–96.

Coccioni, R., and Galeotti, S., 2003, Deep-water benthic foraminiferal events from the Massignano Eocene/Oligocene boundary stratotype section and point (central Italy): Biostratigraphic, paleoecologic, and paleoceanographic implications, *in* Prothero, D.R., Ivant, L.C., and Nesbitt, E., eds., From Greenhouse to Icehouse: The Marine Eocene-Oligocene Transition: New York, Columbia University Press, p. 438–452.

Coccioni, R., Monaco, P., Monechi, S., Nocchi, M., and Parisi, G., 1986, The Eocene/Oligocene boundary at Massignano (Ancona, Italy): Biostratigraphy based on calcareous nannofossils and planktonic foraminifera: Bulletin of Liaison and Information, International Geological Correlation Program Project 196, v. 6, p. 37–44.

Coccioni, R., Monaco, P., Monechi, S., Nocchi, M., and Parisi, G., 1988, Biostratigraphy of the Eocene/Oligocene boundary at Massignano (Ancona, Italy), *in* Premoli Silva, I., Coccioni, R., and Montanari, A., eds., The Eocene/Oligocene Boundary in the Marche-Umbria Basin (Italy): International Subcommission on Paleogene Stratigraphy Special Publication: Ancona, Industrie Grafiche Fratelli Aniballi, p. 59–80.

Coccioni, R., Moretti, E., Nesci, O., Savelli, D., Tramontana, M., Veneri, F., and Astracedi, M., 1997, Carta Geologica con Itinerari Escursionistici—Parco Naturale del Conero: Firenze, Società Elaborazioni Cartografiche.

Coccioni, R., Basso, D., Brinkhuis, H., Galeotti, S., Gardin, S., Monechi, S., and Spezzaferri, S., 2000, Marine biotic signals across a late Eocene impact layer at Massignano, Italy: Evidence for long-term environmental perturbations?: Terra Nova, v. 12, p. 258–263, doi: 10.1046/j.1365-3121.2000.00305.x.

Colacicchi, R., and Passeri, L., 1981, Mesozoic pelagic sequences and their relations to carbonate platform margin in Umbria-Marche Apennine, *in* Ricci Lucchi, F., eds., Excursion Guidebook with Contributions on Sedimentology of Some Italian Basins; Second European Regional Meeting: Perugia, University of Perugia, Geological Institute, p. 57–76.

Coxall, H.K., Wilson, P.A., Pälike, H., Lear, C.H., and Backman, J., 2005, Rapid stepwise onset of Antarctic glaciation and deeper calcite compensation in the Pacific Ocean: Nature, v. 433, p. 53–57, doi: 10.1038/nature03135.

Dall'Antonia, B., Bossio, A., and Guernet, C., 2003, The Eocene/Oligocene boundary and the psychrospheric event in the Tethys as recorded by deep-sea ostracods from the Massignano global boundary stratotype section and point, central Italy: Marine Micropaleontology, v. 48, p. 91–106, doi: 10.1016/S0377-8398(02)00163-9.

de Baar, H.J.W., and de Jong, J.T.M., 2001, Distribution, sources and sinks of iron in seawater, *in* Turner, D.R., and Hunter, K.A., eds., The Biogeochemistry of Iron in Seawater: International Union for Pure and Applied Chemistry Series in Analytical and Physical Chemistry and Environmental Systems, Volume 7: Chichester, John Wiley and Sons, p. 123–254.

de Baar, H.J.W., de Jong, J.T.M., Bakker, D.C.E., Löscher, B.M., Veth, C., Bathmann, U., and Smetacek, V., 1995, Importance of iron for plankton blooms and carbon dioxide drawdown in the Southern Ocean: Nature, v. 373, p. 412–415, doi: 10.1038/373412a0.

de Baar, H.J.W., Boyd, P.W., Coale, K.H., Landry, M.R., Tsuda, A., Assmy, P., Bakker, D.C.E., Bozec, Y., Barber, R.T., Brzezinski, M.A., Buesseler, K.O., Boye´, M., Croot, P.L., Gervais, F., Gorbunov, M.Y., Harrison, P.J., Hiscock, W.T., Laan, P., Lancelot, C., Law, C.S., Levasseur, M., Marchetti, A., Millero, F.J., Nishioka, J., Nojiri, Y., van Oijen, T., Riebesell, U., Rijkenberg, M.J.A., Saito, H., Takeda, S., Timmermans, K.R., Veldhuis, M.J.W., Waite, A.M., and Wong, C.-S., 2005, Synthesis of iron fertilization experiments: From the Iron Age in the Age of Enlightenment: Journal of Geophysical Research, v. 110, p. C09S16, doi: 10.1029/2004JC002601.

DeConto, R.M., and Pollard, D., 2003, Rapid Cenozoic glaciation of Antarctica induced by declining atmospheric CO_2: Nature, v. 42, p. 1245–1249.

de Jong, J.T.M., den Das, J., Bathmann, U., Stoll, M.H.C., Katnner, G., Nolting, R.F., and de Baar, H.J.W., 1998, Dissolved iron at subnanomolar levels in the Southern Ocean as determined by shipboard analysis: Analytica Chimica Acta, v. 377, p. 113–124, doi: 10.1016/S0003-2670(98)00427-9.

Dercourt, J., Zonenshain, L.P., Ricou, L.-E., Kazmin, V.G., Le Pichon, X., Knipper, A.L., Grandjacquet, C., Sbortshikov, I.M., Geyssant, J., Lepvrier, C., Pechersky, D.H., Boulin, J., Sibuet, J.-C., Savostin, L.A., Sorokhtin, O., Westphal, M., Bazhenov, M.L., Lauer, J.P., and Biju-Duval, B., 1986, Geological evolution of the Tethys belt from the Atlantic to the Pamirs since the Lias: Tectonophysics, v. 123, p. 241–315, doi: 10.1016/0040-1951(86)90199-X.

Dercourt, J., Ricou, L.E., and Vrielynck, B., 1993, Atlas Tethys Paleoenvironmental Maps: Paris, Gauthiers-Villars, p. 1–307.

Dercourt, J., Gaetani, M., Vrielynck, B., Barrier, E., Biju-Duval, B., Brunet, M.F., Cadet, J.P., Crasquin, S., and Sandulescu, M., 2000, 24 maps, Peri-Tethys Atlas Palaeogeographical maps: Paris, France, Commision de la Carte Géologique du Monde (CCGM), 269 p.

Diester-Haass, L., and Zahn, R., 1996, Eocene-Oligocene transition in the Southern Ocean: History water mass circulation and biological productivity: Geology, v. 24, p. 163–166.

Diester-Haass, L., Robert, C., and Chamley, H., 1996, The Eocene-Oligocene preglacial-glacial transition in the Atlantic sector of the Southern Ocean (ODP Site 690): Marine Geology, v. 131, p. 123–149, doi: 10.1016/0025-3227(95)00174-3.

Dupont-Nivet, G., Krijgsman, W., Langereis, C.G., Abels, H.A., Dai, S., and Fang, X., 2007, Tibetan Plateau aridification linked to global cooling at the Eocene-Oligocene transition: Nature, v. 445, p. 635–638, doi: 10.1038/nature05516.

Edgar, K.M., Wilson, P.A., Sexton, P.F., and Suganuma, Y., 2007, No extreme bipolar glaciation during the main Eocene calcite compensation shift: Nature, v. 488, doi: 10.1038/nature06053.

Ehrmann, W.U., 1997, Smectite concentrations and crystallinities: Indications for Eocene age of glaciomarine sediments in the CIROS-1 drill hole, McMurdo Sound, Antarctica, *in* Ricci, C.A., ed., The Antarctic Region, Geological Evolution and Processes: Siena, Italy, Museo Nazionale dell'Antartide, p. 771–780.

Farley, K.A., Montanari, A., Shoemaker, E.M., and Shoemaker, C.S., 1998, Geochemical evidence for a comet shower in the late Eocene: Science, v. 280, p. 1250–1253, doi: 10.1126/science.280.5367.1250.

Gaetani, M., Dercourt, J., and Vrielynck, B., 2003, The Peri-Tethys Programme: Achievements and results: Episodes, v. 26, p. 79–93.

Gaiero, D.M., 2007, Dust provenance in Antarctic ice during glacial periods: From where in southern South America?: Geophysical Research Letters, v. 34, p. L17707, doi: 10.1029/2007GL030520.

Gardin, S., Spezzaferri, S., Basso, D., and Coccioni, R., 1999, Calcareous nannofossil and planktonic foraminiferal response to a meteorite impact in the Umbria-Marche Massignano section (northeastern Apennines, Italy): European Union Geosciences Strasbourg Meeting: Journal of Conference Abstracts, v. 4, p. 271.

Gonzalvo, C., and Molina, E., 1992, Estudio cuantitativo de los foraminiferos planctònicos en el estratotipo del limite Eoceno/Oligoceno en Massignano (Appeninos, Italia): Geogaceta, v. 12, p. 64–67.

Gourlan, A.T., Meynadier, L., and Allègre, C.J., 2008, Tectonically driven changes in the Indian Ocean circulation over the last 25 Ma: Neodymium isotope evidence: Earth and Planetary Science Letters, v. 267, p. 353–364, doi: 10.1016/j.epsl.2007.11.054.

Gradstein, F.M., Ogg, J.G., and Smith, A.G., 2004, A Geologic Time Scale: Cambridge, Cambridge University Press, 589 p.

Harzhauser, M., Piller, W.E., and Steininger, F.F., 2002, Circum-Mediterranean Oligo-Miocene biogeographic evolution—The gastropods' point of view: Palaeogeography, Palaeoclimatology, Palaeoecology, v. 183, p. 103–133, doi: 10.1016/S0031-0182(01)00464-3.

Hempton, M.R., 1987, Constraints on Arabia plate motion and extensional history of the Red Sea: Tectonics, v. 6, p. 687–705, doi: 10.1029/TC006i006p00687.

Hessami, K., Koyi, H.A., Talbot, C.J., Tabasi, H., and Shabanian, E., 2001, Progressive unconformities within an evolving foreland fold-thrust belt, Zagros Mountains: Journal of the Geological Society of London, v. 158, p. 969–981, doi: 10.1144/0016-764901-007.

Hilgen, F.J., and Kuiper, K.F., 2009, this volume, A critical evaluation of the numerical age of the Eocene-Oligocene boundary, *in* Koeberl, C., and Montanari, A., eds., The Late Eocene Earth—Hothouse, Icehouse, and Impacts: Geological Society of America Special Paper 452, doi: 10.1130/2009.2452(09).

Hotinski, R.M., and Toggweiler, J.R., 2003, Impact of a Tethyan circumglobal passage on ocean heat transport and "equable" climates: Paleoceanography, v. 18, 1007 p., doi: 10.1029/2001PA000730.

Hu, X., Jansa, L., and Sarti, M., 2005, Mid-Cretaceous oceanic red beds in the Umbria-Marche Basin, central Italy: Constraints on paleoceanography and paleoclimate: Palaeogeography, Palaeoclimatology, Palaeoecology, v. 233, p. 163–186, doi: 10.1016/j.palaeo.2005.10.003.

Huber, M., and Sloan, L.C., 2001, Heat transport, deep waters, and thermal gradients: Coupled simulation of an Eocene greenhouse climate: Geophysical Research Letters, v. 28, p. 3481–3484, doi: 10.1029/2001GL012943.

Huber, H., Koeberl, C., King, D.T., Petruny, L.W., and Montanari, A., 2002, Effects of bioturbation through the late Eocene impactoclastic layer near Massignano, Italy, *in* Buffetaut, E., and Koeberl, C., eds., Geological and Biological Effects of Impact Events: Berlin, Springer-Verlag, p. 197–216.

Hyland, E., Murphy, B., Varela, P., Marks, K., Colwell, L., Tori, F., Monechi, S., Cleaveland, L., Brinkhuis, H., van Mourik, C.A., Coccioni, R., Bice, D., and Montanari, A., 2009, this volume, Integrated stratigraphy and astrochronologic calibration of the Eocene-Oligocene transition in the Monte Cagnero section (Northeastern Apennines, Italy): A potential parastratotype for the Massignano GSSP, *in* Koeberl, C., and Montanari, A., eds., The Late Eocene Earth—Hothouse, Icehouse and Impacts: Geological Society of America Special Paper 452, doi: 10.1130/2009.2452(19).

Ivany, L.C., Lohmann, K.C., Hasiuk, F., Blake, D.B., Glass, A., Aroson, R.B., and Moody, R.M., 2008, Eocene climate record of a high southern latitude continental shelf: Seymour Island, Antarctica: Geological Society of America Bulletin, v. 120, p. 659–678.

Johnson, C.C., Barron, E.J., Kauffman, E.G., Arthur, M.A., Fawcett, P.J., and Yasuda, M.K., 1996, Middle Cretaceous reef collapse linked to ocean heat transport: Geology, v. 24, p. 376–380, doi: 10.1130/0091-7613(1996)024<0376:MCRCLT>2.3.CO;2.

Jovane, L., 2006, Magnetostratigrafia di Successioni Sedimentarie Marine Deposte in Concomitanza di Alcuni tra i Principali Eventi Climatici del Paleogene [Ph.D. thesis]: Bologna, Università degli Studi di Bologna–Alma Mater Studiorum, Dipartimento di Fisica, 146 p. (Available from: http://paleomag.ucdavis.edu/luigi/JovanePhD.pdf.)

Jovane, L., Florindo, F., and Dinarès-Turell, J., 2004, Environmental magnetic record of paleoclimate change from the Eocene-Oligocene stratotype section, Massignano, Italy: Geophysical Research Letters, v. 31, p. L15601, doi: 10.1029/2004GL020554.

Jovane, L., Florindo, F., Sprovieri, M., and Pälike, H., 2006, Astronomic calibration of the late Eocene/early Oligocene Massignano section (central Italy): Geochemistry, Geophysics, Geosystems, v. 7, p. Q07012, doi: 10.1029/2005GC001195.

Jovane, L., Sprovieri, M., Florindo, F., Acton, G., Coccioni, R., Dall'Antonia, B., and Dinarès-Turrell, J., 2007a, Eocene-Oligocene paleoceanographic changes in the stratotype section, Massignano, Italy: Clues from rock magnetism and stable isotopes: Journal of Geophysical Research, v. 112, p. B11101, doi: 10.1029/2007JB004963.

Jovane, L., Florindo, F., Coccioni, R., Dinarès-Turell, J., Marsili, A., Monechi, S., Roberts, A.P., and Sprovieri, M., 2007b, The middle Eocene climatic optimum (MECO) event in the Contessa Highway section, Umbrian Apennines, Italy: Geological Society of America Bulletin, v. 119, p. 413–427, doi: 10.1130/B25917.1.

Katz, M.E., Miller, K.G., Wright, J.D., Wade, B.S., Browning, J.V., Cramer, B.S., and Rosenthal, Y., 2008, Stepwise transition from the Eocene greenhouse to the Oligocene icehouse: Nature Geoscience, v. 1, p. 329–334, doi: 10.1038/ngeo179.

Kennett, J.P., and Shackleton, N.J., 1976, Oxygen isotopic evidence for development of psychrosphere 38 Myr ago: Nature, v. 260, p. 513–515, doi: 10.1038/260513a0.

Kennett, J.P., and Stott, L.D., 1990, Proteus and proto-Oceanus: Ancestral Paleogene oceans as revealed from Antarctic stable isotopic results: ODP Leg 113, *in* Barker, P.F., Kennett, J.P., O'Connell, S., and Pisias, N.G., et al., Proceedings of the Ocean Drilling Program, Scientific Results, Volume 113: College Station, Texas, Ocean Drilling Program, p. 865–880.

Lanci, L., Lowrie, W., and Montanari, A., 1996, Magnetostratigraphy of the Eocene/Oligocene boundary in a short drill-core: Earth and Planetary Science Letters, v. 143, p. 37–48, doi: 10.1016/0012-821X(96)00136-7.

Lanci, L., and Lowrie, W., 1997, Magnetostratigraphic evidence that 'tiny wiggles' in the oceanic magnetic anomaly record represent geomagnetic paleointensity variations: Earth and Planetary Science Letters, v. 148, p. 581–592, doi: 10.1016/S0012-821X(97)00055-1.

Lanci, L., Lowrie, W., and Montanari, A., 1998, Stratigrafia magnetica ad alta risoluzione del limite Eocene-Oligocene nella successione Umbro-Marchigiana: Atti della Accademia Nazionale dei Lincei, Rendiconti di Fisica dell'Accademia dei Lincei, v. 9, p. 109–129.

Langenhorst, F., and Clymer, A.K., 1996, Characteristics of shocked quartz in late Eocene impact ejecta from Massignano (Ancona, Italy): Clues to shock conditions and source crater: Geology, v. 24, p. 487–490, doi: 10.1130/0091-7613(1996)024<0487:COSQIL>2.3.CO;2.

Laskar, J., Robutel, P., Joutel, F., Gastineau, M., Correia, A., and Levrard, B., 2004, A long term numerical solution for the insolation quantities of the Earth: Astronomy & Astrophysics, v. 428, p. 261–285, doi: 10.1051/0004-6361:20041335.

Lawver, L.A., and Gahagan, L.M., 2003, Evolution of Cenozoic seaways in the circum-Antarctic region: Palaeogeography, Palaeoclimatology, Palaeoecology, v. 198, p. 11–37, doi: 10.1016/S0031-0182(03)00392-4.

Lear, C.H., Elderfield, H., and Wilson, P.A., 2000, Cenozoic deep-sea temperatures and global ice volumes from Mg/Ca in benthic foraminiferal calcite: Science, v. 287, p. 269–272, doi: 10.1126/science.287.5451.269.

Lear, C.H., Rosenthal, Y., Coxall, H.K., and Wilson, P.A., 2004, Late Eocene to early Miocene ice sheet dynamics and the global carbon cycle: Paleoceanography, v. 19, p. PA4015, doi: 10.1029/2004PA001039.

Lear, C.H., Bailey, T.R., Pearson, P.N., Coxall, H.K., and Rosenthal, Y., 2008, Cooling and ice growth across the Eocene-Oligocene transition: Geology, v. 36, p. 251–254, doi: 10.1130/G24584A.1.

Livermore, R., Nankivell, A., Eagles, G., and Morris, P., 2005, Paleogene opening of Drake Passage: Earth and Planetary Science Letters, v. 236, p. 459–470, doi: 10.1016/j.epsl.2005.03.027.

Livermore, R., Hillenbrand, C.-D., Meredith, M., and Eagles, G., 2007, Drake Passage and Cenozoic climate: An open and shut case?: Geochemistry, Geophysics, Geosystems, v. 8, p. Q01005, doi: 10.1029/2005GC001224.

Löscher, B.M., de Baar, H.J.W., de Jong, J.T.M., Veth, C., and Dehairs, F., 1997, The distribution of Fe in the Antarctic Circumpolar Current: Deep-Sea Research, Part II–Topical Studies in Oceanography, v. 44, p. 143–187, doi: 10.1016/S0967-0645(96)00101-4.

Lowrie, W., and Lanci, L., 1994, Magnetostratigraphy of Eocene-Oligocene boundary sections in Italy: No evidence for short subchrons within chrons

12R and 13R: Earth and Planetary Science Letters, v. 126, p. 247–258, doi: 10.1016/0012-821X(94)90110-4.

Luyendyk, B.P., Forsyth, D., and Phillips, J.D., 1972, Experimental approach to the paleocirculation of the oceanic surface waters: Geological Society of America Bulletin, v. 83, p. 2649–2663, doi: 10.1130/0016-7606(1972)83[2649:EATTPO]2.0.CO;2.

Lyle, M., Gibbs, S., Moore, T.C., and Rea, D.K., 2008, Late Oligocene initiation of the Antarctic Circumpolar Current: Evidence from the South Pacific: Geology, v. 35, p. 691–694.

Marchegiani, L., Bertotti, G., Cello, G., Deiana, G., Mazzoli, S., and Tondi, E., 1999, Pre-orogenic tectonics in the Umbria-Marche sector of the Afro-Adriatic continental margin: Tectonophysics, v. 315, p. 123–143, doi: 10.1016/S0040-1951(99)00277-2.

Martini, E., 1971, Standard Tertiary and Quaternary calcareous nannoplankton zonation, *in* Farinacci, A., ed., Proceedings of the Second Planktonic Conference, Rome 1970: Rome, Tecnoscienza, v. 2, p. 739–785.

Mattias, P., Crocetti, G., Barrese, E., Montanari, A., Coccioni, R., Farabollini, P., and Parisi, E., 1992, Caratteristiche mineralogiche e lithostratigrafiche della sezione eo-oligocenica di Massignano (Ancona, Italia) comprendente il limite Scaglia Variegata-Scaglia Cinerea: Studi Geologici Camerti, v. 12, p. 93–103.

McClay, K.R., Whitehouse, P.S., Dooley, T., and Richards, M., 2004, 3D evolution of fold and thrust belts formed by oblique convergence: Marine and Petroleum Geology, v. 21, p. 857–877, doi: 10.1016/j.marpetgeo.2004.03.009.

Merico, A., Tyrell, T., and Wilson, P.A., 2008, Eocene/Oligocene ocean de-acidification linked to Antarctic glaciation by sea-level fall: Nature, v. 452, p. 979–982.

Miller, K.G., Wright, J.D., and Fairbancks, R.G., 1991, Unlocking the ice house: Oligocene-Miocene oxygen isotopes, eustasy and margin erosion: Journal of Geophysical Research, v. 96, p. 6829–6848, doi: 10.1029/90JB02015.

Miller, K.G., Wright, J.D., and Browning, J.V., 2005a, Visions of ice sheets in a greenhouse world: Marine Geology, v. 217, p. 215–231, doi: 10.1016/j.margeo.2005.02.007.

Miller, K.G., Kominz, M.A., Browning, J.V., Wright, J.D., Mountain, G.S., Katz, M.E., Sugarman, P.J., Cramer, B.S., Christie-Blick, N., and Pekar, S.F., 2005b, The Phanerozoic record of global sea-level change: Science, v. 310, p. 1293–1298, doi: 10.1126/science.1116412.

Miller, K.G., Browning, J.V., Aubry, M.-P., Wade, B.S., Katz, M.E., Kulpecz, A.A., and Wright, J.D., 2008, Eocene-Oligocene global climate and sea level changes: St. Stephens Quarry, Alabama: Geological Society of America Bulletin, v. 120, p. 34–53, doi: 10.1130/B26105.1.

Molina, E., Gonzalvo, C., and Keller, G., 1993, The Eocene-Oligocene planktic foraminiferal transition: Extinctions, impacts and hiatuses: Geological Magazine, v. 130, p. 483–499.

Monechi, S., Buccianti, A., and Gardin, S., 2000, Biotic signals from nannoflora across the iridium anomaly in the Upper Eocene of the Massignano section: Evidence from statistical analysis: Marine Micropaleontology, v. 39, p. 219–237, doi: 10.1016/S0377-8398(00)00022-0.

Montanari, A., 1988, Geochemical characterization of volcanic biotites from the Upper Eocene–Upper Miocene pelagic sequence of the northeastern Apennines, *in* Premoli Silva, I., Coccioni, R., and Montanari, A., eds., The Eocene/Oligocene Boundary in the Marche-Umbria Basin (Italy): International Subcommission on Paleogene Stratigraphy Special Publication: Ancona, Industrie Grafiche Fratelli Aniballi, p. 209–227.

Montanari, A., 1990, Geochronology of the terminal Eocene impacts; an update, *in* Sharpton, V.L., and Ward, P.D., eds., Global Catastrophes in Earth History: Geological Society of America Special Publication 247, p. 607–616.

Montanari, A., and Bice, D.M., 1986, Paleomagnetic analysis across the Eocene-Oligocene boundary at Massignano: Correlation with the classic sequence of Contessa: Bulletin de Liaison et Informations, International Geological Correlation Program Project 196, v. 6, p. 44–57.

Montanari, A., and Koeberl, C., 2000, Impact stratigraphy—The Italian record: Springer, Heidelberg, Lecture Notes in Earth Sciences, v. 93, p. 384.

Montanari, A., and Odin, G.S., 1989, Eocene-Oligocene boundary at Massignano (Ancona, Italy): Potential boundary stratotype, *in* 28th International Geological Congress Abstracts: Washington, D.C., Abstract 2, p. 453.

Montanari, A., Deino, A.L., Drake, R.E., Turrin, B.D., De Paolo, D.J., Odin, G.S., Curtis, G.H., Alvarez, W., and Bice, D.M., 1988, Radioisotopic dating of the Eocene-Oligocene boundary in the pelagic sequence of the northeastern Apennines, *in* Premoli Silva, I., Coccioni, R., and Montanari, A., eds., The Eocene/Oligocene Boundary in the Marche-Umbria Basin (Italy): International Subcommission on Paleogene Stratigraphy Special Publication: Ancona, Industrie Grafiche Fratelli Aniballi, p. 195–208.

Montanari, A., Deino, A., Coccioni, R., Langenheim, V.E., Capo, R., Monechi, S., 1991, Geochronology, Sr isotopes analysis, magnetostratigraphy, and plankton stratigraphy across the Oligocene-Miocene boundary in the Contessa section (Gubbio, Italy): Newsletter on Stratigraphy, v. 23, p. 151–180.

Montanari, A., Asaro, F., Michel, H.V., and Kennett, J.P., 1993, Iridium anomalies of late Eocene age at Massignano (Italy), and ODP 689B (Maud Rise, Antarctic): Palaios, v. 8, p. 420–437, doi: 10.2307/3515017.

Nocchi, M., Parisi, G., Monaco, P., Monechi, S., and Madile, M., 1988, Eocene and early Oligocene micropaleontology and paleoenvironments in SE Umbria, Italy: Palaeogeography, Palaeoclimatology, Palaeoecology, v. 67, p. 181–244, doi: 10.1016/0031-0182(88)90154-X.

Oberhänsli, H., and Hsü, K.J., 1986, Paleocene-Eocene paleoceanography, *in* Hsü, K.J., ed., Mesozoic and Cenozoic Oceans: American Geophysical Society Geodynamics Monograph 15, p. 85–100.

Oberli, F., and Meier, M., 1991, Age of Eocene-Oligocene boundary in the Marche-Umbria Basin, Italy, by high resolution U/Th-Pb dating: TERRA Abstracts, v. 3, p. 286.

Odin, G.S., 1986, Radiometric data across the Eocene-Oligocene boundary in the Apennines: Bulletin de Liaison et Informations, International Geological Correlation Program Project 196, v. 6, p. 58–62.

Odin, G.S., and Montanari, A., 1988, The Eocene-Oligocene boundary at Massignano (Ancona, Italy): A potential boundary stratotype, *in* Premoli Silva, I., Coccioni, R., and Montanari, A., eds., The Eocene-Oligocene Boundary in the Marche-Umbria Basin (Italy): International Subcommission on Paleogene Stratigraphy Special Publication: Ancona, Industrie Grafiche Fratelli Aniballi, p. 253–263.

Odin, G.S., and Montanari, A., 1989, Age radiométrique et stratotype de la limite Éocène-Oligocène: Comptes Rendus de l'Académie des Sciences Paris, v. 309, p. 1939–1945.

Odin, G.S., Billiou, D., Renard, M., and Létolle, R., 1986, Etude des isotopes stables dans les carbonates de la coupe de Massignano (Ancone, Italie): Caractérisation de la limite Eocène/Oligocène: Bulletin de Liaison et Informations, International Geological Correlation Program Project 196, v. 6, p. 63–64.

Odin, G.S., Clauser, S., and Renard, M., 1988a, Sedimentological and geochemical data on the Eocene-Oligocene boundary at Massignano (Apennines, Italy), *in* Premoli Silva, I., Coccioni, R., and Montanari, A., eds., The Eocene-Oligocene Boundary in the Marche-Umbria Basin (Italy): International Subcommission on Paleogene Stratigraphy Special Publication: Ancona, Industrie Grafiche Fratelli Aniballi, p. 175–186.

Odin, G.S., Guise, P., Rex, D.C., and Kreuzer, H., 1988b, K-Ar and $^{39}Ar/^{40}Ar$ geochronology of late Eocene biotites from the northeastern Apennines, *in* Premoli Silva, I., Coccioni, R., and Montanari, A., eds., The Eocene-Oligocene Boundary in the Marche-Umbria Basin (Italy): International Subcommission on Paleogene Stratigraphy Special Publication: Ancona, Industrie Grafiche Fratelli Aniballi, p. 239–245.

Odin, G.S., Montanari, A., Deino, A., Drake, R., Gruise, P., Kreuzer, H., and Rex, D.C., 1991, Reliability of volcano-sedimentary biotite ages around the Eocene/Oligocene boundary: Chemical Geology, v. 86, p. 203–224.

Okada, H., and Bukry, D., 1980, Supplementary modification and introduction of code numbers to the low-latitude coccolith biostratigraphic zonation (Bukry, 1973; 1975): Marine Micropaleontology, v. 5, p. 321–325, doi: 10.1016/0377-8398(80)90016-X.

Orue-Extebarria, X., and Lamolda, M.A., 1985, Paleobiogeographic characteristics of the Basco-Cantabric basin during the Paleogene: Revue de Micropaléontologie, Paris, v. 27, p. 257–265.

Pagani, M., Zachos, J.C., Freeman, K.H., Tipple, B., and Bohaty, S., 2005, Marked decline in atmospheric carbon dioxide concentrations during the Paleogene: Science, v. 309, p. 600–603, doi: 10.1126/science.1110063.

Parisi, G. and Coccioni, R., 1988, Deep-water benthic Foraminifera at the Eocene-Oligocene boundary in the Massignano section (Ancona, Italy), *in* Premoli Silva, I., Coccioni, R., and Montanari, A., eds., The Eocene/Oligocene Boundary in the Marche-Umbria Basin (Italy): International Subcommission on Paleogene Stratigraphy Special Publication: Ancona, Industrie Grafiche Fratelli Aniballi, p. 97–109.

Pearson, P.N., and Palmer, M.R., 2000, Atmospheric carbon dioxide concentrations over the past 60 million years: Nature, v. 406, p. 695–699, doi: 10.1038/35021000.

Pearson, P.N., Ditchfield, P.W., Singano, J., Harcourt-Brown, K.G., Nicholas, C.J., Olsson, R.K., Shackleton, N.J., and Hall, M.A., 2001, Warm tropical sea surface temperatures in the Late Cretaceous and Eocene epochs: Nature, v. 413, p. 481–487, doi: 10.1038/35097000.

Pearson, P.N., van Dongen, B.E., Nicholas, C.J., Pancost, R.D., Schouten, S., Singano, J.M., and Wade, B.S., 2007, Stable warm tropical climate through the Eocene Epoch: Geology, v. 35, p. 211–214, doi: 10.1130/G23175A.1.

Persico, D., and Villa, G., 2004, Eocene-Oligocene calcareous nannofossils from Maud Rise and Kerguelen Plateau (Antarctica): Palaeoecological and palaeoceanographic implications: Marine Micropaleontology, v. 52, p. 153–179, doi: 10.1016/j.marmicro.2004.05.002.

Pierrard, O., Robin, E., Rocchia, R., and Montanari, A., 1998, Extraterrestrial Ni-rich spinel in Upper Eocene sediments from Massignano, Italy: Geology, v. 26, p. 307–310, doi: 10.1130/0091-7613(1998)026<0307:ENRSIU>2.3.CO;2.

Popov, S.V., and Stolyarov, A.S., 1996, Paleogeography and anoxic environments of the Oligocene–early Miocene Paratethys: Israel Journal of Earth Sciences, v. 45, p. 161–167.

Popov, S.V., Akhmet'ev, M.A., Zaporozhets, N.I., Voronina, A.A., and Stolyarov, A.S., 1993, Evolution of eastern Paratethys in the late Eocene–early Miocene: Stratigraphy and Geological Correlation, v. 1, p. 10–39.

Popov, S.V., Rögl, F., Rozanov, A.Y., Steininger, F.F., Shcherba, I.G., and Kovac, M., 2004, Lithological-paleogeographic maps of Paratethys—10 maps Late Eocene to Pliocene: Frankfurt, Courier Forschungsinstitut Senckenberg, v. 250, 46 p., 10 maps.

Poulsen, C.J., Seidov, D., Barron, E.J., and Peterson, W.H., 1998, The impact of paleoceanographic evolution on the surface oceanic circulation and the marine environment within the mid-Cretaceous Tethys: Paleoceanography, v. 13, p. 546–559, doi: 10.1029/98PA01789.

Poulsen, C.J., Barron, E.J., Arthur, M.A., and Peterson, W.H., 2001, Response of the mid- Cretaceous global oceanic circulation to tectonic and CO_2 forcings: Paleoceanography, v. 16, p. 576–592, doi: 10.1029/2000PA000579.

Premoli Silva, I., and Jenkins, D.G., 1993, Decision on the Eocene-Oligocene boundary stratotype: Episodes, v. 16, p. 379–382.

Premoli Silva, I., Coccioni, R., and Montanari, A., eds., 1988, The Eocene/Oligocene Boundary in the Marche-Umbria Basin (Italy): International Subcommission on Paleogene Stratigraphy Special Publication: Ancona, Industrie Grafiche Fratelli Aniballi, 268 p.

Premoli Silva, I., Montanari, A., and Coccioni, R., 1989, Intercorrelation of time scales: Example of multidisciplinary study from Umbria-Marche pelagic sequences (Italy) at the Eocene-Oligocene transition, *in* 28th International Geological Congress Abstracts: Washington, D.C., Abstracts 2, p. 639.

Premoli Silva, I., Nicora, A., Arnaud Vanneau, A., Budd, A.F., Camoin, G.F., and Masse, J.-P., 1995, Paleobiogeographic evolution of shallow-water organisms from the Aptian to the Eocene in the western Pacific, *in* Haggerty, J.A., Premoli Silva, I., Rack, F., and McNutt, M.K., Proceedings of the Ocean Drilling Program, Scientific Results, Volume 144: College Station, Texas, Ocean Drilling Program, p. 887–893.

Prothero, D.R., 1994, The Eocene-Oligocene Transition: Paradise Lost: New York, Columbia University Press, 291 p.

Prothero, D.R., and Berggren, W.A., eds., 1992, Eocene-Oligocene Climatic and Biotic Evolution: Princeton, New Jersey, Princeton University Press, 568 p.

Prothero, D.R., Ivany, L., and Nesbitt, E., 2003, From Greenhouse to Icehouse: The Marine Eocene-Oligocene Transition: New York, Columbia University Press, 541 p.

Quilty, P.G., 2001, Late Eocene foraminifers and palaeoenvironment, Cascade Seamount, southwest Pacific Ocean: Implications for seamount subsidence and Australia-Antarctica Eocene correlation: Australian Journal of Earth Sciences, v. 48, p. 633–641.

Ravizza, G., and Peucker-Ehrenbrink, B., 2003, The marine $^{187}Os/^{188}Os$ record of the Eocene-Oligocene transition: The interplay of weathering glaciation: Earth and Planetary Science Letters, v. 210, p. 151–165, doi: 10.1016/S0012-821X(03)00137-7.

Raymo, M.E., and Ruddiman, W.F., 1992, Tectonic forcing of the Cenozoic climate: Nature, v. 359, p. 117–122, doi: 10.1038/359117a0.

Rio, D., Premoli Silva, I., and Capraro, L., 2003, *The Geological Time Scale* and the Italian stratigraphic record: Episodes, v. 26, p. 259–263.

Rögl, F., 1998, Palaeogeographic considerations for Mediterranean and Paratethys seaways (Oligocene to Miocene): Annalen des Naturhistorischen Museums in Wien, v. 99A, p. 279–310.

Rögl, F., 1996, Stratigraphic correlation of the Paratethys Oligocene and Miocene: Mitteilungen der Gesellschaft der Geologie- und Bergbaustudenten in Österreich, v. 41, p. 65–73.

Scher, H.D., and Martin, E.E., 2004, Circulation in the Southern Ocean during the Paleogene inferred from neodymium isotopes: Earth and Planetary Science Letters, v. 228, p. 391–405, doi: 10.1016/j.epsl. 2004.10.016.

Schulz, H.M., Bechtel, A., and Sachsenhofer, R.F., 2005, The birth of the Paratethys during the early Oligocene: From Tethys to an ancient Black Sea analogue?: Global and Planetary Change, v. 49, p. 163–176, doi: 10.1016/j.gloplacha.2005.07.001.

Shackleton, N.J., and Kennett, J.P., 1975, Paleotemperature history of the Cenozoic and the initiation of Antarctic glaciation: Oxygen and carbon isotope analyses, *in* Kennett, J.P., Houtz, R.E., et al., DSDP Sites 277, 279, and 281: Initial Reports of the Deep Sea Drilling Project, Volume 29: Washington, D.C., Government Printing Office, p. 743–755.

Sharland, P.R., Archer, R., Casey, D.M., Davies, R.B., Hall, S.H., Heward, A.P., Horbury, A.D., and Simmons, M.D., 2001, Arabian Plate Sequence Stratigraphy: Manama, Bahrain, GeoArabia Special Publication, v. 2, 371 p.

Sloan, L.C., and Rea, D.K., 1996, Atmospheric pCO_2 of the early Eocene: A general circulation modeling sensitivity study: Palaeogeography, Palaeoclimatology, Palaeoecology, v. 119, p. 275–292, doi: 10.1016/0031-0182(95)00012-7.

Spezzaferri, S., Basso, D., and Coccioni, R., 2002, Late Eocene planktonic foraminiferal response to an extraterrestrial impact at Massignano GSSP (northeastern Apennines, Italy): Journal of Foraminiferal Research, v. 32, p. 188–199, doi: 10.2113/0320188.

Stanley, S.M., 1995, New horizons for paleontology, with two examples: The rise and the fall of the Cretaceous Supertethys and the cause of the modern ice age: Journal of Paleontology, v. 69, p. 999–1007.

Sunda, W.G., 2001, Bioavailability and bioaccumulation of iron in the sea, *in* Turner, D.R., and Hunter, K.A., eds., The Biogeochemistry of Iron in Seawater: International Union for Pure and Applied Chemistry Series in Analytical and Physical Chemistry and Environmental Systems, Volume 7: Chichester, John Wiley and Sons, p. 41–84.

Thomas, D.J., 2004, Evidence for deep-water production in the North Pacific Ocean during the early Cenozoic warm interval: Nature, v. 430, p. 65–68, doi: 10.1038/nature02639.

Thomas, D.J., Bralower, T.J., and Jones, C.E., 2003, Neodymium isotopic reconstruction of late Paleocene–early Eocene thermohaline circulation: Earth and Planetary Science Letters, v. 209, p. 309–322, doi: 10.1016/S0012-821X(03)00096-7.

Thomas, E., and Gooday, A.J., 1996, Cenozoic deep-sea benthic foraminifers: Tracers for changes in oceanic productivity?: Geology, v. 24, p. 355–358, doi: 10.1130/0091-7613(1996)024<0355:CDSBFT>2.3.CO;2.

Tori, F., and Monechi, S., 2007, Paleoecological and paleoclimatical significance of calcareous nannofossils at the E/O transition: New data from the Cagnero and Massignano sections, Umbria-Marche area (Italy), *in* Montanari, A., Koeberl, C., Coccioni, R., and Hilgen, F., eds., The Late Eocene Earth: Hot House, Ice House, and Impacts—Abstract with Program and Field Trip Guide: Ancona, Italy, Geological Society of America Penrose Conference, Osservatorio Coldigioco, Frontale di Aprio (Italy), p. 77.

Van Mourik, C., and Brinkhuis, H., 2005, The Massignano Eocene-Oligocene golden spike section revisited: Stratigraphy, v. 2, p. 13–30.

Verducci, M., and Nocchi, M., 2004, Middle to late Eocene main planktonic foraminiferal events in the Central Mediterranean area (Umbria-Marche Basin) related to paleoclimatic changes: Neues Jahrbuch für Geologie und Palaontologie Abhandlungen, v. 234, p. 361–413.

Verosub, K.L., and Roberts, A.P., 1995, Environmental magnetism: Past, present and future: Journal of Geophysical Research, v. 100, no. B2, p. 2175–2192, doi: 10.1029/94JB02713.

Via, R.K., and Thomas, D.J., 2006, Evolution of Atlantic thermohaline circulation: Early Oligocene onset of deep-water production in the North Atlantic: Geology, v. 34, no. 6, p. 441–444, doi: 10.1130/G22545.1.

Vivier, F., Maier-Reimer, E., and Tyler, R.H., 2004, Simulations of magnetic fields generated by the Antarctic Circumpolar Current at satellite altitude: Can geomagnetic measurements be used to monitor the flow?: Geophysical Research Letters, v. 31, p. L10306, doi: 10.1029/2004GL019804.

Vonhof, H.B., Wijbrans, J., and Smit, J., 1995, The Popigai impact crater: $^{39}Ar/^{40}Ar$ dating and its expression in the $^{87}Sr/^{86}Sr$ record of the Massignano section, *in* Montanari, A., and Coccioni, R., eds., The effects of impacts on the atmosphere and biosphere with regard to short- and long-term changes: Abstracts and field trips: Ancona, Italy, European Science Foundation, p. 163–164.

Vonhof, H.B., Smit, J., Brinkhuis, H., Montanari, A., and Nederbragt, A.J., 2000, Global cooling accelerated by early late Eocene impacts?: Geology, v. 28, p. 687–690, doi: 10.1130/0091-7613(2000)28<687:GCABEL>2.0.CO;2.

Yilmaz, Y., 1993, New evidence and model on the evolution of the southeast Anatolian orogen: Geological Society of America Bulletin, v. 105, p. 251–271, doi: 10.1130/0016-7606(1993)105<0251:NEAMOT>2.3.CO;2.

Zachos, J.C., and Kump, L.R., 2005, Carbon cycle feedbacks and the initiation of Antarctic glaciation in the earliest Oligocene: Global and Planetary Change, v. 47, p. 51–66, doi: 10.1016/j.gloplacha.2005.01.001.

Zachos, J.C., Stott, L.D., and Lohmann, K.C., 1994, Evolution of early Cenozoic marine temperatures: Paleoceanography, v. 9, p. 353–387, doi: 10.1029/93PA03266.

Zachos, J.C., Quinn, T.M., and Salamy, K.A., 1996, High resolution (10^4 years) deep sea foraminiferal stable isotope records of the Eocene-Oligocene climate transitions: Paleoceanography, v. 11, p. 251–266, doi: 10.1029/96PA00571.

Zachos, J.C., Pagani, M., Sloan, L., Thomas, E., and Billups, K., 2001, Trends, rhythms, and aberrations in global climate 65 Ma to Present: Science, v. 292, p. 686–693, doi: 10.1126/science.1059412.

Manuscript Accepted by the Society 16 September 2008

Printed in the USA

The Geological Society of America
Special Paper 452
2009

Climate threshold at the Eocene-Oligocene transition: Antarctic ice sheet influence on ocean circulation

Kenneth G. Miller
James D. Wright
Department of Earth and Planetary Sciences, Rutgers University, Piscataway, New Jersey 08854, USA

Miriam E. Katz
Department of Earth and Environmental Sciences, Rensselaer Polytechnic Institute, Troy, New York 12180, USA

Bridget S. Wade
Department of Geology & Geophysics, Texas A&M University, College Station, Texas 77840, USA

James V. Browning
Benjamin S. Cramer
Department of Earth and Planetary Sciences, Rutgers University, Piscataway, New Jersey 08854, USA

Yair Rosenthal
Institute of Marine & Coastal Sciences and Department of Earth and Planetary Sciences, Rutgers University, New Brunswick, New Jersey 08901, USA

ABSTRACT

We present an overview of the Eocene-Oligocene transition from a marine perspective and posit that growth of a continent-scale Antarctic ice sheet (25×10^6 km^3) was a primary cause of a dramatic reorganization of ocean circulation and chemistry. The Eocene-Oligocene transition (EOT) was the culmination of long-term (10^7 yr scale) CO_2 drawdown and related cooling that triggered a 0.5‰–0.9‰ transient precursor benthic foraminiferal $\delta^{18}O$ increase at 33.80 Ma (EOT-1), a 0.8‰ $\delta^{18}O$ increase at 33.63 Ma (EOT-2), and a 1.0‰ $\delta^{18}O$ increase at 33.55 Ma (oxygen isotope event Oi-1). We show that a small (~25 m) sea-level lowering was associated with the precursor EOT-1 increase, suggesting that the $\delta^{18}O$ increase primarily reflected 1–2 °C of cooling. Global sea level dropped by 80 ± 25 m at Oi-1 time, implying that the deep-sea foraminiferal $\delta^{18}O$ increase was due to the growth of a continent-sized Antarctic ice sheet and 1–4 °C of cooling. The Antarctic ice sheet reached the coastline for the first time at ca. 33.6 Ma and became a driver of Antarctic circulation, which in turn affected global climate, causing increased latitudinal thermal gradients and a "spinning up" of the oceans that resulted in: (1) increased thermohaline circulation and erosional pulses of Northern Component Water and Antarctic Bottom Water; (2) increased deep-basin ventilation, which caused a decrease in oceanic residence time, a decrease in deep-ocean acidity, and a deepening of the calcite compensation depth (CCD); and (3) increased diatom diversity due to intensified upwelling.

Keywords: Antarctica, sea level, oxygen isotopes, Oligocene, Eocene.

Miller, K.G., Wright, J.D., Katz, M.E., Wade, B.S., Browning, J.V., Cramer, B.S., and Rosenthal, Y., 2009, Climate threshold at the Eocene-Oligocene transition: Antarctic ice sheet influence on ocean circulation, *in* Koeberl, C., and Montanari, A., eds., The Late Eocene Earth—Hothouse, Icehouse, and Impacts: Geological Society of America Special Paper 452, p. 169–178, doi: 10.1130/2009.2452(11). For permission to copy, contact editing@geosociety.org.

INTRODUCTION: THE BIG CHILL

The Eocene-Oligocene transition is the most profound oceanographic and climatic change of the past 50 m.y. Cooling began in the middle Eocene and culminated in the major earliest Oligocene $\delta^{18}O$ increase (oxygen isotope event Oi-1, ca. 33.55 Ma). The Eocene-Oligocene transition is the largest of three Cenozoic deep-sea benthic foraminiferal $\delta^{18}O$ increases (Fig. 1), which also include the middle Miocene (ca. 14.8 Ma), when a permanent (i.e., dry-based) Antarctic ice sheet developed (Shackleton and Kennett, 1975), and the late Pliocene (ca. 2.6 Ma), when Northern Hemisphere ice volume increased (Shackleton et al., 1984) ("the Ice Ages"). Most studies agree that the earliest Oligocene $\delta^{18}O$ increase (Oi-1; 33.55 Ma) signaled the beginning of the icehouse Earth, with large ice sheets on Antarctica (Miller et al., 1991, 2005a, 2005b; Zachos et al., 1996). The 33.55 Ma event is marked by a 1.0‰ increase in deep-sea benthic foraminiferal $\delta^{18}O$ throughout the Atlantic, Pacific, Indian, and Southern Oceans (Corliss et al., 1984; Coxall et al., 2005; Keigwin, 1980; Kennett and Shackleton, 1976; Miller et al., 1987; Savin et al., 1975; Shackleton and Kennett, 1975; Zachos et al., 2001) (Fig. 1). A latest Eocene precursor 0.5‰ $\delta^{18}O$ increase has been documented in the deep Pacific (Coxall et al., 2005) and Gulf Coast paleoshelf (Miller et al., 2008a; Katz et al., 2008), yielding an overall increase of ~1.5 ‰ across the Eocene-Oligocene transition (Fig. 2).

Deep-sea benthic foraminiferal $\delta^{18}O$ records reflect changes in both temperature and $\delta^{18}O_{seawater}$ due to ice-volume changes, complicating interpretations of the Eocene-Oligocene transition.[1] Over the past 50 m.y., long-term (10^7 yr) deep sea benthic foraminiferal $\delta^{18}O$ values have increased by ~5‰ (Miller et al., 1987, 2005a) (Fig. 1). Assuming that ice-volume changes can only explain ~2.0‰ of the total increase (assuming 0.11‰ per 10 m sea-level change; Fairbanks and Matthews, 1978; see following for discussion of this calibration), more than half of the long-term increase must be attributed to an ~12 °C cooling.

[1]Though local fractionation resulting from evaporation/precipitation changes affects the surface ocean, it has a minimal effect on deep-sea $\delta^{18}O$ records, which reflect ice volume and deep-sea temperature (which in turn reflect high-latitude surface temperature).

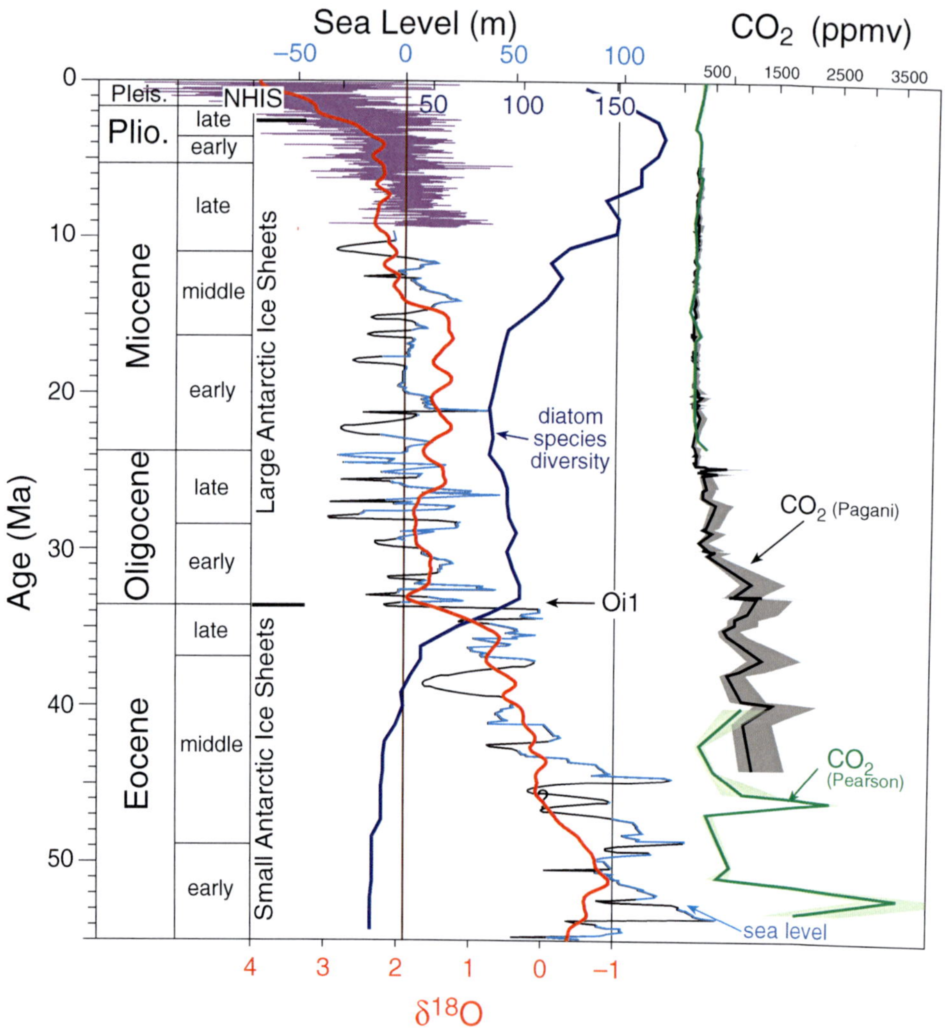

Figure 1. Comparison of the global sea level (blue indicates intervals constrained by data; black indicates estimated lowstands; Kominz et al., 2008), benthic foraminiferal $\delta^{18}O$ values (reported to *Cibicidoides* and smoothed to remove periods shorter than 1 m.y.), diatom diversity curve (Katz et al., 2005), and atmospheric CO_2 estimates derived from alkenones (Pagani et al., 1999, 2005; stippled pattern—error) and boron isotopes (Pearson and Palmer, 2000; stippled pattern—error). NHIS—Northern Hemisphere ice sheets. The benthic foraminiferal curve is the Atlantic synthesis of Miller et al. (2005a; http://geology.rutgers.edu/miller.shtml), which is smoothed to remove periods shorter than 1 m.y.

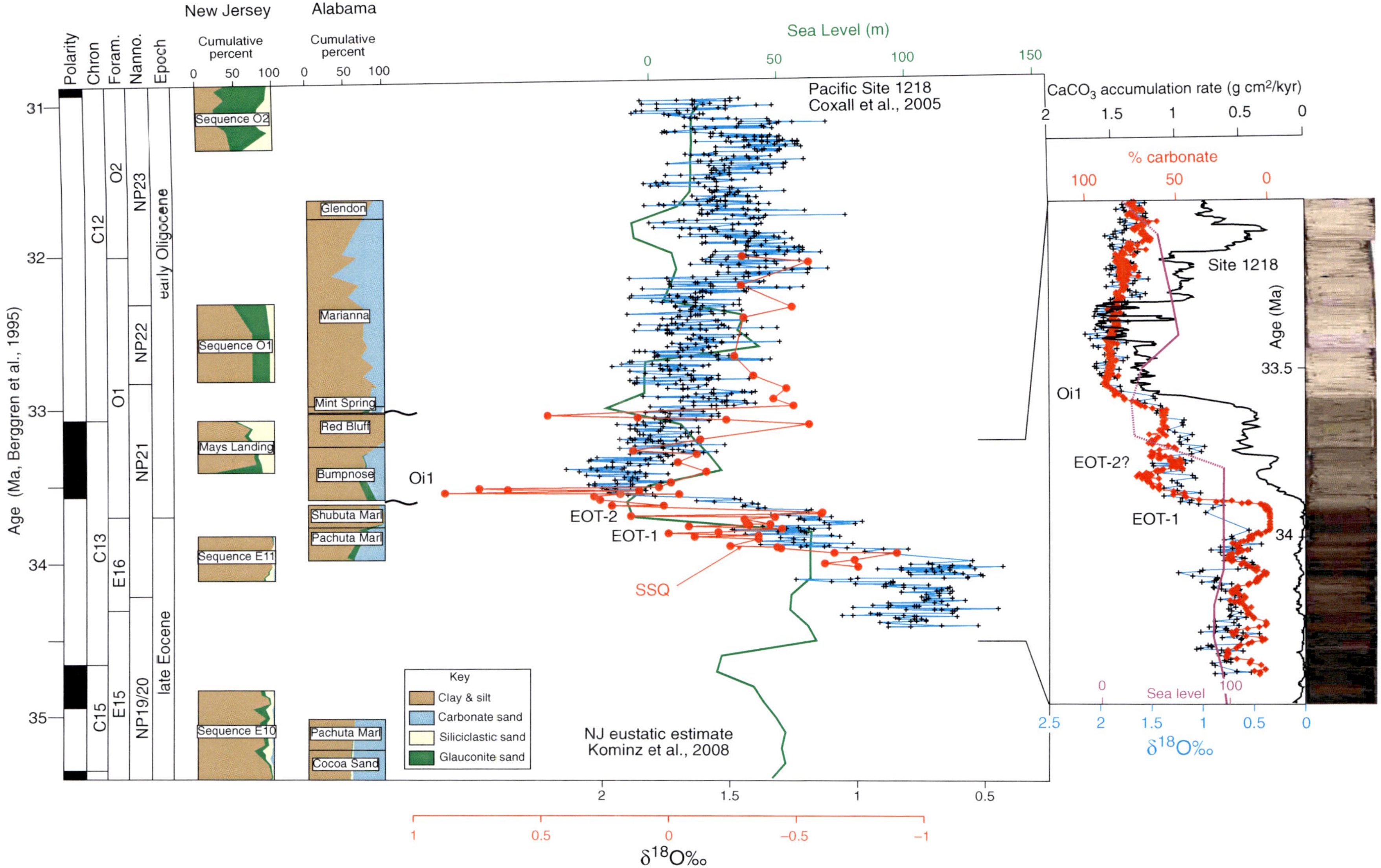

Figure 2. Comparison of continental margin records from New Jersey and Alabama (Miller et al., 2008a), global sea-level estimates (Kominz et al., 2008), and benthic foraminiferal $\delta^{18}O$ records from deep Pacific ODP Site 1218 (Coxall et al., 2005) and St. Stephens Quarry (SSQ), Alabama (Katz et al., 2008). The isotopes from SSQ provide first-order correlations to the sequences and sequence boundaries at SSQ. Close-up of Site 1218 oxygen isotopes at right includes carbonate percent and mass accumulation rate (AR) data (Coxall et al., 2005) and also shows two large drops associated with the EOT-1 and Oi-1 oxygen isotope increases, which are reflected in the core photograph at right (light color—carbonate rich, dark—carbonate poor). Figure was modified after Miller et al. (2008a).

Benthic foraminiferal Mg/Ca data show 12 °C of cooling over the last 50 m.y. (Lear et al., 2000), supporting this interpretation. On shorter time scales (10^5–10^6 yr), distinguishing the relative contributions of ice versus temperature on $\delta^{18}O$ records is difficult. As a result, considerable controversy has surrounded hypotheses about the cause(s) of the latest Eocene through earliest Oligocene $\delta^{18}O$ increases, ranging from early studies that attributed them entirely to deep-water cooling (and hence, high-latitude surface-water cooling; Shackleton and Kennett, 1975), to recent studies that ascribe the increase largely to ice-sheet expansion (Tripati et al., 2005). This latter interpretation requires continental ice storage that is ~1.5 times larger than modern ice sheets in Antarctica and the Northern Hemisphere, and it predicts a 150 m global sea-level (eustatic) fall. The extent of bipolar glaciation has been disputed (Edgar et al., 2007; Katz et al., 2008; Miller et al., 2008a) and is discussed in detail later.

The Eocene-Oligocene transition was associated with a dramatic oceanographic reorganization and the largest climatic cooling of the Cenozoic. There was a major increase in ocean productivity (see summary in Berger, 2007), a very large drop in the calcite compensation depth (CCD) (Van Andel, 1975; Coxall et al., 2005; Rea and Lyle, 2005), and pulses of strongly eroding Antarctic bottom water (see summaries in Kennett [1977] and Wright and Miller [1996]) and Northern Component Water (see summaries in Tucholke and Mountain, 1979; Miller and Tucholke, 1983; and Miller, 1992). Paleontological evidence for cooling includes the development of psychrospheric ("cold-loving") ostracods (Benson, 1975), deep-sea and shelf benthic foraminiferal extinctions and appearances (e.g., Miller et al., 1992; Thomas, 1992), extinctions in tropical planktonic and larger foraminifera (Adams et al., 1986; Keller et al., 1992; Pearson et al., 2008; Wade and Pearson, 2008), and the decline of thermophilic calcareous nannoplankton (Aubry, 1992; Dunkley Jones et al., 2008). Terrestrial cooling is indicated by pollen changes (e.g., New Jersey; Owens et al., 1988) and a mammalian turnover (e.g., England; Hooker et al., 2004). Regional aridification was associated with terrestrial cooling (e.g., the Himalayas; Dupont-Nivet et al., 2007; Zanazzi et al., 2007).

Various hypotheses linking these Eocene-Oligocene changes have been postulated. Until recently, the predominant hypothesis has been that the undocking of Australia from Antarctica and the opening of the Drake Passage led to development of the Antarctic Circumpolar Current. This thermally isolated Antarctica from the relatively warm water coming from the Subtropical Gyre, thereby cooling the continent (Kennett, 1977). The role of the Antarctic Circumpolar Current and thermal isolation has recently been both affirmed (Exon et al., 2004) and challenged (Huber et al., 2004), leaving the role of the gateway uncertain. Alternatively, a coupled global climate-ocean-ice model of the Eocene-Oligocene transition (DeConto and Pollard, 2003) has suggested that pervasive glaciation of Antarctica was triggered by a drop of atmospheric CO_2 below a critical threshold of 2.8 times pre-anthropogenic levels. The DeConto and Pollard (2003) model also suggested that opening of the Drake Passage could only have had a role in triggering the glaciation in the context of atmospheric CO_2 already near the critical threshold, although the increase in biological productivity, stimulated by the initiation/increase of the Antarctic Circumpolar Current, might have indirectly contributed to lowering atmospheric pCO_2 (Scher and Martin, 2006). Though the role of gateways versus CO_2 remains unresolved, it is clear that cooling in Antarctica resulted in the development of a large ice sheet across the Eocene-Oligocene transition. We hypothesize here that when the Antarctic ice sheet grew above a critical threshold, extending beyond the coastline for the first time, it began to influence ocean and climate changes.

Here, we synthesize published $\delta^{18}O$ records spanning the Eocene-Oligocene transition (Fig. 1) and compare them with records of sea level (Miller et al., 2005a; Kominz et al., 2008), passive margin sequences (Miller et al., 2005a, 2008a), CO_2 (Pagani et al., 1999, 2005; Pearson and Palmer, 2000), diatom diversity (Katz et al., 2005), and deep-sea carbonate accumulation (Coxall et al., 2005). We distinguish temperature from ice-volume effects by comparing $\delta^{18}O$ records (Figs. 1 and 2) with: (1) a global sea-level estimate that was derived from backstripping[2] New Jersey coastal plain core holes (Miller et al., 2005a; Kominz et al., 2008) and from evaluation of sequence stratigraphic records from St. Stephens Quarry (SSQ), Alabama (Miller et al., 2008a); and (2) by comparison with published Mg/Ca records from Deep Sea Drilling Project (DSDP) Site 522 (Lear et al., 2000), Ocean Drilling Program (ODP) Site 1218 (Lear et al., 2004), and SSQ (Katz et al., 2008).

ICE VOLUME VERSUS TEMPERATURE CHANGES AND THE EOCENE-OLIGOCENE TRANSITION

The Precursor (EOT1 and EOT2) Increases

Katz et al. (2008) named the transient precursor $\delta^{18}O$ increase (33.8 Ma) at Sites 1218 (Coxall et al., 2005), 522 (Zachos et al., 1996), and SSQ (Miller et al., 2008a; Katz et al., 2008) "EOT-1" for "Eocene-Oligocene transition event 1." EOT-1 is associated with an extinction in planktonic foraminifera (the *Turborotalia cerroazulensis* lineage) (Miller et al., 2008a; Pearson et al., 2008) that has been dated at 33.80 Ma (Berggren et al., 1995). At SSQ, EOT-1 is manifested as a 0.9‰ $\delta^{18}O$ increase in neritic (~75 m paleodepth) benthic foraminifera and 2.5 °C cooling as determined by Mg/Ca paleothermometry (Katz et al., 2008). A similar cooling (2.0 °C) at deep-sea Sites 522 and 1218 determined by Mg/Ca is associated with benthic foraminiferal $\delta^{18}O$ increases of ~0.5‰. This suggests that a $\delta^{18}O_{seawater}$ increase of ~0.4‰ (SSQ) to 0.1‰ (522 and 1218) was associated with EOT-1 (Katz et al., 2008). A $\delta^{18}O_{seawater}$ change of 0.1‰–0.4‰ would have resulted from an ~10–35 m sea-level fall using calibrations discussed herein later.

[2]This is a method that progressively removes the effects of compaction, loading (either Airy or two-dimensional flexural loading), and thermal subsidence from margin records. The residual is the result of eustatic and nonthermal tectonic changes. In the absence of nonthermal tectonism, it provides a eustatic estimate.

Sea-level records across EOT-1 are not clear, in part because the published sea-level estimates (Miller et al., 2005a; Kominz et al., 2008) are not sensitive to small changes on short time scales (<1 m.y.). EOT-1 falls near a long hiatus in New Jersey, and, given the temporal resolution, it is likely that this hiatus is a concatenation of EOT-1 and Oi-1. Thus, the sea-level estimates based on New Jersey sequences (Miller et al., 2005a; Kominz et al., 2008) do not reflect this drop. Published sequence stratigraphic analyses at SSQ did not reveal a sea-level fall associated with the EOT-1 increase measured directly on the core (Miller et al., 2008a); however, a glauconite bed occurs at the same level as the EOT-1 increase at SSQ. This glauconite bed may represent a previously undetected unconformity and associated hiatus (i.e., glauconite sands often overlie sequence boundaries in the coastal plain because lowstand deposits are largely missing; Miller et al., 2005a). The SSQ sections are not sensitive to sea-level variations that are less than ~25 m because they were deposited in relatively deep shelf waters (~75 m; Miller et al., 2008a). As noted by Katz et al. (2008), a global sea-level fall at this time is supported by a correlative erosional event in the Priabonian type section (Brinkhuis and Visscher, 1995) in inner neritic carbonate facies most sensitive to small sea-level changes. We follow Katz et al. (2008) in interpreting EOT-1 as primarily a global cooling event of 2 °C and minor sea-level fall (~25 m).

Katz et al. (2008) noted a second earliest Oligocene (ca. 33.63 Ma) ~0.8‰ $\delta^{18}O$ increase, termed EOT-2, at SSQ (with no detectible hiatus), Site 522, and Site 1218. Mg/Ca data are lacking for this event, and its significance in terms of temperature versus ice volume remains unclear. The EOT-2 $\delta^{18}O$ increase occurs within a sequence at SSQ (within the Shubuta Marl) in a section with no marked lithologic or sequence stratigraphic changes and no evidence of an unconformity as expected from a sea-level fall (Miller et al., 2008a). However, EOT-2 occurs in the upper part of the sequence associated with a shallowing-upward section, and estimates from Sites 1218 and 522 indicate a minor change in the $\delta^{18}O_{seawater}$ with EOT-2 (Katz et al., 2008). Thus, though we interpret EOT-2 as primarily a temperature drop, a sea-level component cannot be precluded.

The Oi-1 Increase

The greenhouse-to-icehouse transition culminated in the earliest Oligocene Oi-1 event (33.55 Ma), which is associated with an ~1‰ $\delta^{18}O$ increase in benthic foraminifera at SSQ and DSDP Site 522 and ODP Sites 744 and 1218. Global sea-level estimates derived from New Jersey backstripping (Figs. 1 and 2) show that a eustatic fall of 55 m occurred at ca. 33.5 Ma (Kominz and Pekar, 2001; Miller et al., 2005a; Kominz et al., 2008), correlated with the earliest Oligocene Oi-1 $\delta^{18}O$ increase. This decrease is equivalent to the growth of an Antarctic ice sheet of 25 × 10^6 km^3, similar to the amount stored in East Antarctica today. The 55 m fall explains 0.55‰–0.66‰ of a global change in seawater $\delta^{18}O$ using the sea-level calibration of 0.10‰/10 m derived for the Oligocene (Pekar et al., 2002), the calibration of 0.11‰/10 m derived for the Pleistocene (Fairbanks and Matthews, 1978), or the calibration of 0.12‰/10 m of Katz et al. (2008) for the Eocene-Oligocene. This indicates that the remaining ~0.34‰–0.45‰ of the increase at 33.55 Ma was due cooling of ~1.5–2 °C.

Uncertainties in the sea-level and/or $\delta^{18}O$ calibration potentially range from ~0.055‰ to 0.12‰/10 m (Miller et al., 1987; Pekar et al., 2002; Katz et al., 2008). The outside limits for any sea-level or $\delta^{18}O$ calibration are provided by the upper limit for freezing of ice sheets (~–17‰) and the lower limit of the modern Antarctic ice sheet (~–40‰), yielding a minimum calibration of 0.055‰/10 m (Miller et al., 1987) and a maximum of 0.12‰/10 m (Katz et al., 2008). The Pleistocene calibration is 0.11‰/10 m (Fairbanks and Matthews, 1978). Pekar et al. (2002) estimated an Oligocene calibration of 0.10 ± 0.02‰/10 m, similar to the value of 0.1‰/10 m obtained from climate models (DeConto and Pollard (2003). Katz et al. (2008) determined a calibration of 0.12‰/10 m for the Eocene-Oligocene transition. Thus, empirical calibration suggests that the average sea-level $\delta^{18}O$ calibration for the Cenozoic is 0.11 ± 0.1‰ (Fairbanks and Matthews, 1978; Pekar et al., 2002; Katz et al., 2008).

There are significant uncertainties in the sea-level estimate. Lowstand deposits are generally not represented in the coastal plain records used to construct the eustatic estimate (Miller et al., 2005a; Kominz et al., 2008). By conducting two-dimensional backstripping, Kominz and Pekar (2001) were able to estimate decreases during lowstand and thus reconstruct a nearly complete sea-level cycle for the earliest Oligocene sea-level lowering that correlates with Oi-1. Still, the upper limit of the eustatic fall could have been as high as 70 m. These estimates of a 55–70 m eustatic change do not take into account the isostatic response of the oceanic lithosphere to the change in weight of the overlying water (hydroisostasy). The actual water-volume change from transferring ocean water to glacial ice is ~33% higher than the eustatic change, assuming full isostatic compensation is attained. Hence, the 55–70 m eustatic change corresponds to a volume-equivalent change in the depth/thickness of ocean water of ~82–105 m (correcting for isostatic loading and assuming full compensation). Thus, the sea-level record only places moderate constraints on the ice-volume component of the $\delta^{18}O$ increase: anywhere from 55% to 100% of the 1.0‰ increase can be explained by a sea-level fall of 55–105 m.

Mg/Ca records place additional constraints on the $\delta^{18}O_{seawater}$ changes associated with the Oi-1 increase, though deep-sea Mg/Ca records may be overprinted by other effects (Lear et al., 2004; Katz et al., 2008). Both deep Atlantic DSDP Site 522 and deep Pacific ODP Site 1218 show no cooling or slight warming at Oi-1 time according to Mg/Ca paleothermometry (Lear et al., 2000, 2004). This observation led Tripati et al. (2005) to attribute the bulk of the 1.5‰ $\delta^{18}O$ increase associated with the Eocene-Oligocene transition (i.e., both the EOT-1 and Oi-1 increase) to growth of ice sheets. However, the ~1.2 km drop in the CCD that occurred at the Eocene-Oligocene transition (van Andel, 1975; Coxall et al., 2005; Rea and Lyle, 2005) may have caused changes in carbonate ion activity that masked an estimated 2 °C cooling in deep-sea Mg/Ca records (Lear et al., 2004). At SSQ, the Mg/Ca record supports 2 °C of cooling associated with the ~1.0‰ Oi-1 $\delta^{18}O$ increase, suggesting a 0.55‰ $\delta^{18}O_{seawater}$ change.

Thus, we conclude that the Eocene-Oligocene transition occurred in three major steps: (1) a precursor EOT-1 $\delta^{18}O$ increase at 33.80 Ma that was caused by a 2 °C deep-water cooling and a minor sea-level fall (~25 m) followed by a partial return to pre-event values; (2) a deep-water cooling and minor sea-level fall associated with the EOT-2 increase at 33.63 Ma; and (3) a deep-water cooling of 2 °C and a major sea-level fall of 80 ± 25 m at Oi-1 time (33.45 Ma).[3] This implies the development of an Antarctic ice sheet that was as large or larger than that of present day. It does not require large Northern Hemisphere ice sheets. Edgar et al. (2007) similarly concluded that large Antarctic ice sheets predated large Northern Hemisphere ice sheets. However, it is likely given evidence from Arctic drilling (St. John, 2008) that at least Greenland-sized ice sheets existed during the Eocene and during times of peak glaciation such as Oi-1 (see following).

IMPLICATIONS OF A LARGE ANTARCTIC ICE SHEET

The cause of the earliest Oligocene oxygen isotope increase and the role of CO_2 (e.g., Pagani et al., 2005) versus changing ocean gateways (Kennett, 1977) have been hotly contested. The long-term $\delta^{18}O$ trend parallels atmospheric CO_2 estimates (Pagani et al., 1999, 2005), linking long-term (10^7 yr) global cooling to lower CO_2 (Fig. 1). Published modeling results (DeConto and Pollard, 2003) emphasized the importance of CO_2 on cryospheric development, where gateways played a role only in adjusting the critical pCO_2 threshold for Antarctic glaciation. These modeling studies suggest that a decrease in atmospheric CO_2 below a critical threshold (2.4–2.8 times pre-anthropogenic levels [280 ppm] for models with open and closed Drake Passage) in the earliest Oligocene could have triggered the growth of a large Antarctic ice sheet (Fig. 1), and sea-level estimates (Miller et al., 2005a; Kominz et al., 2008) that require a near modern-sized Antarctic ice sheet developed at ca. 33.55 Ma. DeConto et al. (2007) linked the growth of the Antarctic ice sheet to a CO_2 fall below this threshold, which is roughly consistent with the observed CO_2 estimates (Fig. 1). Though it is likely that there were small (5–15 × 10^6 km^3), ephemeral ice sheets during the greenhouse world of the Late Cretaceous to Eocene (Miller et al., 2005b), the earliest Oligocene saw the first continent-sized ice sheet since the Permian (ca. 280 Ma). Recent studies point toward much older Northern Hemisphere glaciation than was previously thought (e.g., middle Eocene; Moran et al., 2006; Eldrett et al., 2007), but the restriction of physical evidence for Northern Hemisphere glaciation (see summary in Wright and Miller, 1996) necessarily relegates it to a relatively small role (e.g., similar to the modern-day 5–7 m sea-level equivalent stored in Greenland) until the last 2.6 m.y.

We suggest that the Antarctic ice sheet became a driver of, not just a response to, climate change beginning in the earliest Oligocene, causing increased meridional thermal gradients and a "spinning up" of the oceans (greater thermohaline and wind-driven circulation). Enhanced wind circulation due to katabatic winds resulted in increased wind shear/curl and hence increased circum-Antarctic upwelling. In addition, increased salt rejection due to ice-shelf and sea-ice freezing led to enhanced formation of Antarctic Bottom Water (AABW). Combined, these factors led to increased ventilation of the deep ocean, a decrease in residence time, and increased Southern Ocean productivity as discussed by Berger (2007). During the Eocene-Oligocene transition, the Antarctic ice sheet reached sufficient proportions to influence the ocean adjacent to the continent, resulting in increased sea ice and deep-water production. As first noted by Kennett and Shackleton (1976) and recently emphasized by models (DeConto et al., 2007), the formation of extensive sea ice likely played a role in changing deep-water circulation. There is considerable evidence that the Eocene-Oligocene transition was associated with the greatest change in deep-ocean circulation of the past 50 m.y. (Fig. 3). Thermohaline circulation dramatically increased in the early Oligocene (Fig. 3) with pulses of Southern Component Water (SCW; analogous to modern AABW; Kennett, 1977; Wright and Miller, 1996) and Northern Component Water (NCW; analogous to modern North Atlantic Deep Water) (Miller and Tucholke, 1983; Miller, 1992; Davies et al., 2001). An erosional event in the southern oceans caused hiatuses in numerous early Oligocene southern ocean cores and has been attributed to an increase in SCW (Kennett, 1977; Wright and Miller, 1996). The high supply of SCW is also indicated by the very high $\delta^{18}O$ values measured at high southern latitudes (Kennett and Stott, 1990), indicative of a "cold spigot" (Miller, 1992; Miller et al., 2008b). Seismic profiles and hiatus distributions document a coeval early Oligocene pulse of widespread erosion that cut unconformities associated with Horizon Au (Fig. 3; Mountain and Miller, 1992; Tucholke and Vogt, 1979) and reflector R4 (Miller and Tucholke, 1983) in the North Atlantic. This erosion correlates with high $\delta^{13}C$ values in the North Atlantic following the 33.55 Ma Oi-1 event (Fig. 3; Miller, 1992), supporting seismic and hiatus evidence that indicates a NCW source in the earliest Oligocene. Northern Hemisphere high latitudes cooled concomitantly with polar cooling in the Southern Hemisphere, making the North Atlantic a viable source region for deep water (albeit briefly) in the earliest Oligocene (see also Via and Thomas, 2006; Davies et al., 2001; Abelson et al., 2007).

A dramatic deepening of the CCD occurred across the Eocene-Oligocene transition (van Andel, 1975; Coxall et al., 2005; Rea and Lyle, 2005). Both the 33.80 (EOT-1) and 33.55 Ma (Oi1) $\delta^{18}O$ increases were associated with increases in carbonate content and mass accumulation rates in the deep Pacific (Coxall et al., 2005), and there appears to be a third, smaller increase associated with EOT-2 (Fig. 2). This indicates

[3]These estimates include errors in the eustatic estimate plus those due to water loading; e.g., the eustatic estimate for Oi-1 is 55–70 m, with water thickness increase of 55 m (i.e., no loading) to 105 m (i.e., full compensation).

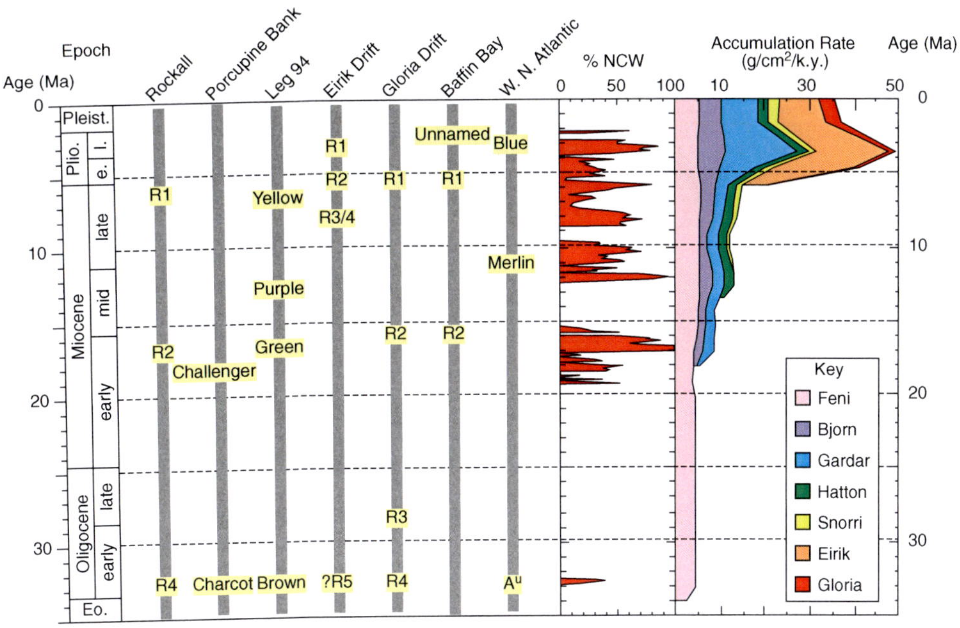

Figure 3. Age of North Atlantic seismic reflectors showing widespread erosional unconformities at ca. 34 Ma, ca. 20–16 Ma, ca. 13–10 Ma, and ca. 5 Ma; W.N.—Western North. Estimated flux of Northern Component Water (NCW) was derived using $\delta^{13}C$ records from the North Atlantic, Southern, and Pacific Oceans (Wright and Miller, 1996). Accumulation rates of northern North Atlantic drifts modified after Wright and Miller (1996). Note that increase in total drift accumulation corresponds to detection of NCW using $\delta^{13}C$ records.

that the CCD deepening occurred in two to three steps tightly coupled to the oxygen isotopic increases. We also note that although the drop in $\%CaCO_3$ at EOT-1 was as large as or larger than at Oi-1 time (Fig. 2), the carbonate accumulation rate data indicate that it was the latter event that was more significant. The deepening of the CCD can be attributed to one of the following: (1) an increase in deep-basin carbonate deposition at the expense of shallow carbonates, likely caused by sea-level fall (shelf-basin fractionation); (2) increased weathering of silicate rocks; (3) a global intensification of carbonate export to the deep sea, possibly caused by increased continental input to the oceans, or (4) a decrease of deep-ocean residence time.

Previous studies have attributed the CCD drop to a glacio-eustatic fall and a shift in shelf-basin fractionation (Coxall et al., 2005; Merico et al., 2008). We argue that shelf-basin fractionation cannot be invoked to explain the CCD drop because: (1) the sea-level fall associated with Oi-1 (33.55 Ma) was too small (80 ± 25 m) to account for the large deepening of the CCD (i.e., numerous Cenozoic sea-level events of this magnitude [Miller et al., 2005a] had little effect on the CCD); and (2) there was a CCD drop associated with EOT-1 despite that fact that there was only a small (~25 m) sea-level fall at that time. Though weathering may have increased in the earliest Oligocene (Robert and Kennett, 1997) and contributed to a global increase in carbonate production, global Oligocene sedimentation rates were low, and it is doubtful that input changes can explain the CCD drop. Rea and Lyle (2005) similarly argued convincingly that the CCD drop could not be entirely due to shelf-basin shift and suggested that a sudden increase in weathering and erosion rates would be unlikely to account for the change, thus implicating variations in deep-sea preservation. We argue that the links among deep-sea temperature, an increase in thermohaline circulation, and the drop in the CCD are manifest as to its cause: increased thermohaline circulation and deep-basin ventilation caused a decrease in oceanic residence time and a decrease in deep-ocean acidity, allowing carbonate to be preserved at greater depth and produce a deeper CCD.

One reason that modeling studies (Merico et al., 2008) have favored shelf-basin fractionation for the CCD drop and ignored the importance of deep-sea ventilation is because they attempted to explain not only the CCD drop but also high global $\delta^{13}C$ values (e.g., Zachos et al., 1996). Carbon isotopic values began to increase at Oi-1 time, and high values lasted for ~1 m.y. (Zachos et al., 1996). Models showed that the CCD drop could be explained by an increase in organic carbon burial, an increase in silica over calcareous plankton deposition, or a shelf-basin fractionation, but only a transient event such as sea-level change could explain the transitory $\delta^{13}C$ peak (Merico et al., 2008). We note that the high $\delta^{13}C$ values must reflect a change in the input or burial of organic carbon relative to carbonate carbon and that such million-year-scale $\delta^{13}C$ cycles of ~1‰ typify the Oligocene to Miocene (Miller and Fairbanks, 1985). We argue that the high carbon isotopic values were not coupled directly to the CCD drop, and the $\delta^{13}C$ increase was due to a million-year-scale increase in organic carbon export relative to carbonate (as first noted by Zachos et al., 1996), and the CCD drop was due to increased deep-sea residence time.

Antarctic cooling coupled with tropical temperature stability (Pearson et al., 2007) implies a sharp increase in latitudinal thermal gradients (Shackleton and Kennett, 1975) that caused increased atmospheric circulation (Dupont-Nivet et al., 2007) and ocean upwelling (Berger, 2007). EOT-2 (33.63 Ma) corresponds to a major drop in nannofossil diversity in Tanzanian cores (Dunkley Jones et al., 2008), which is interpreted as a response to surface-water eutrophication. The increased latitudinal and vertical thermal gradients caused a "spinning up" of the oceans and increased wind-driven upwelling and mesoscale eddy turbulence that contributed to greater nutrient recycling, which in turn stimulated a rapid diversification of diatoms (Falkowski et al., 2004a, 2004b; Katz et al. 2004; Finkel et al., 2005, 2007). As summarized by Katz et al. (2005), marine diatoms are favored over other eukaryotic phytoplankton by conditions with high turbulent mixing and thus are related to increased water-column stratification and latitudinal thermal gradients. Diatom species diversity remarkably mirrors the deep-sea $\delta^{18}O$ record across the Eocene-Oligocene transition (Fig. 1), as expected if deep-sea temperatures were also related to latitudinal thermal gradients and water-column stratification.

We note that diatom diversity increased not only at the Eocene-Oligocene transition but also in the middle Miocene and late Miocene–Pliocene (Fig. 1), when diversification included more bloom-forming diatoms (requiring high nutrient levels); these were also intervals of increased latitudinal thermal gradients and enhanced global circulation (Wright and Miller, 1996). Our comparisons (Fig. 1) indirectly link silica productivity to these intervals of global cooling and increased latitudinal thermal gradients.

As hypothesized by DeConto and Pollard (2003), atmospheric CO_2, which declined below a critical threshold in the earliest Oligocene, may have propelled the climate system into a new state, one with a large Antarctic ice sheet. We suggest the continent-sized ice sheet provided positive feedback that was felt throughout the climate system, probably through the profound effects of sea ice on ocean circulation and chemistry. Though there are large uncertainties in atmospheric CO_2 estimates (Fig. 1), it appears that the broad-scale trends in Cenozoic CO_2 and the long-term (10^7 yr scale) sea-level record (Fig. 1) track each other. The first-order trends of decreasing atmospheric CO_2 and falling sea level appear to have been coupled since 50 Ma, strongly implying a link through changes in ocean crust production rates (Larson, 1991; Miller et al., 2005a). The global long-term cooling of ~12 °C associated with the lowering of CO_2 can only explain 12 m of steric sea-level fall, and therefore the bulk of the long-term fall of ~100 m (Kominz et al., 2008) must have been due to: (1) long-term ice growth and isostatic compensation (~50 m); and (2) lower ocean crust production rates (~40 m). Higher estimates of long-term fall (e.g., Müller et al., 2008) would require greater reduction in ocean crust production rates. The cause of the drop in CO_2 is less clear, and it may be attributed to lower degassing due to lower ocean crust production rates and/or possibly to higher continental weathering rates. The correlation between long-term CO_2 and long-term sea level (Fig. 1) suggests that they are linked through ocean crust production rates.

This evaluation of the Eocene-Oligocene transition provides perspective on the evolution of Cenozoic climate. Oxygen isotope shifts that occur through the Eocene-Oligocene transition (Fig. 1), in the middle Miocene (Fig. 1), and in the mid-Pliocene overshoot estimates of CO_2 and sea-level change, suggesting amplification within the climate system. This implies that CO_2 was not the only driver of Cenozoic climate change, and it leaves open the possibility that CO_2 changes were augmented by a positive feedback mechanism for climate change (e.g., large Antarctic ice sheets, increasing latitudinal thermal gradients, and ocean circulation, resulting in additional CO_2 drawdown and cooling). These events mark major changes in the climate state that were amplified by processes other than changes in CO_2, as first suggested in a prescient paper by Berger (1982). In the case of these three events, it appears that the development of continental-scale ice sheets was an important factor in the amplification of the climate system: (1) in the Eocene-Oligocene transition, the development of a continental-scale ice sheet on Antarctica resulted in major feedbacks through thermohaline circulation and ocean upwelling; (2) in the middle Miocene, the development of a permanent (i.e., dry-based) ice sheet in Antarctica further increased latitudinal thermal gradients (Wright and Miller, 1996); and (3) in the Pliocene, the growth of large Northern Hemisphere ice sheets exerted another amplification, perhaps through a reduction in NCW.

ACKNOWLEDGMENTS

We thank C. Koeberl, D. Hodell, and an anonymous reviewer for comments. This work was supported by National Science Foundation (NSF) grants EAR-06-06693 (Miller) and OCE-06-23256 (Katz, Miller, Wade, Wright).

REFERENCES CITED

Abelson, M., Agnon, A., and Almogi-Labina, A., 2007, Indications for control of the Iceland plume on the Eocene-Oligocene "greenhouse-icehouse" climate transition: Earth and Planetary Science Letters, v. 265, p. 33–48, doi: 10.1016/j.epsl.2007.09.021.

Adams, C.G., Butterlin, J., and Samanta, B.K., 1986, Larger foraminifera and events at the Eocene-Oligocene boundary in the Indo–West Pacific region, *in* Pomerol, C., and Premoli Silva, I., eds., Terminal Eocene Events: Amsterdam, Elsevier, p. 237–252.

Aubry, M.-P., 1992, Late Paleogene calcareous nannoplankton evolution: A tale of climatic deterioration, *in* Prothero, D., and Berggren, W.A., The Eocene-Oligocene Climatic and Biotic Changes: Princeton, New Jersey, Princeton University Press, p. 272–309.

Benson, R., 1975, The origin of the psychrosphere as recorded in changes of deep-sea ostracode assemblages: Lethaia, v. 8, p. 69–83, doi: 10.1111/j.1502-3931.1975.tb00919.x.

Berger, W.H., 1982, Deep-sea stratigraphy: Cenozoic climate steps and the search for chemoclimatic feedback, *in* Einsele, G., and Seilacher, A., eds., Cyclic and Event Stratification: New York, Springer, p. 121–157.

Berger, W.H., 2007, Cenozoic cooling, Antarctic nutrient pump, and the evolution of whales: Deep-Sea Research, Part II—Topical Studies in Oceanography, v. 54, p. 2399–2421, doi: 10.1016/j.dsr2.2007.07.024.

Berggren, W.A., Kent, D.V., Swisher, C.C., III, and Aubry, M.-P., 1995, A revised Cenozoic geochronology and chronostratigraphy, *in* Berggren, W.A., Kent, D.V., and Hardenbol, J., eds., Geochronology, Time Scales and Global Stratigraphic Correlations: A Unified Temporal Framework for an Historical Geology: Society for Sedimentary Geology (SEPM) Special Publication 54, p. 129–212.

Brinkhuis, H., and Visscher, H., 1995, The upper boundary of the Eocene series: A reappraisal based on dinoflagellate cyst biostratigraphy and sequence stratigraphy, *in* Berggren, W.A., Kent, D.V., and Hardenbol, J., eds., Geochronology, Time Scales and Global Stratigraphic Correlations: A Unified Temporal Framework for an Historical Geology: Society for Sedimentary Geology (SEPM) Special Publication 54, p. 295–304.

Corliss, B.H., Aubry, M.-P., Berggren, W.A., Fenner, J.M., Keigwin, L.D., Jr., and Keller, G., 1984, The Eocene/Oligocene boundary event in the deep sea: Science, v. 226, p. 806–810, doi: 10.1126/science.226.4676.806.

Coxall, H.K., Wilson, P.A., Pälike, H., Lear, C.H., and Backman, J., 2005, Rapid stepwise onset of Antarctic glaciation and deeper calcite compensation in the Pacific Ocean: Nature, v. 433, p. 53–57, doi: 10.1038/nature03135.

Davies, R., Cartwright, J., Pike, J., and Line, C., 2001, Early Oligocene initiation of North Atlantic Deep Water formation: Nature, v. 410, p. 917–920, doi: 10.1038/35073551.

DeConto, R., and Pollard, D., 2003, Rapid Cenozoic glaciation of Antarctica induced by declining atmospheric CO_2: Nature, v. 421, p. 245–249, doi: 10.1038/nature01290.

DeConto, R., Pollard, D., and Harwood, D., 2007, Sea ice feedback and Cenozoic evolution of Antarctic climate and ice sheets: Paleoceanography, v. 22, p. PA3214, doi: 10.1029/2006PA001350.

Dunkley Jones, T., Bown, P.R., Pearson, P.N., Wade, B.S., Coxall, H.K., and Lear, C.H., 2008, Major shifts in calcareous phytoplankton assemblages through the Eocene-Oligocene transition of Tanzania and their implications for low-latitude primary production: Paleoceanography, v. 23, PA4204, doi: 10.1029/2008PA001640.

Dupont-Nivet, G., Krijgsman, W., Langereis, C.G., Abels, H.A., Dai, S., and Fang, X., 2007, Tibetan Plateau aridification linked to global cooling at the Eocene-Oligocene transition: Nature, v. 445, p. 635–638, doi: 10.1038/nature05516.

Edgar, K.M., Wilson, P.A., Sexton, P.F., and Suganuma, Y., 2007, No extreme bipolar glaciation during the main Eocene calcite compensation shift: Nature, v. 448, p. 908–911, doi: 10.1038/nature06053.

Eldrett, J.S., Harding, I.C., Wilson, P.A., Butler, E., and Roberts, A.P., 2007, Continental ice in Greenland during the Eocene and Oligocene: Science, v. 446, p. 176–179.

Exon, N.F., Kennett, J.P., and Malone, M.J., 2004, Leg 189 synthesis: Cretaceous-Holocene history of the Tasmanian gateway, *in* Exon, N.F., Kennett, J.P., and Malone, M.J., et al., Proceedings of the Ocean Drilling Program, Scientific Results, Volume 189: College Station, Texas, Ocean Drilling Program, p. 1–37.

Fairbanks, R.G., and Matthews, R.K., 1978, The marine oxygen isotopic record in Pleistocene coral, Barbados, West Indies: Quaternary Research, v. 10, p. 181–196, doi: 10.1016/0033-5894(78)90100-X.

Falkowski, P.G., Katz, M.E., Knoll, A., Quigg, A., Raven, J.A., Schofield, O., and Taylor, M., 2004a, The evolutionary history of eukaryotic phytoplankton: Science, v. 305, p. 354–360, doi: 10.1126/science.1095964.

Falkowski, P.G., Schofield, O., Katz, M.E., van de Schootbrugge, B., and Knoll, A., 2004b, Color wars: Why is the land green and the ocean red?, *in* Thierstein, H., and Young, J.R., eds., Coccolithophores—From Molecular Processes to Global Impact: Berlin, Springer-Verlag, p. 429–453.

Finkel, Z.V., Katz, M.E., Wright, J.D., Schofield, O.M.E., and Falkowski, P.G., 2005, Climatically driven evolutionary change in the size of diatoms over the Cenozoic: Proceedings of the National Academy of Sciences of the United States of America, v. 102, p. 8927–8932, doi: 10.1073/pnas.0409907102.

Finkel, Z.V., Sebbo, J., Feist-Burkhardt, S., Irwin, A.J., Katz, M.E., Schofield, O.M.E., Williams, J., Young, J., and Falkowski, P.G., 2007, A universal driver of macroevolutionary change in the size of marine phytoplankton over the Cenozoic: Proceedings of the National Academy of Sciences of the United States of America, v. 104, no. 51, p. 20,416–20,420, doi: 10.1073/pnas.0709381104.

Hooker, J.J., Collinson, M.E., and Siller, N.P., 2004, Eocene-Oligocene mammalian faunal turnover in the Hampshire Basin, UK: Calibration to the global time scale and the major cooling event: Journal of the Geological Society of London, v. 161, p. 161–172, doi: 10.1144/0016-764903-091.

Huber, M., Brinkhuis, H., Stickley, C.E., Döös, K., Sluijs, A., Warnaar, J., Schellenberg, S.A., and Williams, G.L., 2004, Eocene circulation of the Southern Ocean: Was Antarctica kept warm by subtropical waters?: Paleoceanography, v. 19, p. PA4026, doi: 10.1029/2004PA001014.

Katz, M.E., Finkel, Z., Grzebyk, D., Knoll, A.H., and Falkowski, P.G., 2004, Evolutionary trajectories and biogeochemical impacts of marine eukaryotic phytoplankton. 2004: Annual Review of Ecology Evolution and Systematics, v. 35, p. 523–556, doi: 10.1146/annurev.ecolsys.35.112202.130137.

Katz, M.E., Wright, J.D., Miller, K.G., Cramer, B.S., Fennel, K., and Falkowski, P.G., 2005, Biological overprint of the geological carbon cycle: Marine Geology, v. 217, p. 323–338, doi: 10.1016/j.margeo.2004.08.005.

Katz, M.E., Miller, K.G., Wright, J.D., Wade, B.S., Browning, J.V., Cramer, B.S., and Rosenthal, Y., 2008, Stepwise transition from greenhouse to icehouse climates: Nature Geoscience, v. 1, p. 329–334, doi: 10.1038/ngeo179.

Keigwin, L.D., 1980, Palaeoceanographic change in the Pacific at the Eocene-Oligocene boundary: Nature, v. 287, p. 722–725, doi: 10.1038/287722a0.

Keller, G., MacLeod, N., and Barrera, E., 1992, Eocene-Oligocene faunal turnover in planktic foraminifera, and Antarctic glaciation, *in* Prothero, D.R., and Berggren, W.A., eds., Eocene-Oligocene Climatic and Biotic Evolution: Princeton, New Jersey, Princeton University Press, p. 218–244.

Kennett, J.P., 1977, Cenozoic evolution of Antarctic glaciation, the circum-Antarctic Ocean, and their impact on global paleoceanography: Journal of Geophysical Research, v. 82, p. 3843–3860, doi: 10.1029/JC082i027p03843.

Kennett, J.P., and Shackleton, N.J., 1976, Oxygen isotopic evidence for the development of the psychrosphere 38 Myr ago: Nature, v. 260, p. 513–515, doi: 10.1038/260513a0.

Kennett, J., and Stott, L., 1990, Proteus and Proto-Oceanus: Ancestral Paleogene oceans as revealed from Antarctic stable isotopic results, ODP Leg 113, *in* Barker, P., and Kennett, J., et al., Proceedings of the Ocean Drilling Program, Scientific Results, Volume 113: College Station, Texas, Ocean Drilling Program, p. 865–880.

Kominz, M.A., and Pekar, S.F., 2001, Oligocene eustasy from two-dimensional sequence stratigraphic backstripping: Geological Society of America Bulletin, v. 113, p. 291–314, doi: 10.1130/0016-7606(2001)113<0291:OEFTDS>2.0.CO;2.

Kominz, M.A., Browning, J.V., Miller, K.G., Sugarman, P.J., Misintseva, S., and Scotese, C.R., 2008, Late Cretaceous to Miocene sea-level estimates from the New Jersey and Delaware coastal plain coreholes: An error analysis: Basin Research, v. 20, p. 211–226, doi: 10.1111/j.1365-2117.2008.00354.x.

Larson, R.L., 1991, Geological consequences of superplumes: Geology, v. 19, p. 963–966, doi: 10.1130/0091-7613(1991)019<0963:GCOS>2.3.CO;2.

Lear, C.H., Elderfield, H., and Wilson, P.A., 2000, Cenozoic deep-sea temperatures and global ice volumes from Mg/Ca in benthic foraminiferal calcite: Science, v. 287, p. 269–272, doi: 10.1126/science.287.5451.269.

Lear, C.H., Rosenthal, Y., Coxall, H.K., and Wilson, P.A., 2004, Late Eocene to early Miocene ice sheet dynamics and the global carbon cycle: Paleoceanography, v. 19, p. 4015, doi: 10.1029/2004PA001039.

Merico, A., Tyrrell, T., and Wilson, P.A., 2008, Eocene/Oligocene ocean de-acidification linked to Antarctic glaciation by sea-level fall: Nature, v. 452, p. 979–982, doi: 10.1038/nature06853.

Miller, K.G., 1992, Middle Eocene to Oligocene stable isotopes, climate, and deep-water history: The Terminal Eocene Event?, *in* Prothero, D., and Berggren, W.A., eds., Eocene-Oligocene Climatic and Biotic Evolution: Princeton, New Jersey, Princeton University Press, p. 160–177.

Miller, K.G., and Fairbanks, R.G., 1985, Oligocene to Miocene carbon isotope cycles and abyssal circulation changes, *in* Sundquist, E.T., and Broecker, W.S., eds., The Carbon Cycle and Atmospheric CO_2: Natural Variations Archean to Present: American Geophysical Union, Geophysical Monograph 32, p. 469–486.

Miller, K.G., and Tucholke, B.E., 1983, Development of Cenozoic abyssal circulation south of the Greenland-Scotland Ridge, *in* Bott, M.H.P., Saxov, S., Talwani, M., and Thiede, J., eds., Structure and Development of the Greenland-Scotland Ridge: NATO Conference Series: New York, Plenum Press, p. 549–589.

Miller, K.G., Fairbanks, R.G., and Mountain, G.S., 1987, Tertiary oxygen isotope synthesis, sea level history, and continental margin erosion: Paleoceanography, v. 2, p. 1–19, doi: 10.1029/PA002i001p00001.

Miller, K.G., Wright, J.D., and Fairbanks, R.G., 1991, Unlocking the ice house: Oligocene-Miocene oxygen isotopes, eustasy, and margin erosion: Journal of Geophysical Research, v. 96, p. 6829–6848, doi: 10.1029/90JB02015.

Miller, K.G., Katz, M.E., and Berggren, W.A., 1992, Cenozoic deep-sea benthic foraminifera: A tale of three turnovers, *in* Takayanagi, Y., and Saito, T., eds., Studies in Benthic Foraminifera, Proceedings of the Fourth International Symposium on Benthic Foraminifera, Sendai, 1990 (Benthos '90): Tokyo, Japan, Tokai University Press, p. 245–248.

Miller, K.G., Kominz, M.A., Browning, J.V., Wright, J.D., Mountain, G.S., Katz, M.E., Sugarman, P.J., Cramer, B.S., Christie-Blick, N., and Pekar, S.F., 2005a, The Phanerozoic record of global sea-level change: Science, v. 310, p. 1293–1298, doi: 10.1126/science.1116412.

Miller, K.G., Wright, J.D., and Browning, J.V., 2005b, Visions of ice sheets in a greenhouse world: Marine Geology, v. 217, p. 215–231, doi: 10.1016/j.margeo.2005.02.007.

Miller, K.G., Browning, J.V., Aubry, M.-P., Wade, B.S., Katz, M.E., Kulpecz, A.A., and Wright, J.D., 2008a, Eocene-Oligocene global climate and sea-level changes: St. Stephens Quarry, Alabama: Geological Society of America Bulletin, v. 120, p. 34–53, doi: 10.1130/B26105.1.

Miller, K.G., Wright, J.D., Katz, M.E., Browning, J.V., Cramer, B.S., Wade, B.S., and Mizintseva, S.F., 2008b, A view of Antarctic ice-sheet evolution from sea-level and deep-sea isotope changes during the Late Cretaceous–Cenozoic, *in* Cooper, A.K., Barrett, P.J., Stagg, H., Storey, B., Stump, E., Wise, W., and the 10th International Symposium on Antarctic Earth Sciences team, eds., Antarctica: A Keystone in the Changing World: Proceedings of the 10th International Symposium on Antarctic Earth Sciences: Washington, D.C., National Academy of Sciences Press, p. 55–70.

Moran, K., Backman, J., Brinkhuis, H., Clemens, S.C., Cronin, T., Dickens, G.R., Eynaud, F., Gattacceca, J., Jakobsson, M., Jordan, R.W., Kaminski, M., King, J., Koc, N., Krylov, A., Martinez, N., Matthiessen, J., McInroy, D., Moore, T.C., Onodera, J., O'Regan, M., Pälike, H., Rea, B., Rio, D., Sakamoto, T., Smith, D.C., Stein, R., St John, K., Suto, I., Suzuki, N., Takahashi, K., Watanabe, M., Yamamoto, M., Farrell, J., Frank, M., Kubik, P., Jokat, W., and Kristoffersen, Y., 2006, The Cenozoic palaeoenvironment of the Arctic Ocean: Nature, v. 441, p. 601–605, doi: 10.1038/nature04800.

Mountain, G.S., and Miller, K.G., 1992, Seismic and geologic evidence for early Paleogene deep-water circulation in the western North Atlantic: Paleoceanography, v. 7, p. 423–439, doi: 10.1029/92PA01268.

Müller, R.D., Sdrolias, M., Gaina, C., Steinberger, B., and Heine, C., 2008, Long-term sea-level fluctuations driven by ocean basin dynamics: Science, v. 319, p. 1357–1362, doi: 10.1126/science.1151540.

Owens, J.P., Bybell, L.M., Paulachok, G., Ager, T.A., Gonzalez, V.M., and Sugarman, P.J., 1988, Stratigraphy of the Tertiary sediments in a 945-foot-deep core hole near Mays Landing in the southeastern New Jersey coastal plain: U.S. Geological Survey Professional Paper 1484, p. 1–39.

Pagani, M., Arthur, M.A., and Freeman, K.H., 1999, Miocene evolution of atmospheric carbon dioxide: Paleoceanography, v. 14, p. 273–292, doi: 10.1029/1999PA900006.

Pagani, M., Zachos, J., Freeman, K.H., Bohaty, S., and Tipple, B., 2005, Marked change in atmospheric carbon dioxide concentrations during the Oligocene: Science, v. 309, p. 600–603, doi: 10.1126/science.1110063.

Pearson, P.N., and Palmer, M.R., 2000, Atmospheric carbon dioxide concentrations over the past 60 million years: Nature, v. 406, p. 695–699, doi: 10.1038/35021000.

Pearson, P.N., van Dongen, B.E., Nicholas, C.J., Pancost, R.D., Schouten, S., Singano, J.M., and Wade, B.S., 2007, Stable warm tropical climate through the Eocene epoch: Geology, v. 35, p. 211–214, doi: 10.1130/G23175A.1.

Pearson, P.N., McMillan, I., Wade, B.S., Dunkley Jones, T., Coxall, H.K., Bown, P.R., and Lear, C.H., 2008, Extinction and environmental change across the Eocene-Oligocene boundary in Tanzania: Geology, v. 36, p. 179–182, doi: 10.1130/G24308A.1.

Pekar, S.F., Christie-Blick, N., Kominz, M.A., and Miller, K.G., 2002, Calibration between eustatic estimates from backstripping and oxygen isotopic records for the Oligocene: Geology, v. 30, p. 903–906, doi: 10.1130/0091-7613(2002)030<0903:CBEEFB>2.0.CO;2.

Rea, D.K., and Lyle, M.W., 2005, Paleogene calcite compensation depth in the eastern subtropical Pacific; Answers and questions: Paleoceanography, v. 20, doi: 10.1029/2004PA001064.

Robert, C., and Kennett, J.P., 1997, Antarctic continental weathering changes during Eocene-Oligocene cryosphere expansion: Clay mineral and oxygen isotope evidence: Geology, v. 25, p. 587–590, doi: 10.1130/0091-7613(1997)025<0587:ACWCDE>2.3.CO;2.

Savin, S.M., Douglas, R.G., and Stehli, F.G., 1975, Tertiary marine paleotemperatures: Geological Society of America Bulletin, v. 86, p. 1499–1510, doi: 10.1130/0016-7606(1975)86<1499:TMP>2.0.CO;2.

Scher, H.D., and Martin, E.E., 2006, Timing and climatic consequences of the opening of Drake Passage: Science, v. 312, p. 428–430, doi: 10.1126/science.1120044.

Shackleton, N.J., and Kennett, J.P., 1975, Paleotemperature history of the Cenozoic and the initiation of Antarctic glaciation, oxygen and carbon isotope analyses in DSDP Sites 277, 279, and 281, *in* Kennett, J.P., and Houtz, R.E., et al., Initial Reports of the Deep Sea Drilling Project, Volume 29: Washington, D.C., U.S. Government Printing Office, p. 743–755.

Shackleton, N.J., Backman, J., Zimmerman, H., Kent, D.V., Hall, M.A., Roberts, D.G., Schnitker, D., Baldauf, J.G., Desprairies, A., Homrighausen, R., Huddlestun, P., Keene, J.B., Kaltenback, A.J., Krumsiek, K.A.O., Morton, A.C., Murray, J.W., and Westberg-Smith, J., 1984, Oxygen isotope calibration of the onset of ice-rafting and history of glaciation in the North Atlantic region: Nature, v. 307, p. 620–623, doi: 10.1038/307620a0.

St. John, K., 2008, Cenozoic ice-rafting history of the central Arctic Ocean: Terrigenous sands on the Lomonosov Ridge: Paleoceanography, v. 23, p. PA1S05, doi: 10.1029/2007PA001483.

Thomas, E., 1992, Middle Eocene–late Oligocene bathyal benthic foraminifera (Weddell Sea): Faunal changes and implications for ocean circulation, *in* Prothero, D.R., and Berggren, W.A., eds., Late Eocene–Oligocene Climatic and Biotic Evolution: Princeton, New Jersey, Princeton University Press, p. 245–271.

Tripati, A., Backman, J., Elderfield, H., and Ferretti, P., 2005, Eocene bipolar glaciation associated with global carbon cycle changes: Nature, v. 436, p. 341–346, doi: 10.1038/nature03874.

Tucholke, B.E., and Mountain, G.S., 1979, Seismic stratigraphy, lithostratigraphy and paleosedimentation patterns in the North American Basin, *in* Talwani, M., Hay, W., and Ryan, W.B.F., eds., Deep Drilling Results in the Atlantic Ocean: Continental Margins and Paleoenvironment, Volume 3: Maurice Ewing Symposium Series: Washington, D.C., American Geophysical Union, p. 58–86.

Tucholke, B.E., and Vogt, P.R., 1979, Western North Atlantic: Sedimentary evolution and aspects of tectonic history, *in* Tucholke, B.E., and Vogt, P.R., et al., eds., Initial Reports of the Deep Sea Drilling Project, Volume 43: Washington, D.C., U.S. Government Printing Office, p. 791–825.

van Andel, T.H., 1975, Mesozoic/Cenozoic calcite compensation depth and global distribution of calcareous sediments: Earth and Planetary Science Letters, v. 26, p. 187–194, doi: 10.1016/0012-821X(75)90086-2.

Via, R.K., and Thomas, D.J., 2006, Evolution of Atlantic thermohaline circulation: Early Oligocene onset of deep-water production in the North Atlantic: Geology, v. 34, p. 441–444.

Wade, B.S., and Pearson, P.N., 2008, Planktonic foraminiferal turnover, diversity fluctuations and geochemical signals across the Eocene/Oligocene boundary in Tanzania: Marine Micropaleontology, v. 68, p. 244–255.

Wright, J.D., and Miller, K.G., 1996, Control of North Atlantic Deep Water circulation by the Greenland-Scotland Ridge: Paleoceanography, v. 11, p. 157–170, doi: 10.1029/95PA03696.

Zachos, J.C., Quinn, T.M., and Salamy, S., 1996, High resolution (10^4 yr) deep-sea foraminiferal stable isotope records of the earliest Oligocene climate transition: Paleoceanography, v. 11, p. 251–266.

Zachos, J., Shackleton, N., Revenaugh, J., Pälike, H., and Flower, B., 2001, Climate response to orbital forcing across the Oligocene-Miocene boundary: Science, v. 292, p. 274–278, doi: 10.1126/science.1058288.

Zanazzi, A., Kohn, J.J., MacFadden, B.J., and Terry, D.O., Jr., 2007, Large temperature drop across the Eocene-Oligocene transition in central North America: Nature, v. 445, p. 639–642, doi: 10.1038/nature05551.

Manuscript Accepted by the Society 16 September 2008

Printed in the USA

The Geological Society of America
Special Paper 452
2009

Refined correlation of the UK Late Eocene–Early Oligocene Solent Group and timing of its climate history

J.J. Hooker
Department of Palaeontology, Natural History Museum, Cromwell Road, London, SW7 5BD, UK

S.T. Grimes
School of Earth, Ocean and Environmental Sciences, University of Plymouth, Plymouth, PL4 8AA, UK

D.P. Mattey
M.E. Collinson
N.D. Sheldon*
Department of Geology, Royal Holloway University of London, Egham, Surrey, TW20 0EX, UK

ABSTRACT

The dominantly nonmarine Solent Group of the Hampshire Basin, southern England, spans the Late Eocene and earliest Oligocene. It contains a rich biota, including mammals and charophytes, but contains few fully marine, time-diagnostic intervals. Correlation with standard marine successions is provided here using biostratigraphically significant mammals and charophytes via interdigitation of strata with those containing marine zonal indicators or via magnetostratigraphy in other parts of Europe. This allows the re-identification of recently described normal polarity intervals in the Solent Group to magnetochrons C16n, 15n, 13r1n, and 13n, and improved calibration of the Solent Group to the geomagnetic polarity time scale (GPTS). Consistent with this calibration, interpreted Solent Group sea-level changes can be related to those recognized to be global in deep-marine successions in central Italy. The major sea-level fall associated with polar ice buildup at the Oligocene isotope event Oi-1 near the beginning of the Oligocene is shown to be represented in the Hamstead Member of the Bouldnor Formation by the falling stage systems tract followed by a major hiatus before the next maximum flooding surface. This hiatus encompasses the large European mammalian faunal turnover known as the "Grande Coupure." Improved calibration to the GPTS allows better dating of climate proxies, such that recently described summer freshwater temperatures can be shown to track global fluctuations across the Eocene-Oligocene transition.

Keywords: biostratigraphy, mammals, sea-level change, paleosols.

*Present address: Department of Geological Sciences, University of Michigan, 2534 CC Little Bldg., 1100 N. University Ave., Ann Arbor, Michigan 48109, USA.

Hooker, J.J., Grimes, S.T., Mattey, D.P., Collinson, M.E., and Sheldon, N.D., 2009, Refined correlation of the UK Late Eocene–Early Oligocene Solent Group and timing of its climate history, *in* Koeberl, C., and Montanari, A., eds., The Late Eocene Earth—Hothouse, Icehouse, and Impacts: Geological Society of America Special Paper 452, p. 179–195, doi: 10.1130/2009.2452(12). For permission to copy, contact editing@geosociety.org.

INTRODUCTION

The Solent Group (Insole and Daley, 1985) is a succession of largely continental strata of Late Eocene (Priabonian) to earliest Oligocene (Rupelian) age, originally informally named the Fluvio-marine Formation by Forbes (1853, 1856). It is divided in ascending order into the Headon Hill, Bembridge Limestone, and Bouldnor Formations. The Headon Hill Formation is further subdivided into nine members, and the Bouldnor Formation is divided into three members (Insole and Daley, 1985; Fig. 1). The base of the Solent Group is characterized by a major sea-level fall equated with the TA3.6/4.1 sequence boundary (SB) of Haq et al. (1987) by Plint (1988), which is usually taken to coincide with the end of the Bartonian (late Middle Eocene). The

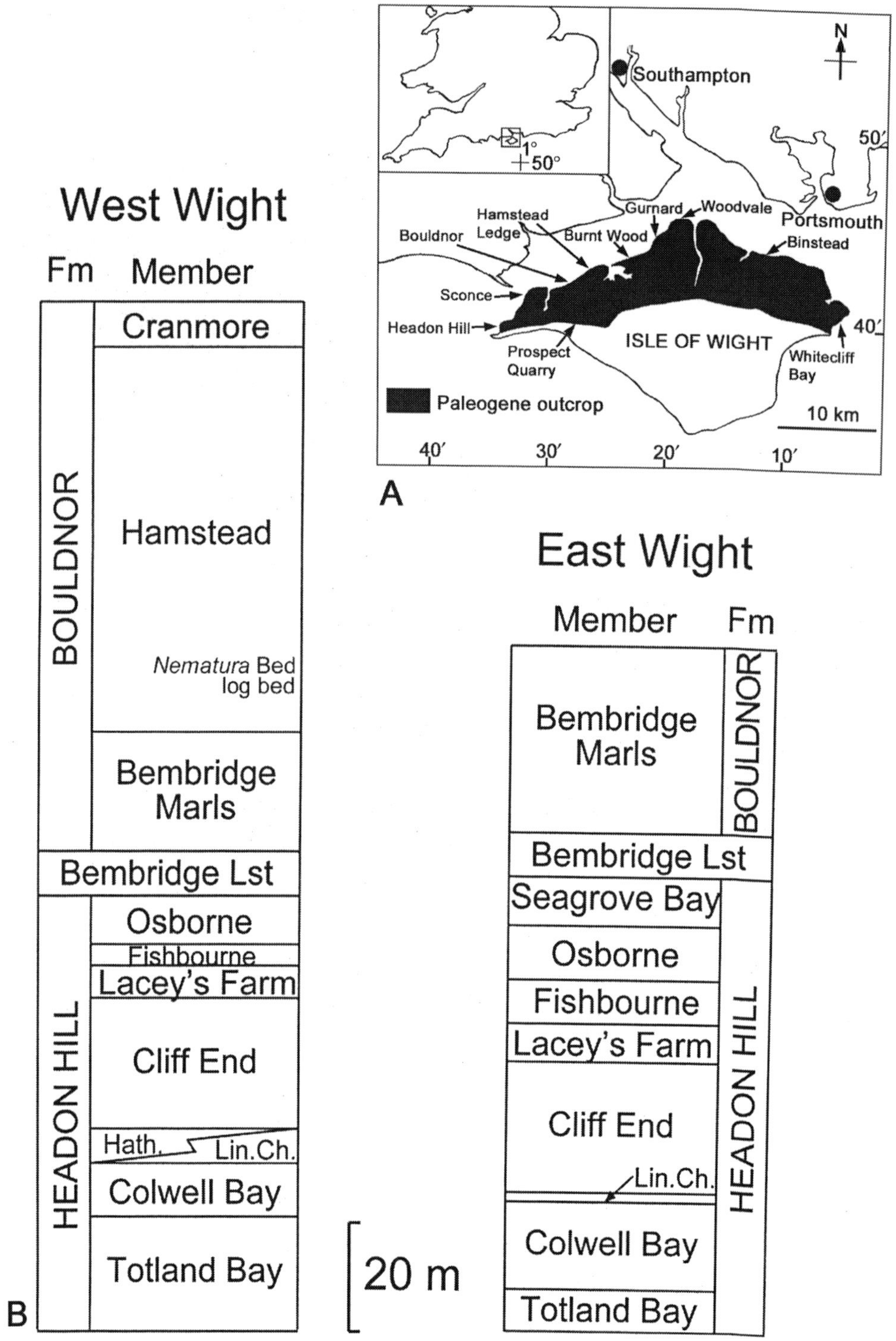

Figure 1. (A) Map of the Isle of Wight, showing the Paleogene outcrop and locations of localities mentioned in the text. (B) Lithostratigraphy of the Solent Group at formational and member levels in western and eastern parts of the Isle of Wight. Thicknesses vary, but those shown are judged to be broadly representative of their areas. There are no implications as to correlation from relative positions on the two logs or for the nature of contacts between units.

highest strata (Cranmore Member) contain calcareous nannoplankton markers indicative of zone NP23 (Martini, 1972) of the early Rupelian (Early Oligocene). The only intermediate horizon with global marine indicators is the Brockenhurst Bed at the base of the Colwell Bay Member, which contains zones NP19–20 (Aubry, 1985, 1986).

The Solent Group sediments are dominated by muds, with generally subordinate limestones and sands. These were deposited on a coastal plain, which fluctuated from lagoonal to lacustrine/palustrine to fluvial and was rarely open marine (e.g., Daley, 1999). They contain a diverse and well-preserved biota dominated by molluscs, vertebrates (especially mammals), charophytes, and vascular plants. The mammals and the charophytes provide good correlation with other continental deposits in Europe (Hooker, 1987, 1992; Hooker et al., 1995, 2004; Riveline, 1984; Sille et al., 2004) but not directly with the marine realm. A major mammalian faunal turnover, known as the "Grande Coupure" (Stehlin, 1910) is recorded in the Hamstead Member of the Bouldnor Formation and has been calibrated to Oligocene isotope event Oi-1 via a chain of biostratigraphic markers and sea-level changes in the Paris and Belgian basins and Italy (Hooker et al., 2004).

A recent magnetostratigraphic and sequence stratigraphic study (Gale et al., 2006) added more parameters to the correlation potential of the Solent Group. Gale et al. (2006, their Fig. 3) identified seven numbered sequences between the base of the Solent Group and the lower part of the Bouldnor Formation, mainly in the Whitecliff Bay section (Isle of Wight), although they did not equate these with the numbered sequences of Haq et al. (1987) or Hardenbol et al. (1998). In Whitecliff Bay, they identified two normal polarity zones, the Headon Hill normal polarity zone and the Bembridge normal polarity zone, which they equated with chrons C15n and C13n, respectively. The Headon Hill normal polarity zone spans the Colwell Bay to lower Cliff End Members of the Headon Hill Formation, while the Bembridge normal polarity zone runs from the upper part of the Bembridge Limestone Formation to the lower part of the Bembridge Marls Member of the Bouldnor Formation. Because of their identification of the Bembridge normal polarity zone with C13n, Gale et al. (2006) equated the base of their sequence 5 with the major sea-level fall associated with the Oi-1 glaciation (and also with the Eocene-Oligocene boundary, which predates it by ~100 k.y.: Fig. 4 in van Mourik, 2006, paper 4). The onset of Oi-1 was judged to be represented by conglomeratic channel fills in the Seagrove Bay Member near the top of the Headon Hill Formation in eastern Isle of Wight and by an implied unconformity at the base of the Bembridge Limestone in western Wight (Gale et al., 2006, their Fig. 5). These correlations conflicted with the earlier identification of Oi-1 in the Solent Group succession at the base of the *Nematura* Bed in the Hamstead Member of the Bouldnor Formation by Hooker et al. (2004). Hooker et al. (2007) argued that on the basis of their 2004 correlation, the Bembridge normal polarity zone could not represent C13n, but fitted temporally most closely with subchron C13r1n (also referred to as C13n2 by Brinkhuis and Visscher [1995] and as short polarity excursion I by Premoli-Silva et al. [1988]).

This paper examines these dating issues in more detail and provides evidence for more resolved correlation of the entire Solent Group to the geomagnetic polarity time scale (GPTS). This involves recalibrating the magnetochrons of Gale et al. (2006). It is then possible to calibrate more precisely the isotope-derived temperature values of Grimes et al. (2005).

NEW CORRELATION

Reinterpretation of Sequence 5 Systems Tracts of Gale et al. (2006)

The factor that appears to have influenced Gale et al. (2006) most in selecting the base of their sequence 5 as representing Oi-1 was the degree of incision, which was judged greater than for any other of the Solent Group sequence boundaries that they studied. They based their estimate of incision, and thus sea-level fall, on the thickness of the lowstand systems tract (LST), which followed earlier practice (Gale et al., 2002). Their LST in sequence 5 in Whitecliff Bay spanned the upper part of the Seagrove Bay Member (from the channel fill deposits upward) and the Bembridge Limestone, a thickness of 14 m. The Seagrove Bay Member is not present on Headon Hill, and the highest parts of the Bembridge Limestone are not exposed there (Hooker et al., 1995), so Gale et al.'s (2006) estimate of LST there was 10+ m.

In Whitecliff Bay, Gale et al. (2006) identified the base of the Bembridge Marls Member as the transgressive surface (TS) and the Bembridge Oyster Bed, only 1.5 m higher, as the maximum flooding surface (MFS). The close proximity of these two surfaces compared to that between the SB and the TS is anomalous. It would be understandable if there were a major unconformity at the TS. Such an unconformity was postulated by Hooker et al. (2004) and used by Gale et al. (2006) in support of their correlation of the Bembridge normal polarity zone with C13n on temporal duration grounds. The boundary, however, is accepted herein to be conformable: the basal Bembridge Marls bed (Daley and Edwards, 1990) rests on a mud-cracked surface of the underlying top Bembridge Limestone bed (see also Gale et al., 2007, p. 687). To have the TS and MFS so closely juxtaposed would suggest that the sediments were accumulating far inland or that there was a considerable gradient, making the TS young rapidly updip. In fact, a brackish interval, the *Corbicula* shell bed, overlying a freshwater muddy interval resting on an erosion surface, is located lower in the sequence within the Bembridge Limestone middle muds (Daley and Edwards, 1990). It is probable that this horizon was not identified as the TS by Gale et al. (2006) because the Bembridge Limestone succession above the middle muds reverts to freshwater deposition up to the top of the formation, and therefore it appeared a minor event. However, clear evidence of intraformational folding in the Bembridge Limestone middle muds is presented here (Fig. 2). In a westerly direction from Whitecliff Bay, the *Corbicula* bed and suprajacent beds of the middle muds are cut out at the base of the upper limestone and then reappear west of Sconce. Similar folding continued late

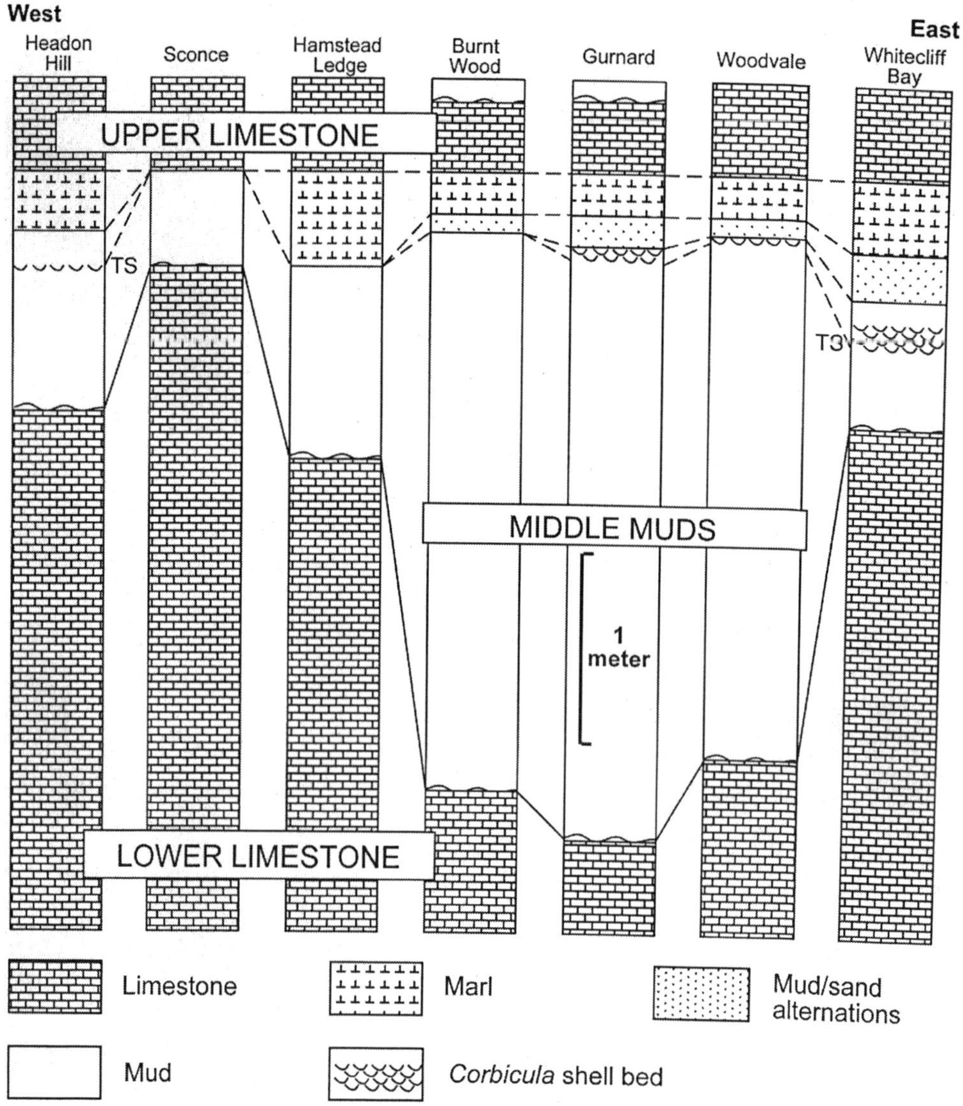

Figure 2. Lithic logs of the middle muds unit of the Bembridge Limestone Formation and adjacent parts of the upper and lower limestone units at localities across the Isle of Wight. Wedging out of higher subdivisions of the middle muds demonstrates intraformational folding prior to deposition of the upper limestone. The transgressive surface (TS) is at the *Corbicula* shell bed. The middle muds include all the lithologies except limestone. The Woodvale log is from Daley and Edwards (1990), the thickness of the upper limestone on Headon Hill is from Bristow et al. (1889), and the rest are from personal observation (Hooker). N.B., the sand/mud unit, is absent from Headon Hill through facies change, not erosion.

during deposition of the upper limestone prior to deposition of Bembridge Marls in the west, such that at Burnt Wood, the Bembridge Marls rest on middle muds of the Bembridge Limestone (Bristow et al., 1889, their Fig. 55; Daley and Edwards, 1971). The sub–upper limestone folding is interpreted to have caused gentle uplift over the whole area, shifting the depositional environment from brackish back to freshwater like that of the lower limestone. The TS of Gale et al.'s (2006) sequence 5 is therefore re-identified as occurring within the Bembridge Limestone at the *Corbicula* shell bed.

The sequence 5 SB was identified by Gale et al. (2006, their Fig. 5) at the base of channels within the Seagrove Bay Member below the Bembridge Limestone in Whitecliff Bay (and other eastern sites) and at the base of the Bembridge Limestone in western Wight, where the Seagrove Bay Member is absent. Although not explicit in their Figure 5, Gale et al. (2006) implied an unconformity between the Osborne Member (highstand systems tract [HST] of sequence 4) and the Bembridge Limestone in west Wight by classifying the latter as late LST of sequence 5. The upper Seagrove Bay Member of eastern sections was classified by Gale et al. as early LST of sequence 5, and on this basis it would be represented in the west by a hiatus. However, there is no evidence for a hiatus between the Osborne Member and the Bembridge Limestone in western sections, where overbank mottled clays pass up almost imperceptibly into lacustrine/palustrine non-mottled clays/marls. In fact, it fits well with Gale et al.'s (2006) characterization of sequences in the Solent Group as having a high sedimentation rate and conformable or near-conformable SBs. This raises the question: with what does the upper Seagrove Bay Member correlate in the west? The question can be answered by means of mammalian biostratigraphy.

An important mammalian dispersal event characterizes the lower Bembridge Limestone on Headon Hill (traced to the lowest 0.5 m, i.e., bed 1 of Daley and Edwards, 1990; Hooker,

1993, personal observ.): the appearance of nine species not recorded below this horizon in the UK (Hooker et al., 1995, 2004). These species dispersed into the UK from France and Germany (Hooker et al., 1995). Of note are the perissodactyls *Plagiolophus minor* and *Palaeotherium medium* and the xiphodont artiodactyl *Xiphodon gracilis*. This event is not just local to Headon Hill, since *P. minor* occurs in the Bembridge Limestone also at Prospect Quarry (lower limestone) and Whitecliff Bay (middle muds), and *P. medium* occurs also at Prospect Quarry (lower limestone) and Sconce (middle muds), although not at the base of the formation. The Osborne Member, which directly underlies the Bembridge Limestone in the west, records *Plagiolophus annectens* at Headon Hill and Thorness Bay; this species is absent from the lower Bembridge Limestone and only reappears fleetingly in the upper Bembridge Limestone (Hooker et al., 1995). An important fauna was collected in the nineteenth century at the eastern site of Binstead in the Seagrove Bay Member. The horizon was recorded as being in the upper 2 m of a sandy unit (Mantell, 1854) now identified with the Seagrove Bay Member; this is supported by the characteristic sandstone matrix adhering to most of the museum specimens. The fauna includes *Plagiolophus annectens*, *Palaeotherium muehlbergi*, *P. curtum*, *P. magnum*, and the xiphodont *Dichodon*. There is no sign of *Plagiolophus minor*, *Palaeotherium medium*, or *Xiphodon gracilis* in this fauna (Hooker, 1992; Hooker, 2008, personal observ.). This argues for correlation of the upper part of the Seagrove Bay Member with higher parts of the Osborne Member in the west. Given that the SB between the Osborne Member and Bembridge Limestone in western sections is conformable, strata during the sea-level fall (the falling stage systems tract [FSST]: see Plint and Nummedal, 2000) of sequence 4 should be preserved. It is likely therefore that this is represented in the east by the Seagrove Bay Member, where gradients steepened and channels cut back into the Isle of Wight monocline, which was already rising in the Middle Eocene (Gale et al., 1999), to produce sandy and rudaceous fills.

Relocation of the SB and TS of sequence 5 to the base of the Bembridge Limestone and the *Corbicula* shell bed of the middle muds, respectively, reduces the LST thickness to 4 m in Whitecliff Bay and 5 m on Headon Hill. This is less than nearly all of the LSTs of the other Solent Group sequences (Gale et al., 2006, their Table 1). Consequently, it does not indicate the deepest incision value and therefore does not in this sense support the positioning of Oi-1 in the Seagrove Bay Member.

Calibration of Mammalian Biostratigraphy to Marine Zonations and Magnetostratigraphy

On biostratigraphic grounds, the Grande Coupure is securely correlated with Oi-1. Early post–Grande Coupure (MP21) mammal localities are recognizable by the persistence of a number of European pre–Grande Coupure endemics alongside newcomers from Asia (Hooker, 2009). Such a case is illustrated by the mammal bed of Hoogbutsel, eastern Belgium (see de Bruijn and Ünay, 1989; Smith, 2003, 2004a, 2004b, 2006; Smith and van den Hoek Ostende 2006, for fauna) at the base of the Borgloon Formation, which represents the LST (Steurbaut, 1992) or more likely the early TST (Hooker et al., 2004) of sequence TA4.4 of Haq et al. (1987). The immediately underlying Neerrepen Sand, which represents the highstand (HST) of sequence TA4.3 (Steurbaut, 1992), and its time equivalent in north Belgium, the Watervliet Clay both contain the last occurrence of the dinoflagellate *Areosphaeridium diktyoplokum*, which consistently occurs within chron C13n, the end of the LST of TA4.4, and the onset of Oi-1 (Brinkhuis and Visscher, 1995; Eldrett et al., 2004). In particular, the $\delta^{18}O$ shift marking Oi-1 has been recognized in marine bivalves, fish otoliths, and benthic foraminifera between the Watervliet Clay and St. Huibrechts Hern Formation (= Grimmertingen Sand of earlier authors) below and the Bilzen Formation above in Belgium (De Man et al., 2004). Another early post–Grande Coupure fauna is that of Santpedor in the Ebro Basin, Spain, where a large species of *Plagiolophus* (probably *P. major* or *P. fraasi*) and *Palaeotherium* (both typical pre–Grande Coupure taxa) co-occur with the post–Grande Coupure (MP21) newcomers *Atavocricetodon atavus* and *Pseudoltinomys gaillardi* (Arbiol and Saez, 1988). The strata involved (Artés Formation) are within a normal polarity zone identified as C13n (Barberà et al., 2001) in a long succession of continental strata with magnetochrons well calibrated by mammals and extending up into the Miocene. In the Hampshire Basin, the Grande Coupure occurs within the Hamstead Member (Figs. 1 and 3). The highest pre–Grande Coupure fauna (MP20) occurs in the log bed (top lower Hamstead Member), immediately below the *Nematura* bed. The lowest post–Grande Coupure fauna occurs some 5 m higher in nonmarine upper Hamstead Member strata following a brackish interval consisting of the *Nematura* Bed plus higher beds with decalcified specimens of the bivalve *Polymesoda*. A major hiatus resulting from erosion associated with the Oi-1 glaciation sea-level fall is judged to occur at the base of the *Nematura* Bed (see following).

Lower parts of the Solent Group are more difficult to correlate despite the occurrence of NP19–20 markers at the base of the Colwell Bay Member in Whitecliff Bay (Aubry, 1986). Zone NP19-20 represents a relatively long time span, lasting nearly 2 m.y. and spanning chrons C16n to early C13r (Gradstein et al., 2004). On Headon Hill, the Colwell Bay Member is not fully marine, so it contains no calcareous nannoplankton, but it is sandwiched between mammal faunas referable to Paleogene reference level MP17 below in the Totland Bay Member and MP18 above in the Linstone Chine (*Microchoerus* Bed) and Hatherwood Limestone Members (Hooker, 1992). The Totland Bay Member also belongs to the *Gyrogona tuberosa* charophyte zone (Riveline, 1984). In the northern Aquitaine Basin, marine and continental strata interdigitate. Here, both MP17 and MP18 reference levels occur at successive levels in the Argiles à *Palaeotherium* at Baby and Civrac, respectively (Ringeade, 1987; Dubreuilh, 1989). Lateral to the MP17 level at Baby, there is a unit called the Calcaire Lacustre de Plassac, which in turn passes laterally into the upper part of the Marnes à *Ostrea cucullaris* with the *Gyrogona*

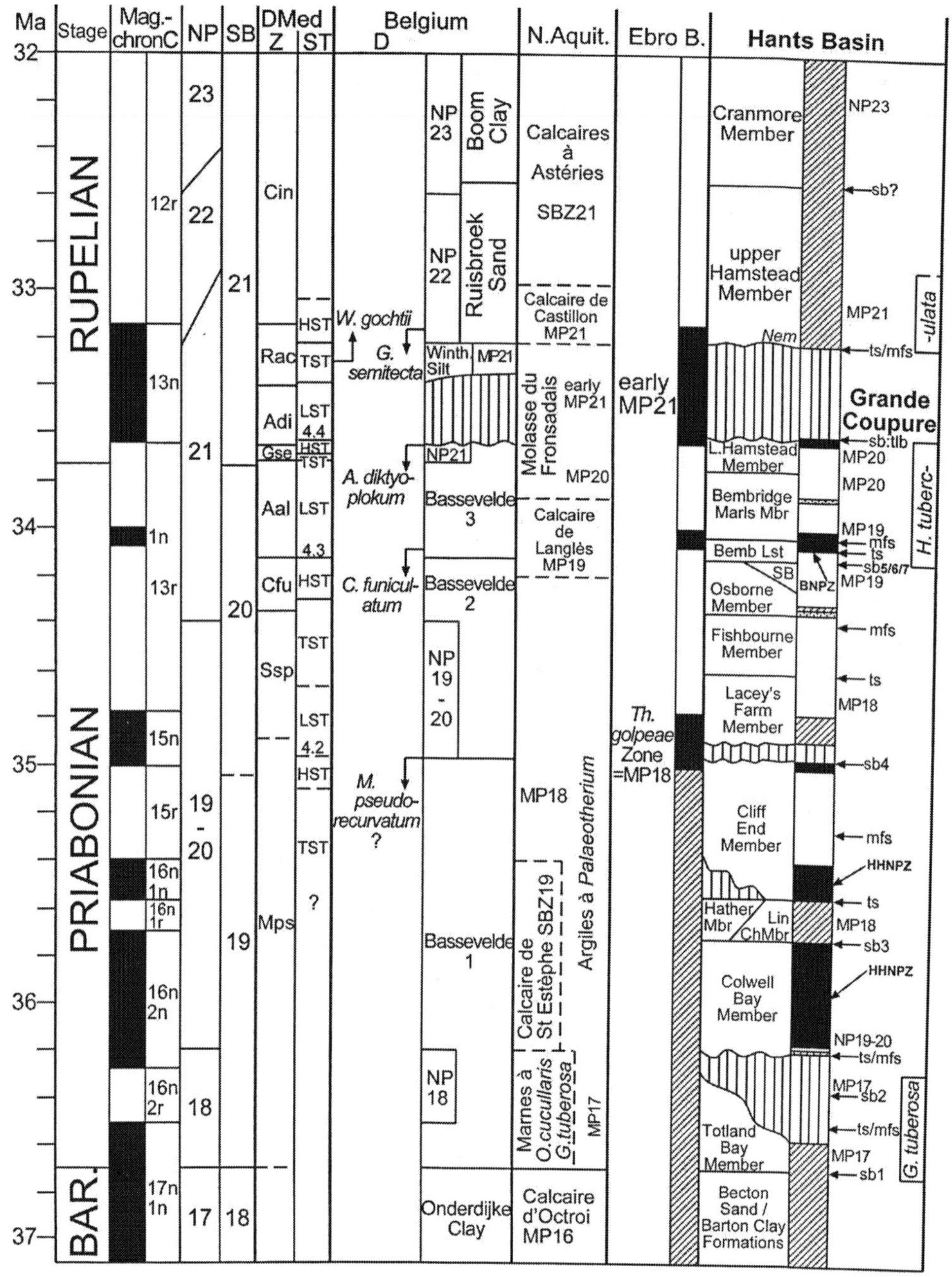

Figure 3. Correlation chart of the Hampshire Basin Solent Group and top Becton Sand and Barton Clay Formations with the geomagnetic polarity time scale (GPTS) via biostratigraphy, magnetostratigraphy, and sequence stratigraphy in the Ebro Basin, Spain, and the Aquitaine Basin, France, Belgium, and Italy. The first column is age in Ma, mainly following Gradstein et al. (2004), but modified above 34 Ma to the astronomically tuned time scale of Coxall et al. (2005). Columns 2–4 show standard stages and Cenozoic magnetochrons (Gradstein et al., 2004), with the addition of subchron C13r1n (see text). Columns 5–6 show standard calcareous nannoplankton (NP; Berggren et al., 1995) and shallow benthic foraminifera (SB; Serra-Kiel et al., 1998) biozones. Columns 7–8 show Mediterranean dinoflagellate (DMed) biozones (Z), and lowstand, transgressive, and highstand systems tracts (LST, TST, HST) derived from them, for northern and central Italy and identified with the Haq et al. (1987) numbered TA sequences (Brinkhuis and Biffi, 1993; Brinkhuis, 1994; Brinkhuis and Visscher, 1995). Columns 9–10 show the main Belgian succession with first and last occurrences of dinoflagellates, calcareous nannoplankton zones (NP), and a Paleogene mammalian reference level (MP) (Vandenberghe et al., 2003). Column 11 shows the succession in the northern Aquitaine Basin with SB and charophyte zones and Paleogene mammalian reference levels (Riveline, 1984; Ringeade, 1987; Dubreuilh, 1989). Column 12 shows magnetochrons and mammalian biostratigraphy in the Ebro Basin, Spain (Barberà et al., 2001). Columns 13–15 show the succession in the Hampshire Basin (west to the left, east to the right), with magnetic polarity intervals from Gale et al. (2006), surfaces of sequences 1–5/6/7 (modified from Gale et al., 2006), and tlb (see text), MP levels from Hooker (1992), NP zones from Aubry (1986) and Martini (1972), and charophyte zones (*G. tuberosa* and *H. tuberculata*) from Riveline (1984) and Sille et al. (2004). Vertical lines indicate hiatuses; stipple indicates intermediate polarity signal; oblique lines indicate intervals without paleomagnetic data. Additional abbreviations: BAR—Bartonian; BNPZ—Bembridge normal polarity zone; Hather—Hatherwood; HHNPZ—Headon Hill normal polarity zone; L—lower; Lin Ch—Linstone Chine; Lst—Limestone; Mbr—Member; mfs—maximum flooding surface; SB—Seagrove Bay Member; sb—sequence boundary; *Nem*—*Nematura* Bed; tlb—top log bed; Winth—Wintham.

tuberosa charophyte zone at Blaye (Riveline, 1984; Dubreuilh, 1989). Above the Marnes à *Ostrea cucullaris* at Blaye and below the MP18 level at Civrac, there lies the marine Calcaire de St. Estèphe, which yields *Nummulites fabianii* (Dubreuilh, 1989, Ringeade, 1987), an index species for Shallow Benthic Zone (SBZ) 19 of Serra-Kiel et al. (1998). Although overlaps between nummulite faunas and calcareous nannoplankton floras are uncommon, overlap does occur in the Biarritz area, where the range of *N. fabianii* coincides with most but not the highest parts of NP19–20 (Mathelin and Sztrákos, 1993). Moreover, according to these authors, a hiatus exists above this, cutting out most or all of the *Turborotalia cerroazulensis cunialensis* zone (P16), indicating that SBZ19 equates with the highest part of P15 and thus only the lower part of the zonal interval NP19–20, which precedes chron C15n (Hardenbol et al., 1998). Recent reclassification of subspecies of *T. cerroazulensis* as distinct species and revision of planktonic foraminiferal zones do not affect the important first occurrence of *T. cunialensis* (Berggren and Pearson, 2005). This means that the Colwell Bay Member record of NP19–20 belongs to the lower part of the zonal interval: the part that correlates with SBZ19 (Fig. 3).

In the Ebro Basin, Spain, the succession of continental strata that calibrates the early post–Grande Coupure mammal fauna of Santpedor to C13n continues downward to another normal polarity zone identified as C15n (Barberà et al., 2001). This too is associated with mammals at Rocafort de Queralt, where a limited fauna is dominated by *Ectropomys exiguus* (Anadón et al., 1987). *E. exiguus* also occurs at nearby San Cugat de Gavadons in a more extensive fauna (Agusti et al., 1987). Both faunas are referred to the local *Theridomys golpeae* zone and equated with MP19–20 (Barberà et al., 2001). However, none of its listed taxa is diagnostic of these levels, and most range from at least MP18 to MP20. In contrast, *T. golpeae* itself occurs also at Civrac, France (Vianey-Liaud and Ringeade, 1993), associated with diagnostic MP18 taxa such as *Palaeotherium magnum girondicum* and *P. medium perrealense* (Ringeade, 1987). *T. golpeae* also occurs at Ste. Croix in the Alès Basin, France (Vianey-Liaud and Ringeade, 1993), associated also with diagnostic MP18 taxa including *P. muehlbergi thaleri* and *P. magnum girondicum* (Remy, 1985). It has not been found associated with diagnostic MP19 or MP20 taxa, and therefore the *T. golpeae* Zone is equated here with MP18, which can thus, in turn, be calibrated to chron C15n (Fig. 3). In the Solent Group, MP18 taxa occur as low as the Linstone Chine Member (*Isoptychus pseudosiderolithicus*) and Hatherwood Limestone Member (*P. muehlbergi thaleri*), and as high as the Lacey's Farm Limestone Member (*P. curtum villerealense*; Hooker, 2008, personal observ.), so C15n should occur somewhere within this interval.

Thus, three key calibrations can be made between Solent Group mammal faunas and the marine record (Fig. 3). First, the Grande Coupure (MP20/21) is directly correlated to both Oi-1 (Belgium) and Chron C13n (Spain). Second, MP17 and 18 bracket SBZ19, showing that the Colwell Bay Member record of NP19–20 falls in the lower part of this zonal interval. Third, the local Spanish *T. golpeae* Zone, which is shown to represent MP18, calibrates in the Ebro Basin to chron C15n, indicating that in the Hampshire Basin, this magnetochron should be present within the span of the Linstone Chine to Lacey's Farm Members.

Recalibration of Solent Group Magnetochrons

Recognition that the NP19–20 record in the Colwell Bay Member in Whitecliff Bay belongs to the lower part of the zonal interval equivalent to SBZ19 means that at least the lower part of the Headon Hill normal polarity zone should represent chron C16n, not C15n (Figs. 3 and 4). The Headon Hill normal polarity zone continues up into the Cliff End Member, although there is an interval of sands identified with higher parts of the Colwell Bay Member (Daley, 1999) where no polarity signal was obtained (Gale et al., 2006, their Fig. 3: 90–100 m depth). As C16n is in two parts separated by a short reversed interval (C16n1r), it is possible that this reversal is hidden within the sandy interval lacking paleomagnetic data in Whitecliff Bay and that C16n1n and C16n2n are both represented by the Headon Hill normal polarity zone.

The association of MP18 with C15n (see previous) indicates that this chron should occur somewhere within the span of the Linstone Chine to Lacey's Farm Members. Candidates are therefore the upper part of the Headon Hill normal polarity zone or the single normally polarized sample at the top of the Cliff End Member (Gale et al., 2006, their Fig. 3). This normally polarized sample at the top of the Cliff End Member is succeeded by ~2 m of the lower part of the Lacey's Farm Member, which does not display a polarity signal, and which also crosses the SB at the base of sequence 4. This single sample could therefore represent part of a longer, largely unrecorded normal polarity interval identifiable with C15n. The most consistent overall pattern in fact is that the Headon Hill normal polarity zone represents all of C16n and that the normal sample at the top of the Cliff End Member belongs to C15n (Fig. 3). This is supported by the order of sequences (see following).

Although originally recognized in Whitecliff Bay (Insole and Daley, 1985), the Linstone Chine Member was later re-identified as the basal part of the Cliff End Member by Daley (1999). However, another sand unit included by Daley (1999) in the upper part of the Colwell Bay Member, at the base of which lies the sequence 3 SB, closely resembles the Linstone Chine Member in west Wight, which there represents the same position in sequence 3 (Gale et al., 2006, their Fig. 5). This sand unit is re-identified as the Linstone Chine Member in Figure 3 herein.

It has been shown (Hooker et al., 2007) that the Bembridge normal polarity zone is too old to be correlative to C13n and too young to represent C15n and that it fits chronologically best with C13r1n (Fig. 3), recognized at Massignano, Contessa Highway, Italy, and Ocean Drilling Program (ODP) Sites 522 and 1090 (South Atlantic) (Premoli-Silva et al., 1988; Tauxe et al., 1984; Channell et al., 2003). C13r1n is not recorded on most global charts (e.g., Gradstein et al., 2004), and Lowrie and Lanci (1994) and Lanci et al. (1996) were unable to relocate this subchron either at Massignano or in a short drill core taken nearby (called the MASSICORE). However, sample spacing in the vicinity of the Eocene-Oligocene boundary in the MASSICORE was 15–20 cm, and at Massignano spacing was ~35 cm, which could exceed the thickness representing the duration of C13r1n (based originally on a single sample at Massignano), raising the possibility that it might have been missed. The depositional rate in the MASSICORE was ~10 m/m.y. for the interval C13n–15n inclusive. At ODP Site 522, it was only ~1.8 m/m.y. for the same interval (calculated from Tauxe et al., 1984, their Fig. 7; Gradstein et al., 2004). Despite this low accumulation rate, one or two normal polarity intervals at or just below the Eocene-Oligocene boundary were recorded as single samples (Tauxe et al., 1984).

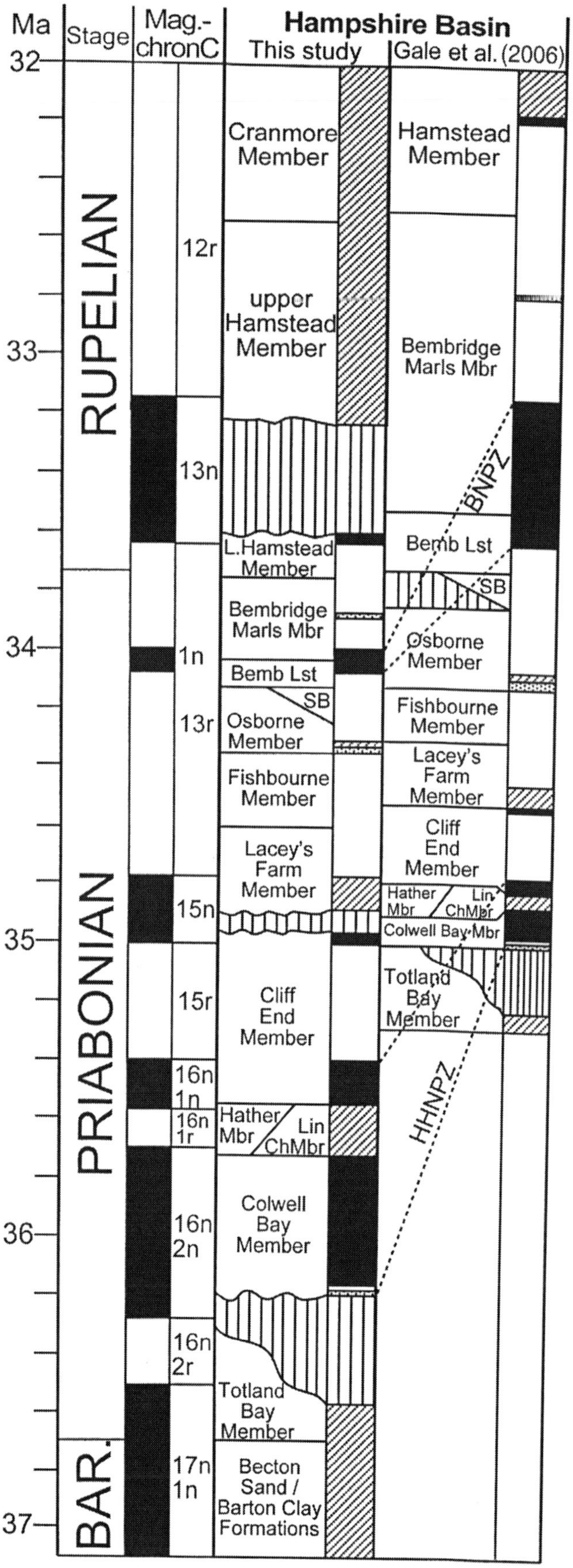

Figure 4. The revised correlation herein shown next to that of Gale et al. (2006, 2007). Dashed tie lines link the Bembridge (BNPZ) and Headon Hill (HHNPZ) normal polarity zones. Abbreviations and log symbols follow those of Figure 3. See text and Figures 2–3 for basis of the new correlation. N.B. Gale et al. (2006) used the Berggren et al. (1995) time scale, but this differs little from that used herein (see Fig. 3) with respect to the positioning and spans of the normally polarized magnetochrons. Calibration of that part of Gale et al.'s correlation above the Bembridge normal polarity zone and below the Headon Hill normal polarity zone to the new correlation is based on their calibration to 400 k.y. cycles (Gale et al., 2006, their Fig. 8). SB—Seagrove Bay Member.

In ODP Site 1090, C13r1n was represented by multiple samples, consistent with a depositional rate in C13r of 38 m/m.y., more than one order of magnitude greater than in Site 522. In ODP Site 1090, the duration of the subchron was estimated to be 79 k.y. (Channell et al., 2003). In contrast, rates in the upper part of the Solent Group may have been as high as 192 m/m.y. according to possible varves in the Bembridge Marls (Collinson, 1983), i.e., more than two orders of magnitude higher than in Site 522, and more than one order of magnitude greater than in the MASSICORE. One could therefore reasonably expect C13r1n to be represented in the Solent Group by multiple samples. Using the Bembridge Marls varve rate as a maximum, the thickness of ~5 m for the Bembridge normal polarity zone would only represent a duration of ~26 k.y. The 79 k.y. duration of C13r1n calculated from ODP Site 1090 implies an average depositional rate for the C13r interval in the Solent Group (Lacey's Farm to lower Hamstead members) of 63 m/m.y., one-third of that postulated from possible varves in the Bembridge Marls. The 1.9 m.y. between the beginning of the Priabonian and the end of chron C15n is represented in west Wight (the more complete succession) by 60–70 m of Solent Group sediments, which (accepting at least two hiatuses) represents an average sedimentation rate of 32–37 m/m.y., thus much less than for the C13r interval. This indicates a marked ascending increase in sedimentation rate through the Solent Group. The Solent Group sediments representing the 250 k.y. interval between C13n and C13r1n, according to the Gradstein et al. (2004) time scale, are ~25 m thick, indicating a rate of 100 m/m.y. for this part of the Solent Group. According to the astronomically tuned time scale of Coxall et al. (2005; see also Fig. 4 in van Mourik, 2006, paper 4) for the terminal Eocene and Oligocene (used in Fig. 3), this same interval is ~350 k.y. long, indicating a depositional rate of 71 m/m.y. for this part of the Solent Group. This suggests that the postulated Bembridge Marls varves represent faster sedimentation over shorter intervals than the average for this time. It is clear though that the Bembridge normal polarity zone covers too short a time span to represent C13n, which lasted nearly half a million years. Lateral thickness variation between Massignano and the MASSICORE (Lanci et al., 1996) suggests that either different rates of sedimentation or, more likely, small unconformities could be responsible for cutting out C13r1n at these sites.

The association between the Grande Coupure and C13n (see previous) indicates that this normal polarity interval should occur somewhere in the Hamstead Member. A single normally polarized sample was recorded by Gale et al. (2006, their Fig. 3) 8 m above the base of the Hamstead Member, which translates approximately to a short distance below the *Nematura* Bed. It is likely, therefore, that this represents the beginning of C13n, the higher parts of which may be missing in the intervening hiatus at the base of the *Nematura* Bed (Hooker et al., 2004; Fig. 3 herein; see following). No paleomagnetic indicators have been recorded higher in the succession, so it is not known whether higher parts of C13n are represented in the *Nematura* Bed and higher strata.

Thus, the calibration of mammal faunas to the marine record allows the identification of chrons C16n, 13r1n, 13n, and tentatively 15n in the Solent Group. Although 15n and 13n are represented only by single samples, their positions are consistent with the biostratigraphy. The Bembridge normal polarity zone in terms of likely duration fits C13r1n much better than C13n. Figure 4 shows the revised correlation herein next to that of Gale et al. (2006, 2007).

Correlation of Solent Group Sequences with Those in Italy and Belgium That Are Equated with Those of Haq et al. (1987)

Top Log Bed Sequence Boundary

Deep-marine successions in central Italy, with standard calcareous nannoplankton (NP) and planktonic foraminiferal (P) zones and magnetochrons, have been correlated with shallow-marine successions in northern Italy by means of dinoflagellates (Brinkhuis and Biffi, 1993; Brinkhuis and Visscher, 1995). Sequence stratigraphy in both areas has been documented using both dinoflagellates as environmental indicators and the sedimentology, and calibration has been established with the sequence stratigraphic scale of Haq et al. (1987) by Brinkhuis and Visscher (1995). Using recognition of many of Brinkhuis and Biffi's (1993) dinoflagellate zones in the Belgian Basin (Stover and Hardenbol, 1994; Vandenberghe et al., 2003) and linking these to mammalian events in the Belgian, Paris, and Hampshire basins, it has been possible to recognize the TA4.3/4.4 SB in the Hampshire Basin (Hooker et al., 2004). This SB occurs at the base of the *Nematura* Bed (top of the log bed) in the Hamstead Member (Figs. 1 and 3). It is marked by an erosion surface overlain by a 1-m-thick interval of ripple-bedded sand/clay alternations with brackish biota, notably the molluscs *Stenothyra* (formerly *Nematura*) and *Polymesoda* (Bristow et al., 1889; Munt et al., 1994), the ostracod *Hemicyprideis* (Keen, 1971), and marine dinoflagellates (Liengjarern et al., 1980), comprising the *Nematura* Bed. The *Nematura* Bed is the most saline, laterally persistent interval between the Bembridge Oyster Bed, nearly 30 m below, and the Cranmore Member, ~55 m above. Most of the intervening strata contain a distinctly freshwater biota, notably the gastropods *Viviparus*, *Lymnaea*, and *Australorbis* (Bristow et al., 1889; Paul, 1988), the ostracods *Moenocypris*, *Cypridopsis*, and *Candona* (Keen, 1972), macro-plants, and algal cysts (Collinson, 1983; Collinson et al., 1993; Collinson and Cleal, 2001). The *Nematura* Bed is thus identified as the maximum flooding surface (MFS). The underlying bed, the log bed, is a muddy sand showing cross-bedding, contorted bedding (Fig. 5), and a loaded base, with ball and pillow structures (Hooker et al., 2004). This shows that dewatering was taking place penecontemporaneously and points to rapid deposition. This bed contains coniferous logs, some large, which sometimes have a coating on their upper surface of *Polymesoda* shells from the overlying *Nematura* Bed, showing that they were projecting from the sediment surface when the *Nematura* Bed began to be deposited. A reworked clast containing rootlets was recorded from the base of the *Nematura* Bed (Hooker et al., 2004). The log bed also contains drifts of seeds (*Potamogeton* and *Stratiotes*), dicotyledonous and monocotyledonous leaves, the gastropods *Australorbis* and *Lymnaea*, and the bones and teeth of mammals. It was therefore clearly deposited in a shallow subaqueous freshwater environment and is not a paleosol (contra Gale et al., 2007, p. 687). The log bed is penetrated by vertical or near-vertical sand-filled burrows, which appear to be penecontemporaneous, since, in some cases, their tops are truncated by superposed sand bodies (Fig. 5).

One might be tempted to interpret the log bed to represent the LST, with a transgressive surface overlying it at the base of the *Nematura* Bed. However, several lines of evidence argue against this interpretation:

1. The estimated amount of sea-level fall at Oi-1 (70 m: Pekar et al., 2002; 55 m: Miller et al., 2008; 67 m: Katz et al., 2008) would mean significant retreat of the coastline and emergence of the sediments of the very shallow freshwater body represented by the log bed. The LST would therefore only be preserved far downdip from the location of the log bed.
2. Evidence of emergence of the top of the log bed is provided by the reworked rooted clast in the *Nematura* Bed.
3. All the mammals in the log bed are present in underlying beds of the Hamstead Member (Hooker, 2009), indicating minimal time between deposition of these lower beds and the log bed. In contrast, ~5 m higher in the succession, there is a typical MP21 post–Grande Coupure mammal fauna with few pre–Grande Coupure survivors, unlike the earlier post–Grande Coupure Hoogbutsel Mammal Bed in Belgium. Thus, the mammal succession points to a significant hiatus between the log bed and higher strata, but not between the log bed and underlying strata.
4. A partially disarticulated skeleton of an immature individual of the artiodactyl mammal *Anoplotherium* has been collected from the log bed (Hooker, 2007). Its unfused articular epiphyses have in some cases separated from their apophyses and lie close by in the sediment, and in other cases, they have remain attached, indicating incomplete decomposition of the animal's cartilaginous tissues at time of burial. Full decomposition of the nonbony parts of such a skeleton, even under water, should not have taken longer than one year. For a bed that does not exceed 1 m

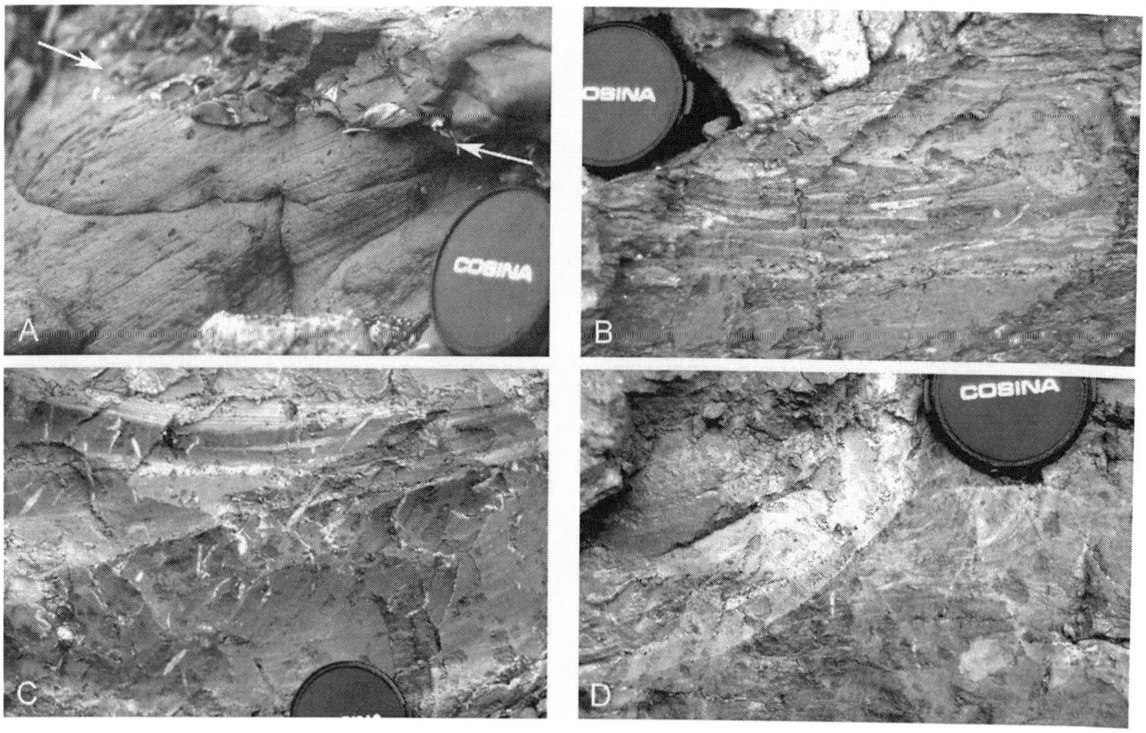

Figure 5. Sedimentary structures in the log bed at the top of the lower Hamstead Member, Bouldnor, Isle of Wight. (A) Cross-bedded muddy sand of log bed truncated by *Nematura* Bed, the base of which is indicated by arrows. (B) Recumbent folds in finely laminated muddy sand cut by sand-filled channel. (C) Sand-filled burrows, some shown to be truncated by overlying laminated sand. (D) Contorted bedding in finely laminated muddy sand cut by steep-sided, cross-bedded, sand-filled channel. Lens cap measures 50 mm in diameter.

thickness, deposition could have taken place in as little as one decade. This is very rapid and does not match the pattern of gentle flooding in other LSTs like the lower Bembridge Limestone Formation and Lacey's Farm Member.

5. The loaded base of the log bed, which penetrates the underlying muds, indicates no hiatus between the two, since the dewatering affected both. The absence of a loaded base to the also sandy *Nematura* Bed indicates a hiatus between it and the log bed.

We conclude therefore that the log bed represents the FSST, as the sea level fell in response to polar ice buildup at Oi-1. Since the log bed is overlain by the *Nematura* Bed, which represents the MFS of the overlying sequence, the LST and TST of this sequence are missing in the intervening hiatus.

Sequences 5–7

Gale et al. (2006, their Fig. 4) did not recognize the top log bed SB but recorded two others (6 and 7) above sequence 5 in Whitecliff Bay in the Bembridge Marls and Hamstead Members (although the Hamstead Member is not recorded as occurring in Whitecliff Bay, and their log may incorporate part of the succession at Hamstead). These sequences were recognized on the basis of gastropod faunas and clay mineralogy. Thus, sequence 6 begins ~4 m above the MFS of sequence 5 in Daley's (1973) bed VI, on the strength of presence of illitic clay and, by comparison with preceding sequences, the implication of a freshwater environment. However, only ~1 m higher, there is a limestone containing brackish corbiculid bivalves well within the sequence 6 LST, indicating the persistence above the Bembridge Oyster Bed of brackish conditions at least to this level (see Daley, 1972). This is paralleled at Hamstead Ledge in the west where brackish molluscs persist up to bed X, an approximate correlative of the top of Whitecliff Bay bed VI (Daley, 1972, 1973). This argues against the existence of Gale et al.'s sequence 6, so that this part of the succession is best combined with the underlying sequence 5.

Sequence 7 begins with the Black Band at the base of the Hamstead Member. The Black Band is a laterally persistent, 40-cm-thick organic mud overlying a bed with "nodules." The possibility of the representation of a terrestrial surface (Forbes, 1856) was expanded by Pomerol (1989) into the hypothesis of a major hiatus representing the sea-level fall attributed now to the Oi-1 glaciation. Hooker et al. (2004) argued against the base of the Black Band as being a SB on the basis of the identity of sediments above and below the Black Band and the presence of similar rooted organic horizons in the underlying Bembridge Marls. Though the "nodules" at the base of the Black Band effervesce slightly when HCl is applied, they are not carbonate nodules sensu Machette (1985), but rather they are clay concretions, formed in situ rather than transported in, given the lack of disturbance fabrics in thin sections of the concretions that could be associated with erosive events (e.g., Retallack, 2005a). The lack of even stage I carbonate development (Machette, 1985) further suggests either a very short formation time for the Black Band paleosol, or that the associated clay concretions are unrelated to pedogenic processes. Thus, there is no evidence to suggest a separate SB associated with the Black Band. Sequences 5, 6, and 7 of Gale et al. (2006) should therefore be amalgamated into one third-order sequence, here termed "sequence 5/6/7."

The SB at the base of sequence 5/6/7 is likely to equate with that of Haq et al.'s (1987) TA4.3. In Italy, the latter coincides with the base of Brinkhuis and Visscher's (1995) Aal dinoflagellate zone, which is shown within chron C13r1n at Massignano. In Whitecliff Bay, however, the sequence 5/6/7 SB is in the reversed interval slightly below C13r1n (Fig. 3). The apparent contradiction may be explained by the expanded estimate of the extent of the subchron in the Massignano section, which was recorded from a single sample.

According to Brinkhuis (1994), the MFS of sequence TA4.3 in northern Italy is within dinoflagellate zone Gse, just after the Eocene-Oligocene boundary at 33.73 Ma on Coxall et al.'s (2005) astronomically tuned time scale. It is, thus, younger than the equivalent MFS identified in the Hampshire Basin at the Bembridge Oyster Bed (see Fig. 3). The equivalent MFS of the Upper Pachuta–Shubuta Sequence in Saint Stephen's Quarry, Alabama, USA, lies at an intermediate position, coinciding with the precursor to the Oi-1 event (Miller et al., 2008; = EOT-1 of Katz et al., 2008) at 33.9 Ma (Coxall et al., 2005). The reason for the differences in timing of the MFS of this sequence in different sections is unclear, although TA4.3 in northern Italy is complicated at Bressana by the inclusion of two parasequences (Brinkhuis and Visscher, 1995).

Sequences 1–4

The next SB down in the Solent Group succession belongs to sequence 4 and is at the base of the Lacey's Farm Member (Fig. 3). In Italy, the equivalently positioned SB, identified as the base of Haq et al.'s (1987) TA4.2, is situated just below the base of Brinkhuis and Biffi's (1993) Ssp dinoflagellate zone and near the top of their underlying Mps zone. In central Italy, an interval barren of dinocysts spans Chron C15n and a small part of C13r, and the Mps-Ssp zonal boundary lies somewhere within it. It is logical on superposition grounds to equate sequence 4 with sequence TA4.2 as recognized in Italy by Brinkhuis and Visscher (1995). The presence of MP18 mammal taxa in the Lacey's Farm Member and the association of MP18 with C15n in Spain (see previous discussion), combined with the order of sequences, support the identification of the normally polarized sample at the top of the Cliff End Member in Whitecliff Bay as representing part of C15n. Solent Group sequence 4 may correlate with the Bassevelde 2 sequence in Belgium (Vandenberghe et al., 2003), since there is a possible upper limit of the Mps zone at the top of the underlying Bassevelde 1 sequence.

Sequences 1–3 extend down to the Haq et al. (1987) TA3.6/4.1 SB at the base of the Solent Group (Plint, 1988). Lack of a record here in central Italy makes calibration difficult. The SBs at the bases of sequences 2 and 3 recorded on Headon Hill and in Whitecliff Bay show fluctuations between lagoonal and lacustrine/palustrine systems and so are likely to represent relatively minor sea-level fluctuations. It is therefore possible that they are not recognizable in most marine successions. The re-identification here of the Headon Hill normal polarity zone as chron C16n suggests that sequences 1–3 may together equate with TA4.1 of Haq et al. (1987).

Summary

Thus, a range of sedimentological, taphonomic, and faunal evidence supports the presence of a major hiatus below the *Nematura* Bed and points to interpretation of the log bed as belonging to the FSST as Oi-1 polar glacial ice formed. The SB between the log bed and the *Nematura* Bed is identified with that of TA4.3/4.4 of Haq et al. (1987), according to correlation with Belgium and Italy (Stover and Hardenbol, 1994; Brinkhuis and Visscher, 1995). Although the Eocene-Oligocene boundary is recognized by the extinction of the planktonic foraminiferal family Hantkeninidae, naturally absent from these continental strata, its approximate position should be intermediate between this SB and the top of the Bembridge normal polarity zone. It should be closer to the former than to the latter according to Coxall et al.'s (2005) astronomically tuned time scale. The boundary may therefore lie high in the Bembridge Marls Member or low in the lower Hamstead Member. Identification of the TA4.3 and 4.2 SBs at the bases of the Bembridge Limestone Formation and Lacey's Farm Member, respectively, is consistent with the biostratigraphically constrained magnetochrons and the order of sequences.

CALIBRATION OF CLIMATE-RELATED PARAMETERS

Isotope-Derived Temperature Values

Grimes et al. (2005) obtained temperatures from $\delta^{18}O$ values in multiple proxies in the Solent Group. By combining those from rodent tooth enamel that mineralized at a constant body temperature with those from pulmonate gastropod (*Lymnaea*) shells, charophyte gyrogonites, and fish otoliths, which belonged to freshwater-dwelling biota, they produced values that were ice-volume independent. These were considered to be slight overestimates but close to the temperatures that existed in the shallow freshwater bodies during the summer months (for methods, see Lindars et al., 2001; Grimes et al., 2003). Grimes et al. (2005) accurately positioned Oi-1 in their succession of temperature values by direct link to the Grande Coupure. However, because of the paucity of marine biostratigraphic indicators lower in the Solent Group succession, calibration of the other values relied largely on assumed constant sedimentation rate. The more refined correlation herein, facilitated by combined biostratigraphy, magnetostratigraphy, and sequence stratigraphy, allows us to adjust the calibration of the temperature values in the lead-up to Oi-1 to the global time scale (Fig. 6). The result is that those of the Lacey's Farm and Osborne Members and Bembridge Limestone Formation shift to younger positions, closer both to one another and to that of the lower Hamstead Member and further from that of the Totland Bay Member. This reflects the increase in sedimentation rates higher in the succession.

In recently published work, recognized climatic events in the marine realm across the Eocene-Oligocene transition have multiplied and, through a combination of $\delta^{18}O$, Mg/Ca, and sea-level studies, in certain cases, temperature and ice-volume have been apportioned (Miller et al., 2008; Pearson et al., 2008; Lear et al., 2008; Katz et al., 2008). These authors have thus calculated that up to half of the $\delta^{18}O$ shift at Oi-1 represents ice-volume increase, whereas at EOT-1 (= the "precursor" event and "step 1") just before the Eocene-Oligocene boundary, most of the isotope shift is attributed to temperature. The role of temperature, not just ice-volume, is also supported by evidence for fluctuations in sea-surface marine temperatures from dinoflagellate assemblages (Brinkhuis and Biffi, 1993; Fig. 3 *in* van Mourik, 2006, paper 4). Although with a much less dense record, it is now possible to match up the Solent Group paleotemperatures with some of these marine events. One of the most useful oceanic sections in

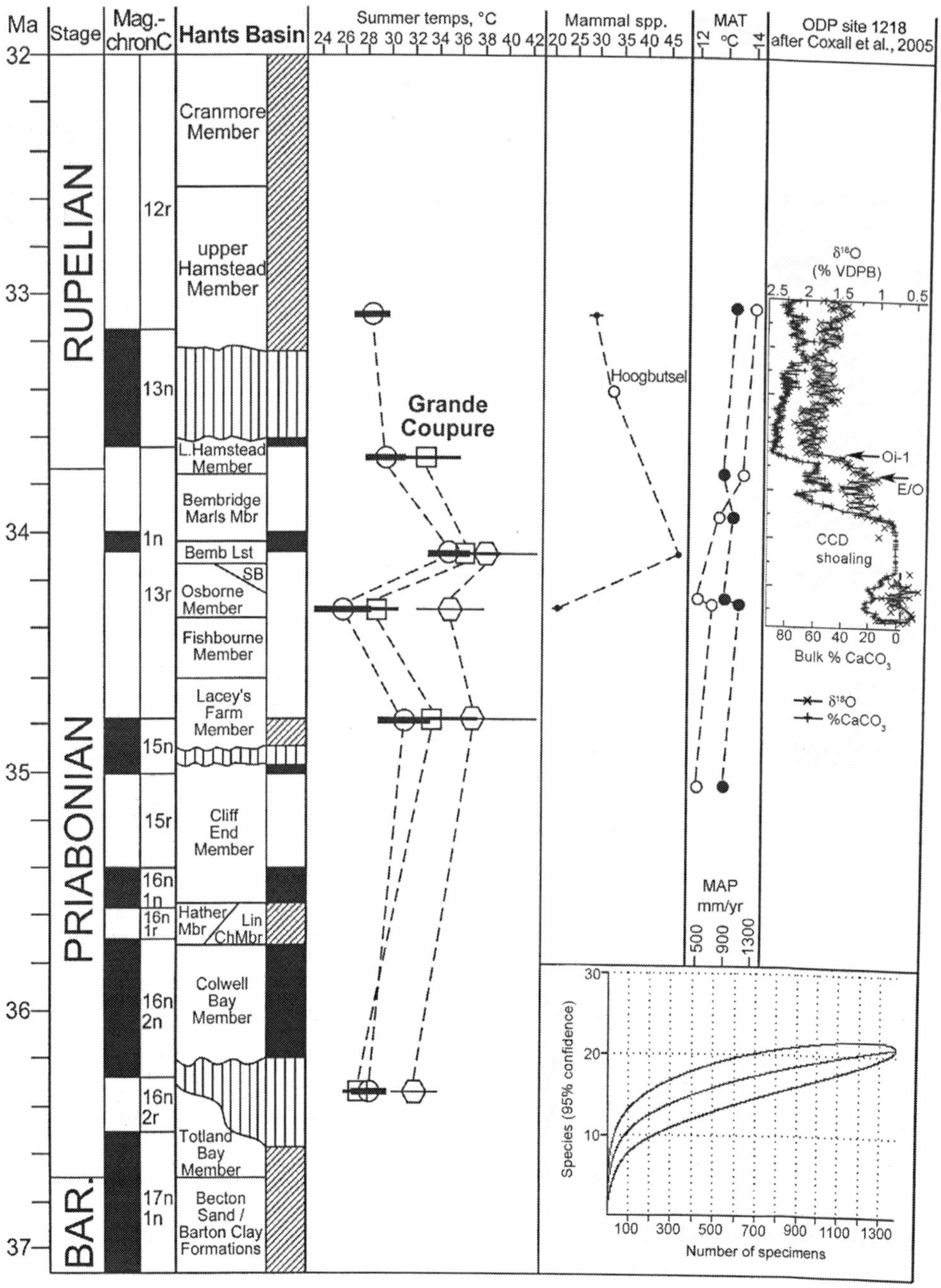

Figure 6. Hampshire Basin oxygen isotope–derived summer temperatures (Grimes et al., 2005), mammal species richness (Hooker et al., 2004; herein), and whole-rock geochemistry-derived temperature (solid circles) and precipitation (hollow circles) values from paleosols (Sheldon et al., this volume.; Sheldon, this volume) calibrated to the geomagnetic polarity time scale (GPTS) following Figure 3, and to the $\delta^{18}O$ and $CaCO_3$ curves at Ocean Drilling Program (ODP) Site 1218 (Coxall et al., 2005). Error bars on summer temperature plots are propagated 2σ standard errors (Grimes et al., 2005). The column "Mammal spp." shows numbers of mammal species at different levels in the Solent Group (solid circles) plus the Belgian site of Hoogbutsel (hollow circle). Error bars on mammal species richness plots (solid) are 95% (inset; Hooker et al., 2004); the Hoogbutsel plot has not been treated statistically. Uncertainties for the paleosol proxy data are ±182 mm/yr and ±4 °C. MAP—mean annual precipitation; MAT—mean annual temperature. Inset is rarefaction curve for mammal species richness in the Osborne Member. VPDB—Vienna Peedee belemnite; CCD—calcite compensation depth.

this context is the low-latitude Pacific ODP Site 1218, where an essentially uninterrupted and biostratigraphically and magnetostratigraphically well-calibrated succession shows a sequence of events leading up to and across Oi-1 (Coxall et al., 2005). Coxall et al.'s (2005) succession extends down to C15r. We follow the identification of the lowest normal interval as C15n by Lanci et al. (2005), rather than as C13r1n by Coxall et al. (2005), as it is the better match with the latter's time scale and with the low-latitude extinction of the NP19–20 marker *Discoaster saipanensis*. Thus, comparing the two independently dated successions (Fig. 6 herein; Coxall et al., 2005, their Fig. 3a): (1) the warming peak in the Bembridge Limestone coincides with the warming interval in Site 1218 around 34 Ma, where the calcite compensation depth (CCD) shoaled; (2) the cooling in the lower Hamstead Member that just precedes the Grande Coupure coincides in Site 1218 with the second stage (itself complicated by the recognition of another small step preceding Oi-1, namely EOT-2, by Katz et al., 2008) of the two-step increase in the CCD and with the positive shift in $\delta^{18}O$ values leading directly to Oi-1; (3) cool temperatures continuing into the upper Hamstead Member (post–Grande Coupure) fit with the maintenance of more positive $\delta^{18}O$ values in Site 1218 through the early part of the Oligocene; and (4) the cooling low in the Osborne Member coincides with a brief positive excursion in benthic foraminiferal $\delta^{18}O$ preceding the shoaling event in Site 1218 (Coxall et al., 2005, their Fig. 3a; Miller et al., 2008, their Fig. 8: "34.1" event), named the "late Eocene event" by Katz et al. (2008).

The close coincidence of the isotope-derived temperature points in the Solent Group (Grimes et al., 2005) with CCD and $\delta^{18}O$ shifts in the low-latitude Pacific (Coxall et al., 2005) suggests that the Solent Group record indeed reflects global temperature fluctuations. The fact that post–Bembridge Limestone temperatures are less cool than those of the preceding Osborne Member may reflect the lesser effect of temperature decrease than of ice-volume increase at Oi-1 compared to earlier in the Eocene-Oligocene transition. Alternatively, it may simply mean that the lower Hamstead plot slightly predates the time of the coolest Oi-1 values. Intriguingly, however, the "late Eocene event" has been tentatively attributed by Katz et al. (2008) to ice buildup, not temperature, because of a sea-level fall-related hiatus at Saint Stephen's Quarry (Miller et al., 2008), plus no significant change in Mg/Ca at ODP 1218 (Lear et al., 2004). However, the hiatus at Saint Stephen's Quarry is a major one and may not simply be attributable to eustatic sea-level fall, since Oi-1 had less erosive impact there (Miller et al., 2008, their Fig. 8). The Osborne Member values clearly show a fall in temperature at this time.

Fluctuations in Mammalian Species Richness

Species richness of Solent Group mammals across the Eocene-Oligocene transition (Fig. 6) essentially tracks the isotope-derived summer temperature values of Grimes et al. (2005). The highest count (47) is that of the Bembridge Limestone lower fauna (Hooker et al., 1995, 2004). That of the post–Grande Coupure upper Hamstead Member fauna is 29 (Hooker et al., 2004; Hooker, 2009). The fauna of the lower part of the Osborne Member consists of 21 species (Fig. 6). The differences between these three faunas are shown by rarefaction analysis to be statistically significant (Fig. 6, inset; Hooker et al., 2004, their Fig. 2). Faunas between the lower part of the Bembridge Limestone and the upper Hamstead Member are still imperfectly known. The mammal fauna of the post–Grande Coupure Hoogbutsel Mammal Bed of Belgium, which predates that of the upper Hamstead Member (see previous), consists of 33 species (Bruijn and Ünay, 1989; Smith, 2003, 2004a, 2004b, 2006; Smith and van den Hoek Ostende, 2006). This is likely to be a reliable count, as it is based on large collections. Species richness of these faunas therefore supports the summer temperature fluctuations of Grimes et al. (2005).

Paleosols

Various methods have been developed for reconstructing paleoclimatic conditions using both physical and chemical traits of paleosols. In particular, two types of paleosol horizons have proven useful when present, the Bk horizon (Retallack, 2005b) and the Bw/Bt horizon (Sheldon et al., 2002). As discussed already, the rare carbonate present in paleosols of the Solent Group represents either a very immature stage of development or represents carbonate unrelated to pedogenesis (Sheldon et al., this volume), and so is not useful for either estimates of paleoprecipitation (Retallack, 2005b) or paleotemperature (Dworkin et al., 2005). In addition, while much of the nonmarine Solent Group shows evidence of some pedogenic alteration in the form of drab-haloed root traces, there are relatively few paleosols developed beyond the Entisol-like stage, and thus, they also lack Bw/Bt horizons. However, there are a few sufficiently developed paleosols (Fig. 6) spanning the Eocene-Oligocene transition interval to be useful for paleoclimatic reconstruction (Sheldon, this volume; Sheldon et al., this volume) using whole-rock geochemistry proxies for paleoprecipitation (Sheldon et al., 2002) and paleotemperature (Sheldon et al., 2002; Sheldon, 2006). These are well calibrated to the time scale. Reconstructed mean annual temperatures are nearly invariant (Fig. 6) across the transition interval from ca. 35 to 33 Ma, consistent with results from various North American sites (Sheldon and Retallack, 2004). In contrast, reconstructed mean annual precipitation values increase across the Eocene-Oligocene transition in the Hampshire Basin and decrease in both continental interior (Nebraska) and coastal plain (Oregon) sites in North America (Sheldon et al., this volume; Sheldon, this volume; Sheldon and Retallack, 2004). This suggests that some elements of Eocene-Oligocene transition climate variation may be strongly regionally variable in the continental realm.

The paleosol samples are from overbank facies in the Cliff End, Osborne, Bembridge Marls, and upper Hamstead members and from lacustrine facies in the Black Band at the base of the lower Hamstead Member. Of these, those from the Osborne and upper Hamstead members approximate horizons from which isotope-derived summer temperatures and significant mammal faunas were obtained. Summer temperatures at these levels are within error of each other and are thus consistent with the paleosol-derived MATs. However, the intervening summer warming recorded in the Bembridge Limestone, which is supported by a rich mammal fauna, has no paleosols suitable for study. Therefore, the invariance of the paleosol-derived MATs may partly reflect sampling resolution.

CONCLUSIONS

Integration of mammalian faunas with marine and other continental biozonal markers, with sea-level changes, and with the Oi-1 event through interdigitation of relevant European strata or direct relationship to identified magnetochrons has allowed us to refine the correlation of the dominantly nonmarine Solent Group to the geomagnetic polarity time scale. As a result, it has been possible to calibrate normal magnetic intervals documented by Gale et al. (2006) to magnetochrons C16n, 15n, 13r1n, and 13n (Fig. 3). The extent of incision at the base of sequence 5 based on lowstand thickness is judged here to be small and therefore does not provide evidence at this level for the degree of sea-level fall expected at Oi-1. The lowstand thickness modification results from recognition that the transgressive surface is at the *Corbicula* shell bed in the middle muds (Fig. 2) rather than at the top of the Bembridge Limestone Formation and that, according to mammalian biostratigraphy, the Seagrove Bay Member of the Headon Hill Formation belongs to sequence 4, not sequence 5. The log bed at the top of the lower Hamstead Member of the Bouldnor Formation is shown to have been deposited rapidly and to be overlain after a considerable hiatus by the *Nematura* Bed (Fig. 5). The log bed is interpreted as the falling stage systems tract recording the beginning of the sea-level fall in the buildup to the Oi-1 glaciation. The hiatus between the log bed and the *Nematura* Bed encompasses the Grande Coupure, which is coincident with Oi-1. The Eocene-Oligocene boundary is constrained to lie between the Bembridge normal polarity zone and the log bed; so, it is likely to be either high in the Bembridge Marls Member or low in the lower Hamstead Member.

The now more resolved correlation means more accurate time calibration of the oxygen isotope–derived summer temperatures (Grimes et al., 2005) and of paleosol-derived MATs and MAPs of the Solent Group. Fluctuations in the former can be seen to coincide with those in global temperature and ice volume, recognized by shifts of the oceanic CCD, sea level, $\delta^{18}O$, and Mg/Ca in benthic foraminifera, and dinoflagellate floral composition (Coxall et al., 2005; Brinkhuis and Biffi, 1993; Fig. 3 *in* van Mourik, 2006, paper 4; Lear et al., 2000, 2004, 2008; Katz et al., 2008), and these fluctuations are tracked by species richness of mammal faunas. Evidence from paleosols indicates a gradual increase in mean annual precipitation from ca. 35 Ma to ca. 33 Ma but no specific shifts associated with Oi-1 or fluctuations in other proxies. Mean annual temperatures from paleosols show no significant fluctuations, but data are lacking at the time of the Bembridge Limestone warming at ca. 34 Ma (Sheldon et al., this volume). The summer cooling at Oi-1 in southern England appears to have been relatively minor (Grimes et al., 2005), consistent with other evidence that shows that up to half of the $\delta^{18}O$ benthic foraminiferal positive shift at Oi-1 records polar ice buildup (Katz et al., 2008).

ACKNOWLEDGMENTS

We thank The National Trust, John and Marion Smith, and the Isle of Wight Countryside Management Service for access to sites; Andy Currant, Bryony Green, Elaine Lindars, John Quayle, Nick Sille, Mike Smith, David and Alison Ward, and Andy Yule for help with UK field work; Denver Fowler, Brian Gasson, John Quayle, Mike Smith, Steve Sweetman, and Andy Yule for donating specimens to the Natural History Museum, London; David Harrison (Harrison Institute, Sevenoaks), Kurt Heissig (Bayerische Staatssammlung für Paläontologie und historische Geologie), Elmar Heizmann (Staatliche Museum für Naturkunde, Stuttgart), Marguerite Hugueney (Faculté des Sciences, Lyon), Martin Munt (Isle of Wight County Museum Services), David Norman (Sedgwick Museum, Cambridge), Marc Godinot, Claire Sagne, and Pascal Tassy (Muséum National d'Histoire Naturelle, Paris), and Monique Vianey-Liaud (Université Montpellier II) for access to museum collections; and Steve Bohaty, Henk Brinkhuis, Lluis Cabrera, Miguel Garces, and Bridget Wade for helpful discussion. Comments on an earlier version by Henk Brinkhuis and Etienne Steurbaut have improved the paper.

REFERENCES CITED

Agustí, J., Anadón, P., Arbiol, S., Cabrera, L., Colombo, F., and Sáez, A., 1987, Biostratigraphical characteristics of the Oligocene sequences of northeastern Spain (Ebro and Campins Basins): Münchner Geowissenschaftliche Abhandlungen, ser. A, v. 10, p. 35–42.

Anadón, P., Vianey-Liaud, M., Cabrera, L., and Hartenberger, J.-L., 1987, Gisements à vertébrés du paléogène de la zone orientale du bassin de l'Ebre et leur apport a la stratigraphie: Paleontología i Evolució, v. 21, p. 117–131.

Arbiol, S., and Sáez, A., 1988, Sobre la edad oligocénica inferior del yacimiento de Santpedor (Cuenca del Ebro, provincia de Barcelona): Acta Geològica Hispànica, v. 23, p. 47–50.

Aubry, M.-P., 1985, Northwestern European Paleogene magnetostratigraphy, biostratigraphy, and paleogeography: Calcareous nannofossil evidence: Geology, v. 13, p. 198–202, doi: 10.1130/0091-7613(1985)13<198:NEPMBA>2.0.CO;2.

Aubry, M.-P., 1986, Paleogene calcareous nannoplankton biostratigraphy of northwestern Europe: Palaeogeography, Palaeoclimatology, Palaeoecology, v. 55, p. 267–334, doi: 10.1016/0031-0182(86)90154-9.

Barberà, X., Cabrera, L., Marzo, M., Parés, J.M., and Agusti, J., 2001, A complete terrestrial Oligocene magnetobiostratigraphy from the Ebro Basin, Spain: Earth and Planetary Science Letters, v. 187, p. 1–16, doi: 10.1016/S0012-821X(01)00270-9.

Berggren, W.A., and Pearson, P.N., 2005, A revised tropical to subtropical Paleogene planktonic foraminiferal zonation: Journal of Foraminiferal Research, v. 35, p. 279–298, doi: 10.2113/35.4.279.

Berggren, W.A., Kent, D.V., Swisher, C.C., III, and Aubry, M.-P., 1995, A revised Cenozoic geochronology and chronostratigraphy, *in* Berggren, W.A., Kent, D.V., Aubry, M.-P., and Hardenbol, J., eds., Geochronology, Time Scales, and Global Stratigraphic Correlation: SEPM (Society for Sedimentary Geology) Special Publication 54, p. 129–212.

Brinkhuis, H., 1994, Late Eocene to Early Oligocene dinoflagellate cysts from the Priabonian type-area (Northeastern Italy): Biostratigraphy and paleoenvironmental interpretation: Palaeogeography, Palaeoclimatology, Palaeoecology, v. 107, p. 121–163, doi: 10.1016/0031-0182(94)90168-6.

Brinkhuis, H., and Biffi, U., 1993, Dinoflagellate cyst stratigraphy of the Eocene/Oligocene transition in central Italy: Marine Micropaleontology, v. 22, p. 131–183, doi: 10.1016/0377-8398(93)90007-K.

Brinkhuis, H., and Visscher, H., 1995, The upper boundary of the Eocene Series: A reappraisal based on dinoflagellate cyst biostratigraphy and sequence stratigraphy, *in* Berggren, W.A., Kent, D.V., Aubry, M.-P., and Hardenbol, J., eds., Geochronology, Time Scales, and Global Stratigraphic Correlation: SEPM (Society for Sedimentary Geology) Special Publication 54, p. 295–304.

Bristow, H.W., Reid, C., and Strahan, A., 1889, The Geology of the Isle of Wight (2nd ed.): London, Memoirs of the Geological Survey of England and Wales, 349 p.

Channell, J.E.T., Galeotti, S., Martin, E.E., Billups, K., Scher, H.D., and Stoner, J.S., 2003, Eocene to Miocene magnetostratigraphy, biostratigraphy, and chemostratigraphy at ODP Site 1090 (sub-Antarctic South Atlantic): Geological Society of America Bulletin, v. 115, p. 607–623, doi: 10.1130/0016-7606(2003)115<0607:ETMMBA>2.0.CO;2.

Collinson, M.E., 1983, Palaeofloristic assemblages and palaeoecology of the Lower Oligocene Bembridge Marls, Hamstead Ledge, Isle of Wight: Botanical Journal of the Linnean Society, v. 86, p. 177–205, doi: 10.1111/j.1095-8339.1983.tb00725.x.

Collinson, M.E., and Cleal, C., 2001, Late Middle Eocene–early Oligocene (Bartonian-Rupelian) and Miocene palaeobotany of Great Britain, *in* Cleal, C.J., Thomas, B.A., Batten, D.J., and Collinson, M.E., eds., Mesozoic and Tertiary Palaeobotany of Great Britain: Peterborough, Joint Nature Conservation Committee, Geological Conservation Review Series, no. 22 p. 227–274.

Collinson, M.E., Singer, R.L., and Hooker, J.J., 1993, Vegetational change in the latest Eocene of southern England, *in* Planderová, E., Konzálová, M., Kvacek, Z., Sitár, V., Snopková, P., and Suballyová, D., eds., Paleofloristic and Paleoclimatic Changes during Cretaceous and Tertiary: Proceedings of the International Symposium September 14–20, 1992: Bratislava, Geologicky Ustav Dionyza Stúra, p. 81–87, plates 20–25.

Coxall, H.K., Wilson, P.A., Pälike, H., Lear, C.H., and Backman, J., 2005, Rapid stepwise onset of Antarctic glaciation and deeper calcite compensation in the Pacific Ocean: Nature, v. 433, p. 53–57, doi: 10.1038/nature03135.

Daley, B., 1972, Macroinvertebrate assemblages from the Bembridge Marls (Oligocene) of the Isle of Wight, England, and their environmental significance: Palaeogeography, Palaeoclimatology, Palaeoecology, v. 11, p. 11–32, doi: 10.1016/0031-0182(72)90035-1.

Daley, B., 1973, The palaeoenvironment of the Bembridge Marls (Oligocene) of the Isle of Wight, Hampshire: Proceedings of the Geologists' Association, v. 84, p. 83–93.

Daley, B., 1999, Palaeogene sections in the Isle of Wight: A revision of their description and significance in the light of research undertaken over recent decades: Tertiary Research, v. 19, p. 1–69.

Daley, B., and Edwards, N., 1971, Palaeogene warping in the Isle of Wight: Geological Magazine, v. 108, p. 399–405.

Daley, B., and Edwards, N., 1990, The Bembridge Limestone (late Eocene), Isle of Wight, southern England: A stratigraphical revision: Tertiary Research, v. 12, p. 51–64.

de Bruijn, H., and Ünay, E., 1989, Petauristinae (Mammalia, Rodentia) from the Oligocene of Spain, Belgium, and Turkish Thrace, *in* Black, C.C., and Dawson, M.R., eds., Papers on Fossil Rodents in Honour of Albert Elmer Wood: Natural History Museum of Los Angeles County Science Series, no. 33, p. 139–145.

De Man, E., Ivany, L., and Vandenberghe, N., 2004, Stable oxygen isotope record of the Eocene-Oligocene transition in the southern North Sea Basin: Positioning of the Oi-1 event: Netherlands Journal of Geosciences, v. 83, p. 193–197.

Dubreuilh, J., 1989, Synthèse paléogéographique et structurale des dépôts fluviatiles tertiaires du nord du Bassin d'Aquitaine; passage aux formations palustres, lacustres et marines: Documents du Bureau de Recherches Géologiques et Minières, v. 172, 461 p.

Dworkin, S.I., Nordt, L.C., and Atchley, S.C., 2005, Determining terrestrial paleotemperatures using the oxygen isotopic composition of pedogenic carbonates: Earth and Planetary Science Letters, v. 237, p. 56–68, doi: 10.1016/j.epsl.2005.06.054.

Eldrett, J.S., Harding, I.C., Firth, J.V., and Roberts, A.P., 2004, Magnetostratigraphic calibration of Eocene-Oligocene dinoflagellate cyst biostratigraphy from the Norwegian-Greenland Sea: Marine Geology, v. 204, p. 91–127, doi: 10.1016/S0025-3227(03)00357-8.

Forbes, E., 1853, On the Fluvio-marine Tertiaries of the Isle of Wight: Quarterly Journal of the Geological Society, London, v. 9, p. 259–269.

Forbes, E., 1856, On the Tertiary Fluvio-marine Formation of the Isle of Wight: Memoirs of the Geological Survey of Great Britain, and of the Museum of Practical Geology, Sheet Memoir no. 10 (old series), p. i–xviii, 1–162.

Gale, A.S., Jeffery, P.A., Huggett, J.M., and Connolly, P., 1999, Eocene inversion history of the Sandown Pericline, Isle of Wight, southern England: Journal of the Geological Society, London, v. 156, p. 327–339, doi: 10.1144/gsjgs.156.2.0327.

Gale, A.S., Hardenbol, J., Hathway, B., Kennedy, W.J., Young, J.R., and Phansalkar, V., 2002, Global correlation of Cenomanian (Upper Cretaceous) sequences: Evidence for Milankovitch control on sea level: Geology, v. 30, p. 291–294, doi: 10.1130/0091-7613(2002)030<0291:GCOCUC>2.0.CO;2.

Gale, A.S., Huggett, J.M., Pälike, H., Laurie, E., Hailwood, E.A., and Hardenbol, J., 2006, Correlation of Eocene-Oligocene marine and continental records: Orbital cyclicity, magnetostratigraphy and sequence stratigraphy of the Solent Group, Isle of Wight, UK: Journal of the Geological Society, London, v. 163, p. 401–415, doi: 10.1144/0016-764903-175.

Gale, A., Huggett, J., and Laurie, E., 2007, Discussion on the Eocene-Oligocene boundary in the UK: Journal of the Geological Society, London, v. 164, p. 686–688.

Gradstein, F.M., Ogg, J.G., and Smith, A.G., eds., 2004, A Geologic Time Scale 2004: Cambridge, Cambridge University Press, 589 p.

Grimes, S.T., Mattey, D.P., Hooker, J.J., and Collinson, M.E., 2003, Paleogene paleoclimate reconstruction using oxygen isotopes from land and freshwater organisms: The use of multiple paleoproxies: Geochimica et Cosmochimica Acta, v. 67, p. 4033–4047, doi: 10.1016/S0016-7037(03)00173-X.

Grimes, S.T., Hooker, J.J., Collinson, M.E., and Mattey, D.P., 2005, Summer temperatures of late Eocene to early Oligocene freshwaters: Geology, v. 33, p. 189–192, doi: 10.1130/G21019.1.

Haq, B.U., Hardenbol, J., and Vail, P.R., 1987, Chronology of fluctuating sea levels since the Triassic: Science, v. 235, p. 1156–1167, doi: 10.1126/science.235.4793.1156.

Hardenbol, J., Thierry, J., Farley, M., Jacquin, T., de Graciansky, P.-C., and Vail, P.R., 1998, Mesozoic and Cenozoic sequence chronostratigraphic framework of European basins, *in* de Graciansky, P.-C., Hardenbol, J., Jacquin, T., and Vail, P.R., eds., Mesozoic and Cenozoic Sequence Stratigraphy of European Basins: SEPM (Society for Sedimentary Geology) Special Publications 60, p. 3–13, 763–781.

Hooker, J.J., 1987, Mammalian faunal events in the English Hampshire Basin (late Eocene–early Oligocene) and their application to European biostratigraphy: Münchner Geowissenschaftliche Abhandlungen, ser. A, v. 10, p. 109–116.

Hooker, J.J., 1992, British mammalian paleocommunities across the Eocene-Oligocene transition and their environmental implications, *in* Prothero, D.R. and Berggren, W.A., eds., Eocene-Oligocene Climatic and Biotic Evolution: Princeton, Princeton University Press, p. 494–515.

Hooker, J.J., 2007, Bipedal browsing adaptations of the unusual Late Eocene–earliest Oligocene tylopod *Anoplotherium* (Artiodactyla, Mammalia): Zoological Journal of the Linnean Society, v. 151, p. 609–659, doi: 10.1111/j.1096-3642.2007.00352.x.

Hooker, J.J., 2009, The "Grande Coupure" in the Hampshire Basin, UK: Taxonomy and stratigraphy of the mammals on either side of this major Paleogene faunal turnover: Micropalaeontological Society Special Publication (in press).

Hooker, J.J., Collinson, M.E., Van Bergen, P.F., Singer, R.L., de Leeuw, J.W., and Jones, T.P., 1995, Reconstruction of land and freshwater palaeoenvironments near the Eocene-Oligocene boundary, southern England: Journal of the Geological Society, London, v. 152, p. 449–468, doi: 10.1144/gsjgs.152.3.0449.

Hooker, J.J., Collinson, M.E., and Sille, N.P., 2004, Eocene-Oligocene mammalian faunal turnover in the Hampshire Basin, UK: Calibration to the global time scale and the major cooling event: Journal of the Geological Society, London, v. 161, p. 161–172, doi: 10.1144/0016-764903-091.

Hooker, J.J., Collinson, M.E., Grimes, S.T., Sille, N.P., and Mattey, D.P., 2007, Discussion on the Eocene-Oligocene boundary in the UK: Journal of the Geological Society, London, v. 164, p. 685–688, doi: 10.1144/0016-76492006-098.

Insole, A., and Daley, B., 1985, A revision of the lithostratigraphical nomenclature of the late Eocene and early Oligocene strata of the Hampshire Basin: Southern England: Tertiary Research, v. 7, p. 67–100.

Katz, M.E., Miller, K.G., Wright, J.D., Wade, B.S., Browning, J.V., Cramer, B.S., and Rosenthal, Y., 2008, Stepwise transition from the Eocene greenhouse to the Oligocene icehouse: Nature Geoscience, v. 1, p. 329–334, doi: 10.1038/ngeo179.

Keen, M.C., 1971, A palaeoecological study of the ostracod *Hemicyprideis montosa* (Jones and Sherborn) from the Sannoisian of NW Europe: Bulletin du Centre de Recherches de Pau, v. 5, supplement, p. 523–543.

Keen, M., 1972, The Sannoisian and some other Upper Palaeogene Ostracoda from north-west Europe: Palaeontology, v. 15, p. 267–325.

Lanci, L., Lowrie, W., and Montanari, A., 1996, Magnetostratigraphy of the Eocene/Oligocene boundary in a short drill-core: Earth and Planetary Science Letters, v. 143, p. 37–48, doi: 10.1016/0012-821X(96)00136-7.

Lanci, L., Parés, J.M., Channell, J.E.T., and Kent, D.V., 2005, Oligocene magnetostratigraphy from equatorial Pacific sediments (ODP Sites 1218 and 1219, Leg 199): Earth and Planetary Science Letters, v. 237, p. 617–634, doi: 10.1016/j.epsl.2005.07.004.

Lear, C.H., Elderfield, H., and Wilson, P.A., 2000, Cenozoic deep-sea temperatures and global ice volumes from Mg/Ca in benthic foraminiferal calcite: Science, v. 287, p. 269–272, doi: 10.1126/science.287.5451.269.

Lear, C.H., Rosenthal, Y., Coxall, H.K., and Wilson, P.A., 2004, Late Eocene to early Miocene ice sheet dynamics and the global carbon cycle: Paleoceanography, v. 19, p. PA4015, doi: 10.1029/2004PA001039.

Lear, C.H., Bailey, T.R., Pearson, P.N., Coxall, H.K., and Rosenthal, Y., 2008, Cooling and ice growth across the Eocene-Oligocene transition: Geology, v. 36, p. 251–254, doi: 10.1130/G24584A.1.

Liengjarern, M., Costa, L.I., and Downie, C., 1980, Dinoflagellate cysts from the Upper Eocene–Lower Oligocene of the Isle of Wight: Palaeontology, v. 23, p. 475–499.

Lindars, E.S., Grimes, S.T., Mattey, D.P., Collinson, M.E., Hooker, J.J., and Jones, T.P., 2001, Phosphate $\delta^{18}O$ determination of modern rodent teeth by direct laser fluorination: An appraisal of methodology and potential application to palaeoclimate reconstruction: Geochimica et Cosmochimica Acta, v. 65, p. 2535–2548, doi: 10.1016/S0016-7037(01)00606-8.

Lowrie, W., and Lanci, L., 1994, Magnetostratigraphy of Eocene-Oligocene boundary sections in Italy: No evidence for short subchrons within chrons 12R and 13R: Earth and Planetary Science Letters, v. 126, p. 247–258, doi: 10.1016/0012-821X(94)90110-4.

Machette, M.N., 1985, Calcic soils of the southwestern United States, *in* Weide, D.L., ed., Soils and Quaternary Geology of the South West United States: Geological Society of America Special Paper 203, p. 1–21.

Mantell, G., 1854, Geological Excursions Round the Isle of Wight and along the Adjacent Coast of Dorsetshire: Illustrative of the Most Interesting Geological Phenomena and Organic Remains, Third edition: London, H. Bohn, p. xiii + 428.

Martini, E., 1972, Die Gattung Eosphaeroma (Isopoda) im europäischen Alttertiär: Senckenbergiana Lethaea, v. 53, p. 65–79.

Mathelin, J.-C., and Sztràkos, K., 1993, L'Eocène de Biarritz (Pyrénées Atlantiques, SW France): Stratigraphie et paléoenvironnement, *in* Monographie des Foraminifères: Cahiers de Micropaléontologie (N.S.), v. 8, 5–182.

Miller, K.G., Browning, J.V., Aubry, M.-P., Wade, B.S., Katz, M.E., Kulpecz, A.A., and Wright, J.D., 2008, Eocene-Oligocene global climate and sea-level changes: St. Stephens Quarry, Alabama: Geological Society of America Bulletin, v. 120, p. 34–53, doi: 10.1130/B26105.1.

Munt, M.C., Daley, B., and Barker, M.J., 1994, Recent borehole data from the Hamstead Member (Bouldnor Formation, early Oligocene) at Cowes, Isle of Wight: Tertiary Research, v. 15, p. 49–51.

Paul, C.R.C., 1988, The molluscan faunal succession in the Hatherwood Limestone Member (Upper Eocene), Isle of Wight, England: Tertiary Research, v. 10, p. 147–162.

Pearson, P.N., McMillan, I.K., Wade, B.S., Dunkley Jones, T., Coxall, H.K., Bown, P.R., and Lear, C.H., 2008, Extinction and environmental change across the Eocene-Oligocene boundary in Tanzania: Geology, v. 36, p. 179–182, doi: 10.1130/G24308A.1.

Pekar, S.F., Christie-Blick, N., Kominz, M.A., and Miller, K.G., 2002, Calibration between eustatic estimates from backstripping and oxygen isotopic records for the Oligocene: Geology, v. 30, p. 903–906, doi: 10.1130/0091-7613(2002)030<0903:CBEEFB>2.0.CO;2.

Plint, A.G., 1988, Global eustasy and the Eocene sequence in the Hampshire Basin, England: Basin Research, v. 1, p. 11–22, doi: 10.1111/j.1365-2117.1988.tb00002.x.

Plint, A.G., and Nummedal, D., 2000, The falling stage systems tract: Recognition and importance in sequence stratigraphic analysis, *in* Hunt, D., and Gawthorpe, R.L., eds., Sedimentary Responses to Forced Regressions: Journal of the Geological Society, London, Special Publication 172, 1–17.

Pomerol, C., 1989, Stratigraphy of the Palaeogene: Hiatuses and transitions: Proceedings of the Geologists' Association, v. 100, p. 313–324.

Premoli-Silva, I., Orlando, M., Monechi, S., Madile, M., Napoleone, G., and Ripepe, M., 1988, Calcareous plankton biostratigraphy and magnetostratigraphy at the Eocene-Oligocene transition in the Gubbio area, *in* Premoli-Silva, I., Coccioni, R., and Montanari, A., eds., The Eocene-Oligocene Boundary in the Marche-Umbria Basin (Italy): Ancona, Italy, International Subcommission on Paleogene Stratigraphy, International Union of Geological Sciences (IUGS) Commission on Stratigraphy, p. 137–161.

Remy, J.-A., 1985, Nouveaux gisements de mammifères et reptiles dans les Grès de Célas (Eocène sup. du Gard). Etude des Palaeothériidés (Perissodactyla, Mammalia): Palaeontographica A, v. 189, p. 171–225.

Retallack, G.J., 2005a, Earliest Triassic claystone breccias and soil erosion crisis: Journal of Sedimentary Research, v. 75, p. 663–679, doi: 10.2110/jsr.2005.055.

Retallack, G.J., 2005b, Pedogenic carbonate proxies for amount and seasonality of precipitation in paleosols: Geology, v. 33, p. 333–336, doi: 10.1130/G21263.1.

Ringeade, M., 1987, Séquences mammaliennes en Aquitaine: Corrélations avec les zones à mammifères classiques, les zones à charophytes et à ostracodes: Münchner Geowissenschaftliche Abhandlungen, ser. A, v. 10, p. 189–196.

Riveline, J., 1984, Les gisements à charophytes du Cénozoïque (Danien à Burdigalien) d'Europe occidentale. Lithostratigraphie, biostratigraphie, chronostratigraphie: Bulletin d'Information des Géologues du Bassin de Paris (Mém. hors Série), v. 4, p. 1–523.

Serra-Kiel, J., et al., 1998, Larger foraminiferal biostratigraphy of the Tethyan Paleocene and Eocene: Bulletin de la Société Géologique de France, v. 169, p. 281–299.

Sheldon, N.D., 2006, Quaternary glacial-interglacial climate cycles in Hawaii: The Journal of Geology, v. 114, p. 367–376, doi: 10.1086/500993.

Sheldon, N.D., 2009, this volume, Nonmarine records of climatic change across the Eocene-Oligocene transition, *in* Koeberl, C., and Montanari, A., eds., The Late Eocene Earth—Hothouse, Icehouse, and Impacts: Geological Society of America Special Paper 452, doi: 10.1130/2009.2452(15).

Sheldon, N.D., and Retallack, G.J., 2004, Regional paleoprecipitation records from the late Eocene and Oligocene of North America: The Journal of Geology, v. 112, no. 4, p. 487–494, doi: 10.1086/421076.

Sheldon, N.D., Retallack, G.J., and Tanaka, S., 2002, Geochemical climofunctions from North America soils and application to paleosols across the Eocene-Oligocene boundary in Oregon: The Journal of Geology, v. 110, no. 6, p. 687–696, doi: 10.1086/342865.

Sheldon, N.D., Mitchell, R.L., Collinson, M.E., and Hooker, J.J., 2009, this volume, Eocene-Oligocene transition paleoclimatic and paleoenvironmental record from the Isle of Wight (UK), *in* Koeberl, C., and Montanari, A., eds., The Late Eocene Earth—Hothouse, Icehouse, and Impacts: Geological Society of America Special Paper 452, doi: 10.1130/2009.2452(16).

Sille, N.P., Collinson, M.E., Kucera, M., and Hooker, J.J., 2004, Evolution within the charophyte genus *Harrisichara*, late Paleogene, southern England; environmental and biostratigraphic implications: Palaeogeography, Palaeoclimatology, Palaeoecology, v. 208, p. 153–173, doi: 10.1016/j.palaeo.2004.03.002.

Smith, R., 2003, Les vertébrés terrestres de l'Oligocène inférieur de Belgique (Formation de Borgloon, MP21): Inventaire et interprétation des données actuelles: Colloquios de Paleontologia (vol. ext.) v. 1, p. 647–657.

Smith, R., 2004a, Insectivores (Mammalia) from the earliest Oligocene (MP21) of Belgium: Netherlands Journal of Geosciences, v. 83, p. 187–192.

Smith, R., 2004b, Nouveau nyctithère (Mammalia, Lipotyphla) de l'Oligocène inférieur de Belgique (Formation de Borgloon, MP 21): Bulletin de l'Institut Royal des Sciences Naturelles de Belgique (Sciences de la Terre), v. 74, supplement, p. 197–204.

Smith, R., 2006, *Sigenyctia oligocaena* n. gen. n. sp., nyctithere (Mammalia, Lipotyphla) de l'Oligocène inférieur de Belgique (Formation de Borgloon,

MP21): Bulletin de l'Institut Royal des Sciences Naturelles de Belgique (Sciences de la Terre), v. 76, p. 131–136.

Smith, R., and van den Hoek Ostende, L.W., 2006, A new heterosoricid shrew from the lowermost Oligocene of Europe: Acta Palaeontologica Polonica, v. 51, p. 381–384.

Stehlin, H.G., 1910, Remarques sur les faunules de Mammifères des couches Éocènes et Oligocènes du Bassin de Paris: Bulletin de la Société Géologique de France, ser. 4, v. 9, p. 488–520.

Steurbaut, E., 1992, Integrated stratigraphic analysis of lower Rupelian deposits (Oligocene) in the Belgian Basin: Annales de la Société Géologique de Belgique, v. 115, p. 287–306.

Stover, L.E., and Hardenbol, J., 1994, Dinoflagellates and depositional sequences in the Lower Oligocene (Rupelian) Boom Clay Formation, Belgium: Bulletin de la Société Belge de Géologie, v. 102, p. 5–77.

Tauxe, L., Tucker, P., Petersen, N.P., and LaBreque, J.L., 1984, Magnetostratigraphy of Leg 73 sediments, *in* Hsu, K.J., LaBreque, J.L., et al., eds., Initial Reports of the Deep Sea Drilling Project, Volume 73: Washington, D.C., Government Printing Office, p. 609–621.

Vandenberghe, N., Brinkhuis, H., and Steurbaut, E., 2003, The Eocene/Oligocene boundary in the North Sea area: A sequence stratigraphic approach, *in* Prothero, D.R., Ivany, L.C., and Nesbitt, E.A., eds., From Greenhouse to Icehouse. The Marine Eocene-Oligocene Transition: New York, Columbia University Press, p. 420–437.

van Mourik, C.A., 2006, The Greenhouse-Icehouse Transition: A Dinoflagellate Perspective: Stockholm, Stockholm University, Meddelanden från Stockholms Universitets Institution för Geologi och Geokemi 327, p. 1–32 + four individually paginated papers.

Vianey-Liaud, M., and Ringeade, M., 1993, La radiation des Theridomyidae (Rodentia) hypsodontes à l'Eocène supérieur: Geobios, v. 26, p. 455–495, doi: 10.1016/S0016-6995(06)80230-5.

Manuscript Accepted by the Society 16 September 2008

Printed in the USA

The Geological Society of America
Special Paper 452
2009

Biostratigraphy and paleoclimatology of the Eocene-Oligocene boundary section at Toadstool Park, northwestern Nebraska, USA

Alessandro Zanazzi*
Department of Geological Sciences, University of South Carolina, Columbia, South Carolina 29208, USA

Matthew J. Kohn*
Department of Geosciences, Boise State University, Boise, Idaho 83725, USA

Dennis O. Terry Jr.*
Department of Earth and Environmental Science, Temple University, Philadelphia, Pennsylvania 19122, USA

ABSTRACT

The stratigraphic section located at Toadstool Park, northwestern Nebraska, preserves a detailed sedimentological and fossil record of the Eocene-Oligocene boundary, yet chronologic control is limited to sparse ash dates and magnetostratigraphy. To improve the chronologic control of the Toadstool Park sediments, we examined the fossil collection at the University of Nebraska State Museum and recorded the stratigraphic level of some important biochronologic events. All stratigraphic levels were determined relative to the Upper Purplish White layer (UPW), a distinctive ash that is chemically and mineralogically correlated to an ash dated at 34.6 ± 0.1 Ma. The small artiodactyls *Leptomeryx speciosus*, *Leptomeryx mammifer*, and *Leptomeryx yoderi* co-occur at –34 ± 2 m, implying that the early Chadronian–middle Chadronian boundary (35.7 Ma) is located at this level. The first occurrence of *L. mammifer* (i.e., the middle Chadronian–late Chadronian boundary, 34.8 Ma) is located at –18 ± 2 m, and the first occurrence of the camel *Poebrotherium wilsoni* (34.2 Ma) is located at 0 ± 2 m. The first occurrences of two other artiodactyls (*Hypertragulus calcaratus* and *Leptomeryx evansi*), which define the Chadronian-Orellan boundary (33.9 Ma), are located approximately at the UPW, ~5 m lower than in the nearby type sections of Wyoming. This apparent diachroneity may reflect sampling bias in Nebraska, i.e., loose material washed down from higher outcrops. The first occurrences of the dwarfed oreodont *Miniochoerus affinis* and of the rodent *Eumys elegans* (i.e., late early Orellan, 33.6 Ma) are located at +8 ± 2 m and +11 m, respectively. Finally, the first occurrence of *M. gracilis* (i.e., early late Orellan, 33.2 Ma) is located at +13 ± 2 m.

These time points, along with magnetostratigraphy and $^{40}Ar/^{39}Ar$ dates obtained for this section, were used to refine the age model for a previously published stable isotope record of fossil bone CO_3. The record is augmented here with 59 bone samples from federal lands near Toadstool Park spanning the critical time period of 33.8–33.4 Ma. The

*Zanazzi, corresponding author: azanazzi@geol.sc.edu. Kohn: mattkohn@boisestate.edu. Terry: doterry@temple.edu.

Zanazzi, A., Kohn, M.J., and Terry, D.O., Jr., 2009, Biostratigraphy and paleoclimatology of the Eocene-Oligocene boundary section at Toadstool Park, northwestern Nebraska, USA, *in* Koeberl, C., and Montanari, A., eds., The Late Eocene Earth—Hothouse, Icehouse, and Impacts: Geological Society of America Special Paper 452, p. 197–214, doi: 10.1130/2009.2452(13). For permission to copy, contact editing@geosociety.org.

$\delta^{18}O$ values in the complete data set show a step to higher values (from 23.0‰ ± 0.2‰ to 24.5‰ ± 0.2‰) at ca. 34 Ma, synchronous with the marine record. In combination with unchanged enamel compositions, this increase in bone CO_3 $\delta^{18}O$ translates into a mean annual temperature drop of 7.1 ± 3.1 °C across the boundary. The time of onset of the climate change coincides with an increase in sedimentation rate from 19 to 37 m/m.y., confirming the strong role of climate on continental erosion and sediment yield.

Keywords: Eocene-Oligocene, biostratigraphy, paleoclimatology, stable isotopes.

INTRODUCTION

The fine-grained volcaniclastic deposits of the White River sequence in the western United States record faunal, depositional, and paleoclimatic changes during the late Eocene and early Oligocene. They are one of the most productive fossil mammal terrains in the world. Research on vertebrate and invertebrate paleontology, sedimentology, stratigraphy, and paleopedology of this sequence has provided crucial information on late Paleogene paleoclimates and on mammalian evolution (e.g., Prothero and Berggren, 1992; Terry et al., 1998). White River exposures are present in Colorado, Wyoming, and North Dakota, where they are treated as a formation, and in Nebraska and South Dakota, where they are considered a group. The most famous area of White River sediments is located in southwestern South Dakota and is known as the "Big Badlands" (Fig. 1).

Despite extensive study elsewhere, the section located along the Pine Ridge escarpment at Toadstool Park in northwestern Nebraska (Fig. 1) is particularly important because it preserves a continuous sedimentological and fossil record of the Eocene-Oligocene boundary (33.9 ± 0.1 Ma; all ages are reported relative to the time scale of Gradstein et al., 2004), a period characterized by one of the most pronounced episodes of Cenozoic climate change (e.g., Miller et al., 1987; Zachos et al., 2001). In the Big Badlands, the boundary coincides with an unconformity marked by removal of most of chron C13n (33.7–33.2 Ma), and by an absence of early Oligocene (Orellan North American Land Mammal Age [NALMA]; 33.9–32 Ma) taxa known from Wyoming and Nebraska (Prothero and Whittlesey, 1998). In contrast, the stratigraphic interval corresponding to the late Eocene–early Oligocene is present at Toadstool Park and is represented by the Big Cottonwood Creek Member of the Chadron Formation and by the Orella Member of the Brule Formation (Terry and LaGarry, 1998).

In spite of the great significance of the White River sections of northwestern Nebraska, chronostratigraphic control remains problematic for two reasons. First, interpretations of biostratigraphic zonation in this region have been coarse owing to the lack of modern systematics of some of the White River mammals and to the sketchy stratigraphic information that accompanies many of the fossils collected from this area, which are now housed at the University of Nebraska State Museum. In many (but by no means all) cases, collection notes for these specimens specify only the alphabetically labeled stratigraphic interval (e.g., Chadron B, Orella A) based on the obsolete zonation of Schultz and Stout (1955) and not the level within the zone. The substantial thickness of each zone (10–25 m, representing ~500 k.y.) does not allow the use of these fossil specimens for high-resolution biostratigraphy. Second, previous tephrostratigraphic studies exhibit some inconsistencies. For instance, the most distinctive ash layer of the area, the Upper Purplish White layer (UPW) of Schultz and Stout (1955), has long been correlated with the 5 Tuff (ca. 33.8 Ma) near Douglas, Wyoming (Evanoff et al., 1992). Larson and Evanoff (1998), however, showed that the 5 Tuff in Douglas is geochemically distinct from the UPW of northwestern Nebraska. These authors indicated that the UPW is more similar to the 3c Tuff of Douglas and Ash J (ca. 34.6 Ma) of Flagstaff Rim, Wyoming.

All these problems and inconsistencies regarding the chronostratigraphy of the White River Group in Nebraska complicate the possibility of conducting paleoclimatic and paleontological studies with high temporal resolution. The main purpose of this paper is therefore to present detailed biostratigraphic information for the Toadstool Park geologic section that considerably improves

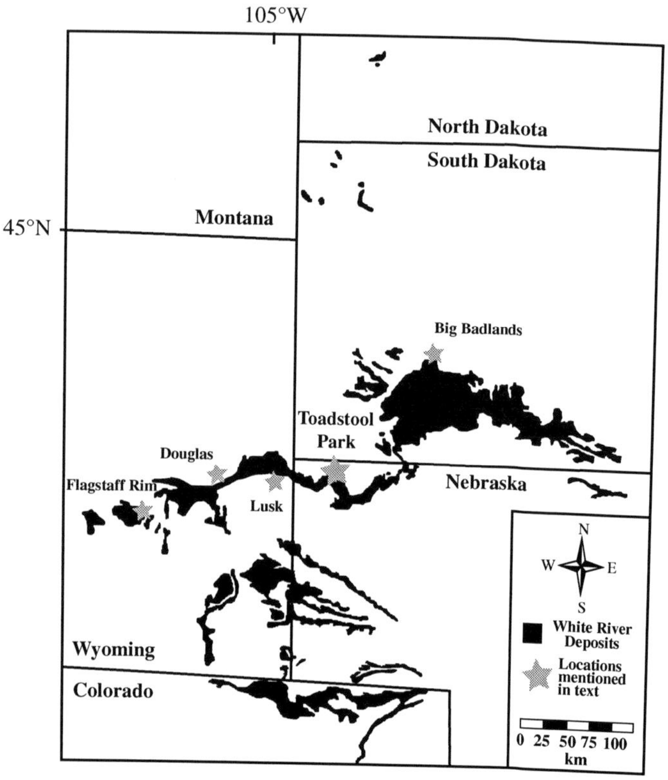

Figure 1. Map showing the deposits of the White River Group and the locations mentioned in the text (modified from Terry, 2001).

the chronologic control of these sediments. These data are critical for temporal placement of the high-resolution fossil bone stable isotope record of the Eocene-Oligocene boundary from Toadstool Park (Zanazzi et al., 2007). To complete this record, we also present here new fossil bone isotope compositions spanning a key time interval (from 33.8 to 33.4 Ma) characterized by sparse data in the record of Zanazzi et al. (2007). These new data allow a better understanding of the timing relationship between continental and oceanic climate change of the Eocene-Oligocene boundary and are used to investigate the relationship between climate change and sedimentation rates in the North American midcontinent.

BACKGROUND

Stratigraphy

Regional uplift during the Laramide orogeny forced the retreat of the Cretaceous Western Interior Seaway, resulting in the subareal exposure and weathering of the Early Cretaceous Pierre Shale, which unconformably underlies the White River Group. White River sediments are composed of volcaniclastic and siliciclastic sediments deposited within fluvial, lacustrine, and eolian environments. The volcanogenic material was reworked by rivers that flowed east-southeast across the study area (Swinehart et al., 1985) and by eolian processes. The White River Group is overlain by the late Oligocene to early Miocene Arikaree Group.

Lithostratigraphy

The majority of research on the White River Group of northwestern Nebraska is based on the stratigraphic classification of Schultz and Stout (1955) (Fig. 2). These authors divided the White River Group of northwestern Nebraska into the Chadron and Brule Formations. The Chadron Formation, previously named Titanotherium Beds after the discovery in these rocks of the remains of large ungulates called titanotheres or brontotheres, was further subdivided into informal units termed A, B, and C. The boundaries between these units were defined at prominent marker beds that consisted of paleosols, volcanic ashes, gypsum,

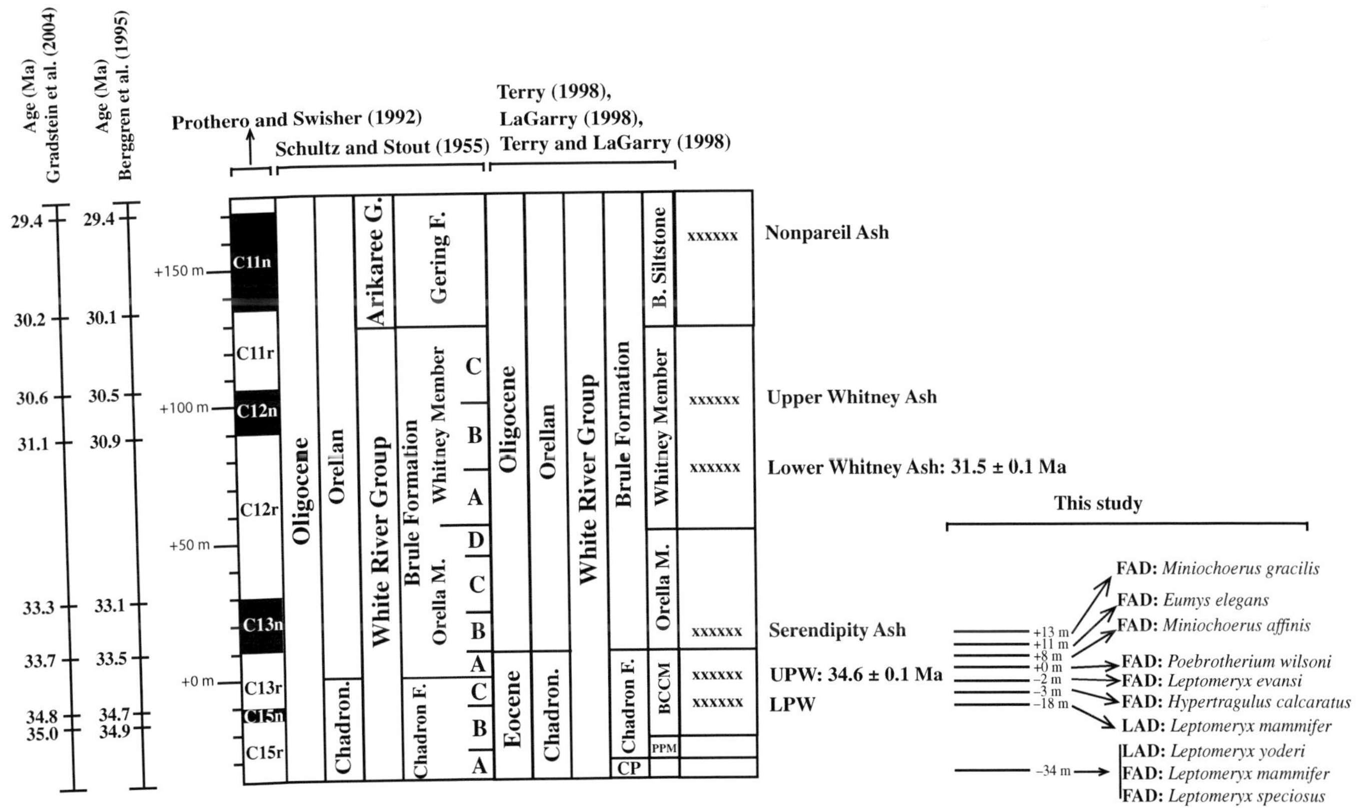

Figure 2. Stratigraphic section of Toadstool Park (Nebraska). The Upper Purplish White layer (UPW) has been correlated with Ash J of Flagstaff Rim, the age of which is from Obradovich et al. (1995). The age of the Lower Whitney Ash is the average of the ages obtained by Deino (1996, personal commun. *in* LaGarry, 1998). The ages of both ashes are corrected by 0.65% according to Kuiper et al. (2008). On the left, the ages of the chron boundaries are reported according to the time scales of both Gradstein et al. (2004) and Berggren et al. (1995). Abbreviations: LPW—Lower (= second) Purplish White layer; UPW—Upper Purplish White layer; CP—Chamberlain Pass Formation; PPM—Peanut Peak Member; BCCM—Big Cottonwood Creek Member; FAD—first appearance datum; LAD—last appearance datum. Stratigraphic level of important marker horizons (relative to the UPW): LPW = –10 m; Serendipity Ash = +17 m; Lower Whitney Ash = +77 m; Upper Whitney Ash = +100 m; Nonpareil Ash = +160 m.

and limestones (Terry and LaGarry, 1998). These marker beds were named by Schultz and Stout (1955) "purplish white layers." Schultz and Stout (1955) described Chadron A (lower Chadron), which is incised into a weathered zone of the Cretaceous Pierre Shale (the Interior Paleosol Complex), as a brilliantly colored conglomeratic and/or sandy unit. Chadron A is overlain by Chadron B (middle Chadron), which was described as the most fossiliferous part of the formation. Chadron B, composed of greenish gray clay with interbedded green channel sandstones, was further subdivided into four units (B^1, B^2, B^3, B^4), the boundaries of which were defined on the basis of three "purplish white layers" (the Third, Fourth, and Fifth). The boundary between Chadron B and the greenish gray clay of Chadron C (upper Chadron) was placed at the Lower Purplish White layer.

The Upper Purplish White (UPW) layer, a volcanic ash of regional extent, was assigned by Schultz and Stout (1955) as the boundary between the Chadron and Brule Formations. The Brule Formation was divided into the Orella and Whitney Members. The Orella Member was further subdivided, analogous to the Chadron Formation, into Orella A, Orella B, Orella C, and Orella D based on the presence of paleosol complexes, nodular beds, and other distinctive marker horizons. Orella A (lower part, lower Orella) and Orella B (upper part, lower Orella) are divided by a prominent nodular bed ("lower nodules"), and both consist of greenish gray to brownish buff claystones but are easily distinguished from each other by prevalent sheet sands and numerous rows of nodular siltstone and sandstone concretions in Orella B. Orella C (middle Orella) consists of laminated brownish buff silt and clay, and it is marked at its base by another nodular bed ("upper nodules") and by a thick channel fill (the Toadstool Park Channel Complex) that represents one of several episodes of fluvial downcutting and backfilling within the Orella Member (Wells et al., 1994). Orella C is capped by a prominent erosional step (the Bench), which separates it from the overlying Orella D (upper Orella). Orella D consists of laminated to massive, brownish buff silt and clay with several key nodular beds (Bed Z, Bed Y, and Bed X) of local correlative value.

The next two higher units (Whitney Member and "Brown Siltstone" member of the Brule) have received less paleontological and sedimentological scrutiny, but they are important because they host dated ashes. The Whitney Member was subdivided by Schultz and Stout (1955), similar to the Chadron Formation and Orella Member, into Whitney A (lower Whitney), Whitney B (middle Whitney), and Whitney C (upper Whitney). These units are uniformly massive pinkish buff siltstones and silty claystones and are mutually separated by two prominent ashes: the Lower and Upper Whitney ashes.

Schultz and Stout's stratigraphic classification of the Chadron and Brule Formations was revised by LaGarry (1998), Terry (1998), and Terry and LaGarry (1998), who placed the boundaries at distinctive changes in lithology (Fig. 2). These authors abandoned the A, B, and C subdivisions of the Chadron Formation. They correlated Chadron A with the Chamberlain Pass Formation, and the lower part of Chadron B with the Peanut Peak Member of the Chadron Formation of South Dakota. The Chamberlain Pass Formation also includes the Interior Paleosol Complex of Schultz and Stout (1955). In addition, the upper two-thirds of the Chadron Formation (Chadron B and Chadron C) and the lowermost part of the Orella Member of the Brule Formation (Orella A) were unified into a newly defined member of the Chadron Formation, the Big Cottonwood Creek Member of Terry and LaGarry (1998). According to these authors, the difference between this member and the underlying bluish-green, gray, and olive claystones of the Peanut Peak Member lies in its variegated color and higher component of silt, unaltered vitric volcaniclastic rocks, lacustrine limestone, pedogenic calcrete, and marl interbeds. The upper boundary of the Big Cottonwood Creek Member was assigned to the first occurrence of sheet sandstones, which corresponds with a lithologic change from pedogenically modified green, red, and pink volcaniclastic silty claystones to the thinly interbedded and less pedogenically modified brown, orange, and tan volcaniclastic clayey siltstones and sheet sandstones. This revision shifted the boundary between the Chadron and Brule Formations upwards by 9–11 m relative to the Upper Purplish White layer (UPW, the previously assigned boundary; Schultz and Stout, 1955). LaGarry (1998) also revised the boundary between the Orella and Whitney Members of the Brule Formation, placing it 2.5–5 m below the position previously proposed by Schultz and Stout (1955), the X or White Bed.

Tephrostratigraphy

Intense volcanism throughout western North America deposited numerous thin, widespread ash beds in the White River Group. These intercalated ash beds represent excellent local and regional chronostratigraphic markers and have been used for stratigraphic positioning of fossil localities, magnetic polarity zones, and depositional features despite their substantial distance from the source, which makes the dating of the ashes either directly or via tephrostratigraphy rather difficult. Table 1 lists the ages and correlations of the White River ashes present in the locations mentioned in this paper.

The most prominent ash beds present in the Toadstool Park section are, in order of increasing stratigraphic level: the Lower Purplish White layer (−10 m from the UPW) and the UPW in the Big Cottonwood Creek Member of the Chadron Formation, the Serendipity Ash (+17 m from the UPW) in the Orella Member of the Brule Formation, the Lower (+77 m from the UPW) and Upper (+100 m from the UPW) Whitney ashes in the Whitney Member of the Brule Formation, and the Nonpareil Ash (+160 m from the UPW) in the "Brown Siltstone" member of the Brule Formation (Fig. 2). Larson and Evanoff (1998) described the Chadron and Orella tuffs as primarily rhyolitic to rhyodacitic. The Lower Whitney Ash is rhyodacitic, whereas the Upper Whitney Ash and the Nonpareil Ash are dacitic. With regard to grain size, the Orella tuffs are finer than those of the underlying Chadron Formation and overlying Whitney Member. Based on the ages, chemical and mineralogical compositions, and trend in grain size changes of the White River ashes, Larson and Evanoff

TABLE 1. AGE AND CORRELATIONS OF THE WHITE RIVER ASHES IN THE LOCATIONS MENTIONED IN THE TEXT

	Wyoming			Nebraska
NALMA	**Flagstaff Rim**	**Douglas**	**Lusk**	**Toadstool Park**
Arikareean				Nonpareil Ash
Whitneyan				Upper Whitney Ash Lower Whitney Ash, **31.47 ± 0.06 Ma (2)***
Orellan		7 Tuff White Tuff		Serendipity Ash
		Black tuffs	Black tuffs	
Chadronian		5 Tuff, **33.81 ± 0.02 Ma (1)*** 4 Tuff		
	Ash J, **34.58 ± 0.11 Ma (1)***	3C Tuff	?Persistent White layer (PWL)	?Upper Purplish White layer (UPW)
	Ash G +73	3B Tuff		
	Ash G +36	3A Tuff		
	Ash F	2 Tuff		
				Lower Purplish White layer
	Ash B +6C	1 Tuff		
	Ash B, **35.64 ± 0.14 Ma (1)***	0 Tuff		

Note: Data are modified from Larson and Evanoff (1998). NALMA—North American Land Mammal Age. Ages of the ashes are normalized to the time scale of Gradstein et al. (2004). References: (1) Obradovich et al. (1995); (2) Deino (1996, personal commun. *in* LaGarry, 1998).
*Ages of the ashes are corrected by 0.65% according to Kuiper et al. (2008).

(1998) suggested that the source of the volcanogenic material was likely the Great Basin in eastern Nevada and western Utah.

The Lower and Upper Whitney ashes and the Nonpareil Ash were dated in northwestern Nebraska using $^{40}Ar/^{39}Ar$ techniques by Swisher and Prothero (1990). Unfortunately, the dates obtained by these authors are considered inaccurate, likely because of excess ^{40}Ar (Prothero and Emry, 2004). Subsequently, however, the Lower Whitney Ash was redated, yielding a $^{40}Ar/^{39}Ar$ age of 31.47 ± 0.06 Ma (Deino, 1996, personal commun. *in* LaGarry, 1998; corrected by 0.65% according to Kuiper et al., 2008). Among the other ashes, the UPW was correlated via phenocryst assemblages with Ash J of Flagstaff Rim, the 3c Tuff of Douglas, and the Persistent White layer of Lusk, Wyoming (Larson and Evanoff, 1998). Ash J was dated by Obradovich et al. (1995) and yielded an age of 34.58 ± 0.11 Ma ($^{40}Ar/^{39}Ar$ on sanidine, corrected by 0.65% according to Kuiper et al., 2008).

Magnetostratigraphy

The results of the magnetostratigraphic studies conducted on the Toadstool Park section were summarized by Prothero and Swisher (1992) and Prothero (1996a) and are shown in Figure 2. The section presented by Prothero and Swisher (1992) is the most complete and shows five intervals of normal polarity and six intervals of reversed polarity (Fig. 2.5 *in* Prothero and Swisher, 1992). These authors correlated the normal intervals in the Chadron Formation (between −14 and −10 m relative to the UPW) and Orella Member of the Brule Formation (+9 to +28 m relative to the UPW) with chron C15n (35.0–34.8 Ma) and chron C13n (33.7–33.3 Ma), respectively. The long reversed interval spanning the upper Orella Member and lower Whitney Member (+43 to +90 m relative to the UPW) was interpreted to correspond with chron C12r (33.3–31.1 Ma), and the normal interval around the Upper Whitney Ash (+90 to +106 m relative to the UPW) was correlated with chron C12n (31.1–30.6 Ma). According to these correlations, the reversed interval at the base of the section (below −14 m relative to the UPW) corresponds to chron C15r (35.4–35.0 Ma), and the reversed intervals in the Big Cottonwood Creek–Orella members (−10 to +9 m relative to the UPW) and in the upper Whitney Member (+106–137 m relative to the UPW) correspond to chron C13r (34.8–33.7 Ma) and chron C11r (30.6–30.2 Ma), respectively. In the uppermost part of the section, the long normal interval in the "Brown Siltstone" member (+137 to +171 m relative to the UPW), which includes the Nonpareil Ash, corresponds to chron C11n (30.2–29.5 Ma). These correlations do not take into account a short normal interval present in Orella C (+40 to +43 m relative to the UPW; see Figure 2.5 *in* Prothero and Swisher, 1992). This interval is also ignored in this paper because it is based on only one significant site with no thermal demagnetization and therefore likely represents a spurious overprint that was not removed prior to analysis (Prothero and Swisher, 1992).

Paleoclimatology

The Eocene-Oligocene boundary is considered one of the most pronounced episodes of climate change of the Cenozoic and is often described as the transition from a "greenhouse" to an "icehouse" world. Research conducted on marine sediments indicates the presence of global extinctions and turnovers in planktic and benthic foraminifera, calcareous nannoplankton, diatoms, and mollusks (Aubry, 1992; Baldauf, 1992; Keller et al., 1992; Thomas, 1992; Hansen et al., 2004). In addition, oxygen isotope records of benthic foraminifera show an abrupt, ~1‰ increase in $\delta^{18}O$ across the boundary, which reflects 3–4 °C of cooling of oceanic bottom waters combined with a substantial growth of the ice sheet on Antarctica (e.g., Zachos et al., 2001). The most recent, highest resolution marine isotope records of the Eocene-Oligocene boundary show that the increase in $\delta^{18}O$ occurred from 34.0 to 33.5 Ma in two 40-k.y.-long steps separated by a 200-k.y.-long intermediate plateau (Coxall et al., 2005; Lear et al., 2008).

In North America, paleoclimate research on the Eocene-Oligocene boundary mainly includes paleosol, sedimentological, paleontological, leaf morphology, and stable isotopes studies. Based on the depth of the calcic (Bk) horizon, which is sensitive to precipitation, Retallack (1992) inferred a decrease in mean annual precipitation (MAP) from more than 1000 mm/yr in the late Eocene to 500–900 mm/yr in the early Oligocene at Badlands National Park. In addition, based on the root traces observed in the paleosols, Retallack (1983, 1992) inferred a change from dry forests in the late Eocene to dry woodlands and wooded grasslands in the early Oligocene. These results were subsequently confirmed by Terry (2001) at Toadstool Park. Based on changes in soil structure, root traces, and geochemical trends, he inferred a change from humid, forested conditions in the late Eocene to more seasonal, semiarid, open conditions in the early Oligocene. Terry's MAP estimates for the late Eocene (855 ± 282 mm/yr) and early Oligocene (739 ± 282 mm/yr) are in general agreement with those of Sheldon and Retallack (2004) and Retallack (2007) from paleosols of western Nebraska (from 752 ± 156 to 628 ± 156 mm/yr).

An increase in aridity is generally consistent with sedimentological and paleontological studies. River systems changed from meandering and anastomosing in the late Eocene to braided with large width-to-depth ratios in the early Oligocene (Clark et al., 1967; Emry, 1992; Evanoff et al., 1992). With respect to faunas, Evanoff et al. (1992) observed a decrease in shell size of gastropods across the Eocene-Oligocene boundary. Late Eocene land snails from Douglas are large-shelled, tropical forms similar to the modern species found in the southern Rocky Mountains or central Mexican Plateau, and they seem to indicate a mean annual temperature (MAT) of 16.5 °C and a MAP of 450 mm/yr. These large-shelled forms were subsequently replaced in the early Oligocene by small-shelled, drought-tolerant taxa indicative of open woodlands with a pronounced dry season, similar to the modern forms of Baja California. In conformity with these results, Hutchison (1982, 1992) observed a rapid decline in aquatic reptiles and amphibians (salamanders, crocodilians, and pond turtles) across the boundary in western North America but only a modest decrease in terrestrial forms. He interpreted these observations as reflecting a major increase in aridity but only a small decrease in MAT. Unlike the records of ectotherms, the vast majority (62 out of 70) of mammalian lineages persisted through the Eocene-Oligocene climatic event with no observable response (Prothero and Heaton, 1996).

Some studies suggest a large drop in MAT across the Eocene-Oligocene boundary. Based on leaf morphology in the Pacific Northwest, Wolfe (1978) inferred a drop in MAT across the Eocene-Oligocene boundary of ~12–13 °C over a period of less than 1 m.y., although later studies in Oregon suggested a smaller MAT decline (~8 °C; from ~20 °C to 12 °C), similar to that observed in other North American sites (e.g., Alaska, Rocky Mountains) (Wolfe, 1992, 1994). Wolfe (1992) also inferred a large increase in temperature seasonality (MART, mean annual range of temperature) across the boundary from ~10 °C to ~25 °C. Retallack et al. (2004) provided new age calibrations for Oregon fossil floras and suggested more protracted (~6 m.y.) cooling, although data presented by Sheldon and Retallack (2004) are consistent with an abrupt MAT drop of 5–6 °C across the Eocene-Oligocene boundary in Oregon.

The stable isotope study of Zanazzi et al. (2007) indicates a large MAT drop across the Eocene-Oligocene boundary in the North American midcontinent. This study combined in vivo isotope compositions of tooth enamel and pedogenically altered compositions of fossil bones. Tooth enamel, because of its low porosity and high crystallinity, is resistant to isotopic exchange, so after burial and fossilization, it retains its biological $\delta^{18}O$ signal (see summary of Kohn and Cerling, 2002). This signal is temperature-independent because mammals are homeothermic, and it is instead directly correlated with rainwater composition. In contrast to tooth enamel, bones become isotopically altered during fossilization by soil water on likely time scales of 20–50 k.y. due to their higher porosity, lower crystallinity, and higher content of organic matter. In this context, bone CO_3 $\delta^{18}O$ becomes dependent on both soil water composition and temperature (Kohn and Law, 2006). Zanazzi et al. (2007) therefore used tooth enamel $\delta^{18}O$ to track changes in water composition and used this information to infer temperature changes from the $\delta^{18}O$ of fossil bones. In addition to temperature changes, major trends in aridity can be investigated from the carbon isotope composition of fossil bones. As indicated by Kohn and Law (2006), the $\delta^{13}C$ of fossil bones reflects the composition of soil CO_2, which, in turn, depends on plant organic matter $\delta^{13}C$ and penetration of atmospheric CO_2 (Cerling and Quade, 1993). Because C4 plants were not yet abundant in the late Eocene and early Oligocene (Cerling et al., 1997), bone $\delta^{13}C$ principally reflects water stress and canopy structure, and it shows increases with increasing aridity.

Based on the ~1.7‰ increase in mean bone $\delta^{18}O$ and on the insignificant change in mean enamel $\delta^{18}O$ across the boundary, Zanazzi et al. (2007) calculated a temperature drop of 8.2 ± 3.1 °C (from 21.0 ± 10.1 °C to 13.1 ± 9.5 °C), similar to estimates from leaf morphology analysis (Wolfe, 1992, 1994). Sparse data in the critical time interval 33.7–33.3 Ma, however, did not allow a precise comparison between the timing of onset of the continental versus oceanic isotopic shift. This study also indicates a higher Oligocene (~25.8 °C) versus Eocene (~21.9 °C) MART; this increase in MART is minor, however, compared to the one inferred by Wolfe (1992). Finally, in contrast to the paleosols and sedimentological studies, the indistinguishable Oligocene versus Eocene average bone $\delta^{13}C$ value suggests no major change in aridity across the boundary (Zanazzi et al., 2007).

Materials and Methods

The study site is located in the Toadstool Park area, ~24 km north of Crawford (northwestern Nebraska, Fig. 1). The overall biostratigraphy and biochronology of the Chadronian and Orellan NALMAs for the White River sequence were reviewed by Prothero and Emry (2004). To obtain a better chronostratigraphic framework for the geologic sections at Toadstool Park, we examined collections housed at the University of Nebraska State Museum (UNSM) and recorded the stratigraphic level of important faunal events. Fossils used for biostratigraphy were collected from the park area itself or within a few kilometers. Note that some key locations, particularly for the lower sections of the Chadron Formation, are on private lands that are no longer accessible to scientific study. Many fossils have poor associated stratigraphic information, but collection notes associated with each sample allow reconstruction of stratigraphic locations of a large subset to within ±1–2 m. Using only well-located samples, we obtained key datum levels for the late early Chadronian–middle Chadronian boundary, middle Chadronian–late Chadronian boundary, Chadronian-Orellan boundary, late early Orellan, and early late Orellan (Fig. 2).

Prothero and coworkers have long criticized the UNSM collections as providing too little stratigraphic information for biochronologic comparisons to type sections in Wyoming (collections housed in the American Museum of Natural History; e.g., Prothero and Emry, 1996, 2004; Prothero and Whittlesey, 1998). We agree that stratigraphic levels for most of the USNM samples are indeed coarse, but many fossils are quite precisely located, in some cases, better than in Wyoming, at least for the few artiodactyls that we examined. Considering proximity to volcanic sources, sedimentation rates in Wyoming were probably higher than in Nebraska, at least for the Chadron Formation. So, in principle, the Wyoming sections might permit higher resolution records through the late Eocene. However, collecting at Toadstool was logistically much easier and allowed us to place our work within existing and concurrent paleopedological and geochemical studies (Terry, 2001; Grandstaff and Terry, 2009).

In our study of the UNSM collections, we mainly identified species, but for *Leptomeryx evansi* and *Leptomeryx speciosus*, we additionally observed the development of a fold on the lower molars (*Palaeomeryx* fold; Heaton and Emry, 1996) and assigned a "*Palaeomeryx* score" from zero (*Palaeomeryx* fold absent) to four (*Palaeomeryx* fold exceptionally well developed). For *L. speciosus*, *Leptomeryx yoderi*, and *Leptomeryx mammifer*, we also measured the length and width of anterior and posterior selenes and scored the development of an accessory labial cusp (buccal column; Storer, 1984) from zero (absent) to four (exceptionally well developed). We additionally examined oreodont species, which are still organized in the UNSM collections according to the obsolete taxonomy of Schultz and Falkenbach (1956, 1968), who divided the lineage into a multitude of invalid genera and species. To correctly identify species, we made several careful measurements on the skulls of the *Miniochoerus* specimens and compared them with the revised classification of Stevens and Stevens (1996).

With respect to the fossil bone stable isotope record, we improved the data set of Zanazzi et al. (2007) by analyzing 59 additional bone samples collected at submeter stratigraphic resolution from federal lands adjacent to the park by agreement with the National Forest Service. These samples cover the critical stratigraphic interval located at +6 to +16 m relative to the UPW (the climate transition is centered at ~+10 m). The stratigraphic position of these samples was measured with a high-precision Jacob's staff relative to the UPW. Uncertainty in stratigraphic placement of in situ samples was ±0.1 m. Some samples were collected from float, but always within 1–2 m from the tops of flat ridges, so their stratigraphic placement uncertainties are no worse than ±1 m, and typically ±0.5 m. For inferred sedimentation rates of ~20–40 m/m.y., this implies chronologic resolution of, at worst, ±50 k.y. In the laboratory, small fragments of bones were removed with a slow-speed diamond-wafering saw. The fragments were finely powdered with an agate mortar and pestle and treated with H_2O_2 to remove organic contaminants and with an acetic acid–calcium acetate buffer to remove any labile carbonate components that might be adsorbed onto phosphate crystallites (Koch et al., 1997). Small aliquots (1–3 mg) of the samples were then dissolved in phosphoric acid and analyzed by an ISOCARB automated carbonate system attached to a VG-OPTIMA isotope-ratio mass spectrometer housed at the Department of Geological Sciences, University of South Carolina. All samples analyzed in this study were run in duplicate. Results are reported in per mil (‰) with the conventional δ notation relative to the Vienna standard mean ocean water (VSMOW, $\delta^{18}O$) and Vienna Peedee belemnite (VPDB, $\delta^{13}C$) standards. Precision of the analyses is typically better than ±0.2‰ ($\pm1\sigma$). The phosphate standard NBS 120c yielded an average $\delta^{13}C$ value of –6.1‰ and an average $\delta^{18}O$ value of 28.6‰.

RESULTS

Biostratigraphy

Late Early Chadronian–Middle Chadronian Boundary (35.7 Ma)

Because it is common, abundant, and distinctive, the most useful taxon to define the late early Chadronian–middle Chadronian boundary is the tiny deer-like ruminant *Leptomeryx*. Heaton and Emry (1996) provided the most detailed analysis of the group. In the early Chadronian, the only species present is *L. yoderi*. At the beginning of the middle Chadronian, the genus split into two species: the larger *L. mammifer* (derived by a gradual anagenetic size increase from *L. yoderi*) and *L. speciosus*. The last appearance datum (LAD) of *L. yoderi* and the first appearance data (FAD's) of *L. mammifer* and *L. speciosus* define the boundary between the early and middle Chadronian.

L. yoderi and *L. mammifer* are easily distinguished by the size of their teeth (see Tables 3, 4, 5, 6 *in* Heaton and Emry,

1996). *L. speciosus* is similar in size to *L. yoderi* but can be distinguished by the shape of a small cusp (entoconulid) on the last lower molar. In *L. yoderi*, the entoconulid is long and narrow, and the anterior and posterior ridges extending from its peak slope downward at about the same angle when viewed lingually. In contrast, in *L. speciosus*, the entoconulid is more rounded and pronounced, and it slopes steeply downward posteriorly (Fig. 3). According to Storer (1984) an accessory labial cusp (buccal column) tends to be larger and more common in *L. speciosus* compared to *L. yoderi*. Samples from UNSM indicate that the three *Leptomeryx* species (i.e., *L. yoderi*, *L. speciosus*, and *L. mammifer*) co-occur at 34 ± 2 m below the UPW (Table A1), implying that the late early Chadronian–middle Chadronian boundary occurs at this level in the Toadstool Park area.

Middle Chadronian–Late Chadronian Boundary (34.8 Ma)

One biostratigraphic event that can be used to define the middle Chadronian–late Chadronian boundary is the LAD of *Leptomeryx mammifer* (Prothero and Emry, 1996). Based on the measurements shown in Table A1, *L. mammifer* last occurs at ~18 ± 2 m below the UPW, suggesting that the middle Chadronian–late Chadronian boundary occurs at this level in the Toadstool Park area.

Chadronian–Orellan Boundary (33.9 Ma)

The Chadronian-Orellan boundary (33.9 Ma, late chron C13r) has long been defined by the LAD of the huge ungulates called brontotheres. However, the LAD of the brontotheres is an impractical biostratigraphic marker because this taxon is not very common and because its LAD was subsequently found to be highly variable among different White River locations (Prothero and Whittlesey, 1998). Consequently, Prothero and Whittlesey (1998) defined the Chadronian-Orellan boundary based on the FAD of the distinctive small ruminant *Hypertragulus calcaratus*, which appears in the White River Group with no late Chadronian predecessors. Prothero and Whittlesey (1998) also indicated other biostratigraphic events useful for identifying the beginning of the Orellan, including the FAD of *Leptomeryx evansi*, very abundant in all early Orellan localities, and the FAD of the camel *Poebrotherium wilsoni*.

L. evansi evolved from the late Chadronian *L. speciosus* and can be recognized by the *Palaeomeryx* fold and crenulations on the lower molars (Fig. 3) (Heaton and Emry, 1996). *H. calcaratus* differs from *L. evansi* in having a strong cingulum, in lacking the *Palaeomeryx* fold and crenulations on the lower molars, and in having more conical cusps on the lower molars (Fig. 4). With respect to other camelid species, *P. wilsoni* is morphologically similar to its predecessor *P. eximium* but can be distinguished by the longer diastema between P1 and P2 (Prothero, 1996b).

The FAD's of *H. calcaratus*, *L. evansi*, and *P. wilsoni* are located ~7 m above the Persistent White Layer (PWL) at Lusk (ca. 34.5 Ma, assuming linear sedimentation rate and PWL = Ash J; ca. 33.7 assuming linear sedimentation rate and PWL = 5 Tuff) and 7 m above the 5 Tuff at Douglas (ca. 33.7 Ma, assuming linear sedimentation rate) (Prothero and Whittlesey, 1998). Toadstool samples housed at UNSM indicate a mix of *L. speciosus* and *L. evansi* a few meters below and at the UPW, but a preponderance of *L. evansi* starting at ~2 m above the UPW (Table A2). The stratigraphically lowest *H. calcaratus* samples at Toadstool were collected 3 m below the UPW (Table A3). Therefore, if the correlations between the White River

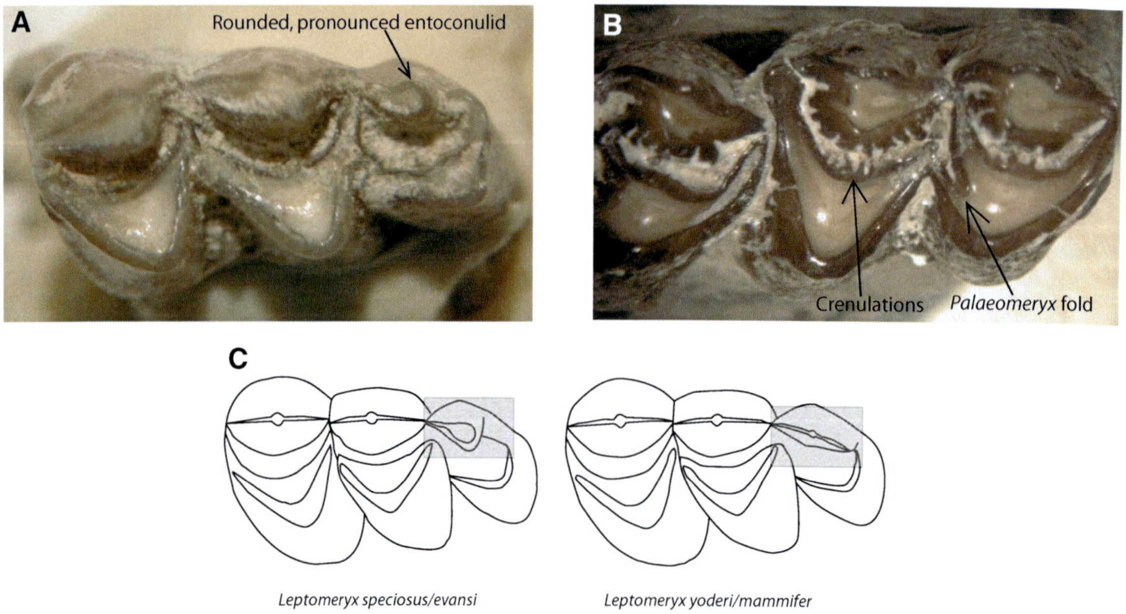

Figure 3. (A) Right third lower molar (m3) of *Leptomeryx speciosus* (sample 123975; –9 m relative to the UPW). (B) Left second lower molar (m2) of *Leptomeryx evansi* (sample 64406; Orella A). (C) Diagram showing the different configurations of the entoconulid in *L. yoderi/mammifer* and *L. speciosus/evansi*.

ashes proposed by Larson and Evanoff (1998) are correct (i.e., UPW = PWL = 3c Tuff = Ash J; Table 1), our study indicates that the first occurrence of these *H. calcaratus* and *L. evansi* could be diachronous by as much as ~1 m.y. among different White River locations. A diachronous biostratigraphic boundary is supported by the first occurrence of *P. wilsoni* at the UPW in northwestern Nebraska (sample 2513–60). However, for two reasons, we do not favor biochronologic diachroneity. First, camelids are not very abundant, and the published FAD of *P. wilsoni* in Wyoming may underestimate the age of its true first occurrence. Second, leptomerycid fossils that we examined in the UNSM collections may have washed down from higher outcrops. More studies and more intense sampling are required to resolve this issue, and we conservatively assign the Chadronian-Orellan boundary to 2 ± 5 m above the UPW.

Late Early Orellan

The late early Orellan can be defined by the FAD's of the artiodactyl *Miniochoerus affinis* and of the rodent *Eumys elegans* (Prothero and Whittlesey, 1998). *M. affinis* belongs to a lineage of oreodonts significantly smaller than the more common *Merycoidodon. M. affinis* first appears at ~17 m above the PWL at Lusk (Prothero and Whittlesey, 1998). This level corresponds to an age of ca. 34.3 Ma (assuming linear sedimentation rate and PWL = Ash J) or ca. 33.5 Ma (assuming linear sedimentation rate and PWL = 5 Tuff). Our measurements for a sample collected 8 m above the UPW (p1–p4 length = 62.5 mm [left]; m1–m3 length = 36.1 mm [left]; p1–p4 lengths = 28.0 mm [right], 27.8 mm [left]) indicate that the first occurrence of *M. affinis* corresponds to the base of chron C13n (ca. 33.7 Ma). The first occurrence of the common species of rodent *E. elegans*, which appears at Lusk 27 m above the PWL, is located at Toadstool Park at ~11 m above the UPW, slightly above the base of chron C13n (ca. 33.6 Ma) (Schultz and Stout, 1955).

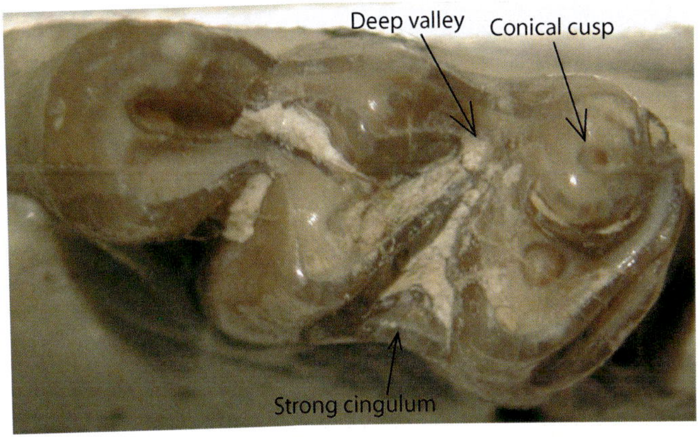

Figure 4. Right third lower molar (m3) of *Hypertragulus calcaratus* (sample 124013; –2.7 m relative to the UPW). This taxon can be distinguished from *Leptomeryx* by the strong cingulum (i.e., ridge) on the lingual side of the m3s and by the prominent, conical cusps on the metaconid of the lower molars. A deep valley between the entoconid and metaconid of the lower molars is also well developed in this taxon.

Early Late Orellan

Early late Orellan strata are correlative with the earliest part of chron C12r and are marked by only a few biostratigraphic events. One of them is the FAD of the dwarfed oreodont *Miniochoerus gracilis*, which appears ~27 m above the PWL (ca. 34.1 Ma, assuming a linear sedimentation rate and PWL = Ash J; ca. 33.3 Ma, assuming a linear sedimentation rate and PWL = 5 Tuff) at Lusk and ~57 m above the 5 Tuff (ca. 33.0 Ma, assuming linear sedimentation rate) at Douglas. This event is used to identify the beginning of the early late Orellan (Prothero and Whittlesey, 1998). The skull to snout lengths of four samples housed at the UNSM (10–28–8-35, 114–5-8–36, 111–5-8–36, 1012–3-8) indicate that *M. gracilis* (samples 114–5-8–36 and 111–5-8–36) first occurs at ~13 m above the UPW at Toadstool Park, in the middle of chron C13n (ca. 33.5 Ma).

Sedimentation Rates

Ages and stratigraphic levels of the chron boundaries, volcanic ashes, and bioevents are reported in Table 2 and plotted in Figure 5. We regressed data, accounting for errors both in stratigraphic location and in age (York, 1966), for data subsets within the Chadron and Brule Formations. For all regressions, R^2 values are >0.99, and slopes differ at >99% confidence. The standard errors for the predicted ages are ~250 k.y., similar to the error of the age model for the highest resolution Eocene-Oligocene marine record (214 k.y.; Coxall et al., 2005). The intersection of the two regression lines occurs at +10–11 m, which corresponds exactly with the lithostratigraphic change from the Chadron to Brule Formations. The linearity of the age models for the Chadron

TABLE 2. AGE AND STRATIGRAPHIC LEVEL OF THE TIME POINTS PRESENTED IN THE TEXT

Event	Stratigraphic level (m)*	Age (Ma)†
LAD *L. yoderi*; FAD *L. mammifer*, *L. speciosus* (1)	–34	35.8
Base chron C15n (2)	–14	35.0
LAD *L. mammifer* (1)	–18	34.8
Base chron C13r (2)	–10	34.8
$^{40}Ar/^{39}Ar$ date on UPW (3)	0	34.6
FAD *P. wilsoni* (1)	0	34.2
FAD *H.calcaratus* (1)	2	33.9
FAD *L. evansi* (1)	4	33.9
Base chron C13n (2)	9	33.7
FAD *Eumys elegans* (1, 4)	11	33.6
FAD *Miniocherus gracilis* (1)	13	33.5
Base chron C12r (2)	28	33.3
$^{40}Ar/^{39}Ar$ date on Lower Whitney Ash (5)	77	31.5
Base chron C12n (2)	90	31.1
Base chron C11r (2)	106	30.6
Base chron C11n (2)	137	30.2
Base chron C10r (2)	171	29.5

Note: References: (1) this study; (2) Prothero and Swisher (1992); (3) Obradovich et al. (1995); (4) Schultz and Stout (1955); (5) Deino (1996, personal commun. *in* LaGarry, 1998). Abbreviations: LAD—last appearance datum; FAD—first appearance datum.

*Stratigraphic level is measured relative to the Upper Purplish White (UPW) layer.

†Time scale of Gradstein et al. (2004).

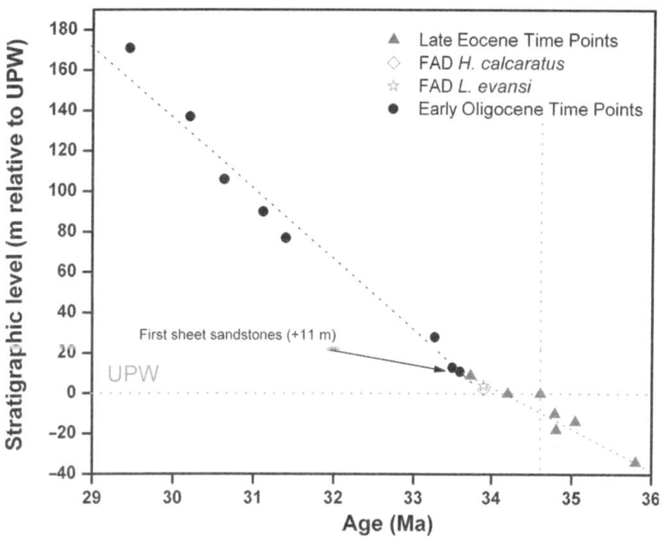

Figure 5. Stratigraphic level of the time points mentioned in the text plotted versus age (time scale of Gradstein et al., 2004). The time points define linear sedimentation rates in the Chadronian and Orellan with a slope of 19 and 37 m/m.y., respectively. The change in sedimentation rate occurs ~11 m above the Upper Purplish White layer (UPW), coincident with the lithostratigraphic boundary between the Chadron and Brule Formations (sensu Terry and LaGarry, 1998), which is marked by the appearance of sheet sandstones.

and Brule Formations, their mutual intersection at a clear lithostratigraphic change, and lack of evidence of any depositional hiatus in the outcrops provide persuasive stratigraphic evidence for continuous sedimentation in the Toadstool Park section.

The equations of the regression models are as follows:

$$Age = 34.22 \pm 0.11 - 0.0525 \pm 0.0088 \cdot Strat.Level, \quad (1)$$

$$Age = 33.96 \pm 0.36 - 0.0270 \pm 0.0030 \cdot Strat.Level, \quad (2)$$

where Equation 1 applies when stratigraphic level is <+11m, and Equation 2 applies when stratigraphic level is ≥+11 m, and errors are at 95% confidence. This model was used to determine the age of the fossil bone samples. Inverse regressions (stratigraphic level versus age) indicate that sedimentation rate increased from 19 ± 3 m/m.y. in the Chadron to 37 ± 4 m/m.y. in the Brule Formation.

Stable Isotopes

The results of the stable isotope analyses of the fossil bone samples analyzed in this study are reported in Table A4 and shown in Figure 6. These samples span a time interval from 33.8 to 33.4 Ma and complete the data set of Zanazzi et al. (2007). The δ values show a fairly large within-level variability (±1‰–2‰). This within-level variability is comparable to that of paleosol $CaCO_3$ compositions (e.g., Koch et al., 2003) and may simply derive from the high variability in the factors that control isotope compositions in continents, such as MAT, MAP, MART, and precipitation seasonality. Given this scatter, we binned the whole data set into Eocene (i.e., pre–33.85 Ma) and Oligocene (i.e. post–33.50 Ma) values to investigate the climate change of the Eocene-Oligocene boundary. With respect to $\delta^{13}C$, Eocene values range from −9.81‰ to −6.31‰ and show an average (±2 s.e.) of −7.82‰ ± 0.16‰ (n = 76). Oligocene $\delta^{13}C$ values show a larger range (from −9.92‰ to −5.67‰) and average −7.57‰ ± 0.11‰ (n = 169). Statistical analysis indicates a marginal difference in Eocene versus Oligocene average $\delta^{13}C$ ($p = 0.014$, t-test; $p = 0.007$ Mann-Whitney test). With respect to $\delta^{18}O$, Eocene values range from 21.23‰ to 26.23‰ and show an average value of 23.02‰ ± 0.22‰ (n = 73). Oligocene values range from 20.68‰ to 26.95‰ and average 24.50‰ ± 0.19‰ (n = 169). Statistical tests indicate a very significant difference in Eocene versus Oligocene average $\delta^{18}O$ ($p < 0.001$, t-test; $p < 0.001$ Mann-Whitney test). In terms of variability, a Brown-Forsythe test indicates a higher Oligocene versus Eocene $\delta^{18}O$ variance ($p = 0.0006$). In contrast, Eocene and Oligocene $\delta^{13}C$ variances are indistinguishable (p = 0.745, Brown-Forsythe test). Linear regressions of the Eocene $\delta^{13}C$ and $\delta^{18}O$ data subsets versus time do not resolve significant trends ($\delta^{13}C$: $p = 0.083$, $R^2 = 0.04$; $\delta^{18}O$: $p = 0.305$, $R^2 = 0.0148$). In contrast, regressions of the Oligocene $\delta^{13}C$ and $\delta^{18}O$ values versus time are significant ($\delta^{13}C$: $p < 0.001$; $\delta^{18}O$: $p = 0.013$). Correlations are, however, very weak ($\delta^{13}C$: $R^2 = 0.09$; $\delta^{18}O$: $R^2 = 0.0361$). Similarly, regressions of Eocene and Oligocene $\delta^{18}O$ versus $\delta^{13}C$ are significant (Eocene: $p < 0.001$, Oligocene: $p = 0.007$), but the amount of variation in bone $\delta^{18}O$ explained by its linear relationship with $\delta^{13}C$ is very low (Eocene: $R^2 = 0.156$; Oligocene: $R^2 = 0.0429$).

DISCUSSION

Eocene-Oligocene Climate Change

The stable isotope data presented here permit a comparison of the continental and marine records. Our $\delta^{18}O$ data indicate that, within age uncertainties, the isotope shift in the North American midcontinent was synchronous with that in the oceans, both of which started ca. 34 Ma and ended ca. 33.5 Ma (Fig. 6). These results suggest therefore a strong ocean-continent coupling during episodes of abrupt climate change such as that of the Eocene-Oligocene boundary. These new data also allow a refinement of the temperature estimates proposed by Zanazzi et al. (2007). Most notably, the MAT drop that occurred across the boundary can be quantified by binning the oxygen isotope compositions according to the marine record (i.e., comparing data older than 33.85 Ma versus younger than 33.50 Ma). Given the increase in Oligocene versus Eocene average bone CO_3 $\delta^{18}O$ ($\Delta\delta^{18}O$ = 1.5‰ ± 0.3‰), considering a change in rainwater $\delta^{18}O$ of 0‰ ± 0.5‰ (Zanazzi et al., 2007), and assuming the same temperature dependence for the carbonate component of apatite as for calcite ($\Delta\varepsilon_{CaCO_3\text{-}H_2O}$ = −0.21‰/°C; Kim and O'Neil, 1997), the calculated MAT drop across the boundary is equal to 7.1 ± 3.1 °C. The largest source of error for this estimate lies in the estimated shift for rainwater

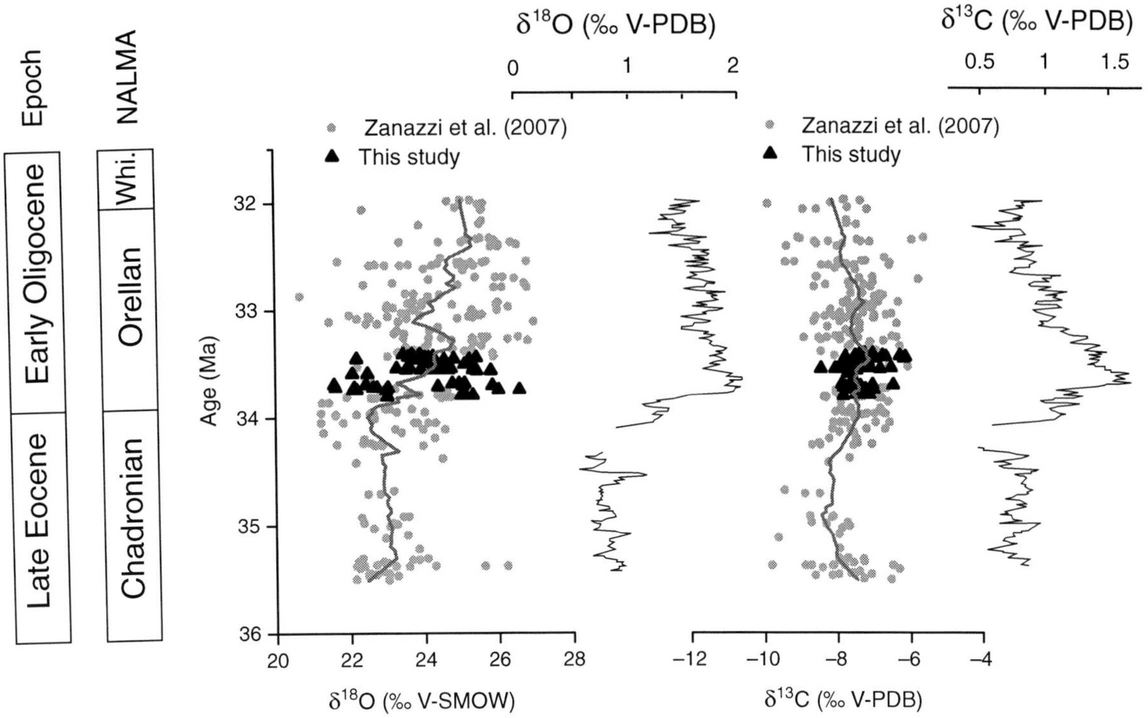

Figure 6. High-resolution bone carbonate $\delta^{18}O$ and $\delta^{13}C$ plotted versus age (time scale of Gradstein et al., 2004). The dark gray curves were obtained by exponential smoothing the raw data. The black curves refer to the marine record of Coxall et al. (2005). Abbreviations: NALMA—North American Land Mammal Ages; Whi—Whitneyan.

$\delta^{18}O$. This MAT drop is much larger than those estimated from isotope records of coastal sites, which indicate little or no temperature change across the Eocene-Oligocene boundary (Kohn et al., 2004, 2008; Grimes et al., 2005).

In addition to the temperature drop, values for Eocene and Oligocene MAT can also be calculated. Given the equation relating rainwater $\delta^{18}O$ to enamel CO_3 $\delta^{18}O$ in modern horses ($\delta^{18}O_{rainwater} = 1.16 \pm 0.08[\delta^{18}O_{enamel\ CO_3}] - 36.4 \pm 1.6‰$; Kohn and Fremd, 2007) and using the average value for *Mesohippus* enamel $\delta^{18}O$ (24.3‰ ± 0.5‰; Zanazzi et al., 2007), an average value for rainwater $\delta^{18}O$ can be calculated (−8.2‰ ± 2.0‰). This value for rainwater $\delta^{18}O$ can be used to calculate Eocene and Oligocene MAT by using the average Eocene and Oligocene bone CO_3 $\delta^{18}O$, the offset between bone CO_3 and paleosol $CaCO_3$ $\delta^{18}O$ ($\delta^{18}O_{paleosolCaCO_3} = [\delta^{18}O_{boneCO_3}] - 2.2‰ \pm 0.6‰$; Kohn and Law, 2006), and the temperature dependence of the fractionation between calcite and water $\delta^{18}O$ (Kim and O'Neil, 1997). Calculated values for MAT are 20.5 ± 10.0 °C (Eocene) and 13.6 ± 9.5 °C (Oligocene); in this case, the largest source of error lies in the estimated rainwater $\delta^{18}O$, the error of which in turn derives from the calibration for modern horses.

In addition to the lower MAT, the Oligocene was also likely characterized by larger temperature variability relative to the Eocene, as indicated by the larger Oligocene versus Eocene $\delta^{18}O$ variance. In terms of aridity, the $\delta^{13}C$ data presented here do not essentially change the conclusions of Zanazzi et al. (2007). Because CO_2 in the atmosphere undergoes rapid exchange with the surface of the ocean, which in turn exchanges carbon with the deep ocean, the slightly higher Oligocene versus Eocene average bone $\delta^{13}C$ likely reflects the similar increase in the $\delta^{13}C$ of atmospheric CO_2 inferred from the marine record (Coxall et al., 2005). Our data therefore indicate no major change in aridity in the North American midcontinent across the boundary. This result may seem inconsistent with most of the research previously conducted in this area. However, Terry (2007) suggested that discrepancies between paleosol and stable isotope data can be resolved if sedimentation rates are taken into consideration. More specifically, the weak development of the paleosols in the Orella Member, which has been interpreted as reflecting dry conditions, may simply derive from the increase in sedimentation rate that was observed across the boundary. Increased sedimentation rates influence soil formation by reducing the amount of time available for pedogenesis, allowing preservation of a greater proportion of parent materials and resulting therefore in weaker soils.

Links between Climate Change and Sedimentation Rate

Three factors govern rates of regional erosion and associated rates of terrigenous sedimentation in continental interiors. Crustal deformation creates elevated terrains that provide potential energy to rivers and glaciers, the main agents of erosion, whereas tectonic subsidence provides accommodation space. In this volcanically influenced depositional system, the rate and style of volcanism determine the amount of pyroclastic material available for transport and deposition. Finally, climate controls the abundance and temporal distribution of the erosive agents, the efficiency of pyroclastic transport, and the type, density, and distribution of vegetation that protects Earth's surface from erosion. The data presented here indicate that the increase in sedimentation rate at Toadstool occurs, within uncertainty, synchronously or slightly delayed relative to the increase in bone and foraminifera $\delta^{18}O$; i.e., with the global climate change of the Eocene-Oligocene boundary. The correspondence between the change in sedimentation rate and the Eocene-Oligocene climatic event suggests a stronger role of climate relative to tectonics and volcanism. This finding supports the conclusion of recent studies that have attributed the major increase in Cenozoic sedimentation rates to climate (Peizhen et al., 2001; Molnar, 2001, 2004; Harris and Mix, 2002).

The increased rate of sedimentation following the cooling and (perhaps) drying event at the Eocene-Oligocene boundary is consistent with the relationship of Langbein and Schumm (1958). According to these authors, the specific sediment yield, in continental climates, increases with increasing MAP up to a peak, which is followed by a decline when MAP increases further. Langbein and Schumm (1958) indicated that this peak in sediment yield corresponds to a MAP of ~300 mm. The existence of this peak can be explained by the nonlinear, complex relationship between rainfall and erosion. When MAP is low (<300–400 mm), an increase in MAP causes an increase in erosion rate because the effect of rainfall erosivity prevails over the resistance effect of vegetation. In contrast, when MAP is higher (>400 mm), forest cover starts to increase rapidly, and vegetative protection begins to exceed rainfall erosivity (Jiongxin, 2005). The studies of Retallack (1992) and Terry (2001) on the paleosols of South Dakota and Nebraska clearly indicate, for both the late Eocene and early Oligocene, MAP estimates higher than the peak of Langbein and Schumm (1958). Although not directly supported by our isotope data, a small decrease in MAP across the boundary such as that inferred by Retallack (1992) and Terry (2001) should cause an increase in sediment yield, in agreement with our observations. Either alternatively, or in addition, climate seasonality may have increased Oligocene erosion rates. Equable and uniform Eocene climates, in which landscapes had attained equilibrium configuration, may have changed to Oligocene climates characterized by frequent and abrupt changes in temperature, precipitation, and vegetation that prevented fluvial systems from establishing equilibrium rates (Peizhen et al., 2001). The higher variability in Oligocene versus Eocene isotope compositions in both our terrestrial and oceanic records indeed suggests a more variable Oligocene versus Eocene climate.

Although our data support a temporal and therefore presumably causal link between climate change and sedimentation rates at Toadstool Park, other tectonic and volcanic processes such as localized high rates of basin subsidence and accumulation, increased siliciclastic influx from individual Laramide uplifts, and increase in volcaniclastics from the Great Basin region of Nevada and Utah may also have played an important role in causing the observed increase in sedimentation rate across the Eocene-Oligocene boundary. Further study of sedimentation rates, sedimentary facies, and localized tectonics in different locations within the White River Group would therefore be useful to confirm the links among Eocene-Oligocene climate change, erosion rates, and sediment yield.

CONCLUSIONS

Detailed biostratigraphy for the Eocene-Oligocene boundary section of the Toadstool Park area (northwestern Nebraska) based on collections at the University of Nebraska State Museum improves chronologic control of these sediments. Specifically, fossils of the small artiodactyls *Leptomeryx speciosus*, *L. mammifer*, and *L. yoderi* co-occur at 34 ± 2 m below the Upper Purplish White layer (UPW) and mark the early Chadronian–middle Chadronian boundary (35.7 Ma). The last occurrence of *L. mammifer* (34.8 Ma) is located 18 ± 2 m below the UPW. The first occurrences of *Hypertragulus calcaratus* (another small artiodactyl) and *L. evansi* (both of which define the Chadronian-Orellan boundary at 33.9 Ma) are located approximately at the UPW, considerably lower than in Wyoming. Stratigraphically low occurrences may reflect collections bias. Finally, the dwarfed oreodonts *Miniochoerus affinis* and *M. gracilis* first occur 8 ± 2 m and 13 ± 2 m above the UPW, respectively, and mark the late early Orellan and early late Orellan.

These time points were combined with magnetostratigraphy and chronostratigraphy to construct an age model for fossil bone samples collected from these strata. Carbon and oxygen isotope compositions of these samples, along with the data of Zanazzi et al. (2007), indicate a temperature drop of 7.1 ± 3.1 °C and no major change in aridity across the boundary. In addition, the isotope shift in bone $\delta^{18}O$ in the North American midcontinent occurs synchronously with that recorded in the oceans and coincides with an increase in sedimentation rate from 19 to 37 m/m.y. These findings therefore suggest a strong ocean-continent coupling during periods of abrupt global change and a causal relationship among climate, continental erosion, and sediment yield.

ACKNOWLEDGMENTS

The project was funded by National Science Foundation (NSF) grant ATM 0400532 to Kohn. Additional support for this research was provided by the Travel Fund for Vertebrate Paleontology within the Department of Geological Sciences, University of South Carolina. The comments and suggestions of Nathan Sheldon and Donald Prothero greatly contributed to the improvement of this paper. The authors thank the National Forest Service and Barbara Beasley for permitting collection of fossils, Eric Tappa for facilitating analyses, and Mike Voorhies and George Corner for access to and help negotiating the University of Nebraska State Museum's collections.

APPENDIX

TABLE A1. MEASUREMENTS AND OBSERVATIONS ON *LEPTOMERYX MAMMIFER, YODERI*, AND *SPECIOSUS* CONDUCTED AT THE UNIVERSITY OF NEBRASKA STATE MUSEUM

Sample ID	Stratigraphic level (m)*	Tooth type	Length of ant. selene (mm)	Length of two selenes (mm)	Width of anterior selene (mm)	Width of posterior selene (mm)	Tooth length (mm)	Buccal column†	Entoconulid§	Diagnosis
123706	–34.5	M1/M2	4.7	N.A.	7.7	8	10.5	2	N.A.	L. mammifer
123707	–34.5	M3	5	10.8	8.5	7.9	12.7	1	N.A.	L. mammifer
123708	–34.5	M1/M2	5	N.A.	7.8	8.2	10.2	2	N.A.	L. mammifer
123709	–34.5	M3	3.9	7.4	6.2	5.5	11.1	0	Y/M	L. mammifer
123710	–34.5	M3	3	6	4.8	4.3	8.5	3	Y/M	L. yoderi
123710	–34.5	M2	3.4	N.A.	4.8	4.9	6.2	1	N.A.	L. yoderi
123710	–34.5	M1	2.9	N.A.	4.3	4.6	6.5	2	N.A.	L. yoderi
123711	–34.5	M3	N.A.	N.A.	N.A.	N.A.	N.A.	N.A.	S	L. speciosus
123712	–34.5	M3	N.A.	N.A.	N.A.	N.A.	N.A.	N.A.	S	L. speciosus
123718	–34.5	M3	N.A.	N.A.	N.A.	N.A.	N.A.	N.A.	S	L. speciosus
123722	–34.5	M3	3.4	6.2	4.7	4.1	9.7	0	S	L. speciosus
123726	–34.5	M3	N.A.	N.A.	N.A.	N.A.	N.A.	N.A.	S	L. speciosus
123731	–34.5	M3	3.7	3.6	5	4.2	10.5	0	S	L. speciosus
123737	–34.5	Upper Molar	N.A.	N.A.	N.A.	N.A.	N.A.	N.A.	N.A.	L. mammifer
123765	–34.5	M3	N.A.	N.A.	N.A.	N.A.	N.A.	N.A.	S	L. speciosus
123861	–34.5	M3	5.2	9.9	7.4	6.9	13.6	1	Y/M	L. mammifer
123862	–34.5	M1/M2	4.1	N.A.	6.9	5.6	8.5	2	N.A.	L. mammifer
123865	–34.5	M1/M2	4.3	N.A.	7.2	7.5	8.1	2	N.A.	L. mammifer
123889	–25.3	M3	3.3	3	4.2	4	8.8	N.A.	S	L. speciosus
123890	–25.3	M3	3.6	6.9	4.7	4.3	9.4	N.A.	S	L. speciosus
124156	–24.7	M3	N.A.	N.A.	N.A.	N.A.	N.A.	N.A.	S	L. speciosus
123783	–18.3	M1/M2	4.8	N.A.	7.1	7.7	10.5	3	N.A.	L. mammifer
123975	–9.1	M3	N.A.	N.A.	N.A.	N.A.	N.A.	4	S	L. speciosus
123976	–9.1	M3	N.A.	N.A.	N.A.	N.A.	N.A.	4	S	L. speciosus
123989	–9.1	M3	4.1	3.4	N.A.	4.1	10.1	4	S	L. speciosus

*Stratigraphic level is measured relative to the Upper Purplish White (UPW) layer.
†Buccal column scores: 1—none, 2—very small, 3—large, 4—very large.
§Entoconulid: Y—*L. yoderi*; M—*L. mammifer*, S—*L. speciosus*.

TABLE A2. OBSERVATIONS RELATIVE TO *PALAEOMERYX* FOLD AND CRENULATIONS CONDUCTED ON THE *LEPTOMERYX* SAMPLES AT THE UNIVERSITY OF NEBRASKA STATE MUSEUM

Sample ID	Stratigraphic level (m)*	Tooth type	*Palaeomeryx* fold score[†]
123793	–6.7 to –5.8	M1/M2	0
123793	–6.7 to –5.8	M2	0
123793	–6.7 to –5.8	M3	0
123796	–6.7 to –5.8	M2	0
123796	–6.7 to –5.8	M3	0
123800	–6.7 to –5.8	M2	0
123800	–6.7 to –5.8	M3	0
Average score (–6.7 m to –5.8 m)			**0**
123873	–2.7	M1	0
123873	–2.7	M2	0
123873	–2.7	M3	0
123873	–2.7	M1	0
123873	–2.7	M2	0
123873	–2.7	M3	0
Average score (–2.7 m)			**0**
123856	–1.5	M2	0
123856	–1.5	M3	0
123868	–1	M3	0
64301	–0.3	M1	2
124465	0	M1	0
63497	1	M3	0
63496	1	M1/M2	2
64297	1 to 1.2	M3	1
64297	1 to 1.2	M2	1
64297	1 to 1.2	M1	1
Average score (–1.5 m to 1.2 m)			**0.7**
124450	1.5 to 3	M1	3
124451	1.5 to 3	M2	0
124451	1.5 to 3	M3	0
124452	1.5 to 3	M1	1
124452	1.5 to 3	M2	1
124452	1.5 to 3	M3	1
124453	1.5 to 3	M1	0
124453	1.5 to 3	M2	0
124453	1.5 to 3	M3	0
124454	1.5 to 3	M1	2
124454	1.5 to 3	M2	1
124455	1.5 to 3	M1	3
124455	1.5 to 3	M2	2
124457	1.5 to 3	M3	2
124458	1.5 to 3	M3	3
124460	1.5 to 3	M1	3
124460	1.5 to 3	M2	3
124461	1.5 to 3	M2	0
124461	1.5 to 3	M3	0
124462	1.5 to 3	M1	0
124462	1.5 to 3	M2	1
Average score (1.5 m to 3 m)			**1.2**
124413	3	M1	4
124542	3	M1	2
124542	3	M2	1
124542	3	M3	2
124543	3	M1	3
124543	3	M2	1
124544	3	M1	1
124545	3	M2	3
124547	3	M3	0
124548	3	M1/M2	2
124551	3	M2	2
124551	3	M3	2
124553	3	M2	1
124553	3	M3	1
124554	3	M2	1
124554	3	M3	1
124563	3	M1/M2	3
124566	3	M1/M2	4
124568	3	M1	1
124568	3	M2	1
124569	3	M1	2
124569	3	M2	1
124581	3	M3	1
124582	3	M3	0
124622	3	M3	1
124629	3	M3	3
Average score (3 m)			**1.7**
63272	4.6	M3	3
63272	4.6	M2	2
63279	4.6	M1	4
64525	1.8 to 6.1	M1	1
64539	1.8 to 6.1	M2	1
64539	1.8 to 6.1	M1	1
64546	1.8 to 6.1	M1	4
64366	1.8 to 6.1	M3	3
64366	1.8 to 6.1	M3	4
64366	1.8 to 6.1	M3	2
64365	1.8 to 6.1	M3	1
64365	1.8 to 6.1	M2	1
64365	1.8 to 6.1	M1	3
Average score (1.8 m to 6.1 m)			**2.3**

(*continued*)

TABLE A2. OBSERVATIONS RELATIVE TO *PALAEOMERYX* FOLD AND CRENULATIONS CONDUCTED ON THE *LEPTOMERYX* SAMPLES AT THE UNIVERSITY OF NEBRASKA STATE MUSEUM (*continued*)

Sample ID	Stratigraphic level (m)*	Tooth type	*Palaeomeryx* fold score†
63477	5.5	M3	0
63477	5.5	M2	0
63477	5.5	M1	0
64542	6.1	M3	2
64549	6.1	M3	4
64549	6.1	M2	2
64643	6.1	M3	1
64643	6.1	M2	1
64643	6.1	M1	1
124583	6.1	M1	4
124583	6.1	M2	3
124641	6.1 to 7	M3	2
124645	6.1 to 7	M3	1
64322	4 to 6.4	M3	2
64330	4 to 6.4	M1/M2(?)	1
64331	4 to 6.4	M1	2
64332	4 to 6.4	M2	2
64332	4 to 6.4	M1	3
64337	4 to 6.4	M3	1
Average score (4 m to 7 m)			**1.7**
60607	7	M3	1
60607	7	M2	1
60607	7	M1	2
64294	7	M3	1
64294	7	M2	2
64294	7	M1	2
60638	7	M1	2
Average score (7 m)			**1.6**
60639	7.6	M2	1
63913	7.6	M3	4
60639	7.6	M1	2
60640	7.6	M2	1
60640	7.6	M1	2
60643	7.6	M2	4
60643	7.6	M1	3
60644	7.6	M2	1
60644	7.6	M1	2
60652	7.6	M1	1
60657	7.6	M3	3
60667	7.6	M2	4
124409	8.2	M1	2
124409	8.2	M2	2
Average score (7.6 m to 8.2 m)			**2.3**

Sample ID	Stratigraphic level (m)*	Tooth type	*Palaeomeryx* fold score†
64710	Orella A (0 to 10)	M1	3
60753	Orella A (0 to 10)	M2	2
60859	Orella A (0 to 10)	M3	1
60859	Orella A (0 to 10)	M2	2
60864	Orella A (0 to 10)	M3	2
60870	Orella A (0 to 10)	M2	2
60870	Orella A (0 to 10)	M1	1
60888	Orella A (0 to 10)	M3	3
60888	Orella A (0 to 10)	M2	2
60888	Orella A (0 to 10)	M1	2
63030	Orella A (0 to 10)	M2	4
63030	Orella A (0 to 10)	M1	2
63045	Orella A (0 to 10)	M2	2
63045	Orella A (0 to 10)	M1	2
63507	Orella A (0 to 10)	M3	1
63507	Orella A (0 to 10)	M2	1
63652	Orella A (0 to 10)	M2	2
63654	Orella A (0 to 10)	M2	3
63654	Orella A (0 to 10)	M1	3
63662	Orella A (0 to 10)	M3	4
63662	Orella A (0 to 10)	M2	2
63913	Orella A (0 to 10)	M2	3
64383	Orella A (0 to 10)	M2	2
64383	Orella A (0 to 10)	M1	2
64385	Orella A (0 to 10)	M1	3
64405	Orella A (0 to 10)	M3	3
64405	Orella A (0 to 10)	M2	2
64406	Orella A (0 to 10)	M3	4
64406	Orella A (0 to 10)	M2	2
64542	Orella A (0 to 10)	M2	1
64542	Orella A (0 to 10)	M1	3
65396	Orella A (0 to 10)	M1	3
Average score (Orella A)			**2.3**

*Stratigraphic level is measured relative to the Upper Purplish White (UPW) layer.
†*Palaeomeryx* fold scores: 1—absent, 2—very small, 3—large, 4—very large.

TABLE A3. SAMPLES OF *HYPERTRAGULUS CALCARATUS* IDENTIFIED AT THE UNIVERSITY OF NEBRASKA STATE MUSEUM AND THEIR STRATIGRAPHIC PLACEMENT

Sample ID	Stratigraphic level (m)*	Tooth type
123871	–3.0	M2–M3
123874	–2.7	M1–M2
124013	–2.7	M3
63499	0.6	M1/M2
63495	0.9	M1–M3
124456	1.5 to 3	M1–M3
124570	3.0	M1–M2
124562	3.0	M1/M2
124575	3.0	M1–M3
124577	3.0	M3
124576	3.0	M3
124573	3.0	M1–M3
124564	3.0	M1/M2
124555	3.0	M3
124549	3.0	M1/M2
124552	3.0	M2–M3
124578	3.0	M3
124556	3.0	M1–M2?
124599	3 to 4.5	P4–M3
63280	5.5 to 7.9	M1–M2
64298	9.4	P3–M2
64543	6.1	M3–M2
63911	7.0 to 8.2	M3–M1
63284	4.0 to 5.5	M3–P4
63286	4.0 to 5.5	M1–M3
64544	6.1	N.A.
64344	4.0 to 6.4	N.A.
64545	6.1	N.A.
63283	4.0 to 6.4	N.A.
64354	1.8 to 6.1	N.A.
64329	4.0 to 6.4	N.A.
64343	4.0 to 6.4	N.A.
64326	4.0 to 6.4	N.A.
64335	4.0 to 6.4	N.A.
65237	6.1	N.A.
65246	6.1	N.A.
64228	6.1	N.A.
64355	1.8 to 6.1	N.A.
125414	4.0	N.A.
125423	5.5	N.A.

*Stratigraphic level is measured relative to the Upper Purplish White (UPW) layer.

TABLE A4. CARBON AND OXYGEN ISOTOPE COMPOSITION OF THE FOSSIL BONE SAMPLES ANALYZED IN THIS STUDY

Sample ID	Stratigraphic level (m)*	Age (Ma)†	$\delta^{13}C$ (‰, VPDB)	$\delta^{18}O$ (‰, VSMOW)
TD6-05	6.1	33.80	–7.85	23
TD6-03-1	6.3	33.79	–7.15	25.3
TD6-03-2	6.3	33.79	–7.35	24.99
TD6-03-3	6.3	33.79	–7.51	25.1
TD6-11-1	7.3	33.74	–7.46	22.09
TD6-11-2	7.3	33.74	–7.03	26.56
TD6-11-3	7.3	33.74	–7.34	25.99
TD6-11-4	7.3	33.74	–7.25	22.97
TD6-11-5	7.3	33.74	–7.75	22.18
TD6-12-1	7.8	33.72	–7.68	22.12
TD6-12-2	7.8	33.72	–7.76	22.73
TD6-12-3	7.8	33.72	–7.55	23.02
TD6-12-4	7.8	33.72	–7.84	22.62
TD6-12-5	7.8	33.72	–7.49	21.62
TD6-08	8	33.71	–7.39	24.36
TD6-09-1	8.1	33.70	–6.49	25.83
TD6-09-2	8.1	33.70	–7.06	25.03
TD6-06-1	8.3	33.69	–7.89	21.58
TD6-06-2	8.3	33.69	–7.64	22.43
TD6-04-1	8.5	33.68	–7.52	24.75
TD6-04-2	8.5	33.68	–7.52	25.08
TD6-04-3	8.5	33.68	–7.63	24.93
TD6-07-1	10.2	33.59	–7.69	22.07
TD6-07-2	10.2	33.59	–7.73	22.47
TD6-10-1	10.8	33.56	–7.69	25.78
TD6-10-2	10.8	33.56	–7.49	25.4
TD6-16-2-1	11	33.55	–6.85	24.22
TD6-16-2-2	11	33.55	–7.52	23.51
TD6-27	11	33.55	–8.08	24.22
TD6-29-1	11	33.55	–8.48	24.37
TD6-29-2	11	33.55	–7.69	25.3
TD6-30	11.1	33.55	–7.67	24.65
TD6-31	11.1	33.55	–8.01	23.9
TD6-14-1	11.3	33.54	–7.25	24.71
TD6-14-2	11.3	33.54	–7.49	24.23
TD6-13	11.4	33.54	–7.19	23.25
TD6-19	11.5	33.54	–6.55	25.33
TD6-17	11.9	33.53	–7.15	23.6
TD6-18	11.9	33.53	–7.88	24.55
TD6-28	12.4	33.51	–7.8	24.32
TD6-26-1	12.8	33.50	–7.43	25.08
TD6-26-2	12.8	33.50	–7.62	25.1
TD6-21	13.4	33.49	–7.18	24.15
TD6-24	14.3	33.46	–7.45	23.88
TD6-35-1	14.4	33.46	–7.22	24.53
TD6-35-2	14.4	33.46	–7.59	25.2
TD6-25-2	14.6	33.45	–7.32	22.18
TD6-34-1	14.6	33.45	–7.47	24.79
TD6-34-2	14.6	33.45	–6.2	23.73
TD6-22	14.7	33.45	–7.01	24.09
TD6-23-1	15	33.44	–6.7	25.39
TD6-23-2	15	33.44	–6.16	24.01
TD6-33	15.2	33.44	–7.8	24.11
TD6-20-1	15.4	33.43	–6.79	23.9
TD6-20-2	15.4	33.43	–6.31	23.56
TD6-20-3	15.4	33.43	–7.53	24.23
TD6-15-1	16	33.41	–7.4	23.42
TD6-15-2	16	33.41	–7.28	23.66
TD6-15-3	16	33.41	–7.06	23.88

*Stratigraphic level is measured relative to the Upper Purplish White (UPW) layer.
†Time scale of Gradstein et al. (2004)

REFERENCES CITED

Aubry, M.P., 1992, Late Paleogene calcareous nannoplankton evolution: A tale of climatic deterioration, *in* Prothero, D.R., and Berggren, W.A., eds., Eocene-Oligocene Climatic and Biotic Evolution: Princeton, New Jersey, Princeton University Press, p. 272–309.

Baldauf, J.G., 1992, Middle Eocene through early Miocene diatom floral turnover, *in* Prothero, D.R., and Berggren, W.A., eds., Eocene-Oligocene Climatic and Biotic Evolution: Princeton, New Jersey, Princeton University Press, p. 310–326.

Berggren, W.A., Kent, D.V., Swisher, C.C., III, and Aubry, M.P., 1995, A Revised Cenozoic Geochronology and Chronostratigraphy: Society for Sedimentary Geology (SEPM) Special Publication 54, p. 129–212.

Cerling, T.E., and Quade, J., 1993, Stable carbon and oxygen isotopes in soil carbonates, *in* Swart, P.K., Lohmann, K.C., McKenzie, J., and Savin, S., eds., Climate Change in Continental Isotopic Records: American Geophysical Union Geophysical Monograph 78, p. 217–231.

Cerling, T.E., Harris, J.M., MacFadden, B.J., Leakey, M.G., Quade, J., Eisenmann, V., and Ehleringer, J.R., 1997, Global vegetation change through the Miocene-Pliocene boundary: Nature, v. 389, p. 153–158, doi: 10.1038/38229.

Clark, J., Beerbower, J.R., and Kietze, K.K., 1967, Oligocene Sedimentation, Stratigraphy, Paleoecology, and Paleoclimatology in the Big Badlands of South Dakota: Fieldiana Geology Memoir 5, 158 p.

Coxall, H.K., Wilson, P.A., Pälike, H., Lear, C.H., and Backman, J., 2005, Rapid stepwise onset of Antarctic glaciation and deeper calcite compensation in the Pacific Ocean: Nature, v. 433, p. 53–57, doi: 10.1038/nature03135.

Emry, R.J., 1992, Mammalian range zones in the Chadronian White River Formation at Flagstaff Rim, Wyoming, *in* Prothero, D.R., and Berggren, W.A., eds., Eocene-Oligocene Climatic and Biotic Evolution: Princeton, New Jersey, Princeton University Press, p. 106–115.

Evanoff, E., Prothero, D.R., and Lander, R.H., 1992, Eocene-Oligocene climatic change in North America: The White River Formation near Douglas, east-central Wyoming, *in* Prothero, D.R., and Berggren, W.A., eds., Eocene-Oligocene Climatic and Biotic Evolution: Princeton, New Jersey, Princeton University Press, p. 116–130.

Gradstein, F., Ogg, J., and Smith, A., 2004, A Geologic Time Scale 2004: Cambridge, U.K., Cambridge University Press, 610 p.

Grandstaff, D.E., and Terry, D.O., 2009, Rare earth element composition of Paleogene vertebrate fossils from Toadstool Geologic Park, Nebraska, USA: Applied Geochemistry (in press).

Grimes, S.T., Hooker, J.J., Collinson, M.E., and Mattey, D.P., 2005, Temperatures of late Eocene to early Oligocene freshwaters: Geology, v. 33, p. 189–192, doi: 10.1130/G21019.1.

Hansen, T.A., Kelley, P.H., and Haasl, D.M., 2004, Paleoecological patterns in molluscan extinctions and recoveries: Comparison of the Cretaceous-Paleogene and Eocene-Oligocene extinctions in North America: Palaeogeography, Palaeoclimatology, Palaeoecology, v. 214, p. 233–242.

Harris, S.E., and Mix, A.C., 2002, Climate and tectonic influences on continental erosion of tropical South America, 0–13 Ma: Geology, v. 30, p. 447–450, doi: 10.1130/0091-7613(2002)030<0447:CATIOC>2.0.CO;2.

Heaton, T.H., and Emry, R.J., 1996, Leptomerycidae, *in* Prothero, D.R., and Emry, R.J., eds., The Terrestrial Eocene-Oligocene Transition in North America: Cambridge, U.K., Cambridge University Press, p. 581–608.

Hutchison, J.H., 1982, Turtle, crocodilian, and champsosaur diversity changes in the Cenozoic of the north-central region of western United States: Palaeogeography, Palaeoclimatology, Palaeoecology, v. 37, p. 149–164, doi: 10.1016/0031-0182(82)90037-2.

Hutchison, J.H., 1992, Western North America reptile and amphibian record across the Eocene/Oligocene boundary and its climatic implications, *in* Prothero, D.R., and Berggren, W.A., eds., Eocene-Oligocene Climatic and Biotic Evolution: Princeton, New Jersey, Princeton University Press, p. 451–463.

Jiongxin, X., 2005, Precipitation-vegetation coupling and its influence on erosion on the Loess Plateau, China: Catena, v. 64, p. 103–116, doi: 10.1016/j.catena.2005.07.004.

Keller, G., MacLeod, N., and Barrera, E., 1992, Eocene-Oligocene faunal turnover in planktic foraminifera, and Antarctic glaciation, *in* Prothero, D.R., and Berggren, W.A., eds., Eocene-Oligocene Climatic and Biotic Evolution: Princeton, New Jersey, Princeton University Press, p. 218–244.

Kim, S.T., and O'Neil, J.R., 1997, Equilibrium and nonequilibrium oxygen isotope effects in synthetic carbonates: Geochimica et Cosmochimica Acta, v. 61, p. 3461–3475, doi: 10.1016/S0016-7037(97)00169-5.

Koch, P.L., Tuross, N., and Fogel, M.L., 1997, The effects of sample treatment and diagenesis on the isotopic integrity of carbonate in biogenic hydroxylapatite: Journal of Archaeological Science, v. 24, p. 417–429, doi: 10.1006/jasc.1996.0126.

Koch, P.L., Clyde, W.C., Hepple, R.P., Fogel, M.L., Wing, S.L., and Zachos, J.C., 2003, Carbon and oxygen isotope records from paleosols spanning the Paleocene-Eocene boundary, Bighorn Basin, Wyoming, *in* Wing, S.L., et al., eds., Causes and Consequences of Globally Warm Climates in the Early Paleogene: Geological Society of America Special Paper 369, p. 49–64.

Kohn, M.J., and Cerling, T.E., 2002, Stable isotope compositions of biological apatite, *in* Kohn, M.J., Rakovan, J., and Hughes, J.M., eds., Phosphates: Geochemical, geobiological, and materials importance: Reviews in Mineralogy and Geochemistry, v. 48, p. 455–488.

Kohn, M.J., and Fremd, T.J., 2007, Tectonic controls on isotope compositions and species diversification, John Day Basin, central Oregon: PaleoBios, v. 27, p. 48–61.

Kohn, M.J., and Law, J.M., 2006, Stable isotope chemistry of fossil bone as a new paleoclimate indicator: Geochimica et Cosmochimica Acta, v. 70, p. 931–946, doi: 10.1016/j.gca.2005.10.023.

Kohn, M.J., Josef, J.A., Madden, R., Kay, R., Vucetich, G., and Carlini, A.A., 2004, Climate stability across the Eocene-Oligocene transition, southern Argentina: Geology, v. 32, p. 621–624, doi: 10.1130/G20442.1.

Kohn, M.J., Zanazzi, A., and Josef, J.A., 2008, Stable isotopes of fossil teeth and bones at Gran Barranca as a monitor of climate change and tectonics, *in* Madden, R.H., Vucetich, M.G., Carlini, A.A., and Kay, R.F., eds., The Paleontology of Gran Barranca: Evolution and Environmental Change through the Middle Cenozoic of Patagonia: Cambridge, U.K., Cambridge University Press (in press).

Kuiper, K.F., Deino, A., Hilgen, F.J., Krijgsman, W., Renne, P.R., and Wijbrans, J.R., 2008, Synchronizing rock clocks of Earth history: Science, v. 320, p. 500–504, doi: 10.1126/science.1154339.

LaGarry, H.E., 1998, Lithostratigraphic revision and redescription of the Brule Formation, White River Group, western Nebraska, *in* Terry, D.O., LaGarry, H.E., and Hunt, R.M., eds., Depositional Environments, Lithostratigraphy, and Biostratigraphy of the White River and Arikaree Groups (Late Eocene to Early Miocene, North America): Geological Society of America Special Paper 325, p. 63–91.

Langbein, W.B., and Schumm, S.A., 1958, Yield of sediment in relation to mean annual precipitation: Eos (Transactions, American Geophysical Union), v. 39, p. 1076–1084.

Larson, E.E., and Evanoff, E.E., 1998, Tephrostratigraphy and source of the tuffs of the White River sequence, *in* Terry, D.O., LaGarry, H.E., and Hunt, R.M., eds., Depositional Environments, Lithostratigraphy, and Biostratigraphy of the White River and Arikaree Groups (Late Eocene to Early Miocene, North America): Geological Society of America Special Paper 325, p. 1–14.

Lear, C.H., Bailey, T.R., Pearson, P.N., Coxall, H.K., and Rosenthal, Y., 2008, Cooling and ice growth across the Eocene-Oligocene transition: Geology, v. 36, p. 251–254, doi: 10.1130/G24584A.1.

Miller, K.G., Fairbanks, R.G., and Mountain, G.S., 1987, Tertiary oxygen isotope synthesis, sea level history, and continental margin erosion: Paleoceanography, v. 2, p. 1–19, doi: 10.1029/PA002i001p00001.

Molnar, P., 2001, Climate change, flooding in arid environments, and erosion rates: Geology, v. 29, p. 1071–1074, doi: 10.1130/0091-7613(2001)029 <1071:CCFIAE>2.0.CO;2.

Molnar, P., 2004, Late Cenozoic increase in accumulation rates of terrestrial sediment: How might climate change have affected erosion rates?: Annual Review of Earth and Planetary Sciences, v. 32, p. 67–89, doi: 10.1146/annurev.earth.32.091003.143456.

Obradovich, J.D., Evanoff, E., and Larson, E.E., 1995, Revised single-crystal laser-fusion $^{40}Ar/^{39}Ar$ ages of Chadronian tuffs in the White River Formation of Wyoming: Geological Society of America Abstracts with Programs, v. 27, no. 3, p. A-77.

Peizhen, Z., Molnar, P., and Downs, W.R., 2001, Increased sedimentation rates and grain sizes 2–4 Myr ago due to the influence of climate change on erosion rates: Nature, v. 410, p. 891–897, doi: 10.1038/35073504.

Prothero, D.R., 1996a, Magnetic stratigraphy of the White River Group in the high plains, *in* Prothero, D.R., and Emry, R.J., eds., The Terrestrial Eocene-Oligocene Transition in North America: Cambridge, U.K., Cambridge University Press, p. 262–277.

Prothero, D.R., 1996b, Camelidae, *in* Prothero, D.R., and Emry, R.J., eds., The Terrestrial Eocene-Oligocene Transition in North America: Cambridge, U.K., Cambridge University Press, p. 591–633.

Prothero, D.R., and Berggren, W.A., 1992, Eocene-Oligocene Climatic and Biotic Evolution: Princeton, New Jersey, Princeton University Press, 568 p.

Prothero, D.R., and Emry, R.J., 1996, Summary, *in* Prothero, D.R., and Emry, R.J., eds., The Terrestrial Eocene-Oligocene Transition in North America: Cambridge, U.K., Cambridge University Press, p. 646–664.

Prothero, D.R., and Emry, R.J., 2004, The Chadronian, Orellan, and Whitneyan North American Land Mammal Ages, *in* Woodburne, M.O., ed., Late Cretaceous and Cenozoic Mammals of North America: Biostratigraphy and Geochronology: New York, Columbia University Press, p. 156–168.

Prothero, D.R., and Heaton, T.H., 1996, Faunal stability during the early Oligocene climatic crash: Palaeogeography, Palaeoclimatology, Palaeoecology, v. 127, p. 257–283, doi: 10.1016/S0031-0182(96)00099-5.

Prothero, D.R., and Swisher, C.C., III, 1992, Magnetostratigraphy and geochronology of the terrestrial Eocene-Oligocene transition in North America, *in* Prothero, D.R., and Berggren, W.A., eds., Eocene-Oligocene Climatic and Biotic Evolution: Princeton, New Jersey, Princeton University Press, p. 46–73.

Prothero, D.R., and Whittlesey, K.E., 1998, Magnetic stratigraphy and biostratigraphy of the Orellan and Whitneyan land-mammal "ages" in the White River Group, *in* Terry, D.O., LaGarry, H.E., and Hunt, R.M., eds., Depositional Environments, Lithostratigraphy, and Biostratigraphy of the White River and Arikaree Groups (Late Eocene to Early Miocene, North America): Geological Society of America Special Paper 325, p. 39–61.

Retallack, G.J., ed., 1983, Late Eocene and Oligocene Paleosols from Badlands National Park, South Dakota: Geological Society of America Special Paper 193, 82 p.

Retallack, G.J., 1992, Paleosols and changes in climate and vegetation across the Eocene/Oligocene boundary, *in* Prothero, D.R., and Berggren, W.A., eds., Eocene-Oligocene Climatic and Biotic Evolution: Princeton, New Jersey, Princeton University Press, p. 382–398.

Retallack, G.J., 2007, Cenozoic paleoclimate on land in North America: The Journal of Geology, v. 115, p. 271–294, doi: 10.1086/512753.

Retallack, G.J., Orr, W.N., Prothero, D.R., Duncan, R.A., Kester, P.R., and Ambers, C.P., 2004, Eocene-Oligocene extinction and paleoclimatic change near Eugene, Oregon: Geological Society of America Bulletin, v. 116, p. 817–839, doi: 10.1130/B25281.1.

Schultz, C.B., and Falkenbach, C.H., 1956, Miniochoerinae and Oreonetinae, two new subfamilies of oreodonts: Bulletin of the American Museum of Natural History, v. 109, p. 377–482.

Schultz, C.B., and Falkenbach, C.H., 1968, The phylogeny of the oreodonts, Parts 1 and 2: Bulletin of the American Museum of Natural History, v. 139, p. 1–498.

Schultz, C.B., and Stout, T.M., 1955, Classification of Oligocene sediments of Nebraska: Bulletin of the University of Nebraska State Museum, v. 4, p. 17–52.

Sheldon, N.D., and Retallack, G.J., 2004, Regional paleoprecipitation records from the late Eocene and Oligocene of North America: The Journal of Geology, v. 112, p. 487–494, doi: 10.1086/421076.

Stevens, M.S., and Stevens, J.B., 1996, Merycoidodontinae and Miniochoerinae, *in* Prothero, D.R., and Emry, R.J., eds., The Terrestrial Eocene-Oligocene Transition in North America: Cambridge, U.K., Cambridge University Press, p. 498–573.

Storer, J.E., 1984, Fossil mammals of the Southfork local fauna (early Chadronian) of Saskatchewan: Canadian Journal of Earth Sciences, v. 21, p. 1400–1405.

Swinehart, J.B., Souders, V.L., Degraw, H.M., and Diffendal, R.F., 1985, Cenozoic paleogeography of western Nebraska, *in* Flores, R.M., and Kaplan, S., eds., Cenozoic Paleogeography of the West Central United States: Rocky Mountain Section, Society of Economic Paleontologists and Mineralogists, Rocky Mountain Paleogeography Symposium 3, p. 209–229.

Swisher, C.C., III, and Prothero, D.R., 1990, Single crystal $^{40}Ar/^{39}Ar$ dating of the Eocene-Oligocene transition in North America: Science, v. 249, p. 760–762, doi: 10.1126/science.249.4970.760.

Terry, D.O., 1998, Lithostratigraphic revision and correlation of the lower part of the White River Group: South Dakota to Nebraska, *in* Terry, D.O., LaGarry, H.E., and Hunt, R.M., eds., Depositional Environments, Lithostratigraphy, and Biostratigraphy of the White River and Arikaree Groups (Late Eocene to Early Miocene, North America): Geological Society of America Special Paper 325, p. 15–37.

Terry, D.O., 2001, Paleopedology of the Chadron Formation of northwestern Nebraska: Implications for paleoclimate change in the North American mid-continent across the Eocene-Oligocene boundary: Palaeogeography, Palaeoclimatology, Palaeoecology, v. 168, p. 1–38, doi: 10.1016/S0031-0182(00)00248-0.

Terry, D.O., 2007, The Eocene-Oligocene transition of northwestern Nebraska, USA: Bolides, glaciers, or none of the above?: Boulder, Colorado, Geological Society of America, Penrose Conference Abstracts with Program, p. 75.

Terry, D.O., and LaGarry, H.E., 1998, Big Cottonwood Creek Member: A new member of the Chadron Formation in northwestern Nebraska, *in* Terry, D.O., LaGarry, H.E., and Hunt, R.M., eds., Depositional Environments, Lithostratigraphy, and Biostratigraphy of the White River and Arikaree Groups (Late Eocene to Early Miocene, North America): Geological Society of America Special Paper 325, p. 117–141.

Terry, D.O., LaGarry, H.E., and Hunt, R.M., eds., 1998, Depositional Environments, Lithostratigraphy, and Biostratigraphy of the White River and Arikaree Groups (Late Eocene to Early Miocene, North America): Geological Society of America Special Paper 325, 216 p.

Thomas, E., 1992, Middle Eocene–late Oligocene bathyal benthic foraminifera (Weddell Sea): Faunal changes and implications for ocean circulation, *in* Prothero, D.R., and Berggren, W.A., eds., Eocene-Oligocene Climatic and Biotic Evolution: Princeton, New Jersey, Princeton University Press, p. 245–271.

Wells, W.B., Terry, D.O., and Nixon, D.A., 1994, Multiple cut and fill paleovalley sequences within Oligocene sediments (White River Group) of northwestern Nebraska: Geological Society of America Abstracts with Programs, v. 26, no. 7, p. A-492.

Wolfe, J.A., 1978, A paleobotanical interpretation of Tertiary climates in the Northern Hemisphere: American Scientist, v. 66, p. 694–703.

Wolfe, J.A., 1992, Climatic, floristic, and vegetational changes near the Eocene-Oligocene boundary in North America, *in* Prothero, D.R., and Berggren, W.A., eds., Eocene-Oligocene Climatic and Biotic Evolution: Princeton, New Jersey, Princeton University Press, p. 421–436.

Wolfe, J.A., 1994, Tertiary climatic changes at middle latitudes of western North America: Palaeogeography, Palaeoclimatology, Palaeoecology, v. 108, p. 195–205, doi: 10.1016/0031-0182(94)90233-X.

York, D., 1966, Least-squares fitting of a straight line: Canadian Journal of Physics, v. 44, p. 1079–1086.

Zachos, J.C., Pagani, M., Sloan, L., Thomas, E., and Billups, K., 2001, Trends, rhythms, and aberrations in global climate 65 Ma to present: Science, v. 292, p. 686–693, doi: 10.1126/science.1059412.

Zanazzi, A., Kohn, M.J., MacFadden, B.J., and Terry, D.O., 2007, Large temperature drop across the Eocene-Oligocene transition in central North America: Nature, v. 445, p. 639–642, doi: 10.1038/nature05551.

Manuscript Accepted by the Society 16 September 2008

Printed in the USA

The Geological Society of America
Special Paper 452
2009

The Australo-Antarctic Gulf and the Auversian facies shift

Brian McGowran*
School of Earth & Environmental Sciences, Mawson DX 650 313, The University of Adelaide, Adelaide, SA 5005, Australia

ABSTRACT

Three time lines through the neritic stratigraphic record distributed around the northern margin of the Australo-Antarctic Gulf (AAG) mark three fundamental shifts in global environments collectively comprising the Auversian facies shift. The three lines are: (1) the beginning: the Khirthar transgression and the onset of neritic carbonate accumulation in the Bartonian Age (preceding onset of the Middle Eocene climatic optimum [MECO]); (2) the midlife change (Bartonian-Priabonian transition): the shift from carbonate-rich to carbonate-poor, higher-nutrient environments under estuarine circulation, causing widespread dysaerobia culminating in opaline silicas; and (3) the Eocene-Oligocene = Priabonian-Rupelian boundary and glaciation during oxygen isotope event Oi-1, with return of improved ventilation in neritic environments and resumption of carbonate accumulation. Meanwhile, it was warm and very wet at ~60°S. In developing a scenario for the death of the AAG, the birth of the Southern Ocean, and the transition from Paleogene greenhouse Earth to Neogene icehouse Earth, the neritic record of the northern margin is more in accord with the "Dinocyst biogeographic hypothesis" than with the "Tasman gateway hypothesis."

Keywords: Eocene stratigraphy, southern Australia, Australo-Antarctic Gulf, Southern Ocean, MECO, Bartonian, Priabonian.

INTRODUCTION

The Australo-Antarctic Gulf (AAG) formed as seafloor spreading propagated to the southeast from the Indian Ocean, between two fragments of disintegrating Gondwanaland. During the first 50 m.y., spreading was slow, and the AAG widened slowly. Fifty million years ago, India collided with Asia, global spreading systems were rearranged, Australia-Antarctica separation accelerated, and by Oligocene times, the AAG was being absorbed into the new Southern Ocean.

The AAG has been the site of several controversies. Terrestrial vertebrate animal fossils posed intractable questions in the battles over continental drift. How could Triassic tetrapods get to Antarctica (in pre–AAG times)? How could marsupials get to Australia? How indeed could therians get out of Australia? Then, as Australia/Antarctica separated, the Southern Ocean grew, and so too did the Circum-Antarctic Current and the polar ice caps. Apropos of the latter: was it the insulating effect of the Circum-Antarctic Current or was it the collapse of carbon dioxide levels? Huber and Nof (2006) suggested that the question, "what caused Oligocene cooling?" may be approached more profitably by asking, "what made the Eocene warm, and why did it stop?" For such questions of how we chart the greenhouse-to-icehouse transition and how we explain it, the AAG and its environs remain highly pertinent, as Ocean Drilling Program (ODP) Leg 189 has demonstrated.

This paper summarizes the stratigraphic record of the neritic realm in southern Australia—the north shore of the AAG.

*brian.mcgowran@adelaide.edu.au.

McGowran, B., 2009, The Australo-Antarctic Gulf and the Auversian facies shift, *in* Koeberl, C., and Montanari, A., eds., The Late Eocene Earth—Hothouse, Icehouse, and Impacts: Geological Society of America Special Paper 452, p. 215–240, doi: 10.1130/2009.2452(14). For permission to copy, contact editing@geosociety.org.

Emphasis is on the stratigraphic, sometimes at the expense of the sedimentological and the paleobiological, and emphasis is on the three-part slice of biogeohistory—the sudden expansion of carbonates and coals in the Bartonian, the profound environmental shift into the Priabonian, and the making of the modern environment in the Rupelian. However, as stratigraphers, we are always looking sideways, chronologically speaking, and we should never forget that our neat ad hoc and local or regional explanations have a habit of merging into a still larger canvas. I begin at this broader level before scrutinizing the record in the AAG.

There is no record of bolide impact in the Bartonian-Priabonian of the AAG—nor, however, has there been a coherent and sustained search.

WHAT IS THE AUVERSIAN FACIES SHIFT?

Amongst the turbulent progress in our understanding of the later Paleogene in the 1970s–2000s, chronological resolution and correlation advancing hand in hand with construction of new geochemical time series (Shackleton and Kennett, 1975), I acknowledge two conceptual insights. One was identification and naming of the major paleoceanographic transition—the Auversian facies shift of Berger and Wefer (1996) and Berger (2007); the other insight involved insinuating nutrient supply into the sedimentological mindset, which had previously been dominated by depth, temperature, salinity, and so on, resulting in the trophic resource continuum of Hallock (1987), Hallock et al. (1991), and Boersma et al. (1998).

The Eocene Epoch has long been perceived as a time of fundamental change at all levels, endogenic to exogenic in the old physico-geographic categories. The levels extend from the deep crust (the death of Tethys closing the low-latitude and circumglobal warm ring; the opening of the Southern Ocean producing the cold ring), through the resulting paleoceanographic and paleoclimatic changes (the "greenhouse" to "icehouse" transition), to the biosphere, with its replacement of the archaic by more modern lineages and biotas. Berger and Wefer (1996) and Berger (2007) focused on the intermediate, paleoceanographic levels in naming the Auversian facies shift. (The Auversian itself is a venerable but redundant stage, being coextensive with the Bartonian, so the name can serve as an appellation for a fundamental shift in the exogenic realm.) The point of the Auversian facies shift was to contrast the earlier oceanic record, which was geochemically less differentiated, and which had more mixed and variable oceanic facies, with the later (especially Oligocene) oceanic record, with its more clearly sequestered oceanic facies and its clean, "boring" carbonates dominating extensive tracts of the global deep-sea floor.

For Berger and Wefer, the Auversian facies shift was of several million years' duration centered roughly on 40 Ma, i.e., at or near the Bartonian-Priabonian boundary. It soon became clear during the early days of the Deep Sea Drilling Project that the later Eocene carbonate record in oceanic facies was inferior to the Oligocene because it was more patchy, riddled with hiatuses, and displayed abundant signs of carbonate corrosion. Such observations, quantified and synthesized, yielded the trajectory of the calcite compensation depth (CCD) from consistently shallow until the later Eocene, when it plunged by kilometers into the early Oligocene (van Andel et al., 1975). The large and sudden drop in the CCD (Rea and Lyle, 2005) was intimately linked to ice buildup on Antarctica (Coxall et al., 2005). Adding to the sense of major geohistorical transition responding to deep-crustal shifts, there is the sharp upward turn in the strontium-isotopic curve (McArthur and Howarth, 2004).

The Auversian facies shift was an oceanic concept begging, as it were, for scrutiny in the neritic and terrestrial realms. I attempted this scrutiny not in response to the Auversian facies shift exposition but to stratigraphic patterns emerging when comparing Indian Ocean stratigraphy with circum–Indian Ocean stratigraphy (McGowran, 1977, 1978b). The central theme of cause-and-effect was simple:

1. There was a three-part tectonic succession shaped by two clusters of events: early Eocene collisions in Tethys and cessation of Tasman and Coral Sea spreading (earlier), and rapid Australia-Antarctica separation within reorganized late middle Eocene ocean spreading (later);
2. an extraordinarily well displayed circum–Indian Ocean marine transgression in the late middle Eocene (the *Khirthar transgression*) seemed to be a tectono-eustatic response to the new crustal spreading regime; and
3. several diverse patterns in the fossil record fitted this physical pattern rather well, inviting the inference of detectable physical impact on biohistory (McGowran, 1990, 1991).

Thus, the world was cooler and regressive following the collisions and the death of Tethys, and evolutionary activity on several indicators was muted. However, following the invigorated spreading in the new tectonic regime, there was evidence of warming and invigorated evolutionary activity—pelagic, neritic, and terrestrial. Things physical and biospheric changed markedly around the late middle Eocene and the Khirthar transgression, at or close to the Lutetian-Bartonian Stage boundary. In the pelagic realm, benthic foraminifera underwent a massive faunal change, and planktonic foraminifera sharply increased their rates of extinction and speciation. In the neritic realm, there were sharp increases in diversity (molluscs) and in biogeographic expansion toward the poles (photosymbiotic foraminifera). In the terrestrial realm (floras, mammalian faunas), there were parallel expansions in taxic evolution and in warming indicators.

These changes challenged the standard scenario of a steady deterioration in global climate from the Early Eocene climatic optimum (EECO) to the end-Eocene chill (Oi-1). There should have been some signal from the pelagic realm confirming the importance of the Khirthar transgression in the neritic record. In due course it emerged—MECO, the middle Eocene climatic optimum (Bohaty and Zachos, 2003). We would now assert that the Auversian facies shift can usefully be broadened—beyond the pelagic realm—and stratigraphically sharpened—beginning at about the base of the Bartonian Stage, where there is a rapid succession from reorganized seafloor spreading, through a major marine transgression, to a sharp reversal of the general global cooling and a series of biospheric effects seen in the fossil record.

Meanwhile, the availability, supply, and delivery of nutrients, their movement through the exogenic system and the reactive lithosphere, increasingly penetrated our consciousness, not least through the elegant conceptualization in the trophic resource continuum (Fig. 1). The trophic resource continuum stimulated a way of comparing the effects of nutrient shifts in different communities, such as the open-ocean planktonic foraminifera and the warm-water neritic "larger" foraminifera (Hallock et al., 1991). These authors made a double point—that parallel patterns in evolutionary development suggested a common factor in changes in nutrient flux to surface waters; and that apparently sudden shifts between "warm" and "cool" biotas were not so much that as an enhanced reaction to changes in nutrient flux. Thus, oligotrophic/eutrophic may well be more heuristic than warm/cool.

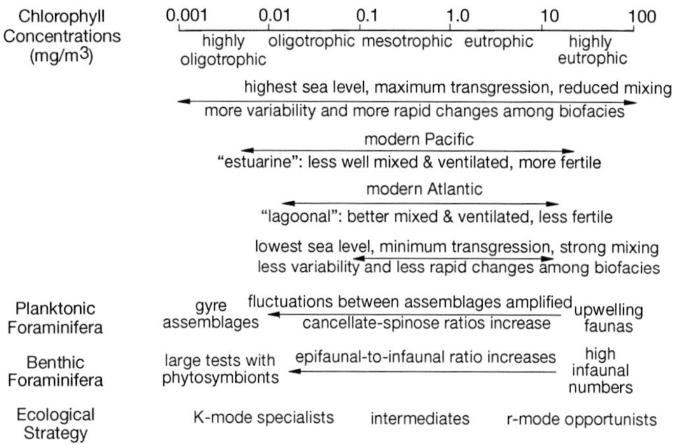

Figure 1. The trophic resource continuum (TRC), a visualization of nutrients in changing oceanographic states (Hallock, 1987; Hallock et al., 1991, 1998; McGowran, 2005). Double-headed arrows suggest expansion-contraction rather than diachronous shifts to left or right.

Chronofaunas and Turnover

These points can be broadened by the chronological comparisons in Figure 2. A skeptic may well murmur "post hoc ergo propter hoc": do not derive cause and effect from temporal correlations and ordinations. Even so, the drive might be inferred from the crustal shift from the Laramide to the Himalayan tectonic regime. The global climate embraces both MECO and Oi-1.

Students of the North American land mammal succession have long identified the Duchesnian Age as a time of enhanced faunal turnover from the long-lived Eocene chronofauna to the even longer-lived White River chronofauna (references in Woodburne, 2004, 2007). The chronofaunal boundary is drawn sharply within the Duchesnian and within chron C18n.1n by Prothero (2004), who was concerned with demonstrating that the mammalian chronofaunal change was not matched by a climatic event, since the cooling at the middle-late Eocene (Bartonian-Priabonian) boundary was distinctly younger, within C17n.1n. Although numerous other biotic changes could be matched with that cooling, the mammalian turnover was due, Prothero concluded, to intrinsic biological factors. However, the chronofaunal boundary may well have been coeval with a major nutritional shift in the neritic and pelagic realms (see following).

Hottinger (1982, 1997) recognized a similar punctuating effect in the long-lived (while evolving within themselves) large, photosymbiotic foraminiferal communities of the Tethyan neritic realm, which were subsequently (McGowran and Li, 2001) named the early Paleogene and Oligocene-Miocene chronofaunas interrupted by the impoverished Priabonian chronofauna. These forms occupied the oligotrophic reaches of the trophic resource continuum. In the shallow-benthic zonal succession (SBZ) (Serra-Kiel et al., 1998), the major depletion of the most prominent clades (especially *Alveolina* and large *Nummulites*) occurred in the upper part of SBZ17, correlated with upper chron C18–lower C17. Also prominent, there is the extinction in the pelagic realm of the two planktonic foraminiferal genera, *Acarinina* and *Morozovelloides*, which are considered to have lived uppermost in the water column and which possessed algal photosymbionts (Norris, 1996). They too occupied the oligotrophic reaches of the trophic resource continuum. The extinction of this group (with the exception of the lingering *A. collactea*) was sudden, at chron C17n.3n (Sexton et al., 2006), an event marking the zone E13-E14 boundary (Berggren and Pearson, 2005). The possibly simultaneous removal of the dominating photosymbiotic and oligotrophic components from the pelagic and neritic realms implies a major shift in the nutritional regime at the global scale.

In the deep ocean, benthic foraminiferal (chrono)faunas persisted for 10–20 m.y. with turnover times of 2–3 m.y. (Thomas et al., 2000). Figure 2 makes the point that the Auversian facies shift signals a time of enhanced change.

Tethyan Succession of Neritic Biofacies

Hottinger (1982, 1997) emphasized the magnitude of the change in the oligotrophic foraminiferal communities in the Tethyan neritic realm at the termination of the early Paleogene chronofauna at the end of the middle Eocene. Nebelsick et al. (2005) made the same point in a different way (Fig. 3). They employed the concept of "facies dynamics" to define and interpret spatial and temporal changes in neritic carbonate biofacies in the circum-Alpine region. Definitions were constructed by recognizing 14 major facies types. Middle Eocene major facies types are dominated by ecologically partitioned larger foraminifera, plotted as seven biofacies. Four of those major facies types had disappeared by late Eocene time, in keeping with the major extinctions in these oligotrophic organisms, while the algal-dominated major facies types expanded their ranges and diversity and the bryozoans appeared. The replacement of oligotrophs by

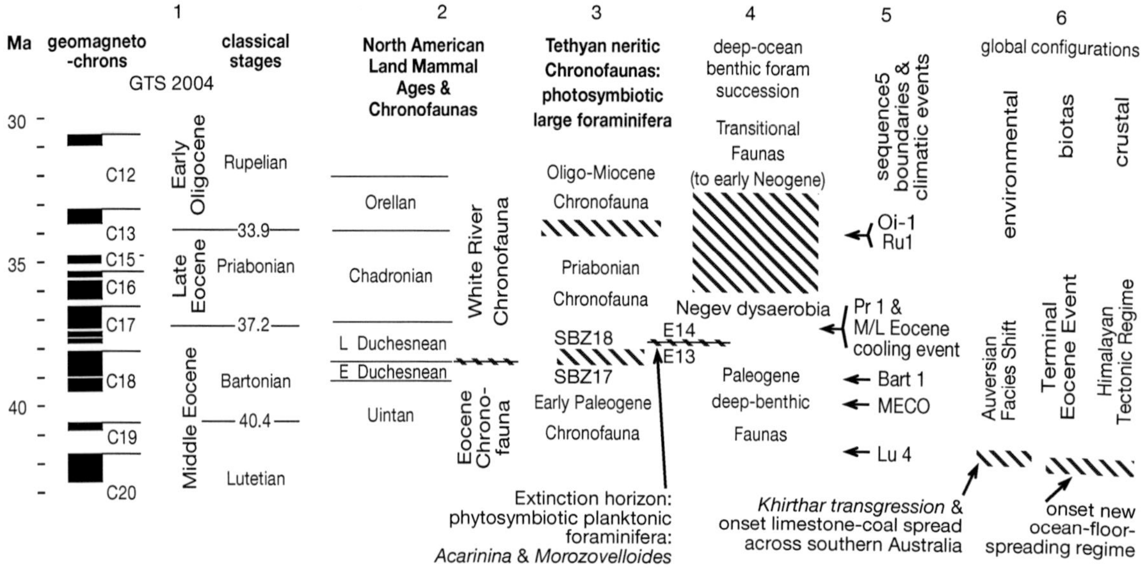

Figure 2. Correlations in the late Lutetian–early Rupelian interval from the terrestrial, neritic, and pelagic realms, adapted from McGowran (2005, his Figures 6.11, 6.12). The point is to emphasize (1) the sense of long-term stability punctuated by short-term turnover, and (2) chronological parallels between the three environmental realms. The mammalian, neritic foraminiferan, and deep-ocean foraminiferan "fauna" or "chronofauna" has each persisted throughout the Eocene, until Bartonian time. Underlying the environmental and biotic shifts, the new tectonic regime is exemplified by invigorated seafloor spreading in the Australian-Antarctic Gulf (and soon its death), the major stratigraphic event of the Khirthar transgression, and onset of extensive neritic carbonate accumulations. 1—A geologic time scale 2004 (GTS 2004; Gradstein et al., 2004). 2—Woodburne (2004, 2007); chronofaunal boundary from Prothero (2004). 3—Hottinger (1997), McGowran and Li (2001); shallow benthic zones 17, 18 (SBZ17–18; Serra-Kiel et al., 1998) occur between 3 and 4, E14-E13 boundary, planktonic foraminiferal zones, and extinction horizon, Berggren and Pearson (2005), Sexton et al. (2006); 4—Two successional benthic foraminiferal faunas (chronofaunas), Thomas et al. (2000); Negev dysaerobia, Barbieri et al. (2003). 5—Middle-late Eocene boundary cooling event, Prothero (2004); MECO (MECO, Bohaty and Zachos, 2003). 6—see text.

algae is a strong signal of a general increase in the supply of nutrients to the Tethyan neritic realm in general, and their expansion in the Priabonian has been noted previously (Hallock et al., 1991). In addition, the foraminifera do not subsequently recover—indeed the orthophragminids (*Discocyclina* etc.) also disappear; the main survivors into the Oligocene are the non-photosymbiotic smaller miliolids, and the bryozoans expand in the early Oligocene. Thus, we have a chronological sandwich—the Priabonian neritic record is a perceptibly higher-nutrient assembly of biofacies interpolated between the oligotrophic Bartonian and the Rupelian. The Bartonian assemblages are ancient insofar as they are clearly different from those present today, whereas the early Oligocene biofacies found in the studied region are modern, all being found in Holocene environments, a point emphasized by Nebelsick et al. (2005). The neritic pattern is *congruent* with the pelagic Auversian facies shift. The essential point taken here from the Nebelsick et al. survey is that traditional environmental parameters such as depth, temperature, light, or salinity simply do not suffice to explain this pattern of generalized but comprehensive stratigraphic distribution. A widespread increase in fertility in the neritic realm is the key, a point now extended to the pelagic realm by the likely coeval extinction of oligotrophic planktonic foraminifera (see previous).

THE AUSTRALO-ANTARCTIC GULF

During the Mesozoic breakup of eastern Gondwanaland and the dispersal of its fragments, sedimentary troughs developed and filled in the rift between the future Antarctica and Australia as the nascent ocean intervened. As the spreading ridge in the ocean to the west (Southeast Indian Ridge) propagated from west to east, that ocean opened in the same direction, and the AAG grew very slowly for ~50 m.y. Things changed dramatically in the middle Eocene, ca. 50–40 Ma, tectonically, oceanographically, and climatically. About 42 Ma, the AAG was at latitudes of 55–60°S (perhaps 25° beyond today's position) and was still only a few hundred kilometers wide after ~50 m.y. It was either closed to the SW Pacific to the southeast or connected by a very narrow and very shallow strait—the South Tasman Saddle between Tasmania and the South Tasman Rise. The AAG was floored by oceanic crust. At about this time, the Southeast Indian Ridge changed from slow spreading to rapid spreading, and the spreading occurred in a more northerly direction than before. This was a major component of a global change in plate-tectonic pattern and movement, brought about by collisions, especially the onset of welding the Indian continent to Asia (i.e., McGowran, 1989a, 1991; Veevers, 2000).

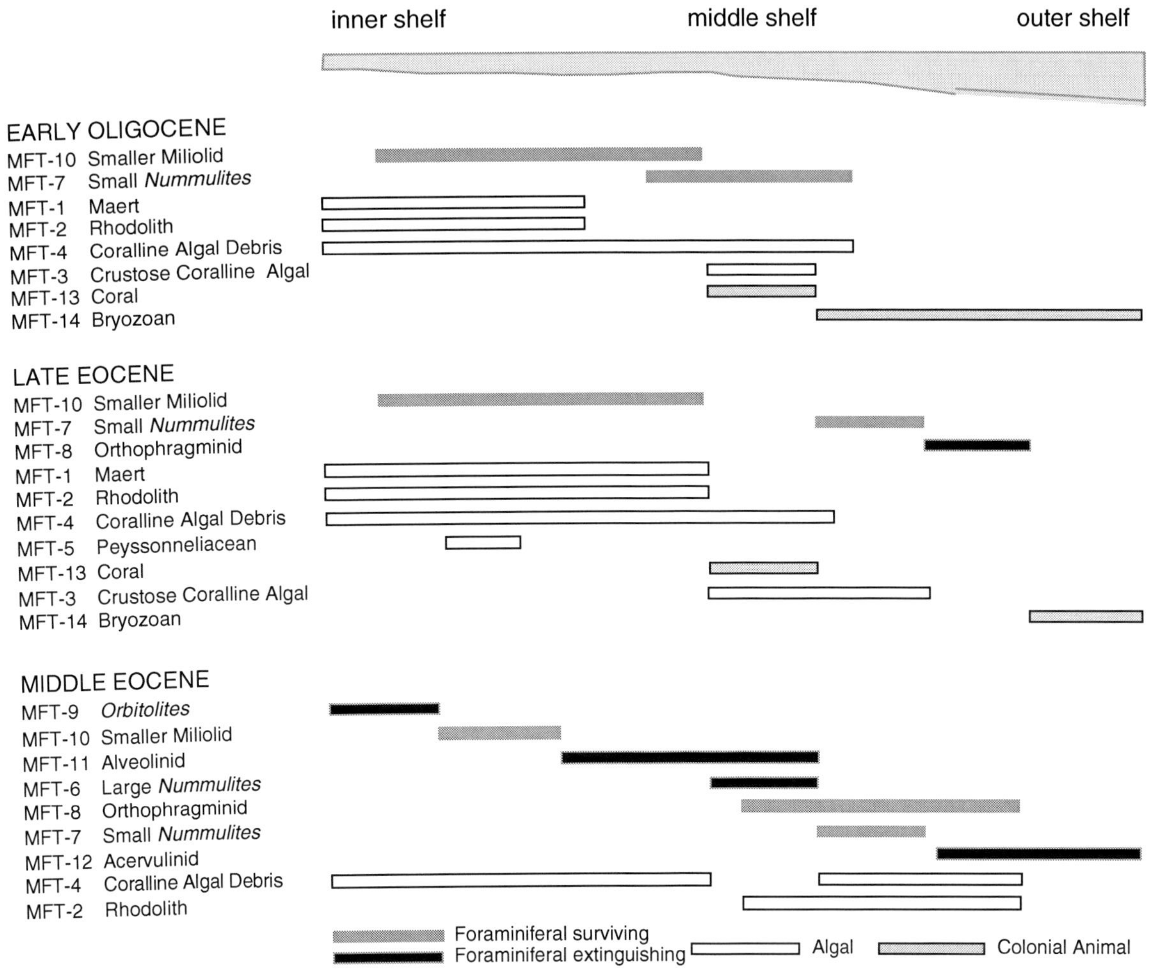

Figure 3. Spatial and temporal depiction of biofacies successions in the Tethyan neritic realm (adapted from Nebelsick et al., 2005). MFT—major facies type. In the Bartonian, the oligotrophic larger foraminifera are spread through the photic zone (MFT 9, 11, 6, 7, [12]) but undergo comprehensive extinction (7 excepted), to be replaced not by their kind or analogues of their kind but by more eutrophic algal communities in the Priabonian. This implies a major shift in the nutritional regime between the Bartonian and the Lutetian. The change to early Oligocene faunas and facies is also profound in that they are truly modern, whereas their predecessors are not.

One outcome in this region was the subsidence and opening of the Tasmania-Antarctica gap—and the death of the AAG as such, as it became subsumed in the rapidly growing Southern Ocean (Lawver and Gahagan, 2003; Cande and Stock, 2004).

Our access to the actual record of the northern neritic reaches of the AAG and its terrestrial surrounds is by scattered and patchy outcrop (due to little deformation, little deep dissection, little subsequent removal of the extensive regolith, shielding by lithified and cherty carbonates), and by economic drilling, but not generally in a culture of detailed stratigraphic drilling and sampling for biostratigraphy. Oceanic drilling occurred, once, just outside the mouth of the AAG (Dee Sea Drilling Project [DSDP] Site 264, Naturaliste Plateau) and just inside in the far southeast (the DSDP Leg 29 sites). Although the Leg 29 holes within the AAG contributed rather little, the leg itself gave rise to a pioneering oxygen-isotopic series (Shackleton and Kennett, 1975) and to one of the two main theories of southern cooling (Kennett et al., 1974). There have been two more recent forays within the AAG. Several holes in the western part of the Great Australian Bight intersected strata of interest here (Li et al., 2003a, 2003b). In the southeast, the results of ODP Leg 189 are of considerable interest (Exon et al., 2004a, 2004b). Dredging during a rare marine survey yielded a suite of samples from the eastern Great Australian Bight (Davies et al., 1989; McGowran, 1991). Basins, locations, and sites are shown on Figure 4.

P.G. Quilty (2008, personal commun.) has pointed out that "drilling along the Antarctic margin has been disappointing in not providing data from the other side of the AAG and that, except for information from well west of the AAG, only patchy information, not readily placed in the timeframe discussed here, has been recovered" (e.g., Quilty et al., 2000; Quilty, 2001).

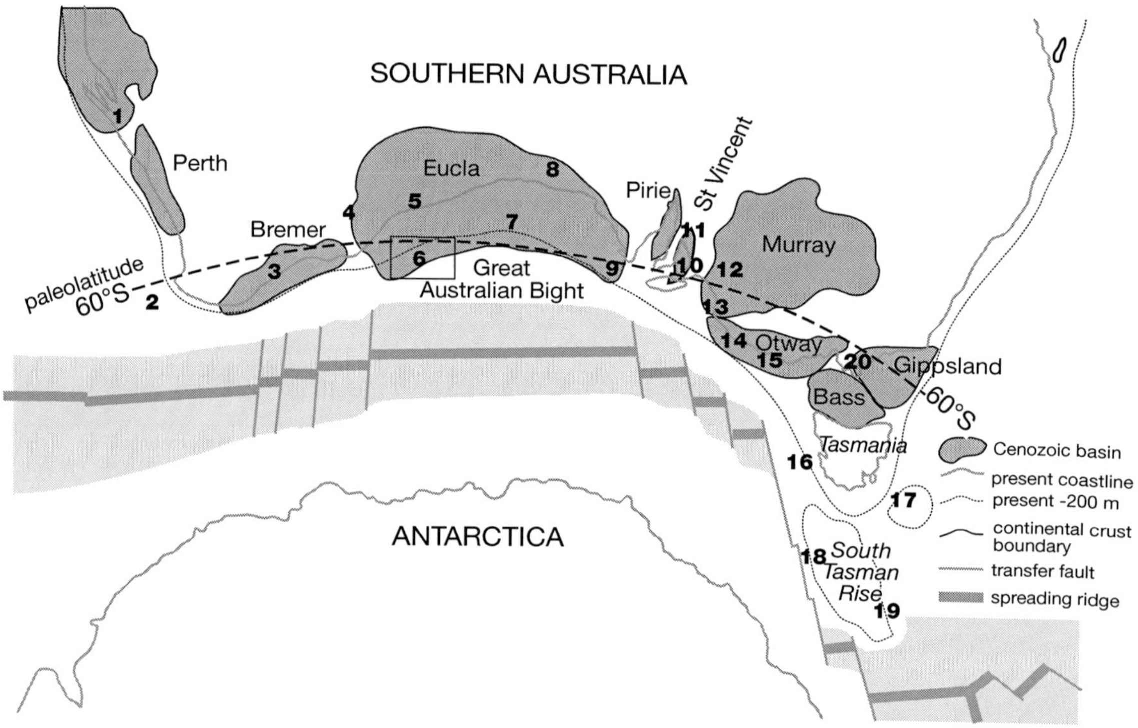

Figure 4. The Australo-Antarctic Gulf (AAG) and localities in and around it. 1—Siliceous Priabonian, southern Carnarvon Basin (Haig and Mory, 2003). 2—Lutetian carbonate, Deep Sea Drilling Project (DSDP) Site 264, Naturaliste Plateau (McGowran, 1977, 1978a). 3—Bremer Basin (Gammon et al., 2003a). 4—Eucla Basin, western estuaries (de Broekert and Sandiford, 2005; Clarke et al., 1996). 5—Eucla Basin Eyre #1 and Gambanga #1 (Fig. 9). 6—Ocean Drilling Program (ODP) 182 sites and Jerboa #1 (Li et al., 2003a, 2003b). 7—Potoroo #1 (Fig. 9). 8—Eucla Basin and eastern estuaries (Hou et al., 2003, 2006). 9—Marine surveying and dredging (Davies et al., 1989; McGowran, 1991). 10, 11—St. Vincent Basin, eastern and northern (Alley and Lindsay, 1995; McGowran and Alley, 2008). 12, 13—Western Murray Basin (Alley and Lindsay, 1995). 14—Gambier Platform, western Otway Basin (Li et al., 2000). 15—Otway Basin (Gallagher and Holdgate, 2000; Holdgate and Gallagher, 2003). 16—ODP Site 1168, west of Tasmania. 17—ODP Site 1172, East Tasman Plateau (Exon et al., 2004a, 2004b; Warnaar, 2006). 18—ODP Sites 1169, 1170, South Tasman Rise. 19—ODP Site 1171, South Tasman Rise (Exon et al., 2004a, 2004b; Pekar et al., 2005). 20—Latrobe Valley and western Gippsland Basin (Holdgate, 2003). General stratigraphy-centered reviews: McGowran et al. (2004), Holdgate and Clarke (2000).

BIOSTRATIGRAPHIC AND CHRONOSTRATIGRAPHIC FRAMEWORK FOR THE AUSTRALO-ANTARCTIC GULF

Foraminiferal Biostratigraphy

In recent decades, foraminiferal biostratigraphy has driven correlation and age determination (McGowran et al., 2004), supported and sometimes contradicted by nannofossil biostratigraphy (Shafik, 1981, 1982, 1993, 1995; Shafik and Idnurm, 1997; Waghorn, 1989; Kamp et al., 1990). The most important recent contribution to biostratigraphy has been the dinocyst biostratigraphy of Leg 189 integrated with magnetostratigraphy (Brinkhuis et al., 2003a, 2003b). Extension of correlations and age control into marginal marine and terrestrial facies has relied on palynology, which, however, has proven rather refractory to advances in resolution and precision (Partridge and Dettmann, 2003). At high southern latitudes well to the west of the Gulf, Huber and Quillévéré (2005) revised the foraminiferal-biostratigraphic and magnetostratigraphic succession. To the east of the gulf in New Zealand, which has a much richer tradition of Cenozoic biostratigraphy than does southern Australia, Morgans et al. (2004) updated and calibrated the Paleogene. Both compilations are tied to the paleomagnetic time scale, which is the backbone of modern Cenozoic geochronology, but has not been, until recently, significant in AAG geochronology. To clarify the discussion here, Figure 5 displays some of the lithostratigraphic nomenclature in southern Australia.

Figure 6 displays the geochronological framework for the northern shores of the AAG. The growth of our understanding of the foraminiferal-biostratigraphic succession has been very similar to the growth of the classical stage succession in NW Europe—scattered outcrops of mostly neritic and commonly inner neritic facies necessitating a piecemeal accumulation and ordination of events (Parr, 1947; Glaessner, 1951; Carter, 1958a, 1958b; Wade, 1964; Ludbrook and Lindsay, 1969; McGowran

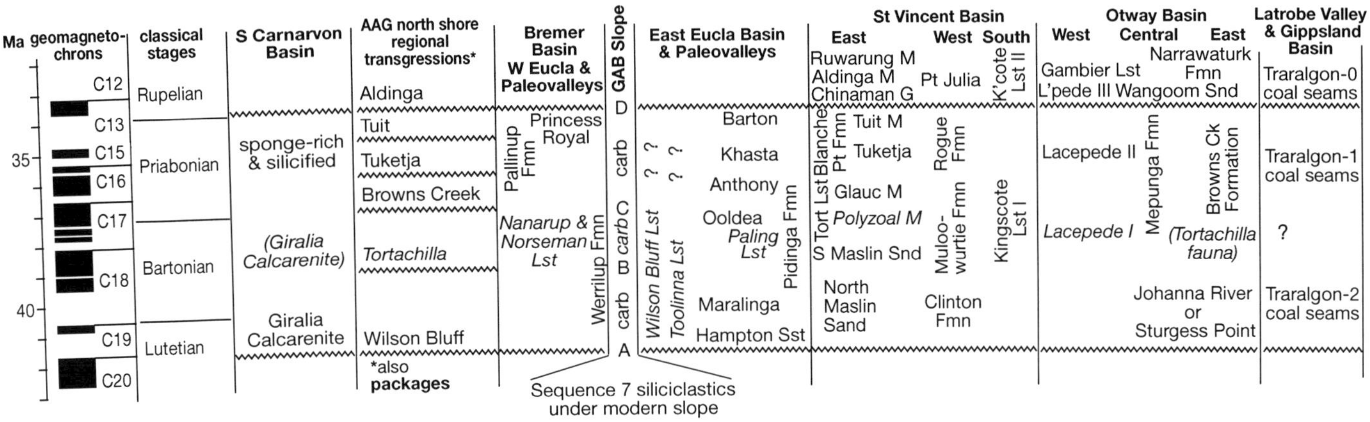

Figure 5. Lithostratigraphic nomenclature and lithologies for southern Australia—the north and east flank of the Australo-Antarctic Gulf (AAG)—in a rough guide to units mentioned in text. Sources are as mentioned in text and for Figure 4. The carbonates of the Tortachilla package are italicized to highlight the most extensive Bartonian spread of neritic carbonates just prior to the onset of the carbonate-poor environments of the Priabonian. GAB—Great Australian Bight; Fmn—Formation; Lst—limestone; Sst—sandstone; Snd—sand; M—member. K'cote—Kingscote; L'pede—Lacepede; G—Gully; Tort—Tortachilla; Pt—Point; Glauc—glauconitic. East Eucla Basin: query marks indicate doubt as to securely dated carbonates in the Priabonian.

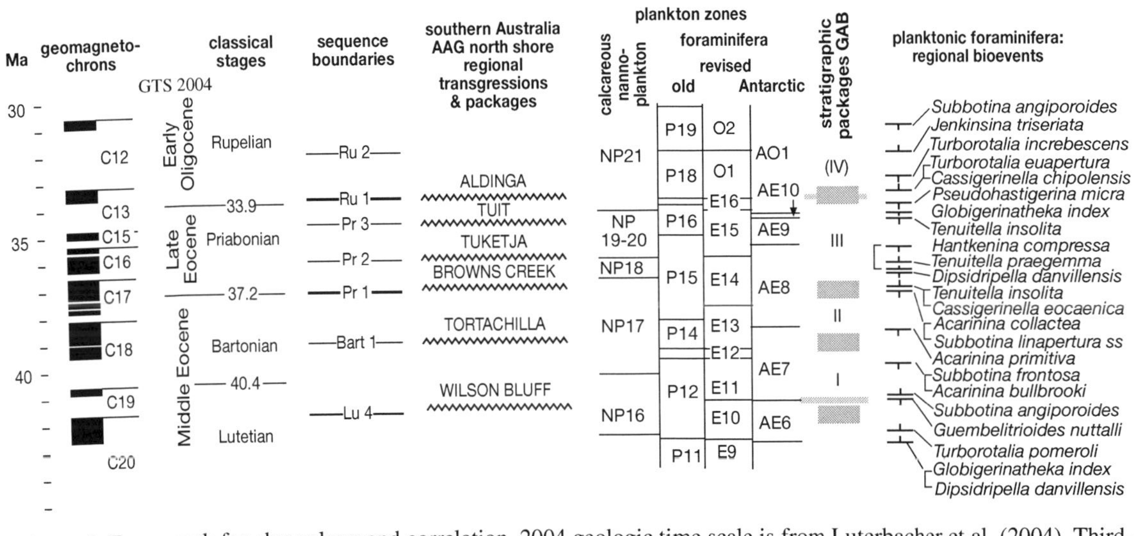

Figure 6. Framework for chronology and correlation. 2004 geologic time scale is from Luterbacher et al. (2004). Third-order sequences are from Hardenbol et al. (1998). Biostratigraphic framework for low-latitude and high-austral oceanic facies is from Berggren and Pearson (2005) and Huber and Quillévéré (2005). Regional bioevents in planktonic foraminifera are from numerous sources, including Li et al. (2003a, 2003b), McGowran et al. (2004), and Pearson et al. (2006). Stratigraphic packages in Great Australian Bight Basin (GAB) are from Li et al. (2003a).

et al., 1971; McGowran, 1978a, 1989a; Lindsay and McGowran, 1986; Li et al., 2003a, 2003b; McGowran et al., 1997a, 2004). Biozonation was attempted (e.g., Ludbrook and Lindsay, 1969), but the Eocene component was unstable and did not prosper; efforts were deemed better invested in the ordination of events and their attempted correlation with the low-latitude "standard" zones (McGowran, 1978a). Unfortunately, this permitted the inference that we actually had the standard zones in the AAG, which was largely untrue.

Meanwhile, Cenozoic regional stages (Singleton, 1941) also did not prosper, although the most valuable survey of the molluscan succession was set in terms of regional stages (Darragh, 1985). In the present context, the Aldingan Stage is an unfortunate example of how regional stages fall into disrepair and disuse. The Aldingan Stage was deemed unusable by Ludbrook and Lindsay (1966) and revised by them to extend from the base of the Tortachilla Limestone to the top of the Aldinga Member of the Port Willunga Formation in the St. Vincent Basin (Fig. 5).

In present-day terms, this means that the stage extends from the maximum flooding surface of Bart-1, across four third-order unconformities and sequence boundaries and the Eocene-Oligocene boundary and the Oi-1 downcut, to the Ru-1 maximum flooding surface. There is no coherent or imaginable use for the Aldingan Stage that would not obfuscate any discourse beyond recovery; nor has anyone seen fit to repair or replace the unfortunate entity.

Marine Transgressions, Stratal Packages, and Third-Order Sequences

Yet another approach was made during the 1970–1980s, not coincidentally, during the gestation of sequence stratigraphy: a pattern of regional transgressions emerged that seemed not to be broadly diachronous, at least not within the chronological resolutions available (McGowran, 1989a). The regional transgressions have proved to be useful—always with the proviso that their near-synchrony is not to be taken for granted. Four developments have arisen from scrutiny of the regional transgressions—allostratigraphic packaging, third-order sequences, ecostratigraphy, and biogeography and species turnover.

As to allostratigraphic packaging (e.g., Li et al., 2003a), a sustained stratigraphic integration could well produce rigorous and useful regional stages from such packages (that it has not done is due largely to the lack of intense interest, economic or scientific, in refining the Cenozoic stratigraphic record of tectonically relatively tranquil southern Australia, in contrast to turbulent New Zealand, where regional geology and mapping are hugely indebted to biostratigraphy).

As to the obvious comparison with the putatively global third-order "sequences" (McGowran et al., 1992), the present situation on the northern margin of the AAG, discussed later, consists of six sequences, Lutetian-4 to Rupelian-1; there are six recognized and named regional marine transgressions, Wilson Bluff to Aldinga; and there is enough support for approximate correlation to render anything but a tight one-to-one correlation quite unlikely. (I add a chastened footnote of heuristic, Popperian conjecture and refutation here. There were five transgressions but six sequence boundaries, a mismatch driving me back to the evidence and forcing the acknowledgment that I had missed a transgression, now named the Browns Creek.)

The third outcome from the pattern of regional transgressions is ecostratigraphic. The section exposed at Maslin and Aldinga Bays in the St. Vincent Basin (Fig. 7) displays fluctuations of tens of percent between the dominating groups of the benthic foraminifera, the epifaunal Cibicididae and infaunal Uvigerinidae plus Bolivinidae in the late Eocene, and a pronounced drop in the infauna in the early Oligocene (McGowran and Beecroft, 1986a; McGowran et al., 1992). The strongest fluctuations occur in the highly siliceous Blanche Point Formation, and they invite comparison and hopefully correlation with other sections, as discussed here later. It would seem that the packages based on marine transgressions and ecostratigraphy work, stratigraphically (McGowran et al., 1992, 2004), and it would seem too that they give powerful regional confirmation to the third-order sequences. However, there are ambiguities and uncertainties to be discussed.

The fourth outcome from the pattern of regional transgressions is biogeographic and speciational (McGowran and Beecroft,

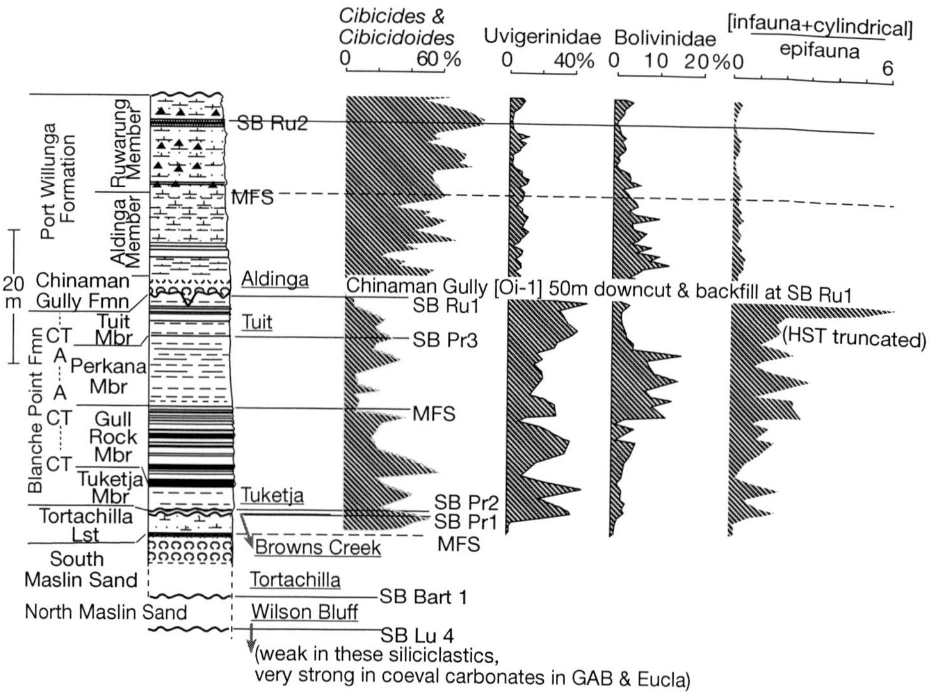

Figure 7. Bartonian-Rupelian section at Maslin Bay and Aldinga Bay, eastern St. Vincent Basin, the most thoroughly studied section in the AAG (but lacking geomagnetics). This figure was modified after McGowran and Beecroft (1986a) and McGowran et al. (2004). SB—sequence boundary, labeled after Hardenbol et al. (1998). Wilson Bluff transgression is not to scale; in this basin, the package consists of the North Maslin Sand with brackish patches, but also the rich Maslin Bay flora. The Browns Creek package (see Fig. 10) is reduced here to a few centimeters' thickness. At left: silica mineralogy in shorthand: CT—opal-CT in transgressive tract; A—opal-A in highstand (HST); MFS—maximum flooding surface. GAB—Great Australian Bight Basin.

1986b; Moss and McGowran, 1993, 2003). The major change in benthic foraminiferal species overturn is at the Eocene-Oligocene boundary, which is also the level of biogeographic contraction in the regional neritic record, a contraction that reverses during the Oligocene development of faunas. Biogeography is discussed more broadly in the next section.

Biogeography and Marginal and Terrestrial Palynobiostratigraphy and Paleomagnetics

Terrestrial palynomorph successions have long since been correlated with the marine record (McGowran et al., 1971). The zoning of the southeastern Australian sequences (Stover and Partridge, 1973; Stover and Evans, 1973) has been tested for many years, but little advance has been made in chronological resolution (Partridge and Dettmann, 2003). Palynological zonation commonly shows the lower *Nothofagidites asperus* zone in contact with the early Eocene *Proteacidites asperopolus* zone (e.g., Partridge and Dettmann, 2003), but it does not demonstrate an actual stratal record of 5 m.y. or more between constrained dates in the two zones. The lower *N. asperus* zone is characterized by the incoming of abundant *Nothofagidites* spp., sometimes known as the *Nothofagus* revolution, in which the extant tropical subgenus *Brassospora* is prominent—clearly a major environmental and ecological event in the terrestrial realm and equally clearly deserving of stronger chronological constraints on its configuration.

A central theme of the present paper is the facies change (and its environmental implications) between the Bartonian carbonate-rich and the Priabonian carbonate-poor packages. However, the middle *Nothofagidites asperus* zone, which is defined by the range of *Triorites magnificus* and characterized by rather strong taxic overturn at or near its boundaries, and which is discernible along the north flank of the AAG to the far southwest in the Bremer Basin (Stover and Partridge, 1982), at present spans the Tortachilla to Tuit packages inclusive (Fig. 8). We need more clarity and resolution than that to distinguish Priabonian strata from Bartonian strata in terrestrial palynofacies. The lack of temporal resolution makes for considerable frustration in scrutinizing data-rich and comprehensive surveys, especially the major review by Macphail et al. (1994). A dinocyst (microplankton) succession closely parallels the pollen zones (Partridge and Dettmann, 2003).

ODP Leg 189 palynology has made four major contributions—(1) a Priabonian succession and correlation of dinocyst events, (2) Priabonian paleomagnetic calibration within the AAG, (3) Bartonian-Priabonian biogeographic contrasts within and outside the AAG at its eastern end, and (4) a glimpse for the first time of sediments of early middle Eocene age, just outside the AAG on the South Tasman Rise (ODP 1171) and East Tasman Rise (ODP 1172) (Brinkhuis et al., 2003a, 2003b; Sluijs et al., 2003; Röhl et al., 2004; Warnaar, 2006; Pekar et al., 2005.) Figure 7 displays the succession of bioevents and their paleomagnetic ties at Site 1168, the only within-AAG site.

Depending on one's viewpoint, biogeography has been most informative about paleoclimatic shifts, and conversely—but not in circularity, for consilience is in play—paleoclimatic shifts have been insightful on biogeography (e.g., McGowran et al., 1997b, 2000, 2004). In a major advance, rich dinocyst assemblages could be quantified environmentally in various ways (Sluijs et al., 2003; Röhl et al., 2004; Warnaar, 2006). Thus, the percent endemic and bipolar cysts is an endemic versus nonendemic (including cosmopolitan) ratio. Meanwhile, two major taxic groups are considered to be largely heterotrophic and autotrophic, respectively, so that high gonyaulacoid/peridinioid (G/P) values may signal relatively oligotrophic (and more offshore) conditions, and conversely, low G/P values may signal eutrophic (and more inshore) conditions under a brackish lid.

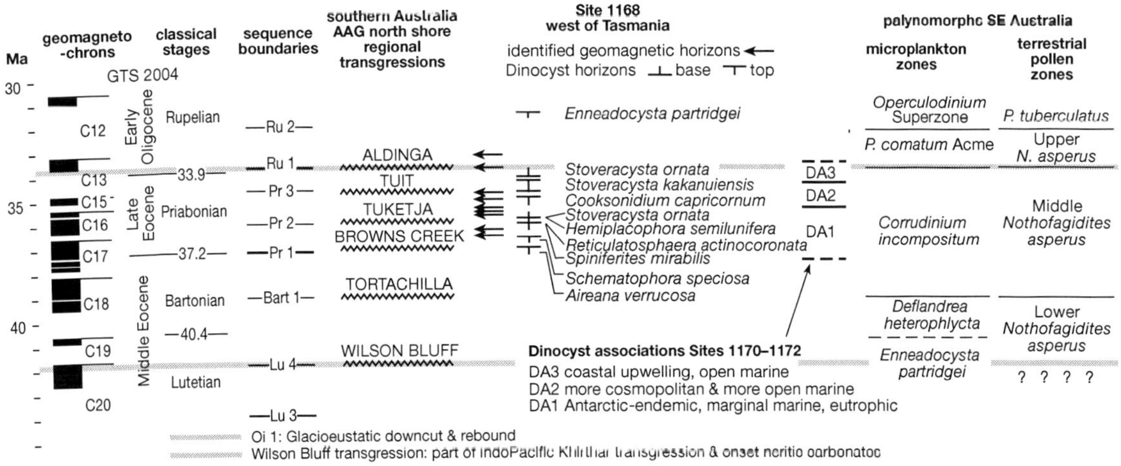

Figure 8. Bartonian-Rupelian palynobiostratigraphy. Far right—microplankton (dinocyst) and terrestrial pollen zones (from Partridge and Dettmann, 2003). Center—dinocyst events, geomagnetic calibration, and dinocyst facies succession and interpretation from Ocean Drilling Program (ODP) Site 189 (from Brinkhuis et al., 2003a, 2003b). (The latter is shown slightly differently in Figure 14.)

STRATIGRAPHIC CONFIGURATIONS, NORTH FLANK OF AUSTRALO-ANTARCTIC GULF

Lutetian-Bartonian Correlations

Before the Khirthar Transgression

Biostratigraphic investigations of all facies revealed a break in the succession of microfossil assemblages between late early Eocene and late middle Eocene, the so-called "Lutetian gap" through the AAG (McGowran et al., 2004). Within the AAG, in the Great Australian Bight the oldest sediments drilled in ODP 182 are Sequence 7, siliciclastics perhaps of early middle Eocene age (Li et al., 2003a).

De Broekert and Sandiford (2005) emphasized the great extent of the "inset valleys" surrounding and encroaching into the Eucla and Great Australian Bight and draining from the cratons into the AAG, where much of the area was subsequently drowned by the Wilson Bluff transgression. (Inset valleys have previously been called deep leads, an old gold-prospecting term, and paleochannels.) Stimulating their formation was a lowered geomorphic base level and increased stream gradients during epeirogenic uplift, which (I would add) preceded the accelerated widening of the AAG by seafloor spreading.

Outside the AAG in the west, a pelagic carbonate section on the Naturaliste Plateau cuts out at about the onset of the Wilson Bluff neritic accumulation (McGowran, 1978a). At and near the head of the AAG in the southeast, there are two sections bridging the "Lutetian gap": ODP 1171 on the South Tasman Rise and ODP 1172 on the East Tasman Plateau (Exon et al., 2004a). At Site 1171, almost 500 m of carbonate-poor silts and clays accumulated in inner neritic environments during the Ypresian-Lutetian, mostly in water <50 m depth, and for the upper Lutetian part, in <20 m depth. Ages were controlled by dinocyst events and geomagnetics; however, Wei (2004) recorded the last occurrence of the nannofossil *Discoaster kuepperi* at Sites 1171 and 1172, giving a very early Lutetian age. The combination of sustained epeirogenic subsidence and sustained siliciclastic accumulation supports the de Broekert-Sandiford scenario in the far west.

Wilson Bluff Package

The base of the marine later Eocene in southern Australia was firmly dated by correlation with the planktonic foraminiferal zone P12 or E11 (in the terminology in Fig. 6) (McGowran and Lindsay, 1969). The assemblage occurs in Eyre #1 and Gambanga #1 (Fig. 9) at the base of the largest Paleogene body of extratropical neritic carbonates on the planet. Four decades later, including ODP and other offshore drilling in the bight (Fig. 8) and dredging of the shelf recovering calcareous samples (Davies et al., 1989; McGowran, 1991), this level remains the floor to later middle Eocene stratal accumulation in the AAG (Li et al., 2003a, 2003b); nor is there any earlier carbonate in the easterly parts of the AAG (Holdgate and Gallagher, 2003; Exon et al., 2004a).

It has also long been known that the onset of this sedimentary accumulation event was coeval with carbonates accumulating on the western and northern Australian continental margins (the latter in Papua New Guinea). On the western margin, Haig et al. (1997) and Haig and Mory (2003) confirmed that the Giralia Calcarenite begins at this level, and Haig and Mory emphasized the preceding gap in the record and major erosional episode, which is our Lutetian gap. In turn, these configurations are part of the still greater Khirthar transgression, the major event in the Indo-Pacific region (Nagappa, 1959; McGowran, 1977, 1978b). It follows that explanations for the onset of carbonate accumulation in merely regional terms will fall short, as in the intuitively attractive notion that the southern continental margin drifted into carbonate-friendly latitudes after Australia/Antarctica separated. (The same argument as applied widely to terrestrial-floral change and inferred paleoclimatic change is similarly inadequate.)

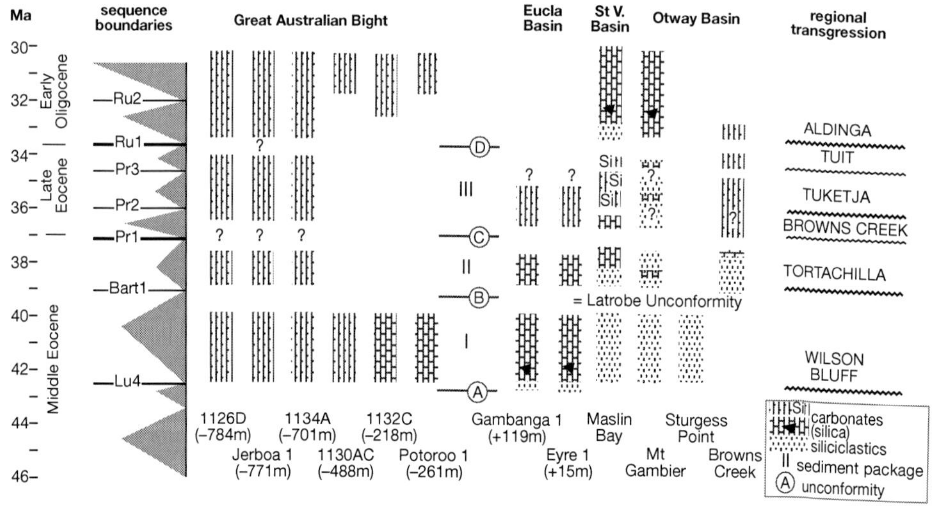

Figure 9. Stratigraphic columns illustrating widespread and meaningful hiatuses (adapted from Li et al., 2003a; McGowran et al., 2004). Hiatuses A–D bounding packages I–III were identified in the Ocean Drilling Program (ODP) program in the bight and are shown to correlate with land-based sections; non-identification of the Tuit package is due to sampling failure.

Another attractive notion (to this author) was the observation that there was an early middle Eocene carbonate section on the Naturaliste Plateau at the portal to the AAG (McGowran, 1978), although this section cuts out just as the AAG neritic carbonates (and Khirthar generally) begin, inducing notions of shelf/basin fractionation. To complete this scenario, perhaps the new oceanic spreading regime drove the transgression tectono-eustatically, although we do know too, from upward-deepening biofacies especially, but not only, in the Great Australian Bight (Li et al., 2003a), that passive-margin basins were undergoing rapid subsidence in response to the reinvigorated spreading.

The Wilson Bluff planktonic foraminiferal assemblage included *Globigerinatheka index* and others, *Acarinina primitiva*, *Acarinina bullbrooki*, *Subbotina frontosa*, *Planorotalites pseudoscitula*, *Turborotalia pomeroli*, *Dipsidripella danvillensis*, *Guembelitrioides nuttalli*, among others (McGowran and Lindsay, 1969; Fig. 3 *in* McGowran, 1989a; Li et al., 2003a, 2003b). It is a somewhat reduced version of the coeval assemblage in the Carnarvon Basin (Haig et al., 1997; Haig and Mory, 2003). In the "old" P zones, stable for many years, the lowest planktonic foraminiferal faunas fall within P12. In the Berggren and Pearson (2005) revision, the few specimens of *Guembelitrioides nuttalli* (previously *"Globigerinoides" higginsi*) become important because the top of this species defines the top of zone E10 and is reported to fall within chron C19r. In terms of the Antarctic zonation (Huber and Quillévéré, 2005), we are around the zone AE6/AE7 boundary, defined by first appearance of *Subbotina angiporoides*.

Benthos in the neritic realm included *Crespinella chapmani, Maslinella chapmani*; these species also occurred at the (neritic) base of the section in the ODP Great Australian Bight sites, but rapid subsidence forced upward-deepening and their replacement by bathyal species such as *Nuttallides truempyi* and *Oridorsalis umbonatus* (Li et al., 2003a).

Calcareous nannofossil biostratigraphy is consistent with this (Shafik, *in* Davies et al., 1989). The lowest assemblage contains *Cyclicargolithus reticulatus*, *Chiasmolithus grandis*, *C. solitus*, *Discoaster barbadiensis*, and *Neococcolithus dubius*. Relying on the absence of *Reticulofenestra scissura* (which then comes in), Shafik bracketed the assemblage by the two close events, base *C. reticulatus* and base *R. scissura*, and correlated this narrow interval with about the middle of the Martini zone NP16 or near the base of the Okada-Bukry zone CP14.

Wei (2004) unwittingly confirmed the Khirthar transgression and base of the Wilson Bluff package with calcareous nannofossil dates from the inner-neritic clays and silts in the ODP Sites 1170, 1171, and 1172 by identifying the first appearance of glauconite in each case between the first appearance of *Reticulofenestra reticulata* and *Reticulofenestra umbilicus*, virtually the same narrow interval at about the middle of zone NP16. He interpreted this event at around the chron C19-C18 boundary as the opening of the gateway to shallow-water circulation. Much more cogently, it is part of the Khirthar transgression.

Although we lack geomagnetic correlations, the biostratigraphic evidence consistently indicates indirect correlation of the onset of the Wilson Bluff transgression and package with chron C19r. This is just above the third-order sequence boundary Lu 4, as shown for example in the 2004 geological time scale (Luterbacher et al., 2004, their Figure 20.1), and it is close to the Lutetian-Bartonian boundary within zones NP16, E11, and lower AE7.

The Wilson Bluff transgression was accompanied by a warming of the surficial ocean, the main evidence for which is an expansion to higher latitudes of warm-water organisms, especially the oligotrophic larger benthic foraminifera. They island-hopped down the Ninetyeast Ridge and down the Western Australian margin to the Carnarvon Basin (*Asterocyclina, Discocyclina, Nummulites, Alveolina*; Haig and Mory, 2003), against the oceanic gyre, on an inferred proto–Leeuwin Current hugging the continental margin (McGowran et al., 1997b). The current entered the AAG, but these larger foraminifera seemingly did not. This was the first of several warm incursions similar to the later middle Eocene, seen also in New Zealand in the Bortonian stage (e.g., McGowran et al., 2004, their Fig. 8). Wilson Bluff equivalents on the East Tasman Plateau are inner-neritic, carbonate-poor claystones that display strong fluctuations in the dinocyst G/P ratio. At the putatively oligotrophic peak falling within chron C18r, the %(endemic + bipolar) dinocysts decreases sharply, allowing "tentative" identification of MECO (Röhl et al., 2004, their Figure 14; Fig. 5 *in* Warnaar, 2006), but since we now have the warm peak in three facies, with two different microfossil groups and oxygen isotopes, all within the same identified or correlated geomagnetic interval, the consilience is strong enough to cancel the "tentative" at these high southern latitudes.

Accordingly, we have solid supporting evidence for the postulated very warm and wet conditions in the environs of the AAG (Macphail et al., 1994). As it happens, the largest Cenozoic brown coal body on the planet is in the Latrobe Valley facing the Gippsland Basin (the Traralgon-2 body), and smaller but still substantial bodies occur within the AAG drainage (Holdgate and Clarke, 2000; Holdgate et al., 2000, 2009; Holdgate, 2003). Being in the lower *Nothofagidites asperus* zone, they are part of the Wilson Bluff package. At Site 1171 on the South Tasman Rise, MK Macphail (*in* Exon et al., 2004a) found the expected dominance of gymnosperms, especially Araucariaceae, in the early Eocene microfloras, indicating a warm, wet but weakly seasonal climate, which was confirmed by dinocyst indications of sea-surface temperatures. The subsequent increase in the (angiosperm) *Nothofagus* rain forest was taken to indicate that cooler and uniformly wet climates characterized the middle to late Eocene. Even so, the rain forests bordering the AAG were in the meso- to megathermal range at ~60°S (references in Hill [2004] and Martin [2006]).

Tortachilla Package

Carbonates belonging to this package are more extensive than those in the preceding package, which were restricted to the Eucla and GA Bight Basins—in this package, they extend into the Bremer, St. Vincent, Murray, and Otway Basins, and further from the shelf edge in the Eucla Basin. The planktonic foraminiferal assemblage includes *Globigerinatheka index*, *Acarinina collactea*, *A. primitiva* (rare), *Dipsidripella danvillensis*, *Subbotina linaperta*, and *Turborotalia increbescens*.

The warm-water, oligotrophic larger benthic foraminifera appear within the AAG for the first time on the inferred proto–Leeuwin Current. Although the phytosymbiotic orbitoid *Asterocyclina* reached the Bremer Basin, but no further into the AAG, warm- and shallow-water forms spread from the Bremer (Quilty, 1981) to the eastern Otway Basin—*Linderina glaessneri*, *Maslinella chapmani*, *Halkyardia bartrumi*, *Pseudopolymorphina carteri*, *Wadella rotaliformis*, *Sherbornina atkinsoni*, and *Crespinina kingscotensis*. The neritic carbonate biotas attained peak diversities in echinoids, mollusks, and bryozoans, among many others, especially as displayed in the Tortachilla Limestone in the St. Vincent Basin. As previously for MECO, this warm event is also confirmed by a sharp decrease in the %(endemic + bipolar) dinocysts as found in chron C18n.1n at ODP 1172 (Röhl et al., 2004, their Figure 14; Fig. 4 *in* Warnaar, 2006).

The Wilson Bluff–Tortachilla package boundary is controlled biostratigraphically in SE Australia, including the Gippsland Basin, and falls at the Lower-Middle *Nothofagidites asperus* zone boundary. It is also the locus of a gentle, short-lived but defined and widespread deformation feature giving rise to the Latrobe unconformity, sensu stricto (e.g., Holdgate and Gallagher, 2003; McGowran et al., 2004). In the Great Australian Bight, Li et al. (2003a) could trace a hiatus into deeper water; this hiatus is coeval with the Latrobe unconformity, inviting a common cause in short-lived tectonism.

Priabonian Correlations

Recognizing Browns Creek, Tuketja, and Tuit Packages

As observed already, fluctuations in relative abundance of benthic taxa can attain tens of percent. There are several nonquantified examples of this effect where it becomes a disjunct distribution. For example, Figure 10 of Lindsay (1985) shows three species, *Linderina glaessneri*, *Maslinella chapmani*, and *Halkyardia bartrumi*, as occurring in the Tortachilla and the Aldinga packages but not through the late Eocene. The same is seen in some planktonics, particularly *Globigerinatheka index*, *Cassigerinella eocaenica*, and *Chiloguembelina ototara*, which tend to alternate, probably in response to trophic resource continuum shifts in the mixed layer from more oligotrophic to more eutrophic, respectively (McGowran, 1987). In neritic facies, this Lazarus-like effect could be ascribed to neritic environmental effects and explains our tendency to rely on assemblages instead of ranges for age determination. However, two planktonic species, *Dipsidripella danvillensis* and *Turborotalia increbescens*, display the same disjunction above a bathyal water column in Jerboa #1 (McGowran, 1991, his Fig. 6). Furthermore, the general top of *D. danvillensis* (previously *aculeata*) in New Zealand and more widely in the Eocene ocean is at the lower level (McGowran et al., 2001). This is strong evidence that local explanations of neritic phenomena ignoring evidence from the pelagic realm will fall short (see following).

The backbone for Priabonian correlations in the AAG is shown in Figures 10 and 11. First (Fig. 10), the section at Browns Creek (Holdgate and Gallagher, 2003): the Tortachilla package is established in a very thin but unambiguous interval at the top of the Johanna River Formation based on the planktonic foraminifera shown together with *Linderina* and *Maslinella*. In the next unit, siliciclastics fine upward into *Spirocolpus* clays. The *Notostrea* greensand marks a new transgression and a succession of banded gray marls, which terminate at a poorly fossiliferous carbonaceous unit that marks the top of several abundant Eocene species. Note too the disjunct pattern of *Dipsidripella danvillensis*, seen also in the bathyal (see previous section). The Eocene species are still present in the recurring *Spirocolpus* clays and marls, which have been erosionally truncated.

There is enough lithofacies and biofacies/biostratigraphic evidence here both to pick third-order sequence boundaries and to correlate them with the "global" sequence boundaries Pr-1, Pr-2, and Pr-3. Correlation of Browns Creek with Maslin/Aldinga Bay 600 km away makes the case much stronger (Fig. 11), as follows.

The Tortachilla package is coeval and ecologically consistent between the sections. At Maslin Bay, the upper member of the Tortachilla Limestone (Reynolds, 1953; Buonaiuto, 1979; James and Bone, 2000) is quite different from the oligotrophic main unit: it is a very thin, glauconitic, *Spirocolpus*-rich facies, and its shift in the eutrophic direction is reinforced by a shift from a highly diverse bryozoan fauna to one dominated by *Celleporaria* (Hageman et al., 2003). It should have formational status; sequence, stage, and subseries boundaries pass between the two ecologically disparate "members," and the upper unit has much in common with the overlying facies, as James and Bone also observed. The *Spirocolpus* clays at Browns Creek represent its coeval and facies counterpart in a siliciclastically expanded section. The richly glauconitic Tuketja Member is likewise both ecologically equivalent to and coeval with the *Notostrea* greensand (*Hantkenina primitiva*, *Isthmolithus recurvus*, *Aturia clarkei*, and the *Pseudohastigerina micra* "spike"). The Maslin/Aldinga section then becomes richly siliceous (opal-CT in the transgressive tract; opal-A in the highstand tract) in an oceanographically restricted basin, whereas Browns Creek is open to the AAG (contrast the total plankton numbers) and consists of banded marls. *Globigerinatheka index* disappears at Aldinga Bay after peak "b" at level 1 but continues at Browns Creek up to level 2. The Tuit transgression is signaled by the return of opal-CT at Aldinga Bay and by the *Spirocolpus* facies in both sections, and it is confirmed by the surely coeval "spikes" in aragonitic-benthic Robertinacea and planktonic *Chiloguembelina ototara,* respectively. Some part of the Priabonian is missing at the top: less than half the last and briefest of the three third-order Priabonian sequences.

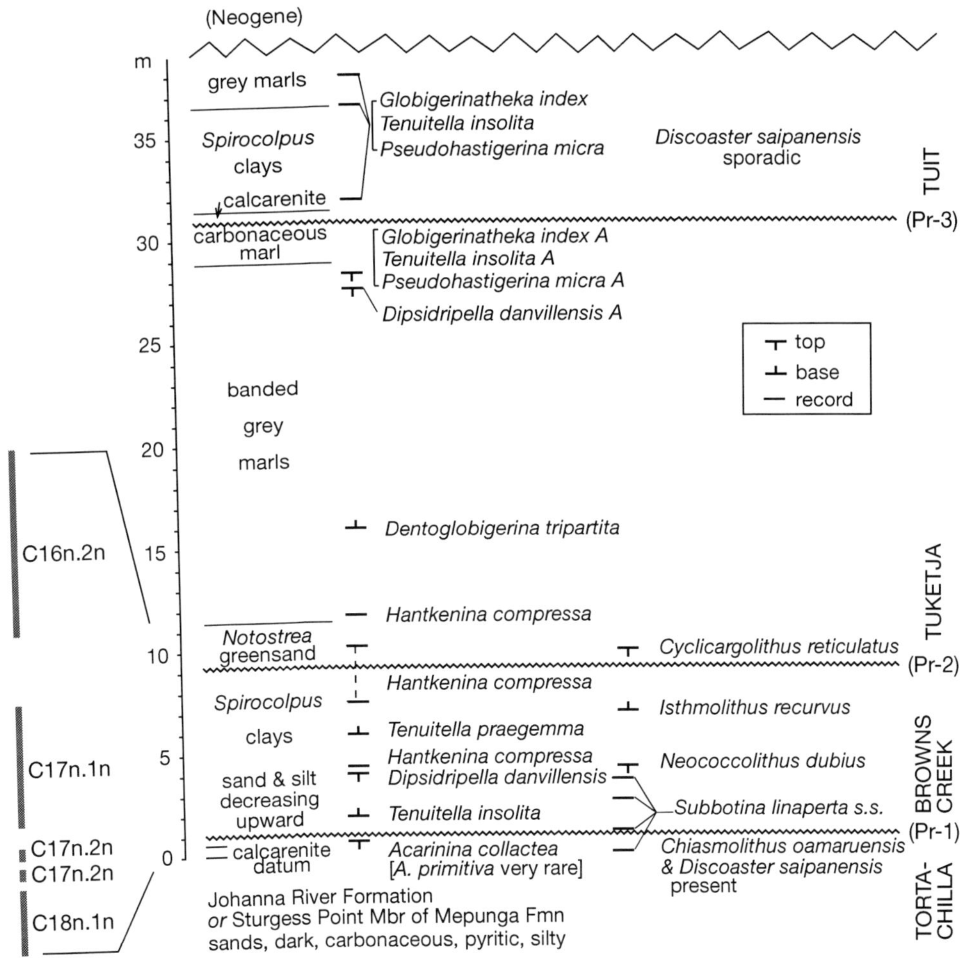

Figure 10. Biostratigraphic succession at Browns Creek, eastern Otway Basin. Calcareous nannofossils are after Shafik (1995). Planktonic foraminifera are from McGowran (1978a, 1989a), unpublished observations, and various sources. Geomagnetic succession and identifications at left are after Shafik and Idnurm (1997), who also presented an alternative interpretation (which they preferred) that was based on a very few scattered specimens of *Isthmolithus recurvus* well below the base shown here (and unexplained), and that entailed rejection of planktonic foraminiferal events (especially top *Acarinina collactea*) as having chronological value. A, abundant.

From this comparison: (1) there are parallels in lithofacies and biofacies between sections 600 km apart in the north neritic realm of the AAG; (2) three sequence boundaries are strongly identified and equally strongly matched with the putatively global sequence boundaries Pr-1, Pr-2 and Pr-3; and (3) whether as transgressions (McGowran, 1989a), packages (Li et al., 2003a), or third-order sequences (McGowran et al., 2004), these divisions can be used chronostratigraphically in integration with biostratigraphy and ecostratigraphy.

Carnarvon, Bremer, and Eucla Basin Stratigraphy

The limestones of the later Eocene were regarded as of late Eocene age, then, as worked progressed and the biostratigraphic significance of the acarininids *A. primitiva* and *A. collactea* was recognized, as of late-middle to late Eocene age. As the extent of other and especially siliceous facies became appreciated, the presence of late Eocene neritic carbonates became moot.

In the Carnarvon Basin, an isolated, weathered remnant of a siliceous unit apparently is younger than the Giralia Calcarenite (Haig and Mory, 2003). The relationship parallels the siliceous-above-carbonate pattern in the Bremer Basin. There, the Nanarup carbonates occur in the Tortachilla package (McGowran, 1989a), are overlain by the Pallinup, are rich in spongolites and spiculites, and carry a highly diverse assemblage of sponges (Gammon et al., 2003a, 2003b, 2000b; Pisera and Bitner, 2007). In paleovalleys flanking the western Eucla Basin, the limestone makes up the Norseman (Tortachilla transgression), and the silica is in the Princess Royal Spongolite (Tuketja transgression) (Clarke et al., 1996). The sedimentological model in these studies for the Priabonian involves silica-rich sediments inshore and calcareous sediments offshore. In the eastern Eucla Basin, Hou et al. (2003, 2006) produced a more detailed but plausible stratigraphic scenario wherein the furthest-reaching carbonate (Paling) is Tortachilla, and succeeding strata seem to fit the regional transgression model based primarily on the packages in the St. Vincent Basin. Here too, the perception is of offshore carbonates giving way to noncarbonates inshore.

The identification of late middle Eocene carbonates (Tortachilla) is secured primarily on the presence of the acarininids *A. primitiva* and *A. collactea*. The late Eocene is less secure, unless significant late Eocene species (Fig. 6) are identified: taking the presence of *Globigerinatheka index* in the absence of *Acarinina* as late Eocene is inferior as evidence to the actual presence of late Eocene species. It is also noteworthy that most of

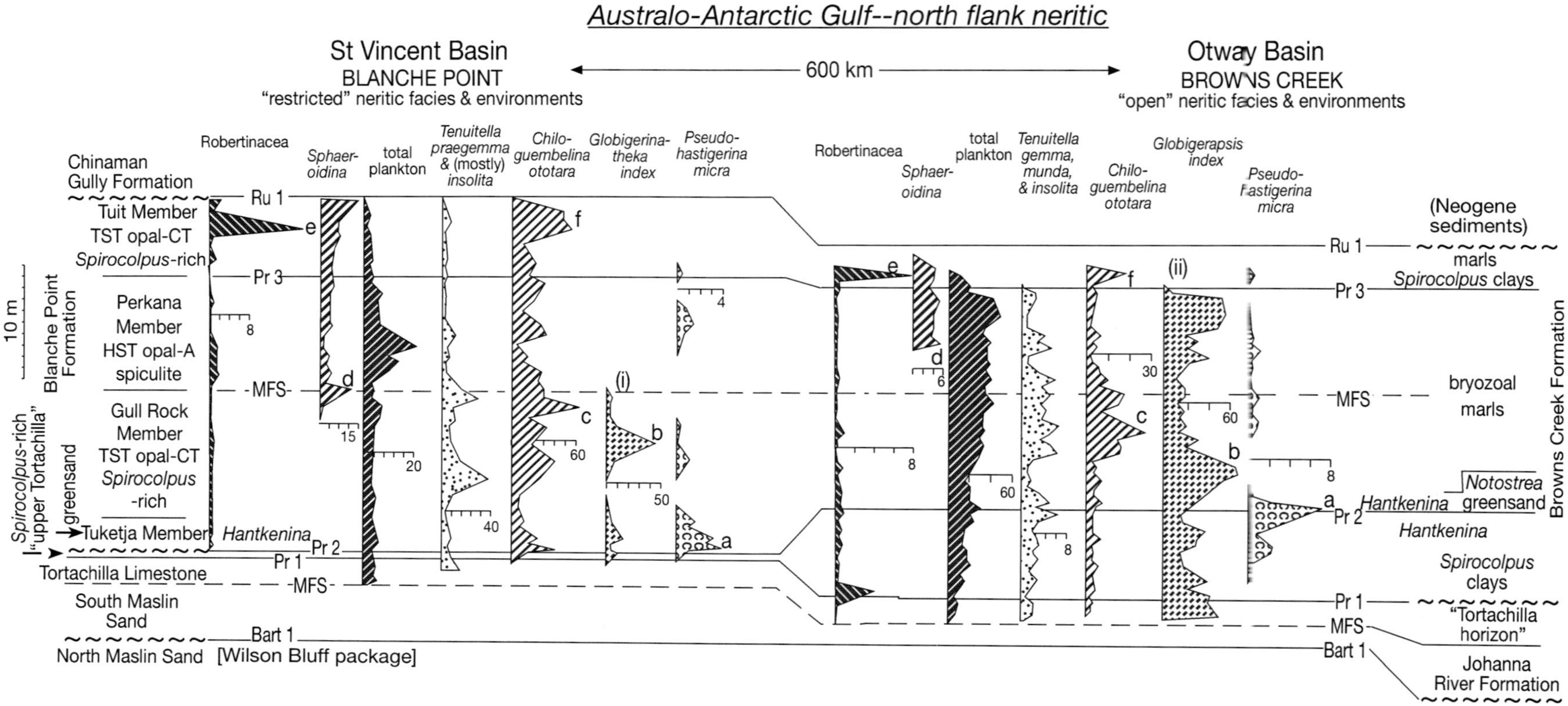

Figure 11. Comparison of biofacies succession at Maslin-Aldinga Bays and Browns Creek, 600 km apart (10 and 15 on Fig. 4), both in the north neritic area of the Australo-Antarctic Gulf, the former in a basin of relatively restricted circulation and a poor planktonic microfossil record, the latter in more open conditions and a better record. This diagram emphasizes strong parallels in biofacies or ecostratigraphy and in interruptions defining packages. Note changes in scale in the relative abundances of foraminifera. Points a to f suggest coeval ecoshifts. Points (i) and (ii) indicate allochronous top to *Globigerinatheka index*. The main difference is the rich opal-A/opal-CT alternation at Maslin Bay and its absence at Browns Creek. Although a strong tendency to dysaerobia under an estuarine circulatory tendency was very widespread in the Australo-Antarctic Gulf, silica fixation had a western tendency. This figure was modified after McGowran et al. (1992). HST—highstand systems tract; TST—transgressive systems tract; MFS maximum flooding surface.

the age-determined limestone is Tortachilla. Certainly, the neritic carbonate facies are much contracted in the Priabonian compared to the Bartonian; it is less certain whether there is any secure late Eocene carbonate to support the notion of coeval silica-carbonate facies change in the neritic record. Bathyal carbonates of Priabonian age are not questioned (McGowran, 1991; Li et al., 2003a), nor are the neritic marls at Browns Creek (Fig. 10). The offshore carbonates cited to support the coeval silica-carbonate pattern in the western Eucla Basin (Gammon et al., 2000) and in the St. Vincent Basin (James and Bone, 2000) may be Priabonian, but these correlations are not supported by evidence. Extensive removal of Priabonian strata during the Oi-1 fall in sea level (see following) was not clearly recognized until recently.

Eocene-Oligocene Boundary and Early Rupelian

Chinaman Gully Downcut

The St. Vincent Basin, centrally placed in the AAG neritic realm, has the clearest record of the Eocene-Oligocene boundary. The silica-rich and dysaerobic facies are cut through to ~50 m depth, and the backfill is made up of coarse to fine siliciclastics (Chinaman Gully Formation) on both margins of the basin. Removal of section at the same time by comparable amounts has been inferred for the western Otway Basin (Li et al., 2000). Coal swamps resumed their accumulation in several basins from the St. Vincent Basin to the southeast. To the west, no neritic or nonmarine strata have been recognized to the mouth of the AAG and beyond. The Chinaman Gully downcut and erosional surface are responsible for our generally patchy Priabonian record in southern Australia.

Aldinga Package

In the St. Vincent Basin, marine sedimentation soon resumed (the Aldinga transgression) in facies contrasting strongly with the latest Eocene. The contrast is seen most succinctly in the epifaunal/infaunal ratio of the neritic foraminifera (Fig. 7). Significant planktonic foraminifera (Lindsay, 1967, 1981, 1985; Lindsay and McGowran, 1986) include *Cassigerinalla eocaenica* (soon accompanied by *C. chipolensis*), *Dipsidripella danvillensis*, *Jenkinsina triseriata*, *Tenuitella praegemma*, *Turborotalia ampliapertura*, *T. increbescens*, *T. euapertura*, *Subbotina angiporoides*, "*S. linaperta*," and *Globoturborotalita labiacrasata.* (Sporadic records of *Globigerinatheka index*, given considerable weight by Lindsay, are now considered to have been reworked from the soft late Eocene strata during downcutting.) There is a brief but sharp warm interval, inferred to be a rebound after the Oi-1 chill, marked by the brief return of the warmer-water benthics of the Tortachilla assemblage: *Halkyardia*, *Linderina*, *Maslinella*, and *Crespinina*.

Neritic carbonate accumulation commenced over a wide area in the Otway Basin early in the early Oligocene, but it has not been dated with resolution and precision (Gallagher et al., 1999; Gallagher and Holdgate, 2000; Holdgate and Gallagher, 2003, their Figure 10.8; McGowran et al., 2004, their Fig. 10). The first carbonates have been studied in most detail in the western sector, the Gambier Embayment or Gambier Platform (Li et al., 2000), where the Gambier Limestone is one of the purest and temporally most persistent neritic carbonates facing the neritic Southern Ocean. With higher planktonic foraminiferal numbers than the Aldinga member, the Gambier Limestone contains the same suite plus *Globoturborotalita brevis*.

The base of the Gambier Limestone has the configuration of a "carbonate flood" across the shelf on the rebound from the Oi-1 fall in sea level. The critical interval is found only in the subsurface and has not been recovered at high-resolution standards, but it is apparent that carbonate accumulation in the St. Vincent Basin, the Gambier Platform, and Otway Basin was established by very early in chron C12 and that water depths of 50–100 m were attained and sustained quite rapidly. Bathyal calcareous facies of very early Oligocene age are known from the eastern Great Australian Bight but only from grab samples (McGowran, 1991).

Mentions thus far of glaciation Oi-1 have been confident in identifying the effects of same. An earlier attempt to locate the isotopic signature (Kamp et al., 1990) in SE Australia foundered in dating uncertainties. However, the isotopic signature of Oi-1 has been found, and constrained by paleomagnetic identification, in ODP 1128 in the Great Australian Bight (Mallinson et al., 2003a, 2003b). In this section, carbonate percentages increase across the Eocene-Oligocene boundary, the $\delta^{18}O$ values shift by about 2‰, and there is a 20-m-thick nannofossil chalk during Chron C13n.

The most striking facies change found in the ODP Leg 189 drilling sites is the onset of calcareous sedimentation (e.g., Exon et al., 2004, their Fig. 4). In each case, the Upper Eocene is marine but essentially noncalcareous; the Lower Oligocene is foraminiferal/nannofossil ooze/chalk. For the ODP 189 sites, Wei (2004) used calcareous nannofossils to date the change from siliciclastics to pelagic carbonates at ca. 31 Ma, high in chron C12r. The facies changes in oceanic drilling in the west and southeast could still be coeval with the Gambier flood on continental crust and may be regarded as the same event.

DISCUSSION

Summary of Stratigraphy and Generalizations

The following generalizations are pertinent (Fig. 12).

The stratigraphic record divides the upper Lutetian to lower Rupelian into six packages that match six putatively global third-order sequences. Carbonates are prominent in the Bartonian and Rupelian packages. The Wilson Bluff and Aldinga carbonates include chert-carbonate facies, but the Tortachilla, the most extensive carbonate, does not. In strong contrast, the biofacies and lithofacies of the three Priabonian packages, silts and silicas with minor carbonates, imply high fertility and dysaerobic conditions. Neritic carbonates taken as Priabonian may indeed be of that age, but their age is uncertain, and their extent assuredly is much less than in the Bartonian or Rupelian.

Marginal to the marine environments, extensive coal-forming swamps developed. The Traralgon coals in the Gippsland Basin

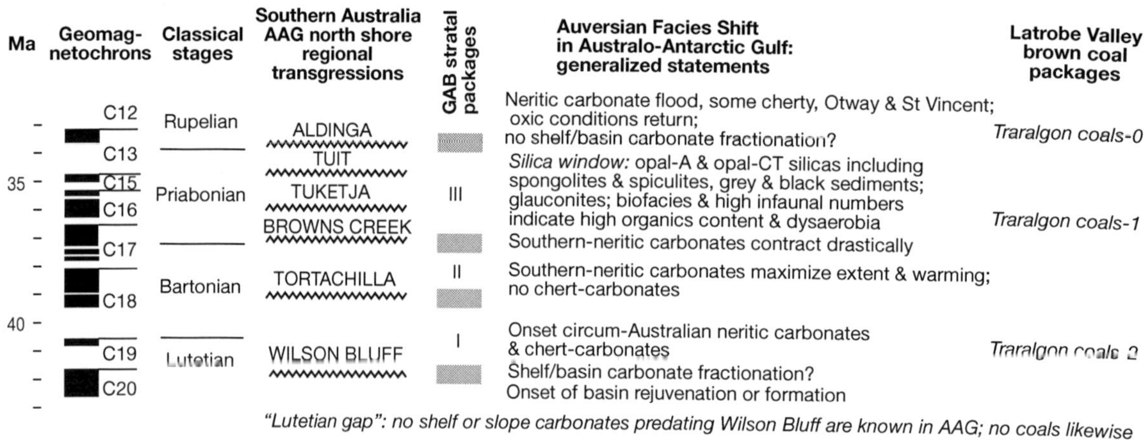

Figure 12. Summary statements capturing stratigraphic generalizations on the Australo-Antarctic Gulf. GAB—Great Australian Bight.

attain ~350 Gt. The Priabonian Traralgon-1 may be the biggest Cenozoic coal body on the planet. It may be that of the six packages, only the Tortachilla lacks substantial coals; correlations are not sufficiently rigorous to be more positive.

These strata together comprise a second-order cycle. The beginning is sudden, since sediments of early or mid-Lutetian age are not identified until ODP 189. The end of the cycle is the general regression and restricted record of marine strata in the mid-Oligocene, before the next major transgression in the Chattian. This cycle matches the second-order, Hardenbol cycle extending from the late Lutetian through the Rupelian.

The contrasts between the neritic *facies* of warm-ring western Tethys (Fig. 3) and the developing cold-ring AAG are obvious. However, equally apparent are the *stratigraphic* similarities seen in parallel fertility signals from the Bartonian into Priabonian and thence into the Rupelian.

Paleoceanographic and Paleoclimatic Context

MECO, CCD, and Bartonian Carbonates

It has been clear for three decades that the Khirthar transgression was a fundamental turning-point in southern Australian geohistory (e.g., McGowran, 1979) and that there was a pronounced movement poleward by warm-water organisms. There have been more general indicators of significant change around the Lutetian-Bartonian boundary, not so much in sequence-stratigraphic patterns, but rather in glimpses of climatic changes preceding the Oligocene cooling (e.g., Browning et al., 1996), and there is a long-recognized lack of signal in the oceanic record of this major change in the neritic record. (This was due in part to inadequate sampling of the oceanic record, but also to an absence of challenge from sequence stratigraphy.) However, then the transient MECO was discovered in the Southern Ocean (Bohaty and Zachos, 2003). The MECO lasted about half a million years in the lower part of chron C18 and in the lower part of zone AE7 (Huber and Quillévéré, 2005). It has been identified in the Northern Hemisphere (Jovane et al., 2007). Recent scrutiny indicates that the MECO builds up to a peak near the chron C18r-C18n boundary (Steven Bohaty, 2008, personal commun.).

The aforementioned lack of oceanic record had its origins in the shallow but poorly resolved trajectory of the CCD prior to the early Oligocene plunge. Resolution was poor because the Bartonian-Priabonian interval was poor in drilled oceanic sections, but more recent drilling has made it possible to reconstruct the later Eocene CCD in the eastern Pacific (Tripati et al., 2005). Two propositions dating from the 1970s have been reinvigorated: (1) Carbonates accumulate in the neritic realm at the expense of the pelagic realm, where the CCD shoals. Contracting shelf areas reverses the balance. (2) Contractions in the neritic realm by an episode of glaciation and rapid deepening of the CCD are linked. With high-quality sections through pelagic carbonates and high-resolution techniques, Tripati et al. could reconstruct CCD fluctuations to become proxies for transient glaciations, of which there are four in the initial study (Fig. 13).

With reconstructions of atmospheric carbon dioxide added, we have a scenario of rapid feedback and environmental shift. It becomes clearer as tools of chronological resolution, correlation, and age determination improve. Zachos et al. (2008) and Thomas (2008) captured this sense of impending breakthrough, but Thomas also warned that correlations are not yet robust enough to disentangle the feedbacks. Meanwhile, I consider some implications for the AAG (Figs. 13 and 14).

Tripati et al. (2005) concluded that the CCD steadily deepened, with early, transient glacials, as global cooling progressed through most of the Lutetian. Pekar et al. (2005) took sequence identification on the South Tasman Rise (ODP 1171) further: synchrony of their Ypresian and Lutetian sequence boundaries with determinations in the Northern Hemisphere implied a global control and glacioeustasy "in this supposedly ice-free world." The final plunge in chron C19 in the Tripati et al. reconstruction was ascribed to the most intensive of the transient glaciations. The spectacular reversal and shoaling of the CCD

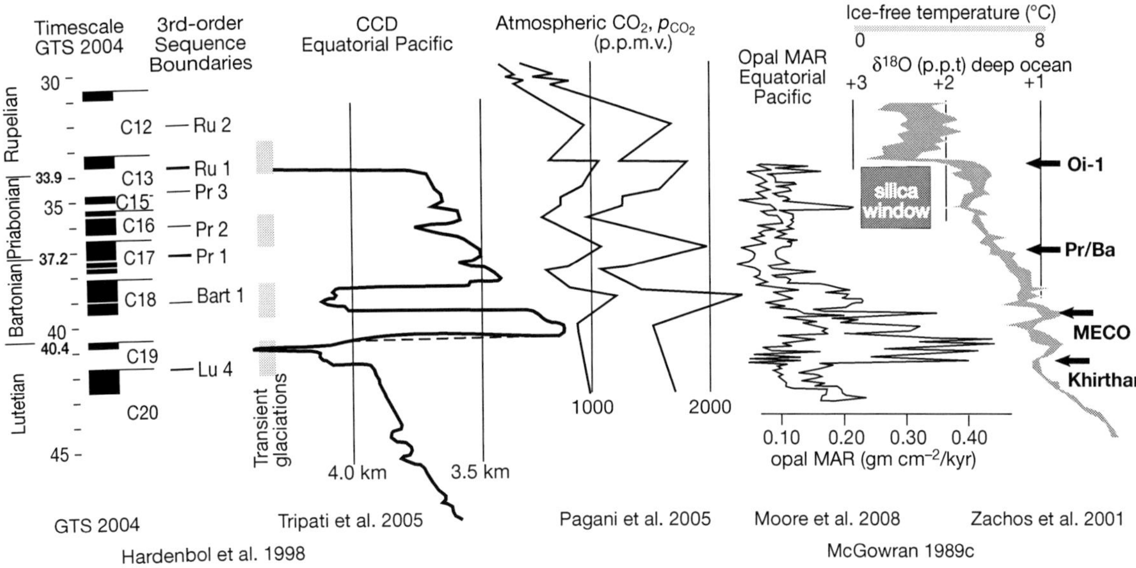

Figure 13. Auversian facies shift, Bartonian-Rupelian, in paleoceanographic setting. Reconstruction of calcite compensation depth (CCD), eastern Pacific, and inferred transient glaciations are from Tripati et al. (2005). Reconstruction of atmospheric CO_2 is from Zachos et al. (2008) after Pagani et al. (2005) (lines define error envelope). Biogenic opal, equatorial Pacific, is from Moore et al. (2008); MAR—mass accumulation rate. Silica window is from McGowran (1989c). Bottom water oxygen-isotopic curve was adapted from Zachos et al. (2001). Four gray bars emphasize in succession the Khirthar transgression, the MECO, the Bartonian-Priabonian boundary (Pr/Ba), which is a major environmental and facies shift in the AAG, and glaciation Oi-1.

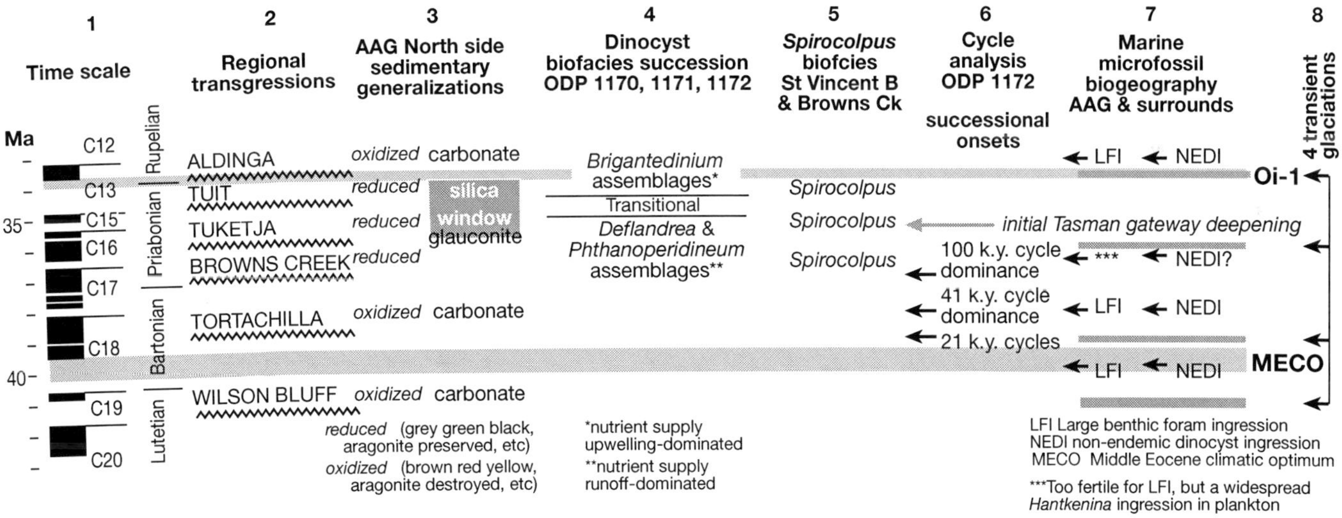

Figure 14. Environmental chronicle and scenario for the last days of the AAG. (1, 2) Time scale and regional packages and transgressions, from previous figures. (3) Characteristics of five stratigraphic packages simplified to carbonate-rich and oxidized on the one hand and carbonate-poor and reduced on the other. (4) Dinocyst biofacies succession in three (extra-AAG) holes (from Sluijs et al., 2003, 2005) (shown slightly differently in Fig. 8). (5) Three *Spirocolpus*-rich horizons at Maslin-Aldinga Bays, indicating high organics and inferred dysaerobia; the middle horizon is missing at Browns Creek, as is silica window. (6) Milankovitch cycle dominance changing across the Bartonian-Priabonian shift, Ocean Drilling Program (ODP) Site 1172 (just outside the AAG) (from Röhl et al., 2004). (7) Biogeographic horizons indicating warming. LFI—large benthic foraminiferal records argued to be real horizons, not preservational remnants of ranges (from McGowran et al., 2001, 2004). (Note absence from Priabonian—deemed too estuarine and fertile for these sophisticated oligotrophs.) NEDI—nonendemic dinocyst ingressions at ODP Site 1172 (from Röhl et al., 2004; Warnaar, 2006). (8) Four transient glaciations inferred from high-resolution analysis of oceanic carbonates (from Tripati et al., 2005) (Fig. 13). Gray arrow: horizon of initial deepening of Tasman gateway (Stickley et al., 2004), some 2 m.y. after the Priabonian shift to estuarine dysaerobia and some 2 m.y. before any cogent proto–Antarctic Circumpolar Current, thereby not supporting a simple link between tectonic shifts and biogeography.

happened at the time (approximately, at least) of the Khirthar transgression and the first coal swamps (since the Ypresian) and preceded the MECO. Perhaps, then, we may state that certain things happened in this order: (1) the glaciation culminated the long-term trend; (2) the Khirthar–Wilson Bluff transgression was forced tectono-eustatically, thereby (3) causing warming, terminating the glaciation, and forming abundant neritic space and carbonate-friendly environments; and (4) this reversal of cooling culminated in MECO.

Meanwhile, the accumulation of peats and carbonates had to affect the carbon dioxide balances in atmosphere and ocean in this "greenhouse" world, which still lacked the strong buffers that developed in the Neogene. The MECO did not last; there was a second strong plunge in the CCD and inferred transient glaciation, and a second strong rebound. Correlational uncertainties notwithstanding, the two shoalings appear to match with the Wilson Bluff and Tortachilla carbonate packages. One might speculate on cause and effect among shelf/basin carbonate fractionation, massive coal accumulation, and fluctuations in greenhouse-level atmospheric CO_2, but our correlations are not yet robust enough.

Mysterious Priabonian: AAG Climates and Terrestrial Floras

Through the Priabonian, the CCD deepens somewhat but at shallower levels than during the Lutetian, implying that shelf carbonates are more influential globally (if not in the AAG). The plunge in CCD at Oi-1 is confirmed, but it highlights the absence of Priabonian plunge and reversal. One feature still to be explained is the overall environmental contrast between the Bartonian and Priabonian both regionally, as in the AAG, and more generally, as in the Tethyan neritic record. There was a shift in the trophic resource continuum in the eutrophic direction. Trophic resource continuum theory would suggest a contraction at both ends in the Priabonian vis-à-vis the Bartonian—this is still to be clarified.

It has long been held with confidence that rain forests were extensive in southern Australia in the Paleocene and Eocene, with tendencies toward drying becoming apparent in the Oligocene. An ecological plant classification for these communities recognizes three thermal response groups (with mean air temperatures), megatherm (>24 °C), mesotherm (>14 °C, <20 °C), and microtherm (<12 °C), linked by intermediate categories. So long as there is not undue limiting of light and moisture, these categories correspond roughly to tropical, subtropical/warm-temperate, and cool/cold-temperate, respectively (e.g., Macphail et al., 1994). Surveys of the macrobotanical and palynological records agree on the points in the following paragraph. (Several good entries into the subject are available in Hill [1994, including several chapters], Hill [2004], Macphail et al. [1994], and Martin [2006].)

In the Paleocene, conditions were warming up and very wet, and there were rain forests in inland Australia. In very broad but broadly acceptable terms, there were wet-conifer forests in the Paleocene. The early Eocene can be characterized as the time of maximum development of nonseasonal megathermal rain forests in Australia. This was in the coastal and lowland areas, including the northern shores of the AAG, and this was the time of the EECO, of two or more hyperthermals (Zachos et al., 2008), and of pronounced deep weathering (e.g., McGowran, 1989b).

The second set of assemblages is usually lumped together as middle-late Eocene, the implication being a time of some 15 m.y., ca. 49–34 Ma, a large slice to generalize over. I repeat two comments. One, paleobotanists have tended to assume more or less continuity from the early Eocene, following palynology and industrial stratigraphy, whereas I have asserted that almost half of this time is unproven in the strata (until ODP 1171, as noted here previously). The Lutetian gap would tend to exaggerate the already strong contrast in the floras. Two, palynology does not resolve the four stratigraphic packages within the middle *N. asperus* zone, which bestrides the stage boundary. Even so, the middle-late Eocene can be characterized as the time of maximum development of nonseasonal mesothermal rain forest. The most striking feature of the floral assemblages is the rapid and widespread expansion of *Nothofagus*, mostly in the species group *Brassospora*. This is the "*Nothofagus* revolution."

The third set of assemblages is the latest Eocene–earliest Oligocene. There are indications that this was a biological turning point in the history of southern Australian vegetation (Macphail et al., 1994)—the onset of the "modern," as in other parts of the biosphere. Thus, we have a deep-time sweep of wet-conifer, megathermal, and mesothermal floral communities, broadly paralleling the global warming to the early Eocene, cooling somewhat into the middle and late Eocene, and finishing at the end of the Eocene with the start of something new. We need now to look more carefully at the major environmental changes within the "nonseasonal mesothermal" times. The key is to imagine a circulating body of water that can be at any scale from a puddle to an ocean, and that is prone to flipping from lagoonal (surface-water inflow, deep-water outflow) to estuarine (surface-water outflow, deep-water inflow) circulation. Development of a barrier would lead to hypersalinity (lagoonal or anti-estuarine) or anoxia (estuarine).

Modern circulation in southern Australia is lagoonal, driven by high evaporation rates compared to runoff from rain and rivers; with minor exceptions, marine environments are well aerated, and oxygen-minima zones are thin and weak, or absent, as is upwelling at the continental margin. The north shore of the AAG was very different when rainfall was very much higher. Surface-water outflow could well have formed brackish lids, seriously inhibiting gas exchange with marine waters and bottom muds. A microfossil record should preserve the ensuing biofacies patterns, the key ability of which would be to distinguish (even imperfectly) epifauna from infauna, as is well known in micropaleontology. Brackishness and dysaerobia strongly discourage the calcifying members of the normal marine neritic biota, but fertility levels and food burial would be high and inviting to opportunists to colonize in high numbers, such as the shallow-infaunal gastropod *Spirocolpus* and the burrowing crustacean *Thalassinoides*. However, siliceous sponges now enter the picture to the extent of becoming rock-forming, as spiculites and spongolites. McGowran and Beecroft (1986a) argued for the AAG (as named subsequently) and these notions of estuarine circulation, brackish water lid and stratification, and biofacies signals of dysaerobia in the Blanche Point Formation at the geographic scale of the St. Vincent Basin. A broader stratigraphic context was already clear—to the west, the spiculites and spongolites of the Eucla and Bremer Basins were coeval; to the east in New Zealand, there was the "deep-water" Oamaru Diatomite; and the concept of a chronologically defined silica window was extended to the eastern Pacific margin. All of these silicas could be constrained to within one radiolarian zone and two Exxon sequences (McGowran, 1989c). In present terms, the silica window is confined to two packages and sequences (Figs. 13 and 14).

This account distinguishes the seemingly well-ventilated Bartonian (and Rupelian) AAG from the Priabonian AAG with its problems, but we still need a reason the shift to estuarine mode occurred *within* the times of extensive mesotherm rain forest. Among other attributes, the AAG and its surrounds fostered the development of extensive swamps and the accretion of world-class coal bodies. The global benthic $\delta^{18}O$ curve offers a clue (Fig. 13). Inferred bottom temperatures decreased in the Lutetian, reversed and fluctuated at MECO, and then fell again in the Priabonian. Ice-free temperatures, post–MECO, were ~5–7 °C in the Bartonian and ~4–6 °C in the Priabonian. This, translated to a warm gulf at high southern latitudes, was far from a chill (with a chill's accompanying tendency to aridity): it was more likely to enhance the rain belts and the winter fog blankets. The high-latitude rainforests, adapted to these conditions (Hill, 2004), probably flourished even more and delivered still more nutrient to the estuarine gulf.

At ODP 1172 (ETP), a latest Bartonian section of pronounced condensation (and/or hiatus) truncates the Bartonian trend of increased Ca intensity and increased G/P ratio in dinocysts, which together give an onshore-to-offshore trend (within the neritic realm) (Röhl et al., 2004). The interval extends from the base of polarity zone C17n.1n to the top of C17n.3n. There was a return to shallow eutrophic conditions, including high dominance by *Deflandrea*, and Röhl et al. (2004) showed a well-marked Priabonian transgression at about the Browns Creek level. Meanwhile, the clay mineral suite, almost exclusively smectite through most of the middle Eocene (Robert, 2004), showed increasing amounts of kaolinite (as from the Tortachilla level) and suggested a still further increase in precipitation, weathering, and runoff due to climatic warming. It is worth recording too that cyclostratigraphic studies of high-frequency micropaleontological and geochemical data indicate shifts in dominance from 21 k.y. to 41 k.y. to 100 k.y. cycles, the latter of which was established by the early Priabonian (Fig. 14).

We come to the Priabonian dinocyst biofacies (Sluijs et al., 2003, 2005; Röhl et al., 2004). Biofacies and biogeography are overlapping categories; the former is emphasized here, and the latter is discussed in the next section. There is a consistent succession of assemblages in the three ODP sections 1170, 1171, and 1172, which span the time of initiation of the Tasman gateway (Fig. 14). The cysts of genera such as *Deflandrea*, *Vozzhennikovia*, and *Phthanoperidineum* were interpreted as representing shallow-marine, heterotrophic dinoflagellates, here indicating deltaic systems and organic-rich facies. Thus, the assemblages dominated by them were interpreted as responding to and thereby signaling a nutrient supply dominated by runoff. The wet *Nothofagus*-dominated forests were very close. *Nothofagus* is much more common in the Priabonian coals than in the Bartonian coals (Barbara Wagstaff, 2008, personal commun.; Holdgate et al., 2009). Blooms of *Brigantedineum* on the other hand are well known from upwelling regions. The *Brigantedineum* assemblage replacing the *Deflandrea-Vozzhennikovia-Phthanoperidineum* assemblage implies that upwelling-dominated nutrient sources replaced runoff-dominated sources. This dinocyst succession, reconstructed with the evolution of the Tasman gateway in mind, is also a spectacular biofacies signal of the decline of the AAG and its estuarine circulation late in the Priabonian.

Biogeography and Paleocirculation: Death of AAG and Birth of Southern Ocean

Thus far, I have referred to biogeography in a paleoclimatic context, namely, the penetration by tropical-type benthic foraminifera to the far southern reaches of the AAG north shore, indicating warming pulses in three packages, and peaks in the proportion of nonendemic dinocyst taxa, which imply much the same thing. The two quite independent lines of evidence seem to be congruent (Fig. 14) in what is commonly called an "integrated" or "multiproxy" approach, but is simply a powerful and satisfying *consilience* (McGowran, 2005, ch. 8). The forminifera-dinocyst consilience validates MECO in this region. The Maslin Bay flora in the St. Vincent Basin and several coeval floras are considered to be the warmest in southern Australia.

Now we come to scenarios of biogeography in paleogeography and the distinctive contribution from the dinocysts. The Tasman gateway hypothesis, now more than three decades old (Kennett et al., 1974; Kennett and Exon, 2004), begins with Australia/Antarctica separation leading in due course to deepening of the Tasman gateway and to thermal isolation. It is important that warm currents, principally the (proto) East Australia Current (EAC), maintain warmth at high latitudes. With the breakthrough and the onset of the Antarctic Circumpolar Current (ACC), the EAC would be pushed north.

Dinocyst biogeography challenges this scenario. Where chronological resolution obviously is essential for chronology and biofacies change, the ODP Leg 189 palynologists went the other way in assessing the two-part biogeographic problem of the AAG, namely, (1) distinguishing endemic elements, or floras, from cosmopolitan elements; and (2) using those insights to test the integration of the AAG into the Southern Ocean and notions of far-south circulation patterns (M. Huber et al., 2004; Warnaar, 2006).

Their point of departure was the general perception of a high-latitude assemblage of endemic and/or bipolar affinities, also known as the trans-Antarctic flora. Contrasted with this, there is the cosmopolitan flora, also known as tropical, or tropical/subtropical, or Tethyan. This nomenclatural plethora is subsumed in "non-endemic." Next, there is the insouciant assumption of lumping broad swathes of time, based on the impression that deep-ocean temperatures remained much the same during the middle and late Eocene—and that most of the cooling from the warm early Eocene to cool early Oligocene had actually occurred by middle Eocene times. (The identification of the MECO at Site 1172 is an important counterexample to lumping. Discussions, presented earlier herein, about the Bartonian-Priabonian contrasts do not support the assumption of stable temperatures. Even so, we can provisionally accept the ~10 m.y. of lumping.)

The dinocysts in the section at ODP Site 1168 are non-endemic. Almost all are warm-water types, and the Antarctic endemic taxa are conspicuously absent. In stark contrast, the dinocysts at Sites 1170, 1171, and 1172 are mostly Antarctic endemics or bipolar. More than 90% are in this category. Diatoms are conspicuously absent from 1168, whereas abundant, diverse, and well-preserved forms of typically shallow-water aspect are found at Sites 1170, 1171, and 1172.

These contrasts between two assemblages of dinocysts and diatoms, lumped through the sections, are at the core of the results. The addition of other data from the AAG and the southern oceanic regions reinforces the key generalization that there was a sustained marked difference between the surface waters of the AAG and the nearby SW Pacific. This contrast is consistent with a barrier lying between the AAG and the Ross Sea region. The biogeographic argument lying beyond this barrier argument is that the high endemism in the water mass of the Ross Sea region signals sources in the Antarctic region, not in the low-latitude western Pacific.

A point not discussed here is the extensive paleoclimatic-paleocirculation modeling conducted to test the biogeography, and the data supporting it (M. Huber et al., 2004; Warnaar, 2006; Huber and Nof, 2006)). Circulation in the AAG and the nascent Southern Ocean in the light of dinocyst and diatom biogeography is sketched in Figure 15 to contrast with the Tasman gateway hypothesis. For the latter, the key point, strong delivery of heat to Antarctica and its environs on the EAC in the

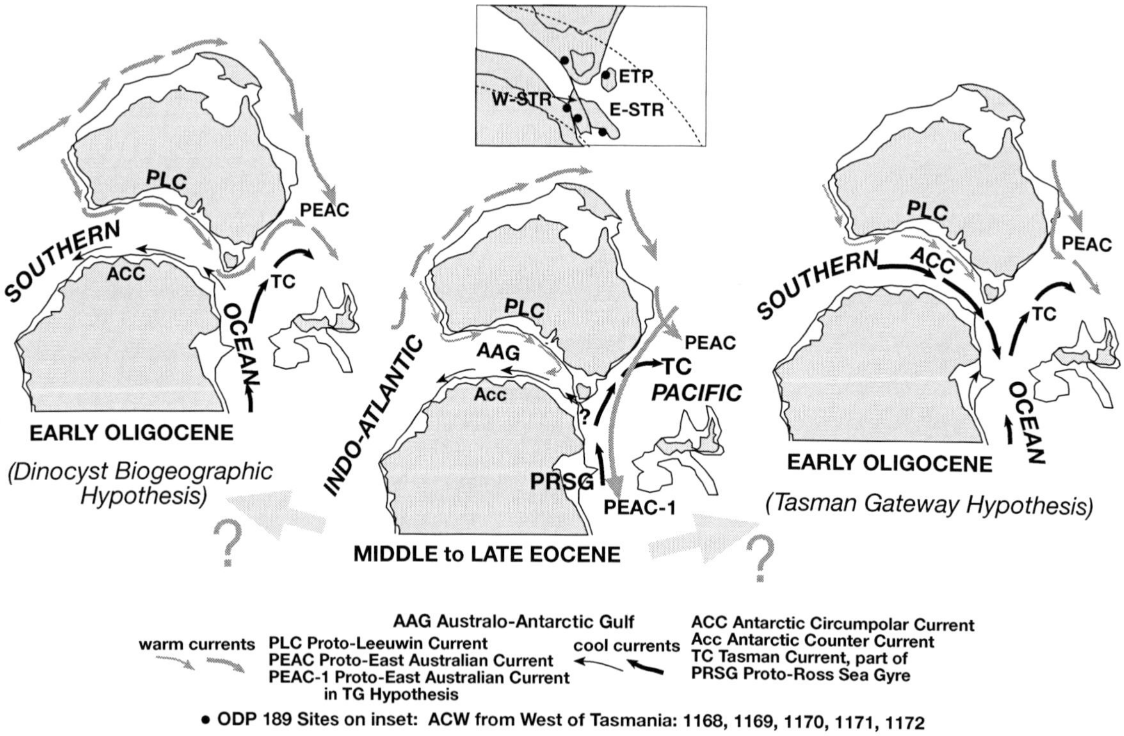

Figure 15. Paleogeographic scenarios of death of the AAG by absorption into the newly formed Southern Ocean. Central map: later Eocene, showing clockwise circulation in the AAG with a significant proto–Leeuwin Current (PLC), and two opposing views of a SW Pacific, *either* a strong Tasman Current (TC, black) as part of a cool proto–Ross Sea Gyre (PRSG), *or* a strong proto–East Australian Current (EAC) delivering heat virtually to Antarctica itself (PEAC-1, gray). Map on the right: Early Oligocene, the Tasman gateway hypothesis (Kennett et al., 1974; Kennett and Exon, 2004). Map on the left: Early Oligocene, the alternative, acknowledging the strong influence of high-latitude provincialism and biofacies in dinocysts and diatoms, here labeled the "Dinocyst biogeographic hypothesis" (i.e., Huber et al., 2004; Warnaar, 2006). Inset: ETP—East Tasman Plateau, W-STR and E-STR—South Tasman Rise, east and west blocks.

Eocene, is shown as an alternative to the cold Tasman Current. However, the main contrast is in the early Oligocene, where general acceptance of the "Dinocyst biogeographic hypothesis" will turn on the robustness and relevance of the biogeographic arguments. As things stand, this hypothesis has a richer biogeographic underpinning and explains rather more than does its rival. Barker et al. (2007) observed the conflicts in these Leg 189 scenarios and offered little hope of a reconciliation—with which one cannot but agree.

The very strong "nonendemic" presence in the dinocyst assemblages in the far east of the AAG gives the proto–Leeuwin Current, always a strongly biogeographic hypothesis, a still stronger transporting role. The current plays little part in the quarantine and chill scenario, and it even has been written out of existence under that sway (Gammon et al., 2003b).

Silica

Stratigraphically speaking, silica occurs broadly as follows. Oceanic silica in chert-carbonates peaks at the EECO and declines steeply through the Lutetian (along with progressive cooling), on the evidence of cherts in space and time in oceanic drilled sections (Muttoni and Kent, 2007). The more pertinent signals here are mass accumulation rates (MAR) of biogenic silica in the equatorial Pacific (Moore et al., 2008). There is a very sharp increase in MAR at the level of the Khirthar transgression in the late Lutetian (Fig. 13). Levels are high but fluctuating into the Bartonian (MECO), and Moore et al. (2008) identified a distinct and steplike decrease in the late Bartonian, just below the top of chron C18n.1n.

In the Wilson Bluff package, the AAG neritic and bathyal carbonates have cherty horizons. Chert-carbonates seem to be absent from the Tortachilla package. In the Priabonian, we have the quite different facies of the spongolites, spiculites, and opal-A/opal-CT rearrangements in the silica window. The Rupelian sees the return of neritic chert carbonates in the Aldinga package.

Oceanic and neritic chert carbonates are two sides of the coin of fertility, where silica acts more like a nutrient than like bicarbonate. The silica window in the AAG is different. It hugs the coasts in protected or barred locales, promoting the association with dysaerobia and grays and blacks in the sediments (McGowran and Beecroft, 1986a). Temperature has little to do with it, except in so far as the Priabonian was wetter, perhaps because it was cooler (doubtful). Gammon et al. (2000) promoted the contrast of Priabonian silicas with textbook situations of silica genesis in much cooler environments, but the important question here concerns the Bartonian-Priabonian contrast and the shift from a more oligotrophic situation to a more eutrophic situation. Our early ideas of estuarine circulation and brackish lids over limited marginal areas are strengthened, but the early problem of the silica window remains—local sedimentological, geochemical, and biofacies explanations do not address the wider picture, to which we now come.

CONCLUDING REMARKS

The late Lutetian to early Rupelian is a particularly critical slice of time in the transformation of the early Paleogene "greenhouse" times to the Neogene "icehouse" times. The long-term cooling of the deep oceans was interrupted briefly by MECO, and this event coincided with the Khirthar transgression and a major Bartonian episode of neritic carbonates in the Indo-Pacific region. This episode marks the beginning of the comprehensive stratigraphic record of the last ~10 m.y. of the AAG. Penecontemporaneously, world-class coal swamps developed, the biggest of which is in the Latrobe Valley in the Gippsland Basin, but is actually distributed along the AAG margin. The gymnosperm-dominated rain forests of the Paleocene and early Eocene were altered by the advent of angiosperms, especially the "southern beech" genus *Nothofagus*.

The Bartonian-Priabonian transition is marked by shrinkage and withdrawal of neritic carbonate facies and expansion of gray-green-black sediments with characteristic dominances by dysaerobic and/or high-nutrient biotas. The "extreme" cases are the silicas—spiculites, spongolites, and opal-A/opal-CT alternations, all of which occur in more marginal, barred, or otherwise protected environments. There is no reason to doubt that the AAG at 60+°S paleolatitude enjoyed warm temperatures, high rainfall, and deep chemical weathering of the hinterland. However, this leaves the Bartonian-Priabonian contrast unexplained. It has long been suggested that estuarine hydrographic conditions (brackish lid, surface-water outflow) were enhanced over areas of the AAG, going well beyond ad hoc explanations of the source and genesis of silicas. The key may be linked to some cooling into the Priabonian and still higher rainfall, supported by much higher *Nothofagus*, in the Priabonian than the Bartonian coals.

Is our problem of the Bartonian-Priabonian transition strictly regional? Prothero et al. (2003) and Prothero (2004) summarized changes in the neritic and terrestrial realms including, in North America, dense tropical forests being replaced by more open forests under less than half the rainfall—coeval with, but hardly consistent with, changes in the AAG. However, there is evidence from the deep ocean for shifts in ventilation and nutrient regimes. In Israel, Mediterranean Tethys, Barbieri et al. (2003) identified a benthic foraminiferal biofacies in basal Priabonian (zone NP18 and lowest NP19–20) bathyal chalks. In contrast to their Assemblages I and III, the intervening Assemblage II is marked by a very strong increase in infauna (from 40%–50% to 70%–80%), an increase in inferred "anaerobic" taxa, and a sharp decline in epifauna. This <2 m.y. interval is immediately above the long-known, deep-water ecostratigraphic event, the decline in *Nuttallides truempyi*, and immediately above the planktonic nutritional event, the extinction of the acarininids. Barbieri et al. (2003) attributed this pattern to oxygen depletion, pointing to a coeval blooming of bolivinids and uvigerinids on the northern side of Mediterranean Tethys (Nocchi et al., 1988). Barbieri et al. (2003, p. 466) concluded: "It is so far impossible to assess whether the faunal perturbation evidenced by Assemblage II originated solely from regional events (e.g., upwelling) or as a faunal response of the global paleoceanographic reorganization related to the changes in origin (high- or low-latitude sources) of deep waters throughout the middle Eocene." We can say now that (1) the latter is strongly favored, and the event is at the deep-benthic chronofaunal turnover (Fig. 2), and (2) the neritic and terrestrial realms are deeply engaged with the pelagic.

The conditions of poor ventilation and dysaerobia, high infaunal numbers, and inferred estuarine circulation in the dying (Priabonian) days of the AAG largely disappear at glaciation Oi-1 and into the Rupelian. Mallinson et al. (2003b) used the record of clay mineralogy in the Great Australian Bight (ODP 1128) to infer a transition from cool and wet conditions over Australia to cool and dry conditions during the Oi-1 glaciation. However, coals of substance still accumulated in the southeastern sector of the continental margin. This was the time of the subsidence of the barrier between the AAG and the SW Pacific. The neritic record has not played a major role in developing the dominating scenario of a proto–Circum-Antarctic Current and growth of a south-polar ice cap, but the record, including the role of the proto–Leeuwin Current, is more in accord with the "Dinocyst biogeographic hypothesis" than with the "Tasman gateway hypothesis."

However cogent our regional scenarios, they remain a subset of global changes in this time interval, for which I urge the adoption of Berger's Auversian facies shift. An example from the neritic of Tethys shows a wholesale increase in available nutrient from the Bartonian into the Priabonian. Coeval with the Bartonian-Priabonian shift in the AAG, and with the extinction of the photosymbiotic planktonic foraminifera in the pelagic realm, it demands a comprehensive explanation.

ACKNOWLEDGMENTS

I thank the convenors, Alessandro Montanari, Christian Koeberl, Frits Hilgen, and Rodolfo Coccioni, for the opportunity to attend the excellently organized and highly stimulating Penrose Conference at Monte Cònero (Ancona) on *The Late Eocene Earth*. Qianyu Li and Pat Quilty reviewed the manuscript encouragingly, carefully, and helpfully. I also thank Steven Bohaty, Aradhna Tripati, Barbara Wagstaff, Pat Quilty, Guy Holdgate, and Qianyu Li for answering questions and providing information. Christian Koeberl displayed editorial forbearance.

REFERENCES CITED

Alley, N.F., and Lindsay, J.M., 1995, Tertiary, *in* Drexel, J.F., and Preiss, W.V., eds., The Geology of South Australia. Volume 2. The Phanerozoic: Adelaide, South Australia Department of Mines and Energy Bulletin, 54, p. 150–217.

Barbieri, R., Benjamini, C., Monechi, S., and Reale, V., 2003, Stratigraphy and benthic foraminiferal events across the middle-late Eocene transition in western Negev, Israel, *in* Prothero, D.R., Ivany, L.C., and Nesbitt, E.A., eds., From Greenhouse to Icehouse: The Marine Eocene-Oligocene Transition, New York, Columbia University Press, p. 453–470.

Barker, P.F., Filipelli, G.M., Florindo, F., Martin, E.E., and Scher, H.D., 2007, Onset and role of the Antarctic Circumpolar Current: Deep-Sea Research, Part II–Topical Studies in Oceanography, v. 54, p. 2388–2398, doi: 10.1016/j.dsr2.2007.07.028.

Berger, W.H., 2007, Cenozoic cooling, Antarctic nutrient pump, and the evolution of whales: Deep-Sea Research, Part II–Topical Studies in Oceanography, v. 54, p. 2399–2421, doi: 10.1016/j.dsr2.2007.07.024.

Berger, W.H., and Wefer, G., 1996, Expeditions into the past: Paleoceanographic studies in the South Atlantic, *in* Wefer, G., Berger, W.H., Siedler, G. and Webb, D.J., eds., The South Atlantic: Present and Past Circulation: Berlin, Springer-Verlag, p. 363–410.

Berggren, W.A., and Pearson, P.N., 2005, A revised tropical to subtropical Paleogene planktonic foraminiferal zonation: Journal of Foraminiferal Research, v. 35, p. 279–298, doi: 10.2113/35.4.279.

Boersma, A., Premoli Silva, I., and Hallock, P., 1998, Trophic models for the well-mixed and poorly mixed warm oceans across the Paleocene/Epoch Epoch boundary, *in* Aubry, M.-P., Lucas, S., and Berggren, W.A., eds., Late Paleocene–Early Eocene Climatic and Biotic Events in the Marine and Terrestrial Records: New York, Columbia University Press, p. 204–213.

Bohaty, S.M., and Zachos, J.C., 2003, Significant Southern Ocean warming event in the late middle Eocene: Geology, v. 31, p. 1017–1020, doi: 10.1130/G19800.1.

Brinkhuis, H., Munsterman, D.K., Sengers, S., Sluijs, A., Warnaar, J., and Williams, G.L., 2003a, Late Eocene to Quaternary dinoflagellate cysts from ODP Site 1168, off western Tasmania, *in* Exon, N.F., Kennett, J.P., and Malone, M.J., et al., Proceedings of the Ocean Drilling Program, Scientific Results, Volume 189: College Station, Texas, Ocean Drilling Program, p. 1–36 (online), doi: 10.2973/odp.proc.sr.189.106.2003.

Brinkhuis, H., Sengers, S., Sluijs, A., Warnaar, J., and Williams, G.L., 2003b, Latest Cretaceous to earliest Oligocene, and Quaternary dinoflagellate cysts from ODP Site 1172, East Tasman Plateau, *in* Exon, N.F., Kennett, J.P., and Malone, M.J., et al., Proceedings of the Ocean Drilling Program, Scientific Results, Volume 189: College Station, Texas, Ocean Drilling Program, p. 1–48.

Browning, J.V., Miller, K.G., and Pak, D.K., 1996, Global implications of lower to middle Eocene sequence boundaries on the New Jersey coastal plain: The icehouse cometh: Geology, v. 24, p. 639–642, doi: 10.1130/0091-7613(1996)024<0639:GIOLTM>2.3.CO;2.

Buonaiuto, M.F., 1979, Late Eocene Molluscs and Related Composite Species from Southern Australia [Ph.D. thesis]: Adelaide, The University of Adelaide, 492 p.

Cande, S.C., and Stock, J.M., 2004, Cenozoic reconstructions of the Australia–New Zealand–South Pacific sector of Antarctica, *in* Exon, N.F., Kennett, J.P., and Malone, M.J., eds., The Cenozoic Southern Ocean: Tectonics, Sedimentation, and Climate Change between Australia and Antarctica: American Geophysical Union Geophysical Monograph 151, p. 5–18.

Carter, A.N., 1958a, Pelagic foraminifera in the Tertiary of Victoria: Geological Magazine, v. 95, p. 297–304.

Carter, A.N., 1958b, Tertiary foraminifera from the Aire District, Victoria: Geological Survey of Victoria Bulletin 55, p. 1–76.

Clarke, J.D.A., Bone, Y., and James, N.P., 1996, Cool-water carbonates in an Eocene paleoestuary, Norseman Formation, Western Australia: Sedimentary Geology, v. 101, p. 213–226.

Coxall, H.K., Wilson, P.A., Pälicke, H., Lear, C.H., and Backman, J., 2005, Rapid stepwise onset of Antarctic glaciation and deep calcite compensation in the Pacific Ocean: Nature, v. 433, p. 53–57, doi: 10.1038/nature03135.

Darragh, T.A., 1985, Molluscan biogeography and biostratigraphy of the Tertiary of southern Australia: Alcheringa, v. 9, p. 83–116.

Davies, H.L., Clarke, J.D.A., Stagg, H.M.J., Shafik, S., McGowran, B., Alley, N.F., and Willcox, J.B., 1989, Maastrichtian and younger sediments from the Great Australian Bight: Canberra, Bureau of Mineral Resources, Geology and Geophysics Report 288, 40 + 14 p.

de Broekert, P., and Sandiford, M., 2005, Buried inset-valleys in the eastern Yilgarn craton, Western Australia: Geomorphology, age, and allogenic control: The Journal of Geology, v. 113, p. 471–493, doi: 10.1086/430244.

Exon, N.F., Kennett, J.P., and Malone, M.J., 2004a, 1. Leg 189 synthesis: Cretaceous-Holocene history of the Tasman Gateway, *in* Exon, N.F., Kennett, J.P., and Malone, M.J. et al., Proceedings of the Ocean Drilling Program, Scientific Results, Volume 189: College Station, Texas, Ocean Drilling Program, p. 1–22.

Exon, N.F., Brinkhuis, H., Robert, C.M., Kennett, J.P., Hill, P.J., and Macphail, M.K., 2004b, Tectono-sedimentary history of uppermost Cretaceous through Oligocene sequences from the Tasmanian region, a temperate

Antarctic margin, *in* Exon, N.F., Kennett, J.P., and Malone, M.J., eds., The Cenozoic Southern Ocean: Tectonics, Sedimentation, and Climate Change between Australia and Antarctica: American Geophysical Union Geophysical Monograph 151, p. 319–344.

Gallagher, S.J., Jonasson, K., and Holdgate, G.R., 1999, Foraminiferal biofacies and palaeoenvironmental evolution of an Oligo-Miocene cool-water carbonate succession in the Otway Basin, southeast Australia: Journal of Micropalaeontology, v. 18, p. 143–168.

Gallagher, S.J., and Holdgate, G.R., 2000, The palaeogeographic and palaeoenvironmental evolution of a Paleogene mixed siliciclastic cool-water succession in the Otway Basin, southeast Australia: Palaeogeography, Palaeoclimatology, Palaeoecology, v. 156, p. 19–50, doi: 10.1016/S0031-0182(99)00130-3.

Gammon, P.R., James, N.P., and Pisera, A., 2000, Eocene spiculites and spongolites in southwestern Australia: Not deep, not polar, but shallow and warm: Geology, v. 28, p. 855–858, doi: 10.1130/0091-7613(2000)28<855:ESASIS>2.0.CO;2.

Gammon, P.R., James, N.P., Clarke, J.D.A., and Bone, Y., 2003a, Sedimentology and lithostratigraphy of Upper Eocene sponge-rich sediments, southern Western Australia: Australian Journal of Earth Sciences, v. 47, p. 1087–1103, doi: 10.1046/j.1440-0952.2000.00835.x.

Gammon, P.R., Bone, Y., Clarke, J.D.A., and James, N.P., 2003b, Upper Eocene greenhouse deposits in southern Australia: Paleoclimatology and paleoceanography, *in* Prothero D.R., Ivany L.C., and Nesbitt, E.A., eds., From Greenhouse to Icehouse: The Marine Eocene-Oligocene Transition: New York, Columbia University Press, p. 107–118.

Glaessner, M.F., 1951, Three foraminiferal zones in the Tertiary of Australia: Geological Magazine, v. 88, p. 273–283.

Gradstein, F.M., Ogg, J.G., and Smith, A.G., eds., 2004, A Geologic Time Scale 2004: Cambridge, Cambridge University Press, 589 p.

Hageman, S.J., Lukasik, J., McGowran, B., and Bone, Y., 2003, Paleoenvironmental significance of *Celleporaria* (Bryozoa) from modern and Tertiary cool-water carbonates of southern Australia: Palaios, v. 18, p. 510–527, doi: 10.1669/0883-1351(2003)018<0510:PSOCBF>2.0.CO;2.

Haig, D.W., and Mory, A.J., 2003, New record of siliceous, marine, later Eocene from Kalbarri, Western Australia: Journal of the Royal Society of Western Australia, v. 86, p. 107–113.

Haig, D.W., Smith, M., and Apthorpe, M.C., 1997, Middle Eocene foraminifera from the type Giralia Calcarenite, Gascoyne Platform, southern Carnarvon Basin, Western Australia: Alcheringa, v. 20, p. 41–68.

Hallock, P., 1987, Fluctuations in the trophic resource continuum: A factor in global diversity cycles?: Paleoceanography, v. 2, p. 457–471, doi: 10.1029/PA002i005p00457.

Hallock, P., Premoli Silva, I., and Boersma, A., 1991, Similarities between planktonic and larger foraminiferal evolutionary trends through Paleogene paleoceanographic changes: Palaeogeography, Palaeoclimatology, Palaeoecology, v. 83, p. 49–64, doi: 10.1016/0031-0182(91)90075-3.

Hardenbol, J., Thierry, J., Farley, M.B., Jacquin, T., de Graciansky, P.-C., and Vail, P.R., 1998, Mesozoic and Cenozoic sequence chronostratigraphic framework of European basins, *in* de Graciansky, P.-C., Hardenbol, J., Jacquin, T., and Vail, P.R., eds., Mesozoic and Cenozoic Sequence Stratigraphy of European Basins: SEPM (Society of Sedimentary Geology) Special Publication 60, p. 3–13.

Hill, R.S., ed., 1994, History of the Australian Vegetation: Cretaceous to Recent: Cambridge, Cambridge University Press, 433 p.

Hill, R.S., 2004, Origins of the southeastern Australian vegetation: Philosophical Transactions Royal Society of London, ser. B, v. 359, p. 1537–1549.

Holdgate, G.R., 2003, Coal: World-class energy reserves without limits, *in* Birch, W.D., ed., Geology of Victoria: Melbourne, Geological Society of Australia (Victoria Division) Special Publication 23, p. 489–516.

Holdgate, G.R., and Clarke, J.D.A., 2000, A review of Tertiary brown coal deposits in Australia—Their depositional factors and eustatic correlations: The American Association of Petroleum Geologists Bulletin, v. 84, p. 1129–1151.

Holdgate, G.R., and Gallagher, S.J., 2003, Tertiary: A period of transition to marine basin environments, *in* Birch, W.D., ed., Geology of Victoria: Melbourne, Geological Society of Australia (Victoria Division) Special Publication 23, p. 289–335.

Holdgate, G.R., Wallace, M., Gallagher, S.J., and Taylor, D.J., 2000, A review of the Traralgon Formation in the Gippsland Basin—A world class brown coal resource: International Journal of Coal Geology, v. 45, p. 55–84, doi: 10.1016/S0166-5162(00)00020-3.

Holdgate, G.R., McGowran, B., Fromhold, T., Wagstaff, B.E., Gallagher, S.J., Wallace, M.W., Sluiter, I.R.K., and Whitelaw, M., 2009, Eocene-Miocene carbon-isotope and floral record from brown coal seams in the Gippsland Basin of southeast Australia: Global and Planetary Change, v. 65, p. 89–103, doi: 10.1016/j.gloplacha.2008.11.001.

Hottinger, L., 1982, Larger foraminifera, giant cells with a historical background: Naturwissenschaft, v. 69, p. 361–371, doi: 10.1007/BF00396687.

Hottinger, L., 1997, Shallow benthic foraminiferal assemblages as signals for depth of their deposition and their limitations: Bulletin de la Societé Géologique de France, v. 168, no. 4, p. 491–505.

Hou, B., Frakes, L.A., Alley, N.F., and Clarke, J.D.A., 2003, Characteristics and evolution of the Tertiary palaeovalleys in the northwest Gawler craton, South Australia: Australian Journal of Earth Sciences, v. 50, p. 111–130.

Hou, B., Alley, N.F., Frakes, L.A., Stoian, L., and Cowley, W.M., 2006, Eocene stratigraphic succession in the Eucla Basin of South Australia and correlation to major regional sea-level events: Sedimentary Geology, v. 183, p. 297–319, doi: 10.1016/j.sedgeo.2005.10.007.

Huber, B.T., and Quillévéré, F., 2005, Revised Paleogene planktonic foraminiferal biozonation for the austral realm: Journal of Foraminiferal Research, v. 35, p. 299–314, doi: 10.2113/35.4.299.

Huber, M., and Nof, D., 2006, The ocean circulation in the Southern Hemisphere and its climatic impacts in the Eocene: Palaeogeography, Palaeoclimatology, Palaeoecology, v. 231, p. 9–28, doi: 10.1016/j.palaeo.2005.07.037.

Huber, M., Brinkhuis, H., Stickley, C.E., Döös, K., Sluijs, A., Warnaar, J., Schellenberg, S.A., and Williams, G.L., 2004, Eocene circulation of the Southern Ocean: Was Antarctica kept warm by subtropical waters?: Paleoceanography, v. 19, p. PA4026, doi: 10.1029/2004PA001014.

James, N.P., and Bone, Y., 2000, Eocene cool-water carbonate and biosiliceous sedimentation dynamics, St. Vincent Basin, South Australia: Sedimentology, v. 47, p. 761–786, doi: 10.1046/j.1365-3091.2000.00315.x.

Jones, J.B., and Fitzgerald, M.J., 1984, Extensive volcanism associated with the separation of Australia and Antarctica: Science, v. 226, p. 346–348, doi: 10.1126/science.226.4672.346.

Jones, J.B., and Fitzgerald, M.J., 1986, Silica-rich layering at Blanche Point, South Australia: Australian Journal of Earth Sciences, v. 33, p. 529–551, doi: 10.1080/08120098608729389.

Jones, J.B., and Fitzgerald, M.J., 1987, An unusual and characteristic sedimentary mineral suite associated with the evolution of passive margins: Sedimentary Geology, v. 52, p. 45–63, doi: 10.1016/0037-0738(87)90016-9.

Jovane, L., Florindo, F., Coccioni, R., Dinarès-Tureli, J., Marsili, A., Monechi, S., Roberts, A.P., and Sprovieri, M., 2007, The MECO event in the Contessa Highway section, Umbrian Apennines, Italy: Geological Society of America Bulletin, v. 119, p. 413–427, doi: 10.1130/B25917.1.

Kamp, P.J.J., Waghorn, D.B., and Nelson, C.S., 1990, Late Eocene–early Oligocene integrated isotope stratigraphy and biostratigraphy for paleoshelf sequences in southern Australia: Palaeogeography, Palaeoclimatology, Palaeoecology, v. 80, p. 311–323, doi: 10.1016/0031-0182(90)90140-3.

Kennett, J.P., and Exon, N.F., 2004, Paleoceanographic evolution of the Tasman Seaway and its climatic implications, *in* Exon, N.F., Kennett, J.P., and Malone, M.J., eds., The Cenozoic Southern Ocean: Tectonics, Sedimentation, and Climate Change between Australia and Antarctica: American Geophysical Union Geophysical Monograph 151, p. 345–367.

Kennett, J.P., Houtz, R.E., Andrews, P.B., Edwards, A.R., Gostin, V.A., Hajos, M., Hampton, M.A., Jenkins, D.G., Margolis, S.V., Ovenshine, A.T., and Perch-Nielsen, K., 1974, Development of the Circum-Antarctic Current: Science, v. 186, p. 144–147, doi: 10.1126/science.186.4159.144.

Lawver, L.A., and Gahagan, L.M., 2003, Evolution of Cenozoic seaways in the circum-Antarctic region: Palaeogeography, Palaeoclimatology, Palaeoecology, v. 198, p. 11–37, doi: 10.1016/S0031-0182(03)00392-4.

Li, Q., McGowran, B., and White, M.R., 2000, Sequences and biofacies packages in the mid-Cainozoic Gambier Limestone, South Australia: Reappraisal of foraminiferal record: Australian Journal of Earth Sciences, v. 47, p. 955–970, doi: 10.1046/j.1440-0952.2000.00824.x.

Li, Q., James, N.P., and McGowran, B., 2003a, Middle and late Eocene Great Australian Bight lithobiostratigraphy and stepwise evolution of the southern continental Australian margin: Australian Journal of Earth Sciences, v. 50, p. 113–128, doi: 10.1046/j.1440-0952.2003.00978.x.

Li, Q., McGowran, B., and James, N.P., 2003b, Eocene-Oligocene planktonic foraminiferal biostratigraphy of Sites 1126, 1130, 1132, and 1134, ODP Leg 182, Great Australian Bight, *in* Hine, A.C., Feary, D.A., and Malone, M.J., et al., Proceedings of the Ocean Drilling Program, Scientific results,

Volume 182: College Station, Texas, Ocean Drilling Program, p. 1–28 (online), doi: 10.2973/odp.proc.sr.182.006.2003.

Lindsay, J.M., 1967, Foraminifera and stratigraphy of the type section of Port Willunga Beds, Aldinga Bay, South Australia: Transactions of the Royal Society of South Australia, v. 91, p. 93–110.

Lindsay, J.M., 1981, Tertiary Stratigraphy and Foraminifera of the Adelaide City area, St. Vincent Basin, South Australia [M.Sc. thesis]: Adelaide, The University of Adelaide, 681 p.

Lindsay, J.M., 1985, Aspects of South Australian foraminiferal biostratigraphy, with emphasis on studies of *Massilina* and *Subbotina*, *in* Lindsay, J.M., ed., Stratigraphy, Palaeontology, Malacology; Papers in Honour of Dr. Nell Ludbrook: Adelaide, D.J. Woolman, Government Printer, Department of Mines and Energy Special Publication 5, p. 187–214.

Lindsay, J.M., and McGowran, B., 1986, Eocene/Oligocene boundary, Adelaide region, South Australia, *in* Pomerol, Ch., and Premoli Silva, I., eds., Terminal Eocene Events: New York, Elsevier Science, p. 165–173.

Ludbrook, N.H., and Lindsay, J.M., 1966, The Aldingan Stage: Quarterly Geological Notes: Geological Survey of South Australia, v. 19, p. 1–2.

Ludbrook, N.H., and Lindsay, J.M., 1969, Tertiary foraminiferal zones in South Australia, *in* Brönnimann, P., and Renz, H.H., eds., Proceedings of the First International Conference on Planktonic Microfossils: Leiden, E.J. Brill, v. 2, p. 366–374.

Luterbacher, H.P., Ali, J.R., Brinkhuis, H., Gradstein, F.M., Hooker, J.J., Monechi, S., Ogg, J.G., Powell, J., Röhl, U., Sanfilippo, A., and Schmitz, B., 2004, The Paleogene Period, *in* Gradstein, F.M., Ogg, J.G., and Smith, A.G., eds., A Geologic Time Scale 2004: Cambridge, Cambridge University Press, p. 384–408.

Macphail, M.K., Alley, N.F., Truswell, E.M., and Sluiter, I.R.K., 1994, Early Tertiary vegetation: Evidence from spores and pollen, *in* Hill, R.S., ed., History of the Australian Vegetation: Cretaceous to Recent: Cambridge, Cambridge University Press, p. 189–261.

Mallinson, D.J., Flower, B.A., Hine, A.C., Brooks, G.R., Garza, R.M., Drexler, T.M., and Leg 182 Scientific Party, 2003a, Data report: Mineralogy and geochemistry of ODP Site 1128, Great Australian Bight, *in* Hine, A.C., Feary, D.J., and Malone, M.J., et al., Proceedings of the Ocean Drilling Program, Scientific Results, Volume 182: College Station, Texas, Ocean Drilling Program, p. 1–17 (online), doi: 10.2973/odp.proc.sr.182.001.2003.

Mallinson, D.J., Flower, B.A., Hine, A.C., Brooks, G.R., and Garza, R.M., 2003b, Paleoclimate implications of high latitude precession-scale mineralogic fluctuations during early Oligocene Antarctic glaciation: The Great Australian Bight record: Global and Planetary Change, v. 39, p. 257–269, doi: 10.1016/S0921-8181(03)00119-X.

Martin, H.A., 2006, Cenozoic climatic change and the development of the arid vegetation in Australia: Journal of Arid Environments, v. 66, p. 533–563, doi: 10.1016/j.jaridenv.2006.01.009.

McArthur, J.M., and Howarth, R.J., 2004, Strontium isotope stratigraphy, *in* Gradstein, F.M., Ogg, J.G., and Smith, A.G., eds., A Geologic Time Scale 2004: Cambridge, Cambridge University Press, p. 96–105.

McGowran, B., 1977, Maastrichtian to Eocene foraminiferal assemblages in the northern and eastern Indian Ocean region: Correlations and historical patterns, *in* Heirtzler, J.R., Bolli, H.M., Davies, T.A., Saunders, J.B., and Sclater J.G., eds., Indian Ocean Geology and Biostratigraphy, Studies Following Deep Sea Drilling Project Legs 22–29: Washington, D.C., American Geophysical Union, p. 417–458.

McGowran, B., 1978a, Early Tertiary biostratigraphy in southern Australia: A progress report, *in* Belford, D.J., and Scheibnerova, V., eds., The Crespin Volume: Essays in Honour of Irene Crespin: Bureau of Mineral Resources of Australia, Geology and Geophysics Department Bulletin 192, p. 83–95.

McGowran, B., 1978b, Stratigraphic record of early Tertiary oceanic and continental events in the Indian Ocean region: Marine Geology, v. 26, p. 1–39, doi: 10.1016/0025-3227(78)90043-9.

McGowran, B., 1979, The Australian Tertiary: Foraminiferal overview: Marine Micropaleontology, v. 4, p. 235–264, doi: 10.1016/0377-8398(79)90019-7.

McGowran, B., 1987, Late Eocene perturbations: Foraminiferal biofacies and evolutionary overturn, southern Australia: Paleoceanography, v. 2, p. 715–728.

McGowran, B., 1989a, The later Eocene transgressions in southern Australia: Alcheringa, v. 13, p. 45–68, doi: 10.1080/03115518908619125.

McGowran, B., 1989b, Silica burp in the Eocene ocean: Geology, v. 17, p. 857–860, doi: 10.1130/0091-7613(1989)017<0857:SBITEO>2.3.CO;2.

McGowran, B., 1989c, Comment: Late Eocene diatomite from the Peruvian coastal desert, coastal upwelling in the eastern Pacific, and Pacific circulation before the terminal Eocene event: Geology, v. 17, p. 957–959, doi: 10.1130/0091-7613(1989)017<0957:CAROLE>2.3.CO;2.

McGowran, B., 1990, Fifty million years ago: American Scientist, v. 78, p. 30–39.

McGowran, B., 1991, Maastrichtian and early Cainozoic, southern Australia: Foraminiferal biostratigraphy, *in* Williams, M.A.J., De Deckker, P., and Kershaw, A.P., eds., The Cainozoic of the Australian Region: Geological Society of Australia Special Publication 18, p. 79–98.

McGowran, B., 2005, Biostratigraphy: Microfossils and geological time: Cambridge, Cambridge University Press, 459 p.

McGowran, B., and Alley, N.F., 2008, Chapter 2, History of the Cenozoic St. Vincent Basin in South Australia, *in* Shepherd, S.A., Bryars, S., Kirkegaard, I., Harbison, P., and Jennings, J.T., eds., Natural History of Gulf St. Vincent: Royal Society of South Australia Occasional Publication 8, p. 13–28.

McGowran, B., and Beecroft, A., 1986a, Foraminiferal biofacies in a silica-rich neritic sediment, late Eocene, South Australia: Palaeogeography, Palaeoclimatology, Palaeoecology, v. 52, p. 321–345, doi: 10.1016/0031-0182(86)90053-2.

McGowran, B., and Beecroft, A., 1986b, Neritic, southern extratropical foraminifera and the terminal Eocene event: Palaeogeography, Palaeoclimatology, Palaeoecology, v. 55, p. 23–34, doi: 10.1016/0031-0182(86)90134-3.

McGowran, B., and Li, Q., 2001, Evolutionary palaeoecology of Cainozoic forminifera: Tethys, IndoPacific, southern Australasia: Historical Biology, v. 15, p. 3–27.

McGowran, B., and Lindsay, J.M., 1969, A middle Eocene planktonic foraminiferal assemblage from the Eucla Basin: Geological Survey of South Australia Quarterly Geological Notes, v. 30, p. 2–10.

McGowran, B., Lindsay, J.M., and Harris, W.K., 1971, Attempted reconciliation of Tertiary biostratigraphic systems, Otway Basin, *in* Wopfner, H., and Douglas, J.G., eds., The Otway Basin in Southeastern Australia: Adelaide and Melbourne, Geological Surveys of South Australia and Victoria, Special Bulletin, p. 273–281.

McGowran, B., Moss, G., and Beecroft, A., 1992, Late Eocene and early Oligocene in southern Australia: Local neritic signals of global oceanic changes, *in* Prothero, D.R., and Berggren, W.A., eds., Eocene-Oligocene Climatic and Biotic Evolution: Princeton, Princeton University Press, p. 178–201.

McGowran, B., Li, Q., and Moss, G.D., 1997a, The Cenozoic neritic record in southern Australia: The biogeohistorical framework, *in* James, N.P., and Clarke, J.D.A., eds., Coolwater Carbonates in Space and Time: Society of Economic Paleontologists and Mineralogists Special Volume 56, p. 185–203.

McGowran, B., Li, Q., Cann, J., Padley, D., McKirdy, D., and Shafik, S., 1997b, Biogeographic impact of the Leeuwin Current in southern Australia since the late middle Eocene: Palaeogeography, Palaeoclimatology, Palaeoecology, v. 136, p. 19–40, doi: 10.1016/S0031-0182(97)00073-4.

McGowran, B., Archer, M., Bock, P., Darragh, T.A., Godthelp, H., Hageman, S., Hand, S.J., Hill, R.S., Li, Q., Maxwell, P.A., McNamara, K.J., Macphail, M., Mildenhall, D., Partridge, A.D., Richardson, J., Shafik, S., Truswell, E.M., and Warne, M., 2001, Australasian palaeobiogeography: The Palaeogene and Neogene record, *in* Wright, A.J., Young, G.C., Talent, J.A., and Laurie, J.R., eds., Palaeobiogeography of Australasian faunas and floras: Australasian Association of Palaeontologists Memoir 23, p. 405–470.

McGowran, B., Holdgate, G.R., Li, Q., and Gallagher, S.J., 2004, Cenozoic stratigraphic succession in southeastern Australia: Australian Journal of Earth Sciences, v. 51, p. 459–496, doi: 10.1111/j.1400-0952.2004.01078.x.

Moore, T.C., Jr., Jarrard, R.D., Lyle, A.O., and Lyle, M., 2008, Eocene biogenic silica accumulation rates at the Pacific equatorial divergence zone: Paleoceanography, v. 23, p. PA2202, doi: 10.1029/2007PA001514.

Morgans, H.E.G., Beu, A.G., Cooper, R.A., Crouch, E.M., Hollis, C.J., Jones, C.M., Raine, J.I., Strong, C.P., Wilson, G.J., and Wilson, G.S., 2004, Chapter 11, Paleogene (Dannevirke, Arnold and Landon Series), *in* Cooper, R.A., ed., The New Zealand Geological Time Scale: Institute of Geological and Nuclear Sciences Monograph 22, p. 125–163.

Moss, G.D., and McGowran, B., 1993, Foraminiferal turnover in neritic environments: The end-Eocene and mid-Oligocene events in southern Australia, *in* Jell, P.A., ed., Palaeontological studies in honour of Ken Campbell: Association of Australasian Palaeontologists Memoir 15, p. 407–416.

Moss, G., and McGowran, B., 2003, Oligocene neritic foraminifera in Southern Australia: Spatiotemporal biotic patterns reflect sequence-stratigraphic environmental patterns, *in* Olson, H., and Leckie, M., eds., Paleobiological, Geochemical, and Other Proxies of Sea Level Change: SEPM (Society of Sedimentary Geology) Special Volume 75, p. 117–138.

Muttoni, G., and Kent, D.V., 2007, Widespread formation of cherts during the early Eocene climatic optimum: Palaeogeography, Palaeoclimatology, Palaeoecology, v. 253, p. 348–362, doi: 10.1016/j.palaeo.2007.06.008.

Nagappa, Y., 1959, Foraminiferal biostratigraphy of the Cretaceous-Eocene succession in the India-Pakistan-Burma region: Micropaleontology, v. 5, p. 145–192, doi: 10.2307/1484208.

Nebelsick, J.H., Rasser, M.W., and Bassi, D., 2005, Facies dynamics in Eocene to Oligocene circumalpine carbonates: Facies, v. 51, p. 197–216, doi: 10.1007/s10347-005-0069-2.

Nocchi, M., Parisi, G., Monaco, P., Monechi, S., and Madile, M., 1988, Eocene and early Oligocene micropaleontology and paleoenvironments in SE Umbria, Italy: Palaeogeography, Palaeoclimatology, Palaeoecology, v. 67, p. 181–244, doi: 10.1016/0031-0182(88)90154-X.

Norris, R.D., 1996, Symbiosis as an evolutionary innovation in the radiation of Paleocene planktic foraminifera: Paleobiology, v. 22, p. 461–480.

Pagani, M., Zachos, J.C., Freeman, K.H., Tipple, B., and Bohaty, S., 2005, Marked decline in atmospheric carbon dioxide concentrations during the Palaeogene: Science, v. 309, p. 600–603, doi: 10.1126/science.1110063.

Parr, W.J., 1947, An Australian record of the foraminiferal genus *Hantkenina*: Proceedings of the Royal Society of Victoria, v. 58, p. 45–47.

Partridge, A.D., and Dettmann, M.E., 2003, Chapter 22.4.2, Plant microfossils, *in* Birch W.D., ed., Geology of Victoria: Melbourne, Geological Society of Australia (Victoria Division) Special Publication 23, p. 639–652.

Pearson, P.N., Olsson, R.K., Huber, B.T., Hemleben, C., and Berggren, W.A., eds., 2006, Atlas of Eocene Planktonic Foraminifera: Fredericksburg, Cushman Foundation for Foraminiferal Research Special Publication 41, 514 p.

Pekar, S.F., Hucks, A., Fuller, M., and Li, S., 2005, Glacioeustatic changes in the early and middle Eocene (51–42 Ma): Shallow-water stratigraphy from ODP Leg 189 Site 1171 (South Tasman Rise) and deep-sea $\delta^{18}O$ records: Geological Society of America Bulletin, v. 117, p. 1081–1093, doi: 10.1130/B25486.1.

Pisera, A., and Bitner, M.A., 2007, The sponge genus *Brachiaster* (Pachastrellidae, Demospongiae) and its first known fossil representative, from the late Eocene of southwestern Australia: Alcheringa, v. 31, p. 365–374, doi: 10.1080/03115510701640629.

Prothero, D.R., 2004, Did impacts, volcanic eruptions, or climate change affect mammalian evolution?: Palaeogeography, Palaeoclimatology, Palaeoecology, v. 214, p. 283–294.

Prothero, D.R., Ivany, L.C., and Nesbitt, E.A., eds., 2003, From greenhouse to icehouse: The marine Eocene-Oligocene transition: New York, Columbia University Press, 541 p.

Quilty, P.G., 1981, Late Eocene benthic Foraminiferida, south coast, Western Australia: Journal of the Royal Society of Western Australia, v. 64, p. 79–100.

Quilty, P.G., 2001, Reworked Paleocene and Eocene foraminifera, MacRobertson Shelf, East Antarctica: Paleoenvironmental implications: Journal of Foraminiferal Research, v. 31, p. 369–384, doi: 10.2113/0310369.

Quilty, P.G., Truswell, E.M., O'Brien, P.E., and Taylor, F., 2000, Paleocene-Eocene biostratigraphy and palaeoenvironments: New data from MacRobertson Shelf and western parts of Prydz Bay East Antarctica: Australian Geological Survey Organization (AGSO) Journal of Australian Geology & Geophysics, v. 17, p. 133–143.

Rea, D.K., and Lyle, M.W., 2005, Paleogene calcite compensation depth in the eastern subtropical Pacific: Answers and questions: Paleoceanography, v. 20, p. PA1012, doi: 10.1029/2004PA001064.

Reynolds, M.A., 1953, The Cainozoic succession of Maslin and Aldinga Bays, South Australia: Transactions of the Royal Society of South Australia, v. 7, p. 114–140.

Roberts, C., 2004, Cenozoic environments in the Tasmanian area of the Southern Ocean (ODP Leg 189): Inferences from bulk and clay mineraology, *in* Exon, N.F., Kennett, J.P., and Malone, M.J., eds., The Cenozoic Southern Ocean: Tectonics, sedimentation, and climate change between Australia and Antarctica, Washington, American Geophysical Union, Geophysical Monograph 151, p. 43–62.

Röhl, U., Brinkhuis, H., Stickley, C.E., Fuller, M., Schellenberg, S.A., Wefer, G., and Williams, G.L., 2004, Sea level and astronomically induced environmental changes in middle and late Eocene sediments from the East Tasman Plateau, *in* Exon, N.F., Kennett, J.P., and Malone, M.J., eds., The Cenozoic Southern Ocean: Tectonics, Sedimentation, and Climate Change between Australia and Antarctica: American Geophysical Union Geophysical Monograph 151, p. 127–152.

Serra-Kiel, J., Hottinger, L., Caus, E., Drobne, K., Ferràndez, C., Jauhri, A.K., Less, G., Pavlovec, R., Pignatti, J., Samso, J.M., Schaub, H., Sirel, E., Strougo, A., Tambareau, Y., Tosquella, J., and Zakrevskaya, E., 1998, Larger foraminiferal biostratigraphy of the Tethyan Paleocene and Eocene: Bulletin de la Société Géologique de France, v. 169, p. 281–299.

Sexton, P.F., Wilson, P.A., and Pearson, P.N., 2006, Palaeoecology of late middle Eocene planktic foraminifera and evolutionary implications: Marine Micropaleontology, v. 60, p. 1–16, doi: 10.1016/j.marmicro.2006.02.006.

Shackleton, N.J., and Kennett, J.P., 1975, Paleotemperature history of the Cenozoic and the initiation of Antarctic glaciation: Oxygen and carbon isotope analyses in DSDP Sites 277, 279, and 281, *in* Kennett, J.P., and Houtz, R.E., et al., Initial Reports of the Deep Sea Drilling Project, Volume 29: Washington, D.C., U.S. Government Printing Office, p. 743–755.

Shafik, S., 1981, Nannofossil biostratigraphy of the *Hantkenina* (foraminiferal) interval in the Upper Eocene of southern Australia: Bureau of Mineral Resources (BMR) Journal of Geology & Geophysics, v. 6, p. 108–116.

Shafik, S., 1983, Calcareous nannofossil biostratigraphy: An assessment of foraminiferal and sedimentation events in the Eocene of the Otway Basin, southeastern Australia: Bureau of Mineral Resources (BMR) Journal of Australian Geology and Geophysics, v. 8, p. 1–17.

Shafik, S., 1993, The significance of some subtle changes in the composition of Late Cretaceous and early Tertiary nannofloras from Australia: International Nannofossil Association (INA) Newsletter, v. 15, p. 93–94.

Shafik, S., 1995, Calcareous microplankton biostratigraphy of the Eocene Browns Creek Clay in the Aire District, Otway Basin of southeastern Australia: An update: Australian Geological Survey Organization (AGSO) Journal of Geology & Geophysics, v. 16, p. 333–344.

Shafik, S., and Idnurm, M., 1997, Calcareous microplankton and polarity reversal stratigraphies of the Upper Eocene Browns Creek Clay in the Otway Basin, southeast Australia: Matching the evidence: Australian Journal of Earth Sciences, v. 44, p. 77–86, doi: 10.1080/08120099708728295.

Sluijs, A., Brinkhuis, H., Stickley, C.E., Warnaar, J., Williams, G.L., and Fuller, M., 2003, Dinoflagellate cysts from the Eocene/Oligocene transition in the Southern Ocean: Results from ODP Leg 189, *in* Exon, N.F., Kennett, J.P., and Malone, M.J., et al., Proceedings of the Ocean Drilling Program, Scientific Results, Volume 189: College Station, Texas, Ocean Drilling Program, p. 1–42 (online), doi: 10.2973/odp.proc.sr.182.104.2003.

Sluijs, A., Pross, J., and Brinkhuis, H., 2005, From greenhouse to icehouse; organic-walled dinoflagellate cysts as paleoenvironmental indicators in the Paleogene: Earth-Science Reviews, v. 68, p. 281–315.

Stickley, C.E., Brinkhuis, H., Schellenberg, S., Sluijs, A., Roehl, U., Fuller, M., Grauert, M., Warnaar, J., and Williams, G.L., 2004, Timing and nature of the deepening of the Tasman Gateway: Paleoceanography, v. 19, p. PA4027, doi: 10.1029/2004PA001022.

Stover, L.E., and Partridge, A.D., 1973, Tertiary and Late Cretaceous spores and pollen from the Gippsland Basin, southeastern Australia: Proceedings of the Royal Society of Victoria, v. 85, p. 237–286.

Stover, L.E., and Partridge, A.D., 1982, Eocene spore-pollen from the Werillup Formation, Western Australia: Palynology, v. 6, p. 69–96.

Thomas, E., 2008, Descent into the icehouse: Geology, v. 36, p. 191–192, doi: 10.1130/focus022008.1.

Thomas, E., Zachos, J.C., and Bralower, T.J., 2000, Deep-sea environments on a warm earth: Latest Paleocene-early Eocene, *in* Huber, B.T., MacLeod, K.G., and Wing, S.L., eds., Warm Climates in Earth History: Cambridge, Cambridge University Press, p. 132–160.

Tripati, A., Backman, J., Elderfield, H., and Ferretti, P., 2005, Eocene bipolar glaciation associated with global carbon cycle changes: Nature, v. 436, p. 341–346, doi: 10.1038/nature03874.

van Andel, Tju. H., Heath, G.R., and Moore, T.C., Jr., 1975, Cenozoic history and paleoceanography of the central Pacific Ocean: Geological Society of America Memoir, 143, p. 1–134.

Veevers, J.J., 2000, Billion-Year Earth History of Australia and Neighbours in Gondwanaland: Sydney, ARC National Key Centre for Geochemical Evolution and Metallogeny of Continents (GEMOC), Department of Earth and Planetary Sciences, Macquarie University, 400 p.

Wade, M., 1964, Application of lineage concept to biostratigraphic zoning based on planktonic foraminifera: Micropaleontology, v. 10, p. 273–290, doi: 10.2307/1484576.

Waghorn, D.B., 1989, Middle Tertiary calcareous nannofossils from Aire District, Victoria: A comparison with equivalent assemblages in South Australia and New Zealand: Marine Micropaleontology, v. 14, p. 237–255, doi: 10.1016/0377-8398(89)90036-4.

Warnaar, J., 2006, Climatological Implications of Australian-Antarctic Separation [Ph.D. thesis]: Utrecht, Universiteit Utrecht, 143 p.

Wei, W., 2004, Opening of the Australia-Antarctica gateway as dated by nannofossils: Marine Micropaleontology, v. 52, p. 133–152, doi: 10.1016/j.marmicro.2004.04.008.

Woodburne, M.O., ed., 2004, Late Cretaceous and Cenozoic Mammals of North America: Biostratigraphy and Geochronology: New York, Columbia University Press, 391 p.

Woodburne, M.O., 2007, Mammal ages: Stratigraphy, v. 3 (2006), p. 229–261.

Zachos, J.C., Pagani, M., Sloan, L., Thomas, E., and Billups, K., 2001, Trends, rhythms, and aberrations in global climate 65 Ma to present: Science, v. 292, p. 686–693, doi: 10.1126/science.1059412.

Zachos, J.C., Dickens, G.R., and Zeebe, R.E., 2008, An early Cenozoic perspective on greenhouse warming and carbon-cycle dynamics: Nature, v. 451, p. 279–283, doi: 10.1038/nature06588.

Manuscript Accepted by the Society 16 September 2008

Printed in the USA

The Geological Society of America
Special Paper 452
2009

Nonmarine records of climatic change across the Eocene-Oligocene transition

Nathan D. Sheldon*
Department of Geological Sciences, University of Michigan, 1100 N. University Ave., Ann Arbor, Michigan 48109, USA

ABSTRACT

The greenhouse-icehouse change across the Eocene-Oligocene transition and associated Oi-1 glaciation event is the most profound climatic change in Earth's recent geological history. Marine reconstructions of the Oi-1 glaciation using foraminiferal $\delta^{18}O$ isotopic compositions suggest that much of the change was associated with Antarctic ice growth rather than climatic change. Nonetheless, some cooling is expected to have occurred on land in addition to drier conditions associated with water tied up in the polar ice caps, and some recent results based on stable isotope analyses of bones support this viewpoint. Nonmarine paleoclimatic conditions (mean annual temperature, mean annual precipitation) may be quantitatively reconstructed using paleosols preserved in continental successions to test this general model. Results from Oregon and Nebraska suggest moderate drying and cooling, not as a stepwise change at the time of the Oi-1 glaciation, but as part of a long-term aridification and cooling event associated in part with emplacement of the Cascade Range. In contrast, intermontane Montana's paleoprecipitation and paleotemperatures fluctuated on short-term (i.e., Milankovitch) time scales but on balance were both essentially unchanged by the Oi-1 glaciation. Results from Europe (UK, Spain) suggest a different pattern characterized by stable (i.e., unchanging) paleotemperatures in both localities and increasingly wet conditions in the UK. Taken together, these results indicate that (1) strongly regionalized climatic change was associated with the Oi-1 glaciation, (2) physiographic position with respect to orographic features played a key role in determining those regional climatic responses to the global event, and (3) there was little or no cooling on land associated with the Oi-1 glaciation.

Keywords: Eocene-Oligocene transition, paleoclimate, paleosols, rain shadow.

INTRODUCTION

In the past, the search for robust nonmarine paleoclimatic and paleoenvironmental records has been hampered by incompleteness (e.g., plant fossil assemblages) or by poor temporal control. Though incompleteness is perhaps inescapable, in many continental basins, a combination of radiometric dating, sequence stratigraphy, biostratigraphy, and magnetostratigraphy can provide temporal control that rivals that of the marine realm. Many of those basins also contain similar sedimentary archives that include alluvial and lacustrine deposits that have been overprinted by fossil soils (paleosols), allowing for the construction

*nsheldon@umich.edu.

Sheldon, N.D., 2009, Nonmarine records of climatic change across the Eocene-Oligocene transition, *in* Koeberl, C., and Montanari, A., eds., The Late Eocene Earth—Hothouse, Icehouse, and Impacts: Geological Society of America Special Paper 452, p. 241–248, doi: 10.1130/2009.2452(15). For permission to copy, contact editing@geosociety.org.

of both spatially and temporally related climate records. Given that paleosols form at Earth's surface, in direct contact with the atmospheric and climatic conditions at the time of their formation, well-dated paleosol records are potentially crucial to understanding global change in Earth's past.

The Eocene-Oligocene transition is the Earth's most recent "greenhouse" to "icehouse" transition (Zachos et al., 2001); it was a complex event, and the main Oi-1 glaciation event is now known to have lagged temporally after the Eocene-Oligocene transition (e.g., Coxall et al., 2005). Because much of the large isotopic excursion at (Zachos et al., 2001) and immediately following (Oi-1 glaciation; Coxall et al., 2005) the Eocene-Oligocene transition was associated with the buildup of ice on Antarctica rather than oceanic temperature change, other methods are also needed to assess the climatic event. Isotopic proxies based on fossil mammals and invertebrates give contrasting results: some authors suggest a significant cooling event (Zanazzi et al., 2007), and others suggest stable, unchanging conditions (Grimes et al., 2003, 2005). Resolution of this disagreement for continental successions is crucial to understanding the climatic dynamics of the whole Oi-1 glaciation event. Long sequences of paleosols in continental successions yield additional lines of evidence by providing quantitative data about mean annual precipitation and mean annual temperature. Results are collected here for the first time from a variety of North American (Oregon, Montana, Nebraska; Sheldon and Retallack, 2004; Retallack, 2007) and European (UK, Spain; Sheldon et al., this volume) sites, and these are used to identify regional and global nonmarine paleoclimatic trends associated with the Oi-1 glaciation.

MATERIALS AND METHODS

Field Sites and Sample Collection

Stratigraphic data and samples were collected from a variety of field sites centered around five areas (Fig. 1): (1) central Oregon near John Day; (2) southwestern Montana near Dillon; (3) northwestern Nebraska near Harrison; (4) Maians (Spain) northwest of Barcelona; and (5) the Isle of Wight (UK). Because the North American stratigraphic successions are typically nearly flat-lying and contain dated units (e.g., tuffs) that can be correlated between sections, stratigraphic superposition can be used to infer the age of a given paleosol. Retallack (2007) has compiled these age relationships for the past 45 m.y., a subset of which (36–33 Ma) is used here. Geological ages of the paleosols in Oregon were inferred from paleomagnetic estimates (Prothero and Rensberger, 1985) and radiometric ages regressed against stratigraphic level (Retallack et al., 2000; Retallack, 2007). Geological ages of paleosols in Montana come from bio- and magnetostratigraphic data (Tabrum et al., 1996) regressed against the composite stratigraphic section of Hanneman and Wideman (1991). Geological ages of paleosols in Nebraska were determined using bio- and magnetostratigraphic data (Prothero and Swisher, 1992; Prothero and Whittlesey, 1998) regressed against stratigraphic sections. The standard errors of the age models are typically ~200 k.y., ranging from ~160 k.y. for the Oligocene of Oregon to ~400 k.y. for the Eocene of Nebraska (Sheldon and Retallack, 2004), if a constant sedimentation rate is assumed between individual age determinations rather than for the succession as a whole, though the sedimentation rates for all three

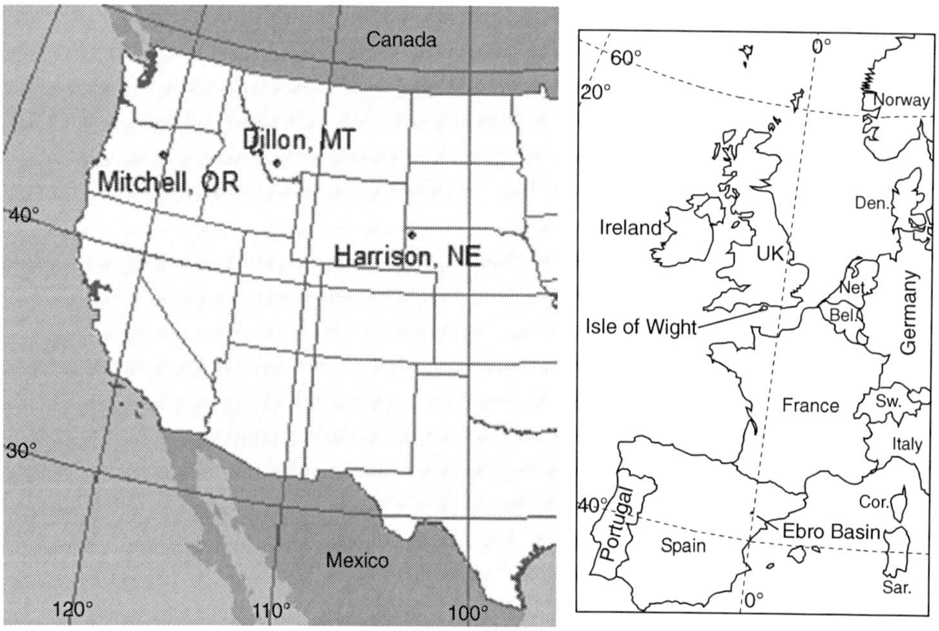

Figure 1. Maps showing the location of all of the field sites in North America and Europe. MT—Montana, NE—Nebraska, OR—Oregon.

areas are nearly constant within a given succession (data fits have $R^2 = 0.95–0.99$; Sheldon and Retallack, 2004).

Age determinations for the Isle of Wight's Solent Group succession are based on a combination of magnetostratigraphy (Gale et al., 2006) and mammalian biostratigraphy, including the Oi-1–coincident European "Grande Coupure" mammalian turnover event (Hooker et al., this volume). Additional age control includes calcareous nannoplankton in some parts of the succession (Hooker et al., this volume), both pre– and post–Eocene-Oligocene transition.

Maians (Spain) is relatively poorly dated in comparison to the other sections owing to a dearth of fossil material, though ongoing work by L. Cabrera, M. Garces, and E. Diaz at Universitat de Barcelona should soon improve the chronology. Their magnetostratigraphic record (L. Cabrera, 2007, personal commun.) is used here to determine the relative ages of the paleosols in the succession by assuming a constant sedimentation rate, though at present, those ages (see record in Results section) should be considered relatively imprecise in comparison to the other sections.

Paleoclimatic Proxies

Paleoprecipitation estimates from paleosols examined here are based on two independent proxies, depth to a calcic horizon, and the degree of chemical weathering. Both approaches yield results consistent with independent estimates of paleoenvironmental conditions from fossil plants and animals, and with each other (Sheldon et al., 2002; Sheldon and Retallack, 2004; Tabor et al., 2004). Soils forming in areas receiving higher mean annual precipitation generally have Bk (carbonate-bearing) horizons lower in their profiles (Retallack, 1991). This relationship, quantified for modern soils (most recently calibrated by Retallack, 2005), has been used in investigations of paleosols (Retallack et al., 2000; Retallack, 2007). Bk depths are measured from the top of a paleosol profile using a tape measure and are corrected for any structural dip. The top of a profile is typically markedly finer grained than the overlying sediments, and other evidence of pedogenesis (e.g., root traces) is used to confirm the location of the profile top. Further, as noted already, paleoprecipitation estimates from Bk horizons have been shown to correspond well (±<200 mm) to estimates from fossil plants (Retallack et al., 2000) and also to estimates from the degree of chemical weathering (Sheldon et al., 2002, this study). The transfer function is configured here to make depth to the Bk horizon (as a positive number) the input variable as follows:

$$\text{Precipitation (mm)} = 137.24 + 6.45D - 0.013D^2, \quad (1)$$

where D is the depth (cm), $R^2 = 0.52$, and the standard error is 147 mm/yr (Retallack, 2005). This relationship can be improved upon substantially for some soil orders (e.g., Vertisols) or textures (clay-rich; Nordt et al., 2006), but, in general, the paleosols discussed within this paper lack vertic features and are too coarse-grained (cf. the Isle of Wight succession is fine-grained enough, but it has no paleosols with Bk horizons) for the more precise functions of Nordt et al. (2006) to be applicable, so the generalized relationship of Retallack (2005) is used instead.

Bk depth collection methods (Sheldon and Retallack, 2004) and data are archived elsewhere (Retallack et al., 2000; Sheldon and Retallack, 2004; Retallack, 2007). Montana Bk depths ($n = 223$) were collected from paleosol-bearing sections at Pipestone Creek, Little Pipestone Creek, McCartney's Mountain, Cook Ranch, Easter Lily, and Mill Point and are presented here (Fig. 2; Retallack, 2007, supplemental data). Nebraska Bk depths ($n = 4$) are from Sheldon and Retallack (2004), and they are drawn from a variety of sites, including some of Terry's (2001) sites. In each case, measured Bk depths were adjusted using the decompaction equations of Sheldon and Retallack (2001) to account for compaction of the paleosols due to burial. Depths of burial may be found in Retallack et al. (2000) and Retallack (2007) but, in some cases, would have resulted in compaction sufficient to give a 200+ mm underestimation of paleoprecipitation if the original Bk depths were uncorrected.

The second paleoclimatic proxy is based on the observation that soils receiving higher mean annual precipitation show a greater degree of chemical weathering than soils receiving lower precipitation. This general relationship is due to two factors: (1) higher precipitation leads to greater potential for runoff and for dissolution of soil mineral matter, and (2) higher precipitation facilitates greater plant productivity, which augments chemical weathering as plants remove nutrients from the soil. Results are consistent with paleoprecipitation estimates from fossil plants (e.g., Wolfe, 1995) where the proxies can be directly compared (e.g., middle Eocene to late Oligocene; Sheldon et al., 2002), and this approach has been widely applied (Tabor et al., 2004; Sheldon, 2005; Hamer et al., 2007a, 2007b). Data on the degree of chemical weathering of Oregon paleosols were compiled elsewhere (Retallack et al., 2000; Sheldon et al., 2002; Retallack, 2007). Paleoprecipitation can be estimated using the chemical index of alteration without potassium (CIA – K; Maynard, 1992) of paleosol B horizons if burial K metasomatism (e.g., Sheldon, 2003) can be eliminated, and the following equation is derived by relating the CIA – K values of modern soils to their measured mean annual precipitation (mm/yr):

$$P = 221.12e^{0.0197(\mathrm{CIA-K})}, \quad (2)$$

where $R^2 = 0.72$, and standard error is 182 mm/yr (Sheldon et al., 2002). For further discussion of the applicability of the proxy, see Sheldon et al. (2002).

Another transfer function relates salinization (S) to mean annual temperature (MAT; °C):

$$\mathrm{MAT} = -18.5S + 17.3, \quad (3)$$

where S is the molar ratio of potash and soda to alumina, $R^2 = 0.37$, and the standard error is 4.4 °C for the relationship (Sheldon et al., 2002). For one paleosol order (Inceptisols), a

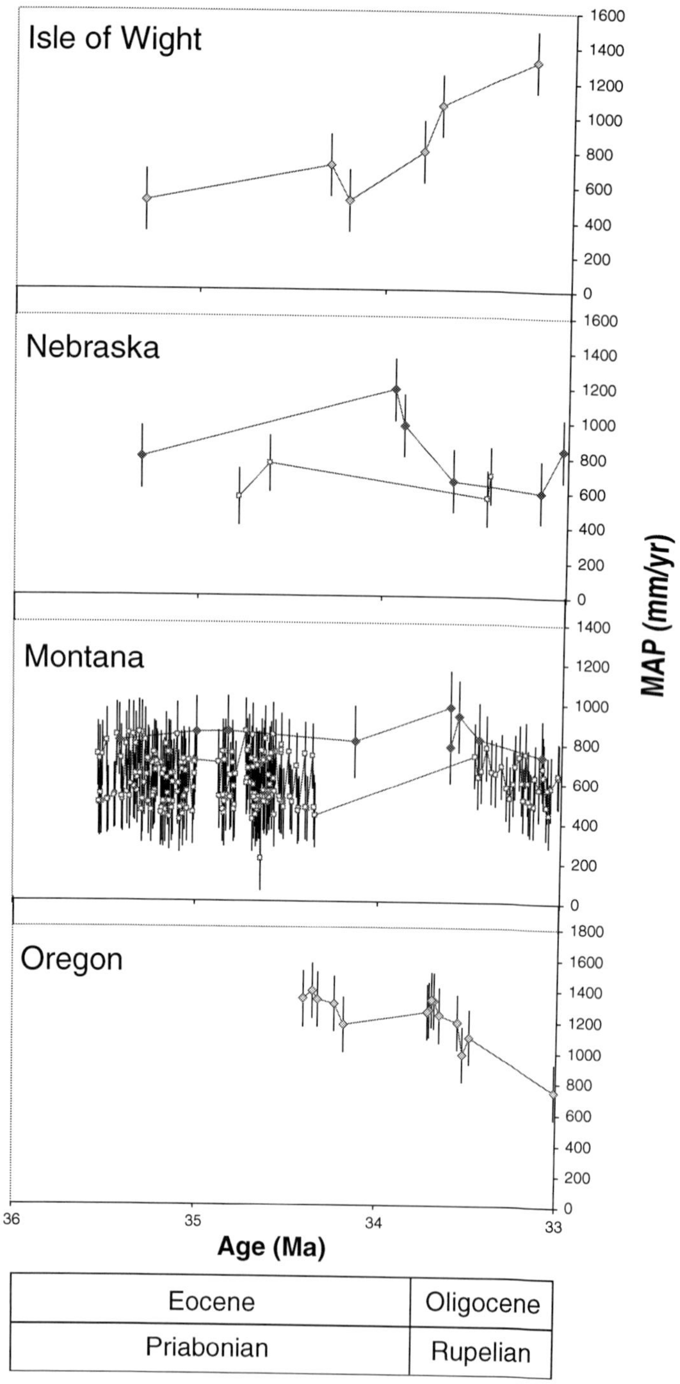

Figure 2. Aggregate nonmarine climatic record from paleosols: mean annual precipitation (MAP). Open symbols in each panel are based on Bk horizon data, and closed symbols are based on whole-rock geochemistry. There are no Bk depth data for Oregon prior to ca. 30.5 Ma because there were no Bk-bearing paleosols (Sheldon and Retallack, 2004). Reconstructed paleoprecipitation values were calculated using Equation 1 for Bk depths and Equation 2 for whole-rock geochemistry.

more precise relationship exists ($R^2 > 0.9$; S.E. = 0.6 °C; Sheldon, 2006b); it relates MAT (°C) to "clayeyness" (C; Al/Si ratio) and was derived from the same data set as Equations 2 and 3:

$$\text{MAT} = 46.9C + 4. \tag{4}$$

Both relationships were applied at Maians, which is dominated by Inceptisol-like paleosols, giving rise to similar results, so this second, more precise climofunction is preferred for that section. Samples for paleoclimatic analysis for the 36–33 Ma time window were collected at the same sites listed for Oregon (n = 15), Montana (v = 9), Nebraska (n = 6), the Isle of Wight (n = 6), and the Ebro Basin (n = 55); additional samples collected to characterize pedogenic processes but not used for paleoclimatic reconstruction (e.g., Table 2 in Sheldon et al., this volume) are discussed and tabulated elsewhere (Retallack et al., 2000; Sheldon and Retallack, 2004; Retallack, 2007).

RESULTS

Figures 2 and 3 show compiled paleoclimatic records for the Eocene-Oligocene transition in Oregon, Montana, Nebraska, and Isle of Wight derived from the paleoprecipitation (Fig. 2; Eqs. 1 and 2) and paleotemperature (Fig. 3; Eq. 3) proxies described in the Materials and Methods section. Whereas the North American records show declining (Oregon and Nebraska) or steady (Montana) paleoprecipitation values, the Isle of Wight record suggests wetter conditions prevailed in the earliest Oligocene relative to the late Eocene (Fig. 2). A previous study of the Oregon succession based on a very limited number of data (Bestland et al., 1997) suggested a stepwise climatic change, but paleotemperature records from North American sites indicate instead gradually cooling conditions across the Eocene-Oligocene transition, and no strong step function coincident with the isotopic shifts observed in the marine realm (Zachos et al., 2001; Coxall et al., 2005; Katz et al., 2008). More recent work in Oregon (Sheldon and Retallack, 2004) has demonstrated that the Eocene-Oligocene decline in North America is part of a long-term (36–27 Ma) cooling event. In the highest-resolution record (Bk-derived record from Montana), Milankovitch-scale climatic change is indicated (Fig. 4), even in the absence of large ice sheets in the late Eocene. Paleotemperature results from the temporally poorly constrained Maians record indicate similar short-term climate fluctuations but no significant change associated with the Eocene-Oligocene transition or Oi-1 glaciation (Fig. 5).

DISCUSSION

Differences among Paleoprecipitation Proxies

There is a systematic, though slight, offset between paleoprecipitation estimates derived from Bk horizon depths (drier) and Bw/Bt horizon geochemistry (Fig. 2). This is due to several well-known factors. First, there is a limit as to how wet conditions may be before carbonate formation becomes limited. That

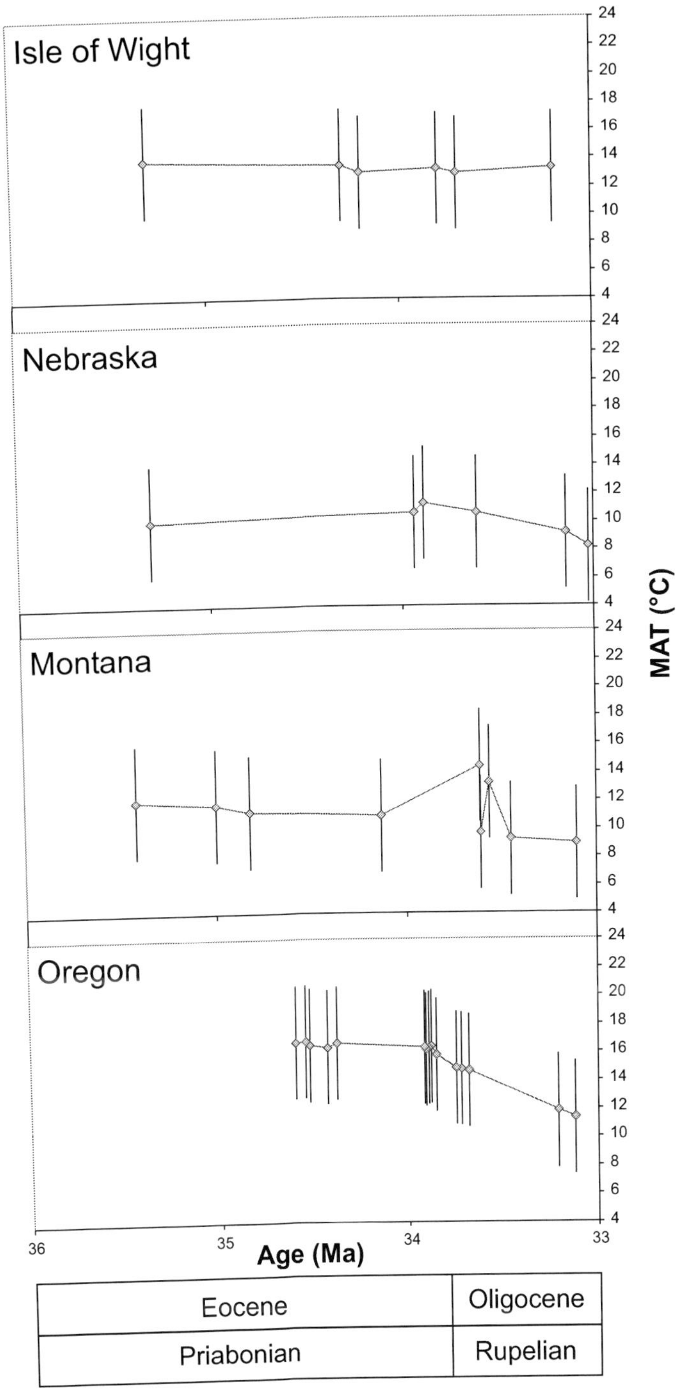

Figure 3. Aggregate nonmarine climatic record from paleosols: mean annual temperature (MAT). All of the reconstructed paleotemperatures were calculated using Equation 3.

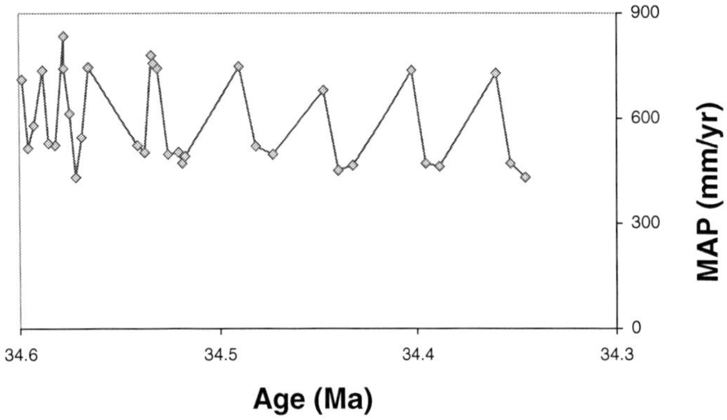

Figure 4. Milankovitch-scale mean annual precipitation (MAP) cyclicity in Montana. A typical climate cycle consists of two paleosols that indicate arid conditions (shallow Bk horizons) followed by one that indicates wetter semiarid or subhumid conditions (deep Bk horizons).

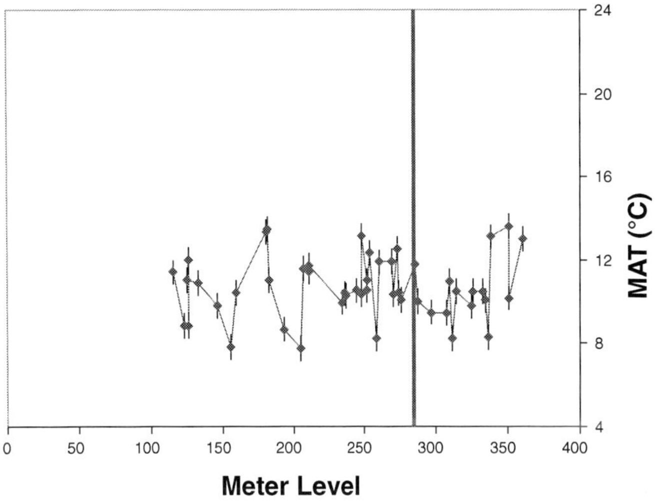

Figure 5. Paleotemperature record from Maians, Spain. MAT—mean annual temperature. Though the stratigraphic level of transition between chrons 13r and 13n is known and calibrated based on biostratigraphy (L. Cabrera, pers. comm.), sedimentation rates have not been precisely determined yet, so the paleoclimatic data are plotted against meter level rather than age as in the other sections. The reconstructed mean Eocene conditions (n = 39; 10.7 °C ± 3.3 °C [2 S.D.]) and Oligocene conditions (n = 16; 10.6 °C ± 3.2 °C [2 S.D.]) are indistinguishable.

isohyet is at 760 mm/yr for North America and 600 mm/yr for Africa currently (Retallack, 2001), though strongly seasonal precipitation regimes (i.e., monsoonal; Sheldon, 2005; Nordt et al., 2006) may allow for the formation of limited (stage I or II; Machette, 1985) amounts of carbonate in places with wetter mean annual conditions. Thus, Bk-bearing paleosols do not necessarily form at all unless the conditions are sufficiently arid. In Nebraska, for example, Bk-bearing paleosols become much more common in the Oligocene relative to the Eocene, and in Oregon, no Bk-bearing paleosols are known until ca. 30.6 Ma (Sheldon and Retallack, 2004). Second, the Bk proxy relies on physical measurements of the depth of the horizon, meaning that proper identification of the top of a given paleosol is necessary, and also that erosion by fluvial or alluvial processes prior to burial could result in lower estimates of paleoprecipitation. Identification of the top of the paleosol is typically based on the uppermost level from which root traces emanate downward, and erosive processes typically act "unevenly," leaving evidence of an erosion surface at the top of a paleosol. Thus, with careful observation, it should be possible to overcome this limitation. Finally, the Bw/Bt horizon proxy works best for paleosols that have limited amounts of pedogenic carbonate and no inherited carbonate (Sheldon et al., 2002), so it should be most reliable for the slightly wetter conditions that suppress carbonate formation. Therefore, (when possible) both proxies together give a more complete picture of the climatic conditions and represent the range of short-term variation rather than some systematic mismatch between the proxies. This is further supported by the fact that the standard error envelopes associated with both proxies overlap (Fig. 2) and that both proxies give results consistent with paleobotanical estimates for the Eocene-Oligocene transition (Sheldon et al., 2002).

Paleoclimatic Spatial Trends

There is a spatial component to the North American paleoclimatic records; both near-coastal Oregon and continental Nebraska are wetter in the Eocene than intermontane Montana, which lies between them on an east-west transect (Fig. 1). By 30 Ma, all three regions received similar amounts of precipitation and would be classified as semiarid (Sheldon and Retallack, 2004). Thus, between 36 and 30 Ma, there was a homogenization of western and midwestern climatic conditions in North America. Part of this change has been attributed to the growth locally of a rain shadow in response to the emplacement of the Cascades to the west of the Oregon sites (Sheldon and Retallack, 2004). Calculations of the rate of the expansion of the rain shadow show that there was significant expansion along the Oregon-Montana transect coincident with the Eocene-Oligocene transition (Retallack, 2007). Thus, the long-term aridification was a function of both water availability, which was lessened by the formation of permanent Antarctic ice sheets (e.g., Dupont-Nivet et al., 2007), and of proximity to an orographic impediment that created a rain shadow in Oregon. In contrast, the low-lying, coastal Hampshire Basin in which the Isle of Wight paleosols were deposited actually became wetter following the Oi-1 glaciation event, again suggesting that the distribution of water masses and their proximity to orographic impedances play a key role in local paleoclimatic responses to global events, because the physiographic setting for the Isle of Wight was otherwise similar to that of Oregon (Sheldon et al., this volume).

The paleotemperature records show a similar spatial pattern with long-term cooling trends in Oregon and Nebraska (Sheldon and Retallack, 2004; Retallack et al., 2004), along with perhaps slight cooling in Montana (Fig. 3). In contrast, both European sites show steady paleotemperatures across the Eocene-Oligocene transition (Figs. 3 and 5). The spatial trends in North America may also be a function of regional orographic effects, where the more "open" depositional systems in Europe are more representative of global conditions, but given that predictions of future climatic change based on general circulation models (GCMs) (IPCC, 2007) suggest strong regional variation in the nature and intensity of climatic change, it is perhaps unsurprising that similar variability was present in the geologic past.

Paleoclimatic Temporal Trends

None of the five sites (Figs. 3 and 5) on two continents shows a stepwise or drastic cooling event; thus, evidence for a significant continental cooling event associated with the Oi-1 glaciation event is equivocal. Among the five records, the Oregon record shows the greatest change in paleoprecipitation, dropping from humid precipitation levels in the Eocene to semiarid precipitation levels by ca. 30.5 Ma (Fig. 2; Sheldon and Retallack, 2004). Much of this drop occurs in the Oligocene and significantly lags the Oi-1 glaciation event. Nonetheless, the contrast between near-coastal sites of long-term drying in Oregon (backarc basin behind a nascent arc) and long-term wetting on the Isle of Wight (coastal plain) indicates a substantial reorganization of the global hydrological cyclical (Wolfe, 1995; Robert and Kennett, 1997), consistent with the formation of permanent ice sheets on Antarctica (e.g., Robert and Kennett, 1997).

The intermontane basins in Maians (Fig. 5) and Montana (Fig. 2) preserve significant short-term (Milankovitch-scale) climatic variation both prior to and following the formation of Antarctic ice sheets but show no overall temporal trend. For example, the average mean annual precipitation for Montana of all Eocene paleoprecipitation estimates, 595 ± 243 (2σ) mm/yr, is indistinguishable from the Oligocene average of 612 ± 172 (2σ) mm/yr. Similarly, there is no difference among paleotemperatures in Spain between the average Eocene value of 10.7 °C ± 3.3 °C (2σ) and the average Oligocene value of 10.6 °C ± 3.2 °C (2σ). Comparable short-term variation has been documented in Oligocene-Miocene paleosols of the central Ebro Basin (Hamer et al., 2007a, 2007b) as in the Eocene-Oligocene Maians section of the eastern Ebro Basin. Discounting the possibility that the Ebro Basin is somehow special, there are two other options, either intermontane basins record short-term variation better than other depositional settings, or intermontane basins can be relatively immune to global climatic changes and instead are driven primarily by intrinsic climatic factors (Sheldon, 2006a), with a more minor role for global drivers (e.g., Milankovitch insolation cyclicity). In the context of this paper and this paleoclimatic event, the latter possibility is more likely, but further work is needed for other paleoclimatic events to discount the former idea.

Comparison to Isotope-Based Records

Two competing hypotheses have been put forward on the basis of isotopic analyses of fossil materials: one asserts no paleotemperature change in continental settings, and one asserts a very significant cooling event. Grimes et al. (2003, 2005) used a multiorganism approach (rodent teeth, fish scales, gastropods, charophytes, and otoliths) to reconstruct paleotemperatures in the Hampshire Basin (Isle of Wight), and they found no change between Eocene and Oligocene summer temperatures using a variety of proxies. This result is consistent with paleotemperature estimates reconstructed using co-occurring paleosols (Fig. 3; Sheldon et al., this volume).

In contrast, Zanazzi et al. (2007, this volume) proposed a significant cooling event (>8 °C) based on an ~1.7‰ isotopic shift in mean bone calcite values ($\delta^{18}O_{CO_3}$) collected from the same Nebraska sites studied by Terry (2001) and discussed in this paper. This is not consistent with results obtained from co-occurring paleosols (Fig. 3; Terry, 2001; Sheldon and Retallack, 2004; Retallack, 2007) or from other studies using a similar methodology on fossils from the Eocene-Oligocene transition in Argentina (Kohn et al., 2004). Zanazzi et al. (2007, this volume) used bone calcite rather than enamel; bone calcite is highly subject to diagenetic alteration (Iacumin et al., 1996), but Zanazzi et al. (2007, this volume) argued that changes to the $\delta^{18}O_{CO_3}$ composition of the diagenetic water should reflect meteoric water values in some way and that relative changes are still useful for interpreting paleotemperatures. There are three problems with these assertions. First, the assumption that early diagenetic waters will change in lock-step with meteoric water is speculative, and the timing of the diagenetic alteration is difficult to constrain in fossil bones. Second, both prior to and following the Eocene-Oligocene transition and Oi-1 glaciation, Zanazzi et al.'s (2007, this volume) data show substantial variation, essentially yielding a cloud spanning the interval, and therefore implying that the way in which the data were binned temporally has an important role in determining how large a $\delta^{18}O_{CO_3}$ and paleotemperature shift one would interpret. They binned the data into pre– and post–33.5 Ma bins and, consequently, argued that the drop in mean annual temperature lagged the marine cooling trend by ~400 k.y. However, the exponentially smoothed curve in their Figure 3 appears to suggest synchroneity instead. If their data are rebinned, and the same paleothermometry method is used (reference 40 in their supplementary information), the magnitude of the $\delta^{18}O_{CO_3}$ would be 2.6‰, equivalent to a 12 °C shift, which is not consistent with any other proxy, marine or nonmarine. Some stratigraphic levels have a >4.0‰ difference (e.g., 7.3 m level; Zanazzi et al., this volume), which would result in massively different paleotemperature estimates from data points that should all yield the same value if the method were viable. Third, although the authors acknowledge that water stress or increasing aridity could contribute to changes in the diagenetic (or meteoric) $\delta^{18}O_{CO_3}$ composition, they rule out both possibilities. Results from paleosols (Terry, 2001; Sheldon and Retallack, 2004; Retallack, 2007, this paper) have consistently shown aridification across the Eocene-Oligocene transition in Nebraska, which suggests that the apparent large temperature drop hypothesized by Zanazzi et al. (2007) was due, at least in part, to changes in water availability. A more modest change, consistent with the paleosol result (Fig. 3), is more reasonable, especially in light of the consistent results obtained by comparing paleosols elsewhere with independent proxies like fossil plants (Sheldon et al., 2002; Retallack et al., 2004). Both Oregon (Retallack et al., 2004) and the Isle of Wight (Sheldon et al., this volume) contain co-occurring paleosols and plant and animal fossils, and neither area shows a large temperature drop. Furthermore, in both areas, multiple proxies indicate similar magnitudes of climatic change (i.e., minimal temperature change) associated with the Eocene-Oligocene transition and the Oi-1 glaciation.

CONCLUSIONS

Regional differences among paleoclimatic records of the Eocene-Oligocene transition and Oi-1 glaciation event suggest both that the consequences of the climatic change were variable, and that other, local factors such as orography are important when considering continental paleoclimate records. North American sites show significant long-term aridification, of which the Oi-1 glaciation was an important part, rather than the lone driving force. Significantly wetter Oligocene conditions on the Isle of Wight indicate that different regions responded differently to the global climatic change and that there was a significant reorganization of the global hydrologic cycle. Unchanging or gradual (and longer-term) paleotemperature changes are consistent with the idea that much of the marine oxygen isotopic shift at the Eocene-Oligocene transition and Oi-1 glaciation was due to the formation of permanent ice sheets on Antarctica (Zachos et al., 2001) rather than to a stepwise or abrupt climate change (Retallack et al., 2004; Sheldon and Retallack, 2004).

ACKNOWLEDGMENTS

Funding in support of this work was obtained from Natural Environment Research Council (NERC) (NE/D000521/1) and the Strategic Research Fund (SRF) at Royal Holloway, University of London, during the author's time there. Neil Tabor and Steve Driese are thanked for thorough and constructive reviews, along with Editor Christian Koeberl, who had further suggestions to enhance clarity.

REFERENCES CITED

Bestland, E.A., Retallack, G.J., and Swisher, C.C., III, 1997, Stepwise climate change recorded in Eocene-Oligocene paleosol sequences from Central Oregon: The Journal of Geology, v. 105, p. 153–172.

Coxall, H.K., Wilson, P.A., Palike, H., Lear, C.H., and Backman, J., 2005, Rapid stepwise onset of Antarctic glaciation and deeper calcite compensation in the Pacific Ocean: Nature, v. 433, p. 53–57, doi: 10.1038/nature03135.

Dupont-Nivet, G., Krijgsman, W., Langereis, C.G., Abels, H.A., Dai, S., and Fang, X.M., 2007, Tibetan Plateau aridification linked to global cooling at the Eocene-Oligocene transition: Nature, v. 445, p. 635–638, doi: 10.1038/nature05516.

Gale, A.S., Huggett, J.M., Palike, H., Laurie, E., Hailwood, E.A., and Hardenbol, J., 2006, Correlation of Eocene-Oligocene marine and continental records: Orbital cyclicity, magnetostratigraphy and sequence stratigraphy on the Solent Group, Isle of Wight, UK: Journal of the Geological Society of London, v. 163, p. 401–415, doi: 10.1144/0016-764903-175.

Grimes, S., Mattey, D., Hooker, J.J., and Collinson, M.E., 2003, Paleogene palaeoclimate reconstruction using oxygen isotopes from land and freshwater organisms: The use of multiple palaeoproxies: Geochimica et Cosmochimica Acta, v. 67, p. 4033–4047, doi: 10.1016/S0016-7037(03)00173-X.

Grimes, S.T., Hooker, J.J., Collinson, M.E., and Mattey, D.P., 2005, Temperatures of late Eocene to early Oligocene freshwaters: Geology, v. 33, p. 189–192, doi: 10.1130/G21019.1.

Hamer, J.M.M., Sheldon, N.D., Nichols, G.J., and Collinson, M.E., 2007a, Late Oligocene–early Miocene palaeosols of distal fluvial systems, Ebro Basin, Spain: Palaeogeography, Palaeoclimatology, Palaeoecology, v. 247, p. 220–235, doi: 10.1016/j.palaeo.2006.10.016.

Hamer, J.M.M., Sheldon, N.D., and Nichols, G.J., 2007b, Global aridity during the early Miocene? A terrestrial paleoclimate record from the Ebro Basin, Spain: The Journal of Geology, v. 115, p. 601–608, doi: 10.1086/519780.

Hanneman, D.H., and Wideman, C.J., 1991, Sequence stratigraphy of Cenozoic continental rocks, southwestern Montana: Geological Society of America Bulletin, v. 103, p. 1335–1345, doi: 10.1130/0016-7606(1991)103<1335:SSOCCR>2.3.CO;2.

Hooker, J.J., Grimes, S.T., Mattey, D.P., Collinson, M.E., and Sheldon, N.D., 2009, this volume, Refined correlation of the UK late Eocene–early Oligocene Solent Group and of its climate history, *in* Koeberl, C., and Montanari, A., eds., The Late Eocene Earth—Hothouse, Icehouse, and Impacts: Geological Society of America Special Paper 452, doi: 10.1130/2009.2452(12).

Iacumin, P., Bocherens, H., Mariotti, A., and Longinelli, A., 1996, Oxygen isotope analyses of co-existing carbonate and phosphate in biogenic apatite: A way to monitor diagenetic alteration of bone phosphate?: Earth and Planetary Science Letters, v. 142, p. 1–6, doi: 10.1016/0012-821X(96)00093-3.

Intergovernmental Panel on Climate Change (IPCC), 2007, The Physical Science Basis: United Nations Environment Programme (UNEP), Geneva, 996 p.

Katz, M.E., Miller, K.G., Wright, J.D., Wade, B.S., Browning, J.V., Cramer, B.S., and Rosenthal, Y., 2008, Stepwise transition from the Eocene greenhouse to the Oligocene icehouse: Nature Geoscience, v. 1, p. 329–334, doi: 10.1038/ngeo179.

Kohn, M.J., Josef, J.A., Madden, R., Vucetich, G., and Carlini, A.A., 2004, Climate stability across the Eocene-Oligocene transition, southern Argentina: Geology, v. 32, p. 621–624, doi: 10.1130/G20442.1.

Machette, M.N., 1985, Calcic soils of the southwestern United States, *in* Weide, D.L., ed., Soils and Quaternary Geology of the Southwest United States: Geological Society of America Special Paper 203, p. 1–21.

Maynard, J.B., 1992, Chemistry of modern soils as a guide to interpreting Precambrian paleosols: The Journal of Geology, v. 100, p. 279–289.

Nordt, L., Orosz, M., Driese, S., and Tubbs, J., 2006, Vertisol carbonate properties in relation to mean annual precipitation: Implications for paleoprecipitation estimates: The Journal of Geology, v. 114, p. 501–510, doi: 10.1086/504182.

Prothero, D.R., and Rensberger, J.M., 1985, Preliminary magnetostratigraphy of the John Day Formation, Oregon, and the North American Oligocene-Miocene boundary: Newsletters in Stratigraphy, v. 15, p. 59–70.

Prothero, D.R., and Swisher, C.C., III, 1992, Magnetostratigraphy and geochronology of the terrestrial Eocene-Oligocene transition in North America, *in* Prothero, D.R., and Berggren, W.A., eds., Eocene-Oligocene Climatic and Biotic Evolution: Princeton, New Jersey, Princeton University Press, p. 46–75.

Prothero, D.R., and Whittlesey, K.E., 1998, Magnetic stratigraphy and biostratigraphy of the Orellan and Whitneyan land mammal "ages" in the White River Group, *in* Terry, D.O., Lagarry, H.E., and Hunt, R.M., eds., Depositional Environments, Lithostratigraphy, and Biostratigraphy of the White River and Arikaree Groups (Late Eocene to Early Miocene), North America: Geological Society of America Special Paper 325, p. 39–61.

Retallack, G.J., 1991, Untangling the effects of burial alteration and ancient soil formation: Annual Review of Earth and Planetary Sciences, v. 19, p. 183–206, doi: 10.1146/annurev.ea.19.050191.001151.

Retallack, G.J., 2001, Soils of the Past (2nd edition): London, Blackwell Science, Ltd., 404 p.

Retallack, G.J., 2005, Pedogenic carbonate proxies for amount and seasonality of precipitation in paleosols: Geology, v. 33, p. 333–336, doi: 10.1130/G21263.1.

Retallack, G.J., 2007, Cenozoic paleoclimate on land in North America: The Journal of Geology, v. 115, p. 271–294, doi: 10.1086/512753.

Retallack, G.J., Bestland, E.A., and Fremd, T.J., 2000, Eocene and Oligocene Paleosols in Central Oregon: Geological Society of America Special Paper 344, 192 p.

Retallack, G.J., Orr, W.J., Prothero, D.R., Duncan, R.A., Kester, P.A., and Ambers, C.P., 2004, Eocene-Oligocene extinction and paleoclimatic change near Eugene, Oregon: Geological Society of America Bulletin, v. 116, p. 817–839, doi: 10.1130/B25281.1.

Robert, C., and Kennett, J.P., 1997, Antarctic continental weathering during the Eocene-Oligocene cryosphere expansion: Clay mineral and oxygen isotopic evidence: Geology, v. 25, p. 587–590, doi: 10.1130/0091-7613(1997)025<0587:ACWCDE>2.3.CO;2.

Sheldon, N.D., 2003, Pedogenesis and geochemical alteration of the Picture Gorge Subgroup, Columbia River Basalt, Oregon: Geological Society of America Bulletin, v. 115, p. 1377–1387, doi: 10.1130/B25223.1.

Sheldon, N.D., 2005, Do red beds indicate paleoclimatic conditions?: A Permian case study: Palaeogeography, Palaeoclimatology, Palaeoecology, v. 228, p. 305–319, doi: 10.1016/j.palaeo.2005.06.009.

Sheldon, N.D., 2006a, Abrupt chemical weathering increase across the Permian-Triassic boundary: Palaeogeography, Palaeoclimatology, Palaeoecology, v. 231, p. 315–321, doi: 10.1016/j.palaeo.2005.09.001.

Sheldon, N.D., 2006b, Quaternary glacial-interglacial climate cycles in Hawaii: The Journal of Geology, v. 114, p. 367–376, doi: 10.1086/500993.

Sheldon, N.D., and Retallack, G.J., 2001, Equation for compaction of paleosols due to burial: Geology, v. 29, no. 3, p. 247–250, doi: 10.1130/0091-7613(2001)029<0247:EFCOPD>2.0.CO;2.

Sheldon, N.D., and Retallack, G.J., 2004, Regional precipitation records from the late Eocene and early Oligocene of North America: The Journal of Geology, v. 112, p. 487–494, doi: 10.1086/421076.

Sheldon, N.D., Retallack, G.J., and Tanaka, S., 2002, Geochemical climofunctions from North American soils and application to paleosols across the Eocene-Oligocene boundary in Oregon: The Journal of Geology, v. 110, p. 687–696, doi: 10.1086/342865.

Sheldon, N.D., Mitchell, R.L., Collinson, M.E., and Hooker, J.J., 2009, this volume, Eocene-Oligocene transition paleoclimate and paleoenvironmental record from the Isle of Wight (UK), *in* Koeberl, C., and Montanari, A., eds., The Late Eocene Earth—Hothouse, Icehouse, and Impacts: Geological Society of America Special Paper 452, doi: 10.1130/2009.2452(16).

Tabor, N.J., Montañez, I.P., Zierenberg, R., and Currie, B.S., 2004, Mineralogical and geochemical evolution of a basalt-hosted fossil soil (Late Triassic, Ischigualasto Formation, northwest Argentina): Potential for paleoenvironmental reconstruction: Geological Society of America Bulletin, v. 116, p. 1280–1293, doi: 10.1130/B25222.1.

Tabrum, A.R., Prothero, D.R., and Garcia, D., 1996, Magnetostratigraphy and biostratigraphy of the Eocene-Oligocene transition, southwestern Montana, *in* Prothero, D.R., and Emry, R.J., eds., The Terrestrial Eocene-Oligocene Transition in North America: New York, Cambridge University Press, p. 278–311.

Terry, D.O., Jr., 2001, Paleopedology of the Chardron Formation, northwestern Nebraska: Implications for paleoclimate in the North American mid-continent across the Eocene-Oligocene boundary: Palaeogeography, Palaeoclimatology, Palaeoecology, v. 168, p. 1–39, doi: 10.1016/S0031-0182(00)00248-0.

Wolfe, J.A., 1995, Paleoclimatic estimates from Tertiary leaf assemblages: Annual Review of Earth and Planetary Sciences, v. 23, p. 119–143, doi: 10.1146/annurev.ea.23.050195.001003.

Zachos, J., Pagani, M., Sloan, L., Thomas, E., and Billups, K., 2001, Trends, rhythms, and aberrations in global climate 65 Ma to present: Science, v. 292, p. 686–693, doi: 10.1126/science.1059412.

Zanazzi, A., Kohn, M.J., McFadden, B.J., and Terry, D.T., 2007, Large temperature drop across the Eocene-Oligocene transition in central North America: Nature, v. 445, p. 639–642.

Zanazzi, A., Kohn, M.J., and Derry, D.T., 2009, this volume, Biostratigraphy and paleoclimatology of the Eocene-Oligocene boundary section at Toadstool Park, northwestern Nebraska, U.S.A., *in* Koeberl, C., and Montanari, A., eds., The Late Eocene Earth—Hothouse, Icehouse, and Impacts: Geological Society of America Special Paper 452, doi: 10.1130/2009.2452(13).

Manuscript Accepted by the Society 16 September 2008

Printed in the USA

The Geological Society of America
Special Paper 452
2009

Eocene-Oligocene transition paleoclimatic and paleoenvironmental record from the Isle of Wight (UK)

Nathan D. Sheldon*
Ria L. Mitchell
Margaret E. Collinson
Geology Department, Royal Holloway, University of London, Egham, Surrey TW20 0EX, UK

Jerry J. Hooker
Palaeontology Department, Natural History Museum, Cromwell Road, London SW7 5BD, UK

ABSTRACT

Four different types of paleosols are recognized in the late Eocene–earliest Oligocene Solent Group (Isle of Wight, UK), representing a patchwork of ecosystems. Weakly developed marsh paleosols (Entisol-like; histic Inceptisol-like) are the most common, and there are relatively fewer, slightly elevated Inceptisol-like and Alfisol-like paleosols present as well. The more developed paleosols allow for a quantitative paleoclimatic reconstruction. The Eocene-Oligocene transition is associated globally with the Oi-1 glaciation event. Some nonmarine sequences show long-term cooling and aridification associated with the glaciation. Reconstructed paleoclimatic conditions using Solent Group paleosols do not; instead, they reflect steady mean annual temperatures and gradually increasing mean annual precipitation. This result is consistent with previous evaluations of floral assemblages, which indicate consistent vegetative covering and niche floral elements spanning the Eocene-Oligocene transition. In contrast, there is a significant change in the mammalian faunas found throughout western Europe (Grande Coupure). The evidence for relatively static climatic conditions is not consistent with the scenario of a climatically driven turnover event for the Grande Coupure, although the impact of increased seasonality cannot be ignored.

Keywords: England, Eocene-Oligocene, paleosols, Isle of Wight, paleoclimate.

INTRODUCTION

The Eocene-Oligocene transition coincides with the first permanent ice sheets of the Cenozoic (oxygen isotope Oi-1 glaciation; e.g., Zachos et al., 2001; Katz et al., 2008) and immediately predates the Grande Coupure ("big break"; Stehlin, 1910), a major mammalian turnover event across Europe (Hooker et al., 2004). Lagoonal and continental Hampshire Basin strata on the Isle of Wight, UK, preserve that faunal turnover (Hooker et al., 2004) and include charophytes (Sille et al., 2004) and other fossils useful for paleotemperature reconstructions (Grimes et al., 2003, 2005). Other paleontological finds include algae, dinoflagellates, foraminifera, mollusks, ostracodes, crustaceans, insects, arachnids, amphibians, reptiles, and fish (Gale et al., 2006).

*Present address: Department of Geological Sciences, University of Michigan, 2534 CC Little Bldg., 1100 N. University Avenue, Ann Arbor, Michigan 48103, USA; nsheldon@umich.edu.

Sheldon, N.D., Mitchell, R.L., Collinson, M.E., and Hooker, J.J., 2009, Eocene-Oligocene transition paleoclimatic and paleoenvironmental record from the Isle of Wight (UK), *in* Koeberl, C., and Montanari, A., eds., The Late Eocene Earth—Hothouse, Icehouse, and Impacts: Geological Society of America Special Paper 452, p. 249–259, doi: 10.1130/2009.2452(16). For permission to copy, contact editing@geosociety.org.

The substrate in which this rich assemblage has been preserved has been subject to relatively less study (e.g., Daley, 1999), and though the presence of paleosols has been previously noted (e.g., Gale et al., 2006), they have not been documented in any meaningful way. The purpose of this paper is to describe those paleosols and to use them to quantitatively estimate the climatic conditions in which they formed.

GEOLOGICAL SETTING

Eocene-Oligocene transition strata on the Isle of Wight are part of the Solent Group, which is composed of three formations and various members (Fig. 1; Daley, 1999). The Solent Group consists of brackish and freshwater clays, paleosols, sands, and thin freshwater limestones; the full succession reaches a maximum thickness of 200 m (Gale et al., 2006). On the basis of laminated, varve-like sediments, Hooker et al. (2005) estimated a maximum sedimentation rate of 0.019 cm/yr for parts of the Bembridge Marls but considerably less for lower parts of the succession (Hooker et al., this volume). Pre– and post–Grande Coupure faunas are separated by ~5 m stratigraphically from the base of at the *Nematura* Bed (Fig. 1). Given the hiatus below the *Nematura* Bed, there is a gap of at least 350 k.y. in the mammalian record of the Grande Coupure at this site (see Hooker et al., this volume). Though the precise level of the Eocene-Oligocene boundary was a source

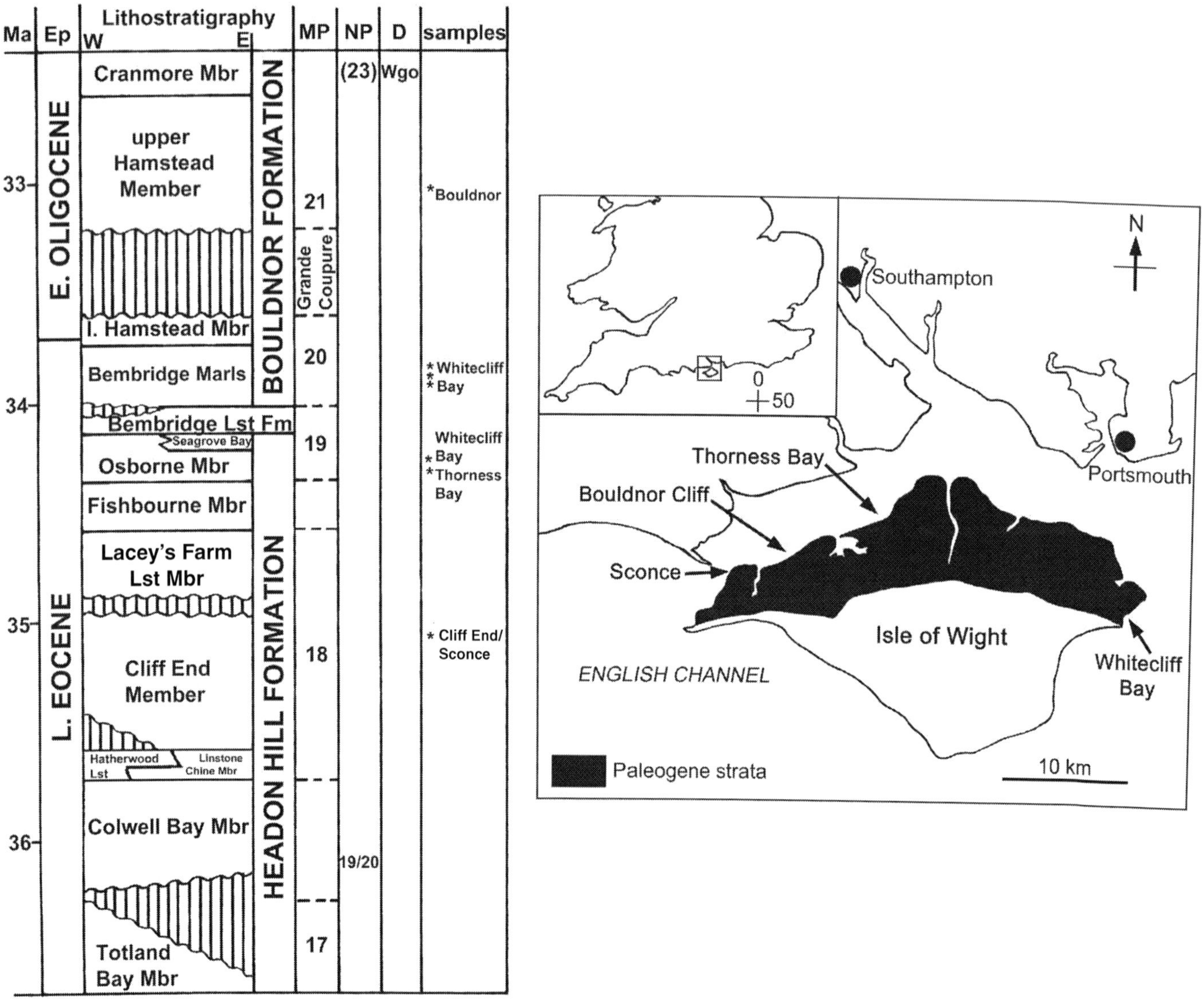

Figure 1. Solent Group stratigraphy and location map. Asterisks represent sample levels of paleosols. For further explanation of the stratigraphic correlation of the Solent Group to other European sequences, see Hooker et al. (this volume). MP—Paleogene mammal level; NP—nannoplankton zone; D—Belgian dinoflagellate zone.

of recent discussion (e.g., Gale et al. [2006] versus Hooker et al. [2004]), this has been resolved (Hooker et al., this volume), and there is general agreement that the Grande Coupure postdates the Eocene-Oligocene boundary and coincides with the initiation of the Oi-1 glaciation (Zachos et al., 2001).

METHODS

Stratigraphic sections were measured and samples (~50 g per sample) were collected from paleosols exposed at each of the sites (Mitchell, 2006). Paleosol profiles were classified with modern soil taxonomic terms (Soil Survey Staff, 1999; Table 1) using observations of Munsell color, reaction with dilute acid, and other field characteristics (Retallack, 2001). However, given diagenetic (e.g., Retallack, 2001; Sheldon and Retallack, 2001) changes to the paleosols, modern taxonomy is not strictly applicable, and therefore "-like" has been added to the end of the taxonomic term most closely relevant to the paleosol type (e.g., Entisol-like). The tops of profiles are identified as the surface from which root traces, rhizoliths, or drab-haloed root traces emanate. Major and trace elements were analyzed at Royal Holloway University London (RHUL) using X-ray fluorescence (XRF); the major-element data are contained in Table 2; trace-element and loss on ignition (LOI) data may be found in Mitchell (2006). Samples were dried in a 60 °C oven and crushed using a Teema Swing Mill. Fused pellets were made by combining 0.8 g of sample with 4.8 g of lithium tetraborate flux and placing the mixture into an 1100 °C furnace; the resulting melts were cast as glass disks and cooled for major-element analysis on the Phillips PW-1480 XRF at RHUL. Pressed pellets were made for analysis of trace-element composition on the same XRF by combining 7 g of sample with 5–10 drops of poly vinyl acetate (PVA) (as needed) and ~25 g of boric acid. The resulting mixture was poured into a mold and subjected to a pressure of 10 tons per square inch for up to 2 min to make the pellet. Analytical uncertainty was 0.01% for the major elements and 1–3 ppm for trace elements (Mitchell, 2006). Relative ages were derived using a combination of stratigraphic superposition and mammal biostratigraphy, following the scheme of Hooker et al. (2004, this volume). Additional paleopedological details may be found in Mitchell (2006).

RESULTS

Paleosol Taxonomy

Four distinctive pedotypes were identified. In comparison with modern soil orders, and in order of progressive maturity, they are Entisol-like, Inceptisol-like, histic Inceptisol-like, and Alfisol-like (Fig. 2; Soil Survey Staff, 1999). Entisol-like paleosols are typically thin, characterized by minimal horizonation (A–C profiles; Fig. 2E) or development, identified primarily on the basis of drab-haloed root traces in the A horizon, and found throughout the sequence. They were not considered sufficiently developed to be used for paleoclimatic reconstruction purposes (Sheldon et al., 2002; Table 1). Inceptisol-like paleosols (Bembridge Marls; Upper Hamstead Member; Cliff End Member; Osborne Member) are moderately developed with moderate horizonation (A-Bw-C profiles; e.g., Fig. 2A), and they typically have larger and more deeply penetrating drab-haloed root traces than Entisols. More well-developed pedotypes are rare, but Alfisol-like and a single histic Inceptisol-like pedotype were also identified. The Alfisol-like paleosols (e.g., Osborne Member) are moderately developed, have strong horizonation (A-Bt-C profiles), and typically have large, deeply penetrating drab-haloed root traces (Fig. 2C), including rare taproot traces. The lone histic Inceptisol-like paleosol is represented by the "Black Band" (Fig. 1; lower Hamstead Member), which is a thin accumulation (decimeter-scale) of organic-rich mud with root traces and carbonate globules immediately beneath the organic O/A horizon. All of the paleosols are clay-rich (>60% based on thin sections), and most of the clay is illite or illite-rich illite-smectite and lesser amounts of kaolinite and chlorite (extensive analysis of the clay minerals of the Solent Group may be found in Gale et al., 2006).

Postburial Alteration of Paleosols

Though these paleosols and their associated sediments have not been deeply buried, and thus show minimal compaction (e.g., Sheldon and Retallack, 2001), they have still been subjected to some postburial alteration. All of the paleosols described herein show evidence of two alteration processes:

TABLE 1. PALEOENVIRONMENTAL INTERPRETATION OF EOCENE-OLIGOCENE PALEOSOLS

Paleosol	Paleoclimate	Former vegetation	Paleotopography	Parent material	Time (yr)*
Entisol-like	Too poorly developed, but probably humid	Marsh, poorly drained	Boggy floodplain at least seasonally inundated	Silts, sands, muds	50–500
Inceptisol-like	Humid (600–1200 mm/yr), seasonally dry	Marsh, moderately drained	Microhighs in the topography	Silts, sands, muds	200–2000
Histic Inceptisol-like	Humid (>1000 mm/yr)	Marsh, poorly drained	Microlows in the topography	Silts and muds	200–2000
Alfisol-like	Humid (600–1200 mm/yr), seasonally dry	Riparian swale woodland	Terraces	Silts and muds	200–2000

*Soil formation times are estimates based on the qualitative degree of development of features such as ped development, horizonation, the presence of relict bedding, etc., and Figure 8-17 of Birkeland (1984), and Retallack (2001). Application of modern taxonomy is limited to the Order level; however, given that many of these paleosols show evidence of hydromorphy and waterlogging, the paleosols could be further classified as Aquents, Aquepts, and Aqualfs.

TABLE 2. GEOCHEMICAL DATA CORRECTED FOR LOSS ON IGNITION

Sample	Strat. pos.*	Horizon	SiO_2	Al_2O_3	Fe_2O_3	MgO	CaO	Na_2O	K_2O	TiO_2	MnO	P_2O_5	MAP†	MAT†	Type
FV1	−145	C	36.34	10.36	4.83	1.25	22.87	0.12	1.88	0.58	0.15	0.05	n/a	n/a	Incpt.
FV2	−100	C	43.99	14.40	6.56	1.76	13.38	0.18	3.35	0.59	0.05	0.08	n/a	n/a	Incpt.
FV3	−85	BC2	44.26	14.71	5.79	1.73	13.84	0.08	3.28	0.60	0.07	0.07	n/a	n/a	Incpt.
FV4	−75	BC	31.49	10.17	5.54	1.04	24.92	0.29	1.68	0.51	0.20	0.08	n/a	n/a	Incpt.
FV5	−65	Bw	50.77	15.79	6.36	1.29	21.79	0.07	2.62	0.83	0.17	0.07	n/a	n/a	Incpt.
FV6	−35	Bw	42.40	13.55	6.38	1.13	16.36	0.04	2.24	0.69	0.24	0.08	512	13.0	Incpt.
NT2	−188	BC	56.34	21.36	12.94	1.90	1.35	0.74	3.15	1.09	0.02	0.01	n/a	n/a	Alfis.
NT3	−165	Bt2	48.77	16.85	9.42	1.87	18.76	0.48	2.57	0.78	0.19	0.14	n/a	n/a	Alfis.
NT4	−125	Bt	55.25	20.91	9.26	1.73	7.40	0.42	3.33	0.95	0.06	0.09	716	13.5	Alfis.
NT5	−120	C	60.42	18.70	6.46	1.77	7.59	0.16	3.21	0.95	0.06	0.11	n/a	n/a	Incpt.
NT7	−65	BC	37.02	12.77	6.56	1.33	19.62	0.18	2.03	0.62	0.14	0.13	n/a	n/a	Incpt.
NT8	−20	Bw	42.18	17.35	9.04	1.47	12.17	0.74	2.60	0.70	0.10	0.16	512	13.0	Incpt.
NT9	−10	A	25.06	9.75	2.79	1.22	30.66	0.09	1.53	0.43	0.19	0.07	n/a	n/a	Incpt.
WB1	−275	R	73.02	9.14	2.49	0.87	5.87	0.46	1.73	0.87	0.05	0.14	n/a	n/a	Incpt.
WB2	−250	R	75.49	9.42	2.90	0.82	3.18	0.19	1.79	0.99	0.04	0.11	n/a	n/a	Incpt.
WB3	−233	C	45.22	14.48	7.29	1.42	13.37	0.11	2.43	0.67	0.14	0.14	n/a	n/a	Incpt.
WB4	−220	C	40.24	13.59	6.13	1.49	17.70	0.10	2.21	0.61	0.19	0.12	n/a	n/a	Incpt.
WB5	−202	BC	62.53	18.01	6.87	1.55	1.02	0.20	3.23	0.91	0.02	0.07	n/a	n/a	Incpt.
WB6	−185	BC	69.64	15.35	5.03	1.34	0.61	0.22	2.77	0.92	0.01	0.08	n/a	n/a	Incpt.
WB7	−175	BC	69.71	14.22	4.43	1.19	0.98	0.22	2.58	0.89	0.01	0.06	n/a	n/a	Incpt.
WB8	−143	BC	62.37	19.00	8.92	1.72	2.85	0.31	3.29	0.85	0.03	0.13	n/a	n/a	Incpt.
WB9	−100	Bt/Bw	57.09	16.65	7.44	1.60	4.68	0.26	2.97	0.77	0.04	0.09	796	13.2	Incpt.
WB10	−25	A	47.13	11.59	5.92	1.51	15.14	0.22	2.69	0.57	0.12	0.13	n/a	n/a	Incpt.
WB11	−10	A	52.88	12.00	5.93	1.59	11.41	0.09	2.99	0.63	0.08	0.16	n/a	n/a	Incpt.
BC1	−37	C	25.84	10.54	46.47	3.76	8.82	0.46	1.46	0.49	1.85	0.38	n/a	n/a	H. Inc.
BC2	−11	Bw	59.68	24.03	7.95	2.20	2.41	1.10	3.60	0.92	0.11	0.23	1058	12.9	H. Inc.
BC3	−1	A/Bw	49.78	19.66	8.12	1.94	3.86	0.79	3.17	0.82	0.06	0.24	n/a	n/a	H. Inc.
BC4	−158	C	63.30	22.86	5.64	1.50	0.65	0.58	3.57	1.31	0.01	0.07	n/a	n/a	Incpt.
BC5	−120	BC	59.19	20.03	9.33	1.37	0.62	0.23	2.94	1.22	0.02	0.11	n/a	n/a	Incpt.
BC6	−63	Bt/Bw	58.84	19.13	9.12	1.24	0.60	0.55	2.98	1.20	0.03	0.09	1316	13.3	Incpt.
BC7	−5	A	64.55	18.76	5.60	1.29	0.56	0.29	2.99	1.34	0.02	0.07	n/a	n/a	Incpt.

*Stratigraphic positions are in cm as measured down from the top of the paleosol.

†MAP (mean annual precipitation; mm yr^{-1}) and MAT (mean annual temperature; °C) are calculated using equations (1) and (2), except for FV6, which used equation (3) of Sheldon et al. (2002) because of CaO contamination. Samples are from the Cliff End Member (Sconce; FV), Osborne Member (Thorness Bay, NT), Bembridge Marls Member (Whitecliff Bay, WB), and lower (BC1-3) and upper (BC4-7) Hamstead Members (Bouldnor; BC).

(1) oxidation of organic matter; and (2) reddening of oxyhydroxides by dehydration (Retallack, 1991).

Preserved organic matter is nearly absent in these paleosols, with the exception of the histic Inceptisol "Black Band" level. Paleosols typically have significantly less organic matter than their modern equivalents due to efficient biotic decomposition as well as burial dissolution (Retallack, 1991). Nonetheless, there is significant evidence for the former presence of organic matter in the form of drab-haloed root traces, which form due to micro-reducing conditions around decomposing organic matter in an otherwise oxidizing setting (Figs. 2B–2C; e.g., Sheldon, 2005). Burial has also significantly reddened most of the paleosols via dehydration of iron oxyhydroxides (e.g., goethite) to hematite (Retallack, 1991).

Factors of Soil Formation

Climate

Though "calcrete nodules" have previously been reported (e.g., Gale et al., 2006), none that fits a classical soil science definition (Machette, 1985) has been found, even at those previously discussed levels. Instead, those "nodules" are unlithified carbonate-rich globules, where at least some of the carbonate comes from detrital shell material rather than from in situ pedogenic processes. The globules do not form in distinctive horizons and are instead randomly arrayed. As a result, widely applied relationships between mean annual precipitation and Bk (carbonate) horizon depth (Retallack, 2005) are not applicable here. Instead, paleoclimatic conditions are estimated on the basis of chemical weathering proxies.

Measurements of the degree of chemical weathering of paleosol Bt and Bw horizons can be related to mean annual precipitation (MAP) and mean annual temperature (MAT) using climofunctions derived from modern soils that give results consistent with independent estimates including plant fossils (Sheldon et al., 2002). The chemical index of alteration without potassium (*CIA-K*) can be related to MAP as follows:

$$\text{MAP (mm)} = 221.2e^{0.0197 CIA\text{-}K}, \quad (1)$$

where $R^2 = 0.72$, standard error is ±182 mm (Sheldon et al., 2002), and $CIA\text{-}K = 100([mAl_2O_3]/[mAl_2O_3 + mCaO + mNa_2O])$ (Maynard, 1992). MAT can be estimated with less accuracy as follows:

$$\text{MAT (°C)} = -18.5S + 17.3, \quad (2)$$

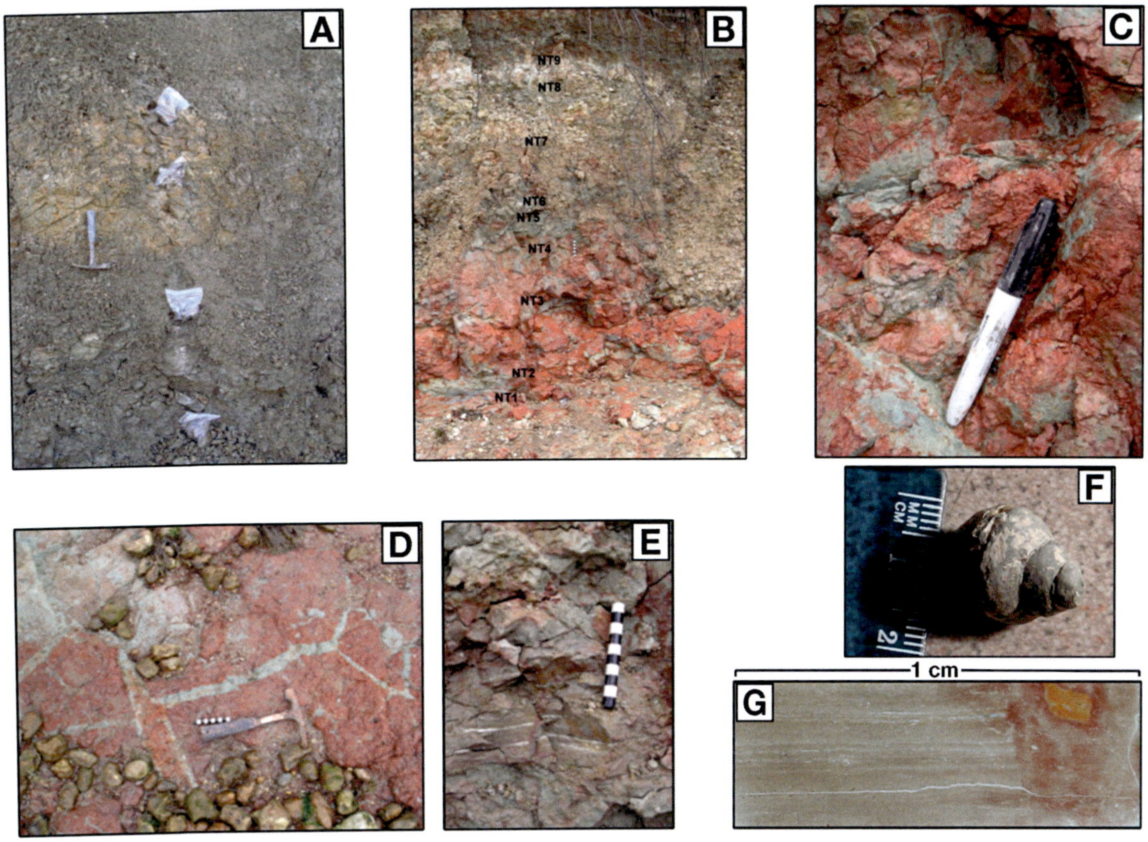

Figure 2. Field photos of Solent Group paleosols. (A) Color change indicates horizonation. (B) Profile through two paleosols in the Osborne Member in Thorness Bay; sample numbers are indicated on the photo. (C) Drab-haloed root traces in an Alfisol-like paleosol in Thorness Bay. The micro-gleyed root-traces formed in response to oxidation of organic matter. (D) Desiccation cracks in mudstone in Thorness Bay. (E) Laminated Entisol-like paleosol in Thorness Bay. (F) Internal mold of freshwater gastropod *Viviparus* from the Cliff End Member at Sconce. (G) Large-format thin section of pedogenic modification of the laminated muds at Thorness Bay; sample from E. Scales are: rock hammer = 30 cm, Sharpie = 12 cm; others have centimeter scales given.

where R^2 = 0.37, standard error (SE) is 4.4 °C, and $S = (mK_2O + mNa_2O)/mAl_2O_3$ (Sheldon et al., 2002). This approach has been widely applied to paleosols of various ages and various types, ranging from the Paleozoic (Tabor et al., 2004; Driese et al., 2005) to the Cenozoic (e.g., Sheldon, 2003; Sheldon and Retallack, 2004; Sheldon, 2005; Sheldon, 2006b; Hamer et al., 2007a, 2007b). Reconstructed MAP values for the Solent Group range from 500 to 1300 mm/yr (±182 mm SE) with a steady MAT of 13–14 °C (±4 °C SE) (Table 2; Fig. 3). Additional evidence of fluctuating climatic conditions comes in the form of desiccation cracks throughout the Solent Group (Fig. 2D). Gale et al. (2006) also found that the clay mineral assemblage was highly variable throughout the Solent Group, with most of the variation taking place between the proportion of illite and smectite, and suggested that the variability indicated changing seasonality.

Organisms

On the basis of nearest living relatives, the flora of the Solent Group represents freshwater marsh, pond, and lake plants and intermittent periods that include both relatively drier floral elements and brackish rather than freshwater conditions (Collinson and Cleal, 2001). During deposition of the Bembridge Marls Member (Fig. 1; Bouldnor Formation), the vegetation has been reconstructed as a *Typha*/*Acrostichum* marsh with patches of open water supporting a rich wetland flora including marginal (e.g., Cyperaceae, Alismataceae, Juncaceae), open-water (e.g., Nymphaeaeceae, Potamogetonaceae, Characeae), and free-floating (e.g., *Azolla*, *Stratiotes*) plants. This vegetation is best represented in the Bembridge Marls at Hamstead Ledge (Collinson, 1983) in western Isle of Wight, but it is also well-documented further east at Gurnard Ledge (Collinson, 1983) and at one horizon in Whitecliff Bay in the east part of the island (Collinson, 1990).

The existence of this freshwater marsh flora is supported by a wide range of evidence, both sedimentological and paleobiological. Plant macrofossils and mesofossils occur in abundance in finely laminated (with couplets, possibly varves), fine-grained, rooted mud rocks and are accompanied by several taxa of freshwater algae (e.g., Sille et al., 2004; fossil resting cysts and cell walls), freshwater planktic and benthic ostracods, fish

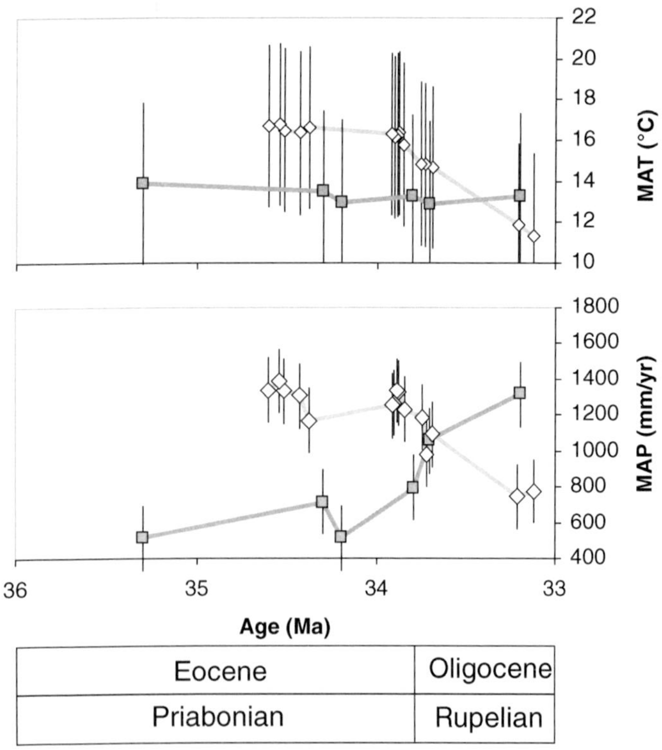

Figure 3. Reconstructed paleoclimate (MAT—mean annual temperature, MAP—mean annual precipitation) based on Solent Group paleosols compared to Oregon. Open symbols represent data from Oregon (Sheldon and Retallack, 2004), and closed symbols represent data from the Solent Group succession.

and turtle remains, and freshwater mollusks, including lymnaeid, planorbid, and viviparid gastropods (Fig. 2F), and occasional unionid bivalves.

In addition, several of the plants from this assemblage are known as fossils from multiple organs, all of which confirm their near identity with living relatives. *Typha* is represented by leaves, fruits, and seeds, and probable pollen (Collinson 1983). *Acrostichum* is represented by foliage, sporangial clusters, sterile hairs, and dispersed sporangia with spores. *Stratiotes* (free-floating monocotyledon) is represented by seeds and leaf-margin teeth. *Azolla* (free-floating water fern) is known from associated and organically connected megaspores and microspore massulae (Collinson 1983). In sum, the Bembridge Marls at Hamstead Ledge yield a rich and diverse biota from the freshwater marshes. This marsh flora also typifies the freshwater parts of the overlying Hamstead Member and, thus, the Bouldnor Formation as a whole.

Many elements of this marsh flora are also recorded in the underlying Bembridge Limestone Formation, rarely at Headon Hill, and in abundance and diversity at Gurnard Ledge (Collinson et al., 1993; Hooker et al., 1995). There are, however, subtle differences between the wetland flora, other biota, and facies in the Bembridge Marls and Bembridge Limestone. Collinson et al. (1993) concluded that while the Bembridge Marls seem to reflect an extensive persistent marshland with some slight fluvial influence and some open waters, the Bembridge Limestone suggests the presence of ponds or lakes with a limited catchment area in comparatively dry surroundings and warm, shallow, clear, calcareous waters including areas of open water and areas with emergent vegetation.

Topographic Relief

There are no indications of significant topography in any of the paleosols, nor on the basis of stratigraphic relationships. There is some very gentle, centimeter-scale ridge-and-swale topography at the top of some of the profiles, but this is likely attributable to shrink-swell behavior of the clays in the Solent Group sediments rather than to any significant (i.e., meter-scale) topography at the time of formation. This is further supported by the paucity of erosional contacts in the Solent Group except at sequence boundaries and at transgressive surfaces (Gale et al., 2006; Hooker et al., this volume). However, given the presence of rooted aquatic plants, floating aquatic plants, and rooted nonaquatic plants, some microtopography must have been present to give rise to the variable water table.

Parent Material

The Solent Group consists of brackish and freshwater clays, sands, and thin freshwater limestones, and the full succession reaching a maximum thickness of 200 m (Gale et al., 2006). Paleosols are most typically parented by freshwater clays, consistent with an alternating brackish and freshwater sequence stratigraphic interpretation (Gale et al., 2006).

Time

Formation times for paleosols are inferred to have been short (Table 1) on the basis of generally weakly developed horizonation, a dearth of large root traces, and relatively high sedimentation rates. In addition, field observations suggest that >90% of the paleosols (or of total pedogenically altered strata) are Entisol-like, indicating short formation times (Retallack, 2001).

Pedogenesis

Weathering Ratios

A number of ratios between major and trace elements have been to shown provide information about both pedogenic processes (Retallack, 2001; Sheldon et al., 2002; Sheldon, 2005; Perez-Huerta and Sheldon, 2006; Sheldon, 2006a, 2006b) and postburial alteration (Retallack, 1991; Sheldon, 2003, 2006a, 2006b). The molecular ratio of Ti to Al is taken as a measure of provenance because the two elements become mobile under very different conditions and weathering environments (Sheldon, 2006a, 2006b). By plotting Ti/Al with depth for each of the paleosols used in the paleoclimatic reconstruction, it is possible to look for evidence of inheritance

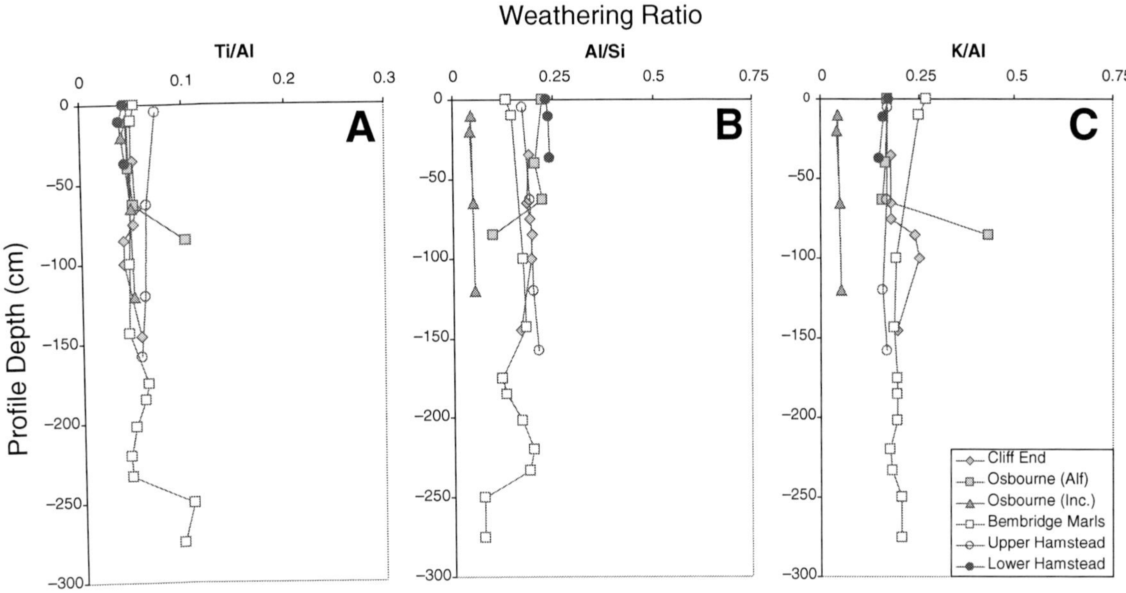

Figure 4. Weathering ratios as a function of depth for the paleosols. The nearly constant Ti/Al (A) and Al/Si (B) values both within and between paleosols indicate a common provenance, and the nearly constant K/Al ratios (C) both within and between paleosols are consistent with minimal postburial alteration or salinization.

or multiple provenances. The constant Ti/Al values with depth both within and among the paleosols (Fig. 4A) suggest a single, shared provenance for all of the paleosols. The molecular ratio of Al to Si is a measure of "clayeyness" (Retallack, 2001; Sheldon, 2006a, 2006b). More intense weathering should result in higher Al/Si ratios or in windborne addition of clay minerals. In general, the Al/Si ratios of the Solent Group paleosols (Fig. 4B) are very consistent both within and between paleosols. The only perceptible shift occurs between the Osborne Inceptisol-like paleosol and the lower Hamstead Inceptisol-like paleosol, which corresponds roughly in time to the Eocene-Oligocene transition, perhaps suggesting a slight increase in chemical weathering coincident with the moderate wetting recorded by the paleosols, though, in general, the degree of "clayeyness" is roughly constant. This result is in contrast to Sheldon et al. (2002), who found declining degrees of chemical weathering across the Eocene-Oligocene transition, though again, any increase in the Solent Group is minor. The most common postburial evidence of metasomatism is the addition of K to paleosols (Maynard, 1992; Fedo et al., 1995; Sheldon, 2003). This typically occurs as K-rich fluids move at contacts between units (Sheldon, 2003), so if metasomatism has occurred, the highest levels of K should be near the top of paleosols. By normalizing K to Al content, it is possible to assess addition of K by metasomatism. As Figure 4C shows, none of the K/Al ratios of the Solent Group paleosols shows any significant up-profile increase, and the values are generally quite consistent among the different types of paleosols, suggesting that they have all undergone similar postburial processes and that metasomatism was not a significant factor.

DISCUSSION

Paleoclimatic Record

Previously published long-term Eocene-Oligocene transition records are primarily from North America (Terry, 2001; Sheldon and Retallack, 2004; Retallack, 2007; Sheldon, this volume, chapter 15). Among those records, the physiographic setting of the Solent Group paleosol sequence is most similar to Oregon, which was relatively close to the ocean and subject to minimal orographic effects until after the Eocene-Oligocene transition, when the emplacement of the Cascade Range resulted in expansion of a rain shadow on its eastern side (Sheldon and Retallack, 2004; Retallack, 2007). The basinal settings are not identical though, since Oregon represents a backarc basin behind the nascent Cascade Range, and the Isle of Wight represents a coastal plain (Hooker et al., this volume). A more thorough discussion comparing and contrasting the different nonmarine records may be found in Sheldon (this volume, chapter 15).

The two sequences show very different paleoclimatic patterns across the Eocene-Oligocene transition (Fig. 3). Oregon experienced cooling and drying paleoclimatic conditions, though not as a step function or associated directly with Oi-1 glaciation (Zachos et al., 2001), but more as a part of longer-term climatic deterioration. In contrast, the Solent Group records relatively steady paleotemperatures accompanied by fairly significant wetting. There are a few possible explanations for these differences. First, both regions were subject to some continental drift: the UK was shifting northward and Oregon was shifting eastward. When combined with the emplacement of the Cascade Range and the newly formed rain shadow (Sheldon and Retallack, 2004), increasing continentality would be expected in Oregon, which would result in falling MAP and MAT values. However, given the stable paleotemperatures indicated by the Isle of Wight succession and the general tendency for more northerly regions to be cooler, this is unlikely to be the whole explanation, because long-term cooling would also be expected there. Second, a transgression could make the coastal plain environment on the Isle of Wight wetter, but all indications are that the Eocene-Oligocene transition was a time falling sea level due to the buildup of ice on Antarctica. Finally, a change to the seasonality of the climate, which would be latitudinally dependent, could result in climatic changes with opposite signs in otherwise relatively similar settings (e.g., Ivany et al., 2000). This final explanation, when combined with the increasing continentality of Oregon, seems to be the most likely explanation. A shift to a Mediterranean climate on the Isle of Wight could have resulted in unchanged paleotemperatures but a net gain in precipitation due to wetter winters. A strongly seasonal climate for the Solent Group has been previously suggested on the basis of a highly variable clay mineral assemblage (Gale et al., 2006).

The Isle of Wight paleoclimatic record is also in contrast to other Eocene-Oligocene paleoclimatic records based on paleosols, because though constant temperatures are also indicated in the Ebro Basin of Spain (Sheldon, this volume, chapter 15) and in Argentina (Kohn et al., 2004), progressively wetter conditions are not known from any other region. This range of responses to the Oi-1 glaciation event is important because it suggests that during relatively rapid climatic change events, different regions will respond differently and that single-region proxy records may not represent global conditions (e.g., Zanazzi et al., 2007). Thus, integration of multiple sites (Sheldon and Retallack, 2004; Sheldon, this volume, chapter 15) is crucial to understanding nonmarine climatic change.

Comparison to Previous Paleontological Results

Mammals

Mammals occur at numerous horizons in the Solent Group, but they are better represented in western (including the Hampshire mainland) than eastern sites. They are diverse, and the most up-to-date lists are by Hooker et al. (2005). Before the major turnover in the earliest Oligocene, nearly all went extinct at family rank, except dormice (Gliridae) and moles (Talpidae), although most are attributable to modern orders. The mammals mainly occur as isolated teeth, jaw fragments, and postcranial bones; articulated skeletons are exceptionally rare. In terms of biostratigraphy, they span European Mammalian reference levels MP17–21 (Fig. 1) and include faunas referable to the entire series; a zonation has also been erected as a more appropriate system for such a continuous sequence of faunas, which is unlike most of Europe, but which has application also outside the UK (Hooker, 1987, 1992). The MP levels and zones are critical to correlation with mammal faunas in continental strata elsewhere in Europe (Hooker et al., this volume). However, they do not operate in other continents or in the marine realm except in cases of co-occurrence with marine zonal indicators or intercalation with units containing them (e.g., the spot occurrence of nannoplankton zones NP19–20 sandwiched between MP17 and 18).

Gradual changes in the mammal faunas provide evidence of evolutionary change (e.g., Hooker, 1991) and of changing environments, broadly indicating opening up of forested habitats (Collinson and Hooker, 1987; Hooker, 1992; Hooker et al., 1995). The greatest change, however, is in the earliest Oligocene, when a major turnover termed the Grande Coupure (Stehlin, 1910) saw extinction of many taxa endemic to "Island Europe" and their replacement or displacement by many newcomers from Asia. This event is recorded between the lower and upper parts of the Hamstead Member and coincides with the Oi-1 glaciation (Hooker et al., 2004).

There are arguments both for climate change and for competition following dispersal as causes of the turnover (Hooker, 2000); however, the relatively steady climatic conditions reconstructed by this study in terms of MAT and MAP and of summer temperatures by fossil-based isotope proxies (Grimes et al., 2005) imply that climatic change had little impact. However, evidence for increased seasonality in the marine realm (e.g., Ivany et al., 2000) suggests that climate may still have played an important role. Ongoing work on Solent Group mammals and associated European mammals (Joomun et al., 2008) and on freshwater gastropods will test the role that seasonal changes may have played in the Grand Coupure. Groups of endemic European mammals that were decimated are: families Palaeotheriidae (horse relatives), Anoplotheriidae, Xiphodontidae (both distant camel relatives), Choeropotamidae (another family of cloven-hoofed mammals; Hooker and Thomas, 2001), Pseudosciuridae (rodents), and Nyctitheriidae (extinct insectivorous relatives of tree-shrews and primates; Hooker, 2001). Apart from one nyctithere tooth fragment, all of these had become extinct in the Hampshire Basin by the Grande Coupure. Families Theridomyidae (another extinct rodent type), Gliridae (dormice), Hyaenodontidae (extinct carnivorous mammals), and Herpetotheriidae (extinct opossum-like marsupials) crossed the Grande Coupure undiminished. Newcomers of Asian origin were rhinoceroses, anthracotheres (extinct hippopotamus relatives), advanced ruminants, hedgehogs, hamsters, beavers, Eomyidae (extinct rodents, sister group to pocket gophers plus kangaroo rats), and bears.

Plants

The variety of paleosol types preserved within the Solent Group suggests a mosaic of ecosystems. As discussed already, many of the floral elements exhibit wetland or marsh affinities. However, there are also a number of plant fossils that indicate drier or more well-drained conditions. For example, two horizons in the lower Hamstead Member contain abundant fossils of the conifer *Quasisequoia couttsiae* (Heer) Kunzmann represented by leafy shoots, seeds, and cone fragments (Hooker et al., 2004). Some of these shoots are woody (with secondary xylem).

Numerous large logs of coniferous wood also occur in the lower Hamstead Member (Hooker et al., 2004). In the early Priabonian Totland Bay Member of the Headon Hill Formation (Fig. 1) at Hordle, Hampshire, in situ tree stumps with *Glyptostroboxylon* wood are associated with foliage, cones, and seeds of *Q. couttsiae* (Fowler et al., 1973; Collinson and Cleal, 2001). The same types of tree stumps also occur at a slightly lower level in the Totland Bay Member at Totland Bay on the Isle of Wight. In the Totland Bay Member at Hordle, these occurrences were interpreted as swamp trees (Fowler et al., 1973), and Plint (1984) interpreted the unit containing them as in-channel point bar deposits of a minor distributary channel. Dispersed pollen records suggest the presence of Taxodiaceae/Cupressaceae throughout the Solent Group (*Inaperturopollenites* pollen). Hooker et al. (2004) concluded that the conifer records in the lower Hamstead Member reflected a vegetation that (together with a distinctive rodent fauna) represented the hinterland brought nearer to the depositional environment by regression during a late highstand systems tract. Palm foliage occurs rarely in the Bembridge Marls, and palm pollen is also present, though it becomes much less common upsection and disappears entirely at the Black Band (basalmost Hamstead Member) (Hooker et al., 2004). From pollen in the Headon Hill Formation and Upper Hamstead Member of the Bouldnor Formation, nonwetland floral elements are recognized (Boulter and Hubbard, 1982; Collinson, 1992) as Eocene deciduous angiosperm forest with water plant associations, though because palynological residues may be transported over long distances, it is difficult to judge how close the trees were to the place in which the pollen was deposited. The other floral elements of interest are rare xeromorphic leaf taxa, namely *Zizyphus zizyphoides* (Bembridge Marls; one specimen) and *Palibinia* (Bembridge Marls and Bembridge Limestone; 10 specimens known) (Collinson et al. 1993; Collinson and Hooker, 2003). These taxa have thick coriaceous microphyllous leaves, and they occur in southern European Oligocene floras in association with an increasing number of sclerophylls. This vegetation is considered by some to be the forerunner of Neogene Mediterranean sclerophyllous vegetation. Taking into account a variety of evidence, Collinson and Hooker (2003) concluded that the vegetation belt in southern Europe, which has common sclerophylls in the early Oligocene, probably reflects the existence of a slightly drier interval within an overall humid warm temperate climate. In the Bembridge Marls and Bembridge Limestone, *Palibinia* is the only leaf taxon consistently associated with the wetland flora. This may indicate sclerophylly, comparable to that of modern *Myrica*, linked to nutrient-poor soils rather than water limitations. Pollen of *Ephedripites* (nearest living relative to *Ephedra*) is present in the Insect Limestone of the Bembridge Marls Member (BMM) and in the Bembridge Limestone Formation (BLF) (Collinson et al., 1993). This is often referred to as an "arid element" in palynofloras. Again though, pollen are known to be transported over long distances, so this lone occurrence is probably not sufficient to argue for significant aridification. Other palynomorphs linked to an early Oligocene shift to a climate with a dry season in southern Europe by Cavagnetto and Anadón (1996), such as grasses, chenopods, Mimosaceae, and *Mahonia*, are not present in the UK succession. For further discussion, see vegetation with sclerophyllous elements in Collinson and Hooker (2003).

Two principal conclusions can be drawn from a comparison between the Solent Group paleosols and fossil flora. First, the mosaic of wet and dry ecosystems was present both in the Eocene and Oligocene with only minor changes in floral composition, and no changes in the paleosol assemblage, suggesting long-term environmental stability across the Eocene-Oligocene transition. Second, the combination of dryland and marsh floral elements and dryland and wetland paleosols is consistent with an Everglades-like collection of ecosystems where microtopographic differences played an important role in controlling the floral composition.

CONCLUSIONS

New observations of Eocene-Oligocene Solent Group paleosols bring to light a number of important results. First, the paleosols are consistent with at least seasonally inundated marsh conditions and with microtopography-controlled drainage patterns. This is consistent with the fossil flora that has been recovered, and with previous interpretations of an Everglades-like assemblage of ecosystems. Second, quantitative reconstruction of the paleoclimatic conditions at the time of the paleosols' formation suggests a long-term shift to wetter conditions (subhumid to humid) associated with the Eocene-Oligocene transition, but no significant cooling event. This result is in contrast to some North American paleosol-bearing sequences, but it is consistent with other European paleosol-bearing sequences. This suggests that care is needed to avoid overinterpretation (i.e., to assume a local sequence represents the global climate) of nonmarine sequences. Third, the relative climate stability seems inconsistent with the Grande Coupure mammalian turnover event having been climatically driven, though increased seasonal climatic variation remains a potential factor.

ACKNOWLEDGMENTS

This research was funded by a Research Strategy Fund grant (Royal Holloway University London) to Sheldon. Mitchell would like to thank M. Thirlwall and C. Manning for X-ray fluorescence (XRF) training. We wish to thank Dennis Terry and Greg Retallack for thorough reviews that improved many of the details in this manuscript, and we thank Christian Koeberl for additional suggestions that helped to improve clarity in places.

REFERENCES CITED

Birkeland, P.W., 1984, Soils and Geomorphology: New York, Oxford University Press, 372 p.

Boulter, M., and Hubbard, R.N.L.B., 1982, Objective palaeoecological and biostratigraphic interpretation of Tertiary palynological data by multivariate statistical analysis: Palynology, v. 6, p. 55–68.

Cavagnetto, C., and Anadón, P., 1996, Preliminary palynological data on floristic and climatic changes during the middle Eocene–early Oligocene of the eastern Ebro Basin, northeast Spain: Review of Palaeobotany and Palynology, v. 92, p. 281–305, doi: 10.1016/0034-6667(95)00096-8.

Collinson, M.E., 1983, Paleofloristic assemblages and paleoecology of the Lower Oligocene Bembridge Marls, Hamstead Ledge, Isle of Wight: Botanical Journal of the Linnean Society, v. 86, p. 177–225, doi: 10.1111/j.1095-8339.1983.tb00725.x.

Collinson, M.E., 1990, Plant evolution and ecology during the early Cainozoic diversification: Advances in Botanical Research, v. 17, p. 1–98, doi: 10.1016/S0065-2296(08)60132-9.

Collinson, M.E., 1992, The early fossil history of Salicaceae: A brief review: Proceedings of the Royal Society of Edinburgh, v. 98B, p. 155–167.

Collinson, M.E., and Cleal, C.J., 2001, Late middle Eocene–early Oligocene (Bartonian-Rupelian) and Miocene paleobotany of Great Britain, *in* Cleal, C.J., Thomas, B.A., Batten D.J., and Collinson, M.E., eds., Mesozoic and Tertiary Palaeobotany of Great Britain: Peterborough, Joint Nature Conservation Committee (JNCC), Geological Conservation Review Series no. 22, p. 185–226.

Collinson, M.E., and Hooker, J.J., 1987, Vegetational and mammalian faunal changes in the Early Tertiary of southern England, *in* Friis, E.M., Chaloner, W.G., and Crane, P.R., eds., The Origins of Angiosperms and Their Biological Consequences: Cambridge, Cambridge University Press, p. 259–304.

Collinson, M.E., and Hooker, J.J., 2003, Paleogene vegetation of Eurasia: Framework for mammalian faunas: Deinsea, v. 10, p. 41–83.

Collinson, M.E., Singer, R.L., and Hooker, J.J., 1993, Vegetational change in the latest Eocene of southern England, *in* Planderová, E., Konzálová, M., Kvacek, Z., Sitár, V., Snopková, P., and Suballyová, D., eds., Paleofloristic and Paleoclimatic Changes during Cretaceous and Tertiary: Bratislava, Dionyz Stur Institute of Geology, p. 81–87, pls. XX–XXV.

Daley, B., 1999, Palaeogene sections on the Isle of Wight. A revision of their description and significance in light of research undertaken over recent decades: Tertiary Research, v. 19, p. 1–69.

Driese, S.G., Nordt, L.C., Lynn, W., Stiles, C.A., Mora, C.I., and Wilding, L.P., 2005, Distinguishing climate in the soil record using chemical trends in a Vertisol climosequence from the Texas coastal prairie, and application to interpreting Paleozoic paleosols in the Appalachian basin: Journal of Sedimentary Research, v. 75, p. 340–353.

Fedo, C.M., Nesbitt, H.W., and Young, G.M., 1995, Unraveling the effects of potassium metasomatism in sedimentary rocks and paleosols, with implications for paleoweathering conditions and provenance: Geology, v. 23, p. 921–924, doi: 10.1130/0091-7613(1995)023<0921:UTEOPM>2.3.CO;2.

Fowler, K., Edwards, N., and Brett, D.W., 1973, In situ coniferous (taxodiaceous) tree remains in the Upper Eocene of southern England: Paleontology, v. 16, p. 205–217.

Gale, A.S., Huggett, J.M., Palike, H., Laurie, E., Hailwood, E.A., and Hardenbol, J., 2006, Correlation of Eocene-Oligocene marine and continental records: Orbital cyclicity, magnetostratigraphy and sequence stratigraphy on the Solent Group, Isle of Wight, UK: Journal of the Geological Society of London, v. 163, p. 401–415, doi: 10.1144/0016-764903-175.

Grimes, S.T., Mattey, D.P., Hooker, J.J., and Collinson, M.E., 2003, Paleogene palaeoclimate reconstruction using oxygen isotopes: Problems and solutions from the use of multiple palaeoproxies: Geochimica et Cosmochimica Acta, v. 67, p. 4033–4047, doi: 10.1016/S0016-7037(03)00173-X.

Grimes, S.T., Hooker, J.J., Collinson, M.E., and Mattey, D.P., 2005, Temperatures of late Eocene to early Oligocene freshwaters: Geology, v. 33, p. 189–192, doi: 10.1130/G21019.1.

Hamer, J.M.M., Sheldon, N.D., and Nichols, G.J., 2007a, Global aridity during the early Miocene? A terrestrial paleoclimate record from the Ebro Basin, Spain: The Journal of Geology, v. 115, p. 601–608, doi: 10.1086/519780.

Hamer, J.M.M., Sheldon, N.D., Nichols, G.J., and Collinson, M.E., 2007b, Late Oligocene–early Miocene palaeosols of distal fluvial systems, Ebro Basin, Spain: Palaeogeography, Palaeoclimatology, Palaeoecology, v. 247, p. 220–235, doi: 10.1016/j.palaeo.2006.10.016.

Hooker, J.J., 1987, Mammalian faunal events in the English Hampshire Basin (late Eocene–early Oligocene) and their application to European biostratigraphy: Münchner Geowissenschaftliche Abhandlungen, v. A10, p. 109–116.

Hooker, J.J., 1991, Two new pseudosciurids (Rodentia, Mammalia) from the English late Eocene, and their implications for phylogeny and speciation: Bulletin of the British Museum (Natural History; Geology), v. 47, p. 35–50.

Hooker, J.J., 1992, British mammalian paleocommunities across the Eocene-Oligocene transition and their environmental implications, *in* Prothero, D.R., and Berggren, W.A., eds., Eocene-Oligocene Climatic and Biotic Evolution: Princeton, New Jersey, Princeton University Press, p. 494–515.

Hooker, J.J., 2000, Paleogene mammals: Crises and ecological change, *in* Culver, S.J., and Rawson, P.F., eds., Biotic Response to Global Change: The Last 145 Million Years: Cambridge, Cambridge University Press, p. 333–349.

Hooker, J.J., 2001, Tarsals of the extinct insectivoran family Nyctitheriidae (Mammalia): Evidence for archontan relationships: Zoological Journal of the Linnean Society, v. 132, p. 501–529, doi: 10.1111/j.1096-3642.2001.tb02473.x.

Hooker, J.J., and Thomas, K.M., 2001, A new species of *Amphirhagatherium* (Choeropotamidae, Artiodactyla, Mammalia) from the late Eocene Headon Hill Formation of southern England and phylogeny of endemic European 'anthracotherioids': Paleontology, v. 44, p. 827–853, doi: 10.1111/1475-4983.00203.

Hooker, J.J., Collinson, M.E., Van Bergen, P.F., Singer, R.L., de Leeuw, J.W., and Jones, T.P., 1995, Reconstruction of land and freshwater palaeoenvironments near the Eocene-Oligocene boundary, S. England: Journal of the Geological Society of London, v. 152, p. 449–468, doi: 10.1144/gsjgs.152.3.0449.

Hooker, J.J., Collinson, M.E., and Sille, N.P., 2004, Eocene-Oligocene mammalian faunal turnover in the Hampshire Basin, UK: Calibration to the global time scale and the major cooling event: Journal of the Geological Society of London, v. 161, p. 161–172, doi: 10.1144/0016-764903-091.

Hooker, J.J., Cook, E., and Benton, M.J., 2005, British Tertiary fossil mammal GCR sites, *in* Benton, M.J., Cook, E., and Hooker, J.J., eds., Mesozoic and Tertiary Fossil Mammals and Birds of Great Britain: Peterborough, Joint Nature Conservation Committee (JNCC), Geological Conservation Review Series no. 32, p. 67–124.

Hooker, J.J., Grimes, S.T., Mattey, D.P., Collinson, M.E., and Sheldon, N.D., 2009, this volume, Refined correlation of the UK late Eocene–early Oligocene Solent Group and of its climate history, *in* Koeberl, C., and Montanari, A., eds., The Late Eocene Earth—Hothouse, Icehouse, and Impacts: Geological Society of America Special Paper 452, doi: 10.1130/2009.2452(12).

Ivany, L.C., Patterson, W.P., and Lohmann, K.C., 2000, Cooler winters as a possible cause of mass extinctions at the Eocene/Oligocene boundary: Nature, v. 407, p. 887–890, doi: 10.1038/35038044.

Joomun, S.C., Hooker, J.J., and Collinson, M.E., 2008, Dental wear variation and diet: An example from Eocene perissodactyls (Mammalia): Palaeogeography, Palaeoclimatology, Palaeoecology, v. 263, p. 92–106, doi: 10.1016/j.palaeo.2008.03.001.

Katz, M.E., Miller, K.G., Wright, J.D., Wade, B.S., Browning, J.V., Cramer, B.S., and Rosenthal, Y., 2008, Stepwise transition from the Eocene greenhouse to the Oligocene icehouse: Nature Geoscience, v. 1, p. 329–334, doi: 10.1038/ngeo179.

Kohn, M.J., Josef, J.A., Madden, R., Kay, R., Vucetich, G., and Carlini, A.A., 2004, Climate stability across the Eocene-Oligocene transition, southern Argentina: Geology, v. 32, p. 621–624, doi: 10.1130/G20442.1.

Machette, M.N., 1985, Calcic soils of the southwestern United States, *in* Weide, D.L., ed., Soils and Quaternary Geology of the Southwest United States: Geological Society of America Special Paper 203, p. 1–21.

Maynard, J.B., 1992, Chemistry of modern soils as a guide to interpreting Precambrian paleosols: The Journal of Geology, v. 100, p. 279–289.

Mitchell, R.L., 2006, An Interpretation of Palaeoclimate from Palaeosols across the Eocene-Oligocene Boundary on the Isle of Wight [M.Sc. thesis]: London, Royal Holloway, University of London, 64 p.

Perez-Huerta, A., and Sheldon, N.D., 2006, Pennsylvanian sea level cycles, nutrient availability, and brachiopod community structure: Palaeogeography, Palaeoclimatology, Palaeoecology, v. 230, p. 264–279, doi: 10.1016/j.palaeo.2005.07.020.

Plint, A.G., 1984, A regressive coastal sequence from the Upper Eocene of Hampshire, southern England: Sedimentology, v. 31, p. 213–225, doi: 10.1111/j.1365-3091.1984.tb01960.x.

Retallack, G.J., 1991, Untangling the effects of burial alteration and ancient soil formation: Annual Review of Earth and Planetary Sciences, v. 19, p. 183–206, doi: 10.1146/annurev.ea.19.050191.001151.

Retallack, G.J., 2001, Soils of the Past (2nd edition): London, Blackwell Science, Ltd., 404 p.

Retallack, G.J., 2005, Pedogenic carbonate proxies for amount and seasonality of precipitation in paleosols: Geology, v. 33, p. 333–336, doi: 10.1130/G21263.1.

Retallack, G.J., 2007, Cenozoic paleoclimate on land in North America: The Journal of Geology, v. 115, p. 271–294, doi: 10.1086/512753.

Sheldon, N.D., 2003, Pedogenesis and geochemical alteration of the Picture Gorge Subgroup, Columbia River Basalt, Oregon: Geological Society of America Bulletin, v. 115, p. 1377–1387, doi: 10.1130/B25223.1.

Sheldon, N.D., 2005, Do red beds indicate paleoclimatic conditions?: A Permian case study: Palaeogeography, Palaeoclimatology, Palaeoecology, v. 228, p. 305–319, doi: 10.1016/j.palaeo.2005.06.009.

Sheldon, N.D., 2006a, Abrupt chemical weathering increase across the Permian-Triassic boundary: Palaeogeography, Palaeoclimatology, Palaeoecology, v. 231, p. 315–321, doi: 10.1016/j.palaeo.2005.09.001.

Sheldon, N.D., 2006b, Quaternary glacial-interglacial climate cycles in Hawaii: The Journal of Geology, v. 114, p. 367–376, doi: 10.1086/500993.

Sheldon, N.D., 2009, this volume, Nonmarine records of climatic change across the Eocene-Oligocene transition, *in* Koeberl, C., and Montanari, A., eds., The Late Eocene Earth—Hothouse, Icehouse, and Impacts: Geological Society of America Special Paper 452, doi: 10.1130/2009.2452(15).

Sheldon, N.D., and Retallack, G.J., 2001, Equation for compaction of paleosols due to burial: Geology, v. 29, no. 3, p. 247–250, doi: 10.1130/0091-7613(2001)029<0247:EFCOPD>2.0.CO;2.

Sheldon, N.D., and Retallack, G.J., 2004, Regional precipitation records from the late Eocene and early Oligocene of North America: The Journal of Geology, v. 112, p. 487–494, doi: 10.1086/421076.

Sheldon, N.D., Retallack, G.J., and Tanaka, S., 2002, Geochemical climofunctions from North American soils and application to paleosols across the Eocene-Oligocene boundary in Oregon: The Journal of Geology, v. 110, p. 687–696, doi: 10.1086/342865.

Sille, N.P., Collinson, M.E., Kucera, M., and Hooker, J.J., 2004, Evolution within the charophyte genus *Harrisichara*, late Paleogene, southern England; environmental and biostratigraphic implications: Palaeogeography, Palaeoclimatology, Palaeoecology, v. 208, p. 153–173, doi: 10.1016/j.palaeo.2004.03.002.

Soil Survey Staff, 1999, Keys to Soil Taxonomy: Blacksburg, Virginia, U.S. Department of Agriculture, 600 p.

Stehlin, H.G., 1910, Remarques sur les faunules de Mammifères des couches Éocènes et Oligocènes du Bassin de Paris: Bulletin de la Société Géologique de France, v. 4, no. 9, p. 488–520.

Tabor, N.J., Montañez, I.P., Zierenberg, R., and Currie, B.S., 2004, Mineralogical and geochemical evolution of a basalt-hosted fossil soil (Late Triassic, Ischigualasto Formation, northwest Argentina): Potential for paleoenvironmental reconstruction: Geological Society of America Bulletin, v. 116, p. 1280–1293, doi: 10.1130/B25222.1.

Terry, D.O. Jr., 2001, Paleopedology of the Chardron Formation, northwestern Nebraska: Implications for paleoclimate in the North American mid-continent across the Eocene-Oligocene boundary: Palaeogeography, Palaeoclimatology, Palaeoecology, v. 168, p. 1–39, doi: 10.1016/S0031-0182(00)00248-0.

Zachos, J., Pagani, M., Sloan, L., Thomas, E., and Billups, K., 2001, Trends, rhythms, and aberrations in global climate 65 Ma to present: Science, v. 292, p. 686–693, doi: 10.1126/science.1059412.

Zanazzi, A., Kohn, M.J., McFadden, B.J., and Terry, D.T., 2007, Large temperature drop across the Eocene-Oligocene transition in central North America: Nature, v. 445, p. 639–642.

Manuscript Accepted by the Society 16 September 2008

Printed in the USA

The Geological Society of America
Special Paper 452
2009

Eocene-Oligocene greenhouse to icehouse transition on a subtropical clastic shelf: The Jackson-Vicksburg Groups of the eastern Gulf Coastal Plain of the United States

Richard H. Fluegeman*
Jeffry D. Grigsby*
Department of Geology, Ball State University, Muncie, Indiana 47306-0475, USA

John V. Hurley*
Department of Earth and Planetary Sciences, MSCO3-2040, 1, University of New Mexico, Albuquerque, New Mexico 87131-0001, USA

ABSTRACT

The Eocene-Oligocene transition in the eastern Gulf Coastal Plain of Alabama and Mississippi occurs within a sequence of marine clastic rocks making up the Jackson Group and the lower Vicksburg Group. The placement and the nature of the Eocene-Oligocene boundary remain controversial after more than 20 yr of detailed study. In Alabama, the Eocene-Oligocene boundary is placed within a condensed section at the contact between the Shubuta Member of the Yazoo Clay and the Red Bluff Clay. In eastern Mississippi, the Eocene-Oligocene boundary is recognized through the use of planktonic foraminifera and calcareous nannofossils as also occurring at the Yazoo–Red Bluff contact, but this surface is considered an unconformity representing a sequence boundary. In the Mossy Grove core of western Mississippi, planktonic foraminifera, in conjunction with radiometrically dated bentonites, place the Eocene-Oligocene boundary within the upper Yazoo Clay.

There is a pronounced turnover in benthic foraminiferal species between the Yazoo Clay of the Jackson Group and the Red Bluff Clay of the Vicksburg Group. This turnover does not occur at the biostratigraphic Eocene-Oligocene boundary within the Yazoo Clay in western Mississippi. Rather, it occurs above the unconformity associated with the base of the Vicksburg Group throughout the Gulf Coastal Plain. This unconformity was produced by a fall in relative sea level, possibly associated with the oxygen isotope event Oi-1. In some sections in the eastern Gulf Coastal Plain (Little Stave Creek, Saint Stephens Quarry, and the Mossy Grove core), the Eocene-Oligocene boundary as defined at the Massignano global stratotype section and point can be recognized. In most sections in the region, however, such as those in

*Fluegeman, corresponding author: rfluegem@bsu.edu. Grigsby: jgrigsby@bsu.edu. Hurley: jvhurley@unm.edu.

Fluegeman, R.H., Grigsby, J.D., and Hurley, J.V., 2009, Eocene-Oligocene greenhouse to icehouse transition on a subtropical clastic shelf: The Jackson-Vicksburg Groups of the eastern Gulf Coastal Plain of the United States, *in* Koeberl, C., and Montanari, A., eds., The Late Eocene Earth—Hothouse, Icehouse, and Impacts: Geological Society of America Special Paper 452, p. 261–277, doi: 10.1130/2009.2452(17). For permission to copy, contact editing@geosociety.org.

the Chickasawhay River valley of Mississippi, subaerial erosion associated with the basal Vicksburg unconformity has removed the latest Eocene and earliest Oligocene (pre–Oi-1) rocks. This means that in most cases in the eastern Gulf Coastal Plain, the Eocene-Oligocene boundary is coincident with a lithostratigraphic and allostratigraphic surface associated with the Oi-1 event rather than the highest occurrence of *Hantkenina*. The Jackson Group—Vicksburg Group contact, with its associated turnovers in benthic foraminifera, is important in identifying the Eocene-Oligocene boundary in the eastern Gulf Coastal Plain. Work currently under way by Charles Betz at Ball State University on stable isotopes in both the Yazoo Clay and Red Bluff Clay will help to further characterize this boundary in the Gulf Coastal Plain and enhance understanding of this important transition.

Keywords: foraminifera, bentonites, stable isotopes, Yazoo Clay, Mississippi.

INTRODUCTION

The Eocene-Oligocene transition in the eastern Gulf Coastal Plain occurs within a sequence of marine clastic rocks (Fig. 1). These rocks include the Eocene-age Moodys Branch Formation and Yazoo Clay, which together make up the Jackson Group, and the Oligocene-age Red Bluff Clay and Forest Hill Formation of the lower Vicksburg Group. In the Gulf Coastal Plain, the Eocene-Oligocene boundary has traditionally been placed at the Jackson-Vicksburg contact (Cooke, 1923), tying the terms "Eocene" and "Oligocene" in the region to lithostratigraphy rather than chronostratigraphy. Jackson and Vicksburg rocks throughout the Gulf Coastal Plain have been of paleontological interest due to the well-preserved fossils they contain. Benthic foraminifera have been of particular interest due to their utility in oil and gas exploration and subsurface correlation (Applin et al., 1925). The works of Cushman (1923, 1935), Howe and Wallace (1933), and Bandy (1949) are four of the more prominent early studies documenting the benthic foraminiferal faunas of the Jackson and Vicksburg Groups in the Gulf Coastal Plain. Despite the fact that these studies of benthic foraminifera were designed to identify key index taxa for the various lithostratigraphic units, a prominent faunal turnover was identified between the Jackson (Eocene) and Vicksburg (Oligocene). Further studies of the faunal turnover across the Jackson-Vicksburg contact were undertaken by Cheetham (1957) and Deboo (1965). Both papers recorded the ranges of foraminifera and other fossils at the Eocene-Oligocene boundary from sections in southwestern Alabama and southeastern Mississippi in order to improve recognition and correlation of the Eocene-Oligocene boundary within the eastern Gulf Coastal

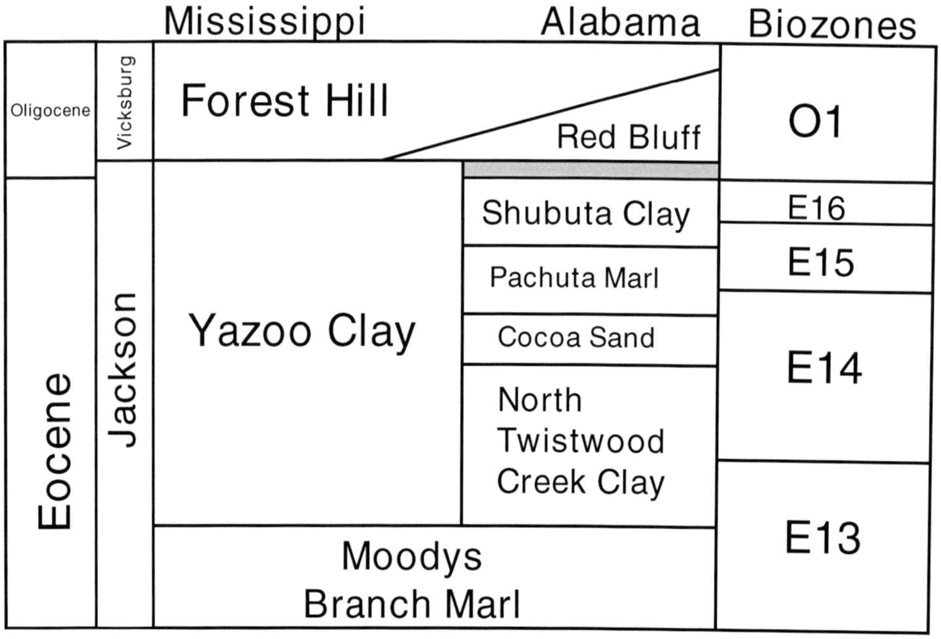

Figure 1. Lithostratigraphic nomenclature for the Jackson and Vicksburg Groups in western Alabama and eastern Mississippi.

Plain. While improving the recognition of the Eocene and Oligocene as chronostratigraphic units within the region, these studies did not facilitate correlation with the European standard sections of the Upper Eocene and Lower Oligocene.

Correlation of the Jackson-Vicksburg sections with the global geochronologic scale is essential if the impact of the developing icehouse world on the Gulf Coast region is to be fully understood. The biostratigraphic work of Mancini (1979), Siesser (1983), and Mancini and Waters (1986) defined the Eocene-Oligocene boundary in the eastern Gulf Coastal Plain in chronostratigraphic terms and correlated it with the global geochronologic scale. In many sections, the Eocene-Oligocene chronostratigraphic boundary is coincident with the Jackson-Vicksburg lithostratigraphic boundary (Mancini and Waters, 1986), but that is not the case in all sections (Fluegeman, 2003; Miller et al., 2008). The biostratigraphic framework established for the Eocene-Oligocene transition in eastern Gulf Coastal Plain is well documented and can be easily correlated with the new biozonation for the Eocene and Oligocene (Berggren and Pearson, 2005).

With an improved biostratigraphic framework, the Eocene-Oligocene transition in the eastern Gulf Coastal Plain has been the focus of many sequence stratigraphic studies. Some of the most important works include Mancini et al. (1987), Baum and Vail (1987, 1988), Loutit et al. (1988), Coleman and Galloway (1990), Dockery (1990), Mancini and Tew (1991), Tew and Mancini (1995), Pasley and Hazel (1995), Mancini (2000), Echols et al. (2003), and Miller et al. (2008). The numerous studies have provided a wealth of data on the Eocene-Oligocene transition in the Gulf Coastal Plain but have generated controversy regarding the nature of this boundary. Fluegeman (2008) provided an overview of some of the key sequence stratigraphic models of this important horizon.

Most of the work on sequence stratigraphy and biostratigraphy has been concentrated in the limited exposures and cores of western Alabama and eastern Mississippi. This region was an area of mixed carbonate and clastic sedimentation during the late Eocene and early Oligocene. It is the purpose of this paper to examine the Eocene-Oligocene record of relative sea-level change on the clastic shelf of western Mississippi.

STUDY AREA

Western Mississippi presents challenges for study of the Eocene-Oligocene transition. The Jackson Group is more than 150 m thick, and yet subtropical weathering has limited most exposures to ~20 m. Additionally, the basal Vicksburg Group in this area is represented by the nonmarine Forest Hill Formation (Fig. 1). The Jackson Group is fossiliferous with abundant planktonic and benthic foraminifera. Additionally, several distinct bentonite beds have been identified in western Mississippi and have provided radiometric ages (Berggren et al., 1992). With the limited exposures typical of the region, the position of any single outcrop within the overall stratigraphic succession cannot be determined.

The Mossy Grove core in Hinds County, Mississippi (Fig. 2), was drilled in an attempt to recover the most complete section across the Eocene-Oligocene transition in eastern Mississippi (Dockery et al., 1991). The continuous core encountered a partial section (15 m) of the Forest Hill Formation, complete sections of the Yazoo Clay (144 m) and the Moodys Branch Formation (3.3 m), and it reached total depth in the Cockfield Formation of the Claiborne Group. Both the Moodys Branch and Yazoo Clay contained abundant foraminifera in the core. A preliminary paleontological report on the planktonic foraminifera in the core (Fluegeman, 1996) placed the Eocene-Oligocene boundary within the Yazoo Clay.

The Yazoo section in the Mossy Grove core is much thicker than coeval sections in eastern Mississippi and western Alabama (Fig. 3). The Yazoo Clay in western Mississippi is not differentiated into distinct members, making correlation with the better-studied sections of eastern Mississippi and western Alabama difficult.

STRATIGRAPHIC SETTING

Biostratigraphy

Planktonic Foraminifera

The planktonic foraminiferal biostratigraphy in the Mossy Grove core was first studied by Fluegeman (1996) and modified by Fluegeman (2003). These studies were based on samples collected at 0.5 m intervals throughout the core and processed in the laboratory for foraminifera. Fluegeman (1996, 2003) used the distribution of planktonic foraminifera to recognize the biozones of Berggren and Miller (1988) and to correlate the Moodys Branch Formation and the Yazoo Clay of the Mossy Grove core to the global geochronologic scale of Berggren et al. (1995). Recent revisions of Eocene planktonic foraminiferal biostratigraphy and taxonomy (Berggren and Pearson, 2005; Pearson et al., 2006) warrant a reexamination of the biostratigraphic data collected in the Mossy Grove core. The distribution of planktonic foraminifera in the Mossy Grove core is shown in Table 1.

Correlations of the Upper Eocene and Lower Oligocene sections with the planktonic foraminiferal biozonation of Berggren and Pearson (2005) are problematic. While planktonic foraminifera are abundant in the Moodys Branch Formation and the Yazoo Clay, key taxa for the recognition of the biozones of Berggren and Pearson (2005) are not present in the eastern Gulf Coastal Plain. The proximity of the shoreline during deposition of the Moodys Branch Formation and the Yazoo Clay likely led to conditions that led to decreased diversity of planktonic foraminifera. The absence of *Morozovelloides crassatus*, *Globigerinatheka semiinvoluta*, and *Globigerinatheka index* from the Moodys Branch and Yazoo Clay requires correlation to the biozonation of Berggren and Pearson (2005) by biostratigraphic proxies. The correlation of the section in the Mossy Grove core with the planktonic foraminiferal biozonation of Berggren and Pearson (2005) and the distribution of key taxa are shown in Figure 4.

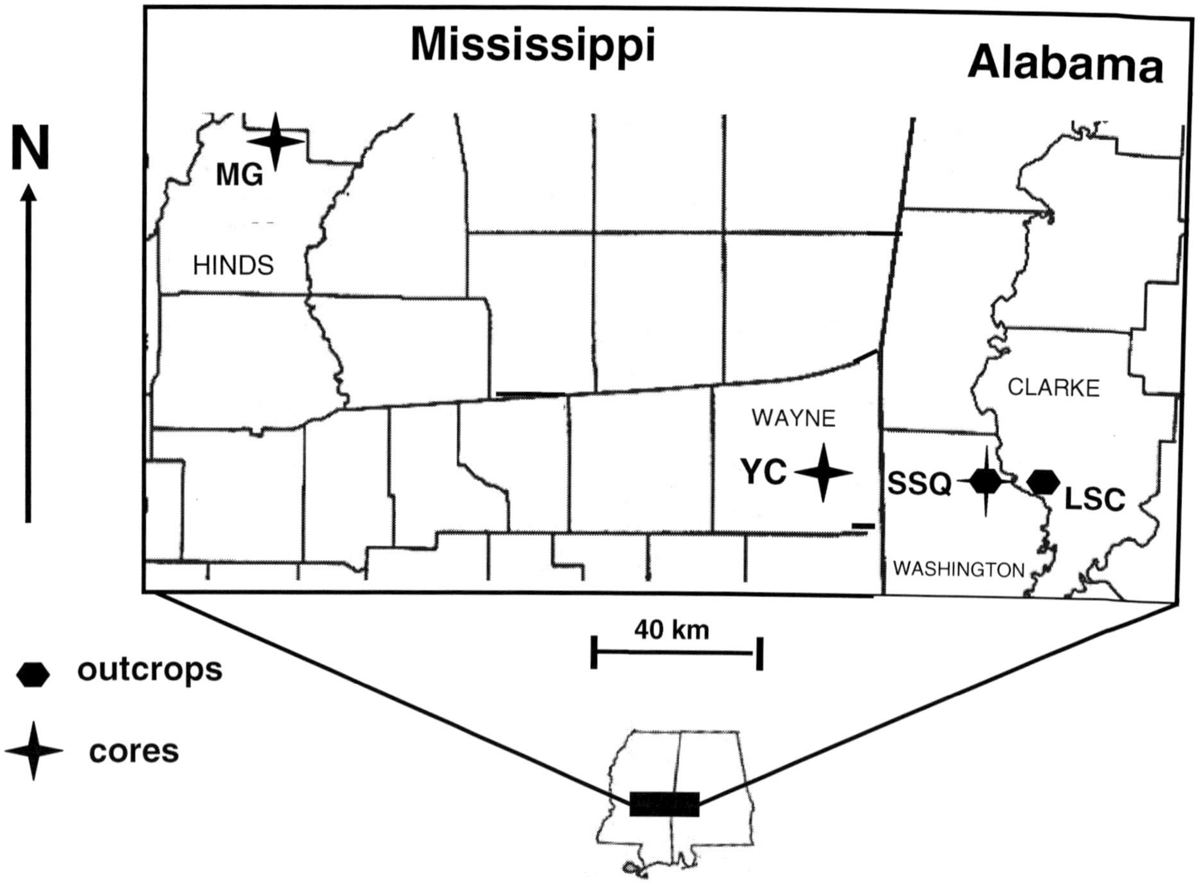

Figure 2. Map of localities mentioned in this study. MG—Mossy Grove core, YC—Mobil #1 Young core, SSQ—St. Stephens Quarry and core, LSC—Little Stave Creek section.

The Moodys Branch Formation and the lower 28 m of the Yazoo Clay in the Mossy Grove core is correlated with the *Morozovelloides crassatus* highest-occurrence zone (zone E13) of Berggren and Pearson (2005). Zone E13 is defined as the biostratigraphic interval represented by the partial range of *Morozovelloides crassatus* between the highest occurrence of *Orbulinoides beckmanni* and the highest occurrence of *M. crassatus*. The recognition of this zone in the Mossy Grove core is problematic because neither *O. beckmanni* nor *M. crassatus* has been identified in the Eocene outcrop belt of the eastern Gulf Coastal Plain. Correlation with biozone E13 was achieved by the use of a biostratigraphic proxy. Seventeen species of planktonic foraminifera were identified from this interval in the Mossy Grove core (Table 1). The presence of *Hantkenina alabamensis, Subbotina gortani*, *Turborotalia increbescens*, and *Acarinina rohri* suggests correlation with biozone E13. The highest occurrence of *Acarinina rohri* is close to the highest occurrence of *Morozovelloides crassatus* (Berggren et al., 2006), and in the Mossy Grove core, the highest occurrence of *Acarinina rohri* (in sample MG408) serves as the approximate upper limit of biozone E13.

The *Globigerinatheka semiinvoluta* highest-occurrence zone (biozone E14) is defined by Berggren and Pearson (2005) as the biostratigraphic interval between the highest occurrence of *Morozovelloides crassatus* and the highest occurrence of *Globigerinatheka semiinvoluta*. In the Mossy Grove core, the base of this biozone can be recognized by the highest occurrence of *Acarinina rohri*. *Globigerinatheka semiinvoluta* does not occur in the Mossy Grove core, nor has it been documented from the Yazoo Clay elsewhere in the eastern Gulf Coastal Plain. The lowest occurrence of *Pseudohastigerina naguewichiensis* (in sample MG330) in the Mossy Grove core is used as a proxy for the upper limit of biozone E14 and the lower limit of the *Globigerinatheka index* highest-occurrence zone (biozone E15).

The *Globigerinatheka index* highest-occurrence zone was defined by Berggren and Pearson (2005) as the biostratigraphic interval between the highest occurrence of *Globigerninatheka semiinvoluta* and the highest occurrence of *Globigerinatheka index*. While recognition of the lower limit of biozone E15 in the Mossy Grove core can be achieved by proxy, correlation of the upper limit of this zone and the base of the *Hantkenina alabamensis* highest-occurrence zone (biozone E16) of Berggren and Pearson (2005) is problematic. Miller et al. (2008) could not separate biozones E15 and E16 in the upper Yazoo Clay of the St. Stephens core, Washington County, southwestern Alabama. In the Mossy

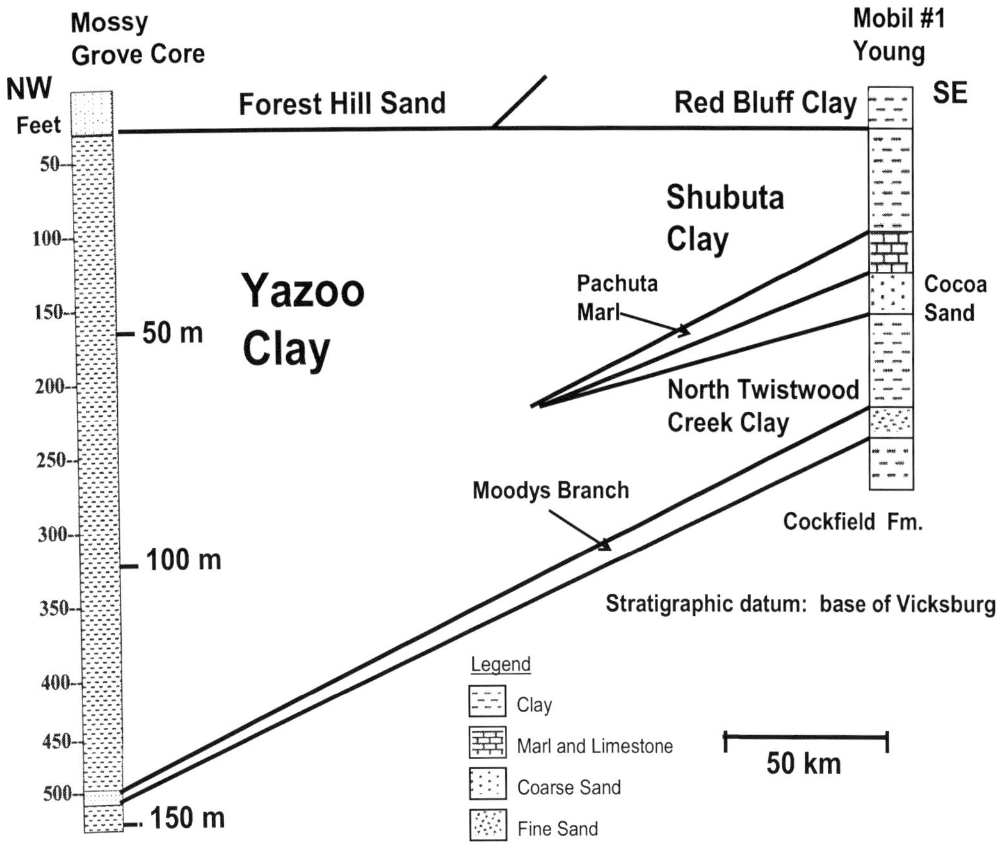

Figure 3. Stratigraphic cross section of the Jackson Group between the Mossy Grove core and the Mobil #1 Young core. The stratigraphic datum is the base of the Vicksburg Group.

Grove core, the overlapping ranges of *Tenuitella gemma* and *Hantkenina alabamensis* serve as a proxy for the recognition of biozone E16. It is recognized that only the upper portion of biozone E16 can be distinguished from biozone E15 in the Mossy Grove core. The interval between sample MG184 and the highest occurrences of *Subbotina yeguaensis*, *Turborotalia cocoaensis*, and *Turborotalia cerroazulensis* in sample MG96 can be correlated with biozone E16.

The upper limit of biozone E16 represents the Eocene-Oligocene boundary (base of the Rupelian stage). The *Hantkenina alabamensis* highest-occurrence zone of Berggren and Pearson (2005) is defined as the partial range of *Hantkenina alabamensis* between the highest occurrence of *Globigerinatheka index* and *H. alabamensis*. The highest occurrence of the family Hantkeninidae serves as a guide event for correlation of the Eocene-Oligocene boundary as defined in the Rupelian global stratotype section and point (GSSP) at Massignano, Italy (Premoli Silva and Jenkins, 1993). *Hantkenina alabamensis* fossils are abundant in the Mossy Grove core and were therefore expected to provide a tool for the recognition of the Eocene-Oligocene boundary. The range of *H. alabamensis* in the Mossy Grove core is depressed. The highest occurrence of *H. alabamensis* is in sample MG120, while *Turborotalia cocoaensis*, *T. cerroazulensis*, and *Subbotina yeguaensis*, three taxa which have a documented shorter range globally (Pearson et al., 2006), are found as high as sample MG96. The joint highest occurrences of *Turborotalia cocoaensis*, *T. cerroazulensis*, and *Subbotina yeguaensis* in sample MG96 serve as an approximation for the upper physical limit of biozone E16, the lower physical limit of biozone O1, and the Eocene-Oligocene series boundary in the Mossy Grove core.

The *Pseudohastigerina naguewichiensis* highest-occurrence zone (biozone O1) is defined as the biostratigraphic interval between the highest occurrence of *Hantkenina alabamensis* and the highest occurrence of *Pseudohastigerina naguewichiensis* (Berggren and Pearson, 2005). In the Mossy Grove core, all of the Yazoo Clay above sample MG96 is correlated to biozone O1. The highest planktonic foraminifera in the Mossy Grove core occur in sample MG52. The Red Bluff Clay of eastern Mississippi and western Alabama, considered a correlative of the overlying Forest Hill Formation of western Mississippi (Echols et al., 2003), has been correlated with biozone O1 (Miller et al., 2008; Fluegeman, 2009). This correlation makes the upper 18 m of the Yazoo Clay early Oligocene in age. If the unconformity at the base of the Forest Hill Formation is in fact related to the global lowering of sea level associated with the oxygen isotope glacial event Oi-1, then the upper Yazoo represents a pre–Oi-1 Oligocene section.

TABLE 1: DISTRIBUTION OF PLANKTONIC FORAMINIFERA IN THE MOSSY GROVE CORE

SAMPLE	*Acarinina rohri*	*Dipsidripella danvillensis*	*Subbotina yeguaensis*	*Subbotina corpulenta*	*Subbotina jacksonensis*	*Cassigerinella eocaenica*	*Pseudohastigerina micra*	*Pseudohastigerina naguewichiensis*	*Turborotalia pomeroli*	*Turborotalia cocoaensis*	*Turborotalia cerroazulensis*	*Turborotalia increbescens*	*Dentoglobigerina tripartita*	*Globoturborotalia ouachitaensis*	*Hantkenina alabamensis*	*Cribrohantkenina inflata*	*Catapsydrax dissimilis*	*Turborotalia ampliapertura*	*Paragloborotalia nana*	*Globigerina officinalis*	*Subbotina gortani*	*Subbotina eocaena*	*Tenuitella gemma*	**Total Planktonics**	**Planktonic:Benthic Ratio**	SAMPLE
MG52																								0	0	MG52
MG54																								0	0	MG54
MG56																				1	1		1	3	0.01010101	MG56
MG58																		2			2	1	2	7	0.023890785	MG58
MG60																							2	2	0.006711409	MG60
MG62							1	1					2					1		2	2	1	3	13	0.045296167	MG62
MG64							2	2					2				1	2	1	1	1	3	2	17	0.060070671	MG64
MG66																		2			2	1	5	10	0.034482759	MG66
MG68																	1	3	4	2	5	2	20	37	0.140684411	MG68
MG70																		3	1		1	1	11	17	0.060070671	MG70
MG72																								0	0	MG72
MG74																					1		2	3	0.01010101	MG74
MG76																								0	0	MG76
MG78							1	2										3	1	1	2	2	9	23	0.083032491	MG78
MG80							1	1										1	1	2	2	1	4	13	0.045296167	MG80
MG82								2										2			3		5	12	0.041666667	MG82
MG84																								0	0	MG84
MG86																		1			1		3	5	0.016949153	MG86
MG88							1	1										1			2	2	4	11	0.038062284	MG88
MG90																				2	3	2	6	13	0.045296167	MG90
MG92																					1		1	2	0.006711409	MG92
MG94																					2	1	2	5	0.016949153	MG94
MG96			1							2	1	2					1	6	2	3	4	1	6	30	0.111111111	MG96
MG98			2	1								1	2					3			2			11	0.038062284	MG98
MG100			2	3			2				2									2	3			15	0.052631579	MG100
MG102																				1	1	2	2	6	0.020408163	MG102
MG104								2												1	1	1	4	9	0.030927835	MG104
MG106			3	2			3	2		1	1	1	2				1	3	2	3	1	4	6	34	0.127819549	MG106
MG108																						1		1	0.003344482	MG108
MG110																							1	1	0.003344482	MG110
MG112							2				1									1		2		3	0.01010101	MG112
MG114																					1		1	2	0.006711409	MG114
MG116																						2		2	0.006711409	MG116
MG118																						2	1	3	0.01010101	MG118
MG120				1			1				1	1	1		1			1				2		9	0.030927835	MG120
MG122			3	2			2	1			2	1	2				1		1	9	2	5	3	32	0.119402985	MG122
MG124																				2	1	3	4	10	0.034482759	MG124
MG126																						1		1	0.003344482	MG126
MG128								1														2		3	0.01010101	MG128
MG130				1							1		1		1	1		1		1	2	3	3	15	0.052631579	MG130

(*continued*)

TABLE 1: DISTRIBU¯ION OF PLANKTONIC FORAMINIFERA IN THE MOSSY GROVE CORE (*continued*)

SAMPLE	*Acarinina rohri*	*Dipsidripella danvillensis*	*Subbotina yeguaensis*	*Subbotina corpulenta*	*Subbotina jacksonensis*	*Cassigerinella eocaenica*	*Pseudohastigerina micra*	*Pseudohastigerina naguewichiensis*	*Turborotalia pomeroli*	*Turborotalia cocoaensis*	*Turborotalia cerroazulensis*	*Turborotalia increbescens*	*Dentoglobigerina tripartita*	*Globoturborotalia ouachitaensis*	*Hantkenina alabamensis*	*Cribrohantkenina inflata*	*Catapsydrax dissimilis*	*Turborotalia ampliapertura*	*Paragloborotalia nana*	*Globigerina officinalis*	*Subbotina gortani*	*Subbotina eocaena*	*Tenuitella gemma*	**Total Planktonics**	**Planktonic:Benthic Ratio**	SAMPLE
MG132			2	1	1					1			2		1			2	1	2	2	1	2	18	0.063829787	MG132
MG134								1					1				2			2	1		1	8	0.02739726	MG134
MG136																						3		3	0.01010101	MG136
MG138							1					1								1	1	1		5	0.016949153	MG138
MG140							1				1													2	0.006711409	MG140
MG142			1	2	1		1				2	1	2	1	1			2	1	2	1	6	4	28	0.102941176	MG142
MG144				1			2	1		1	3	2	3	2	2			1	2	4	3	3	6	36	0.136363636	MG144
MG146			3	3			3	2		1	2	1	4		1			1	1	2	1	7	2	34	0.127819549	MG146
MG148																						3	2	5	0.016949153	MG148
MG150							7				4		1				1			3	1	4	1	11	0.038062284	MG150
MG152							3			1	1	2	4	1	3			5	2	4	2	5	9	42	0.162790698	MG152
MG154																						2	1	3	0.01010101	MG154
MG156																					1	1		2	0.006711409	MG156
MG158																						3		3	0.01010101	MG158
MG160																					1			1	0.003344482	MG160
MG162							2				1													3	0.01010101	MG162
MG164				1				1				1	1				1	1		2	2	8	3	21	0.075268817	MG164
MG166											1		1							1		2		5	0.016949153	MG166
MG168										1		2	2		1			2	1	4	1	4	6	24	0.086956522	MG168
MG170							3				3		4		4	1				3	2	7	3	30	0.111111111	MG170
MG172																						1		1	0.003344482	MG172
MG174							1						1		2					6		5	2	17	0.060070671	MG174
MG176																						2		2	0.006711409	MG176
MG178																						2		2	0.006711409	MG178
MG180							2						3		2							3	2	12	0.041666667	MG180
MG182																				3		1	1	5	0.016949153	MG182
MG184																				2				2	0.006711409	MG184
MG186																								0	0	MG186
MG188																								0	0	MG188
MG190												1	1									2	1	5	0.016949153	MG190
MG192																						1		1	0.003344482	MG192
MG194			1												1					1		2		5	0.016949153	MG194
MG196																				2				2	0.006711409	MG196
MG198													2							1	1	3		7	0.023890785	MG198
MG200										1														1	0.003344482	MG200
MG202				1			3	1				1	1	1	2		1	4	1	2	3	9		30	0.111111111	MG202
MG204																								0	0	MG204
MG206																								0	0	MG206
MG208								1																1	0.003344482	MG208
MG210								2			1													3	0.01010101	MG210

(*continued*)

TABLE 1: DISTRIBUTION OF PLANKTONIC FORAMINIFERA IN THE MOSSY GROVE CORE (*continued*)

SAMPLE	*Acarinina rohri*	*Dipsidripella danvillensis*	*Subbotina yeguaensis*	*Subbotina corpulenta*	*Subbotina jacksonensis*	*Cassigerinella eocaenica*	*Pseudohastigerina micra*	*Pseudohastigerina naguewichiensis*	*Turborotalia pomeroli*	*Turborotalia cocoaensis*	*Turborotalia cerroazulensis*	*Turborotalia increbescens*	*Dentoglobigerina tripartita*	*Globoturborotalia ouachitaensis*	*Hantkenina alabamensis*	*Cribrohantkenina inflata*	*Catapsydrax dissimilis*	*Turborotalia ampliapertura*	*Paragloborotalia nana*	*Globigerina officinalis*	*Subbotina gortani*	*Subbotina eocaena*	*Tenuitella gemma*	**Total Planktonics**	**Planktonic:Benthic Ratio**	SAMPLE
MG212			1				1				1		2		1				1	5	3	4		21	0.075268817	MG212
MG214											1	1	1							6	2	7		18	0.063829787	MG214
MG216																								0	0	MG216
MG218								2												1		1		2	0.006711409	MG218
MG220																								0	0	MG220
MG222								1					1		1					1		3		7	0.023890785	MG222
MG224																						2		2	0.006711409	MG224
MG226																				3	2	4		9	0.030927835	MG226
MG228																						1		1	0.003344482	MG228
MG230				1			1	1			1		1						1	1				7	0.023890785	MG230
MG232																								0	0	MG232
MG234												1								1		1		3	0.01010101	MG234
MG236			1				2	1			2	1	3		2			2	1	2	2	4		23	0.083032491	MG236
MG238			2	3	1		1			1			2		1			3	1	1	1	2		19	0.067615658	MG238
MG240										1										1				1	0.003344482	MG240
MG242			2	2				3		1	1		1		1		1	1		2		11		25	0.090909091	MG242
MG244				2				1				1	3					2		4	1	6		18	0.063829787	MG244
MG246							1				2		2					1		3	2	4		15	0.052631579	MG246
MG248																								0	0	MG248
MG250				1							1									2				4	0.013513514	MG250
MG252												1			1					4	1	3		10	0.034482759	MG252
MG254			1	1						1		1	2		1					2	2	3		14	0.048951049	MG254
MG256																		1		5		2		8	0.02739726	MG256
MG258																				1				1	0.003344482	MG258
MG260			3	4							4				2		3			3		6		25	0.090909091	MG260
MG262													2		1			1		7	4	8		23	0.083032491	MG262
MG264			1															2	1	1	1	2		8	0.02739726	MG264
MG266																				2		5		7	0.023890785	MG266
MG268																				2				2	0.006711409	MG268
MG270													3									3		3	0.01010101	MG270
MG272													2					2		1	1	3		9	0.030927835	MG272
MG274																						1		1	0.003344482	MG274
MG276							1													1				1	0.003344482	MG276
MG278							1															2		3	0.01010101	MG278
MG280											2	2										1		5	0.016949153	MG280
MG282																		2		1				3	0.01010101	MG282
MG284			1	1																2		5		9	0.030927835	MG284
MG286				1					1		1	1	4		3				1	7	3	8		30	0.111111111	MG286
MG288										1										2	1	4		8	0.02739726	MG288
MG290			1																	2		4		7	0.023890785	MG290

(*continued*)

TABLE 1: DISTRIBUTION OF PLANKTONIC FORAMINIFERA IN THE MOSSY GROVE CORE (*continued*)

SAMPLE	*Acarinina rohri*	*Dipsidripella danvillensis*	*Subbotina yeguaensis*	*Subbotina corpulenta*	*Subbotina jacksonensis*	*Cassigerinella eocaenica*	*Pseudohastigerina micra*	*Pseudohastigerina naguewichiensis*	*Turborotalia pomeroli*	*Turborotalia cocoaensis*	*Turborotalia cerroazulensis*	*Turborotalia increbescens*	*Dentoglobigerina tripartita*	*Globoturborotalia ouachitaensis*	*Hantkenina alabamensis*	*Cribrohantkenina inflata*	*Catapsydrax dissimilis*	*Turborotalia ampliapertura*	*Paragloborotalia nana*	*Globigerina officinalis*	*Subbotina gortani*	*Subbotina eocaena*	*Tenuitella gemma*	**Total Planktonics**	**Planktonic:Benthic Ratio**	SAMPLE
MG292																					1	2		3	0.01010101	MG292
MG294																				2				2	0.006711409	MG294
MG296													3		1					11	2	8		25	0.090909091	MG296
MG298			2	4			1		2		1	2	7	1	2		1		1	15	4	12		56	0.229508197	MG298
MG300			4	2							7	4			6		2							53	0.214574899	MG300
MG302				1				2	1				3		1					11	3	9		31	0.115241636	MG302
MG304																				5	1	3		9	0.030927835	MG304
MG306			1	4	1			3		1	1	2	4		1					4	4	7		33	0.123595506	MG306
MG308							2	1	1	1			3		1		1			7	3	3		23	0.083032491	MG308
MG310				1								1	1							3		15		21	0.075268817	MG310
MG312																				1	2	1		4	0.013513514	MG312
MG314																				2				2	0.006711409	MG314
MG316			1	2	1		2	1	2		1	1	4		1					13	3	5		37	0.140684411	MG316
MG318			2	5	2		3	4	2	1	1	1	2	1	5		2		1	21	5	34		92	0.442307692	MG318
MG320			1	8			1			1	2	4	3		4					15		22		61	0.255230126	MG320
MG322																				1		3		4	0.013513514	MG322
MG324																				3				3	0.01010101	MG324
MG326				1			1		3	1	1	2	8		3					12	7	17		55	0.224489796	MG326
MG328				1			2	1				3	7	1	2		1		2	22	8	18		70	0.304347826	MG328
MG330			4				4				1	3	4							11		16		48	0.19047619	MG330
MG332																				2		4		6	0.020408163	MG332
MG334			1																	2	3	4		10	0.034482759	MG334
MG336				7	3		5		2	4	2	1	6		2				1	7	2	9		51	0.204819277	MG336
MG338			2	6	4		3		1			2	4		1					14	4	17		58	0.239669421	MG338
MG340			4	10	2		5		6			4	7		3					4		3		48	0.19047619	MG340
MG342				1			2		1	1		2	4	1			1			17	6	22		58	0.239669421	MG342
MG344			2	4			1						1	2						23	7	19		59	0.244813278	MG344
MG346													2							1	2	3		8	0.02739726	MG346
MG348				2			2						10	1						12	4	24		55	0.224489796	MG348
MG350				12					4			6	8		3					15		21		66	0.282051282	MG350
MG352			1	2					1			2	9	2	1				1	17	6	14		56	0.229508197	MG352
MG354			1	4								1	4		2		2			21	3	11		48	0.19047619	MG354
MG356													5							11	6	12		34	0.127819549	MG356
MG358				3			3		2				2		1				1	3	1	4		20	0.071428571	MG358
MG360			3		1				6			8	5		4									27	0.098901099	MG360
MG362				1			1		1				4		1					5	4	8		25	0.090909091	MG362
MG364																				4				4	0.013513514	MG364
MG366																								0	0	MG366
MG368							1								1					6	2	9		19	0.067615658	MG368
MG370			1	1	2		2						2							1		2		11	0.038062284	MG370

(*continued*)

TABLE 1: DISTRIBUTION OF PLANKTONIC FORAMINIFERA IN THE MOSSY GROVE CORE (*continued*)

SAMPLE	*Acarinina rohri*	*Dipsidripella danvillensis*	*Subbotina yeguaensis*	*Subbotina corpulenta*	*Subbotina jacksonensis*	*Cassigerinella eocaenica*	*Pseudohastigerina micra*	*Pseudohastigerina naguewichiensis*	*Turborotalia pomeroli*	*Turborotalia cocoaensis*	*Turborotalia cerroazulensis*	*Turborotalia increbescens*	*Dentoglobigerina tripartita*	*Globoturborotalia ouachitaensis*	*Hantkenina alabamensis*	*Cribrohantkenina inflata*	*Catapsydrax dissimilis*	*Turborotalia ampliapertura*	*Paragloborotalia nana*	*Globigerina officinalis*	*Subbotina gortani*	*Subbotina eocaena*	*Tenuitella gemma*	**Total Planktonics**	**Planktonic:Benthic Ratio**	SAMPLE
MG372				1					1											4	2	6		14	0.048951049	MG372
MG374			1	1			2								1						2	1		8	0.02739726	MG374
MG376													1							3	1	4		9	0.030927835	MG376
MG378									1			1	5							2		2		11	0.038062284	MG378
MG380			1	3					6			5	6		2					3		3		29	0.10701107	MG380
MG382				1								1	4		1					6	2	4		19	0.067615658	MG382
MG384				1					1		1		2							2		7		14	0.048951049	MG384
MG386			1	1			1		1			1	3	1	1				1	5	4	8		28	0.102941176	MG386
MG388				1			3		3	2	1	2	2							6		3		23	0.083032491	MG388
MG390			2	4			3					3	2		3				2	5		3		27	0.098901099	MG390
MG392			1	3					2	1	2	1	6		4					8	1	7		36	0.136363636	MG392
MG394							2						12		9				1	11	2	8		45	0.176470588	MG394
MG396													1		11					3		5		20	0.071428571	MG396
MG398							1		1				1		8					7	2	9		29	0.10701107	MG398
MG400		2	4	3	2		4						8	2	17					5		6		53	0.214574899	MG400
MG402				1			1					1	3		12		1			7	4	13		43	0.167315175	MG402
MG404				1			1		3						10					12	2	15		44	0.171875	MG404
MG406			1	1			1		1		1	2	4		11				1	12	8	13		56	0.229508197	MG406
MG408	2			2			1								2					1	5	9		20	0.071428571	MG408
MG410	1	1		5	1		2		2			3	6	1	2					3		5		32	0.119402985	MG410
MG412			1	1			3		3	2	1	1	4	1	1					5	2	4		29	0.10701107	MG412
MG414			2	3	1		2		2				6							3		7		26	0.094890511	MG414
MG416			1	4			4		3	1	1	3	8		6					4	5	7		47	0.185770751	MG416
MG418													1							1	2			5	0.016949153	MG418
MG420	1		1		2						4	3	3		2							4		20	0.071428571	MG420
MG422							4		4	1		2	12		3		1			18	5	13		62	0.260504202	MG422
MG424			1	3			2		1			1	8	1	1					6	3	5		32	0.119402985	MG424
MG426			1	2			1		1		1	2	3		1					1		4		17	0.060070671	MG426
MG428																				2				2	0.006711409	MG428
MG430			1	3			1				2	3	1							1		3		14	0.048951049	MG430
MG432							1		1				9		1					12	1	10		35	0.132075472	MG432
MG434				1			2		2	1	1	1	12		4				1	12	2	17		56	0.229508197	MG434
MG436			1				1													8		4		14	0.048951049	MG436
MG438													4							10	1	8		23	0.083032491	MG438
MG440	2	1	1	1	2		4		5		4	1	17	1	5					4		4		53	0.214574899	MG440
MG442			1	1	1		1		1		2		7	1	2					2	1	3		23	0.083032491	MG442
MG444	2		1	4	1		2		2			1	4		2		1			11	1	9		41	0.158301158	MG444
MG446													2							1		2		5	0.016949153	MG446
MG448			1			3	1															2		8	0.02739726	MG448
MG450						9	3		1				1									3		17	0.060070671	MG450

(*continued*)

TABLE 1: DISTRIBUTION OF PLANKTONIC FORAMINIFERA IN THE MOSSY GROVE CORE (*continued*)

SAMPLE	*Acarinina rohri*	*Dipsidripella danvillensis*	*Subbotina yeguaensis*	*Subbotina corpulenta*	*Subbotina jacksonensis*	*Cassigerinella eocaenica*	*Pseudohastigerina micra*	*Pseudohastigerina naguewichiensis*	*Turborotalia pomeroli*	*Turborotalia cocoaensis*	*Turborotalia cerroazulensis*	*Turborotalia increbescens*	*Dentoglobigerina tripartita*	*Globoturborotalia ouachitaensis*	*Hantkenina alabamensis*	*Cribrohantkenina inflata*	*Catapsydrax dissimilis*	*Turborotalia ampliapertura*	*Paragloborotalia nana*	*Globigerina officinalis*	*Subbotina gortani*	*Subbotina eocaena*	*Tenuitella gemma*	**Total Planktonics**	**Planktonic:Benthic Ratio**	SAMPLE
MG452			1			2	2														1	2		8	0.02739726	MG452
MG454				1			1								2					2		5		11	0.038062284	MG454
MG456	1		1	2			4		2	2	1	1	11		1					4		5		35	0.132075472	MG456
MG458							2							3						2		4		11	0.038062284	MG458
MG460	3																					2		5	0.016949153	MG460
MG462				3			1		1			1	3							7	2			17	0.060070671	MG462
MG464																				5				5	0.016949153	MG464
MG466		1	1	4			4		2			2	4		2		1			9		8		38	0.145038168	MG466
MG468																				2				2	0.006711409	MG468
MG470	5	1	2	1					1		2	1								4	3			20	0.071428571	MG470
MG472	3	1	2				1													2				8	0.02739726	MG472
MG474	1		3						1	1			2		1					3	1	4		17	0.060070671	MG474
MG476							2						2							1				5	0.016949153	MG476
MG478	1												1							2		3		8	0.02739726	MG478
MG480	4			3							1				2							4		14	0.048951049	MG480
MG482			1	2			1			1			2									1		8	0.02739726	MG482
MG484									1				1							3				5	0.016949153	MG484
MG486																								0	0	MG486
MG488	1		1	2			1																	5	0.016949153	MG488
MG490			1				1					1			1					1		3		8	0.02739726	MG490
MG492																						2		2	0.006711409	MG492
MG494															1					3		4		8	0.02739726	MG494
MG496	3				1															4		3		11	0.038062284	MG496
MG498	3	1			1															1		2		8	0.02739726	MG498
MG500																						5		5	0.016949153	MG500
MG502							2								1					3		3		8	0.02739726	MG502
MG504				1																5		5		11	0.038062284	MG504
MG506															2					1		2		5	0.016949153	MG506
MG508																				2		3		5	0.016949153	MG508
MG510																								0	0	MG510
MG512	6	3	1	2	3		2								4					7		10		38	0.145038168	MG512

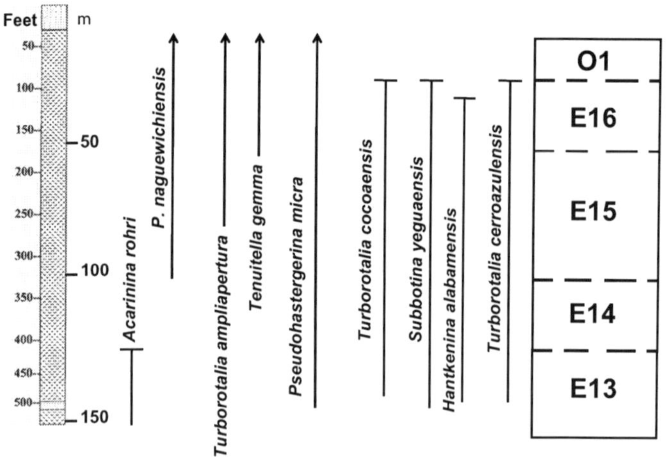

Figure 4. Distribution of key planktonic foraminifera in the Mossy Grove core and correlation to the biozones of Berggren and Pearson (2005).

Benthic Foraminifera

Benthic foraminifera across the Eocene-Oligocene transition in the eastern Gulf Coastal Plain have been studied by Fluegeman (2003) and Fluegeman (2009). There is a pronounced turnover in benthic foraminiferal species between the Yazoo Clay of the Jackson Group and the Red Bluff Clay of the Vicksburg Group. The benthic foraminifera in the Yazoo are typified by species such as *Siphonina jacksonensis*, *Siphonina danvillensis*, *Discorbis cocoaensis*, *Cibicidina walli*, *Uvigerina cocoaensis*, *Bulimina jacksonensis*, *Cibicidoides truncatus*, *Spiroplectammina mississippiensis*, and *Marginulina cocoaensis*. The benthic foraminiferal fauna in the Red Bluff includes *Cibicidoides pippeni*, *Cibicidoides cookei*, *Eponides ellisorae*, *Anomalinoides bilateralis*, *Siphonina advena*, *Hanzawaia mississippiensis*, *Spiroplectammina latior*, *Uvigerina vicksburgensis*, and *Lenticulina vicksburgensis*. Fluegeman (2003) suggested that this turnover was at least in part related to global cooling at this time indicated by Miller et al. (2005) and Zachos et al. (2001). This turnover does not occur at the biostratigraphic Eocene-Oligocene boundary within the Yazoo Clay in western Mississippi. Rather, it occurs above the regional unconformity associated with the base of the Vicksburg Group throughout the Gulf Coastal Plain. This unconformity was produced by a fall in relative sea level, possibly associated with the Oi-1 glaciation of Miller et al. (2005) and Zachos et al. (2001).

In the Mossy Grove core, Fluegeman (2003) also identified benthic foraminiferal assemblages associated with bathymetry. The *Discorbis-Siphonina* assemblage was identified in the Moodys Branch Formation and in the upper portion of the Yazoo Clay. This assemblage is interpreted as representing middle shelf conditions with a depth estimate of 25–50 m (Echols et al., 2003). The lower half of the Yazoo, as well as the interval near the Eocene-Oligocene boundary, is characterized by the *Uvigerina-Bulimina* assemblage. Miller and Lohman (1982) associated this assemblage with outer neritic to upper bathyal conditions and water depths greater than 100–200 m. The *Cibicidoides-Hanzawaia* assemblage was interpreted by Olsson and Wise (1987) as representing outer neritic conditions (60–90 m) on the Eocene New Jersey shelf, and Miller et al. (2008) proposed a similar paleobathymetry for the assemblage in the Yazoo Clay of southwestern Alabama. The distribution of these assemblages is shown in Figure 5.

Bentonite Stratigraphy

The Yazoo Clay in the eastern Gulf Coastal Plain contains several stratigraphically significant bentonite beds. Geochemical analyses of these bentonites indicate a rhyodacite-dacite composition based on the Winchester and Floyd (1977) plot of Zr/TiO_2 against Nb/Y (Grigsby, 1999), and they are considered to have originated in the Trans-Pecos volcanic field of West Texas and the Sierra Madre Occidental volcanic belt of western Mexico. Several of the bentonites have been radiometrically dated (Ghosh, 1972; Berggren et al., 1992; Obradovich et al., 1993).

The bentonite beds of the Yazoo Clay present challenges for stratigraphic correlation. The limited nature of Yazoo outcrops in the eastern Gulf Coastal Plain provides few opportunities to sample bentonite beds. Further, many outcrops have only a few meters of exposure, making it difficult, if not impossible, to determine the stratigraphic position of the bentonite bed within the Yazoo Clay. Cores are important because they can place an individual bentonite bed stratigraphically within the Yazoo Clay, but correlation of individual bentonite beds from locality to locality remains problematic. Biostratigraphy has also been used to determine the relative stratigraphic positions of some of the bentonite beds (Berggren et al., 1992; Fluegeman, 2003), but it does not provide enough resolution to enable lateral correlation of bentonite beds in the region. To date, it is not certain exactly how many distinct bentonite beds exist in the Yazoo Clay in the eastern Gulf Coastal Plain.

Field and laboratory studies of the Yazoo Clay and its bentonite beds have identified six distinct bentonites in Alabama and Mississippi. Ongoing work in the area, as well as studies of sections in Louisiana and Texas, may identify more. The bentonites recognized in this study are identified next. The names used for the individual beds are informal and are derived from localities where they have been previously studied.

Matherville Bentonite

Ghosh (1972) identified a bentonite bed in the North Twistwood Creek Member of the Yazoo Clay from a clay pit 3.2 km east of the town of Matherville, Wayne County, Mississippi. He obtained a K-Ar radiometric age on biotite of 39.5 Ma for the bentonite (reported in Hardenbol and Berggren, 1978; Harland et al., 1990). The bentonite is no longer exposed at this locality, so the stratigraphic position of this bentonite could not be verified. A bentonite collected from the Mississippi Office of Geology #1 Ketler core in Wayne County did contain a bentonite bed within the north Twistwood Creek Clay. This core is 9 km southeast of the Matherville bentonite locality, and it is likely that these two beds are the

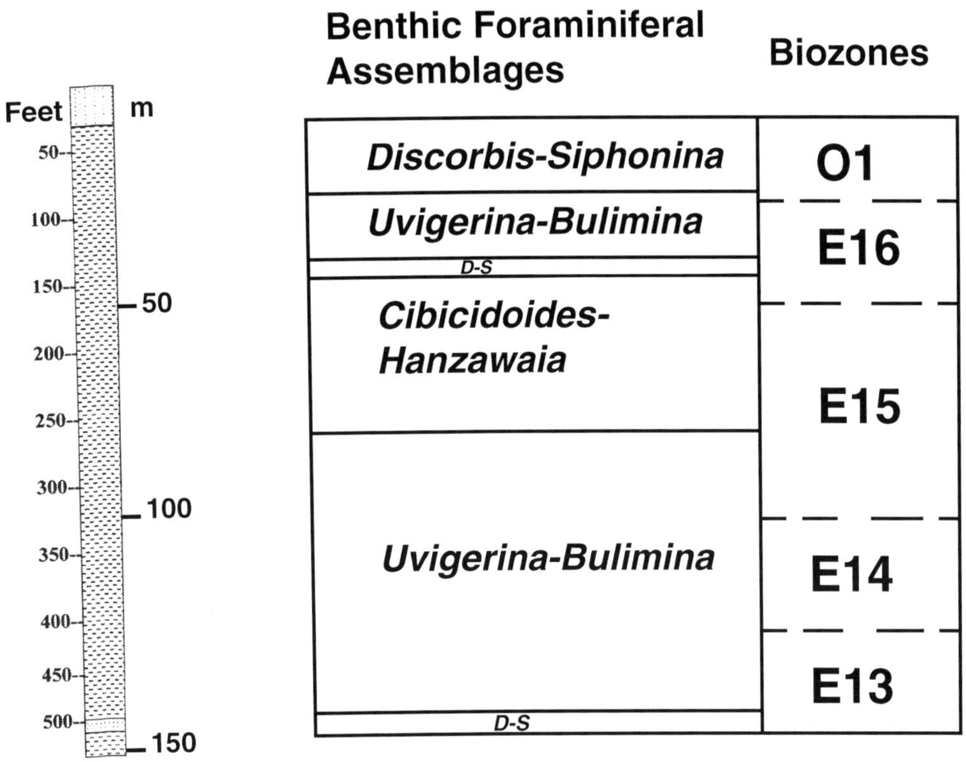

Figure 5. Distribution of benthic foraminiferal assemblages in the Mossy Grove core (modified from Fluegeman, 2003). *D-S—Discorbis-Siphonina.*

same bentonite. The Ketler bentonite is located 7 m below the top of the North Twistwood Creek Clay Member in the core. Micropaleontological examination of samples just above and below the bentonite revealed specimens of *Acarinina rohri*, indicating correlation with biozone E14. Dockery et al. (1994) reported that the Ketler bentonite did not contain sanidine crystals for radiometric dating. To date, this is the only bentonite identified in the North Twistwood Creek Clay Member of the Yazoo Clay.

Mechanicsburg Bentonite

The Mechanicsburg bentonite is named for a bentonite bed exposed in the bank of an unnamed stream in the area SE/4, NE/4, Section 9, T9N, R3W, 1 km north of Mechanicsburg, Yazoo County, Mississippi. This is the "Lower bentonite" from Satartia of Berggren et al. (1992). They reported a $^{40}Ar/^{39}Ar$ age of 34.9 Ma for this bentonite. Planktonic foraminifera obtained from samples collected at this locality included *Subbotina yeguaensis*, *Pseudohastigerina micra*, *Hantkenina alabamensis*, *Turborotalia increbescens*, and *Dentoglobigerina tripartita* among others. The absence of *Pseudohastigerina naguewichiensis* suggests a correlation with biozone E14.

Satartia Bentonite

The Satartia bentonite is named for a bentonite bed exposed along State Highway 3, 0.5 km southeast of Satartia, Yazoo County, Mississippi (SW/4, SE/4, T10N, R3W). The bentonite at this locality was first described by Mellen (1940). This is the "Upper bentonite" from Satartia of Berggren et al. (1992), and they reported a $^{40}Ar/^{39}Ar$ sanidine age of 34.31 Ma. Planktonic foraminiferal assemblages obtained from samples collected below and above the Satartia bentonite include *Pseudohastigerina naguewichiensis* but not *Tenuitella gemma*. This suggests a correlation to biozone E15.

Society Ridge Bentonites

The Society Ridge bentonites were identified by Obradovich et al. (1993) in the Society Ridge core (NW4/SE4, Sec. 24, T7N, R1W), Hinds County, Mississippi. Three bentonites were located 24.6 m (80 ft), 28 m (91 ft), and 29.5 m (96 ft) below the surface. Berggren et al. (1992) reported a mean $^{40}Ar/^{39}Ar$ age on five sanidine samples from the 24.6 m bentonite of 34.3 Ma. A diverse planktonic foraminiferal assemblage from above and below this bentonite includes *Hantkenina alabamensis*, *Subbotina yeguaensis*, *Turborotalia cocoaensis*, and *Tenuitella gemma*, suggesting correlation with biozone E16.

Bentonites in the Mossy Grove Core

The Mossy Grove core represents the longest continuous core through the Yazoo Clay in the eastern Gulf Coastal Plain. Six distinct bentonite beds have been identified in the Yazoo Clay at this locality (Fig. 6). Unlike other bentonite localities, the stratigraphic position of each of the beds is not in dispute.

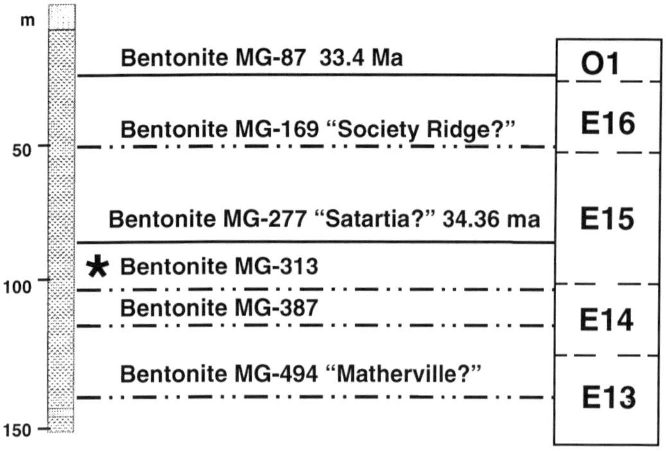

Figure 6. Distribution of bentonite beds in the Mossy Grove core and suggested correlation with other Gulf Coastal Plain bentonites. Solid lines represent radiometrically dated bentonites. Asterisk indicates the location of microspherules in sample MG308.

The bentonites are identified in Figure 6 by their sample number in the core and, where appropriate, a tentative correlation with other bentonite beds. Obradovich et al. (1993) obtained $^{40}Ar/^{39}Ar$ sanidine ages on bentonites MG87 (33.4 Ma) and MG277 (34.56 Ma) in the core. The location of two radiometric dates within the core has enabled the calculation of a sedimentation rate for the upper Yazoo Clay of 7.4 cm/k.y. Using the available radiometric ages for both the Satartia and Mechanicsburg bentonites, a sedimentation rate for the Yazoo Clay of 1.3 cm/k.y. in the outcrop area is indicated.

Bentonite MG87 has not been identified elsewhere in the eastern Gulf Coastal Plain. It has an Oligocene age based on biostratigraphy (this study) and the radiometric date of 33.4 Ma (Obradovich, 1999, personal commun.). Bentonite MG169 has been tentatively correlated with the Society Ridge bentonite. This correlation is based on biostratigraphy, stratigraphic position, and an estimate of the age (33.91 Ma) based on sedimentation rate. The Society Ridge core is 12 km from the Mossy Grove core, and the sections can be directly compared. Using the top of the Moodys Branch Formation as a stratigraphic datum, the Society Ridge bentonite should project into the Mossy Grove core 8.6 m above sample MG169. Allowing for differences in thickness between the two sections (the Moodys Branch alone is 2 m thicker in the Society Ridge core), the correlation of MG169 with the Society Ridge bentonite seems reasonable. Bentonite MG277 has a radiometric age of 34.36 Ma (Obradovich, 1999, personal commun.) and is correlated biostratigraphically with biozone E15. This enables a tentative correlation with the Satartia bentonite.

Based on stratigraphic position and biostratigraphy, bentonite MG315 could be correlated with the Mechanicsburg bentonite. The Mechanicsburg bentonite was determined to be 7.69 m below the Satartia bentonite in the outcrop (Obradovich et al., 1993). In the Mossy Grove core, the distance between bentonites MG277 and MG315 is 11.6 m. Allowing for differences between the two localities, the gap between the two bentonites in the Mossy Grove core is not unreasonable. Bentonite MG315 can be correlated with biozone E14, and this is consistent with the Mechanicsburg bentonite. Bentonite MG315, however, projects into the exposure of Yazoo Clay in the Miss-Lite Clay Pit near Cynthia, Mississippi. Priddy (1960) reported two bentonite beds in this clay pit, but subsequent workers (Dockery and Siesser, 1984; Byerly et al., 1988) did not find bentonite beds. It is possible that bentonite MG 315 and the underlying MG387 bentonite could correlate with one or both of the Cynthia bentonites of Priddy (1960). At present, there is too much stratigraphic uncertainty to propose a tentative correlation for either bentonite MG315 or bentonite MG387.

Bentonite MG494 is in the lower part of the Yazoo Clay in the Mossy Grove core. This part of the Yazoo Clay has been correlated biozone E13. The Matherville bentonite is the only known bentonite in the region that occurs in the lower Yazoo Clay; thus, bentonite MG494 is tentatively correlated with the Matherville bentonite.

Microspherules

Keller (1985) reported a few isolated microspherules in micropaleontological samples collected from an exposure of the Yazoo Clay near Cynthia, Mississippi (the Cynthia clay pit locality of many other workers). Byerly et al. (1988) reexamined Keller's samples from Cynthia and identified more microspherules in sample C. They referred to a "microspherule layer" occurring within the upper Yazoo Clay at this locality. The section exposed in the Cynthia clay pit is stratigraphically below the Society Ridge bentonites (Obradovich et al., 1993). Hazel (1989) resampled the Cynthia clay pit and could not locate a microspherule layer in the section. He suggested that the microspherules might have been the result of modern contamination from a nearby aggregate plant.

In the Mossy Grove core, isolated microspherules have been collected from sample MG308. This sample is within biozone E15 and below bentonite MG277, which has a radiometric date of 34.36 Ma. This is consistent with biostratigraphic and radiometric ages obtained for the North American microtektite strewn field (Glass et al., 1986). Unlike Yazoo microspherules collected from surface exposures, the issue of contamination is minimized in core samples.

SEQUENCE STRATIGRAPHY

The sequence stratigraphy of the Upper Eocene–Lower Oligocene transition in the eastern Gulf Coastal Plain has a complex history of interpretations. While there is general agreement on the number of depositional sequences within the Jackson Group and the lower Vicksburg Group, there are wide differences in the placement of sequence boundaries (Mancini et al., 1987; Baum and Vail, 1987 1988; Loutit et al., 1988; Coleman and Galloway, 1990; Dockery, 1990; Mancini and Tew, 1991;

Tew and Mancini, 1995; Pasley and Hazel, 1995; Mancini, 2000; Echols et al., 2003; Miller et al., 2008). All of these studies are geographically specific, using cores or outcrops within a particular study area of the eastern Gulf Coastal Plain. None of these studies has examined the Eocene-Oligocene transition in the clastic shelf of western Mississippi.

Hurley and Fluegeman (2003) attempted to extract a eustatic signal from the Moodys Branch Formation and the lower Yazoo Clay of the Mossy Grove core. This interval correlates with biozone E13 and depositional sequence TE3.3 of Mancini and Tew and depositional sequence MNT of Miller et al. (2008). To identify a record of fluctuating sea level, they examined benthic $\delta^{18}O$ on *Cibicidoides* through this sequence. Fluctuations in benthic $\delta^{18}O$ are associated with changes in glacioeustasy (Pekar and Miller, 1996). The benthic $\delta^{18}O$ results were verified by Hurley and Fluegman through the use of an independent sea-level proxy: the planktonic:benthic (P:B) ratio (Van der Zwaan et al., 1990). Hurley and Fluegeman (2003) identified a strong correlation between the benthic $\delta^{18}O$ values in the upper part of the depositional sequence and the P:B ratios collected from the same samples. They argued for a role for glacioeustasy in changing sea level from the late Bartonian onward.

Hurley and Fluegeman (2003) looked at only one small part of the Jackson Group in western Mississippi. The Mossy Grove core contains a much longer record through the Eocene-Oligocene transition. The P:B ratios obtained from the Mossy Grove core (including those from Hurley and Fluegeman, 2003) are listed in Table 1. These data, plotted against lithostratigraphy and biostratigraphy in the Mossy Grove core, are shown in Figure 7. An examination of this curve shows three prominent features. First, a sharp decline in P:B values from sample MG380 to sample MG366 may represent the sequence boundary between the sequence MNT and sequence CP of Miller et al. (2008). Second, a gradual increase in P:B ratios to sample MG318 and a P:B ratio of 0.42, the highest P:B value in the entire core, may represent the maximum flooding surface of sequence CP of Miller et al. (2008). Finally, the upper part of the core is characterized by relatively low P:B ratios. This interval is associated biostratigraphically with sequence PS of Miller et al. (2008), but the low P:B ratios may be the result of an increasing sediment influx. Studies of benthic foraminiferal $\delta^{18}O$ isotopes from the upper part of the Mossy Grove core by C. Betz at Ball State University may help to clarify the record of fluctuating sea level of the eastern Gulf Coastal Plain.

The Yazoo Clay of western Mississippi may not fit the sequence stratigraphic model used for sections in western Alabama and eastern Mississippi. The benthic foraminiferal assemblages identified by Fluegeman (2003) indicated outer neritic to upper bathyal conditions. This deep-water setting suggests increasing accommodation space associated with the transgressive systems tract and the maximum flooding surface. Although deep water is indicated by the benthic foraminifera, P:B ratios are more indicative of the filling of the accommodation space associated with a highstand systems tract.

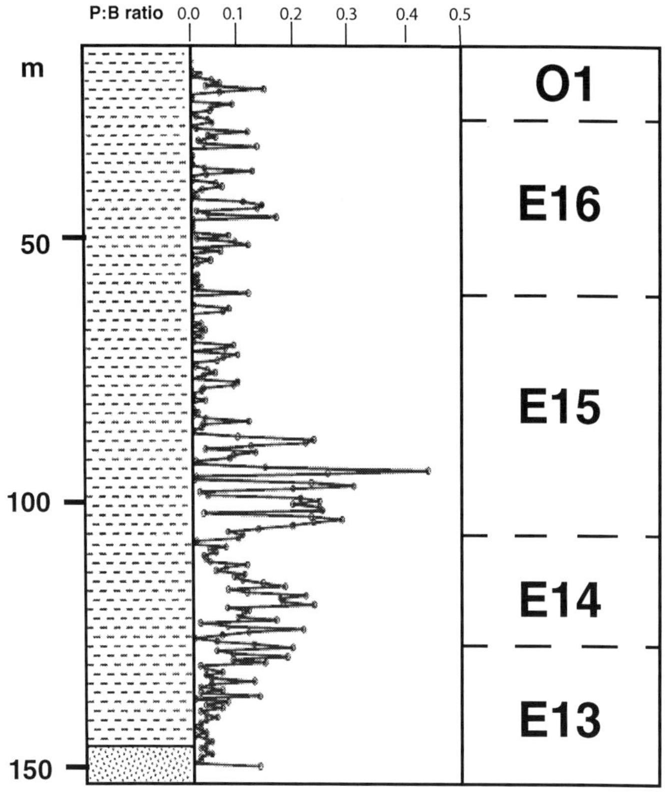

Figure 7. Plot of planktonic:benthic (P:B) ratios in the Mossy Grove core.

Posamentier and Allen (1993) described a relatively deep-water, sand-poor portion of the transgressive systems tract. They referred to the fine-grained sediments deposited seaward of the last clinoforms as the healing phase unit of the transgressive systems tract. Hurley and Fluegeman (2003) interpreted the relatively thick, sand-poor section of the Yazoo Clay between the transgressive surface at the base of the Moodys Branch Formation and the maximum flooding surface as a healing phase unit. Benthic foraminiferal assemblages in this interval are dominated by *Uvigerina*, indicating relatively deep-water deposition. The maximum flooding surface is indicated by the highest relative P:B ratios in the interval. In western Mississippi, the lower and middle Yazoo Clay is a sand-poor unit with benthic foraminiferal assemblages dominated by *Uvigerina* and *Bulimina*. It seems likely that much of the Yazoo represents healing phase units associated with late Eocene and earliest Oligocene transgressive systems tracts.

DISCUSSION

An extensive record of the Eocene-Oligocene transition is preserved in western Mississippi. The Mossy Grove core preserves a continuous record from the late Bartonian through the early Rupelian. Benthic foraminiferal assemblages indicate a relatively deep-water setting for this interval in the western

Mississippi (Fluegeman, 2003) and elsewhere in the Gulf Coastal Plain (Echols et al., 2003; Miller et al., 2008). The record of relative sea-level change in the Mossy Grove core is consistent with results from other workers in the eastern Gulf Coastal Plain for depositional sequences MNT and CP. Additionally, Hurley and Fluegeman argued for the presence of a glacioeustatic signal within sequence MNT based on P:B ratios and benthic foraminiferal $\delta^{18}O$ ratios. P:B ratios in sequence CP of the Mossy Grove core show a similar pattern of fluctuating P:B levels compared to the one identified by Hurley and Fluegeman (2003).

The P:B ratio of depositional sequence PS in the Mossy Grove core differs from the patterns in sequences MNT and CP. No broad increase in values is evident, so it is difficult to identify systems tracts in this interval. P:B values fluctuate widely, suggesting a strong input of clastic sediment in the latest Eocene through the earliest Oligocene. Further evidence of an overall shallowing is seen in the *Hanzawaia-Cibicidoides* benthic foraminiferal assemblages from this part of the core (Fluegeman, 2003). The P:B ratios fluctuate from sample to sample, but the benthic foraminiferal assemblages are more consistent through the core.

The shallowing of the depositional setting in the upper part of the Yazoo Clay is not likely to be associated with the development of the Oi-1 glacial event. A return of the *Uvigerina-Bulimina* assemblage in the latest Eocene and into the earliest Oligocene indicates an increasing bathymetry and a transgression near the beginning of the Oligocene. While this may be a local phenomenon in the section of western Mississippi, it is consistent with observations in the more downdip sections of southwestern Alabama (Mancini and Tew, 1991; Miller et al., 2008). Further studies will help to evaluate the interaction between global and regional processes during the late Eocene and early Oligocene in the production of the sedimentary record of the eastern Gulf Coastal Plain. A better understanding of these processes may enable the resolution of divergent interpretations of the sea-level and paleoecologic record of the latest Eocene and earliest Oligocene in the region.

ACKNOWLEDGMENTS

The writers are indebted to many people for in-depth discussions of the nature of the Eocene-Oligocene boundary in the eastern Gulf Coastal Plain over the years, but especially David T. Dockery, Ernest A. Mancini, Berry H. (Nick) Tew, Ronald J. Echols, and John M. Armentrout. Bridget S. Wade and David Dockery provided helpful reviews of an earlier version of this paper. Acknowledgment is also given to the donors of the Petroleum Research Fund, administered by the American Chemical Society and to the Office of Academic Research and Sponsored Programs at Ball State University for their support of this research.

REFERENCES CITED

Applin, E.R., Ellisor, A.E., and Kniker, H.T., 1925, Subsurface stratigraphy of the coastal plain of Texas and Louisiana: American Association of Petroleum Geologists Bulletin, v. 9, p. 79–122.

Bandy, O.L., 1949, Eocene and Oligocene Foraminifera from Little Stave Creek, Clarke County, Alabama: Bulletins of American Paleontology, v. 32, no. 131, 210 p.

Baum, G.R., and Vail, P.R., 1987, Sequence stratigraphy, allostratigraphy, isotope stratigraphy, and biostratigraphy: Putting it all together in the Atlantic and Gulf Paleogene, *in* McNulty, C.L., and Waterman, A.S., eds., Innovative Biostratigraphic Approaches to Sequence Analysis; New Exploration Opportunities: Houston, Texas, Eighth Annual Research Conference, Gulf Coast Section, Society of Economic Paleontologists and Mineralogists Foundation, p. 15–23.

Baum, G.R., and Vail, P.R., 1988, Sequence stratigraphic concepts applied to Paleogene outcrops, Gulf and Atlantic basins, *in* Wilgus, C.K., Hastings, B.S., Kendall, C.G. St.C., Posamentier, H.W., Ross, C.A., and Van Wagoner, J.C., eds., Sea-Level Changes: An Integrated Approach: Society of Economic Paleontologists and Mineralogists (SEPM) Special Publication 42, p. 309–327.

Berggren, W.A., and Miller, K.G., 1988, Paleogene tropical planktonic foraminiferal biostratigraphy and magnetobiochronology: Micropaleontology, v. 34, p. 362–380, doi: 10.2307/1485604.

Berggren, W.A., and Pearson, P.N., 2005, A revised tropical to subtropical Paleogene planktonic foraminiferal zonation: Journal of Foraminiferal Research, v. 35, p. 279–298, doi: 10.2113/35.4.279.

Berggren, W.A., Kent, D.V., Obradovich, J.D., and Swisher, C.C., III, 1992, Toward a revised Paleogene geochronology, *in* Prothero, D.R., and Berggren, W.A., eds., Eocene-Oligocene Climatic and Biotic Evolution: Princeton, New Jersey, Princeton University Press, p. 29–45.

Berggren, W.A., Kent, D.V., Swisher, C.C., III, and Aubry, M.-P., 1995, A revised Cenozoic geochronology and chronostratigraphy, *in* Berggren, W.A., Kent, D.V., Aubry, M.-P., and Hardenbol, J., eds., Geochronology, Time Scales, and Global Stratigraphic Correlation: Society of Economic Paleontologists and Mineralogists (SEPM) Special Publication 54, p. 129–212.

Berggren, W.A., Pearson, P.N., Huber, B.T., and Wade, B.S., 2006, Taxonomy, biostratigraphy and phylogeny of Eocene *Acarinina*, *in* Pearson, P.N., Olsson, R.K., Huber, B.T., Hemleben, C., and Berggren, W.A., eds., Atlas of Eocene Planktonic Foraminifera: Cushman Foundation Special Publication 41, p. 257–326.

Byerly, G.R., Hazel, J.E., and McCabe, C., 1988, A new late Eocene microspherule in central Mississippi: Mississippi Geology, v. 8, no. 4, p. 1–5.

Cheetham, A.H., 1957, Eocene-Oligocene boundary, eastern Gulf Coast region: Transactions of Gulf Coast Association of Geological Societies, v. 7, p. 89–97.

Coleman, J., and Galloway, W.E., 1990, Sequence stratigraphic analysis of the lower Oligocene Vicksburg Formation of Texas, *in* Armentrout, J.M., Coleman, J.M., Galloway, W.E., and Vail, P.R., eds., Sequence Stratigraphy as an Exploration Tool—Concepts and Practices in the Gulf Coast: Austin, Texas, Eleventh Annual Research Conference: Gulf Coast Section, Society of Economic Paleontologists and Mineralogists (SEPM) Foundation Publication, p. 99–112.

Cooke, C.W., 1923, The correlation of the Vicksburg Group: U.S. Geological Survey Professional Paper 133, p. 1–10.

Cushman, J.A., 1923, The foraminifera of the Vicksburg Group: U.S. Geological Survey Professional Paper 133, p. 11–71.

Cushman, J.A., 1935, Upper Eocene Foraminifera of the Southeastern United States: U.S. Geological Survey Professional Paper 188, 88 p.

Deboo, P.B., 1965, Biostratigraphic Correlation of the Type Shubuta Member of the Yazoo Clay and the Red Bluff Clay with Their Equivalents in Southwestern Alabama: Geological Survey of Alabama Bulletin 80, 84 p.

Dockery, D.T., III, 1990, The Eocene-Oligocene boundary in the northern Gulf—A sequence boundary, *in* Armentrout, J.M., Coleman, J.M., Galloway, W.E., and Vail, P.R., eds., Sequence Stratigraphy as an Exploration Tool—Concepts and Practices in the Gulf Coast: Austin, Texas, Eleventh Annual Research Conference: Gulf Coast Section, Society of Economic Paleontologists and Mineralogists (SEPM) Foundation Publication, p. 141–150.

Dockery, D.T., III, and Siesser, W.A., 1984, Age of the upper Yazoo Formation in central Mississippi: Mississippi Geology, v. 5, p. 1–10.

Dockery, D.T., III, Stover, C.W., Weathersby, P., Stover, C.W., Jr., and Ingram, S.L., 1991, A continuous core through the undifferentiated Yazoo Clay (late Eocene, Jackson Group) of central Mississippi: Mississippi Geology, v. 12, p. 21–27.

Dockery, D.T., III, Thompson, D.E., and Ingram, S.L., 1994, The Mobil-Mississippi Office of Geology Core-Hole Project: Mississippi Geology, v. 15, p. 8–15.

Echols, R.J., Armentrout, J.M., Root, S.A., Fearn, L.B., Cooke, J.C., Rodgers, B.K., and Thompson, P.R., 2003, Sequence stratigraphy of the Eocene/Oligocene boundary interval: Southeastern Mississippi, *in* Prothero, D.R., Nesbitt, E., and Ivany, L., eds., From Greenhouse to Icehouse: The Marine Eocene-Oligocene Transition: New York, Columbia University Press, p. 189–222.

Fluegeman, R.H., 1996, Preliminary paleontological report on the foraminifera of the Mossy Grove core, Hinds County, Mississippi: Mississippi Geology, v. 17, p. 9–15.

Fluegeman, R.H., 2003, Late Eocene–early Oligocene benthic foraminifera in the Gulf Coastal Plain—Regional vs. global influences, *in* Prothero, D.R., Nesbitt, E., and Ivany, L., eds., From Greenhouse to Icehouse: The Marine Eocene-Oligocene Transition: New York, Columbia University Press, p. 283–293.

Fluegeman, R.H., 2008, Late Eocene and early Oligocene benthic foraminiferal paleoecology and sequence stratigraphy in the eastern Gulf Coastal Plain, *in* Demchuk, M.T. and Gary, A., eds., Geologic Problem Solving with Microfossils: Society of Economic Paleontologists and Mineralogists (SEPM) Special Publication no. 93.

Ghosh, P.K., 1972, Use of Bentonites and Glauconites in Potassium-40/Argon-40 Dating in Gulf Coast Stratigraphy [Ph.D. dissertation]: Houston, Texas, Rice University, 136 p.

Glass, B.P., Hall, C.D., and York, D., 1986, $^{40}Ar/^{39}Ar$ laser-probe dating of North American tektite fragments from Barbados and the age of the Eocene-Oligocene boundary: Chemical Geology, v. 59, p. 181–186.

Grigsby, J.D., 1999, Source and correlation of Eocene-Oligocene bentonites of the Gulf Coastal Plain: Geological Society of America Abstracts with Programs, v. 31, no. 7, p. A298.

Hardenbol, J., and Berggren, W.A., 1978, A new Paleogene numerical scale, *in* Cohee, G.V., Glaessner, M.F., and Hedberg, H.D., eds., Contributions to the Geologic Time Scale: American Association of Petroleum Geologists, Studies in Geology, v. 6, p. 213–234.

Harland, W.B., Armstrong, R.L., Cox, A.V., Craig, L.E., Smith, A.G., and Smith, D.G., 1990, A Geologic Time Scale 1989: Cambridge, UK, Cambridge University Press, 263 p.

Hazel, J.E., 1989, Chronostratigraphy of Upper Eocene microspherules: Palaios, v. 4, p. 318–329, doi: 10.2307/3514556.

Howe, H.V., and Wallace, W.E., 1933, Foraminifera of the Jackson Eocene at Danville Landing on the Ouachita, Catahoula Parish, Louisiana: Louisiana Department of Conservation Bulletin, v. 2, p. 127–232.

Hurley, J.V., and Fluegeman, R.H., 2003, Late-middle Eocene glacio-eustasy: Stable isotopes and foraminifera from the Gulf Coastal Plain, *in* Prothero, D.R., Nesbitt, E., and Ivany, L., eds., From Greenhouse to Icehouse: The Marine Eocene-Oligocene Transition: New York, Columbia University Press, p. 223–231.

Keller, G., 1985, Eocene and Oligocene stratigraphy and erosional unconformities in the Gulf of Mexico and Gulf Coast: Journal of Paleontology, v. 59, p. 882–903.

Loutit, T.S., Hardenbol, J., and Vail, P.R., 1988, Condensed sections: The key to age determinations and correlations of continental margin sequences, *in* Wilgus, C.K., Hastings, B.S., Kendall, C.G.St.C., Posamentier, H.W., Ross, C.A., and Van Wagoner, J.C., eds., Sea-Level Changes: An Integrated Approach: Society of Economic Paleontologists and Mineralogists (SEPM) Special Publication 42, p. 309–328.

Mancini, E.A., 1979, Eocene-Oligocene boundary in southwestern Alabama: Transactions of Gulf Coast Association of Geological Societies, v. 29, p. 282–289.

Mancini, E.A., 2000, Sequence stratigraphy and chronostratigraphy of Upper Eocene and Lower Oligocene strata, eastern Gulf Coastal Plain: Transactions of Gulf Coast Association of Geological Societies, v. 50, p. 379–388.

Mancini, E.A., and Tew, B.H., 1991, Relationships of Paleogene stage and planktonic foraminiferal zone boundaries to lithostratigraphic and allostratigraphic contacts in the eastern Gulf Coastal Plain: Journal of Foraminiferal Research, v. 21, p. 48–66.

Mancini, E.A., and Waters, L.A., 1986, Planktonic foraminiferal biostratigraphy of Upper Eocene and Lower Oligocene strata in southern Mississippi and southwestern and south-central Alabama: Journal of Foraminiferal Research, v. 16, p. 24–33.

Mancini, E.A., Tew, B.H., and Waters, L.A., 1987, Eocene-Oligocene boundary in southeastern Mississippi and southwestern Alabama: A stratigraphically condensed section of a type 2 depositional sequence, *in* Ross, C.A., and Haman, D., eds., Timing and depositional history of eustatic sequences: Constraints on seismic stratigraphy: Cushman Foundation for Foraminiferal Research Special Publication 24, p. 41–50.

Mellen, F.F., 1940, Yazoo County Mineral Resources: Mississippi Geological Survey Bulletin 39, 132 p.

Miller, K.G., and Lohman, G.P., 1982, Environmental distribution of Recent benthic foraminifera on the northeast United States continental slope: Geological Society of America Bulletin, v. 93, p. 200–207.

Miller, K.G., Wright, J.D., and Browning, J.V., 2005, Visions of ice sheets in a greenhouse world: Marine Geology, v. 217, p. 215–231, doi: 10.1016/j.margeo.2005.02.007.

Miller, K.G., Browning, J.V., Aubry, M.-P., Wade, B.S., Katz, M.E., Kulpecz, A.A., and Wright, J.D., 2008, Eocene-Oligocene global climate and sea-level changes: St. Stephens Quarry, Alabama: Geological Society of America Bulletin, v. 120, p. 34–53, doi: 10.1130/B26105.1.

Obradovich, J.D., Dockery, D.T., III, and Swisher, C.C., III, 1993, ^{40}Ar-^{39}Ar ages of bentonite beds in the upper part of the Yazoo Formation (Upper Eocene), west-central Mississippi: Mississippi Geology, v. 14, p. 1–9.

Olsson, R.K., and Wise, S.W., 1987, Upper Paleocene to middle Eocene depositional sequences and hiatuses in the New Jersey Atlantic margin, *in* Ross, C., and Haman, D., eds., Timing and Depositional History of Eustatic Sequences: Constraints on Seismic Stratigraphy: Cushman Foundation for Foraminiferal Research Special Publication 24, p. 99–112.

Pasley, M.A., and Hazel, J.E., 1995, Revised sequence stratigraphic interpretation of the Eocene-Oligocene boundary interval, Mississippi and Alabama, Gulf Coast Basin, U.S.A.: Journal of Sedimentary Research, v. B65, p. 160–169.

Pearson, P.N., Olsson, R.K., Huber, B.T., Hemleben, C., Berggren, W.A., and Coxall, H.K., 2006, Overview of Eocene planktonic foraminifera taxonomy, paleoecology, phylogeny, and biostratigraphy, *in* Pearson, P.N., Olsson, R.K., Huber, B.T., Hemleben, C., and Berggren, W.A., eds., Atlas of Eocene Planktonic Foraminifera: Cushman Foundation Special Publication 41, p. 11–28.

Pekar, S., and Miller, K.G., 1996, New Jersey Oligocene "icehouse" sequences (ODP Leg 150X) correlated with global $\delta^{18}O$ and Exxon eustatic records: Geology, v. 24, p. 567–570, doi: 10.1130/0091-7613(1996)024<0567:NJOISO>2.3.CO;2.

Posamentier, H.W., and Allen, G.P., 1993, Variability of the sequence stratigraphic model: Effects of local basin factors: Sedimentary Geology, v. 86, p. 91–109, doi: 10.1016/0037-0738(93)90135-R.

Premoli Silva, I., and Jenkins, G.D., 1993, Decision on the Eocene-Oligocene boundary stratotype: Episodes, v. 16, p. 379–382.

Priddy, R.R., 1960, Madison County Geology: Mississippi Geological Survey Bulletin 88, 123 p.

Siesser, W.G., 1983, Paleogene Calcareous Nannoplankton Biostratigraphy: Mississippi, Alabama, and Tennessee: Mississippi Department of Natural Resources Bureau of Geology Bulletin 125, 61 p.

Tew, B.H., and Mancini, E.A., 1995, An integrated stratigraphic method for paleogeographic reconstruction: Examples from the Jackson and Vicksburg Groups of the eastern Gulf Coastal Plain: Palaios, v. 10, p. 133–153, doi: 10.2307/3515179.

Van der Zwaan, G.J., Jorissen, F.J., and de Stigter, H.C., 1990, The depth dependency of planktonic/benthic foraminiferal ratios: Marine Geology, v. 95, p. 1–16, doi: 10.1016/0025-3227(90)90016-D.

Winchester, J.A., and Floyd, P.A., 1977, Geochemical discrimination of different magma series and their differentiation products using immobile elements: Chemical Geology, v. 20, p. 325–343, doi: 10.1016/0009-2541(77)90057-2.

Zachos, J., Pagani, M., Sloan, L., Thomas, E., and Billups, K., 2001, Trends, rhythms, and aberrations in global climate 65 Ma to present: Science, v. 292, p. 686–693, doi: 10.1126/science.1059412.

Manuscript Accepted by the Society 16 September 2008

The Geological Society of America
Special Paper 452
2009

Reshuffling the cards in the photic zone at the Eocene/Oligocene boundary

Marie-Pierre Aubry*
David Bord
Department of Geological Sciences, Rutgers University, 610 Taylor Road, Piscataway, New Jersey 08854-8066, USA

ABSTRACT

Patterns of diversification and longevity in Paleogene coccolithophorids are analyzed by combining the temporal history of selected genera, families, and orders with the number of discrete morphospecies in them. The coccolithophorids underwent an abrupt mass extinction at the Cretaceous/Paleogene boundary, and a rapid (~1 m.y.-long) global turnover at the Paleocene/Eocene boundary. In contrast, they underwent a diachronous turnover at the Eocene/Oligocene boundary that spread over 6–7 m.y. at mid- and low-latitudes. The turnover included sequential extinctions and speciations of short-lived taxa, beginning slightly before 37 Ma, and losses of taxa that dominated mid- and low latitudes at 34.2 Ma and high latitudes at 32.3 Ma. It is also marked by a few evolutionary appearances, in particular, that of the Family Syracosphaeraceae, which is the most diversified of the living coccolithophorids. Most importantly, the turnover resulted in a shift in the balance between families across several orders, such that families that dominated during the Eocene dwindled during the turnover, and, conversely, families that were little diversified during the Eocene became dominant. Thus, members of Family Coccolithaceae and the genus *Helio-discoaster* typify Eocene communities; members of the Family Calcidiscaceae and the genus *Eu-discoaster* characterize Neogene communities. This shift was accompanied by a decrease in the robustness of coccoliths, suggesting that the Eocene/Oligocene event had a marked effect on the physiology of Eocene coccolithophorids. Bolide impacts and the emplacement of large basaltic provinces provide mechanisms to explain large biotic events. Such mechanisms, however, can be ruled out in the case of the Eocene/Oligocene turnover, which was undoubtedly related to climatic cooling and glaciation. The filtering effect of environmental stress on late Eocene diversity remains to be explained.

Keywords: Coccolithophorids, Paleogene diversity patterns, turnovers, Dominance shift.

*aubry@rci.rutgers.edu.

Aubry, M.-P., and Bord, D., 2009, Reshuffling the cards in the photic zone at the Eocene/Oligocene boundary, *in* Koeberl, C., and Montanari, A., eds., The Late Eocene Earth—Hothouse, Icehouse, and Impacts: Geological Society of America Special Paper 452, p. 279–301, doi: 10.1130/2009.2452(18). For permission to copy, contact editing@geosociety.org.

INTRODUCTION

Once identified as one of the major extinction events of the Phanerozoic (Raup and Sepkoski, 1982), the "terminal Eocene event" (Wolfe, 1978) was subsequently recognized as a significant but progressive turnover in both the marine and terrestrial realms (Prothero and Berggren, 1992). Its status as the main evolutionary event of the Cenozoic after the mass extinction of the Cretaceous/Paleogene boundary was also challenged by the recognition of the significance of the biotic events associated with the Paleocene/Eocene boundary (Aubry et al., 1998). Interestingly, an analysis of diversity of all marine and terrestrial organisms through the Phanerozoic shows greater rates of extinction during the Priabonian (youngest age of the Eocene Epoch) than at any other time during the Cenozoic (Benton, 1995). This analysis did not reveal any diversity change at the family level associated with the Paleocene/Eocene boundary, although it pointed to a middle Eocene (Lutetian) and a late Miocene event. It is thus interesting to revisit the biotic turnover at the Eocene/Oligocene boundary and compare it with the two other main Cenozoic events. The organisms analyzed in this paper in order to decipher changes in diversity patterns at the Eocene/Oligocene boundary are the coccolithophorids. This is a group of marine, unicellular, photosynthetic calcifiers (Haptophytes of the Chromalveolates; Fehling et al., 2007) that have been largely responsible for oceanic primary productivity since at least the Early Cretaceous; they have external skeletal platelets (or coccoliths) that contribute through the deposition of enormous thicknesses of sediments to the sequestration of vast amounts of atmospheric CO_2 (e.g., Hay, 2004; De Vargas et al., 2007). The coccolithophorids have provided critical evidence in the delineation of the major evolutionary moments since their Triassic appearance (most recently: Aubry, 1992a, 1998, 2007; Bown et al., 2004; Bown, 2005). They have registered disruptions in the Earth system with a sensitivity equal to that of planktonic foraminifera (e.g., at the Cretaceous/Paleogene boundary; Bramlette and Martini, 1964; Perch-Nielsen, 1979) and mammals (e.g., at the Paleocene/Eocene boundary; Aubry, 1998).

The objective of this paper is to document the turnover associated with the Eocene/Oligocene boundary, both in terms of timing and the taxonomic levels most affected, and the possible effects of late Eocene–early Oligocene oceanographic, tectonic, and extraterrestrial events are briefly considered. However, the mechanisms that have triggered changes in diversity around the Eocene/Oligocene boundary are not considered here.

MATERIAL AND METHODS

This study builds on a previous analysis of the temporal distribution of Cenozoic families and genera of coccolithophorids (Aubry, 1998), which are revisited here based on recent taxonomic revisions (Aubry, 2009a–i), and the review here is completed by an analysis of diversity patterns at the "species" level within orders, families, and genera (Table 1). The meaning of these diversity patterns requires discussion; the fundamental question is: Do they retrace true biologic diversity through time, and if so, at what taxonomic level? Diversity curves in the coccolithophorids at the species and family levels have been published in recent years (Bown et al., 2004; Bown, 2005), and they have been used to determine (linear) evolutionary trajectories in the planktonic realm through time (e.g., Katz et al., 2004). The patterns obtained by these authors are quite different from those shown here, which gives a measure of the sensitivity of biotic patterns (and their interpretation) to taxonomic concepts. This is not the place to discuss taxonomic differences, but it is important to clarify the nature of the data and the aspect of diversity (e.g., true species richness, morphologic success) being analyzed.

Taxonomy

Although sound in principle, diversity analyses conducted at the species level are unlikely to provide a reliable estimate of true diversity changes through time. A major reason for this is the lack of understanding of the relationships of morphotypes to true species. Molecular biology has revealed that in extant coccolithophores (as in any other group), species may only differ by subtle morphological differences (as in *Calcidiscus*; Geisen et al., 2002). With a few exceptions (Knappertsbusch, 2000; Knappertsbusch et al., 1997), the concept of cryptic species has not been extended to the fossil record, so that a paleontological species may in fact represent the sum of several cryptic species (see De Vargas et al., 2004). Comparative molecular and morphologic studies on planktonic foraminifera (e.g., Huber et al., 1997; De Vargas et al., 2001, 2002; Darling and Wade, 2008) indicate that there is no relationship between high morphologic variability and the numbers of cryptic species, which implies that the "splitter's approach" may not be more suitable than the "lumper's approach" in resolving past biologic diversity. Thus, it is important to acknowledge that, for now, species diversity or species richness for fossil coccolithophorids is not synonymous with true biological diversity; rather, it is an expression of morphologic diversity at any given time, where greater morphologic diversity (= number of morphotypes) probably reflects (although not necessarily) greater biologic diversity (see also De Vargas et al., 2007).

Taxonomy at levels higher than species is rigorous and reproducible when divisions are established based upon the strict construction of coccoliths (Aubry, 1998, 2009a–i). This approach is sound for extant taxa in which the segregation of clades through the techniques of molecular biology overlaps exactly the taxonomic units defined on the basis of morphostructure. There is, however, a difficulty. In addition to the morphostructure, the shape of cells and the arrangement of coccoliths in forming coccospheres are significant characters for the delineation of genera (also supported by molecular biology). Unfortunately, characters as important as dimorphism, varimorphism, and dithecatism are not available to the paleontologist (see also Young et al., 2005). This results in a notable overlap between genus- and family-level classifications in some groups. When the evidence for multiple ancestor-descendant divergence is clear, families are easily delineated. Otherwise families are monogeneric. A multilevel taxonomic approach to diversity is thus highly suitable (see also Benton, 1995).

A dichotomy occurs as to what constitutes an order. Based solely on morphologic/morphostructural evidence, paleontologists have described several orders among coccolithophores (see Aubry [1998] for comparison of the main schemes). In contrast, based mainly on molecular evidence, Edwardsen et al. (2000) recognized only two orders (Isochrysidales and Coccolithales). At the crossroads between these two approaches, a stable order-level classification is being developed (Young et al., 2003; Jordan and Kleijne, 1994; Aubry et al., personal obs.)

In summary, the classification followed in this work (Table 1) at the genus, family, and order levels integrates morphostructural and molecular evidence. Based on objective criteria, it is testable and essentially reflects the different patterns evolved by the coccolithophorids to assemble small rhombohedrons of calcite for the construction of coccoliths. These patterns are well defined and—importantly—do not randomly occur in the stratigraphic record. Some patterns have been dominant in terms of diversification; others have been successful in terms of longevity (see Aubry, 1998). This is analyzed here by combining the temporal history of selected nested genera, families, and orders with a compilation of the number of discrete morphotypes (i.e., paleontological species) in them. The disappearance of an order means that a broad structural pattern is no longer produced; the disappearance of a family or a genus implies that a more restricted structural pattern has disappeared. In a sense, high rank taxonomy in the coccolithophorids mostly retraces the history of calcification in a cellular matrix through time; it expresses conservative phylogenetic traits. In contrast, the extinction of a "species" means that a distinct morphology (i.e., a superficial character such as size, shape, presence of openings, etc.) no longer occurs. When incorporated with high-level taxonomic units, analysis at species-level taxonomy reflects broadly defined morphologic diversity within a given pattern and constitutes an estimate of the success of the pattern.

Stratigraphy and Chronology

The range assigned to a high-level taxon is the sum of the ranges of its component "species." The temporal range of a number of species has been precisely established through astrochronologic calibrations (e.g., Lourens et al., 2004; Raffi et al., 2006; Agnini et al., 2007). The accuracy of these data is offset by the deficiency of our knowledge of the range for other taxa, either because of insufficient geographic coverage, or rare occurrences. The precision achieved for them is at the zonal level, with an uncertainty of up to 4 m.y. The stratigraphic range of the bulk of Cenozoic species is known, however, in terms of relative position within a zone, lowering the age uncertainty on appearances and extinctions to ~±1 m.y. Importantly, the occurrences are well constrained in the vicinity of higher-order chronostratigraphic boundaries.

Rare occurrences raise the problem of bias in fossil diversity caused by differential preservation, as suggested by comparative analysis of the composition in coccolithophores in living communities, sediment traps, and core top sediments (e.g., Andruleit et al., 2004). Recent investigations of sediments with exceptional preservation (Bown et al., 2008) have indicated, however, that Paleogene assemblages are mostly impoverished in holococcoliths. In contrast, unusually high rates of origination often result from the Lagerstätten effect (Benton, 1995). An example of this effect is provided by the late Eocene diversification in the Family Blackitaceae.

The Eocene/Oligocene boundary is formally defined by the lithologic horizon in which the last appearance datum (LAD) of *Hantkenina* spp. is recorded. It is located in the Massignano section, near Ancona, Italy (Odin and Montanari, 1988; Premoli Silva and Jenkins, 1988, see also Van Mourik and Brinkhius, 2005). Its astrochronologic age is 33.79 Ma (Coxall et al., 2005; Pälike et al., 2006) (for details on age estimates and correlation of the Eocene/Oligocene boundary, the reader is referred to Berggren, 2007), which is the age also estimated by Berggren et al. (1995). The latter time scale is used in this study because numerical calibration in the Paleogene time scale is still unstable (the age of the Cretaceous/Paleogene boundary has just been recalibrated; Kuiper et al., 2008), and also because the more recent time scale of Luterbacher et al. (2004) does not provide the means to recalibrate satisfactorily the range of microfossils. Considering the scale of this study, consistency through the Cenozoic time scale is deemed more important than accuracy of numerical ages.

ANATOMY OF THE TURNOVER ASSOCIATED WITH THE EOCENE/OLIGOCENE BOUNDARY

Species Composition of Late Eocene–Early Oligocene Communities

Studies have shown that the Eocene/Oligocene boundary was not associated with a sharp extinction event among the coccolithophorids, but it was bracketed by a long-term (6–7 m.y.) turnover strongly affected by diachrony (Aubry, 1992b). As illustrated at the species level, this complex turnover involved sequential and simultaneous extinctions, as well as sequential evolutionary appearances (Fig. 1). As recorded at low and mid-latitudes, it consisted, in succession, of:

1. a first wave of sequential extinctions (LADs) during Chron C17 (ca. 38–37.1 Ma) (1 in Fig. 1) that mostly affected long-ranging taxa common in early and middle Eocene communities;
2. the sequential evolutionary appearances (FADs) of (mostly) short-ranging taxa (2 in Fig. 1). Although it spread from Chron C17n to Chron C15, this diversification event was mostly centered on 36.8 Ma;
3. the almost simultaneous extinctions (LADs) of the two rosette-shaped discoasters, *Helio-discoaster barbadiensis* and *Helio-discoaster saipanensis*, between 34.3 and 34.2 Ma (Chron C13r) (3 in Fig. 1); and
4. a second wave of sequential extinctions (LADs) during Chron C12 (ca. 32.8 and 30.5 Ma), including the extinction of the *Reticulofenestra umbilicus* group (4 in Fig. 1).

TABLE 1. CLASSIFICATION OF CENOZOIC COCCOLITHOPHORIDS USED IN THIS STUDY
Full references to the taxa cited here can be found in Aubry (2009 a–l)

A—EXTANT ORDERS

Order Isochrysidales Pascher, 1910

Family: Noelarhabdaceae Jerkovic 1970, Young and Bown, 1997
Genera: *Amithalithina, Bekelithella, Crenalithus, Emiliania, Gephyrocapsa, Noelarhabdus, Pyrocyclus, Reticulofenestra, Pseudoemiliania*

Order Coccosphaerales Haeckel, 1894 emend Young and Bown, 1997

Family: Calcidiscaceae Young and Bown, 1997
Genera: *Calcidiscus, Conococcolithus, Cyclolithella, Geminilithella, Hayaster, Oolitothus, Pedinocyclus, Striatococcolithus, Umbilicosphaera.*
Remarks: The proximal shield is monocyclic. The central area is narrow; the outer central cycle is subdued and without massive central structure. *Markalius* is considered the stem genus of this family with coccoliths always much less robust and calcified than in the Family Coccolithaceae. Taxa of the family that are insufficiently described and cannot confidently be assigned to any genus are grouped in this study and labeled "unidentified."

Family: Coccolithaceae Poche 1913 emend Young and Bown 1997
Genera: *Bramletteius, Campylosphaera, Chiasmolithus, Clausibella, Clausicoccus, Coccolithus, Cruciplacolithus, Cruciplectus, Diadelithus, Ectalithus, Sullivania.*
Remarks: *Cruciplacolithus primus*, which evolved during the late Maastrichtian, may be the stem species of this family.

Family: Markaliaceae Aubry (2009a)
Genus: *Markalius*
Remarks: The families Coccolithaceae and Calcidiscaceae were not distinguished by Perch-Nielsen (1985), who considered their genera to have evolved from *Markalius.* Marked structural differences characterize the two families, and *Markalius* is here regarded as ancestral to the Calcidiscaceae.

Family: Stauronerthaceae Aubry (2009a)
Genera: *Catosaura, Stauronertha*
Remarks: Molecular data (Saez et al., 2004; Medlin et al., 2008) demonstrate that the Extant species *Coccolithus neohelis* McIntyre and Bé *in* Hay et al. (1967) and *Coccolithus pelagicus* are genetically distant. This is in conformity with the structural differences exhibited by their coccoliths. Those of *C. neohelis* possess a proximal central cycle that is absent in the placoliths of *C. pelagicus*, as in all the coccoliths assigned to the Family Coccolithaceae. In contrast, this cycle is present in a few early Paleocene (e.g., *Cruciplacolithus filigranus* Mai 2001) and Oligocene taxa (e.g., *Cruciplacolithus tarquinius* Roth and Hay *in* Hay et al., 1967). Consequently, two new genera (*Catosaura, Stauronertha*) and a new family (Stauronerthaceae) are introduced here (Appendix 1).

Incertae sedis
Genera: *Birkelundia, Camuralithus, Coronocyclus, Crassidiscus, Craterolithus, Cryptococcolithus, Hayella*

Order Pontosphaerales Aubry (2009a)

Family: Crepidolithaceae Black 1972
Genus: *Crepidolithus*

Family: Neocrepidolithaceae Aubry (2009a)
Genus: *Neocrepidolithus*

Family: Pontosphaeraceae Lemmermann, 1908
Genera: *Aliculosphaera, Argyrosphaera, Helicosphaerella, Pontosphaera, Scyphosphaera.*

Family: Helicosphaeraceae Black, 1971
Genera: *Helicosphaera*

Order Syracosphaerales Hay, 1977 emend Young et al., 2003

Family: Syracosphaeraceae Lemmermann, 1908
Genera: *Caneosphaera, Coronosphaera, Gaarderia*, Genus 1, *Syracosphaera*
Remarks: The genus *Coronosphaera* is tentatively assigned to this family. Available molecular data are ambiguous (e.g., Medlin et al., 2008; compare figs. 2 and 3).

Family: Umbellosphaeraceae Young and Kleijne *in* Young et al. (2003)
Genus: *Umbellosphaera*

Family: Deutschlandiaceae Kamptner, 1928
Genera: *Calciopappus, Deutschlandia, Michaelsarsia, Ophiaster*

Family: Calciosoleniaceae Kamptner 1927
Genera: *Acanthosolenia, Calciosolenia, Navisolenia*

Family: Rhabdosphaeraceae Lemmermann, 1908
Genera: *Acanthoica, Algirosphaera, Anacanthoica, Cyrtosphaera, Discosphaera, Palusphaera, Rhabdosphaera.*

Family: Blackitaceae Aubry (2009a)
Genera: *Amitha, Aspidorhabdus, Blackites, Cepekiella, Cruxia, Discoturbella, Naninfula, Ommatolithus, Petasus.*

Order Incertae sedis

Family: Braarudosphaeraceae Deflandre, 1947
Genera: *Braarudosphaera, Micrantholithus, Pemma, Pentaster*

Family: Ceraolitholithaceae Norris, 1965
Genera: *Amaurolithus, Ceratolithus*

Incertae sedis
Genera: *Florisphaera, Minylitha*
Remarks: Contiguous stratigraphic range and overall morphologic similarity suggest that *Minylitha* may be ancestral to *Florisphaera* (Gartner, 1992).

(*continued*)

TABLE 1. CLASSIFICATION OF CENOZOIC COCCOLITHOPHORIDS USED IN THIS STUDY (*continued*)
Full references to the taxa cited here can be found in Aubry (2009 a–l).

B—EXTINCT ORDERS

Order Biscutales Aubry (2009a)

Family: Biscutaceae Black 1971
Genus: *Biscutum*

Family: Ellipsolithaceae Aubry (2009)
Genera: *Centumgemina*, *Ellipsolithus*

Family: Prinsiaceae Hay and Mohler 1967
Genera: *Neobiscutum*, *Hornibrookina*, *Prinsius*, *Toweius*

Remarks: *Prinsius* and *Toweius* are assigned to the Family Noelarhabdaceae (Order Isochrysidales) by most authors. However, the proximal cycle is monocyclic in the latter family and bicyclic in the former. The structure of the central area with a prominent outer central cycle and a massive central structure is also different from that in the Noelarhabdaceae.

There is a clear gradation from *Prinsius* to *Toweius*, but coccoliths of the latter genus are considerably derived, with no differentiation of the marginal (shield) elements.

Incertae sedis
Genus: *Hughesius*

Order Predicosphaerales Aubry (2009a)

Family: Predicosphaeraceae Rood, Hay, and Barnard, 1971
Genus: *Predicosphaera*

Family: Praeprinsiaceae Aubry (2009a)
Genera: *Praeprinsius*, *Futyania*

Remarks: Often associated with *Neobiscutum* and *Prinsius*, these two short-ranging and little diversified early Paleocene genera have coccoliths with broad shield elements and an outer central cycle with strongly imbricated elements. These coccoliths are very closely reminiscent of several late Maastrichtian species such as *P. stoveri*, from which they are possibly derived.

Order Watznaueriales Bown 1987

Family: Ellipsagelosphaeraceae Noël 1965
Genus: *Cyclagelosphaera*

Family: Girgisiaceae Aubry (2009a)
Genus: *Girgisia*

Order Chiastozygales Aubry (2009a)

Family: Chiastozygaceae Rood, Hay, and Barnard, 1973
Genus: *Chiastozygus*

Family: Neochiastozygaceae Aubry (2009a)
Genus: *Neochiastozygus*

Family: Neococcolithaceae Aubry (2009a)
Genera: *Neococcolithus*, *Chiphragmalithus*, ? *Isthmolithus*
Remarks: Assignment of *Isthmolithus* to this family is tentative.

Order Zygodiscales Young and Bown 1997 emend Aubry (2009a)

Family: Zeugrhabdotaceae new fam.
Genus: *Placozygus*

Family: Zygodiscaceae Hay and Mohler, 1967
Genera: *Zygodiscus*, *Lophodolithus*. *Helicosphaeroides*

Order Discoasterales Hay, 1977

Family: Biantholithaceae Aubry (2009a)
Genera: *Biantholithus*, Genus B, *Fasciculithus*, *Heliolithus*, *Ilselithina*, *Sphenolithus*.
Remarks: Evolution of *Sphenolithus* from *Biantholithus* (Aubry, 1998) is maintained contra Agnini et al. (2007). Genus B is under description.

The families Facsiculithaceae Hay and Mohler 1967, Heliolithaceae Hay and Mohler 1967, and Sphenolithaceae Deflandre 1952 are under revision and provisionally regrouped here for convenience.

Family: Discoasteraceae Vekshina 1959
Genera: *Catinaster*, *Eu-discoaster*, *Helio-discoaster*

Incertae sedis (Orders)

Family: Goniolithaceae Deflandre, 1957
Genus: *Goniolithus*

Family: Lithostromationaceae Deflandre, 1959
Genera: *Lithostromation*, *Trochoaster*, *Martiniaster*

Family: Rhomboasteraceae Aubry (2009a)
Genera: *Rhomboaster*, *Tribrachiatus*, *Imperiaster*

Family: Triquetrorhabdulaceae Lipps, 1969
Genus: *Triquetrorhabdulus*

Incertae sedis
Family: Lapideacassaceae Bown and Young, 1997
Genera: *Lapideacassis*, *Leesella*

Genera: *Nannotetrina*, *Pseudotriquetrorhabdulus*, *Scampanella*, *Solidipons*, *Tintinnabuliformis*, *Triorbis*, *Trochoastrites*

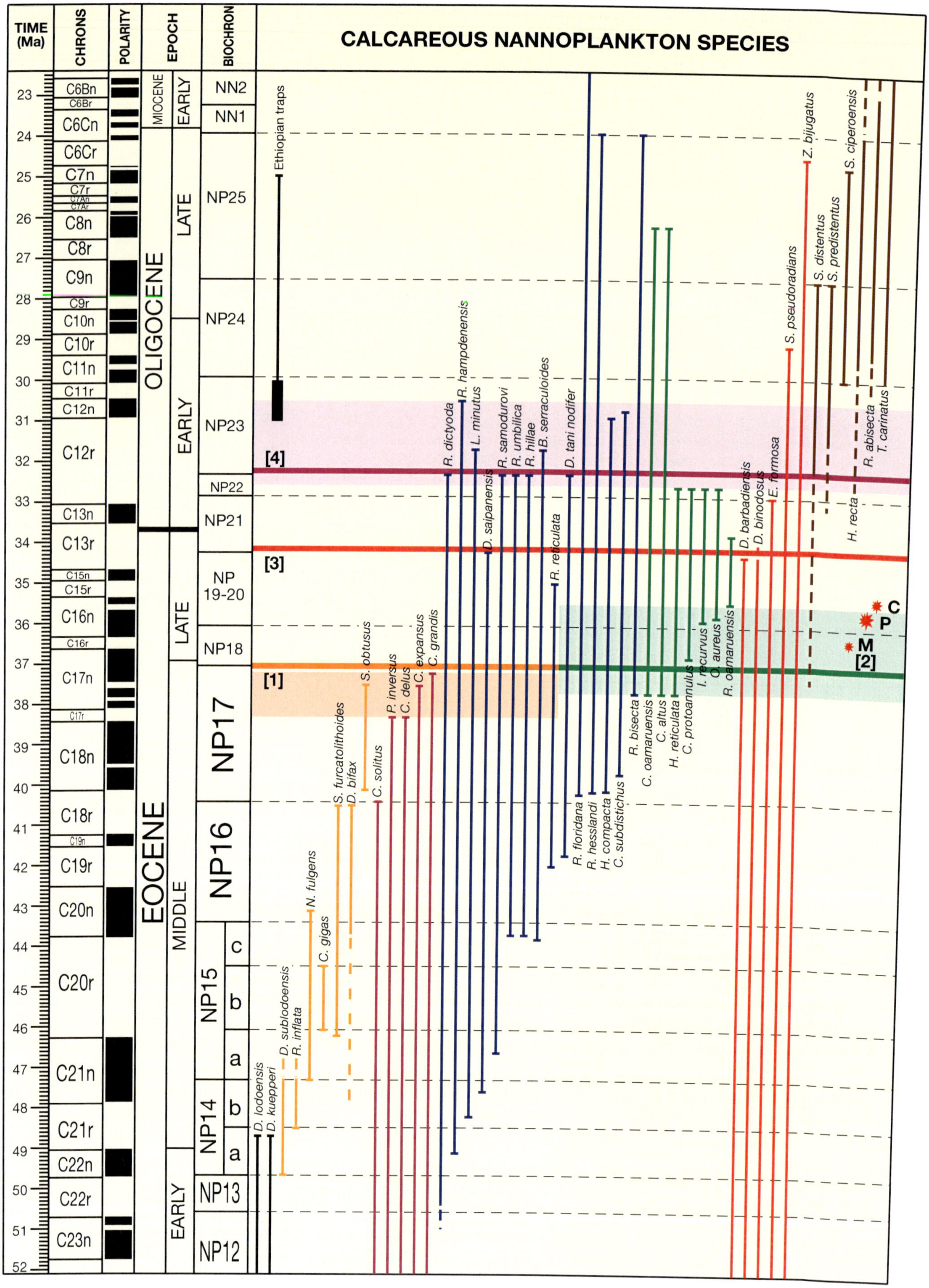

TIME (Ma)
CHRONS
POLARITY
EPOCH
BIOCHRON
CALCAREOUS NANNOPLANKTON SPECIES
23 24 25 26 27 28 29 30 31 32 33 34 35 36 37 38 39 40 41 42 43 44 45 46 47 48 49 50 51 52
C6Bn C6Br C6Cn C6Cr C7n C7r C7An C7Ar C8n C8r C9n C9r C10n C10r C11n C11r C12n C12r C13n C13r C15n C15r C16n C16r C17n C17r C18n C18r C19n C19r C20n C20r C21n C21r C22n C22r C23n
MIOCENE
EARLY
OLIGOCENE
LATE
EARLY
EOCENE
LATE
MIDDLE
EARLY
NN2 NN1 NP25 NP24 NP23 NP22 NP21 NP 19-20 NP18 NP17 NP16 NP15 c b a NP14 b a NP13 NP12
Ethiopian traps
[1]
[2]
[3]
[4]
C
P
M
D. lodoensis
D. kuepperi
D. sublodoensis
R. inflata
N. fulgens
C. gigas
S. furcatolithoides
D. bifax
S. obtusus
C. solitus
P. inversus
C. delus
C. expansus
C. grandis
R. dictyoda
R. hampdenensis
L. minutus
D. saipanensis
R. samodurovi
R. umbilica
R. hillae
B. serraculoides
R. reticulata
D. tani nodifer
R. floridana
R. hesslandi
H. compacta
C. subdistichus
R. bisecta
C. oamaruensis
C. altus
H. reticulata
C. protoannulus
I. recurvus
O. aureus
R. oamaruensis
D. barbadiensis
D. binodosus
E. formosa
S. pseudoradians
Z. bijugatus
S. distentus
S. predistentus
S. ciperoensis
H. recta
R. abisecta
T. carinatus

Two events stand out in this succession. The oldest is the extinction of the two discoasters, 0.5–0.6 m.y. before the Eocene/Oligocene boundary. The youngest is the extinction of the *R. umbilicus* group 1.4 m.y. after the boundary. Both events are highly significant, first because they concern long-ranging taxa, and second because these taxa formed the bulk of Eocene assemblages and were particularly abundant prior to their extinction (Backman, 1987; Aubry, 1992a). There is, however, a major difference between the two events. The youngest event was a global one, even if slightly diachronous (see Berggren et al., 1995). In contrast, the LAD of the two discoasters was a low- and middle-latitude event: *H. saipanensis* disappeared from the high latitudes 1.4 m.y. before the Eocene/Oligocene boundary, and *H. barbadiensis* disappeared 4.6 m.y. earlier (Aubry, 1992b, compare Plate 1 Figure A–B and C–D). The Eocene/Oligocene boundary thus is clearly bracketed by two notable extinction events at low and mid-latitudes, whereas it is followed at high latitudes by the youngest event (Plate 1).

Diachrony affected the geographic distribution of other taxa as well: there is an overlap at mid-latitudes of species that were common at high latitudes but rare to absent at low latitudes (e.g., *Isthmolithus recurvus*, *Chiasmolithus oamaruensis*) and, conversely, species that were common at low latitudes but absent from high latitudes (e.g., *Helicosphaera reticulata*, *Reticulofenestra reticulata*, *Cyclolithella protoannula*).

Figure 1. Anatomy of the evolutionary turnover associated with the Eocene/Oligocene boundary (thick black line at 33.7 Ma; modified from Aubry, 1992a). Time scale is from Berggren et al. (1995). The current ages of the Paleocene/Eocene (55.8 Ma; Storey et al., 2007) and Cretaceous/Paleogene (essentially 66 Ma, Kuiper et al., 2008) boundaries are given for information. The ranges shown are those of the most common taxa encountered in calcareous nannofossil assemblages across the Eocene/Oligocene boundary. The ranges of middle Eocene marker species are also shown to place the Eocene/Oligocene boundary event in a broader temporal context. The ranges are color-coded to readily show the time interval during which they originated. Black (left): species essentially restricted to the early Eocene. Yellow—species restricted to the middle Eocene. Pink—species that evolved during the early Eocene or earlier and became extinct near the middle/late Eocene boundary (37 Ma). Blue—species that evolved during the middle Eocene and became extinct during the Oligocene or later. Green—species that evolved during the late Eocene and became extinct during the Oligocene. Red—species that evolved during the early Eocene or earlier and became extinct during the Oligocene. Violet—Species that evolved after the extinction of the *Reticulofenestra umbilicus* Group. The four critical intervals and moments (1–4; see text) in the Eocene/Oligocene turnover are highlighted as follows: orange shading—first wave of extinctions until the middle/late Eocene boundary (37Ma); orange line marks end of extinctions; green shading—diversification event centered over 36.8 Ma as marked by green line. (For clarity of the drawing, the orange and green lines are aligned, which underlines the approximately contemporaneous extinctions and evolutionary appearances of taxa at the middle/late Eocene boundary.) Red line—extinction (3) of the rosette-shaped discoasters 0.6 m.y. (*H. barbadiensis*) and 0.5 m.y. (*H. saipanensis*) prior to the Eocene/Oligocene boundary (marked by a major extinction event in the planktonic foraminifera), which would be an isolated event except for the fact that these two species constituted a large proportion (possibly as much as 50%) of the late Eocene nannoplankton. Purple shading—second wave of extinctions; the thick horizontal purple line marks the extinction of the *Reticulofenestra umbilicus* group. Star at 35.2–35.5 Ma symbolizes the Chesapeake (C) impact of a bolide (Poag and Aubry, 1995; Lee et al., 2006). The Popigai (P) impact (arctic Siberia; Vishnevsky and Montanari, 1999; Bottomley et al., 1997) is virtually of the same age; the Mistastin (M) impact is considered slightly older (36.4 ± 4 Ma; http://www.unb.ca/passc/ImpactDatabase/NorthAmerica.html). Black box to the left shows the timing of the emplacement of the Ethiopian traps (Touchard et al., 2003).

Temporal Distribution of Genera, Families, and Orders through the Cenozoic

It is helpful to compare different biotic events to determine their respective severity. This is generally done by ranking events according to the number of extinctions or originations or their rates by unit of time (e.g., Roth, 1987; Benton, 1995; Bown, 2005; McGhee et al., 2004). Thus, Bown (2005) determined that the Eocene/Oligocene boundary event ranks sixth of the Cenozoic with regard to rates of extinction of coccolithophorid species. Whereas ranking provides an estimate of the magnitude of diversity changes through time, it ignores the duration and the fabric of evolutionary events. These two important characters are readily accessible by examining the temporal distribution of taxonomic units (Figs. 2A and 2B). The evolutionary events at the generic, family, and order levels shown by the taxo-temporal map are concordant with those reported at the species level by Bown (2005), i.e., (1) increased rates of extinction at the Cretaceous/Paleogene boundary, the Paleocene/Eocene boundary, in the middle Paleocene (60.5 Ma), the latest Miocene (6.5 Ma), and the middle Eocene (45.5 Ma), and (2) increased rates of speciation in the early Danian (63.5 Ma), the middle Paleocene, the late Paleocene (57.5), the early Eocene (51.5 Ma), and the late Miocene (9.5 Ma). The merit of a "map" over a curve is to additionally show the lineages that are involved and the span of time that is concerned for each event. Aubry (1998) previously discussed these events based on an earlier map. It is sufficient here to compare the three events at the Cretaceous/Paleogene, the Paleocene/Eocene, and the Eocene/Oligocene boundaries.

Although the simultaneous Cretaceous extinctions are not shown, the Cretaceous/Paleogene event stands out as a time of abrupt change, when multiple originations occurred at the generic and family levels in most orders. This means that numerous new coccolith patterns evolved after the mass extinction. The Paleocene/Eocene event is also remarkably delineated by extinctions and appearances of genera and families. This is not a mass extinction but a rapid turnover event over ~1 m.y. Over this short duration (~1 m.y.), coccolith patterns were lost (e.g., genera *Fasciculithus*, *Heliolithus*, *Biantholithus*, *Placozygus*, *Ellipsagelosphaera*, Family Neocrepidolithaceae) and others appeared (e.g., Genus *Lophodolithus*; Family Blackitaceae, Calcidiscaceae, Pontosphaeraceae, Rhomboasteraceae).

The Eocene/Oligocene event is of a different nature because it is spread over several million years. A low-diversity

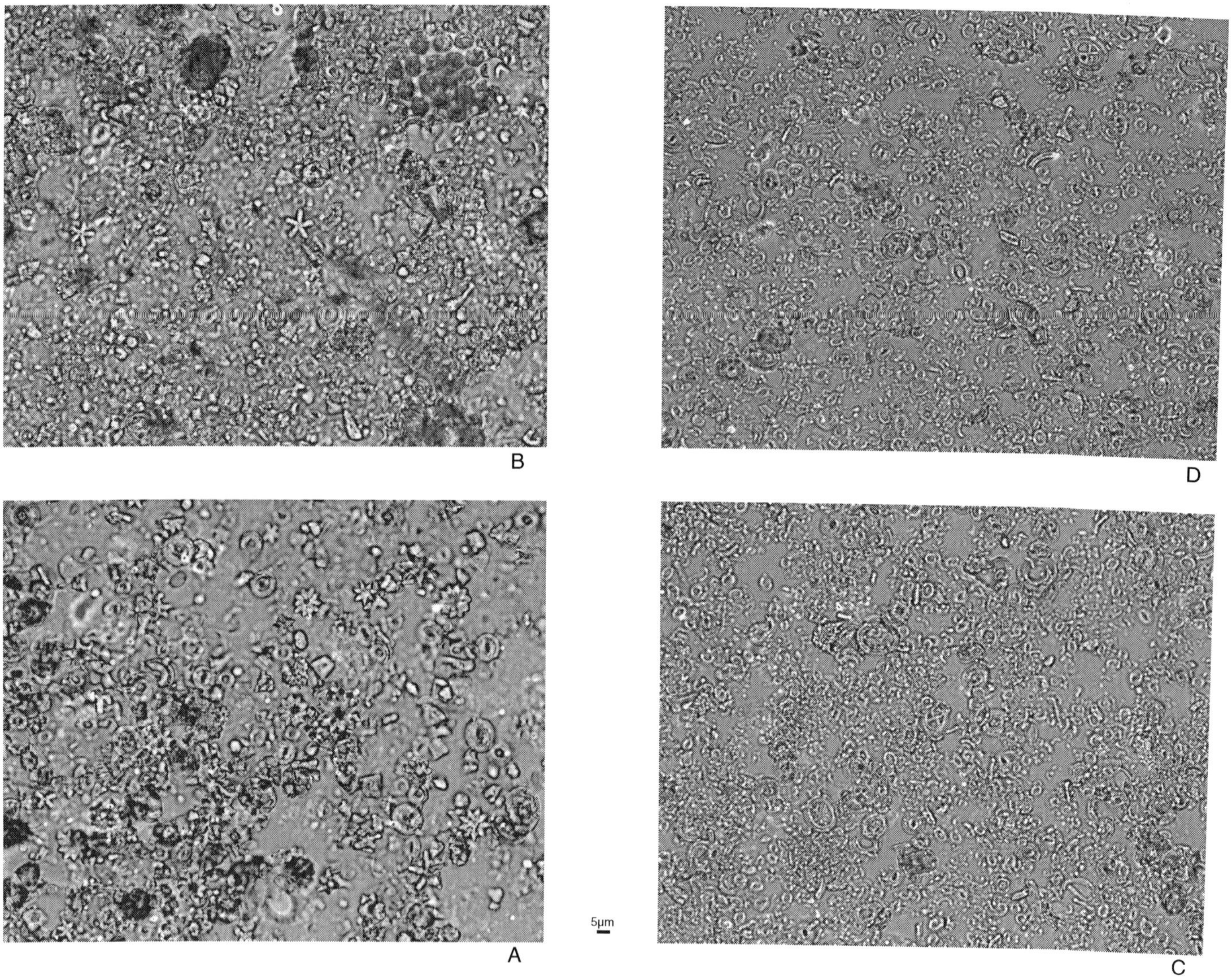

Plate 1. Fossil communities of coccolithophorids across biochronal boundary NP19–20/NP21, i.e., ~0.5 m.y. prior to the Eocene/Oligocene boundary, at a low-latitude site (Indian Ocean Deep Sea Drilling Project [DSDP] Site 219, ~3°S) and a high-latitude site (Southern Pacific DSDP Site 277, 52°17.43′S) (see Corliss et al., 1984; Aubry, 1992a). The extinction of the rosette-shaped discoasters (genus *Heliodiscoaster*) was a massive event (not in terms of diversity but in term of abundance) at low and mid latitudes, but it was a "nonevent" at high latitudes. (A–B) Site 219: (A) sample 219–17–1, 37–42 cm (zone NP19–20) and (B) sample 219–16–6, 137–143 cm (zone NP21). (C–D) Site 277: (C) sample 277–22–3, 136–140 cm (zone NP19–20) and (D) sample 277–22–2, 98–101 cm (zone NP21).

family (Neococcolithaceae) became extinct, and perhaps also the Family Blackitaceae (the record of its short-lived, late Eocene taxa is insufficiently documented). No family with high diversity became extinct, but almost every family was affected, and this occurred in ways that differed among them. Four families underwent a decrease in generic diversity (Table 2). The Family Coccolithaceae was the most severely affected among them, with its first loss of a genus at ca. 37 Ma. From 11 genera before 37 Ma, it was reduced to six genera after ca. 30 Ma, and three genera after ca. 25.5 Ma. Three families (Pontosphaeraceae, Calcidiscaceae, Discoasteraceae,) underwent a rapid turnover marked by the loss of one or two genera, and the gain of a new one. One family (Sphenolithaceae) gained a new genus. Additionally, three families (Helicosphaeraceae, Syracosphaeraceae, Triquetrorhabdulaceae), and several short-lived genera (*Isthmolithus*, *Aliculosphaera*, *Trochastrites*) evolved. It is difficult to document evolutionary changes in the Family Noelarhabdaceae, at least in its earlier history because of its unprecedented morphostructural plasticity, which leads to unsatisfactory taxonomic subdivision (Aubry, 2009, personal obs.). Thus, the genus *Reticulofenestra* is used in a broad sense. There are, however, two major events as shown in Figure 1. The oldest is the extinction of *Reticulofenestra reticulata* (35 Ma at low latitudes; 36.1 Ma at high latitudes), a species that was abundant throughout its range. The youngest event (32.3 Ma) is the loss of all species with a distally opened central area (*R. umbilicus* group; coccoliths possess a proximal grid only).

Morphologic Diversity within High-Level Taxa

The loss of high-level taxa is always significant, but this significance increases with increased morphologic diversity, which is a measure of taxonomic dominance (Wilson, 1992). The Cenozoic coccolithophorids belong to ten well-defined orders (Fig. 3). Extant and fossil taxa of uncertain affinities are regrouped as *Incertae sedis*. All orders originated in the Mesozoic, except the calcifying Isochrysidales (Family Noelarhabdaceae), the earliest fossil record of which is dated at 52.45 Ma (Agnini et al., 2006). Six orders disappeared sequentially during the Cenozoic, where the extant coccolithophorids comprise four orders in addition to several incertae sedis taxa (Young et al., 2003). The Cretaceous/Paleogene and the Paleocene/Eocene boundaries are clearly marked as times of changes in morphologic diversity in all orders, and a previously unidentified decrease in diversity occurred at ca. 40 Ma in all orders. With regard to the Eocene/Oligocene boundary, changes in morphologic diversity are rather diffuse. Most orders show similar but asynchronous trends toward increasing then decreasing diversity over a 10-m.y.-long interval (37–27 Ma). One order (Chiastozygales), already on the decline after the Paleocene/Eocene boundary, disappeared at the Eocene/Oligocene boundary.

Diversity changes at the Eocene/Oligocene boundary are significant at the family and genus level. The balance that prevailed between families during the Eocene became reversed during the Oligocene. This is most notable within the Order Coccolithales. In this order, the Family Coccolithaceae contributed most of the Paleocene and Eocene morphologic diversity (Fig. 4A). In contrast, the Family Calcidiscaceae contributed most of the Neogene diversity (Fig. 4B). The shift in dominance between the two families occurred between 37 and 33 Ma and is part of the Eocene/Oligocene event. The Family Coccolithaceae suffered a progressive loss in diversity beginning at ca. 37 Ma and prolonged until 28.5 Ma. A similar shift occurred in the Order Pontosphaerales between the Genera *Pontosphaera* and *Argyrosphaera* on the one hand, and *Scyphosphaera* and *Helicosphaera* on the other (Fig. 5), and between the genera *Helio-discoaster* and *Eu-discoaster* in the Family Discoasteraceae (Fig. 5). A similar pattern is difficult to establish between the Families Blackitaceae and Rhabdosphaeraceae (Fig. 6). Because of taxonomic limitations (see previous), no pattern can be deciphered in the Family Noelarhabdaceae (Fig. 6).

SIGNIFICANCE OF THE EOCENE/OLIGOCENE TURNOVER

Implications for the Evolution of the Coccolithophorids

The turnover associated with the Eocene/Oligocene boundary is clearly complex. It involved both short- and long-term events. The short-term events include the sudden extinctions of species that (unless the fossil record is irremediably biased) dominated coccolithophorid communities from the middle Eocene to the earliest Oligocene, some at middle and low latitudes, others at all latitudes. The extinctions of *H. barbadiensis* and *H. saipanensis* (the latter at 34.2 Ma), separated by 100,000 years at mid- and low latitudes, and the extinction of the *R. umbilicus* group at 32.8 Ma are the most noticeable short-term events in this turnover. They were preceded, however, by other conspicuous extinctions, such as those of *Reticulofenestra reticulata* (35 Ma), *Chiasmolithus grandis* (ca. 37 Ma), and (in the shallower marine domain) *Pemma* spp. (Pearson et al., 2008). The long-term events include parallel shifts in taxonomic dominance, such as between the families Coccolithaceae and Calcidiscaceae, and the replacement of the genus *Helio-discoaster* by *Eu-discoaster*.

The Eocene/Oligocene boundary event was not only about extinctions. The beginning of the late Eocene (37 Ma) was also marked by a diversification event (Fig. 1). At least nine genera evolved between 37 and 32 Ma, half of them short-lived (<5 m.y.). Among the long-lasting evolutionary events, there were the early Oligocene appearances of the Family Helicosphaeraceae and of the genera *Scyphosphaera* s. st. and *Calcidiscus*, and the late Eocene evolution of the Family Syracosphaeraceae s.l. The end of the turnover is further marked by the evolution of the Family Triquetrorhabdulaceae.

Figure 2 (*on following two pages*). Temporal distribution of coccolithophorid taxa at the order and genus levels. The taxonomic framework (Table 1) used here integrates morphostructural analysis (a methodology explained in Aubry 1998, 2007) and molecular biology (Medlin et al., 2007; Saez et al., 2003, 2004; Aubry et al., 2006, personal obs.). Improved understanding of the morphostructure of some coccoliths (Aubry, 2009 f–i) has led to (1) a reconsideration of earlier generic assignments to families, and (2) a redistribution of families among orders (compare Appendix 10.3 *in* Aubry [1998] and Table 1 herein). The order-level classification follows from a reevaluation of ancestor-descendant relationships between Mesozoic and Cenozoic taxa (Aubry, 2006, personal obs.) and is a major improvement compared to previous classifications. Among the major revisions of ancestor-descendant relationships are: (1) a probable origination of the Family Calcidiscaceae from *Markalius*, and (2) the convergent evolution of *Cyclagelosphaera* and Rhadosphaeraceae coccoliths rather than a divergence of the latter from the former. In the coccolithophorids, dissimilar coccoliths are produced during the equally well-represented haploid and diploid phases. Holococcoliths are considered products of the haploid phase (Geisen et al., 2002), and all fossil holococcoliths have been excluded from this study. Extant genera with no known fossil record were also excluded from this study. *Gladiolithus*, now reported from the Upper Paleocene of Tanzania (Bown et al., 2008), is also excluded. Red thick line—the Cretaceous/Paleogene mass extinction; green shading—the Paleocene/Eocene turnover; violet shading—the Eocene/Oligocene turnover. Orders and families are arranged according to appearance and longevity. For convenience, they are cited in this order in the text rather than alphabetically. Each order is color-coded. Different shades of a color indicate different families (as cited in Table 1). The ancestor to each Cenozoic family is shown (purple color) and included in that family for easy reading. Proper family assignment is given in Table 1.

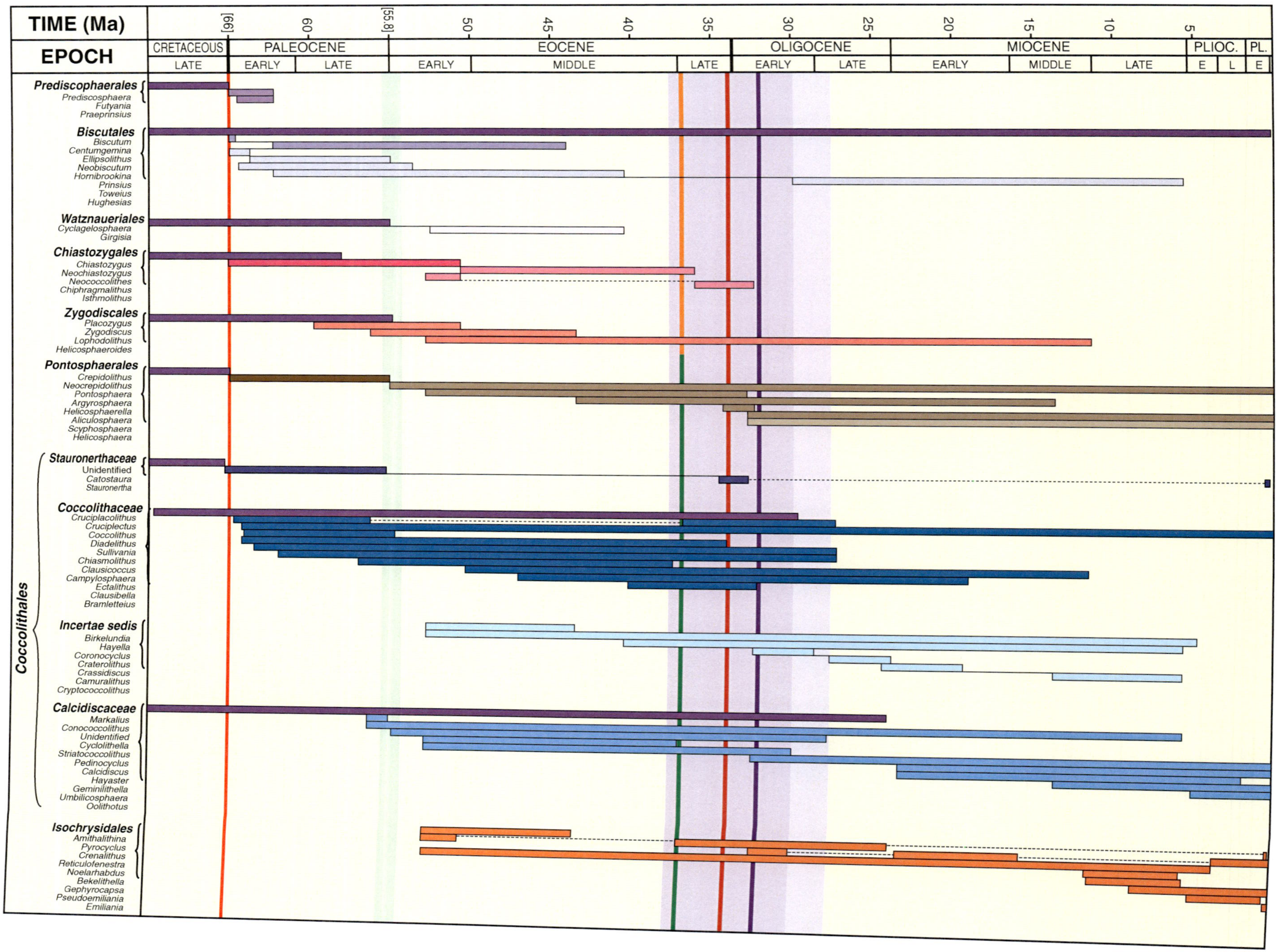

TIME (Ma)
EPOCH
[66]
60
[55.8]
50
45
40
35
30
25
20
15
10
5
CRETACEOUS
PALEOCENE
EOCENE
OLIGOCENE
MIOCENE
PLIOC.
PL.
LATE
EARLY
LATE
EARLY
MIDDLE
LATE
EARLY
LATE
EARLY
MIDDLE
LATE
E
L
E
Prediscophaerales
Prediscosphaera
Futyania
Praeprinsius
Biscutales
Biscutum
Centumgemina
Ellipsolithus
Neobiscutum
Hornibrookina
Prinsius
Toweius
Hughesias
Watznaueriales
Cyclagelosphaera
Girgisia
Chiastozygales
Chiastozygus
Neochiastozygus
Neococcolithes
Chiphragmalithus
Isthmolithus
Zygodiscales
Placozygus
Zygodiscus
Lophodolithus
Helicosphaeroides
Pontosphaerales
Crepidolithus
Neocrepidolithus
Pontosphaera
Argyrosphaera
Helicosphaerella
Aliculosphaera
Scyphosphaera
Helicosphaera
Coccolithales
Stauronerthaceae
Unidentified
Catostaura
Stauronertha
Coccolithaceae
Cruciplacolithus
Cruciplectus
Coccolithus
Diadelithus
Sullivania
Chiasmolithus
Clausicoccus
Campylosphaera
Ectalithus
Clausibella
Bramletteius
Incertae sedis
Birkelundia
Hayella
Coronocyclus
Craterolithus
Crassidiscus
Camuralithus
Cryptococcolithus
Calcidiscaceae
Markalius
Conococcolithus
Unidentified
Cyclolithella
Striatococcolithus
Pedinocyclus
Calcidiscus
Hayaster
Geminilithella
Umbilicosphaera
Oolithotus
Isochrysidales
Amithalithina
Pyrocyclus
Crenalithus
Reticulofenestra
Noelarhabdus
Bekelithella
Gephyrocapsa
Pseudoemiliania
Emiliania

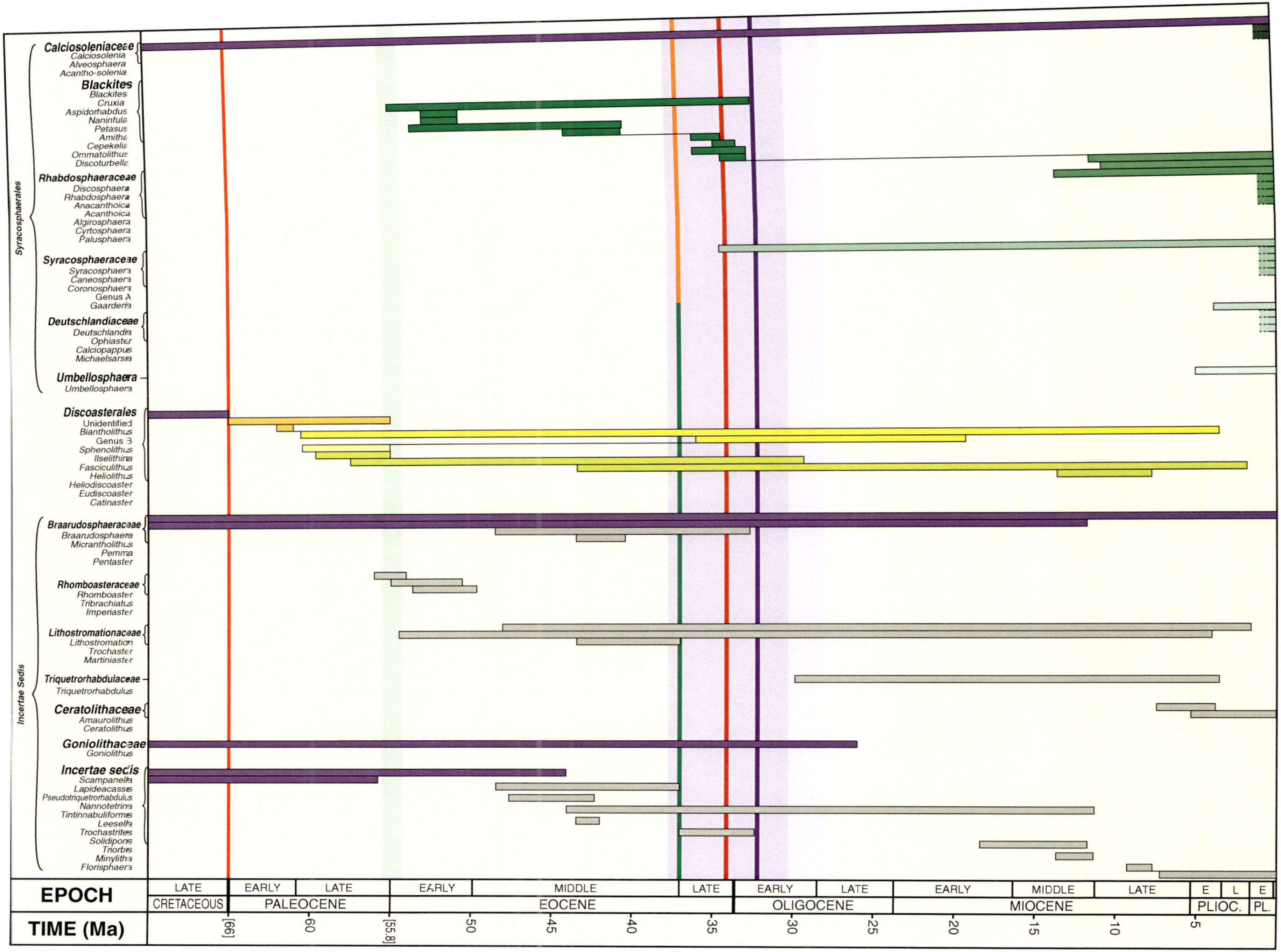

Syracosphaerales
Calciosoleniaceae
Calciosolenia
Alveosphaera
Acantho-solenia
Blackites
Blackites
Cruxia
Aspidorhabdus
Naninfula
Petasus
Amitha
Cepekelia
Ommatolithus
Discoturbella
Rhabdosphaeraceae
Discosphaera
Rhabdosphaera
Anacanthoica
Acanthoica
Algirosphaera
Cyrtosphaera
Palusphaera
Syracosphaeraceae
Syracosphaera
Caneosphaera
Coronosphaera
Genus A
Gaarderia
Deutschlandiaceae
Deutschlandia
Ophiaster
Calciopappus
Michaelsarsia
Umbellosphaera
Umbellosphaera
Discoasterales
Unidentified
Biantholithus
Genus 3
Sphenolithus
Ilselithina
Fasciculithus
Heliolithus
Heliodiscoaster
Eudiscoaster
Catinaster
Incertae Sedis
Braarudosphaeraceae
Braarudosphaera
Micrantholithus
Pemma
Pentaster
Rhomboasteraceae
Rhomboaster
Tribrachiatus
Imperiaster
Lithostromationaceae
Lithostromation
Trochaster
Martiniaster
Triquetrorhabdulaceae
Triquetrorhabdulus
Ceratolithaceae
Amaurolithus
Ceratolithus
Goniolithaceae
Goniolithus
Incertae sedis
Scampanella
Lapideacassis
Pseudotriquetrorhabdulus
Nannotetrina
Tintinnabuliformis
Leesella
Trochastrites
Solidopons
Triorbis
Minylitha
Florisphaera
EPOCH
LATE
EARLY
LATE
EARLY
MIDDLE
LATE
EARLY
LATE
EARLY
MIDDLE
LATE
E
L
E
CRETACEOUS
PALEOCENE
EOCENE
OLIGOCENE
MIOCENE
PLIOC.
PL.
TIME (Ma)
[66]
60
[55.8]
50
45
40
35
30
25
20
15
10
5

TABLE 2. LOSSES (LAST APPEARANCE DATUM [LAD]) AND GAINS (FIRST APPEARANCE DATUM [FAD]) OF GENERA IN DIFFERENT FAMILIES PRIOR TO, DURING, AND AFTER THE EOCENE/OLIGOCENE (E/O) EVENT

Order	Family	No. of genera just prior to the E/O	LADs during the E/O event	FADs during the E/O event	No. of genera after the E/O
Prediscosphaerales		—			
Biscutales		1	0	1 FAD at ~30	2
Watznaueriales		—			
Chiastozygales		1	2	1	0
Zygodiscales		1	0	0	1
Pontosphaerales	**Pontosphaeraceae**	3	2	2	3
	Helicosphaeraceae	0		1	1
Coccosphaerales	Stauronerthaceae	0	0	1	1
	Coccolithaceae	10	4	0	3
			3 at ~27 Ma		
	Incertae sedis	7	1 at ~28 Ma	1	7
				1 at ~30 Ma	
	Calcidiscaceae		2	1	4
			1 at ~27 Ma		
Isochrysidales	**Noelarhabdaceae**		9	1	8
Syracosphaerales	Calciosoleniaceae	3	0	3	3
	Blackitaceae	9	1	0?	0?
	Rhabdosphaeraceae	—			
	Syracosphaeraceae	0		1	1
	Deustchlandiaceae	—			
	Umbellosphaeraceae	—			
Discoasterales	**Sphenolithaceae**	1		1	2
	Discoasteraceae	1	1 at ~28 Ma	1	1
Incertae sedis	Braarudosphaeraceae	3	1	0	2
	Rhomboasteraceae	—			
	Lithostromationaceae	3	1	0	2
	Triquetrorhabdulaceae	0		1 at ~30 Ma	1
	Ceratolithaceae	—			
	Goniolithaceae	1	0	0	1
	Incertae sedis	2	2	1	1

Note: Names of families discussed in the text are in bold face.

Emphasis has been placed on the extinctions at the Eocene/Oligocene boundary. Aubry (1998) pointed to the lack of early Oligocene innovations, and Bown (2005) determined that the rates of speciation associated with the Eocene/Oligocene boundary were at background level. A different picture of the Eocene/Oligocene turnover emerges here. Extinctions were only the most spectacular of three manifestations of the turnover. The shift in dominance at the family and genus levels, and the successful innovations that occurred between 37 and 30 Ma would have the important consequence of transforming typical Eocene communities into typical Neogene communities. Members of the Family Coccolithaceae and the genus *Helio-discoaster* typify Eocene communities; members of the Family Calcidiscaceae and the genus *Eu-discoaster* characterize Neogene ones. The role of the Family Noelarhabdaceae in this narrative is ambiguous, despite the massive event that the extinction of the *R. umbilicus* group represents. Recognition of a diversification event at the Eocene/Oligocene boundary results from both a taxonomic revision prompted by molecular phylogeny (Aubry et al., personal obs.) and documentation of fossil occurrences in well-preserved assemblages (Bown, 2005; Bown et al., 2008). One of the most interesting late Eocene innovations is the introduction of a new coccolith pattern (caneolith). It characterizes the Family Syracosphaeraceae s.l. (sensu Young et al., 2003), which is the most diversified family among extant coccolithophorids.

A morphologic thread connects the taxa involved in the turnover. Taxa characterized by large, robust, heavily calcified coccoliths were eliminated and replaced by taxa with smaller and less robust coccoliths. This is exemplified by the replacement of the Family Coccolithaceae by the Family Calcidiscaceae (Figs. 4A and 4B). The thick coccoliths in the former family consist of several (five), well-developed, interlocking cycles of elements with strongly imbricate elements. In contrast, the coccoliths of

Figure 3. Morphologic diversity ("species richness") at the order level among the Cenozoic coccolithophorids. The reliability of diversity patterns through time is dependent upon the quality of the fossil record. Robust, large coccoliths have greater chance of being preserved than tiny or delicate coccoliths. Thus, in general, coccoliths of the Order Syracosphaerales have a poor fossil record. The sudden Pliocene increase in diversity in this order is artificial and reflects the generally low preservation potential of its Neogene and (to a lesser extent) Paleogene coccoliths. Similarly, its late Eocene–early Oligocene increase in diversity results from the Lägerstatten effect. Diversity among the *Incertae sedis* is also grossly underestimated. This is because all extant taxa without clear phylogenetic affinity and without fossil record are ignored in this study. Orders are color-coded as in Figure 2.

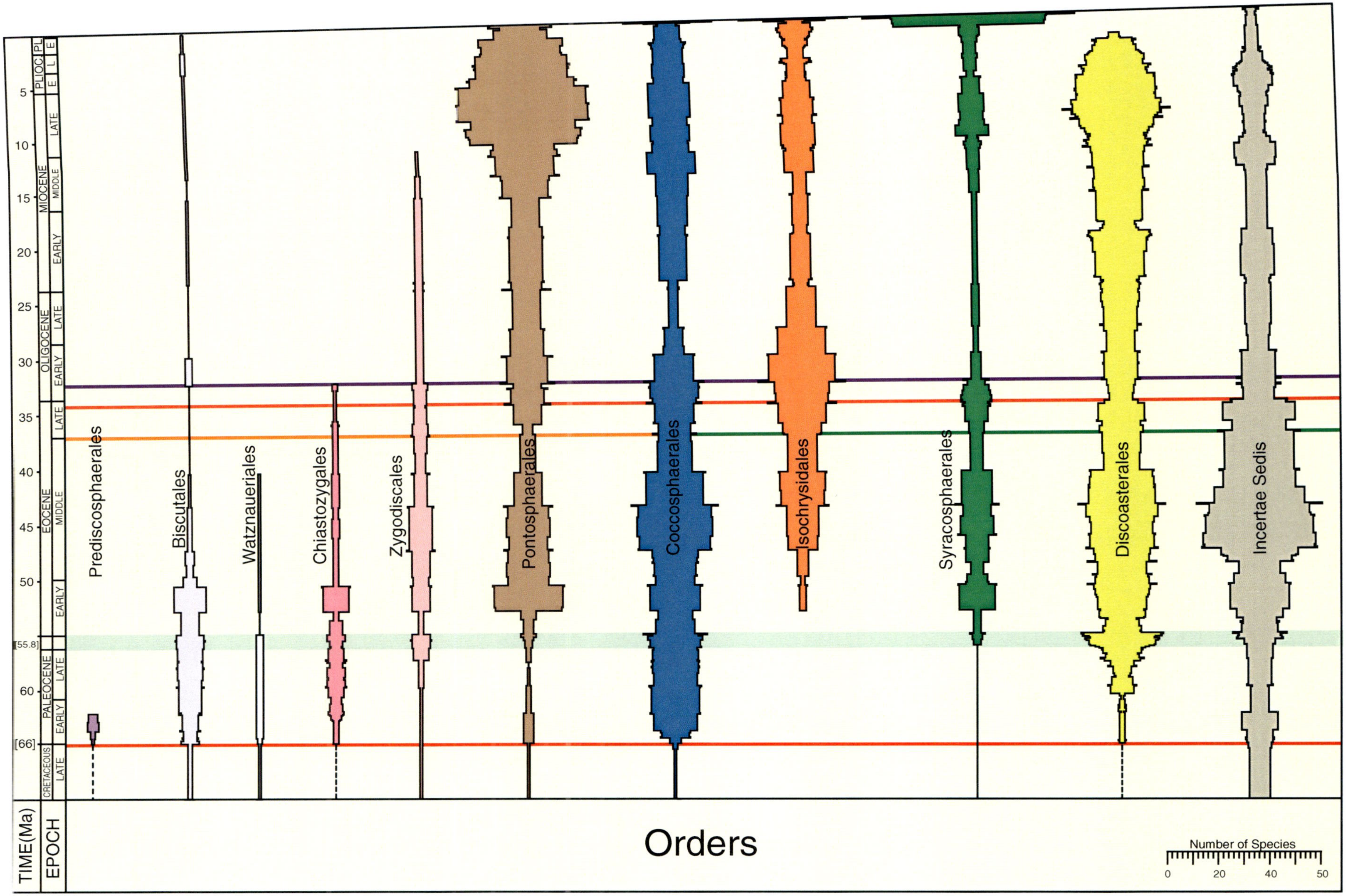

Prediscosphaerales
Biscutales
Watznaueriales
Chiastozygales
Zygodiscales
Pontosphaerales
Coccosphaerales
Isochrysidales
Syracosphaerales
Discoasterales
Incertae Sedis
TIME(Ma)
EPOCH
5
10
15
20
25
30
35
40
45
50
[55.8]
60
[66]
PL
PLIOC
E
L
E
MIOCENE
LATE
MIDDLE
EARLY
OLIGOCENE
LATE
EARLY
EOCENE
LATE
MIDDLE
EARLY
PALEOCENE
LATE
EARLY
CRETACEOUS
LATE
Orders
Number of Species
0
20
40
50

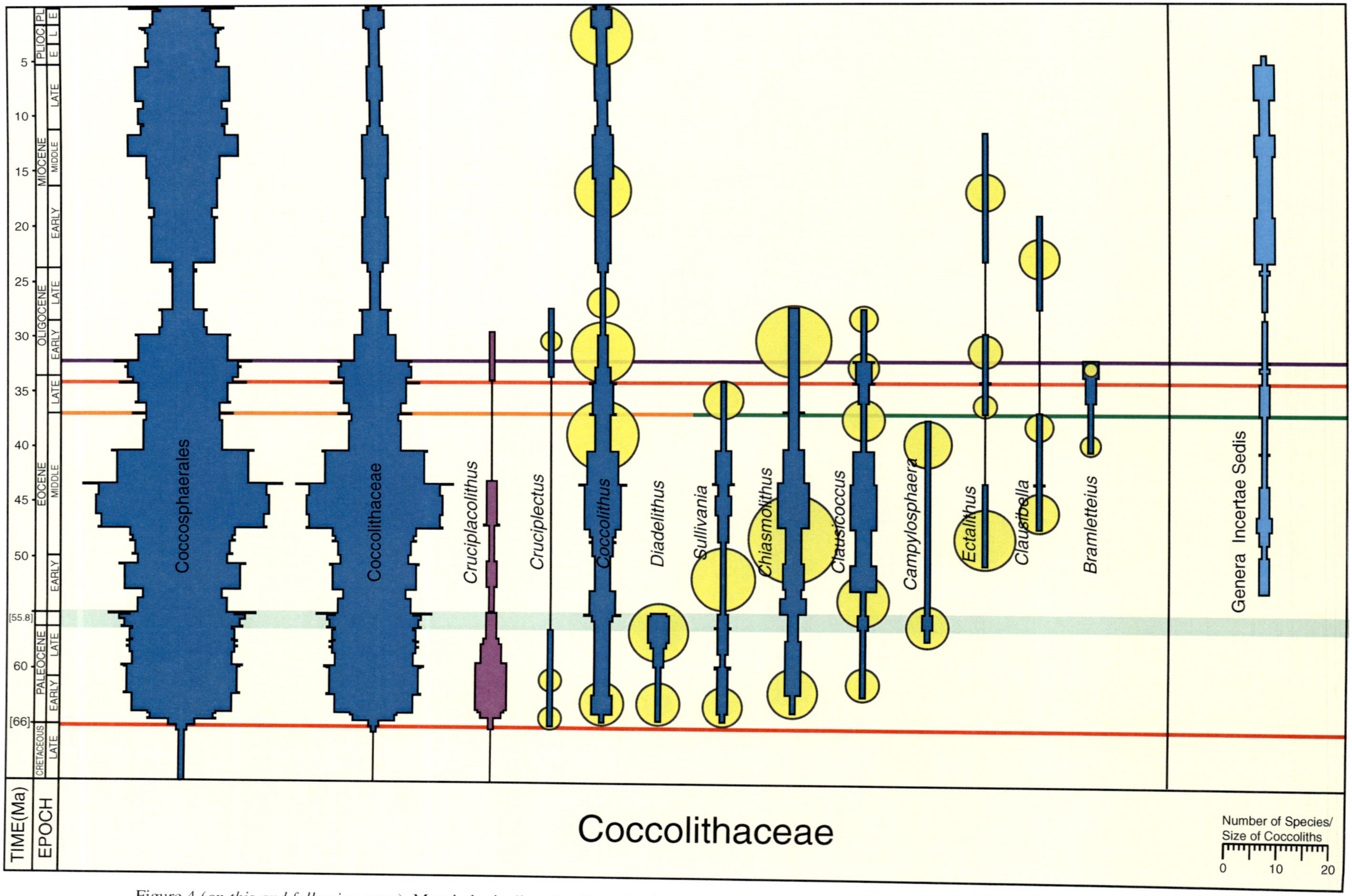

Figure 4 (*on this and following page*). Morphologic diversity ("species richness") at the family and genus levels among the Cenozoic coccolithophorids. (A) Family Coccolithaceae. (B) Family Calcidiscaceae, Family Braarudosphaeraceae. Note that the Family Coccolithaceae makes up the bulk of the Eocene species in the Order Coccolithales, where the Family Calcidiscaceae makes the bulk of the Neogene species. Note the stepwise extinction pattern of genera in the Family Coccolithaceae. Yellow circles denote average size of species in successive genera at different times. Note the general decrease in size across the Eocene/Oligocene boundary, and the much smaller size of the coccoliths in the Family Calcidiscaceae than in the Family Coccolithaceae.

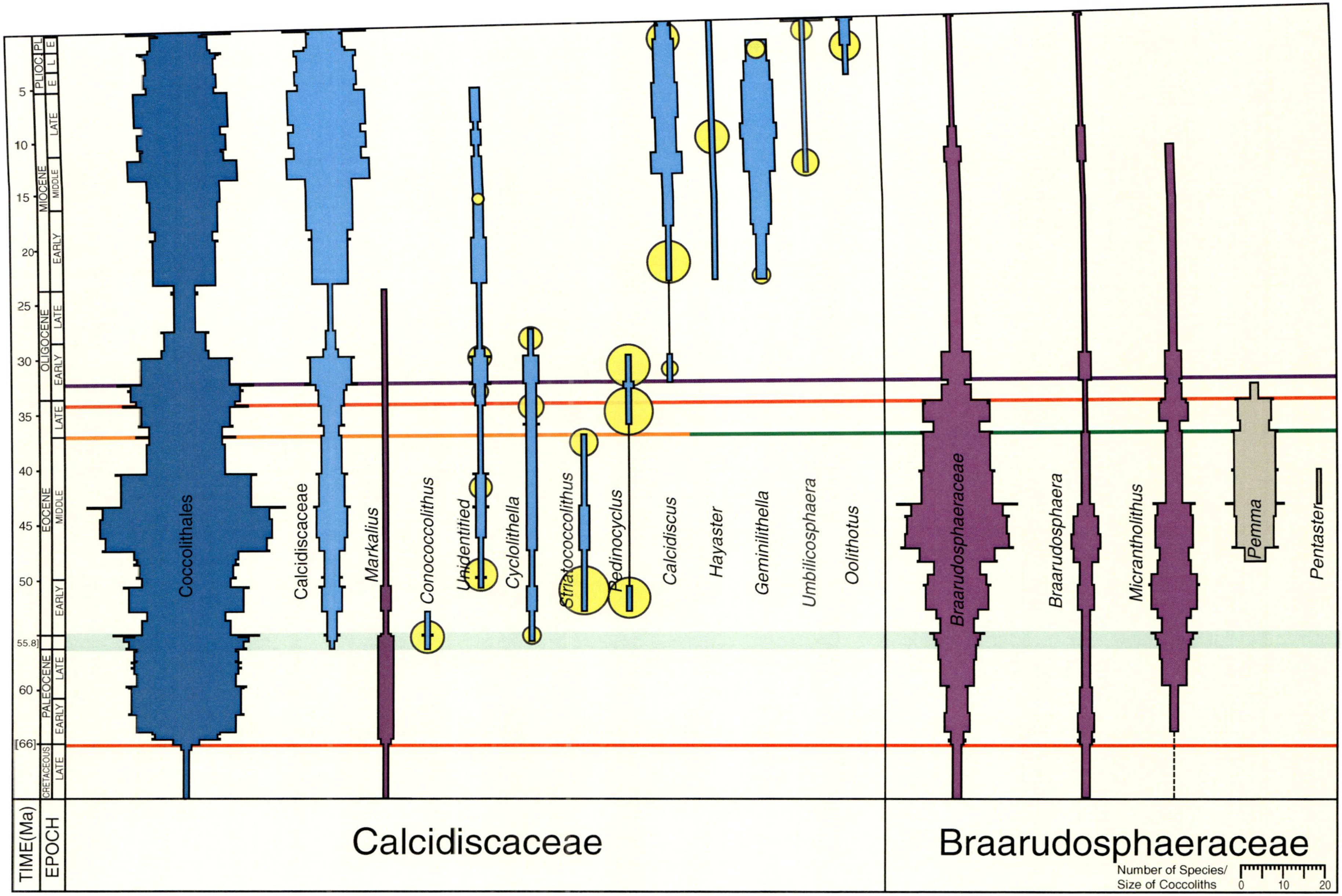

Figure 4 (*continued*).

B

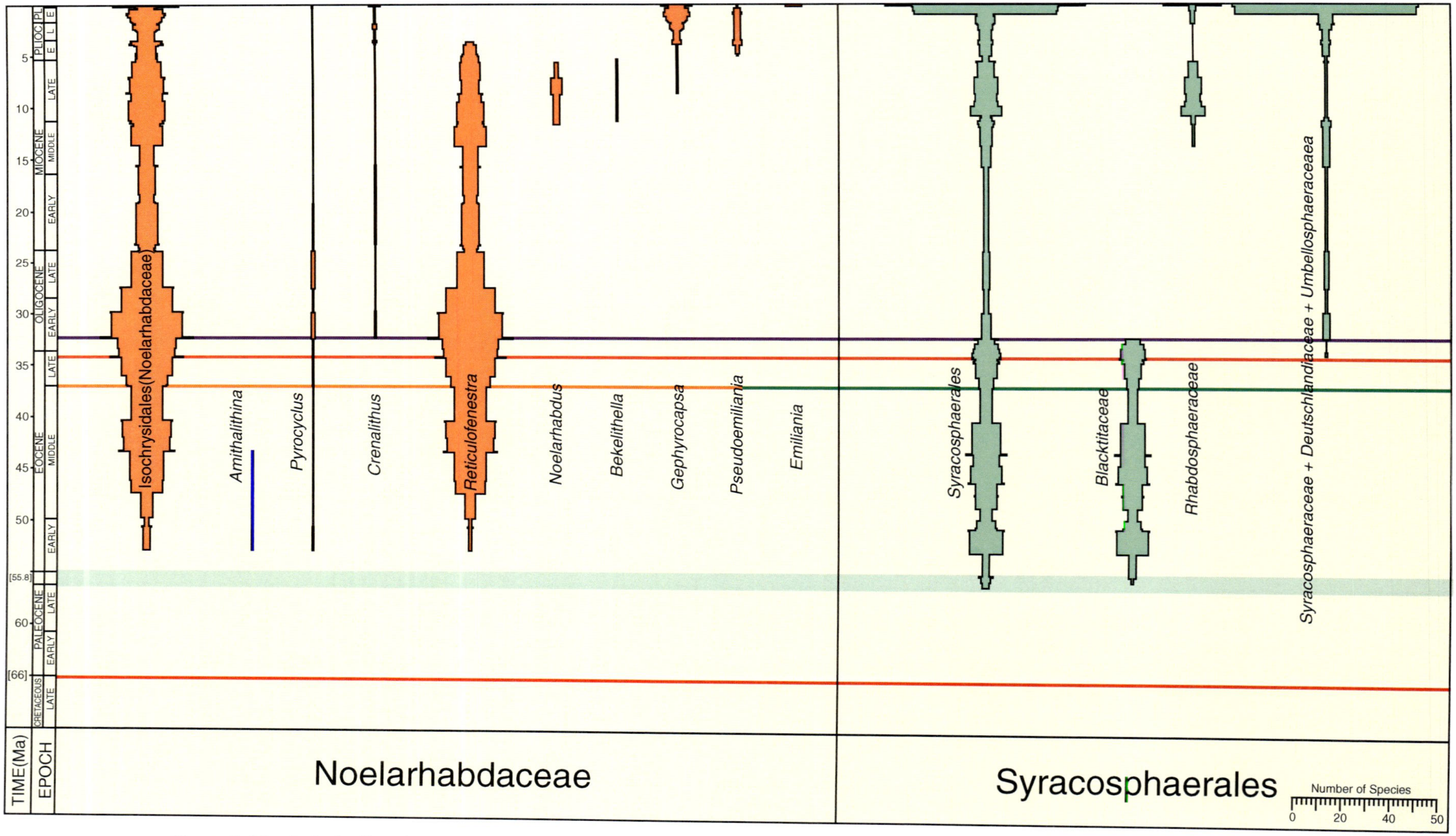

Figure 5. Morphologic diversity ("species richness") at the family and genus levels among the Cenozoic coccolithophorids: Family Noelarhabdaceae; Order Syracosphaerales. The Family Noelarhabdaceae is the only family of the Order Isochrysidales that calcifies. The Order Syracosphaerales also include the Family Calciosoleniaceae, which has a poor fossil record.

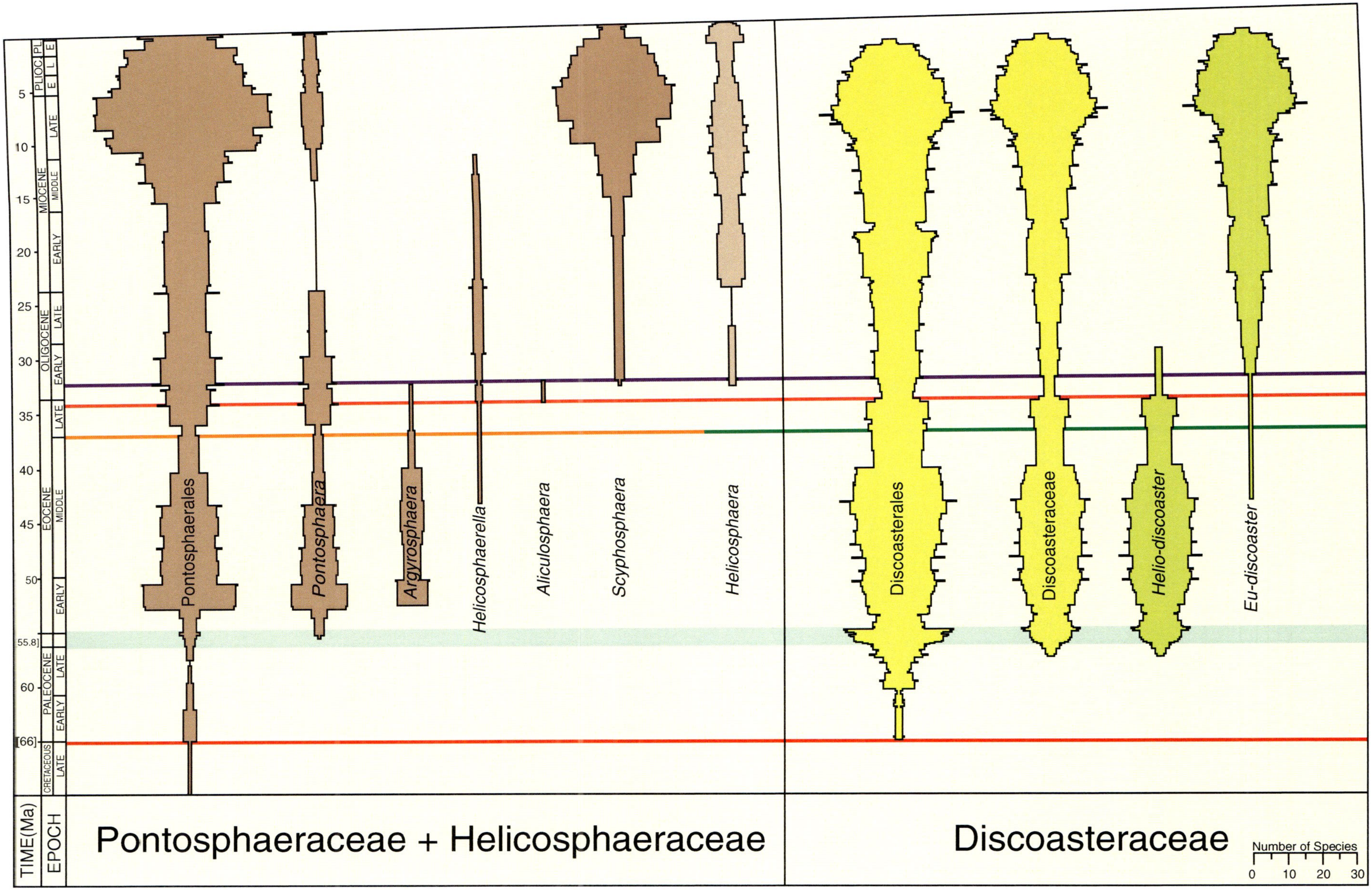

Figure 6. Morphologic diversity ("species richness") at the family level among the Cenozoic coccolithophorids: Family Pontosphaeraceae (brown color) and Helicosphaeraceae (tan color); Family Discoasteraceae. Note the replacement of *Argyrosphaera* by *Scyphosphaera*, and of *Helio-discoaster* by *Eu-discoaster*.

the latter family consist of three thin cycles of poorly imbricate elements. This is exemplified also by the disappearance of the massive rosette-shaped discoasters (genus *Helio-discoaster*) for the benefit of the star-shaped discoasters (genus *Eu-discoaster*), by the extinction of the ornamented pentaliths of *Pemma*, and by the extinction of the large placoliths of the *Reticulofenestra umbilicus* group, and the subsequent dominance of the smaller *R. floridana* and, later, *R. pseudoumbilicus*. The trend toward smaller size that spans the Neogene (Aubry, 2007) was thus initiated during the Eocene/Oligocene turnover. It was a very subtle change, one that would reach its full, albeit delayed, significance in the late Pliocene.

What Triggered the Eocene/Oligocene Turnover?

Large biotic events are attributed to the collision of Earth with cosmic bolides (Raup, 1990; Farley et al., 1998; Rampino, 1999), the emplacement of large basaltic provinces (Courtillot et al., 1999), and, for the largest events, to the simultaneous effects of bolide impacts and volcanism (White and Saunders, 2005). The first two scenarios have been independently proposed to account for the Eocene/Oligocene extinctions, but it would now seem possible to rule out entirely a role for either one (Fig. 1). The three, well-dated late Eocene impacts occurred much too early to have caused the turnover. Conversely, the Ethiopian traps were emplaced subsequent to the turnover.

The nature of the turnover among the coccolithophorids, which combines massive disappearance of taxa, shifts in abundance, and the occurrence of short-lived taxa, suggests evolutionary instability, itself caused by environmental stress and instability. There can be little question thus that the Eocene/Oligocene turnover described here is the response of the coccolithophorids to climatic cooling and the first major glaciation (oxygen isotope event Oi-1) on Antarctica (Miller et al., 1991a, 1991b, 2008; Coxall et al., 2005; Lear et al., 2008; Schouten et al., 2008). This complex, stepwise event is little understood at this time (Thomas, 2008). Whereas the two-step character of the Oi-1 glaciation is now confirmed, the timing of cooling at low latitude is controversial. The biogeographic diachrony of the coccolithophorids between low and high latitudes during the Eocene lends support to essentially warm low latitudes until the end of the Eocene (Pearson et al., 2008). A precise correlation between biotic events in the photic zone and isotopic and other climatic and paleoceanographic proxies is necessary to understand the mechanisms involved in the turnover. It began prior to the Oi-1 glaciation (33.5 Ma) and considerably exceeded its duration. It is reasonable to propose that cooling induced the turnover, as supported by the diachronic nature of the events until 34.2 Ma. Cooling would have been relaxed after the Oi-1 glaciation, however. The unfolding of the turnover beyond this glaciation has two possible explanations. Either intense cooling set an evolutionary response that could not simply be stopped as temperatures warmed; or the turnover was induced by something other than cooling. The question to answer is not simply: What caused the extinctions? A more fundamental question is: What filtering agent caused the redistribution of the roles of families (and genera) during the turnover? Or more simply, what caused the failure of the heavy calcifiers to the benefit of the lighter calcifiers? (We refer here to the amount of calcite used by a cell to produce a single coccolith.) This is a paleophysiological problem to consider (Aubry, 2007; Knoll et al., 2007).

We refrain from discussing any possible factors that may have controlled species extinction, innovations, and shifts in dominance. Changes in nutrient concentrations are often invoked to explain the evolutionary history of the plankton, and it has been hypothesized that increased nutrient concentration in the photic zone as a result of intensification of ocean circulation induced by cooling and ice buildup would explain the Eocene/Oligocene turnover in oceanic plankton (e.g., Aubry, 1992a; Boersma et al., 1987; Bown et al., 2004). The similar response (decrease in size and robustness of coccoliths) across orders to the Eocene/Oligocene turnover and its persistence and intensification through the Neogene suggest that more fundamental mechanisms are involved that affected the physiology of the coccolithophorids.

CONCLUSIONS

This study shows the benefit of a three-tiered approach to diversity analysis around biotic events. (1) The temporal distribution of species that were dominant to common before and after the Eocene/Oligocene boundary shows that the latter was associated with the sequential extinction of long-ranging taxa over a >6 m.y. biotic turnover window at low and mid-latitude, and with two massive events at 34.2 Ma (mid- and low-latitudes only) and 32.3 Ma. (2) The temporal distribution of genera, families, and orders in the vicinity of the Eocene/Oligocene boundary has shown that few structural innovations occurred during the turnover. The most consequential innovation in the long-term—although cryptic at first—was the appearance of the coccoliths characteristic of the Family Syracosphaeraceae s.l., which would achieve success in terms of diversity in the present-day ocean. (3) The analysis of morphologic diversity (= number of paleontological species) within high-level taxa has revealed parallel shifts in dominance within orders and families, most prominently the replacement of the Family Coccolithaceae by the Family Calcidiscaceae, and of the genus *Helio-discoaster* by the genus *Eu-discoaster*. These shifts, which would irremediably change the physiognomy of coccolithophorid communities throughout the Neogene, were associated with the use of lesser amount of calcium carbonate by cells to produce coccoliths.

From a long-term evolutionary perspective, the Eocene/Oligocene turnover was an additional turning point in the evolution of the nannoplankton. The Cretaceous/Paleogene event had resulted in the massive extinction of families without (or with only little) discrimination. The Paleocene/Eocene event had caused the selective extinctions of families. The Eocene/Oligocene event redistributed the contributions of families to diversity in the photic zone.

APPENDIX: TAXONOMIC DESCRIPTIONS (AUBRY)

An effort has generally been made by specialists of the coccolithophorids to avoid the introduction of taxa of higher rank based on indecisive structural differences among coccoliths. Molecular evidence now constitutes an independent means of assessing the significance of these structural differences. It provides firm evidence for confident determination of monophyletism or polyphyletism in a taxon. It also gives clues to the ancestral roots of clades. The following introduction of several new taxa is thus based on morphostructural characters supported by molecular evidence. The introduction of other taxa is based on morphostructure alone. All taxa are thoroughly documented in Aubry (2009, a–i, personal obs.). Genera are described here to avoid the unsatisfactory introduction of invalid names in the figures as nomen nudum. Full references to the taxa cited here can be found in Aubry (2009, a–i, personal obs.).

Order Coccosphaerales Haeckel, 1894 emend. Young and Bown, 1997

Family: Coccolithaceae Poche 1913 emend Young and Bown 1997
Genus *Clausibella* Aubry n. gen.
Type species: *Ericsonia*? *singularis* Perch-Nielsen 1971
Etymology: From clausi (L. [Latin] closed) and bellus (L. beautiful)
Description: Placoliths with a monocyclic distal shield and a bicyclic proximal shield. The central area is filled with elements arranged so as to form a perforated grid. In distal view, no intermediate cycle occurs between the margin (shields) and the central structure. The wall is restricted to proximal position or absent.
Differences: Placoliths of *Pseudoclausicoccus* differ from those of *Clausicoccus* in the absence of an intermediary cycle (wall) between the margin (shields) and the central grid as seen in distal view. In the former genus, the wall is visible only in proximal position. The proximal shield of placoliths of *Hughesius* consists of a single cycle.
Species: *Clausibella singularis* (Perch-Nielsen) Aubry n.c. (=*Ericsonia*? *singularis* Perch-Nielsen 1971, p. 15, pl. 7, figs. 1–3, 5).
Clausibella sp. (Perch-Nielsen) Aubry n.c. (=*Ericsonia cf. E. fenestrata* Perch-Nielsen 1971, p. 15, pl. 7, figs. 1–3, 5).
Clausibella tasmaniae (Edwards and Perch-Nielsen) Aubry n.c. (= *Ericsonia tasmaniae* Edwards and Perch-Nielsen, 1975, pl. 20, figs. 10, 11).
Genus *Cruciplectus* Aubry n. gen.
Type species: *Cruciplacolithus inseadus* Perch-Nielsen 1969
Etymology: From *crux* (L. cross) and *plecto* (L. weave); in reference to the segments that connect the cross bars.
Description: Placoliths with a monocyclic distal shield and a bicyclic proximal shield. The central area is delineated by a narrow outer cycle of elements (the wall), which is leveled with the crest of the margin, and it slopes gently inward in proximal direction. The central opening is spanned by a cross-structure aligned with the major and minor axis of the elliptical coccolith. The cross-bars are connected by short oblique segments.
Differences: Placoliths of this genus differ from those of *Cruciplacolithus* by the presence of connecting segments between the bars of the central structure. Segments and bars consist of small quadrangular elements as in *Cruciplacolithus*. Placoliths of *Sollasites* are very similar, but their proximal shield is monocyclic.
Remarks: Placoliths of *Cruciplectus* may have evolved from *Sollasites* or from *Cruciplacolithus*. Monophyly in the Family Coccolithaceae cannot be established at this time.
Species: *Cruciplectus inseadus* (Perch-Nielsen) Aubry, n.c. (=*Cruciplacolithus inseadus* Perch-Nielsen 1969, p. 58, 59, pl. 1, figs. 5, 6).
Cruciplectus tardus (Roth) Aubry n.c. (= *Sollasites tardus* Roth 1970, p. 845, 846, pl. 3, figs. 2, 3).
Genus *Diadelithus* n. gen.
Type species: *Cyclolithus*? *robustus* Bramlette and Sullivan 1961
Etymology: From *diadelos* (Gr. [Greek] distinctive) and *lithos* (Gr. stone) in reference to the highly modified structure of this member of the Family Coccolithaceae.
Description: Circular placoliths with complex, strongly modified structure. The margin consists of two thick, closely appressed shields that slope steeply in proximal direction. The distal shield is monocyclic; the proximal shield bicyclic. The large, central area cycles strongly overlap the margin distally, largely concealing it.
Differences: Differs from all other genera by its unusual structure.
Remarks: The central cycles are well developed, as in other genera of the Family Coccolithaceae.
Species: *Diadelithus universus* (Wind and Wise) Aubry n.c. (= *Heliolithus universus* Wind and Wise, p. 296, pl. 12, figs. 1–6, pl. 14, figs. 1–3).
Diadelithus robustus (Bramlette and Sullivan) Aubry n.c. (= *Cyclolithus*? *robustus* Bramlette and Sullivan 1961, p. 141, pl. 2, figs. 7a–7c).
Diadelithus fulgidus (Müller) Aubry, n.c. (*Toweius*? *fulgidus* Müller 1974, p. 592, pl. 1, figs. 5–7).
Diadelithus kavaryi (Haq) Aubry, n.c. (= *Cycloplacolithella kavaryi* Haq 1971, p. 20, pl. 5, fig. 13).
Genus *Ectalithus* n. gen.
Type species: *Coccolithus crassus* Bramlette and Sullivan 1961
Etymology: From *ektadios* (Gr. outstretched) and *lithos* (Gr. stone) in reference to the flat, almost horizontally spread shields of the placolith.
Description: Placolith with shields that are flat rather than concave-convex, extending outward at a low angle. A large space separates the two shields, giving thickness to the placoliths. Shield elements are irregularly imbricated clockwise. Sutures between them are straight. The central area is occupied by a prominent cycle (wall) of elements that extends far above the crest of the margin, forming a distal collar. The wall elements are elongate and slope steeply inward and proximally, delineating a central opening of variable diameter.
Differences: The placoliths of *Ectalithos* differ from placoliths in other genera in having flat rather than curved shields, and in having a prominent wall.
Remarks: It is unclear whether the proximal shield is mono- or bicyclic. If monocyclic, it may be necessary to remove the genus from the Family Coccolithaceae and introduce a new family for it.
Species: *Ectalithus crassus* (Bramlette and Sullivan) Aubry n.c. (= *Coccolithus crassus* Bramlette and Sullivan 1961, p. 139, pl. 1, figs. 4a–4d).
Ectalithus simplex (Haq) Aubry n.c. (= *Cycloplacolithella simplex* Haq 1968, p. 25, 26, pl. 8, fig. 8).
Ectalithus tritus (Roth) Aubry n.c. (= *Coccolithus tritus* Roth, 1970, p. 840, pl. 1, fig. 1).
Ectalithus floralis (Wei and Wise) Aubry n.c. (= *Coccolithus pelagicus* [Wallich] Schiller, 1930, subsp. *floralis* Wei and Wise 1992, p. 514, 516, pl. 2, figs. 8, 9).

Family Stauronerthaceae Aubry (2009f)

Genus *Catostaura* n. gen.
Type species: *Cruciplacolithus filigranus* Mai 2001
Etymology: From *Cato* (Gr. beneath) and *Stauro* (Gr. cross); in reference to the proximal location of the central structure.
Description: Concave-convex placolith with a monocyclic distal shield and a bicyclic proximal shield. The central area is strongly depressed in distal view. It is occupied by a composite central structure located in proximal position below the inner contact between the proximal and distal shields and protruding proximally. This structure consists of an outer cycle of elements to which are connected a central cross aligned with the major and minor axes of the elliptical placolith and elements that may fill the openings between the outer cycle and the bars. The

outer cycle consists of elements oriented almost perpendicular to a proximal plane at the basal edge of the central area (where the two shields meet proximally). The bars of the central-cross consist, as in *Cruciplacolithus*, of two parallel rows of subquadrangular elements.
Differences: The main difference with *Cruciplacolithus* is the proximal location and orientation of the central structure. No filling of the central openings is known in *Cruciplacolithus*.
Remarks: Mai (2001) noted the distinct construction of the central area of *C. filigranus* compared to placoliths of *Cruciplacolithus*. It is unclear whether a low wall occurs at the base of the central area, in distal position with respect to the central structure in the generotype.

Ericsonia? *brotzeni* is tentatively assigned to *Eocruciplacolithus*. Perch-Nielsen (1979) noted that the central structure was protruding proximally. In this taxon, the openings delineated by the cross-structure are filled by elements as in *C. filigranus*. The main difference in the latter is the occurrence of a prominent outer central cycle (the wall).
Species: *Catostaura filigrana* (Mai) Aubry n.c. (= *Cruciplacolithus filigranus* Mai 2001, p. 145, pl. 1, figs. 1–6).
Catostaura brotzenii (Perch-Nielsen) Aubry, n.c. (= *Ericsonia*? *brotzenii* Perch-Nielsen 1979, p. 60, 61, pl. 2, figs. 3–6).
Genus *Stauronertha* n. gen.
Type species: *Coccolithus neohelis* McIntyre and Bé, 1967
Etymology: From *Stauro* (Gr. cross) and *nerthe* (Gr. beneath) in reference to the proximal location of the central structure.
Description: Placolith with a monocyclic distal shield and bicyclic proximal shield. The central area is lined by an outer cycle (wall) raised slightly above the crest of the margin. The central opening is spanned by a cross-structure aligned with the major and minor axes of the elliptical placolith. Located in proximal position, it does not rest on the wall (as in *Cruciplacolithus*), but on an additional elliptical cycle located proximally to the inner contact between the proximal and distal shields. The four openings delineated by the bars are filled with delicate elements that connect the bars to the supporting cycle.
Differences: The central structure (additional cycle, cross-bars and filling) in *Stauronertha* is located proximally as in *Eocruciplacolithus*. The two genera are differentiated on the basis of the more protruding character of the central structure in *Eocruciplacolithus* and the occurrence of a distally prominent wall in *Stauronertha*. The bars of the cross-structure are similarly constructed in the two genera.
Species: *Stauronertha neohelis* (McIntyre and Bé) Aubry n.c. (= *Coccolithus neohelis* McIntyre and Bé, 1967, p. 369, fig. 1A–1D).
Stauronertha tarquinius (Hay and Roth *in* Hay et al., 1967) Aubry, n.c. (= *Cruciplacolithus tarquinius* Hay and Roth *in* Hay et al., 1967, p. 446, pl. 6, fig. 8).

Order Pontosphaerales Aubry (2009d)

Family: Pontosphaeraceae Lemmermann, 1908

Molecular evidence shows that *Helicosphaera carteri* and *Scyphosphaera apsteinii* have the same Oligocene ancestral roots. Morphostructural evidence indicates that both evolved from *Pontosphaera*. Helicoliths with bars evolved in the early Eocene and are easily related to *Zygodiscus*. This implies that the genus *Helicosphaera* as typified by *H. carteri* is polyphyletic contra Aubry (1988). Several new genera are introduced here to reconcile molecular and morphostructural evidence (Aubry et al., personal obs.).
Genus: *Aliculosphaera* Aubry n.g.
Type species: *Helicosphaera*? *subantarctica* Edwards and Perch-Nielsen 1975
Etymology: From *alicula* (L. wing); in reference to the shape of the coccolith.
Description: Asymmetrical coccolith, the margin forming a lateral wing. The coccolith is composed of three structural components: an elliptical, ring-shaped, proximal cycle (basal plate) consisting of radiating elements and delineating a large central opening; a marginal cycle (flange) of strongly overlapping elements that first raises almost straight in distal direction and then curves strongly outward. On one side of the coccolith, the flange remains narrow and exhibits a smooth outline, whereas on the other side, it expands in an asymmetrical, roughly triangular wing. On the distal side, the basal plate is covered with small, concentrically arranged elements (distal cover).
Differences: *Aliculosphaera* is closely related to *Helicosphaera* and *Pontosphaera* as indicated by the morphostructure of its coccoliths. It differs from them by the outline of the margin, by the proximal expansion of the flange, and by the distal cover being restricted to the central area.
Species: *Aliculasphaera subantarctica* (Edwards and Perch-Nielsen) Aubry et al., 2008 (= *Helicosphaera subantarctica* Edwards and Perch-Nielsen, 1975, p. 482, pl. 14, figs. 2, 3, 6; pl. 15, figs. 1, 3, 7, 8; pl. 21, figs. 17–20).
Genus *Argyrosphaera* Aubry, Liu, de Vargas and Probert n. gen.
Type species: *Argyrosphaera tubicena* Stradner 1968
Etymology: From *argyris* (Gr. cup).
Description: Lopadoliths are of various morphologies. The basal plate is flat or weakly concave-convex. The margin is thick, often with straight sides; it flares distally and shows a stack pattern in light and electron microscopy.
Differences: The morphologic characters that differentiate lopadoliths of *Argyrosphaera* from those of *Scyphosphaera* are subtle. First, the basal plate is flat or weakly convex distally in *Argyrosphaera*. Second, the margin (the cylindrical part) is thicker in *Argyrosphaera*, and the thickness varies from the base to the top of the lopadolith. Change in focus in light microscopy reveals a stack pattern in the margin that is not seen in lopadoliths of *Scyphosphaera*.
Species: *Argyrosphaera biarritziensis* (Lezaud) Aubry, Liu, de Vargas, and Proberts 2008 (= *Scyphosphaera biarritziensis* Lezaud, 1968, p. 23, pl. 2, figs. 3, 4, 12).
Argyrosphaera columella (Stradner) Aubry, Liu, de Vargas, and Proberts 2008 (= *Scyphosphaera columella* Stradner 1969, p. 416, pl. 88, figs. 1–4, text-fig. 2 [7–9]).
Argyrosphaera decedomurale (Varol) Aubry, Liu, de Vargas, and Proberts 2008 (= *Scyphosphaera decedomurale* Varol, 1991, p. 227, 228, pl. 8, figs. 13–15).
Argyrosphaera expansa (Bukry and Percival) Aubry, Liu, de Vargas, and Proberts 2008 (= *Scyphosphaera expansa* Bukry and Percival, 1971, p. 138, pl. 6, figs. 10–13).
Argyrosphaera olla (Varol) Aubry, Liu, de Vargas, and Proberts 2008 (= *Scyphosphaera olla* Varol, 1989, p. 301, pl. 5, figs. 43, 44).
Argyrosphaera tercisensis (Lezaud) Aubry, Liu, de Vargas, and Proberts 2008 (= *Scyphosphaera tercisensis* Lezaud, 1968, p. 23, pl. 2, figs. 1, 2, 11).
Argyrosphaera tubicena (Stradner) Aubry, Liu, de Vargas, and Proberts 2008 (= *Scyphosphaera tubicena* Stradner 1968, p. 416, pl. 88, figs. 9–12, text-fig. 2).
Argyrosphaera sp. 1 Aubry, Liu, de Vargas, and Proberts 2008 (= *Scyphosphaera apsteinii* Lohmann 1902 *in* Stradner 1969, pl. 89, figs. 1–3, text-fig. 2[1–6]).
Argyrosphaera sp. 2 Aubry, Liu, de Vargas, and Proberts 2008 (= *Scyphosphaera columnella** Stradner 1969 *in* Watkins et al., 1995, pl. 5, fig. 16).
Argyrosphaera sp. 3 Aubry, Liu, de Vargas, and Proberts 2008 (= *Scyphosphaera* cf. *Scyphosphaera galeana* Kamptner 1967 *in* Stradner 1969, p. 416, pl. 88, figs. 1–3).
Argyrosphaera sp. 4 Aubry, Liu, de Vargas, and Proberts 2008 (= *Scyphosphaera tubicena* Stradner, 1969 *in* Watkins et al., 1995, pl. 6, fig. 23).
Genus *Helicosphaerella* Aubry, Liu, de Vargas, and Probert n. g.
Type species: *Helicosphaera seminulum recta* Haq 1966
Description: Helicolith constructed of three structural components: a helicoidal marginal cycle (flange) consisting of strongly overlapping elements; proximally, a helicoidal cycle (basal plate) of radial elements. These two cycles are concealed distally by a cycle of small platelet-like elements arranged concentrically or tangentially to the

meta-pterygal side of the coccolith (see Theodoridis, 1984). The outline may be ovoid or subquadrangular. In the latter case, the flange terminates abruptly and forms a sharp spur.
Differences: The helicoliths of *Helicosphaerella* differ from those of *Helicosphaera* by the asymmetrical (meta-pterygal) distribution of the distal cover. They differ from helicoliths in *Helicosphaeroides* by the absence of a central bar.
Remarks: Presumably, helicoliths with an overlapping truncated flange are derived from helicoliths with a merging wing. The similar development of the distal cover supports such phylogeny.
Species: *Helicosphaerella compacta* (Bramlette and Wilcoxon) Aubry, Liu, de Vargas, and Proberts (= *Helicosphaera compacta* Bramlette and Wilcoxon 1967, p. 105, pl. 6, figs. 5–8).
Helicosphaerella elongata (Theodoridis) Aubry, Liu, de Vargas and Proberts (= *Helicosphaera elongata* Theodoridis 1984, p. 117, pl. 17, figs. 6–9).
Helicosphaerella gertae (Bukry) Aubry, Liu, de Vargas and Proberts (= *Helicosphaera gertae* Bukry 1981, p. 463, pl. 5, figs. 5–13, pl. 6, figs. 1–4).
Helicosphaerella inversaserae (Varol) Aubry, Liu, de Vargas, and Proberts 2008 (= Varol 1991, p. 222, fig. 7/21–24).
Helicosphaerella moorkensis (Steurbaut) Aubry, Liu, de Vargas, and Proberts (= *Helicosphaera moorkensis* Steurbaut 1983, p. 328, pl. 1, figs. 1–9).
Helicosphaerella obliqua (Bramlette and Wilcoxon) Aubry, Liu, de Vargas, and Proberts (= *Helicosphaera obliqua* Bramlette and Wilcoxon 1967, p. 106, pl. 5, figs. 13, 14).
Helicosphaerella perch-nielseniae (Haq) Aubry, Liu, de Vargas, and Proberts (= *Helicopontosphaera perch-nielsenasae* Haq 1971, p. 116, pl. 10, figs. 5, 6; corrected to *perch-nielsenia* by Haq 1973, p. 42).
Helicosphaerella recta (Haq) Aubry, Liu, de Vargas, and Proberts (= *Helicosphaera seminulum recta* Haq 1966, p. 34, pl. 2, fig. 6, pl. 3, fig. 4; = *Helicosphaera recta* (Haq) Jafar and Martini 1975, p. 391).
Helicosphaerella truncata (Bramlette and Wilcoxon) Aubry, Liu, de Vargas, and Proberts (= *Helicosphaera truncata* Bramlette and Wilcoxon, p. 106, pl. 6, figs. 13, 14).

Order Zygodiscales Young and Bown 1997 emend Aubry (2009d)

Family: Zygodiscaceae Hay and Mohler, 1967
Genus *Helicosphaeroides* Aubry, Liu, de Vargas, and Proberts n. g.
Type species: *Helicosphaeroides lophotes* Aubry, Liu, de Vargas, and Proberts n. c.
Etymology: From *Helicosphaera* and *oides* (L. resembling); meaning similar to *Helicosphaera*.
Description: Helicoliths consisting of four structural components: an helicoidal flange of strongly overlapping elements surrounding a basal plate of radial elements delineating a central opening that is spanned by a central bar consisting of two parallel segments. Typically, each end of the bar forms a bevel in the basal plate. These three components are covered distally by a cycle (distal cover) of short tangentially arranged elements.
Differences: The helicoliths of *Helicosphaeroides* differ from helicoliths in other genera in possessing a central bar.
Remarks: The bar may be aligned with the minor or major axis of the helicolith or oblique to the latter. It is always oriented toward the broader pole of the helicolith. Presumably, the helicoliths with an overlapping wing are derived from helicoliths with a merging wing (see Aubry, 1988).
Species: *Helicosphaeroides bramlettei* (Müller) Aubry, Liu, de Vargas, and Proberts n. c. (= *Helicopontosphaera bramlettei* Müller 1970, p. 114, pl. 5, figs. 5, 6).
Helicosphaeroides clarissima (Bown) Aubry, Liu, de Vargas, and Proberts n. c. (= *Helicosphaera clarissima* Bown 2005, p. 30, pl. 11, figs. 26–35).
Helicosphaeroides dinesenii (Perch-Nielsen) Aubry, Liu, de Vargas, and Proberts n.c. (= *Helicopontosphaera dinesenii* Perch-Nielsen 1971, p. 42, pl. 35, figs. 3, 4, pl. 36, figs. 3, 9, 11, pl. 61, figs. 6, 7).
Helicosphaeroides euphratis (Haq) Aubry, Liu, de Vargas, and Proberts n. c. (= *Helicosphaera euphratis* Haq 1966, p. 33, pl. 2, figs. 1, 3).
Helicosphaeroides gartneri (Theodoridis) Aubry, Liu, de Vargas, and Proberts n. c. (= *Helicosphaera gartneri* Theodoridis 1984, p. 129, pl. 15, figs. 8–12).
Helicosphaeroides heezenii (Bukry) Aubry, Liu, de Vargas, and Proberts n. c. (= *Helicopontosphaera heezeni* Bukry 1971, p. 318, pl. 5, figs. 1–5).
Helicosphaeroides intermedia (Martini) Aubry, Liu, de Vargas, and Proberts n. c. (= *Helicosphaera intermedia* Martini 1965, p. 404, pl. 35, figs. 1, 2).
Helicosphaeroides limasera (de Kaenel and Villa) Aubry, Liu, de Vargas, and Proberts n. c. (= *Helicosphaera limasera* de Kaenel and Villa 1996, p. 106, pl. 10, figs. 12, 13).
Helicosphaeroides lophotes (Bramlette and Sullivan) Aubry, Liu, de Vargas, and Proberts n. c. (= *Helicosphaera seminulum lophota* Bramlette and Sullivan 1961, p. 144, pl. 4, figs. 3a, 3b, 4).
Helicosphaeroides neolophotes (Bukry) Aubry, Liu, de Vargas, and Proberts n. c. (= *Helicosphaera neolophota* Bukry *in* Barron et al., 1984, p. 152, pl. 5, figs. 10–21).
Helicosphaeroides papillates (Bukry and Bramlette) Aubry, Liu, de Vargas, and Proberts n. c. (= *Helicopontosphaera papillata* Bukry and Bramlette 1969, p. 133, pl. 2, figs. 1, 2).
Helicosphaeroides parallela (Bramlette and Wilcoxon) Aubry, Liu, de Vargas, and Proberts n. c. (= *Helicosphaera parallela* Bramlette and Wilcoxon 1967, p. 106, pl. 5, figs. 9, 10).
Helicosphaeroides reticulata (Bramlette and Wilcoxon) Aubry, Liu, de Vargas, and Proberts n. c. (= *Helicosphaera reticulata* Bramlette and Wilcoxon 1967, p. 106, pl. 6, fig. 15).
Helicosphaeroides rhomba (Bukry) Aubry, Liu, de Vargas, and Proberts n. c. (= *Helicopontosphaera rhomba* Bukry 1971, p. 320, pl. 5, figs. 6–9).
Helicosphaeroides salebrosa (Perch-Nielsen) Aubry, Liu, de Vargas, and Proberts n. c. (= *Helicopontosphaera salebrosa* Perch-Nielsen 1971, p. 43, pl. 36, figs. 5, 10; pl. 37, fig. 5; pl. 61, figs. 8, 9).
Helicosphaeroides seminulum (Bramlette and Sullivan) Aubry, Liu, de Vargas, and Proberts n. c. (= *Helicosphaera seminulum* Bramlette and Sullivan 1961, p. 144, pl. 4, figs. 1a–1c, 2).
Helicosphaeroides truempyi (Biolzi and Perch-Nielsen) Aubry, Liu, de Vargas, and Proberts n. c. (= *Helicosphaera truempyi* Biolzi and Perch-Nielsen 1982, p. 171, pl. 1, figs. 1–4).
Helicosphaeroides wilcoxonii (Gartner) Aubry, Liu, de Vargas, and Proberts n. c. (= *Helicopontosphaera wilcoxonii* Gartner 1971, p. 110, pl. 2, figs, 2a, 2b, 3).

ACKNOWLEDGMENTS

I (M.-P. Aubry) am most grateful to Christian Koeberl and Alessandro Montanari for their invitation to participate in the Penrose Conference on the Eocene/Oligocene boundary in Ancona and to contribute this paper to the proceedings volume. Discussions with William A. Berggren, Luc Beaufort, Ben Cramer, Claude Hertzberg, Dennis Kent, Hui Liu, Colomban de Vargas, and Bridget Wade are gratefully acknowledged. We thank Hans Egger, Brian McGowran, and John Steinmetz for their thorough review of the manuscript, and Jack Cook (Woods Hole Oceanographic Institute) for drafting Figures 1 and 2. This study was partly supported by a grant to Aubry from PDVSA Petroleos S. A. (Venezuela).

REFERENCES CITED

Agnini, C., Muttoni, G., Kent, D.V., and Rio, D., 2006, Eocene biostratigraphy and magnetic stratigraphy from Possagno, Italy: The calcareous nannofossil response to climatic variability: Earth and Planetary Science Letters, v. 241, p. 815–830, doi: 10.1016/j.epsl.2005.11.005.

Agnini, C., Fornaciaria, E., Raffi, I., Rio, D., Röhl, U., and Westerhold, T., 2007, High-resolution nannofossil biochronology of middle Paleocene to early Eocene at ODP Site 1262: Implications for calcareous nannoplankton evolution: Marine Micropaleontology, v. 64, p. 215–248, doi: 10.1016/j.marmicro.2007.05.003.

Andruleit, H., Rogalla, U., and Stäger, S., 2004, From living communities to fossil assemblages: Origin and fate of coccolithophores in the northern Arabian Sea, *in* Triantaphyllou, M., ed., Advances in the Biology, Ecology, and Taphonomy of Extant Calcareous Nannoplankton: Micropaleontology, v. 50, supplement 1, p. 5–21.

Aubry, M.-P., 1992a, Late Paleogene calcareous nannoplankton evolution: A tale of climatic deterioration, *in* Prothero, D., and Berggren, W.A., eds., Eocene-Oligocene Climatic and Biotic Evolution: Princeton, New Jersey, Princeton University Press, p. 272–309.

Aubry, M.-P., 1992b, Paleogene calcareous nannofossils from Kerguelen Plateau, *in* Wise, S.W., Schlich, R., et al., Proceedings of the Ocean Drilling Program, Scientific Results, Volume 120B: College Station, Texas, Ocean Drilling Program, p. 471–491.

Aubry, M.-P., 1998, Early Paleogene calcareous nannoplankton evolution: A tale of climatic amelioration, *in* Aubry, M.-P., Lucas, S., and Berggren, W.A., eds., Late Paleocene–Early Eocene Climatic and Biotic Events in the Marine and Terrestrial Records: New York, Columbia University Press, p. 158–203.

Aubry, M.-P., 2007, A major Pliocene coccolithophore turnover: Change in morphological strategy in the photic zone, *in* Monechi, S., Rampino, M.R., and Coccioni, R., eds., Large Ecosystem Perturbations: Causes and Consequences: Geological Society of America Special Paper 424, p. 25–51.

Aubry, M.-P., 2009a–i, Handbook on Cenozoic Calcareous Nannoplankton, Volumes 1–9: New York, Micropaleontology Press.

Aubry, M.-P., Lucas, S., and Berggren, W.A., eds., 1998, Late Paleocene–Early Eocene Climatic and Biotic Events in the Marine and Terrestrial Records: New York, Columbia University Press, 513 p.

Backman, J., 1987, Quantitative calcareous nannofossil biochronology of middle Eocene through early Oligocene sediment from DSDP Sites 522 and 523: Abhandlungen Geologische Bundesanstalt, v. 39, p. 21–32.

Benton, M.J., 1995, Diversification and extinction in the history of life: Science, v. 268, p. 52–58, doi: 10.1126/science.7701342.

Berggren, W.A., 2007, Status of the hierarchical subdivision of higher order marine Cenozoic chronostratigraphic units: Stratigraphy, v. 4, p. 99–108.

Berggren, W.A., Kent, D.V., Swisher, C.C., III, and Aubry, M.-P., 1995, A revised Cenozoic geochronology and chronostratigraphy, *in* Berggren, W.A., Kent, D.V., Aubry, M.-P., and Hardenbol, J., eds., Geochronology, Time Scales and Global Stratigraphic Correlations: A Unified Temporal Framework for an Historical Geology: Society of Economic Geologists and Mineralogists Special Volume 54, p. 129–212.

Boersma, A., Premoli-Silva, I., and Shackleton, N., 1987, Atlantic Eocene planktonic foraminiferal paleohydrographic indicators and stable isotope paleoceanography: Paleoceanography, v. 2, p. 287–331, doi: 10.1029/PA002i003p00287.

Bottomley, R.J., Grieve, R., York, D., and Masaitis, V., 1997, The age of the Popigai impact event and its relation to events at the Eocene/Oligocene boundary: Nature, v. 388, p. 365–368, doi: 10.1038/41073.

Bown, P.R., 2005, Calcareous nannoplankton evolution: A tale of two oceans: Micropaleontology, v. 51, p. 299–308, doi: 10.2113/gsmicropal.51.4.299.

Bown, P.R., Lees, J.A., and Young, J.R., 2004, Calcareous nannoplankton evolution and diversity through time, *in* Thierstein, H.R., and Young, J.R., eds., Coccolithophores: From Molecular Processes to Global Impact: Berlin, Springer-Verlag, p. 481–508.

Bown, P.R., Dunkley Jones, T., Lees, J.A., Randell, R.D., Mizzi, J.A., Pearson, P.N., Coxall, H.K., Young, J.R., Nicolas, C.J., Karega, A., Singano, J., and Wade, B.S., 2008, A Paleogene calcareous microfossil konservat-Lagerstätte from the Kilwa Group of coastal Tanzania: Geological Society of America Bulletin, v. 120, p. 3–12.

Bramlette, M.N., and Martini, E., 1964, The great change in calcareous nannoplankton fossils between the Maestrichtian and Danian: Micropaleontology, v. 10, p. 291–322, doi: 10.2307/1484577.

Corliss, B.H., Aubry, M.-P., Berggren, W.A., Fenner, J.M., Keigwin, L.D., Jr., and Keller, G., 1984, The Eocene/Oligocene event in the deep sea: Science, v. 226, p. 806–810, doi: 10.1126/science.226.4676.806.

Courtillot, V., Jaupart, C., Manighetti, I., Tapponnier, P., and Besse, J., 1999, On causal links between flood basalts and continental break up: Earth and Planetary Science Letters, v. 166, p. 177–195, doi: 10.1016/S0012-821X(98)00282-9.

Coxall, H.K., Wilson, P.A., Pälike, H., Lear, C.H., and Backman, J., 2005, Rapid stepwise onset of Antarctic glaciation and deeper calcite compensation in the Pacific Ocean: Nature, v. 433, p. 53–57.

Darling, K.F., and Wade, C.M., 2008, The genetic diversity of planktonic foraminifera and the global distribution of ribosomal RNA genotypes: Marine Micropaleontology, v. 67, p. 216–238, doi: 10.1016/j.marmicro.2008.01.009.

De Vargas, C., Renaud, S., Hilbrecht, H., and Pawlowski, J., 2001, Pleistocene adaptive radiation in *Globorotalia truncatulinoides*: Genetic, morphologic, and environmental evidence: Paleobiology, v. 27, p. 104–125, doi: 10.1666/0094-8373(2001)027<0104:PARIGT>2.0.CO;2.

De Vargas, C., Bonzon, M., Rees, N., Pawlosski, J., and Zaninetti, L., 2002, A molecular approach to biodiversity and biogeography in the planktonic foraminifer *Globigerinella siphonifera* (d'Orbigny), 2002: Marine Micropaleontology, v. 45, p. 101–116, doi: 10.1016/S0377-8398(02)00037-3.

De Vargas, C., Saez, A.G., Medlin, L.K., and Thierstein, H.R., 2004, Super-species in the calcareous plankton, *in* Thierstein, H.R., and Young, J. R., eds., Coccolithophores: From Molecular Processes to Global Impact: Berlin, Springer-Verlag, p. 271–298.

De Vargas, C., Aubry, M.-P., Probert, Y., and Young, J., 2007, Origin and evolution of coccolithophores: A paleonto-genomic approach, *in* Falkowski, P., and Knoll, A., eds., Evolution of Primary Producers in the Sea: Burlington, Massachusetts, San Diego, California, London, UK, Elsevier, p. 251–285.

Edwardsen, B., Eikrem, W., Green, J.C., Andersen, R.A., Moon-van-der Staay, S.Y., and Medlin, L.K., 2000, Phylogenetic reconstructions of the Haptophyta inferred from rRNA sequences and available morphological data: Phycologia, v. 39, p. 19–35.

Farley, K.A., Montanari, A., Shoemaker, E.M., and Shoemaker, C.S., 1998, Geochemical evidence for a comet shower in the late Eocene: Science, v. 280, p. 1250–1253, doi: 10.1126/science.280.5367.1250.

Fehling, J., Stoecker, D., and Baldauf, S.L., 2007, Photosynthesis and the eukaryote tree of life, *in* Falkowski, P., and Knoll, A., eds., Evolution of Primary Producers in the Sea: Burlington, Massachusetts, San Diego, California, London, UK, Elsevier, p. 75–107.

Geisen, M., Billard, C., Boerse, A.T.C., Cros, L., Probert, I., and Young, J.R., 2002, Life-cycle associations involving pairs of holococcolithophorid species: Intra-specific variation or cryptic speciation?: European Journal of Phycology, v. 37, p. 531–550, doi: 10.1017/S0967026202003852.

Hay, W.W., 2004, Carbonate fluxes and calcareous nannoplankton, *in* Thierstein, H.R., and Young, J.R., eds., Coccolithophores: From Molecular Processes to Global Impact: Berlin, Springer-Verlag, p. 509–528.

Huber, B.T., Bijma, J., and Darling, K., 1997, Cryptic speciation in the living planktonic foraminifer *Globigerinella siphonifera* (d'Orbigny): Paleobiology, v. 23, p. 33–62.

Jordan, R.C., and Kleijne, A., 1994. A classification system for living coccolithophores, *in* Winter, A., and Siesser, W.G., eds., Coccolithophores: Cambridge, Cambridge University Press, p. 83–105.

Katz, M.E., Finkel, Z.V., Grzebyk, D., Knoll, A., and Falkowski, P.G., 2004, Evolutionary trajectories and biogeochemical cycles of marine eukaryotic phytoplankton: Annual Review of Ecology Evolution and Systematics, v. 35, p. 523–526, doi: 10.1146/annurev.ecolsys.35.112202.130137.

Knappertsbusch, M., 2000, Morphologic evolution of the coccolithophorid *C. leptoporus* from the early Miocene to Recent: Journal of Paleontology, v. 74, p. 712–730, doi: 10.1666/0022-3360(2000)074<0712:MEOTCC>2.0.CO;2.

Knappertsbusch, M., Cortés, M.Y., and Thierstein, H.R., 1997, Morphologic variability of the coccolithophorid *C. leptoporus* in the plankton, surface sediments, and from the early Pleistocene: Marine Micropaleontology, v. 30, p. 293–317, doi: 10.1016/S0377-8398(96)00053-9.

Knoll, A.H., Bambach, R.K., Payne, J.L., Pruss, S., and Fischer, W.W., 2007, Paleophysiology and end-Permian mass extinction: Earth and Planetary Science Letters, v. 256, p. 295–313, doi: 10.1016/j.epsl.2007.02.018.

Kuiper, K.F., Deino, A., Hilgen, F.J., Krijgsman, W., Renne, P.R., and Wijbrans, J.R., 2008, Synchronizing rock clocks of Earth history: Science, v. 320, p. 500–504, doi: 10.1126/science.1154339.

Lear, C.H., Bailey, T.R., Pearson, P.N., Coxall, H.K., and Rosenthal, Y., 2008, Cooling and ice growth across the Eocene-Oligocene transition: Geology, v. 36, p. 251–254, doi: 10.1130/G24584A.1.

Lee, S.R., Horton, J.W., Jr., and Walker, R.J., 2006, Confirmation of a meteoritic component in impact-melt rocks of the Chesapeake Bay impact structure, Virginia, USA—Evidence from osmium isotopic and PGE systematics: Meteoritics & Planetary Science, v. 41, p. 819–833.

Lourens, L., Hilgen, F., Shackleton, N.J., Laskar, J., and Wilson, D., 2004, The Neogene Period, *in* Gradstein, F.M., Ogg, J., and Smith, A., eds., A Geologic Time Scale 2004: Cambridge, Cambridge University Press, p. 409–440.

Luterbacher, H.P., Ali, J.R., Brinkhuis, H., Gradstein, F.M., Hooker, J.J., Monechi, S., Ogg, J.G., Powell, J., Röhl, U., Sanfilippo, A., and Schmitz, B., 2004, The Paleogene Period, *in* Gradstein, F.M., Ogg, J., and Smith, A., eds., A Geologic Time Scale 2004: Cambridge, Cambridge University Press, p. 384–408.

McGhee, G.R., Jr., Sheehan, P.M., Bottjer, D.J., and Droser, M.L., 2004, Ecological ranking of Phanerozoic diversity crises: Ecological and taxonomic severities are decoupled: Palaeogeography, Palaeoclimatology, Palaeoecology, v. 211, p. 289–297, doi: 10.1016/j.palaeo.2004.05.010.

Medlin, L.K., Saez, A.G., and Young, J.R., 2007, A molecular clock for coccolithophores and implications for selectivity of phytoplankton extinctions across the K/T boundary: Marine Micropaleontology, v. 67, p. 69–86, doi: 10.1016/j.marmicro.2007.08.007.

Miller, K.G., Feigenson, M.D., Wright, J.D., and Clement, B., 1991a, Miocene isotope reference section, Deep Sea Drilling Project Site 608: An evaluation of isotopic and biostratigraphic resolution: Paleoceanography, v. 6, p. 33–52, doi: 10.1029/90PA01941.

Miller, K.G., Wright, J.D., and Fairbanks, R.G., 1991b, Unlocking the ice house: Oligocene-Miocene oxygen isotopes, eustasy, and marine erosion: Journal of Geophysical Research, v. 96, p. 6829–6848, doi: 10.1029/90JB02015.

Miller, K.G., Browning, J.V., Aubry, M.-P., Wade, B.S., Katz, M.E., Kulpecz, A.A., and Wright, J.D., 2008, Eocene-Oligocene global climate and sea-level changes: St. Stephens Quarry, Alabama: GSA Bulletin, v. 120, p. 24–53, doi:10/1130/B26105.1.

Odin, G.S., and Montanari, A., 1988. The Eocene-Oligocene boundary at Massignano Ancona, Italy: A potential stratotype for the Eocene-Oligocene boundary, *in* Premoli-Silva, I., Coccioni, R., and Montanari, A., eds., The Eocene-Oligocene boundary in the Marche-Umbria Basin Italy: Ancona, Italy, International Union of Geological Sciences, Commission on Stratigraphy, International Subcommission on Paleogene Stratigraphy Report, p. 253–263.

Pälike, H., Norris, R.D., Herrle, J.O., Wilson, P.A., Coxall, H.K., Lear, C., Shackleton, N., Tripati, A., and Wade, B.S., 2006, The heartbeat of the Oligocene climate system: Science, v. 314, p. 1894–1898, doi: 10.1126/science.1133822.

Pearson, P.N., McMillan, I.K., Wade, B.S., Jones, T.D., Coxall, H.K., Bown, P.R., and Lear, C.H., 2008, Extinction and environmental change across the Eocene-Oligocene boundary in Tanzania: Geology, v. 36, p. 179–182, doi: 10.1130/G24308A.1.

Perch-Nielsen, K., 1979, Calcareous nannofossil zonation at the Cretaceous/Tertiary boundary in Denmark, *in* Birkelund, T., and Bromley, R.G., eds., Cretaceous-Tertiary Boundary Events (Copenhagen, September 18–24, 1979): Volume 1. The Maastrichtian and Danian of Denmark: Copenhagen, University of Copenhagen, p. 115–135.

Perch-Nielsen, K., 1985, Cenozoic calcareous nannofossils, *in* Bolli, H.M., Saunders, J.B., and Perch-Nielsen, K., eds., Plankton Stratigraphy: Cambridge, Cambridge University Press, p. 427–554.

Poag, C.W., and Aubry, M.-P., 1995, Upper Eocene impactites of the U.S. East Coast: Depositional origins, biostratigraphic framework, and correlation: Palaios, v. 10, p. 16–43, doi: 10.2307/3515005.

Premoli Silva, I., and Jenkins, D.G., 1988, Decision on the Eocene-Oligocene boundary: Episodes, v. 16, p. 379–382.

Prothero, D., and Berggren, W.A., eds., 1992, The Eocene-Oligocene Climatic and Biotic Changes: Princeton, New Jersey, Princeton University Press, p. 272–309.

Raffi, I., Backman, J., Fornaciari, E., Pälike, H., Rio, D., Lourens, L., and Hilgen, F., 2006, A review of calcareous nannofossil astrobiochronology encompassing the past 25 million years: Quaternary Science Reviews, v. 25, p. 3113–3137, doi: 10.1016/j.quascirev.2006.07.007.

Rampino, M.R., 1999, Impact crises, mass extinctions, and galactic dynamics: The case for a unified theory, *in* Dressler, B.O., and Sharpton, V.L., Large Meteorites Impacts and Planetary Evolution II: Geological Society of America Special Paper 339, p. 241–248.

Raup, D.M., 1990, Impact as a general cause of extinction: A feasibility test, *in* Sharpton, V.L., and Ward, P.D., Global Catastrophes in Earth History: Geological Society of America Special Paper 247, p. 27–32.

Raup, D.M., and Sepkoski, J.J., 1982, Mass extinctions in the marine fossil record: Science, v. 215, p. 1501–1503, doi: 10.1126/science.215.4539.1501.

Roth, P., 1987, Mesozoic calcareous nannofossil evolution: Relation to paleoceanographic events: Paleoceanography, v. 2, p. 601–611, doi: 10.1029/PA002i006p00601.

Saez, A.G., Probert, I., Geisen, M., Quinn, P., Young, J.R., and Medlin, L.K., 2003, Pseudo-cryptic speciation in coccolithophores: Proceedings of the National Academy of Sciences of the United States of America, v. 100, p. 7163–7168.

Saez, A.G., Probert, I., Geisen, M., Quinn, P., Young, J.R., and Medlin, L.K., 2004, A review of the phylogeny of Haptophyta, *in* Thierstein, H.R., and Young, J.R., eds., Coccolithophores—From Molecular Processes to Global Impact: Berlin, Springer-Verlag, p. 251–269.

Schouten, S., Eldrett, J., Greenwood, D.R., Harding, I., Baas, M., and Damsté, J.S.S., 2008, Onset of long-term cooling of Greenland near the Eocene-Oligocene boundary as revealed by branched tetraether lipids: Geology, v. 36, p. 147–150, doi: 10.1130/G24332A.1.

Storey, M., Duncan, R.A., and Swisher, C.C., III, 2007, Paleocene/Eocene thermal maximum and the opening of the North Atlantic: Science, v. 316, p. 587–589, doi: 10.1126/science.1135274.

Thomas, E., 2008, Descent into the icehouse: Geology, v. 36, p. 191–192, doi: 10.1130/focus022008.1.

Touchard, Y., Rochette, P., Aubry, M.-P., and Michard, A., 2003, High resolution magnetostratigraphic and biostratigraphic study of Ethiopian traps-related products in Leg 115 Oligocene sediments: Earth and Planetary Science Letters, v. 206, p. 493–508, doi: 10.1016/S0012-821X(02)01084-1.

Van Mourik, C.A., and Brinkhuis, H., 2005, The Massignano Eocene-Oligocene golden spike section revisited: Stratigraphy, v. 2, no. 1, p. 13–30.

Vishnevsky, S., and Montanari, A., 1999, Popigai impact structure (arctic Siberia, Russia): Geology, petrology, geochemistry, and geochronology of glass-bearing impactites, *in* Dressler, B.O., and Sharpton, V.L., eds., Large Meteorite Impacts and Planetary Evolution II: Geological Society of America Special Paper 339, p. 19–59.

White, R.V., and Saunders, A.D., 2005, Volcanism, impact and mass extinction: Incredible or credible coincidences?: Lithos, v. 79, p. 299–316, doi: 10.1016/j.lithos.2004.09.016.

Wilson, E.O., 1992, The Diversity of Life: Cambridge, Massachusetts, The Belknap Press of Harvard University Series, 429 p.

Wolfe, J., 1978, A paleobotanical interpretation of Tertiary climates in the Northern Hemisphere: American Scientist, v. 66, p. 694–703.

Young, J., Geisen, M., Cros, L., Kleijne, A., Probert, I., Sprengel, C., and Ostergaard, J.B., 2003, A guide to extant coccolithophore taxonomy: Journal of Nannoplankton Research, v. 1, special issue, 214 p.

Young, J.R., Geisen, M., and Probert, I., 2005, A review of selected aspects of coccolithophore biology with implications for paleobiodiversity estimation: Micropaleontology, v. 51, supplement 1, p. 267–288, doi: 10.2113/gsmicropal.51.4.267.

Manuscript Accepted by the Society 16 September 2008

The Geological Society of America
Special Paper 452
2009

Integrated stratigraphic and astrochronologic calibration of the Eocene-Oligocene transition in the Monte Cagnero section (northeastern Apennines, Italy): A potential parastratotype for the Massignano global stratotype section and point (GSSP)

Ethan Hyland
Department of Geology, Carleton College, Northfield, Minnesota 55057, USA, and
Osservatorio Geologico di Coldigioco, 62020 Frontale di Apiro, Italy

Brandon Murphy
Stable Isotope Laboratory, University of California, Santa Cruz, California 95064, USA

Phil Varela
Katie Marks
Lauren Colwell
Department of Geology, Carleton College, Northfield, Minnesota 55057, USA, and
Osservatorio Geologico di Coldigioco, 62020 Frontale di Apiro, Italy

Flavia Tori
Simonetta Monechi
Dipartimento di Scienze della Terra dell'Università degli Studi di Firenze, Via La Pira 4, 50121 Firenze, Italy

Laura Cleaveland
Department of Geological Sciences, Brown University, Providence, Rhode Island 02912, USA

Henk Brinkhuis
Laboratory of Palaeobotany and Palynology, Utrecht University, Budapestaan 4, 3584 CD, Utrecht, The Netherlands

Caroline A. van Mourik
Department of Geology and Geochemistry, Stockholm University, S-106 91 Stockholm 9, Sweden

Rodolfo Coccioni
Istituto di Geologia e Centro di Geobiologia, Università degli Studi di Urbino, 61029 Urbino, Italy

David Bice
Department of Geosciences, Pennsylvania State University, University Park, Pennsylvania 16802, USA, and
Osservatorio Geologico di Coldigioco, 62020 Frontale di Apiro, Italy

Alessandro Montanari*
Osservatorio Geologico di Coldigioco, 62020 Frontale di Apiro, Italy

*Corresponding author: sandro.ogc@fastnet.it.

Hyland, E., Murphy, B., Varela, P., Marks, K., Colwell, L., Tori, F., Monechi, S., Cleaveland, L., Brinkhuis, H., van Mourik, C.A., Coccioni, R., Bice, D., and Montanari, A., 2009, Integrated stratigraphic and astrochronologic calibration of the Eocene-Oligocene transition in the Monte Cagnero section (northeastern Apennines, Italy): A potential parastratotype for the Massignano global stratotype section and point (GSSP), *in* Koeberl, C., and Montanari, A., eds., The Late Eocene Earth—Hothouse, Icehouse, and Impacts: Geological Society of America Special Paper 452, p. 303–322, doi: 10.1130/2009.2452(19). For permission to copy, contact editing@geosociety.org.

ABSTRACT

We present the results of integrated biostratigraphic (planktonic foraminifera, calcareous nannofossils, and dinoflagellates), magnetostratigraphic, and cyclostratigraphic analyses of the lower part of Monte Cagnero section (Umbria-Marche Apennines of Italy), a continuous and complete succession of pelagic limestone and marls that provides the means for an accurate and precise astrochronologic calibration of the Eocene-Oligocene transition. This 38.5-m-thick section overlaps the Oligocene section, which, at meter level 188, contains the Rupelian-Chattian boundary corresponding to the O4-O5 planktonic foraminiferal zonal boundary within the upper half of magnetochron C10n. The Eocene-Oligocene boundary at Monte Cagnero, as defined by the last occurrence of hantkeninid planktonic foraminifers (E14-E15 zonal boundary), is found at meter level 114.1, in the upper part of calcareous nannofossil zone CP16a, and very near the Aal-Gse dinocyst zonal boundary. Paleomagnetic analysis has identified all the magnetic reversals from the lower C13r to the lower C12n, precisely overlapping the base of the Oligocene magnetostratigraphic succession and placing the Eocene-Oligocene boundary in the upper part of C13r, in full agreement with the global stratotype section and point (GSSP) at Massignano.

Spectral analysis of calcium carbonate data from bulk samples, collected at 5 cm intervals, indicates that orbital forcing of depositional cycles (i.e., limestone versus marl alternations) is dominant at frequencies corresponding to the theoretical astronomical curves of eccentricity, obliquity, and precessional cycles throughout the studied Eocene-Oligocene transition. Correlation with the astrochronologic time scale allows an age assignment of 33.95 Ma for the Eocene-Oligocene boundary, which is in close agreement with the astrochronologic age for the boundary in the GSSP of Massignano obtained in a similar study by R.E. Brown and colleagues in this volume. Thus, the Monte Cagnero section represents a candidate parastratotype for the Eocene-Oligocene GSSP of Massignano in the eventuality that the oxygen and carbon stable isotope shifts defining the oxygen isotope Oi-1 glaciation will be preferred over the last occurrence of hantkeninids as marker for the boundary, since, at Massignano, the beginning of this isotope shift is barely represented in the uppermost part of the exposed section. The excellent integrated stratigraphic framework of Monte Cagnero presented here will significantly facilitate further high-resolution isotope and paleoecological studies across the time of transition from a hothouse to icehouse Earth.

Keywords: integrated stratigraphy, Eocene-Oligocene boundary, Massignano GSSP, Umbria-Marche Apennines, central Italy.

INTRODUCTION

In the few hundred-thousand-year interval across the Eocene-Oligocene boundary, Earth experienced a change from a "hothouse" climate, or warm, humid period, to an "icehouse" climate, characterized by the expansion of Antarctic ice sheets and consequent accelerated global cooling (e.g., Zachos et al., 2001; van Mourik et al., 2008). This climatic transition was responsible for major changes in both marine and terrestrial biodiversity, and it is considered one of the most pronounced global changes of the Cenozoic Era (e.g., Prothero, 1994).

Extensive pelagic, fossiliferous sedimentary sequences exposed on land that cover this time interval are exceedingly rare around the world. An exception occurs in the Umbria-Marche Basin of the northeastern Apennines of central Italy, which hosts an essentially continuous and complete pelagic carbonate succession covering the earliest Jurassic up to the latest Miocene (e.g., Montanari and Koeberl, 2000). In this Umbria-Marche succession, the Eocene-Oligocene boundary is well preserved within the Scaglia Cinerea Formation in a number of localities. The marly limestones and calcareous marls of the Scaglia Cinerea and the underlying Scaglia Variegata Formations contain a handful of radioisotopically dated biotite-rich volcaniclastic layers (Montanari et al., 1985; Montanari, 1990; Odin et al., 1991), which provide numerical age constraints for detailed stratigraphic studies (e.g., Montanari and Koeberl, 2000).

The Eocene-Oligocene boundary is currently defined as the last occurrence (LO) of the planktonic foraminiferal genus *Hantkenina* (e.g., Premoli Silva et al., 1988), and it is dated at ca. 34 Ma (Miller et al., 2008). The boundary's global stratotype section and point (GSSP) is located at 19 m above the base of the section in the Massignano quarry within the Cònero Regional Park, near the port city of Ancona, Italy (Premoli-Silva and Jenkins, 1993; Fig. 1). Recent studies and current debate within the scientific community suggest that the Eocene-Oligocene boundary may be better defined by the $\delta^{18}O$ and $\delta^{13}C$ isotope excursions occurring between 33.5 and 33.6 Ma, which resulted from the expansion of the Antarctic ice cap (Zachos et al., 2001; Miller et al., 2008), marking the beginning of the present global cold climatic regime often referred to as "icehouse" (Bohaty et al., 2007; van Mourik et al., 2008). However, the Massignano section continues only 4 m beyond the biostratigraphically recognized Eocene-Oligocene boundary, and it does not contain the entirety of the oxygen isotope Oi-1 excursion, which could provide a more correlatable and appropriate epochal boundary worldwide.

While the Massignano section provides limited exposure of Oi-1, the Monte Cagnero section is continuous for ~120 m from the Upper Eocene to the Upper Oligocene. Coccioni et al. (2008) indicated an approximate stratigraphic equivalency between meter level 100 at Monte Cagnero and meter level 0 at Massignano and performed a detailed integrated stratigraphic study of the section from meter level 140 to meter level 225 with the aim of defining the mid-Oligocene Rupelian-Chattian interstage boundary. Here, we extend the stratigraphic study to the lower part of the section, from meter level 146.5 down to meter level 100 with the aim of laying the groundwork for future isotopic investigation and correlating this section with the GSSP of Massignano on the basis of microplankton biostratigraphy (foraminifera, nannofossils, and dinoflagellates), magnetostratigraphy, and astrochronologic cyclostratigraphy using, as a proxy, calcium carbonate content in closely spaced bulk-rock samples.

Location and General Overview of the Section

The section is exposed on the southeastern slope of Monte Cagnero (43°38′50″N, 12°28′05″E of Greenwich, 727 m above sea level [asl]), ~5 km southwest of the town of Urbania (northwestern Marche region; Fig. 1A). The Scaglia Cinerea Formation and the upper portion of the underlying Scaglia Variegata Formation are exposed here over a 100-m-wide front, with a consistent monoclinal bedding dip of ~20°SW. The outcrop is accessible via a steep dirt road departing from the old church of San Paterniano (Fig. 1B). The step morphology of the slope exposure is imparted by the differential erosion between hard marly limestone layers and softer, recessive marly layers (Fig. 2). The section spanning from meter level 140 to meter level 225 measured by Coccioni et al. (2008) stretches along the west side of the exposure, while the section studied herein was measured on the east side of the exposure (Fig. 2). The two sections overlap in a distinctive lithologic interval from meter level 130 to meter level 150, which is composed of a soft, gray and purplish marl up to meter level 142 overlain by a series of hard marly limestone interbedded with softer marly layers, some of which contain volcanosedimentary biotite.

A volcaniclastic layer, which contains unusually large biotite flakes at meter level 145.6, was reported by Coccioni et al. (2008) as being dated to 31.5 ± 0.2 Ma. Another biotite-rich volcaniclastic layer is found at meter level 122.9, and traces of biotite flakes are also found at around meter level 107, where the section is disturbed by a low-angle fault and/or a soft-sediment slump. It is because of this disturbance that we have focused our study on the interval from meter level 108 to meter level 146.5. Below the fault/slump zone, the marly limestone exhibits a pink color down to meter level 102. This interval corresponds to a distinctive and regionally recognizable pink band in the upper part of the Scaglia Variegata Formation. The base of the Massignano GSSP section can be placed two meters below this pink band based on lithology comparisons between the sites. This level would correspond to the base of our Monte Cagnero section, at meter level 100.

MATERIALS AND METHODOLOGY

Micropaleontology

Bulk samples (1–2 kg) were collected at 50 cm intervals throughout the Monte Cagnero section for micropaleontologic analyses. For foraminiferal assemblage analysis, additional samples were collected every 10 cm from meter levels 113–115. Marly and soft marly-limestone samples were easily disaggregated using diluted hydrogen peroxide and Desogen. Planktonic foraminiferal tests from the hard marly-limestone and limestone samples were extracted largely following the procedure described by Lirer (2000), consisting of dissolution in concentrated acetic acid, additional soaking in Desogen for several hours, and immersion in an ultrasonic bath for a few minutes. Samples were subsequently washed through a 40-µm-mesh sieve. The residues were studied with a binocular microscope for their planktonic foraminiferal content.

Nannofossil assemblages were observed on smear slides using ~0.05 g of raw sediment and following standard preparation methods (Hay, 1977). Smear slides were studied under an Orthoplan-Pol Leitz Wetzlar light microscope both under crossed nicols and transmitted light at a magnification of 1560×. Qualitative and semiquantitative analyses were performed on 84 samples by counting 500 specimens with an additional scanning of two traverses (8 mm^2) in order to identify rare species.

For calcareous nannofossils, quantitative analyses were performed in order to recognize first occurrences (FO) and last occurrences (LO), as well as first common occurrence (FCO) and last common occurrence (LCO). FCO and LCO are defined as the beginning and the end of a continuous and consistent pattern. The standard taxonomy of Perch-Nielsen (1985) and zonations of Martini (1970) and Okada and Bukry (1980) are widely used for low- and middle-latitude Eocene-Oligocene biostratigraphic studies, and these were adopted in this paper.

Samples for dinoflagellate biostratigraphic analysis were collected at 1 m intervals from meter level 108 to meter level

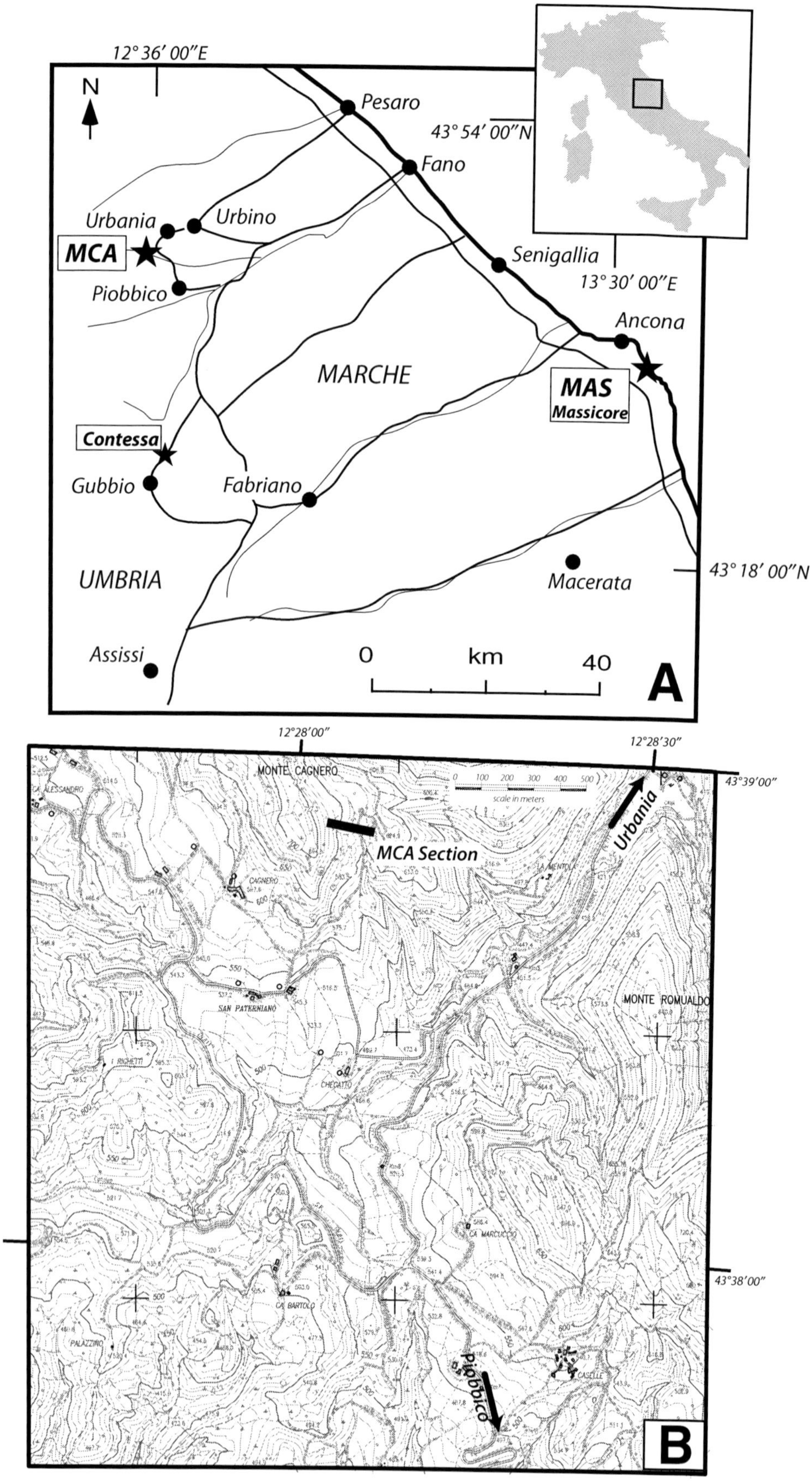

Figure 1. (A) Simplified map of the Umbria-Marche region, with location of the Massignano (MAS) global stratotype section and point (GSSP), Monte Cagnero (MCA), Contessa, and Massicore sections. (B) Topographic map of the Monte Cagnero area and location of the exposed section.

Figure 2. Panoramic view of the Monte Cagnero exposure with meter levels indicated on the left for the section studied by Coccioni et al. (2008), and on the right for the lower part of section studied in this work. Arrow points to a car for scale.

132 and processed using standard palynological techniques (cf. Brinkhuis and Biffi, 1993; van Mourik and Brinkhuis, 2005). Organic walled dinoflagellate cyst (dinocyst) taxonomy follows that cited in Brinkhuis and Biffi (1993) and Fensome and Williams (2004). All material is stored at the Laboratory of Palaeobotany and Palynology, Utrecht, The Netherlands.

Paleomagnetism

Oriented hand-samples were taken at an average of 0.5 m intervals from meter level 108 to meter level 145. The samples were then trimmed and oriented relative to bedding and cored, while maintaining orientation, using a 25-mm-diameter drill press at the Osservatorio Geologico di Coldigioco (Italy), to ~2.5-cm-long cores. Paleomagnetic measurements were carried out using an automatic cryogenic DC-SQuID magnetometer in a magnetically shielded room at Yale University's Department of Geology and Geophysics (New Haven, Connecticut, USA). The core samples were precleaned with liquid nitrogen in a zero field to preferentially reduce multidomain carriers and then treated with low alternating frequency (AF; 0–10 mT, 2 mT increments) and low thermal (TT; 75–200 °C, 25 °C increments) demagnetization processes. Samples were then split into two populations, with at least one core from every stratigraphic level in each; population A–C (n = 51) was further treated with high AF (15–70 mT at 5 mT increments), while population B (n = 41) was treated with high thermal conditions (250–580 °C at increments of between 10 °C and 50 °C, depending on proximity to known unblocking temperatures) in a controlled N_2 environment to prevent the accidental oxidation of weakly magnetized sediments (most samples had magnetizations of $<10^{-6}$ G).

The isolated characteristic remnant magnetization (ChRM) data from these samples were then determined and analyzed using PaleoMag v 3.1b1, where bedding, tilt, and misoriented sample corrections were made and statistical models were applied (least squares models, Fisher statistics, Bingham test, etc., after Fisher, 1953; Zijderveld, 1967; Kirschvink, 1980).

Cyclostratigraphy

We used percentage of calcium carbonate as a proxy for our cyclostratigraphic analysis of the Monte Cagnero section, assuming that the carbonate fraction of these pelagic sediment is predominantly biogenic (i.e., planktonic foraminiferal tests and calcareous nannofossils), while the noncarbonate fraction is made essentially of siliciclastic clay and silt. Samples were collected continuously every 5 cm between meter levels 108 and 146.5. In the laboratory, they were powdered with a brass mortar and pestle and sieved through a 125-µm-grain-size sieve to ensure consistency and complete dissolution when reacted with 10 vol% HCl in a Dietrich-Fruling water calcimeter, which was used to obtain a total $CaCO_3$ content with a precision of ±2%. For each calcimetric analysis, we used ~300 mg of bulk sample. Standards of pure calcium carbonate (Carrara Marble) were run every 20 samples, and water levels in the calcimeter were recalibrated daily to ensure proper calibration.

We conducted spectral analyses of the Monte Cagnero carbonate data in Matlab using algorithms modified from Muller and MacDonald (2000). After first resampling the data to a uniform stratigraphic spacing, we performed fast Fourier transforms (FFTs) of the data (detrended and padded with zeros) in the stratigraphic domain, which yielded a set of spectral peaks representing the stratigraphic wavelengths of the cycles. The spectral power used here is the complex conjugate of the Fourier coefficients, normalized to unit mean power (Muller and MacDonald, 2000). The carbonate data contain an overall trend toward lower carbonate values at the top of the section. This imparts a strong low-frequency spectral peak that obscures spectral peaks at the higher frequencies that may correspond to Milankovitch cycles (precession, obliquity/tilt). This long-term trend was removed by using a rectangular band-pass window that effectively restricted the data to cycles shorter than ~20 m. We evaluated the statistical significance of the remaining spectral peaks by generating a 95% confidence level from a Monte Carlo noise simulation (Muller and MacDonald, 2000). In this approach, 1000 random number sets with the same dimensions as the carbonate data were generated and then run through a FFT to find spectral power at each frequency. The 95% confidence level was estimated from these 1000 spectral results by calculating three times the mean power value for each frequency. Peaks rising above this 95% confidence level were considered to be statistically significant. Prominent, statistically significant peaks related to one another by the same ratios found between Milankovitch cycles are assumed to represent eccentricity, obliquity, and precession, making it possible to correlate meter level to age through the calculation of an average sedimentation rate.

The persistence of the spectral peaks through the stratigraphic section was evaluated through the use of an evolutionary, or sliding window, FFT. In this approach, a subset or window of the data is analyzed using the same FFT; the window then moves up through the time series. The window size was chosen to be ~7 m, which is large enough to enable the detection of a short eccentricity signal given the preexisting sedimentation rate from geochronologic constraints.

Correlations between the carbonate data and astrochronologic time were performed using two general approaches. In the first approach, we compared the detrended carbonate content with a linear combination of the eccentricity, tilt, and precession (ETP; Imbrie et al., 1984) derived from the astronomical variations of Laskar et al. (2004). We modified the ETP slightly by decreasing the contribution of the obliquity, which has the effect of enhancing the longer-period variations due to eccentricity. The section was subdivided into three intervals based on lithologic and spectral characteristics, and each section was stretched until we found the best match with the ETP curve; the correlation was guided in part by trying to match the long-wavelength carbonate variations with the 400 k.y. eccentricity cycle. The preferred correlation provides a time duration and thus sedimentation rate for each of the intervals. This tuning process was constrained by an assumed age for the Eocene-Oligocene boundary (meter level 114.1) of ca. 33.9 Ma, following the work of Brown et al. (this volume), and a radiometric date of 31.5 ± 0.2 Ma for a biotite-rich level at 145.6 m (Coccioni et al., 2008).

In the second approach, we applied band-pass filters to the tuned carbonate time series in order to extract individual presumed orbital signals for comparison with the individual theoretical orbital parameters, which are tied to astrochronologic time. These two approaches also provided a means for correlating the Monte Cagnero section to the Massignano GSSP.

RESULTS

Biostratigraphy

The biostratigraphy of the Monte Cagnero section was initially studied by Baumann and Roth (1969). They established a correlation between planktonic foraminiferal and calcareous nannofossil biozonations for the Eocene-Oligocene transition up to the uppermost Oligocene, and they recognized seven nannoplankton (NP) biozones corresponding to NP 19/20–NP 24.

Revisions by Baumann (1970) and Roth (1970) provided evidence that previous biostratigraphic studies on foraminifera and nannofossils (Subbotina, 1953; Bolli, 1957; Blow and Banner, 1962) were flawed due to facies variation and sedimentation hiatuses. Several other sections of the Umbria-Marche Scaglia succession indicate that the Monte Cagnero section is one of the most extended and continuous sections straddling the Eocene-Oligocene boundary (Baumann, 1970). Further biostratigraphic refinement of the Monte Cagnero section, which covered ~55 m of the Eocene (Scaglia Variegata Formation) and ~30 m of overlying Oligocene (Scaglia Cinerea Formation), was performed by Parisi et al. (1988).

Planktonic Foraminifera

Planktonic foraminifera are abundant in all samples with diverse genera and species typical of late Eocene–early Oligocene low-latitude pelagic environments. Preservation varies from moderate to poor. Specimens often display a slight calcitic overgrowth or infilling. Reworking, which is limited to local bioturbation by *Zoophycos*, *Chondrites*, and *Planolites*, does not affect the biostratigraphic resolution.

Generic and specific concepts adopted here are those of Premoli Silva and Boersma (1989), Spezzaferri (1994), and Pearson et al. (2006). The following main bioevents are identified in the Monte Cagnero section and correlated with the planktonic foraminiferal biozones (from bottom to top): (1) last occurrence (LO) of *Globigerinatheka index* at meter level 107.52; (2) LOs of *Hantkenina alabamensis* and other Hantkeninidae at meter level 114.1; and (3) LO of *Pseudohastigerina naguewichiensis* at meter level 136.0.

We based our biozonal definition on Berggren and Pearson (2006). In agreement with these authors, we used the LO of *G. index* (which they defined as the highest stratigraphic occurrence) to recognize the E15-E16 zonal boundary, the LO of *H. alabamensis* to denote the E16-O1 zonal boundary, and the LO of *P. naguewichiensis* to place the O1-O2 zonal boundary. The Eocene-Oligocene boundary is placed at meter level 114.1, where the Eocene planktonic foraminiferal genus *Hantkenina* becomes extinct (Coccioni et al., 1988; Nocchi et al., 1988; Premoli Silva and Jenkins, 1993).

Calcareous Nannofossils

Calcareous nannofossils of the Monte Cagnero section are abundant, and preservation varies from moderate to poor because of overgrowth and/or etching. Rare Upper Cretaceous reworked specimens have been observed throughout the interval studied, but they never exceed 1%. Otherwise, the assemblages are typical of low- to middle-latitude environments and mainly consist of *Coccolithus pelagicus*, *Cyclicargolithus floridanus*, and *Dictyococcites scrippsae*. Specimens such as *Dictyococcites bisectus*, *Lanternithus minutus*, *Discoaster* spp., *Zygrhablithus bijugatus*, and *Sphenolithus moriformis* fluctuate greatly throughout the section. The studied interval spans from zones CP15 to CP17 and NP19–20 to NP23 as denoted by the standard zonations of Okada and Bukry (1980) and Martini (1970), respectively. The lower part of the section is assigned to zone CP15 (NP19–20), based on the presence of *Isthmolithus recurvus*.

All markers and a few supplementary events have been identified and are reported from the base of the section as follows:

1. The LO of *Cribrocentrum reticulatum*, at meter level 104.50, and the LO of *Discoaster barbadiensis*, at meter level 107.50, together with the LO of *Discoaster saipanensis*, define the marker event for the base of CP16 (Okada and Bukry, 1980). These two events are very close but not simultaneous, as reported by several studies (e.g., Poore et al., 1982; Backman, 1986; Pälike et al., 2005; Zachos et al., 2004). The last occurrence of *D. saipanensis*, at meter 108.5, defines also the base of zone NP21 of Martini (1970).
2. Acme Beginning (AB) of *Clausicoccus obrutus*, characterized by specimens >5.7 μm, is identified at meter level 114.5. This event is proposed here to define the base of the CP 16b subzone instead of the acme end (AE) of *Clausicoccus*, in agreement with Backman (1987). The proposed event is placed immediately after the Eocene-Oligocene boundary in both the Massignano and Monte Cagnero sections, and it shortly predates the inception of the isotopic event Oi-1.
3. The LO of *Calcidiscus protoannulus* was found at meter level 117.5. Although limited to rare occurrences, we observed the first rare occurrence (FRO) of *Helicosphaera ethologa* at meter level 125. The FO of *Sphenolithus akropodus* was observed at meter level 125.5. The continuous, although by no means abundant, distribution of this taxon makes it a good marker for the lower Oligocene at mid- to low latitudes (De Kaenel and Villa, 1996; Monechi et al., 2007).
4. The LO of *E. formosa*, marker for the base of zone NP 22, was observed at meter level 127.50.
5. *Isthmolithus recurvus* is very rare and discontinuously present. The LO is tentatively placed at meter 138, together with the LO of *Reticulofenestra umbilica*, which is the marker of the base of zones NP 23 and CP17. The last occurrence of *Cyclicargolithus abisectus* (>10 μm), is present above meter level 139.
6. *Sphenolithus distentus*, marker of the base of zone NP24, has not been recorded in the upper part of the studied interval.

Dinoflagellates

A new set of Monte Cagnero samples, collected at 1 m intervals from meter level 108–132, mimics the earlier results of Brinkhuis and Biffi (1993), who reported in greater detail on the typical latest Eocene–early Oligocene dinocyst assemblages from this part of the section, and who provided correlations to the Massignano GSSP. Most samples from the interval below meter level 112, which includes a pink band from meter level 102 up to meter level 107.5 overlain by ~4 m of faintly bluish marly limestones, contain rare and poorly preserved dinoflagellate fossils. In general, the samples above this interval are dominated by dinocysts and, to a lesser degree, by bisaccate pollen. The oldest dinocyst zone recognized is zone Cfu from the base of the section to meter level 110, and it is followed by zone Aal up to meter level 113. Dinocyst assemblages in samples from meter levels 114 and 115 pertain to zone Gse, those from meter levels 116–123 pertain to zone Adi, meters 124–129 pertain to zone Rac, and meters 130–132 pertain to zone Cin.

The sparse sampling of this collection does not allow a precise location of dinocyst zonal boundaries better than 1 m. Nevertheless, the dinocyst zones and additional events of Brinkhuis and Biffi (1993), along with the recent modifications proposed by van Mourik and Brinkhuis (2005), have been confirmed. According to this new set of Monte Cagnero samples, it appears that the Eocene-Oligocene boundary, as defined by the LO of hantkeninids at meter level 114.1, falls just above the Aal-Gse zonal boundary, whereas at Massignano, the same dinocyst boundary falls immediately below it. This slight difference, which in terms of true stratigraphic distance translates to a few tens of centimeters, is probably due to the lower resolution of sampling, or to moderate vertical mixing caused by bioturbation, or some combination of the two. Nevertheless, it can be concluded that the dinocyst Aal-Gse zonal boundary essentially approximates the Eocene-Oligocene boundary as defined by the LO of hantkeninids.

Magnetostratigraphy

Nearly all Monte Cagnero samples possess a low-stability overprinting close to the present local field (PLF), probably due to viscous remagnetization (VRM). This field was universally removed (i.e., unblocked) by either controlled-atmosphere thermal demagnetization to around 300 °C (250 °C to 340 °C), or alternating frequency demagnetization to 15 mT, as depicted in the Zijderveld and J/Jo projections (Figs. 3 and 4). An unblocking temperature in this range is indicative of Fe-sulfides, or more specifically, pyrrhotite (Wehland et al., 2005), which is a common alteration mineral in the pelagic sequences of Italy (Freeman, 1986). Some samples also contained a low-stability overprint that had unblocking temperatures between 100 °C and 125 °C, indicative of alteration and weathering to goethite (McElhinny and McFadden, 2000), which is visible as brown staining in the affected cores and hand samples. The primary ChRM in all samples appears to be carried by a component with unblocking temperatures between 570 °C and 580 °C (some as low as

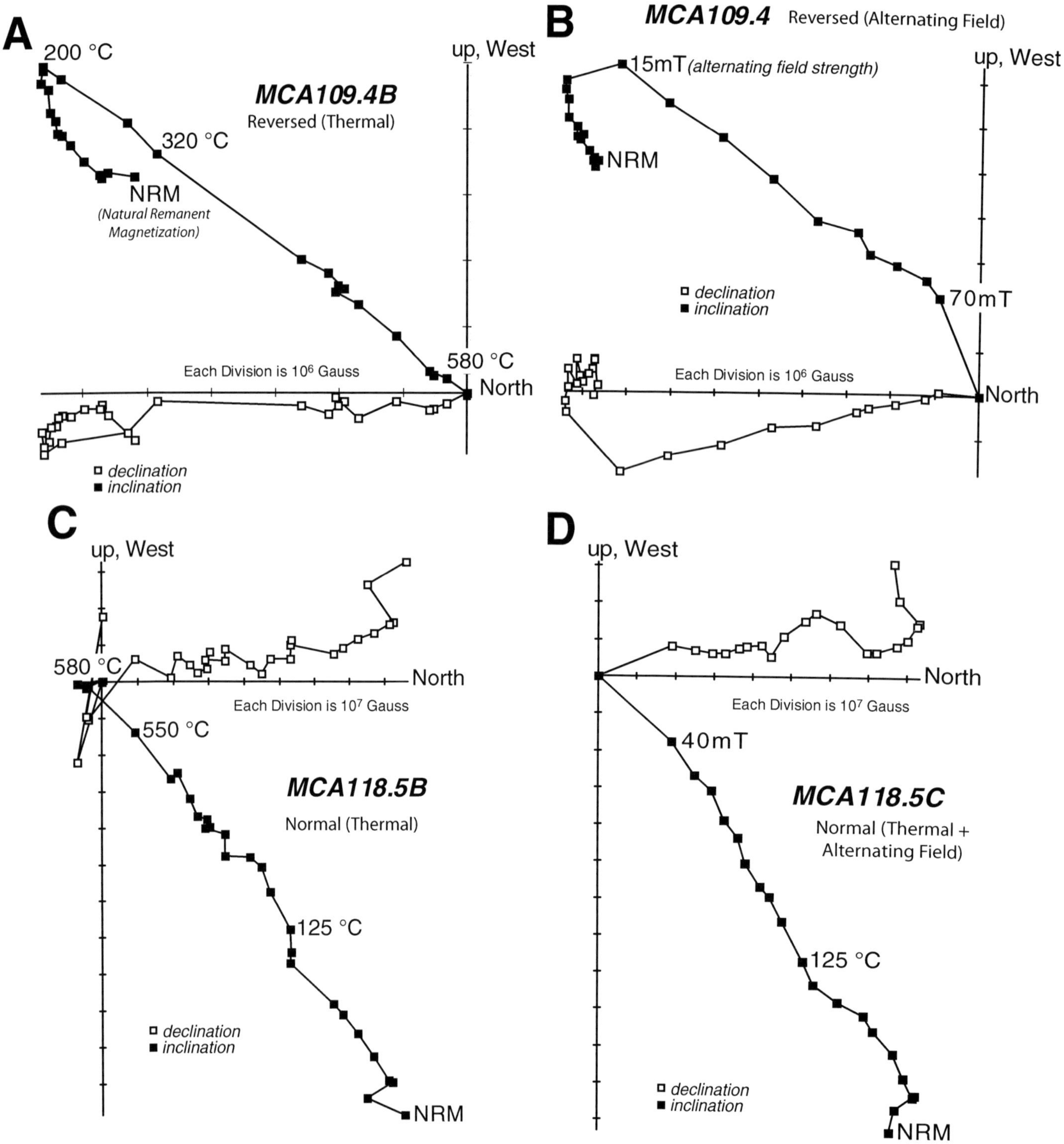

Figure 3. Zijderveld vector diagrams for selected representative Scaglia Cinerea Formation samples: (A) thermal demagnetization of reversed sample 109.4B, (B) alternating frequency demagnetization of reversed sample 109.4, (C) thermal demagnetization of normal sample 118.5B, and (D) alternating frequency demagnetization of normal sample 118.5C. Numbers on graph indicate demagnetizing field in mT or heating temperature in °C. All projections are on vertical E-W planes. Sample numbers correspond to meter levels.

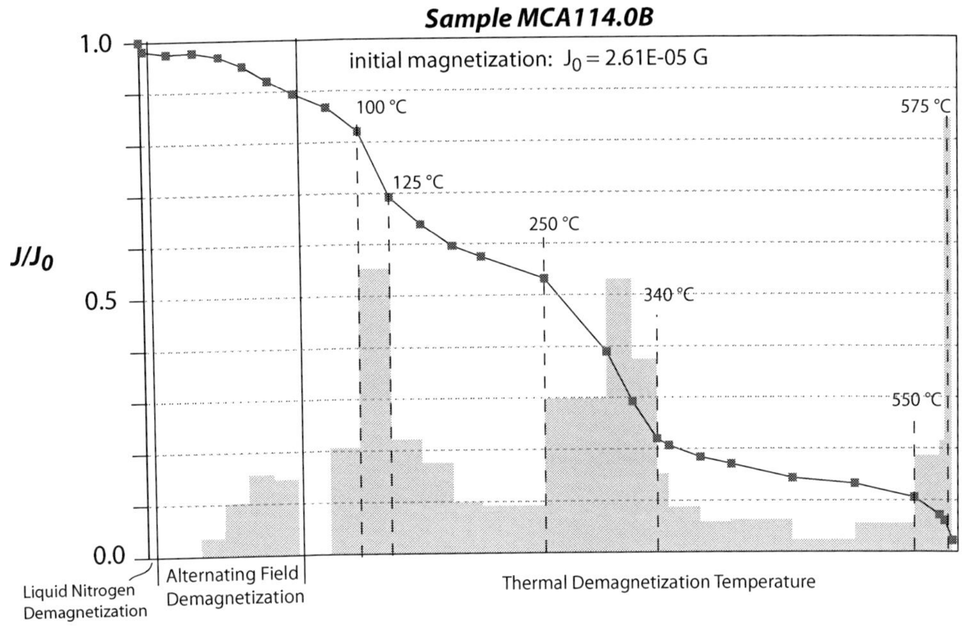

Figure 4. Demagnetization plot for thermal sample 114.0B. Shaded bars indicate magnitude (%) change in magnetization, while points indicate total magnetization of the sample (J/J_0, current/total current). Dotted lines with corresponding temperature markings delineate zones with a high rate of demagnetization, indicative of unblocking points for ferromagnetic minerals. *X*-axis indicates treatment (AF—alternating frequency; thermal) and intensity (AF = 0–10 mT, thermal = 250–580 °C).

550 °C), and coercivities around 40 mT, above which all thermal (TT) samples and some AF samples became directionally unstable, though many AF samples remained stable through the end point at 70 mT. This is indicative of a primary ChRM carrier of (titano) magnetite, with the presence of another higher-stability/coercivity carrier such as hematite (unblocking temperature 675 °C) in some samples (Freeman, 1986).

For both thermal and AF sample groups, the high-stability component was interpreted as the ChRM vector, and least-squares models were forced through the origin as they decayed (Fig. 3), resulting in reverse and normal polarity stable components. Progressive thermal and alternating frequency demagnetization steps defined stable ChRM end-point groups (Fig. 5) of reverse and normal polarity at magnetic intensities of ~10% NRM (<10–6 G). These end-point clusters have a fairly small scatter for both polarity groups (Fig. 5), and a few samples display declinations or inclinations that are intermediate between normal and reverse, meaning that the interpretation of the magnetozones is straightforward.

The ChRM polarity clusters form two nearly antipodal groups (Fig. 5), and the confidence ellipses do not overlap, indicating that the mean directions are statistically different (Fisher, 1953). An F-ratio test reaffirms that they are statistically different at a 95% confidence level (all F-ratios were greater than the proposed null hypothesis; Watson, 1956, 1960). The vector mean directions for the polarity clusters from data set A–C (AF demagnetization) are D = 343°, I = 51° (a95 = 5.9) and D = 169°, I = –54° (a95 = 5.0), while from data set B (TT demagnetization), they are D = 343°, I = 53° (a95 = 7.6) and D = 164°, I = –40° (a95 = 4.0).

When plotted stratigraphically as a local magnetic polarity scale (Fig. 6), the vectors from individual samples within the polarity groups define four distinct magnetozones and at least two shorter reversal periods ("cryptochrons"; e.g., Lanci et al. 1996) in the Monte Cagnero section. Each of the magnetozones is supported by numerous data points and corroborated by both data sets in terms of position and duration, while the smaller polarity periods are supported by either one or two data points in each data set (also coincident in space and time between sets). Resolution of magnetostratigraphy is poor between meters 126 and 145 due to lithologic complications (soft, wet marl), making their interpretation somewhat ambiguous, though data on either side are distinct and appropriate for exact correlations. When the local magnetic polarity scale (LMPS) is tied to known ages within the section, such as the Eocene-Oligocene boundary (as defined by Premoli Silva and Jenkins, 1993), it matches the geomagnetic polarity time scale (GPTS) of Cande and Kent (1995) accurately (Fig. 6).

With the interpreted correlation to the geomagnetic polarity time scale, the outcrop at Monte Cagnero can effectively be linked to other major sections in Italy, such as the Contessa sections, the Massignano Quarry, and the Massicore. These sections are all of different lengths and possess magnetic data of varying resolution, but they can be correlated to each other and the geomagnetic polarity time scale (Fig. 7). When scaled similarly and tied to the Eocene-Oligocene boundary, which has been identified at each outcrop to within 0.5 m precision, the local magnetic polarity time scales and their interpreted chrons are similar in both position (temporal, stratigraphic) and duration (based on individual sedimentation rates constrained by alternate chronologies; Lowrie et al., 1982; Montanari et al., 1985; Premoli Silva et al., 1988), thereby facilitating unambiguous interpretation of Monte Cagnero as contemporaneous, continuous, and correlatable to the other pelagic sequences of the northeastern Apennines.

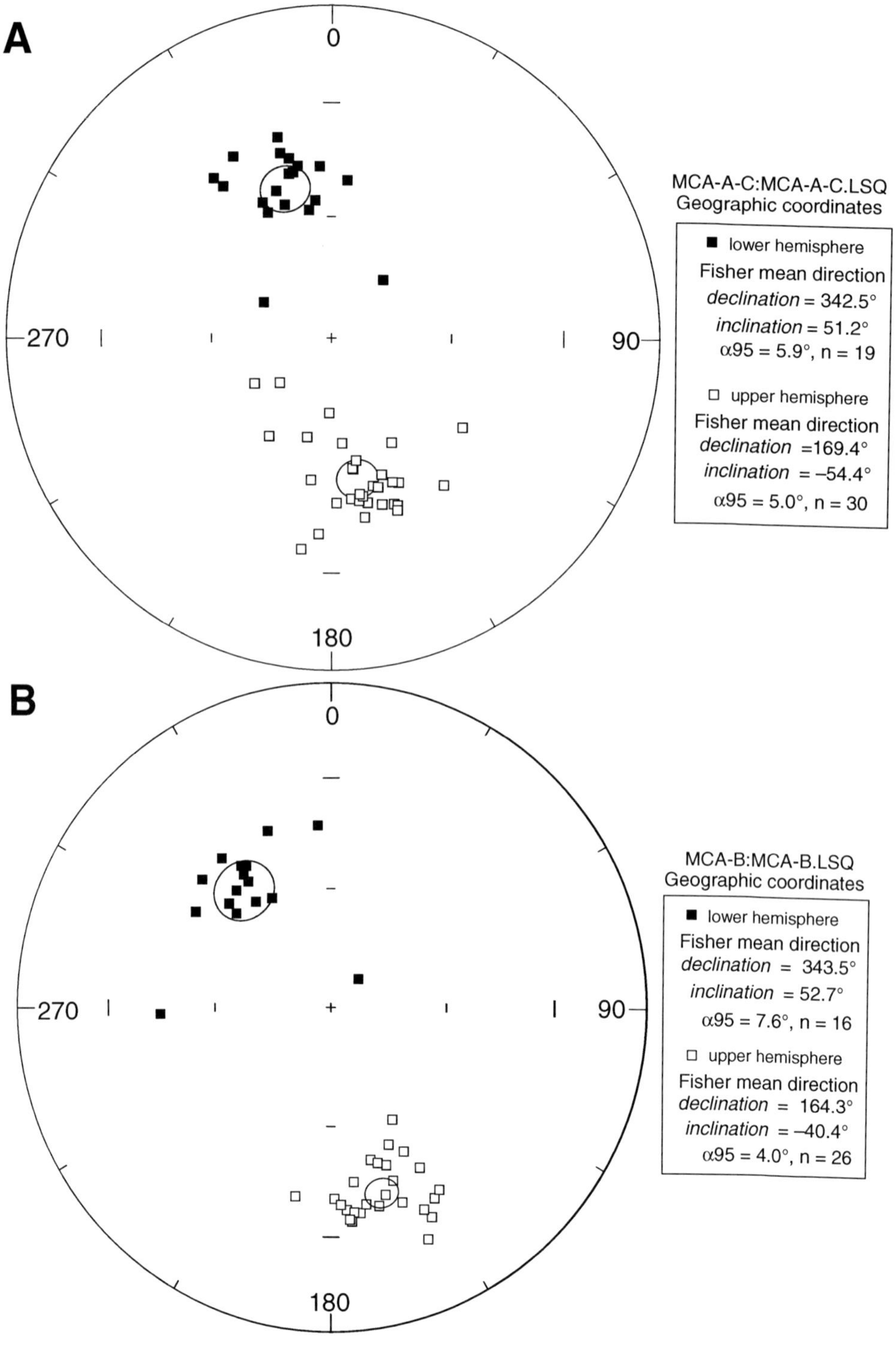

Figure 5. Equal-area stereograms of stable end-point (ChRM) directions for (A) data set MCA-A/C (combined sets A and C), alternating frequency demagnetization, and (B) data set MCA-B, thermal demagnetization. Ellipses indicate Fisher confidence interval least-squares (MCA-A-C.LSQ and MCA-B. LSQ) for end-point groups. All samples have been corrected for simple bedding tilt. MCA—Monte Cagnero.

Cyclostratigraphy and Astrochronologic Calibration

Spectral Peaks: Frequencies and Continuity

A FFT of the detrended carbonate data, viewed in the stratigraphic domain, reveals a number of spectral peaks that rise above the 95% confidence level (Fig. 8A). By simply using a mean sedimentation rate of 12.6 m/m.y. for the whole section, as derived from the radioisotopic age of 31.5 ± 0.2 Ma for the volcaniclastic layer at meter level 145.6 (Coccioni et al., 2008) and the age of 33.9 Ma for the Eocene-Oligocene boundary at meter level 114.1 (Brown et al., this volume), the resulting periods of the spectral peaks include prominent peaks that are generally consistent with Milankovitch cycles (Fig. 8A). An evolutionary FFT of the same data (Fig. 8B) reveals some interesting characteristics of the stationarity and persistence of these signals in the stratigraphic sequence. Some of the sharp, promi-

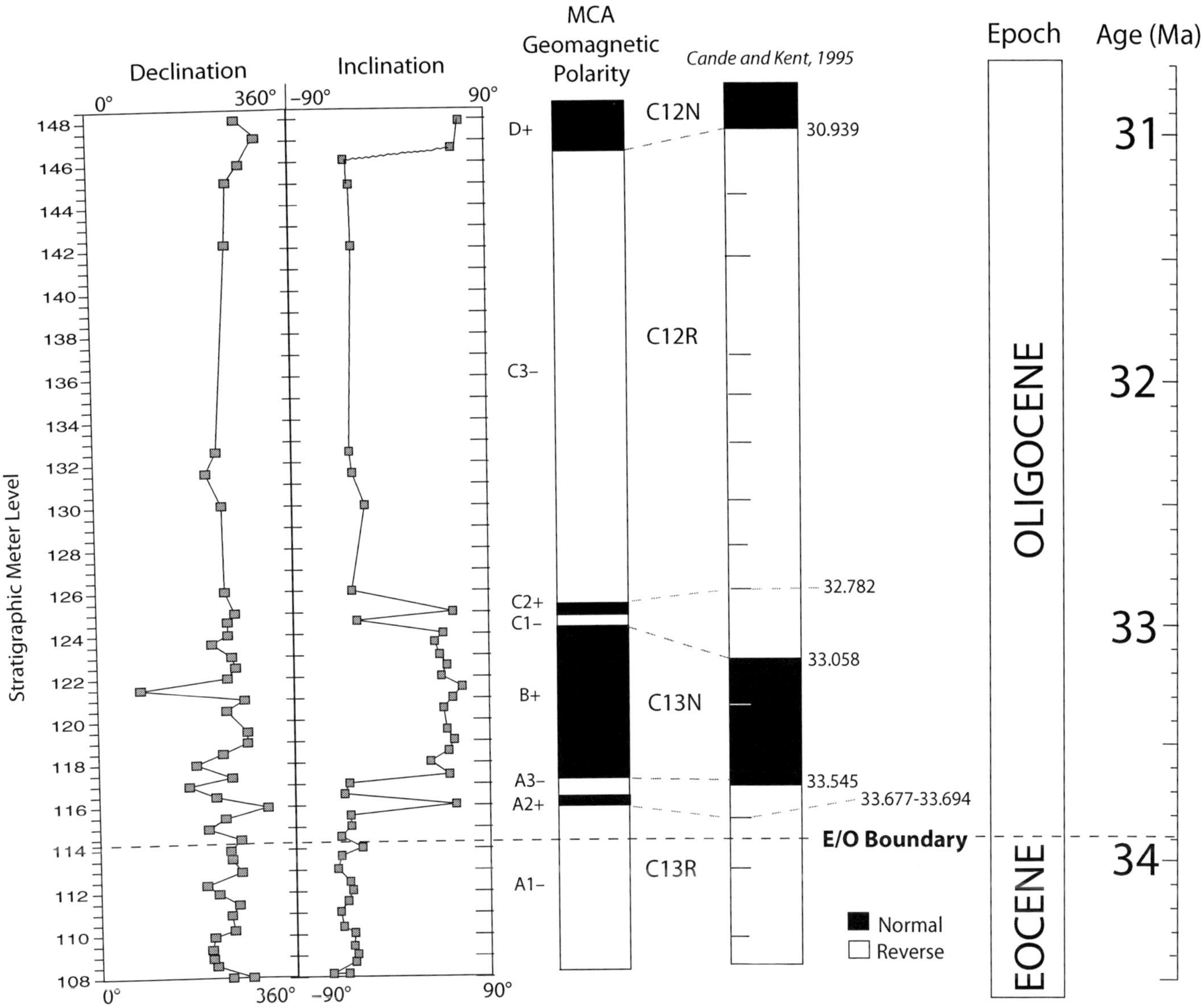

Figure 6. Stratigraphic variation of inclination, the interpreted polarity zones for the Monte Cagnero (MCA) outcrop, and its correlation to the defined magnetozones of the geomagnetic polarity time scale (GPTS) of Cande and Kent (1992, 1995). In the time scale of Cande and Kent (1992, 1995), short bars indicate identified "cryptochrons," and numbers indicate interpolated ages (Ma). E/O—Eocene-Oligocene.

nent spectral peaks present in the FFT of the whole section (Fig. 8A) are, in fact, only present in parts of the section (e.g., spectral peaks with wavelength of 0.41 m and 0.96 m). Other prominent spectral peaks, such as the one with a wavelength of 0.58 m, are more continuous and strong through much of the section, but the wavelength varies through the section, which suggests changes in sedimentation rates during this time period (Fig. 8B). The presumed precessional peaks exhibit a similar nonstationarity when viewed with a shorter window; the longer window shown here (Fig. 8B) allows for enough averaging of cycle lengths to make the signal appear to be more stationary.

If we assume that the ~0.58 m signal represents the orbital obliquity, we see some parallels with the theoretical obliquity—the middle part of the Monte Cagnero section corresponds approximately to the time period when the orbital obliquity is relatively strong (Fig. 8C), and it decreases in relative power toward the bottom and top of the section. Similarly, the spectral peaks that may represent the short eccentricity (in the range of 1.1–1.4 m wavelengths) appear to be stronger near the top and bottom of the section, and the same is true for the short period (95 and 123 k.y.) orbital eccentricity during this time period (Fig. 8C). The longer-period eccentricity (~405 k.y.) shows greater relative spectral power at times when the short eccentricity is weak, and the same appears to be true for the Monte Cagnero section—the peak with a wavelength of ~4.5 m may represent the 400 k.y. cycle of the eccentricity, and it is relatively stronger when the presumed short eccentricity is weaker. In summary, a spectral analysis incorporating a mean sedimentation rate consistent with

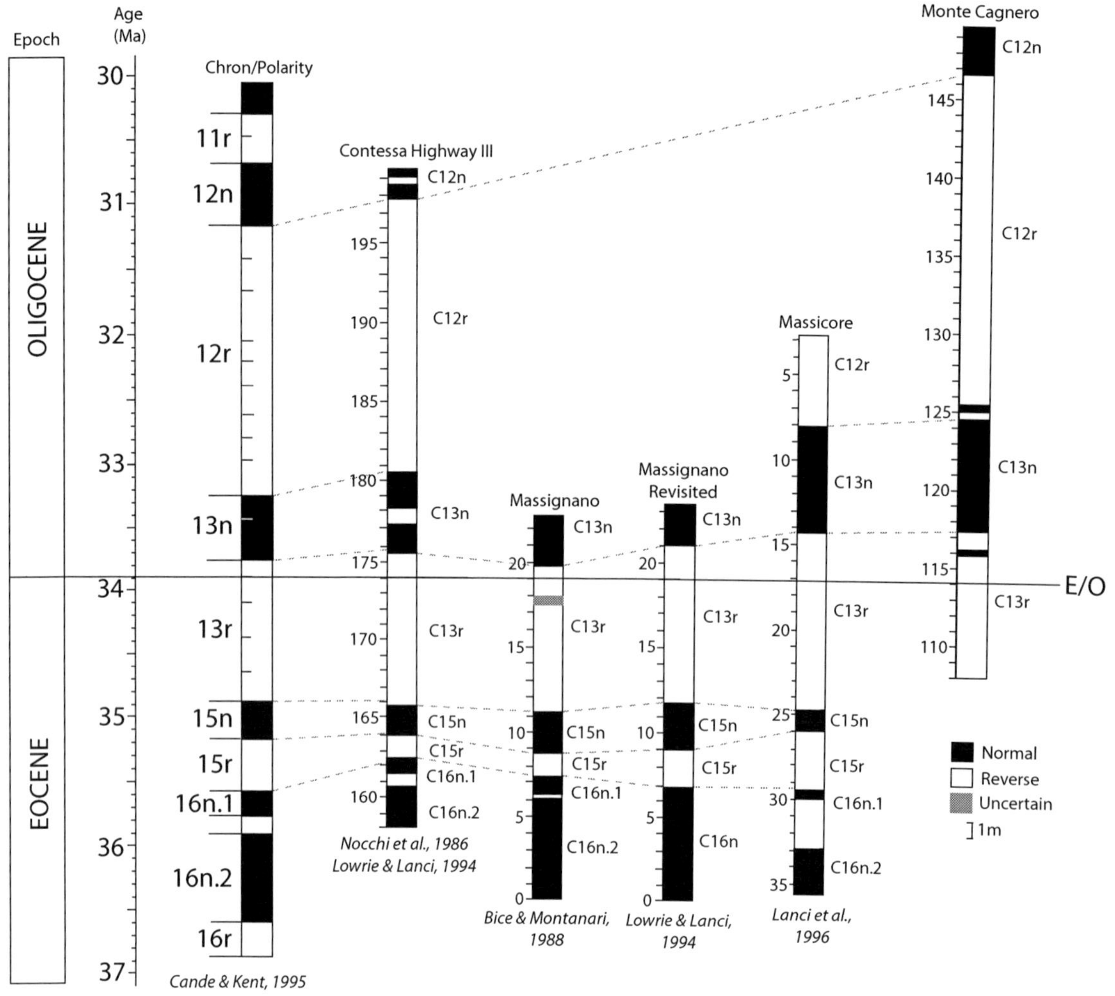

Figure 7. Comparison between geomagnetic polarity time scale (GPTS) of Cande and Kent (1995) and the magnetostratigraphies of the Monte Cagnero (this work), Contessa Highway (Nocchi et al., 1988; Lowrie and Lanci, 1994), Massignano (Bice and Montanari, 1988; Lowrie and Lanci, 1994), and Massicore (Lanci et al., 1996) sections. Time scales are scaled and tied to the Eocene-Oligocene boundary; short bars indicate identified "cryptochrons" in the time scale. E/O—Eocene-Oligocene.

the geochronology indicates that Milankovitch orbital cycles are present in the carbonate data from Monte Cagnero, which raises the possibility of astrochronologic dating of the section. However, as can be seen in the evolutionary FFT (Fig. 8B), the nonstationarity of signals through the section implies shifts in the sedimentation rate that complicate the spectra. Nevertheless, we can resolve some of these complications through a minimal tuning of different intervals of this section, providing an astronomical time scale for the Monte Cagnero section.

Sedimentation Rate Changes

If we consider the spectral peak with a wavelength of ~0.58 m to represent the orbital obliquity, which has an average period for this time of 40 k.y. (Imbrie et al., 1984), we can then infer an average sedimentation rate of ~14.5 m/m.y. for this middle part of the section. This result seems to be consistent with the sedimentation rate based on geochronology (from updated time scales by Kuiper et al., 2008), which, with error bars, permits rates varying from 14.1 to 15.5 m/m.y. This does, however, still suggest that there could be significant changes in sedimentation rate within the section. As it is rare to find an outcrop without at least minor sedimentation rate variations due to circumstances of climate, productivity, and sea-level changes (e.g., Zachos et al., 2001), we consider this to be a significant explanation of variation in the ETP fit; observations of the lithostratigraphy are consistent with changes in the sedimentation rate and lead us to propose three different sedimentation rates for different intervals of the section.

The lower interval, from meter level 108 to 125, is dominated by relatively closely spaced limestone layers ($CaCO_3$ > 75%). On the other hand, the interval between meter levels 125 and 142 is dominated by marls ($CaCO_3$ < 75%), interbedded with

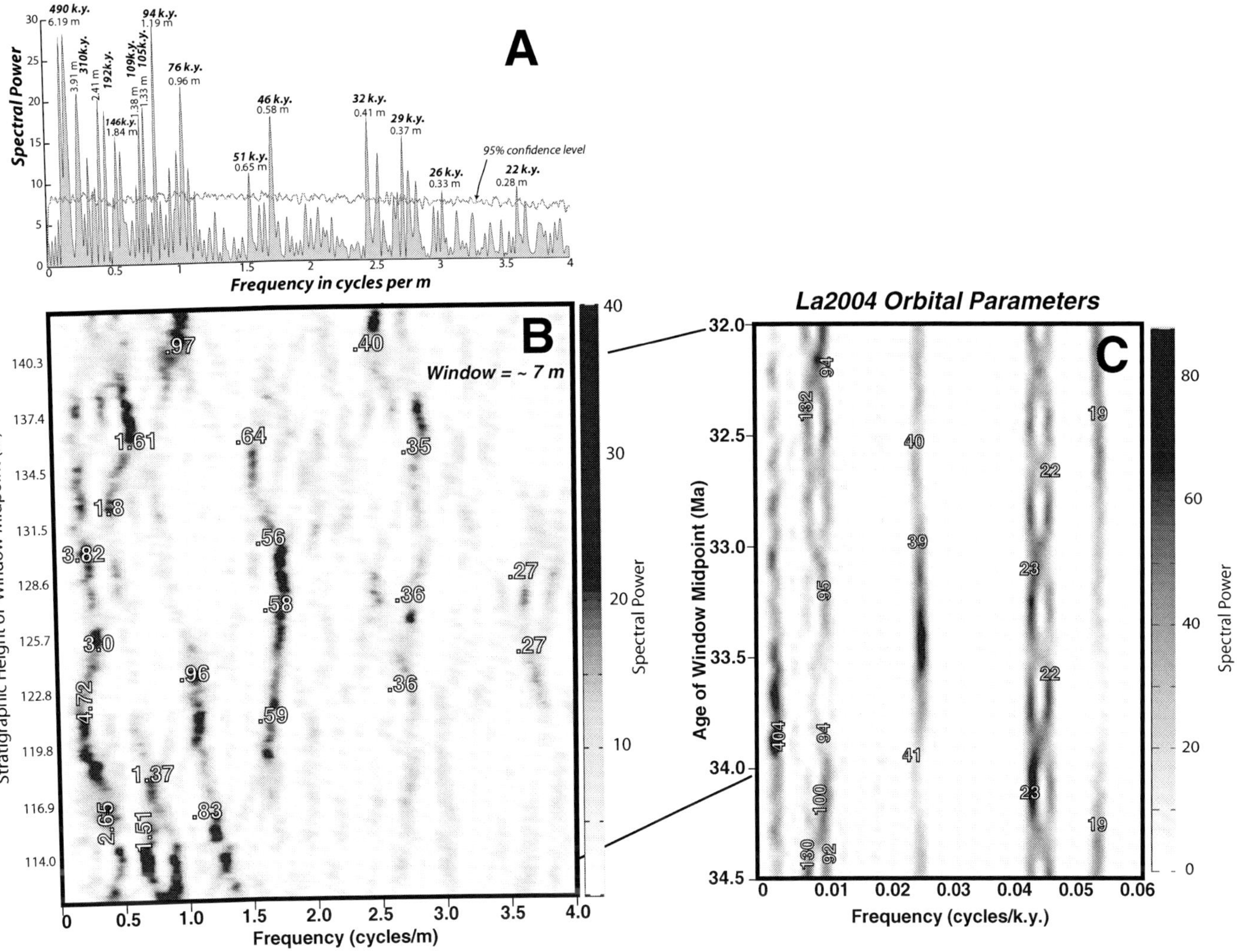

Figure 8. Spectral analysis of the Monte Cagnero carbonate data, from meter level 108 to 146.5. (A) Simple fast Fourier transform (FFT) of the whole section in the stratigraphic domain, with 95% confidence level derived from a Monte Carlo simulation. The stratigraphic wavelengths of significant spectral peaks are labeled; they are converted to time using an initial sedimentation rate of 12.6 m/m.y. (B) Evolutionary FFT with a ~7 m window of the same data set shown in A, showing continuous but nonstationary spectral peak presumed to represent the obliquity and less continuous peaks representing eccentricity. (C) Evolutionary FFT with a window of 500 k.y. of ETP (a linear combination of eccentricity, tilt, and precession) to show the spectral character of the orbital variations during this time period. The ETP here has been modified from Imbrie et al. (1984) by decreasing the contribution of the obliquity by one-half for the purpose of making the precession and eccentricity signals appear more clearly.

rare and thin limestone layers, and it includes a distinctive gray-purple marly unit at the top of this interval. The top 4.5 m of the section are characterized by alternating prominent limestone and recessive marl layers, including four biotite-rich volcano-sedimentary layers (dated by Montanari et al., 1985; revision by Kuiper et al., 2008).

The lower interval includes the Eocene-Oligocene boundary, which has been astrochronologically dated by Brown et al. (this volume) to be 33.9 ± 0.01 Ma. We begin with this constraint and draw a correlation between the stratigraphy and astronomical time by comparing the detrended carbonate data and a 6.8–4.5 m Gaussian band-pass (which should correspond to the long eccentricity cycle) to ETP and the astronomical long eccentricity cycle, extracted from Laskar et al. (2004) using a band-pass filter (Fig. 9). In making correlations between the data and the astronomical curves, we follow Brown et al. (this volume), who used a multiproxy approach to argue that carbonate lows (marls in this case) represent eccentricity and obliquity highs. By performing a simple stretch on the carbonate data, and keeping the Eocene-Oligocene boundary close to 33.9 Ma, we find that the best match occurs when the section spans 1.21 m.y., which would give a sedimentation rate of 14.0 m/m.y. for this lower interval. The match between the carbonate data and the ETP curve is quite good, with the exception of meter levels 115–117, which may

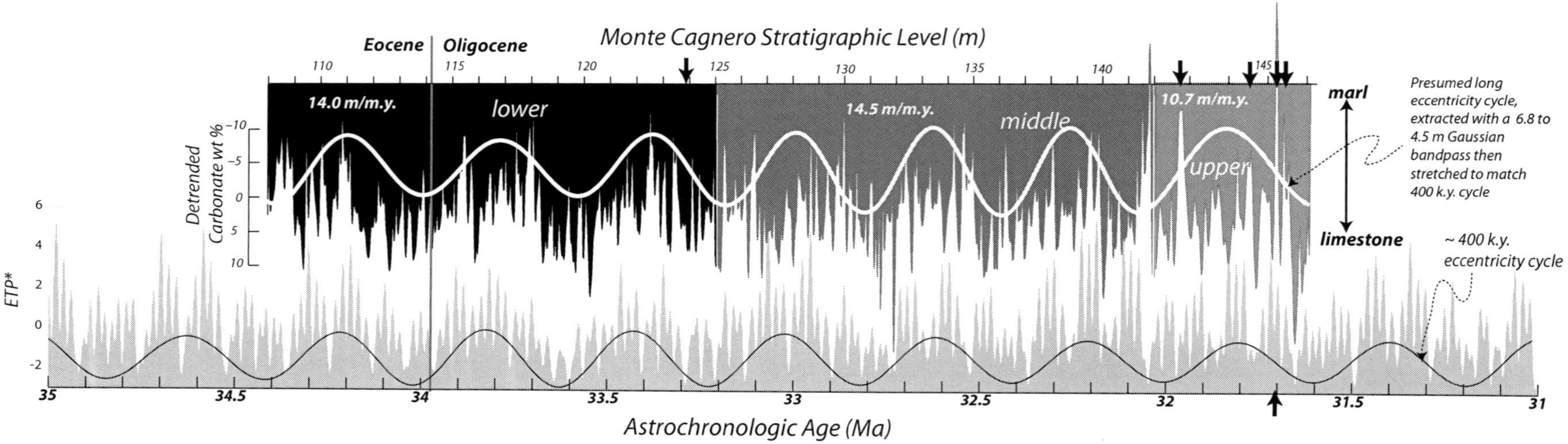

Figure 9. Correlation between the Monte Cagnero detrended carbonate data and the ETP (eccentricity, tilt, precession), which is tied to the astronomical time scale. Carbonate content has been inverted so that higher carbonate values are down and lower values are up; this facilitates the match between marly layers and highs in ETP. The section has been broken into three sections, each of which is stretched to facilitate the best overall match between the carbonate data and the astronomical record. The correlation is consistent with matching the presumed long eccentricity cycle of the carbonate data, extracted with a 6.8–4.5 m Gaussian band-pass filter, with the astronomical long eccentricity, overlain on the ETP. Volcanosedimentary layers are indicated by arrows; the one at 145.6 has been radiometrically dated to 31.7 ± 0.2 Ma. In this correlation, the Eocene-Oligocene boundary is 33.95 Ma, and the dated layer falls very near the radiometric age, at 31.71 Ma.

represent a time of increased sedimentation rate; this interval also corresponds with the disappearance of the strong and persistent obliquity cycle seen in the evolutionary FFT (Fig. 8B), which is also consistent with a substantial change in sedimentation rate. Based on correlations of the magnetic stratigraphy, this time interval represents the onset of a major glaciation, the Oi-1, as in Miller et al. (2008). A simple FFT analysis of this lower interval reveals a set of statistically significant peaks that correspond closely to the expected orbital cycles if we apply a sedimentation rate of 14 m/m.y. (Fig. 10); this is not surprising given the good match between the ETP curve and the data. The best match between the detrended carbonate data and the ETP curve suggests an age for the Eocene-Oligocene boundary of 33.95 Ma (Fig. 9).

The middle section at Monte Cagnero displays a relatively strong and persistent obliquity signal that suggests an average sedimentation rate of 14.5 m/m.y., assuming that the 0.58 m peak shown on Figure 8A represents the 40 k.y. obliquity. We note, however, that this signal is not stationary during this time interval, and the sedimentation rate may rise in the marl-dominated section near the top of this middle interval. If we connect the base of this interval with the top of the lower interval and stretch the data until we find a good fit with the ETP curve (Fig. 9), we find the duration of this interval to be ~1.17 m.y., which yields a mean sedimentation rate of 14.6 m/m.y.

The upper section, extending from meter level 142 to 145.6, is the most problematic in terms of matching its spectral characteristics to the theoretical orbital variations. The cause of this difficulty lies in the occurrence of four volcanosedimentary layers in this interval (Fig. 9), all of which have very low carbonate contents; given the short duration of this interval, these four layers have a significant effect on the analysis of the carbonate record. Nevertheless, if we tie the base of this interval to the top of the middle interval, try to match carbonate highs with ETP lows, match the 6.8–4.5 m band-pass with the 400 k.y. eccentricity, and honor the geochronologic age for the biotite-rich layer at meter level 145.6, we find a reasonable agreement between the broad variations in the carbonate data and ETP (Fig. 9). The preferred correlation shown here gives a duration of 420 k.y. to this interval and thus a mean sedimentation rate of 10.7 m/m.y. At this sedimentation rate, the obliquity cycle would have a stratigraphic wavelength of 0.42 m, and, in fact, there is a similar spectral peak in the evolutionary FFT data near the top of the section (Fig. 8A). The low sedimentation rate for this upper interval also makes sense lithologically; the overall higher carbonate values and more apparent limestone beds correspond to intervals of lower sedimentation rate in other parts of the section.

Correlation with the Massignano Section

The lower portion of the Monte Cagnero section overlaps with the Massignano section, and the carbonate content data from Brown et al. (this volume) provide a means to make a correlation between these two sections based on cyclostratigraphy. The short eccentricity signal, extracted from both data sets using a Gaussian band-pass window, shows remarkable similarity in phase and amplitude and correlates reasonably well with the theoretical eccentricity (Fig. 11). This correlation points to an age of ca. 33.95 for the Eocene-Oligocene boundary, which is consistent with the findings of Brown et al. (this volume) and is also consistent with a correlation between the detrended carbonate data and the astronomical ETP (Fig. 9).

SUMMARY AND CONCLUSIONS

The apparently continuous and complete pelagic section at Monte Cagnero, spanning from the uppermost Eocene through much of the Oligocene, represents an ideal setting for defining an integrated stratigraphy of the interval across the Eocene-Oligocene boundary, and for correlating this section with the Massignano GSSP on the basis of biochronologic, magnetochronologic, and astrochronologic criteria. A stratigraphic synthesis of the Monte Cagnero section (Fig. 12) summarizes many of our key findings.

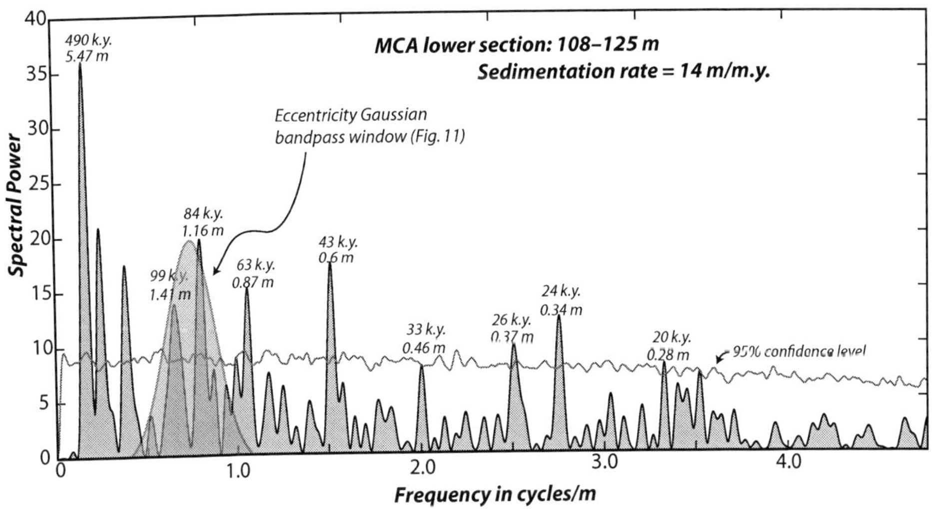

Figure 10. Fast Fourier transform (FFT) of the lower interval, from meter level 108 to 125, showing the stratigraphic wavelength and periods of spectral peaks rising above the 95% confidence level. The labeled periods reflect a sedimentation rate of 14.0 m/m.y., which comes from a comparison of the raw data with the ETP (eccentricity, tilt, precession) in Figure 9; also shown is the band-pass window used to extract the short eccentricity signal of Figure 11.

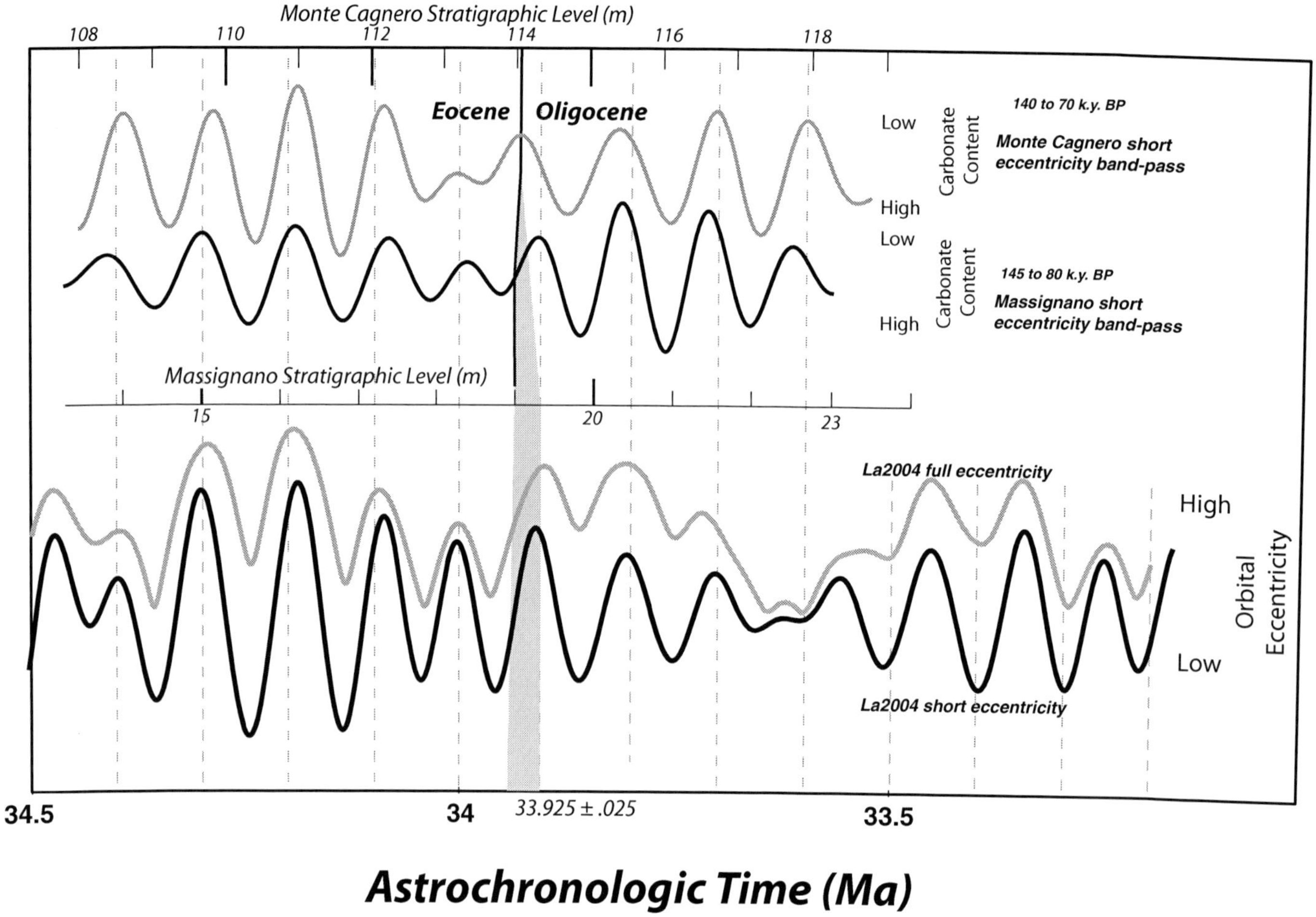

Figure 11. Correlation between the lower Monte Cagnero interval, the upper Massignano interval, and the eccentricity of Laskar et al. (2004). The short eccentricity signals in Monte Cagnero and Massignano are nearly identical and correlate with the astronomical time scale to give an age of the Eocene-Oligocene boundary of 33.95 Ma.

The E15-E16 boundary represents, in this succession, the lowest planktonic foraminiferal boundary sensu Berggren and Pearson (2006), and it is recognized at meter level 107.5 in the vicinity of a disrupted/slumped bed. The Eocene-Oligocene boundary, which corresponds to the E16-O1 zonal boundary marked by the LO of the hantkeninids (P17-P18 of Blow, 1969), is found at meter level 114.1. The O1-O2 zonal boundary, which is placed at the LO of *P. naguewichiensis*, is located at meter level 136 (Fig. 12). The upper part of the studied section, up to meter level 146, overlaps, with zone O2, the mid-upper Oligocene part of the Monte Cagnero exposure studied by Coccioni et al. (2008).

Nannofossil biostratigraphy encompasses zones CP15b through CP17 (NP19/20–NP23), and percentage plots of key species and statistical analyses indicate good correlations when compared with existing studies at Massignano and ODP and DSDP sites (e.g., Premoli Silva et al., 1988; Poore et al., 1982). As for the dinoflagellates, the Monte Cagnero section bears a continuous and complete succession of biozones from zone Cfu up to zone Cin. The Aal-Gse zonal boundary essentially corresponds to the Eocene-Oligocene boundary defined by the LO of hantkeninids at meter level 114.1.

On the basis of an astrochronologic calibration of the lower part of the section (from meter level 108 to meter level 124), the Eocene-Oligocene boundary is dated at 33.95 Ma, which is in close agreement with an age of 33.9 ± 0.01 Ma proposed by Brown et al. (this volume; cf. Pälike et al., 2006), and which is in agreement, within analytical uncertainty, with an age of 33.7 ± 0.4 Ma derived from interpolation between several radioisotopically dated volcanosedimentary layers through the Scaglia

Figure 12. Integrated stratigraphy and astrochronologic time scale for the Monte Cagnero section across the Eocene-Oligocene boundary. Diamonds indicate the location of biotite-rich volcanosedimentary layers. Radioisotopic age of 31.7 ± 0.2 Ma of biotite-rich volcanosedimentary layer at meter level 145.6 is from Coccioni et al. (2008) and Kuiper et al. (2008). The expected location of Oi-1 event in the lower part of chron C13n is from Miller et al. (2008).

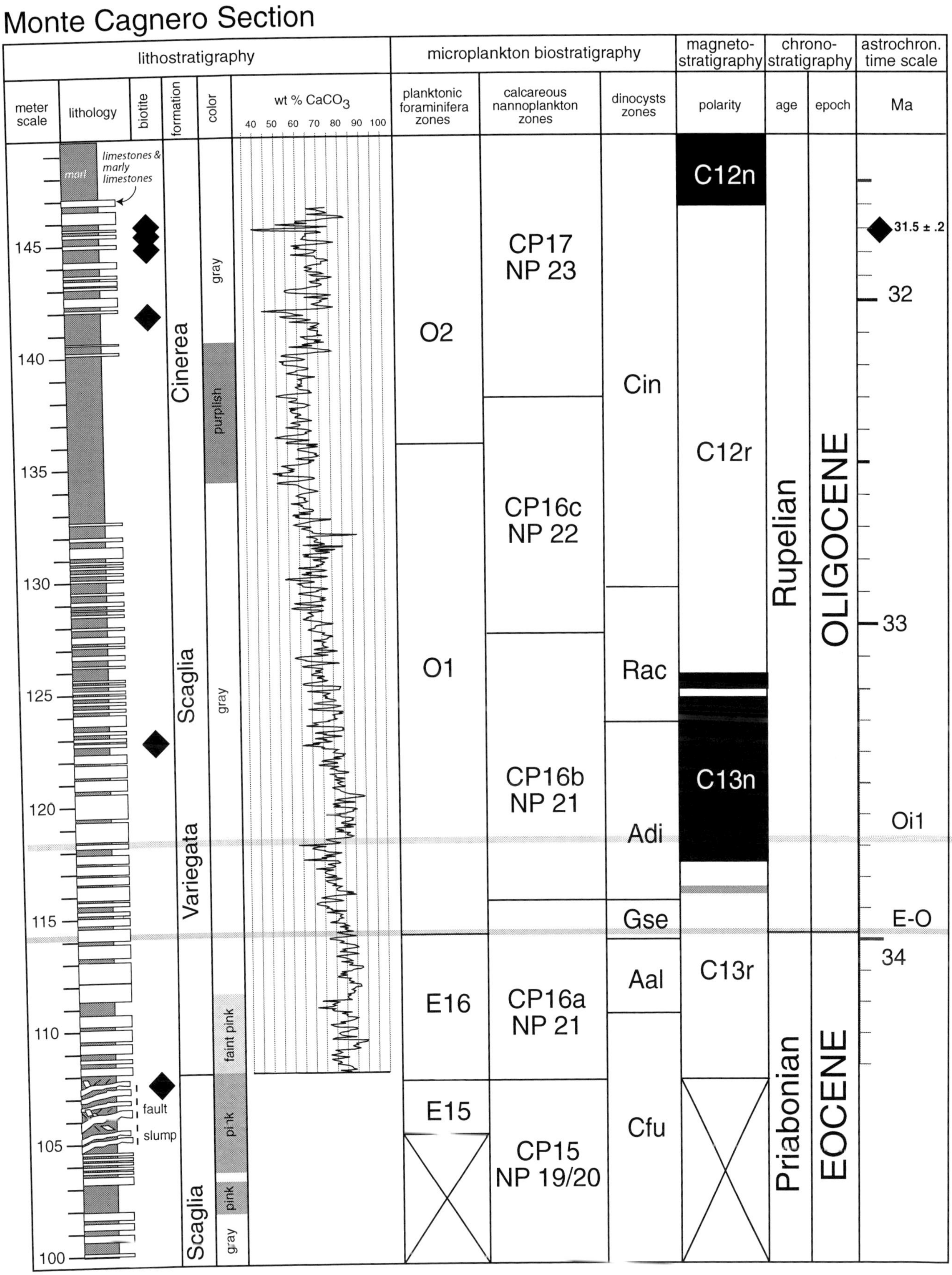
Monte Cagnero Section
lithostratigraphy
microplankton biostratigraphy
magneto-stratigraphy
chrono-stratigraphy
astrochron. time scale
meter scale
lithology
biotite
formation
color
wt % CaCO3
40 50 60 70 80 90 100
planktonic foraminifera zones
calcareous nannoplankton zones
dinocysts zones
polarity
age
epoch
Ma
limestones & marly limestones
marl
fault
slump
145
140
135
130
125
120
115
110
105
100
Cinerea
Scaglia Variegata
Scaglia
gray
purplish
gray
faint pink
pink
pink
gray
O2
O1
E16
E15
CP17 NP 23
CP16c NP 22
CP16b NP 21
CP16a NP 21
CP15 NP 19/20
Cin
Rac
Adi
Gse
Aal
Cfu
C12n
C12r
C13n
C13r
Rupelian
Priabonian
OLIGOCENE
EOCENE
31.5 ± .2
32
33
Oi1
E-O
34

Variegata–Scaglia Cinerea succession in the Umbria-Marche Basin (e.g., Montanari and Koeberl, 2000).

Paleomagnetic data from Monte Cagnero provide an independent correlation with the upper part of the GSSP section at Massignano. An astrochronologic age of 33.75 Ma is obtained for the C13r-C13n boundary (cf. Cande and Kent, 1995; 33.7 Ma) based on a correlation between the astronomical time scale in the form of ETP and a high-resolution $CaCO_3$ profile of the (Fig. 9). The top of C13n, which is not represented in the GSSP at Massignano, is located at meter level 124, and it is dated to an astrochronologic age of 33.23 Ma. The base of C12n at meter level 143.5 has an astrochronologically calibrated age of 31.60 Ma, which is in good agreement with a $^{39}Ar/^{40}Ar$ age of 31.7 ± 0.2 Ma obtained from a biotite-rich volcanosedimentary layer at meter level 145.6 (Coccioni et al., 2008), the astrochronologic age of which is calculated at 31.71 Ma.

According to Miller et al. (2008), the benthic foraminiferal $\delta^{18}O$ shift defining the Oi-1 glaciation is at 33.5 Ma, at the base of C13n. This event is expected to occur, in the Monte Cagnero section, at meter level 118, corresponding to a significant positive $CaCO_3$ shift (Fig. 12) and a probable reduction in the sedimentation rate. According to our time scale, this event occurs at 33.7 Ma, meaning that the event/sequence boundary reported by Miller et al. (2008) is potentially diachronous.

The integrated stratigraphic data and the correlations to the Massignano GSSP section presented in this work render the Monte Cagnero section the most appropriate co-representative for the Umbria-Marche Eocene-Oligocene transition. This section bears continuous and complete biostratigraphic (calcareous plankton, nannofossils, and dinoflagellate cysts), magnetostratigraphic, and astrochronologic records, and it is geochronologically calibrated with a reliable $^{40}Ar/^{39}Ar$ date obtained from a volcaniclastic biotite level. The Monte Cagnero section meets the International Union of Geological Sciences (IUGS) recommendations for a potential designation as a GSSP of the Eocene-Oligocene boundary (e.g., Odin and Montanari, 1988), and it provides exact correlation to the current GSSP at Massignano, thereby qualifying it for designation as a parastratotype. This section and the integrated stratigraphic work presented here invite further research to better calibrate the astrochronologic time scale using proxies in addition to $CaCO_3$, such as stable isotopes, magnetic susceptibility, and quantitative micro-ecopaleontology.

ACKNOWLEDGMENTS

We would like to thank Cameron Davidson (Carleton College), Linda Hinnov (Johns Hopkins University), Andrew Kulpecz (Rutgers University), and Sarah Titus (Carleton College) for their useful comments and involvement in the preparation of this manuscript. We are extremely grateful to David Evans and the Yale University Department of Geology and Geophysics for the use of equipment and supervision on the paleomagnetic analyses. We also thank the Osservatorio Geologico di Coldigioco (Italy), the Coldigioco Research Fund, and the Carleton College Geology Department for funding and supporting this project, and Aaron Fricke, Perry Spector, Sandro Mariani, Paul Kopsick, Gabe Nelson, and Ross Mitchell for their support in field and laboratory work.

REFERENCES CITED

Backman, J., 1986, Accumulation patterns of Tertiary calcareous nannofossils around extinctions: Geologische Rundschau, v. 75, p. 185–196, doi: 10.1007/BF01770187.

Backman, J., 1987, Quantitative calcareous nannofossil biochronology of middle Eocene through early Oligocene sediment from DSDP Sites 522 and 523: Abhandlungen der Geologischen Bundesanstalt, Vienna, v. 39, p. 21–31.

Baumann, P., 1970, Micropaläontologische und stratigraphische Untersuchungen der Obereozänen–Oligozänen Scaglia im zentralen Apennin (Italien): Eclogae Geologicae Helvetiae, v. 63, no. 3, p. 1133–1211.

Baumann, P., and Roth, P.H., 1969, Zonierung des Obereozaens und Oligozaens des Monte Cagnero (Zentralapennin) mit planktonischen Foraminiferen und Nannoplankton: Eclogae Geologicae Helvetiae, v. 62, p. 303–323.

Berggren, W.A., and Pearson, P.N., 2006, Tropical to subtropical planktonic foraminiferal zonation of the Eocene and Oligocene, *in* Pearson, P.N., Olsson, R.K., Huber, B.T., Hemleben, Ch., and Berggren, W.A., eds., Atlas of Eocene Planktonic Foraminifera: Cushman Foundation for Foraminiferal Research Special Publication 41, p. 29–40.

Bice, D.M., and Montanari, A., 1988, Magnetic stratigraphy of the Massignano section across the Eocene-Oligocene boundary: The Eocene-Oligocene boundary in the Marche-Umbria Basin (Italy), *in* Premoli Silva, I., Coccioni, R., and Montanari, A., eds., The Eocene-Oligocene Boundary in the Marche-Umbria Basin (Italy): IUGS 636 Special Publication: Ancona, Italy, Fratelli Aniballi Publishers, v. 2, no. 1, p. 111–117.

Blow, W.H., 1969, Late Middle Eocene to recent planktonic foraminiferal biostratigraphy, *in* Bronnimann, P., and Renz, H.H., eds., Proceedings of the First International Conference of Planktonic Microfossils, Geneva, 1997, v. 1, p. 199–421.

Blow, W.H., and Banner, F.T., 1962, The mid-Tertiary (upper Eocene to Aquitanian) globigerinaceae, *in* Eames, F., Blow, W., Banner, F., and Clarke, W., eds., Fundamentals of mid-tertiary stratigraphical correlation: Cambridge, Cambridge University Press, p. 61–153.

Bohaty, S., Palike, H., and Zachos, J., 2007, Deep-sea stable isotope and cycle stratigraphy of the Eocene-Oligocene transition: Global correlation of records and key problems in ongoing studies, *in* Montanari, A., Koeberl, C., Coccioni, R., and Hilgen, F., eds., Abstracts with Program and Field Guide: The Late Eocene Earth, Penrose Conference (Ancona, Italy, October 3–6), p. 4 (published by the editors, Vienna, Austria).

Bolli, H.M., 1957, Planktonic foraminifera from the Eocene Navet and S. Fernando Formation of Trinidad, British West Indies: Washington, D.C., Smithsonian Institute, U.S. National Museum Bulletin 215, p. 155–172.

Brinkhuis, H., and Biffi, U., 1993, Dinoflagellate cyst stratigraphy of the Eocene/Oligocene transition in central Italy: Marine Micropaleontology, v. 22, p. 131–183, doi: 10.1016/0377-8398(93)90007-K.

Brown, E., Koeberl, C., Montanari, A., and Bice, D.M., 2009, this volume, Evidence for a change in Milankovitch forcing caused by extraterrestrial events at Massignano, Italy: Eocene-Oligocene boundary GSSP, *in* Koeberl, C., and Montanari, A., eds., The Late Eocene Earth—Hothouse, Icehouse, and Impacts: Geological Society of America Special Paper 452, doi: 10.1130/2009.2452(08).

Cande, S., and Kent, D., 1992, A new geomagnetic polarity time scale for the Late Cretaceous and Cenozoic: Journal of Geophysical Research, v. 97, p. 13,917–13,951, doi: 10.1029/92JB01202.

Cande, S., and Kent, D., 1995, Revised calibration of the geomagnetic polarity time scale for the Late Cretaceous and Cenozoic: Journal of Geophysical Research, v. 100, p. 6093–6095, doi: 10.1029/94JB03098.

Coccioni, R., Monaco, P., Monechi, S., Nocchi, M., and Parisi, G., 1988, Biostratigraphy of the Eocene-Oligocene boundary at Massignano (Ancona, Italy), *in* Premoli Silva, I., Coccioni, R., and Montanari, A., eds., The Eocene-Oligocene Boundary in the Marche-Umbria Basin (Italy): IUGS 636 Special Publication: Ancona, Italy, Fratelli Aniballi Publishers, v. 2, no. 1, p. 59–80.

Coccioni, R., Montanari, A., Bellanca, A., Bice, D., Brinkhuis, H., Church, N., Deino, A., Lirer, F., Macalady, A., Maiorano, P., Marsili, A., McDaniel,

A., Monechi, S., Neri, R., Nini, C., Nocchi, M., Pross, J., Rochette, P., Sagnotti, L., Sprovieri, M., Tateo, F., Touchard, Y., Van Simaeys, S., and Williams, G., 2008, Integrated stratigraphy of the Oligocene pelagic sequence in the Umbria-Marche Basin (northeastern Apennines, Italy): A potential GSSP for the Rupelian/Chattian boundary: Geological Society of America Bulletin, v. 120, p. 487–511, doi: 10.1130/B25988.1.

De Kaenel, E., and Villa, G., 1996, Oligocene-Miocene calcareous nannofossil biostratigraphy and paleoecology from the Iberia Abyssal Plain, northeastern Atlantic, *in* Whitmarsh, R.B., Sawyer, D.S., Klaus, A., and Masson, D.G., et al., Proceedings of the Ocean Drilling Program, Scientific Results, Volume 149: College Station, Texas, Ocean Drilling Program, p. 79–145.

Fensome, R.A., and Williams, G.L., 2004, The Lentin and Williams Index of Fossil Dinoflagellates, 2004 Edition: American Association of Stratigraphic Palynologists Contributions Series Number 42, 909 p.

Fisher, R., 1953, Dispersion on a sphere: Proceedings of the Royal Society of London, ser. A, Mathematical and Physical Sciences, v. 217, p. 295–305, doi: 10.1098/rspa.1953.0064.

Freeman, R., 1986, Magnetic mineralogy of pelagic limestones: Geophysical Journal International, v. 85, p. 433–452, doi: 10.1111/j.1365-246X.1986.tb04522.x.

Hay, W., 1977, Calcareous nannofossils, *in* Ramsey, A., ed., Oceanic Micropaleontology II: New York, Academic Press, p. 1055–1200.

Imbrie, J., Hays, J.D., Martinson, D.G., McIntyre, A., Mix, A.C., Morley, J.J., Pisias, N.G., Prell, W.L., and Shackleton, N.J., 1984, Orbital theory of Pleistocene climate: Support from a revised chronology of the marine ^{18}O record, *in* Berger, A., Imbrie, J., Hays, J., Kukla, G., and Saltzman, B., eds., Milankovitch and Climate Part 1: Dordrecht, Netherlands, Reidel Publication Co., p. 267–268.

Kirschvink, J., 1980, The least-squares line and plane and the analysis of palaeomagnetic data: Geophysical Journal of the Royal Astronomical Society, v. 62, p. 699–718.

Kuiper, K., Deino, A., Hilgen, F., Krijgsman, W., Renne, P., and Wijbrans, J., 2008, Synchronizing rock clocks of history: Science, v. 320, p. 500–504, doi: 10.1126/science.1154339.

Lanci, L., Lowrie, W., and Montanari, A., 1996, Magnetostratigraphy of the Eocene-Oligocene boundary in a short drill-core: Earth and Planetary Science Letters, v. 143, p. 37–48, doi: 10.1016/0012-821X(96)00136-7.

Laskar, J., Robutel, P., Joutel, F., Gastineau, M., Correia, A., and Levrard, B., 2004, A long-term numerical solution for the insolation quantities of the Earth: Astronomy & Astrophysics, v. 428, p. 261–285, doi: 10.1051/0004-6361:20041335.

Lirer, F., 2000, A new technique for retrieving calcareous microfossils from lithified lime deposits: Micropaleontology, v. 46, p. 365–369.

Lowrie, W., and Lanci, L., 1994, Magnetostratigraphy of Eocene-Oligocene boundary sections in Italy: No evidence for short subchrons within chrons 12R and 13R: Earth and Planetary Science Letters, v. 126, no. 4, p. 247–258, doi: 10.1016/0012-821X(94)90110-4.

Lowrie, W., Alvarez, W., Napoleone, G., Perch-Nielsen, K., Premoli Silva, I., and Toumarkine, M., 1982, Paleogene stratigraphy in Umbrian pelagic carbonate rocks: The Contessa sections, Gubbio: Geological Society of America Bulletin, v. 93, p. 414–432, doi: 10.1130/0016-7606(1982)93<414:PMSIUP>2.0.CO;2.

Martini, E., 1970, Standard Tertiary and Quaternary calcareous nannoplankton zonation, *in* Farinacci, A., ed., Proceedings of the Second International Conference on Planktonic Microfossils: Rome, Roma Tecnoscienza, v. 2, p. 739–785.

McElhinny, M., and McFadden, P., 2000, Paleomagnetism: Continents and Oceans; International Geophysics Series: San Diego, Academic Press, 386 p.

Miller, K.G., Browning, J.V., Aubry, M., Wade, B.S., Katz, M.E., Kulpecz, A.A., and Wright, J.D., 2008, Eocene-Oligocene global climate and sea level changes: St. Stephens Quarry, Alabama: Geological Society of America Bulletin, v. 120, p. 34–53, doi: 10.1130/B26105.1.

Monechi, S., Riesselman, C.R., and Tori, F., 2007, Global climate reorganization at the Eocene-Oligocene transition: Calcareous nannofossil and benthic geochemical evidence from ODP Site 1263 (Walvis Ridge, South Atlantic), *in* Montanari, A., Koeberl, C., Coccioni, R., Hilgen, F., eds., The Late Eocene Earth: Hot House, Ice House, and Impacts: Abstracts with Program and Field Trip Guide: Geological Society of America, Penrose Conference (Ancona, Italy, October 3–6), p. 51 (published by the editors, Vienna, Austria).

Montanari, A., 1990, Geochronology of the terminal Eocene impacts: An update *in* Sharpton, V., and Ward, P., eds., Global Catastrophes in Earth History: Geological Society of America Special Paper 247, p. 607–616.

Montanari, A., and Koeberl, C., 2000, Impact stratigraphy—The Italian Record: Heidelberg, Springer, Lecture Notes in Earth Sciences, v. 93, 364 p.

Montanari, A., Drake, R., Bice, M.D., Alvarez, W., Curtis, G.H., Turrin, B.D., and DePaolo, D.J., 1985, Radiometric time scale for the Upper Eocene and Oligocene based on K/Ar and Rb/Sr dating of volcanic biotites from the pelagic sequence of Gubbio, Italy: Geology, v. 13, p. 596–599, doi: 10.1130/0091-7613(1985)13<596:RTSFTU>2.0.CO;2.

Muller, R., and MacDonald, G.J., 2000, Ice Ages and Astronomical Causes: Data, Spectral Analysis, and Mechanisms, Volume 9: New York, Springer, 318 p.

Nocchi, M., Parisi, G., Monaco, P., Monechi, S., Madile, M., Napoleone, G., Ripepe, M., Orlando, M., and Bice, D.M., 1986, The Eocene-Oligocene boundary in the Umbrian pelagic sequence, Italy, *in* Pomerol, C., and Premoli Silva, I., eds., Terminal Eocene Events; Developments in Palaeontology and Stratigraphy: Elsevier, v. 9, p. 25–40.

Nocchi, M., Monechi, S., Coccioni, R., Madile, M., Monaco, P., Orlando, M., Parisi, G., and Premoli Silva, I., 1988, The extinction of Hantkeninidae as a marker for recognizing the Eocene-Oligocene boundary: A proposal, *in* Premoli Silva, I., Coccioni, R., and Montanari, A., eds., The Eocene/Oligocene Boundary in the Marche-Umbria Basin (Italy): International Subcommission on Paleogene Stratigraphy Special Publication: Ancona, Italy, Industrie Grafiche Fratelli Aniballi, p. 249–252.

Odin, G., and Montanari, A., 1988, The Eocene-Oligocene boundary at Massignano (Ancona, Italy): A potential boundary stratotype: International Subcommission on Paleogene Stratigraphy Special Publication: IUGS Special Publication: Ancona, Italy, Fratelli Aniballi Publishers, p. 253–263.

Odin, G.S., Montanari, A., Deino, A., Drake, R., Guise, P., Kreuzer, H., and Rex, D.C., 1991, Reliability of volcano-sedimentary biotite ages around the Eocene/Oligocene boundary: Chemical Geology, v. 86, p. 203–224.

Okada, H., and Bukry, D., 1980, Supplementary modification and introduction of code numbers to the low latitude coccolith biostratigraphic zonation (Bukry 1973, 1975): Marine Micropaleontology, v. 5, p. 321–325, doi: 10.1016/0377-8398(80)90016-X.

Pälike, H., Moore, T., Backman, J., Raffi, I., Lanci, L., Parés, J.M., and Janecek, T., 2005, Integrated stratigraphic correlation and improved composite depth scales for ODP Sites 1218 and 1219, *in* Wilson, P.A., Lyle, M., and Firth, J.V., et al., Ocean Drilling Program Scientific Results, Volume 199: College Station, Texas, Ocean Drilling Program, p. 1–41.

Pälike, H., Norris, R., Herrle, J., Wilson, P., Coxall, H., Lear, C., Shackleton, N., Tripati, A., and Wade, B., 2006, The heartbeat of the Oligocene climate cycle: Science, v. 314, p. 1894–1898, doi: 10.1126/science.1133822.

Parisi, G., Guerrera, F., Madile, M., Magnoni, G., Monaco, P., Monechi, S., and Nocchi, M., 1988, Middle Eocene to early Oligocene calcareous nannofossil and foraminiferal biostratigraphy in the Monte Cagnero section, Piobbico (IT), *in* Premoli Silva, I., Coccioni, R., and Montanari, A., eds., The Eocene-Oligocene Boundary in the Marche-Umbria Basin (Italy): International Subcommission on Paleogene Stratigraphy Special Publication: Ancona, Aniballi, p. 119–135.

Pearson, P.N., Olsson, R.K., Hemleben, C., Huber, B.T., and Berggren, W.A., 2006, Atlas of Eocene Planktonic Foraminifera: Cushman Foundation of Foraminiferal Research Special Publication 41, 514 p.

Perch-Nielsen, K., 1985, Cenozoic calcareous nannofossils, *in* Bolli, H.M., Saunders, J.B., and Perch-Nielsen, K., eds., Plankton Stratigraphy: Cambridge, Cambridge University Press, p. 427–554.

Poore, R.Z., Tauxe, L., Percival, S.F., Jr., and LaBrecque, J.L., 1982, Late Eocene–Oligocene magnetostratigraphy and biostratigraphy at South Atlantic DSDP Site 522: Geology, v. 10, p. 508–511, doi: 10.1130/0091-7613(1982)10<508:LEMABA>2.0.CO;2.

Premoli Silva, I., and Boersma, A., 1989, Atlantic Paleogene planktonic foraminiferal bioprovincial indices: Marine Micropaleontology, v. 14, p. 357–371, doi: 10.1016/0377-8398(89)90019-4.

Premoli Silva, I., and Jenkins, D., 1993, Decision on the Eocene-Oligocene boundary stratotype: Episodes, v. 16, p. 379–382.

Premoli Silva, I., Orlando, M., Monechi, S., Madile, A., Napoleone, G., and Ripepe, M., 1988, Calcareous plankton biostratigraphy and magnetostratigraphy at the Eocene-Oligocene transition in the Gubbio area, *in* Premoli Silva, I., Coccioni, R., and Montanari, A., eds., The Eocene-Oligocene Boundary in the Marche-Umbria Basin (Italy): IUGS Special Publication: Ancona, Italy, Fratelli Aniballi Publishers, p. 137–161.

Prothero, D.R., 1994, The Eocene-Oligocene Transition: Paradise Lost: New York, Columbia University Press, 291 p.

Roth, P.H., 1970, Oligocene calcareous nannoplankton biostratigraphy: Eclogae Geologicae Helvetiae, v. 63, p. 799–881.

Spezzaferri, S., 1994, Planktonic foraminiferal biostratigraphy and taxonomy of the Oligocene and Lower Miocene in the oceanic record: An overview: Paleontographia Italica, v. 81, 187 p.

Subbotina, N.N., 1953, Fossil Foraminifers of the U.S.S.R. Globigerinidae, Hantkeninidae, and Globorotalidae: London, Collets, 321 p.

van Mourik, C., and Brinkhuis, H., 2005, The Massignano Eocene-Oligocene golden spike section revisited: Stratigraphy, v. 2, p. 13–30.

van Mourik, C., Lourens, L., Brinkhuis, H., Pälike, H., Hilgen, F., Montanari, A., and Coccioni, R., 2008, From greenhouse to icehouse at the Massignano Eocene-Oligocene GSSP: Implications for the cause, timing and effects of the Oi-1 glaciation, *in* Montanari, A., Koeberl, C., Coccioni, R., and Hilgen, F., eds., Abstracts with Program and Field Guide: The Late Eocene Earth, Penrose Conference (Ancona, October 3–6), p. 78 (published by the editors, Vienna, Austria).

Watson, G., 1956, Analysis of dispersion on a sphere: Monthly Notices of the Royal Society for Astronomy and Geophysics, v. 7, supplement, p. 153–159.

Watson, G., 1960, More significance tests on the sphere: Biometrika, v. 47, p. 87–91.

Wehland, F., Eibl, O., Kotthoff, S., Alt-Epping, U., and Appel, E., 2005, Pyrrhotite pTRM acquisition in metamorphic limestones in the light of microscopic observations: Physics of the Earth and Planetary Interiors, v. 151, p. 107–114.

Zachos, J., Pagani, M., Sloan, L., Thomas, E., and Billups, K., 2001, Trends, rhythms, and aberrations in global climate 65 Ma to the present: Science, v. 292, p. 686–693, doi: 10.1126/science.1059412.

Zachos, J.C., Kroon, D., and Blum, P., and the Leg 208 Scientific Party, 2004, Leg 208: The Early Cenozoic Extreme Climates Transect along Walvis Ridge: Ocean Drilling Program Initial Reports, Volume 208: College Station, Texas, Ocean Drilling Program, 112 p.

Zijderveld, J.D.A., 1967, AC demagnetization of rocks: Analyses of results, *in* Collinson, D.W., Creer, K.M., and Runcorn, S.K., eds., Methods in Paleomagnetism: New York, Elsevier, 257 p.

Manuscript Accepted by the Society 16 September 2008